GeoPlanet: Earth and Planetary Sciences

More information about this series at http://www.springer.com/series/8821

Paweł Rowiński · Artur Radecki-Pawlik
Editors

Rivers—Physical, Fluvial and Environmental Processes

Springer

Editors
Paweł Rowiński
Insitute of Geophysics
Polish Academy of Sciences
Warsaw
Poland

Artur Radecki-Pawlik
Department of Hydraulic Engineering and Geotechnics
Agricultural University of Cracow
Cracow
Poland

The GeoPlanet: Earth and Planetary Sciences Book Series is in part a continuation of Monographic Volumes of Publications of the Institute of Geophysics, Polish Academy of Sciences, the journal published since 1962 (http://pub.igf.edu.pl/index.php).

ISSN 2190-5193 ISSN 2190-5207 (electronic)
GeoPlanet: Earth and Planetary Sciences
ISBN 978-3-319-35502-3 ISBN 978-3-319-17719-9 (eBook)
DOI 10.1007/978-3-319-17719-9

Springer Cham Heidelberg New York Dordrecht London

Softcover reprint of the hardcover 1st edition 2015

Printed on acid-free paper

Springer International Publishing AG Switzerland is part of Springer Science+Business Media (www.springer.com)

Series Editors

Managing Editor

Anna Dziembowska
Institute of Geophysics, Polish Academy of Sciences

Advisory Board

Kon-Kee Liu
Institute of Hydrological
and Oceanic Sciences
National Central University Jhongli
Jhongli, Taiwan

Teresa Madeyska
Research Centre in Warsaw
Institute of Geological Sciences
Warszawa, Poland

Stanisław Massel
Institute of Oceanology
Polish Academy of Sciences
Sopot, Poland

Antonio Meloni
Instituto Nazionale di Geofisica
Rome, Italy

Evangelos Papathanassiou
Hellenic Centre for Marine Research
Anavissos, Greece

Kaja Pietsch
AGH University of Science and
Technology
Kraków, Poland

Dušan Plašienka
Prírodovedecká fakulta, UK
Univerzita Komenského
Bratislava, Slovakia

Barbara Popielawska
Space Research Centre
Polish Academy of Sciences
Warszawa, Poland

Tilman Spohn
Deutsches Zentrum für Luftund
Raumfahrt in der Helmholtz
Gemeinschaft
Institut für Planetenforschung
Berlin, Germany

Krzysztof Stasiewicz
Swedish Institute of Space Physics
Uppsala, Sweden

Ewa Szuszkiewicz
Department of Astronomy
and Astrophysics
University of Szczecin
Szczecin, Poland

Roman Teisseyre
Department of Theoretical Geophysics
Institute of Geophysics
Polish Academy of Sciences
Warszawa, Poland

Jacek Tronczynski
Laboratory of Biogeochemistry
of Organic Contaminants
IFREMER DCN_BE
Nantes, France

Steve Wallis
School of the Built Environment
Heriot-Watt University
Riccarton, Edinburgh
Scotland, UK

Wacław M. Zuberek
Department of Applied Geology
University of Silesia
Sosnowiec, Poland

Piotr Życki
Nicolaus Copernicus Astronomical
Centre
Polish Academy of Sciences
Warszawa, Poland

"So lasting they are, the rivers!" Only think. Sources somewhere in the mountains pulsate and springs seep from a rock, join in a stream, in the current of a river, and the river flows through centuries, millennia. Tribes, nations pass, and the river is still there, and yet it is not, for water does not stay the same, only the place and the name persist, as a metaphor for a permanent form and changing matter...

From "Rivers" by Czesław Miłosz,
Polish poet who was awarded the 1980
Nobel Prize in Literature (translated from
Polish by the Poet and Robert Hass)

Preface

Rivers as we know them are the most interesting creatures on Earth—they are wild and independent, they are friendly and dangerous, and they are fearsome and fragile all at the same time. For us they are charming when working with them as scientists.

In simple terms, rivers are bodies of water with current moving in one general direction. They can vary in size, with smaller versions of rivers being referred to as streams, creeks, creaks, brooks, licks, torrents, gills, flows, burns, or runs. Life on Earth would be very different without rivers. In addition to the life-giving water of rivers, fluvial processes physically transform landscapes. Among all the various land-forming agents, the work of running water is the most ubiquitous. Nearly, every part of the Earth has seen the imprint of fluvial processes at sometime in its geologic past.

It is more the rule than the exception that scientists examine separately either the physical aspects of river systems or its environmental or ecological components. Also, the tradition of writing books on particular aspects of river systems has developed, and therefore, we have an abundance of literature items devoted to sediment transport, water quality problems, and open-channel hydraulics separately. This volume is deliberately interdisciplinary because the issues addressed do not fit into one discipline. The focus of this book is on physical, fluvial, and environmental linkages in river ecosystems, and there is a systematic treatment of the mechanisms behind these linkages. Thus, this book should be very important for professionals from a practical point of view. The research questions posed in this book are analyzed through an interdisciplinary, international, and often case-study approach. The essential feature of this volume is to mobilize a wider range of considerations and sources of information than those used in conventional approaches. Our contributors are world-renowned experts representing as many as **28** research institutions from **15** countries, namely Austria, Canada, Germany, Finland, France, India, Italy, the Netherlands, Norway, Poland, Portugal, Spain, Switzerland, the UK, and the USA. We mention here only the authors of the book chapters, but when the reviewers (from 10 countries) are added to this list, we may claim that we practically cover the expertise in the field from every corner of the globe.

The idea of writing this book originated from discussions with a number of colleagues from all over the world who all complained about the lack of such a state-of-the-art report from which one can learn about various aspects to more fully understand rivers. We believe that the reader of this book will get to know and gain more understanding about the physical phenomena that we deal with in rivers and will see how those phenomena influence fluvial and environmental processes. We want this volume to serve as a practical guide for specialists and students to better understand rivers and have the wealth of knowledge at their fingertips. We will be happy if this single text serves pure scientists, field engineers, researchers, and designers.

It is fair to say that study of the dynamics of rivers has revolutionized in the past few decades, largely through the development of new techniques for measuring turbulent flow and bed morphology in both natural conditions and experimental flumes as well as the development of computational methods. We hope this volume well represents the current state of knowledge. Bertrand Russell in his "A history of Western philosophy" said "Science may set limits to knowledge but should not set limits to imagination." Understanding river systems requires such imagination—they are extremely complex, cause and effect closely interact, and most of the river processes are nonlinear by nature. The consecutive chapters are based on the progress made by river research worldwide.

The first eleven chapters of the book are related to basic physical processes that occur in rivers. The first one prepared by John Fenton is a kind of an introduction to the entire volume; he mainly concentrates on the basics of modeling river flows. Since all river flows are turbulent in nature, the paper of Franca and Brocchini covers the subject of river turbulence from theoretical and experimental viewpoints. The paper of Ferraro and Dey continues this subject but with the emphasis put on formation mechanisms of bedforms and their stability. Another aspect of riverbed changes, namely the characteristics of scouring downstream of low-head hydraulic structures, is given by Pagliara and Palermo. Since some environmental flows can be considered as shallow, the chapter by Uijtewaal discusses the basic physics and ways of modeling of such flows. Three consecutive chapters deal with modeling issues, namely with the problem of one- to three-dimensional flows in rivers (Gąsiorowski et al., Moussa and Cheviron, and Oertel). The next two chapters deal with laboratory experiments; the one of Bung concerns capacities and limits of hydraulic modeling, whereas the chapter by Kozioł and Kubrak covers the turbulence structure in compound channels. The last chapter in this part *Physical Processes* deals with an important issue of uncertainties that are an integral part of modeling and is written by Warmink and Booij. Church opens the second part *Fluvial Processes* with his in-depth discussion of the problems of morphodynamics and morphology of rivers. The principles of sediment transport are given by Ferreira and Hassan. Da Silva provides a review of the present understanding of the kinematics of meandering flow, and its relationship to bed deformation as well as downstream migration and lateral expansion of meander loops. Braiding, as a separate fluvial process, is covered by Surian. Bialik deals with another scale of the process, and he touches the topic of solid particle motion in open channels. The last chapter in this part of the book, provided by Velasco and Ubeda, concerns a completely different topic, namely the problem of soil erosion in watersheds after forest fires. The final part *Environmental Processes* starts with the discussion by Hilderbrand and Utz on the notions of ecological thresholds and resilience that help us understand how lotic ecosystems change. In the next chapter, Gibbins provides an insight about how physical and biological processes are coupled, particularly how flowing water influences benthic organisms. Radecki-Pawlik shows in what way bankfull and dominant discharges influence environmental processes in river channels. The chapter by Aberle and Järvelä concentrates on hydrodynamic processes that govern flow patterns in vegetated channels, while the next three chapters are devoted to the transport of various kinds of pollutants in rivers. Manson and Wallis discuss the problem of stream metabolism; Zaramella et al. focus on pollutant transport and retention processes induced by the transient storage in the dead zones by the riverbed topography and vegetation and by evapotranspiration; and Kalinowska and Rowiński concentrate on the fate of heated water in rivers.

We are very thankful and give our appreciation to those authors who contributed to this volume. We are also grateful to those colleagues who provided constructive reviews of the papers. We are indebted to Anna Dziembowska, Ewelina Brulińska, and Karolina Branicka for spending much time ensuring this book is both professionally produced and published. Finally, we are thankful to our mother institutions: Institute of Geophysics of the Polish Academy of Sciences in Warsaw and Agricultural University in Cracow, for the support of this important work.

Paweł Rowiński
Artur Radecki-Pawlik

List of Reviewers

Hubert Chanson, School of Civil Engineering, The University of Queensland, Brisbane, Australia
Robert Ettema, University of Wyoming, USA
Robert Bialik, Institute of Geophysics Polish Academy of Sciences, Poland
Jarosław Napiórkowski, Institute of Geophysics Polish Academy of Sciences, Poland
John Fenton, Institute of Hydraulic and Water Resources Engineering, Vienna University of Technology, Austria
Steven Wallis, Heriot-Watt University, Edinburgh, UK
Ian Guymer, University of Warwick, UK
Peggy Zinke, SINTEF Energy Research, Division Energy Systems, Water Resources Research Group, Trondheim, Norway
Fabian A. Bombardelli, Department of Civil and Environmental Engineering, University of California, Davis, USA
Ana Maria da Silva, Department of Civil Engineering, Queen's University, Canada
Romuald Szymkiewicz, Department of Hydraulic Engineering, Gdańsk University of Technology, Poland
Subhashish Dey, Department of Civil Engineering, Indian Institute of Technology, Kharagpur
Il Von Seo, Department of Civil and Environmental Engineering, Seoul National University, South Korea
Dubravka Pokrajac, University of Aberdeen, UK
Maciej Zalewski, EUROPEAN REGIONAL CENTRE FOR ECOHYDROLOGY under the auspices of UNESCO, International Institute of Polish Academy of Sciences, Poland
Paul A. Carling, University of Southampton, UK
Blake Tullis, Civil and Environmental Engineering, Utah State University, USA

António José Bento Gonçalves, Geography Department, ICS, University of Minho, Portugal
Michele Palermo, Department of Civil Engineering, University of Pisa, Pisa, Italy
Giuseppe T. Aronica, Department of Civil Engineering, University of Messina, Italy

Contents

Editors and Contributors

About the Editors

Paweł Rowiński is Professor in Earth Sciences and Corresponding Member of the Polish Academy of Sciences. Since 2008 he has been the CEO of the Institute of Geophysics, Polish Academy of Sciences. He is also the chairman of the Board of Directors of Earth and Planetary Research Centre (GeoPlanet) and the chair of the council of provosts at Division III: Mathematics, Physics, Chemistry and Earth Sciences of the Polish Academy of Sciences. He is a member of Leadership Team of Europe Regional Division of IAHR. His research interests and contributions are basically in mathematical modelling of hydrological processes, fluvial hydraulics, river turbulence, pollution and sediment transport in rivers, two-phase flows, chaotic dynamics. He has well over 120 scientific publications and numerous presentations to his credit. He has been a co-author and co-editor of 13 scientific volumes. On May 15, 2015, he was nominated Vice-President of the Polish Academy of Sciences.

Artur Radecki-Pawlik is Professor in Environmental Engineering: River Training, Hydromorphology, Hydraulic Structures. Professional Engineer: RP-Upr.365/91/KR (water reclamation, river training, civil engineering, hydraulic structures). Employed in the Agricultural University of Kraków, Faculty of Environmental Engineering and Geodesy, Poland. Previously employed in CBS and PBW Hydroprojekt Designing Office, Kraków, as a designer. Scientific interests: Hydromorphology and hydrology; River training and hydraulic structures; Sediment transport; Open channel hydraulics; Applied fluvial geomorphology. He was awarded the "Decaban Scholarship" in UBC in Canada. He is an author and co-author of more than 300 papers, projects and professional software as well as 8 books. Also, he is a member of IAHR and Polish Academy of Sciences. Currently he is Editor for the Acta Scientiarum Polonorum series: Environmental Processes (Firmatio Circumiectus). He has a wife Margaret and one son Bartosz. He lives in Kraków.

Contributors

Jochen Aberle received his education in civil engineering from the University of Karlsruhe (TH), with a diploma in 1996 and the Ph.D. title in 2000. In 2001 he was awarded the "Ehrensenator-Huber Preis" by the Department of Civil Engineering, Geo- and Environmental Sciences of Karlsruhe University (TH). After a two-year postdoctoral stay at the National Institute of Water and Atmospheric Research (NIWA) in Christchurch, New Zealand, he joined the Leichtweiss-Institute for Hydraulic Engineering and Water Resources (LWI) at the Technische Universität Braunschweig in 2003 as Research Fellow. In 2008 he was promoted to the head of the LWI Hydraulic Laboratory, and in 2012 he became a full professor at the Norwegian University of Science and Technology (NTNU). Professor Aberle's general research interests are environmental fluid mechanics, sediment transport, cohesive sediment dynamics, and measurement techniques. He authored over 80 journal and conference papers and is currently Associate Editor for the Journal of Hydraulic Research.

Robert J. Bialik has been employed as an Assistant Professor and the leader of the Experimental Hydrodynamics Group at the Institute of Geophysics, Polish Academy of Sciences (PAS) since June 2010. Before that he had been a Ph.D. student at the Department of Hydrology and Hydrodynamics. His particular research experience and achievements relate to Lagrangian modeling of sediment transport in open-channels. Moreover, he has served as an Associate Editor of Acta Geophysica since 2014 and he was an Invited Editor of the Special Issue of GeoPlanet: Earth and Planetary Sciences associated with the 60th Anniversary of the Institute of Geophysics PAS.

Martijn J. Booij is Assistant Professor in Hydrology at the University of Twente, The Netherlands. He holds an MSc degree in Hydrology and Quantitative Water Management from Wageningen University and a Ph.D. degree in Hydrology and Climatology from the University of Twente. His research interests and activities focus on catchment hydrology in general and modelling of hydrological systems, assessment of environmental change impacts and uncertainty analysis in particular. These research themes are reflected in Booij's past and on-going research projects, international collaboration and over 100 scientific publications. He teaches in the area of hydrology and water management and supervises Ph.D. students, MSc theses and BSc theses.

A. Bottacin-Busolin is a Lecturer in Water Engineering in the School of Mechanical, Aerospace and Civil Engineering at the University of Manchester (UK). He has an MSc in Environmental Engineering and a Ph.D. in Civil and Environmental Engineering from the University of Padua (Italy). His research interests involve flow and transport processes in surface and subsurface systems.

Maurizio Brocchini is an expert in the hydrodynamics and morphodynamics of natural water streams. His main area of research is the mathematical and numerical modeling of shallow water flows. He graduated in Theoretical Physics, with full marks and honors, in 1989 at the University of Bologna (Italy). He earned his Ph.D. in Applied Mathematics in 1996 at the University of Bristol (UK), under the tutoring of Prof. D.H. Peregrine. Currently, he is Full Professor of Hydraulics and Fluid Mechanics at the Università Politecnica delle Marche, Ancona, Italy. He is Associate Editor of the Journal of Waterways Ports Coasts and Ocean Engineering, A.S.C.E. He is author/co-author of about 200 peer-reviewed papers, of which 80 appearing on ISI-Scopus listed international journals. He was awarded by the European Community with Marie Curie Fellowship for research in the years 1993–1996.

Daniel B. Bung is a Professor for Hydraulic Engineering at FH Aachen University of Applied Sciences, Germany. He finished his studies on Civil Engineering at Bergische University of Wuppertal, Germany, where he was also awarded his doctoral degree in 2009. In his Ph.D. project he studied flow properties of self-aerated stepped spillway flows by means of experimental work. D.B. Bung was involved in several hydraulic model studies ever since. His main research interests are air-water ffiows, gas transfer as well as experimental and numerical modeling with focus on hydraulic structures. He is also a leadership team member of the prestigious IAHR Hydraulic Structures Technical Committee and reviewer for several international journals.

Bruno Cheviron graduated in Physics and Geophysics and received his Ph.D. in 2004 from the University Paris VI-Pierre and Marie Curie, on the subject of determining vertical water fluxes from soil temperature measurements. His past research activities cover pesticide leaching in groundwater contamination, free-surface flow and erosion modeling, sediment transport and model analysis. He has joined IRSTEA, the French National Research Institute of Science and Technology for Environment and Agriculture, in Montpellier in 2012. He works on the development and automation of the PILOTE crop model, for scenarios of agricultural yield, and on the automation of the hydrological MACRO model, for pesticide fate scenarios. Renewed collaborations target the use of high-precision temperature sensors to follow shallow water and vapor movements in soils.

Mike Church is Professor Emeritus in the University of British Columbia, where he taught hydrology, geomorphology and natural resources management for 38 years. His research is focused on sediment transport and morphodynamic change in rivers, and includes an experimental program on sediment transport. He has made long-term field studies of Peace River and Fraser River, two large rivers in western Canada. He also studies steep mountain creek with a focus on stability and public safety. Mike is a Fellow of the Royal Society of Canada.

Ana Maria Ferreira da Silva is a Professor in the Department of Civil Engineering of Queen's University, Kingston, Canada, where she carries out research in the area of fluvial hydraulics, with a special focus on large-scale river morphology and morphodynamics. She has authored or co-authored over 100 scientific publications, including the 2001 IAHR Monograph Fluvial Processes. She is a recipient of the IAHR Arthur Thomas Ippen Award (2005) and the CSCE Camille A. Dagenais Award (2013). She previously was Chair of the *IAHR Fluvial Hydraulics Committee* (2009–2011), and co-opted member of the IAHR Council (2007–2009). She served as Associate Editor of the ASCE Journal of Hydraulic Engineering (2003–2015) and the Canadian Journal of Civil Engineering (2005–2011). She presently serves as Chair of the IAHR Education and Professional Development Committee and Editorial Board Member of the Journal of Environmental Fluid Mechanics.

Subhasish Dey is a Professor and Head of the Department of Civil Engineering at Indian Institute of Technology Kharagpur. He also holds an *Adjunct Professor* position in Physics & Applied Mathematics Unit at Indian Statistical Institute Kolkata. Besides, he has held numerous visiting professorships, including those at *Universität Stuttgart, Technische Universität Darmstadt, University of Iowa, Technical University of Denmark, Adelaide University, University of Bradford, Tsinghua University, University of Hong Kong, Università di Pisa, Università della Calabria*. He is an *associate editor* of *Journal of Hydraulic Engineering, Journal of Hydraulic Research, Sedimentology, Acta Geophysica, Journal of Hydro-Environment Research, International Journal of Sediment Research* etc. His research interests are applied hydrodynamics, turbulence and sediment transport, in which he has more than thirty years of experience. He is internationally known for his research and acclaimed for his contributions in developing theories and solution methodologies of various problems on hydraulics. He has published more than 130 international journal papers. He is also the author of a book entitled Fluvial Hydrodynamics published by Springer.

He is member of IAHR Fluvial Hydraulics Committee (2014-) and a past-council member of the *World Association for Sedimentation and Erosion Research* (WASER), Beijing (2010–2013). He is a fellow of *Indian Academy of Sciences* (FASc), *The National Academy of Sciences India* (FNASc) and *Indian National Academy of Engineering* (FNAE).

John Fenton grew up on a farm in Australia, did a bachelor's degree and master's degree in civil engineering and a Ph.D. in applied mathematics. He taught at various universities in Australia and New Zealand, initially mathematics, moving back to civil engineering, while his research interests moved from ocean waves to open channel hydraulics. In his retirement he has worked at universities in Germany and Austria.

Domenico Ferraro is a Ph.D. student at the Civil Engineering Department of the University of Calabria, Italy. There he got the Master's Degree in Civil Engineering, with a thesis on "Scour at complex piers", the results of which were published on the ASCE-JHE. He won two Lifelong Learning Programme/Erasmus scholarships, spent at HR Wallingford Ltd (Oxford, UK) and IST (Lisboa, Portugal). He got a post-graduate specialisation as "Expert in integrated and sustainable management of the water-energy cycle in urban drainage systems". As a Ph.D. student in Hydraulics, he is currently working on turbulent phenomena in open channels. During the Ph.D. course, he spent a training research period at IIT-Kharagpur (West Bengal, India).

Rui M.L. Ferreira is Assistant Professor at Instituto Superior Técnico, Universidade de Lisboa. His academic research includes experimental investigation of fluvial and estuarine processes, mostly involving turbulence and sediment transport, development of laboratory instrumentation and mathematical modelling of open-channel flows. Coordinates STAV project—a platform for modelling transient open-channel flows over mobile beds, and DIYPIV project—a non-commercial particle image velocimetry system. Has published more than 100 academic texts, including over 25 journal papers. Has supervised or currently supervises more than 25 researchers, including 6 Ph.D. candidates. Is active in the organisation of scientific events and is a reviewer of journals in the fields of fluid mechanics, hydraulics and geophysics.

Carles Ferrer-Boix is Postdoctoral Research Fellow and Teaching Fellow in the Department of Geography at The University of British Columbia. He received his Ph.D. at the Technical University of Catalonia. His research encompasses diverse issues of fluvial geomorphology, sediment transport and river engineering, with specific attention to mountain streams and the role played by sediment mixtures on river adjustments. He is also interested on the impacts of human activities on riverine ecosystems.

Mário Franca graduated in Civil Engineering (1998) and Hydraulic and Water Resources (MSc, 2002) from the Technical University of Lisbon, and completed his Ph.D. in 2005 at the École Polytechnique Fédérale de Lausanne (EPFL). During his Ph.D. he applied two approaches in the study of the turbulent velocity field in natural gravel-bed rivers: analysis of the mean turbulence characteristics and analysis of particular instantaneous features in the flow. Mário is author of several publications in the field of fluvial hydraulics and works in such subjects as dynamics of gravel-bed channels, gravel augmentation in sediments depleted rivers, sediment transport in suspension, check dams, turbidity currents and lock-exchange density currents. He served as Assistant Professor in the University of Coimbra (invited, 2007–2008) and in the New University of Lisbon (tenure track, 2010–2012), and in the private sector in the periods 1998–2002 and 2008–2010 as hydraulic engineer. He is scientist at the Laboratory of Hydraulic Constructions (EPFL) since 2012.

Dariusz Gąsiorowski is an Assistant Professor at the Faculty of Civil and Environmental Engineering, Gdańsk University of Technology, Poland. He was employed at the Institute of Hydro-Engineering of the Polish Academy of Sciences in Gdańsk, where he obtained a Ph.D. in Civil Engineering. The main area of his activity focuses on mathematical and numerical modelling of flow in open channels, reservoirs and floodplains.

Chris Gibbins is a freshwater ecologist based at the Northern Rivers Institute, School of Geosciences, University of Aberdeen. He has been based at the University for 15 years, undertaking work spanning fundamental and applied ecology. His main interest is how physical habitat conditions influence river ecosystems, especially fish and invertebrates. His work on invertebrates extends from trying to understand the controls on benthic community structure, to how flow forces and sedimentary conditions influence drift. His applied work includes assessing the impacts of flow regulation, the development of environmental flow regimes and the benefits of river restoration.

Marwan Hassan is a Professor of Geomorphology in the Department of Geography, The University of British Columbia, who has worked extensively throughout the world, within small and large basins alike, including major studies completed for the Yangtze, Yellow and Mississippi river basins. His research interests cover broad aspects of the field from basic process-oriented field studies of sediment transport, channel stability and morphology, to physical experimentation of bedload transport processes, to in-depth reviews of the challenges surrounding water in the Middle East. He has pioneered work in particle tracing technologies, and more importantly linking insights to sediment transport theory, and was one of the first to document the significant effect of spawning salmonids on annual bedload sediment budgets, making clear connections between biological and physical processes. He manages a long running field research station in the Coastal Mountains just east of Vancouver, British Columbia, and he directs the brand new, state of the art Mountain Channel Hydraulic Experimental Laboratory at the University of British Columbia. Surprisingly, Marwan has time for other interests and pursuits, including fathering three beautiful children, and a love for history and politics.

Robert Hilderbrand is an Associate Professor at the Appalachian Laboratory, University of Maryland Center for Environmental Science in Frostburg, MD, USA. His main research interests revolve around the conservation, management, and restoration of headwater streams with a particular emphasis on responses to changing land uses and indicators of risk. He lives on a small farm with his wife and two daughters in the central Appalachian Mountains.

Juha Järvelä is a Staff Scientist at the Aalto University School of Engineering, Finland. He received his D.Sc. (Tech.) degree from the Helsinki University of Technology in 2004 and was awarded the Vilamo Foundation's doctoral dissertation award for the thesis "Flow resistance in environmental channels: Focus on vegetation". Subsequently, the Alexander von Humboldt Foundation awarded him a fellowship for post-doctoral research at TU Braunschweig, Germany. Dr. Järvelä's research interests are in the field of environmental hydraulics and ecohydraulics. His scientific contributions deal with flow-vegetation-sediment interactions, with expertise in experimental research with natural plants. Most recently, his research has focused on improving flow resistance models based on physically-based characterisation of complex vegetation using novel remote sensing methods.

Monika B. Kalinowska is an Assistant Professor in the Department of Hydrology and Hydrodynamics, Institute of Geophysics, Polish Academy of Sciences. She obtained a MSc in Physics from the University of Science and Technology in 2003, and a Ph.D. in Geophysics from the Institute of Geophysics, Polish Academy of Sciences, in 2008. In 2002–2003 she worked at CERN (European Organization for Nuclear Research) in ATLAS Experiment by the LHC, working on Online Software of the ATLAS data acquisition (DAQ) system. During her Ph.D. study she developed a River Mixing Model—the two-dimensional numerical model of the spread of passive pollutants in flowing surface water, solving the 2D advection-diffusion equation with the included off-diagonal dispersion tensor components. Her main research interests are modeling of pollution spreading in flowing surface waters and numerical methods in fluid mechanics.

Adam Kozioł is an Assistant Professor at the Department of Water Engineering, Faculty of Civil and Environmental Engineering, Warsaw University of Life Sciences—SGGW, Poland. His research interests cover open-channel flow, hydraulic resistance and turbulence in open-channels and rivers.

Janusz Kubrak is a Lecturer at the Warsaw University of Life Sciences—SGGW. He teaches fluid mechanics and hydraulics at the Department of Hydraulic Engineering. He is currently the head of the Division of Hydraulics. His research activities comprise open channel hydraulics and hydraulic consequences of dam breaks. The author of four books and over 100 papers.

Russell Manson is a Professor of Computational Science and the Director of the Masters program in Computational Science (MSCP) at The Richard Stockton College of New Jersey. He graduated from Strathclyde University, Scotland with a first class honours degree in Civil Engineering in 1989. As an undergraduate summer employee at Hydraulics Research Ltd, Wallingford, England, he worked on the development and application of two and three-dimensional computer models of lake and estuarine cooling water discharges. This led him to read for his Ph.D. in Civil Engineering at the University of Glasgow, Scotland. His thesis was concerned with the development of computational models of flow and transport processes in flooding meandering rivers and was awarded in 1994. His educational passions lie in engineering, science and technology and in particular in computational modeling. Dr. Manson's research is in water and environmental engineering, computational geophysics and natural systems management with particular emphasis on surface water transport processes and hydrology. His work is published in journals such as the Journal of Computational Physics, the International Journal for Numerical Methods in Fluids, Geophysical Research Letters, Environmental Science and

Technology, Water Resources Research, Water Research and Freshwater Biology. He has an vigorous international research agenda with ongoing projects in Scotland, Iceland and Poland.

Andrea Marion, EAP student at UC Berkeley, M.S. Env.Eng.Sc at CalTech, Visiting Professor at Drexel University (USA) and at UC S. Barbara (USA), Visiting Scientist at NIWA (NZ), at the Institute of Geophysics, Polish Academy of Sciences, EU Research Fellow at HR Wallingford (UK) and Delft Hydraulics (NL). Ass. Editor of the Journal of Hydraulic Engineering (ASCE). 30 invited seminars and public lectures. Participant in 31 national and international research projects, 14 times as PI. Member of the National Expert List for Evaluation of Res. (CIVR) in 2005–2006, and of the Italian National Assessment on Research (VQR 2004–2010) in 2012. Coordinator of a 7-team PRIN Project (Research Projects of National Significance). Coordinator of FP7-PEOPLE-2012-ITN HYTECH, 2013–2016.

Roger Moussa is a Research Director at the French National Institute of Agricultural Research in Montpellier, France. He received his Civil Engineering degree from the University Saint-Joseph, Beirut (Lebanon), his Ph.D. degree in Hydrological Sciences from the University of Montpellier (France) in 1991, and his "Habilitation à Diriger des Recherches" from the University of Montpellier in 2003. His research focuses on hydrology, hydraulic, spatially distributed modelling, flood routing, numerical methods, geomorphology and fractals. He participates in many research programs mainly in hydrology, is member of various scientific councils, and is a guest lecturer in many French universities.

Jarosław J. Napiórkowski is Professor of Earth Sciences and Head of the Department of Hydrology and Hydrodynamics at the Institute of Geophysics, Polish Academy of Sciences (IGF). He obtained a MSc in Electronic Engineering (specialization—automatic control) from Warsaw Technical University and Ph.D. from the IGF. His Ph.D. research was concerned with application of Volterra series in dynamic hydrology. During his postdoc at University College Dublin he applied the linear and nonlinear system theory to modelling of flow routing in open channels. He is an expert in modelling of hydrologic processes and in water management. He is the Editor-in-Chief of the international journal Acta Geophysica and former head of the Ph.D. programme at the IGF.

Mario Oertel Full Professor at the Hydraulic Engineering Section, Lübeck University of Applied Sciences, since September 2012. Studied Civil Engineering at the University of Wuppertal in the years 1998–2003 (Dipl. Ing. in August 2003). He was a Research Assistant in 2003–2007, receiving the Ph.D. degree (Dr.-Ing.), summa cum laude, at the Hydraulic Engineering Section, University of Wuppertal, in December 2007. Then he worked as a Senior Researcher over the years 2008–2012, obtaining habilitation (habil.) degree in Hydraulic Engineering and Water Resources Management, University of Wuppertal, in May 2012.

Stefano Pagliara is Professor of Hydraulic Construction at the University of Pisa, Pisa, Italy. He is active in Hydraulic Construction with experience on flood hydraulic and stream rehabilitation. He has published more than 150 scientific papers, mainly on experimental hydraulic, hydraulic constructions and hydrology. His main scientific interests are: localized scour phenomena downstream of hydraulic structures, air-water flow properties on rough beds, design and analysis of eco-friendly structures (block ramps, rock chutes, wood structures, etc.) for river restoration, physical modelling of hydraulic phenomena. He was chair of the IAHR Committee on Hydraulic Structures.

Michele Palermo is Researcher at the University of Pisa, Pisa, Italy. He received the degree in Civil Engineering in 2005 and the Ph.D. in Civil Engineering Sciences and Techniques in 2009 from the University of Pisa. His main scientific interests are: localized scour phenomena downstream of hydraulic structures, air-water flow properties on rough beds, design and analysis of eco-friendly structures (block ramps, rock chutes, stepped gabion weirs, etc.) for river restoration, physical modelling of hydraulic phenomena and hydraulic jump. He is co-author of more than 50 scientific papers.

Nicola Surian is Associate Professor in the Department of Geosciences at the University of Padua (Italy), where he teaches physical geography, geomorphology, and remote sensing. His research interests are primarily focused on fluvial processes (e.g. channel adjustments, sediment transport) and on the application of geomorphological approaches in river management and restoration. Over the last few years he has focused also on channel response to extreme flood events. He has studied different fluvial environments, but a large part of his research has dealt with braided gravel-bed rivers. Author of several papers and book chapters in international journals and monographs.

Romuald Szymkiewicz, born in 1948, is a Professor at the Faculty of Civil and Environmental Engineering, Gdańsk University of Technology, Poland. His research interests focus on hydromechanics and hydraulics, in particular mathematical modelling of unsteady open channel flow, unsteady pipe flow, transport of contaminants and development of the numerical methods. He is an author of more than 90 papers published in Polish and international journals, as well as 10 books and monographs.

Matteo Tregnaghi: Expert in river hydraulics and sediment transport. Visiting scientist at the University of Sheffield (UK) in 2004 and at the University of Auckland (NZ) in 2006. He was awarded his Ph.D. in Civil and Environmental Engineering at the University of Padua in 2008. He was awarded a FP7-PEOPLE-2008 Marie Curie Individual Fellowship at the University of Bradford (UK) in 2009–2011. Senior Post-doc at the University of Padua in 2011–2013 and Guest Professor at the University of Padua in Fluid Mechanics since 2013. Currently Project Manager of the FP7-PEOPLE-2012 Marie Curie Initial Training Network Project 'HYTECH' (2013–2016).

Xavier Úbeda is a Senior Lecturer in the University of Barcelona. He has more than 20 years of experience in the study of fire effects in soil properties, hydrology and soil erosion. He participated in two European projects where the main interest was to understand the importance of the fire intensity in soil physics changes, erosion processes and vegetation regrowth. Thanks to these projects he finished his Ph.D. in Geography in 1998. Since 2000 his main interest is the effects of the prescribed fires on soil properties. Three Spanish projects have sponsored the investigation about the short, medium and long-term effects in soils after prescribed fires.

Wim Uijttewaal is a Professor of Experimental Hydraulics at the faculty of Civil Engineering and Geosciences of Delft University of Technology where he is heading the Environmental Fluid Mechanics Laboratory. He is specialised in shallow flow turbulence and experimental techniques with applications to rivers, but also studies wave impact on structures and interaction of waves and currents with vegetation as well as sediment. He (co-)organised conferences on Shallow Flows, Particle Laden Flows and Ecohydraulics and is active in the Fluid Mechanics section of IAHR. In 2020 he organises the International conference on Fluvial Hydraulics in Delft.

Ryan Utz is an Assistant Professor of water resource management in the Falk School of Sustainability at Chatham University in Pittsburgh, Pennsylvania, USA. His research explores decadal scale changes in physicochemical attributes of streams, the effects of urbanization on water resources, and the efficacy of ecological restoration efforts.

Antonio Velasco graduated in Geography in the University of Barcelona in 2013 and finished his Master's degree in 2014 in "Environmental Management and Land Planning". His work has been focussed on the effects of high intensity fires in the soil hydrology as well as the changes in soil physical properties at a short and long term. He is interested in the role of ashes in the soil system in the post-fire scenarios and the effects of different fire intensities in the regrowth of vegetation and the consequences of post-fire forest management in Mediterranean landscapes.

Steve Wallis graduated from Birmingham University, UK in 1977 with a degree in Civil Engineering, and completed a Ph.D. degree from the same university in 1982 on flow resistance in tidal flows. Between 1983 and 1986 Steve worked at Lancaster University, UK on the Aggregated Dead Zone model for solute transport in rivers. In 1986 he became a Research Fellow at the University of Strathclyde, UK and developed a water quality model of the inner Clyde estuary. In 1988 he was appointed as a Lecturer in Water Engineering at Heriot-Watt University, UK where he continues to undertake research on the mathematical modelling of water movement, solute transport and water quality in rivers, coastal waters and drainage systems. Significant international research collaboration has been undertaken with the Institute of Geophysics in Poland since 2004.

Jord Warmink obtained a MSc in Physical Geography from Utrecht University in 2007, and a Ph.D. from the University of Twente in 2011 working with Professor Suzanne Hulscher. During his Ph.D. research he developed a method to identify uncertainties in (river) models and quantified the uncertainty due to the hydraulic roughness in river models. The research was performed in close cooperation with Deltares in Delft and Jord, cooperating with the University of Osnabruck, Germany, and GEUS in Denmark. During his postdoc, he implemented a model to predict bed form roughness for operational water level predictions together with two Ph.D. students. In 2014 he was appointed assistant professor at the University of Twente.

Mattia Zaramella is a Research Fellow at the Department of Industrial Engineering of the University of Padua and a contracted lecturer in Fluid Mechanics class for the Energy Engineering course. He is an expert of experimental activities with water tracers and of micro-hydropower technologies.

Part I
Physical Processes

Chapter 1
Basic Physical Processes in Rivers

John D. Fenton

Abstract A "one-and-a-half"-dimensional model of a river is developed. It is actually one-dimensional but allows for horizontal curvature using natural curvilinear co-ordinates. The governing long wave equations can be developed with very few limiting approximations, especially using momentum rather than energy. The curvature is then shown to be rarely important and is subsequently ignored. Wave periods, imposed by boundary conditions, are asserted to be fundamental. Long waves have speeds and propagation properties that depend on period, and there is no such thing as a single long wave speed. Examination of dimensionless equations and solution of linearised equations using wave period shows a novel interpretation of terms in the momentum equation: the "kinematic" approximation and wave are misnomers: the approximation lies not in the neglect of inertial terms but is actually a very long period one. The outstanding problem of river modelling, however, is that of resistance to the flow. A large data set from stream-gauging is considered and it is shown that the state of the bed, namely the arrangement of bed grains by previous flows, is more important than actual grain size. A formula for resistance is proposed which contains a parameter representing bed state. As that state is usually changing with flow, one can not be sure what the resistance actually will be. This uncertainty may have important implications for modelling. The momentum principle is then applied also to obstacles such as bridge piers, and a simple approximation gives greater understanding and a practical method for incorporation in river models. Finally, river junctions are considered, and the momentum approach with the very long period approximation shows that they can be modelled simply.

Keywords One-dimensional · Long wave equations · Meandering · Kinematic · Telegraph equation · Resistance · Obstructions · Bridge piers · River junctions

---

J.D. Fenton (✉)
Institute of Hydraulic and Water Resources Engineering, Vienna University of Technology, Karlsplatz 13/222, 1040 Vienna, Austria
e-mail: JohnDFenton@gmail.com

P. Rowiński and A. Radecki-Pawlik (eds.), *Rivers—Physical, Fluvial and Environmental Processes*, GeoPlanet: Earth and Planetary Sciences,
DOI 10.1007/978-3-319-17719-9_1

1.1 Introduction

We follow an approach guided by William of Ockham, a mediaeval British monk and philosopher, who developed the principle known as *Ockham's Razor*. This states that if one can explain a phenomenon without assuming something, there is no ground for assuming it, i.e. that one should always opt for an explanation in terms of the fewest possible number of causes, factors, or variables. This was paraphrased by Albert Einstein as:

> Make things as simple as possible, but not simpler.

Unfortunately we live in a world where complexity is honoured more than simplicity, and the process of peer review for scientific publications contributes to that. The process of modelling, which is what we are doing, also seems to have an in-built one-way filter. One produces a simple model, and if required, it is refined. It is rare for the process to go in the other direction. An aspect of our culture of complexity is that it leads to centralisation and homogenisation and further dominance by the dominant. It is difficult to start competing with all the resources at the disposal of those who have already succeeded. Our scientific world seems to reflect some of the problems of our economic world in the allocation of resources. Maybe we need a culture of Appropriate Technology also in our research, where small, appropriate, and simple would be honoured more than it is.

The word "basic" in the title of this chapter has two contradictory dictionary definitions. One is "to do with the essence, fundamental, or basic truths, concerned with fundamental scientific principles". The second is "constituting or serving as the basis or starting point—a basic set of tools". Concerning the first definition, there are many books and scientific papers concerned with all aspects of the fundamentals of river mechanics, including the subsequent chapters in this book, especially those in *Part 1—Physical Processes*. A single general chapter such as this cannot hope to deal with details in such depth. Instead, it will follow the spirit of the second definition, and describe the simplest approaches and theories—and try to justify their study as providing useful understanding and results. Rather than a review, we will try to present several unusual, possibly unexpected, relatively simple results that might challenge the canonical view, or at least put it in a new light.

One thing that such an approach can do is to develop understanding. We note the comment of Birta and Arbez (2013, p 37):

> The reader is cautioned not to dismiss the issue with the simplistic and naive assertion that the goal is to solve the problem!

A similar more exaggerated view is the opening statement by Hamming (1973) in his book on numerical methods:

> The Purpose of Computing is Insight, not Numbers.

We cannot completely agree with that statement, but we can view it as a reminder. It is hoped that the presentation here will indeed provide insight. However, the simple theories can also be used to provide numbers, useful numbers, and more quickly because of simplicity. Moreover, those numbers will be approximate, and therefore possibly more useful. This is often forgotten in computing in river engineering. The equations that are used may be more-or-less accurate, but often the parameters, inputs, and the modelling of processes such as resistance, are poorly known at best. Accordingly, the results of computations may not be very accurate, but will represent a solution whose accuracy is in accordance with that of the problem that was posed. Too much importance is attached to the use of sophisticated—and expensive—software, when often a simple and conceptual model that reveals the real nature of the problem is adequate. The dominance in hydraulics of large software houses has meant that yet another pre-computer Cottage-Industry has fallen victim to our more modern Industrial Revolution: people in every water office and university once could understood and solve problems, now they are often little more than software operators.

A full model of a river would be three-dimensional and would use the fundamental equations of fluid mechanics, the Navier-Stokes equations in three dimensions, complete with boundary conditions over the irregular boundary, were we able to specify that. Of course, such a model would provide us with little understanding —and very few general answers, such as the all-important question of resistance to motion in a stream. More realistically, we could obtain a numerical solution for an idealised geometry, such as a flow in a smooth rectangular channel around circular arcs, giving results for the primary and secondary flows and wall stresses. Or, in a more general approach we could study the Reynolds-averaged Navier-Stokes equations and obtain theoretical deductions from the nature of terms in the equations. Whereas the equations of fluid mechanics can help with understanding of river processes, none has yet been able to help us when it comes to calculating the resistance to flow in a natural stream. Or in telling us how a flood wave propagates.

We could relax our approach and use a two-dimensional depth-averaged physical and numerical model, applied to rectangular and compound channels with circular arc bends. Or we could even add a mobile bed plus bed transport equation. Our data needs are now much less, however, now we have to use several assumptions and approximations—a bed resistance formula, an assumed longitudinal velocity profile, an assumed variation of a Boussinesq eddy viscosity coefficient, and an assumed formula for the transverse velocity at the free surface. And so we may also obtain good agreement with experiment for longitudinal velocity, passive scalar concentration, and mobile bed changes. Importantly, however, in addition to the assumptions, we have still had to measure and calibrate the approximations used.

In this work we are more concerned with the mechanics of a whole stream, or reach of a stream. We are going to make a simple one-dimensional model, such that we cannot predict velocity distributions, either longitudinal or secondary, and we cannot predict the differential movement of sediment across the section. Although our model will be formally one-dimensional, we will not use the traditional approximation of one dimensional modelling, that all streams are straight. Instead, we will use a curvilinear co-ordinate system in which we incorporate the effects of stream curvature in the horizontal plane, giving a "one-and-a-half-dimensional model". With this model we will be able to describe the nature of the movement of waves and disturbances in the river and to incorporate the effects of obstacles and river junctions and to show what the important quantities are. We will also be able to show where approximations can be made.

Two of the aims of the paper will be seen to have been realised in a negative sense: we show that stream curvature has very little effect on flow and wave propagation, and can usually be ignored. And, we show that the resistance of the bed in a natural stream cannot be predicted accurately at any moment in time, because the arrangement of bed particles is usually important but unknown. We can, however, put bounds on the resistance—and possibly be grateful that the ephemeral nature of the problem tells us that we do not have to model it accurately.

1.2 Momentum and Energy in River Mechanics

Traditionally in open channel hydraulics there has been some confusion between energy and momentum, which has been spelt out by Liggett (1993), who also noted that the Bernoulli equation is a result of conservation of momentum. Fenton (2010) derived the unsteady long wave equations using both momentum and energy. The momentum approach requires only the mechanism of resistance at the boundary. The energy equation requires more coefficients to approximate integrals, as well as ideally requiring the approximation of different forms of energy dissipation in boundary layers, shear layers, separation zones, vortices, and subsequent turbulent decay in the wake.

In spite of the simplicity of momentum, it seems that energy is still widely believed to be an important governing principle in open channel hydraulics. There might be a simple reason for this—that resistance even in momentum formulations is often written in terms of a single dimensionless symbol such as S_{f}, which in itself is not wrong, but what is wrong is that it is described as being the slope of a line representing the total head at a section, the energy grade line. In a momentum equation it comes from the resistance force on the perimeter, and the symbol S_{f} is simply *the ratio of resistance force to gravitational force*. The difference is important in understanding, if only to remind one that we are using momentum, and not energy.

The belief in energy is also bound up with the widespread use of an unnecessarily complicated numerical method for the solution of steady gradually-varied flow problems. The so called "Standard step method" inflicted on students has a simplistic energy derivation but requires the complicated numerical solution of a transcendental equation at each step by iterative means (e.g. Chow 1959, Sect. 10.4). It is much easier, especially for teaching purposes, to take the more general gradually-varied flow equation, a differential equation obtained using momentum, and to use any simple explicit numerical method to solve it. Generally, it is neither necessary nor desirable to use the concept of energy gradient in open channel hydraulics.

Another physical problem better described by momentum is that of any obstacle in a channel such as a bridge pier. The object causes momentum loss to that flow, which takes place immediately at the front, even if it is then gradually diffused through the flow. The force, and hence the momentum loss to the flow, can be simply approximated using a drag force formula, or measured experimentally. In comparison, there are almost no energy losses at the front face of an obstacle, but because of the sudden local diversion of the flow, as the water flows downstream, interacting flows and turbulent flow processes cause energy losses for some distance, difficult to quantify or measure. For these reasons, momentum is to be preferred in analysing the effects of bridge piers and debris in a channel, and for the design of stilling basins with baffle blocks.

Another application where momentum can be used to give a rather simpler description is in the description of river junctions, where momentum is conserved but energy dissipated due to interacting flows.

Each of the above-mentioned problems will be described and analysed below using momentum.

1.3 The One-Dimensional Long Wave Equations for Straight and Curved Channels

In view of our appeal to Ockham's razor above, the modelling of a river or canal by one-dimensional means seems appropriate. It can be done with surprisingly few limiting assumptions or approximations, and can tell us much about wave propagation, both theoretically for the purpose of understanding, and practically, for the purpose of generating numerical approximations. We now describe this.

If a reader were not interested in the details of the derivation, they could skip to Sect. 1.3.4.

1.3.1 Mass and Momentum Conservation Equations

The use of the Reynolds Transport Theorem for an arbitrarily moving and deformable control volume enables us to obtain useful results relatively simply. For a more complete derivation reference can be made to Fenton (2010) for the equations for a straight channel, the traditional approximation, or to Fenton and Nalder (1995) for channels with horizontal curvature.

For an arbitrarily moving control surface and volume, the mass and momentum conservation equations are (e.g. White 2009, Sects. 3.3 and 3.4):

$$\frac{\mathrm{d}}{\mathrm{d}t}\int_{\mathrm{CV}} \rho \,\mathrm{d}\mathscr{V} + \int_{\mathrm{CS}} \rho \,\mathbf{u}_{\mathrm{r}} \cdot \hat{\mathbf{n}}\,\mathrm{d}S = 0, \tag{1.1a}$$

$$\frac{\mathrm{d}}{\mathrm{d}t}\int_{\mathrm{CV}} \rho \,\mathbf{u}\,\mathrm{d}\mathscr{V} + \int_{\mathrm{CS}} \rho \,\mathbf{u}\,\mathbf{u}_{\mathrm{r}} \cdot \hat{\mathbf{n}}\,\mathrm{d}S = \mathbf{T}, \tag{1.1b}$$

where t is time, ρ is the fluid density, $\mathrm{d}\mathscr{V}$ is an element of the control volume CV, $\mathbf{u}_{\mathrm{r}}$ is the velocity of a fluid particle relative to that of the control surface, $\hat{\mathbf{n}}$ is a unit normal vector directed outwards such that $\mathbf{u}_{\mathrm{r}} \cdot \hat{\mathbf{n}}$ is the normal component of fluid velocity relative to the element $\mathrm{d}S$ of the control surface CS, and is that responsible for the transport of any quantity across the surface. $\mathbf{T}$ is the force exerted on the fluid in the control volume by both body and surface forces. In the concise form of Eqs. 1.1a, b, the physical significance of the terms is obvious: $\rho\ \mathrm{d}V$ is an element of mass, integrating over the whole volume gives total mass, and weighting with velocity $\mathbf{u}$ gives momentum. Differentiating with respect to time gives the total rate of change of each inside the control volume. Similarly, $\mathbf{u}_{\mathrm{r}} \cdot \hat{\mathbf{n}}\,\mathrm{d}S$ is the rate of fluid volume leaving via the elemental area, multiplying by ρ gives the mass rate, integrating gives the total rate of mass leaving the control volume while weighting with velocity and integrating gives the net momentum flux.

To evaluate the terms in the conservation Eqs. 1.1a, b for a curved stream we consider a control volume consisting of an elemental slice as shown in Fig. 1.1. We consider a reference axis s which is an arbitrary curve in a horizontal plane along the course of the river, and at any point on this curve there is a local orthogonal curvilinear co-ordinate system (s, n, z), where n is horizontal and transverse to s, and z is vertically upwards, The local radius of curvature of the reference axis is r, such that in an elemental increment Δs the s axis turns through an angle $\Delta\theta$, such

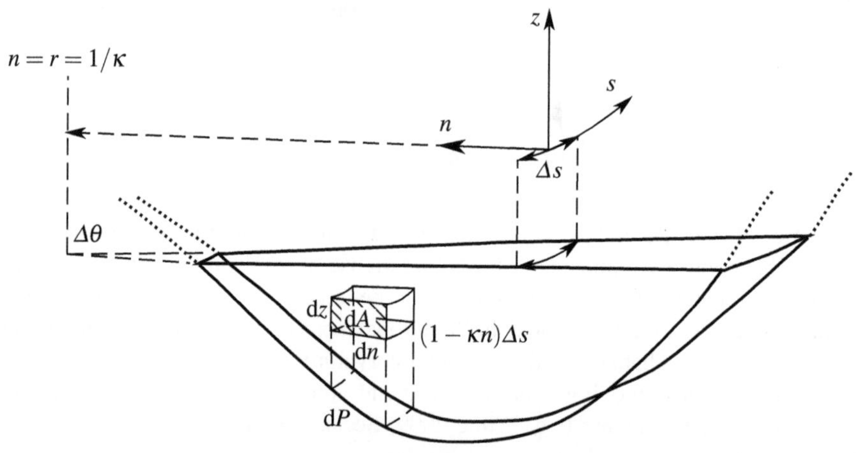

Fig. 1.1 Elemental slice and dimensions with orthogonal curvilinear (s, n, z) coordinate system

that $\Delta s = r\Delta\theta$, where a positive angle (turning anticlockwise around the centre of curvature) corresponds to a positive radius of curvature. The control surface consists of the bed of the river $z = Z(s,n)$ which we assume not to move, two stationary planar faces orthogonal to the streamwise co-ordinate s with an angle $\Delta\theta$ between them, and the free surface $z = \eta(s,n,t)$, which can move and the control surface moves with it. It is convenient to introduce curvature $\kappa = 1/r$, so that an element of volume has dimensions $\Delta s\,(1 - \kappa n)$, Δn and Δz, as shown in the figure. For practical use, the curvature and hence the alignment of the s axis, should represent the stream as a whole, and it will usually be reasonable to choose it so that it follows the course of the river in some average sense, possibly chosen to be the midpoint of the surface of the river at a particular stage, as taken from cross-sections or aerial photographs, or chosen to be the path followed by the deepest part or thalweg, but it is not necessary for it to coincide with either of these features.

Now we consider the contributions to force **T**:

Body force: The only body force acting is that due to gravity. When we use the (s,n,z) co-ordinate system with z vertical and consider only momentum in the s direction, gravity gives no contribution. The way in which it enters is from the next contribution, the pressure.

Pressure forces: The total pressure force on the fluid acting on the control surface is $-\int_{\mathrm{CS}} p\hat{\mathbf{n}}\,\mathrm{d}S$, where p is the pressure, and the negative sign shows that the local force acts in the direction opposite to the outward normal $\hat{\mathbf{n}}$. In this form it is difficult to evaluate for rivers, as the pressure and the non-constant unit vector have to be integrated over all the submerged faces of the control surface. A considerably simpler derivation is obtained if the term is evaluated using Gauss' divergence theorem of vector calculus, replacing the integral over the rather complicated control surface by a volume integral (e.g. Milne-Thomson 1968, Sect. 2.61, Eq. 1.3, but with a sign convention for $\hat{\mathbf{n}}$ opposite to ours), so that here

$$-\int_{\mathrm{CS}} p\,\hat{\mathbf{n}}\,\mathrm{d}S = -\int_{\mathrm{CV}} \nabla p\,\mathrm{d}\mathcal{V}, \tag{1.2}$$

where ∇ is the gradient operator such that ∇p is a vector whose components are the pressure gradients in each direction. Using the result from Batchelor (2000, Appendix 2), in a curvilinear coordinate system, the s component of ∇p at a general point is $(\partial p/\partial s)/(1 - \kappa n)$. However, $d\mathcal{V} = (1 - \kappa n)\Delta s\, dA$, and so we obtain the streamwise contribution

$$-\Delta s \int_A \frac{\partial p}{\partial s}\,\mathrm{d}A,$$

where A is the cross-sectional area.

Resistance: The resistance to motion comes from stress applied at the boundaries, which makes using momentum rather simpler than energy. Here we consider only the contribution from the stress component τ_s in the s direction, for which, until we have to evaluate it, we write the contribution simply as $\int_{\mathrm{Bed}} \tau_s \, \mathrm{d}S$.

Collecting contributions, the s-component of the momentum Eq. 1.1b becomes:

$$\frac{d}{\mathrm{dt}} \int_{\mathrm{CV}} \rho \, u \, d\mathcal{V} + \overline{\int_{\mathrm{CS}} \rho u \, \mathbf{u}_r \cdot \hat{\mathbf{n}} \, \mathrm{d}S} = -\Delta s \int_A \frac{\partial p}{\partial s} \mathrm{d}A + \int_{\mathrm{Bed}} \tau_s \, \mathrm{d}S. \tag{1.3}$$

At the same time we have actually taken a short-term time averaging operation on the whole equation to allow for turbulent fluctuations. If the density is constant, and turbulent fluctuations do not change the geometry, meaning the free surface, the only contribution which is nonlinear is the second one, to which we have added an overbar to indicate the process of taking a short-term mean in time. Provided that the other quantities such as u, p, and τ are time-mean values, the equation then incorporates the effects of turbulence. We have made no other approximations yet.

1.3.2 Approximation of the Equations

Now we evaluate all the integrals in the mass and momentum conservation Eqs. 1.1a and 1.3. The use of a moving control surface simplifies the evaluation of the contributions on the stream bed and on the surface, as on those parts of the possibly-moving control surface the only mass or momentum that cross them is due to inflow or outflow, whether from another stream, rain, or seepage.

1.3.2.1 Mass Conservation

The first term in Eq. (1.1a) is the rate of change of mass inside the elemental control volume. Again using $\mathrm{d}\mathcal{V} = (1 - \kappa n)\,\mathrm{d}A\Delta s$, assuming ρ constant and taking constant quantities outside the integral we obtain

$$\rho \frac{\mathrm{d}\mathcal{V}}{\mathrm{dt}} = \rho \, \Delta s \frac{\partial}{\partial t} \int_A (1 - \kappa n) \, \mathrm{d}A = \rho \, \Delta s \frac{\partial (A - \kappa A \tilde{n})}{\partial t}, \tag{1.4}$$

where A is the cross-sectional area of the channel flow, and $A\tilde{n}$ is the first moment of area of the cross-section about the vertical z axis, such that $\tilde{n}$ is the horizontal position of the centroid, as shown in Fig. 1.2.

Considering the mass rate of flow crossing the boundary $\int_{\mathrm{CS}} \rho \, \mathbf{u}_r \cdot \hat{\mathbf{n}} \, \mathrm{d}S$, on the stationary vertical faces across the flow, $\mathbf{u}_r = \mathbf{u}$, the actual fluid velocity. On the upstream face $\mathbf{u}_r \cdot \hat{\mathbf{n}} = -u$, where u is the s-component of velocity at a point, and

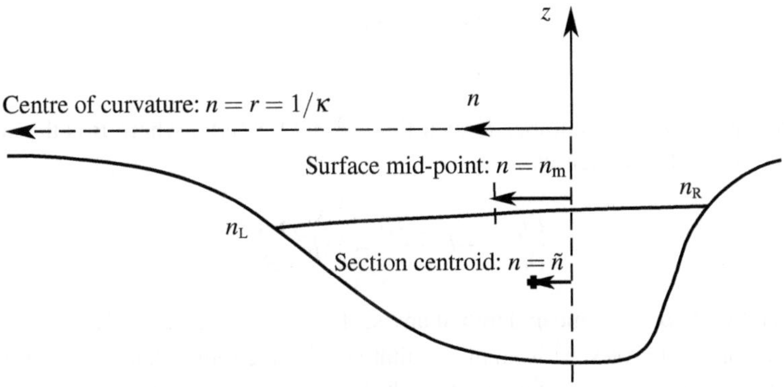

Fig. 1.2 Section of curved river, showing important dimensions and axes

the minus sign is because the velocity is opposite to the outwards normal, which is upstream here. The contribution to the integral is then $-\int_A \rho u \, dA$, which for an incompressible fluid is simply $-\rho Q$, where Q is the volume rate of flow, the discharge. The contribution to the second integral, on the downstream face, is of opposite sign and in general has changed with s. We write it as a Taylor series, giving the combination of the two:

$$-\rho Q|_s + \rho Q|_{s+\Delta s} = \rho \Delta s \frac{\partial Q}{\partial s} + \text{terms like}(\Delta s)^2. \tag{1.5}$$

Finally, an allowance is made for any fluid entering the control volume from rainfall, seepage, or tributaries, with a volume rate i per unit length and density ρ, assumed to be the same as that already in the channel. Its contribution is $-\rho \Delta s i$, negative because it is entering the control volume. Combining this contribution and those of Eqs. 1.4 and 1.5 to Eq. 1.1a, dividing by $\rho \Delta s$ and taking the limit as $\Delta s \to 0$ gives

$$\frac{\partial}{\partial t}(A - \kappa A \tilde{n}) + \frac{\partial Q}{\partial s} = i, \tag{1.6}$$

Unusually in hydraulics this is an exact equation (for a homogeneous fluid). No assumption regarding the flow has had to be made. It suggests that the integral quantities of cross-sectional area A and discharge Q are the most fundamental quantities.

It is simpler if we assume that the free surface is horizontal across the channel. For a straight stream this is usually a good approximation, and in any case it is not clear how in this one-dimensional model we would do otherwise. For curved streams, initially we make the less-restrictive approximation that the free surface is a straight line, whose slope is given by the ratio of the mean local apparent centrifugal acceleration to that of gravity, such that

$$\frac{\partial \eta}{\partial n} = -\frac{(Q/A)^2}{gr}. \tag{1.7}$$

Introducing the Froude number F such that $F^2 = Q^2B/gA^3$, where B is the surface width, this can be written as

$$\frac{\partial \eta}{\partial n} = -F^2 \frac{A/B}{r} = -F^2 \frac{\kappa A}{B}.$$

The quantity A/B is the mean depth at any section. In most rivers, $(A/B)/r$ is small, and also for most rivers F^2 is small, so that usually the cross-channel slope will be very small, and so we will neglect it, such that the elevation η is assumed constant across the stream. If we ever did require a cross-stream super-elevation after computations, we could use Eq. 1.7.

To evaluate $\partial A/\partial t$ and $\partial(A\tilde{n})/\partial t$ in Eq. 1.6 we use a simple geometric argument: consider the laterally-horizontal surface to be raised by a uniform small amount $\delta\eta$. The change in area due to the increment is $\delta A = B\,\delta\eta$. The centroid of the increment is at $n = n_{\mathrm{m}}$, the mid-point of the surface, as shown in Fig. 1.2. Hence, the change in the first moment of area about the z axis is $\delta(A\tilde{n}) = Bn_{\mathrm{m}}\,\delta\eta$. From these two increments we obtain

$$\frac{\partial A}{\partial t} = B\frac{\partial \eta}{\partial t} \quad \text{and} \quad \frac{\partial}{\partial t}(A\tilde{n}) = Bn_{\mathrm{m}}\frac{\partial \eta}{\partial t}. \tag{1.8}$$

Using Eq. 1.6 with these we obtain two versions of the mass conservation equation, one in terms of (A, Q) and for more practical purposes one in terms of (η, Q):

$$(1-\kappa n_{\mathrm{m}})\frac{\partial A}{\partial t} + \frac{\partial Q}{\partial s} = i, \tag{1.9a}$$

$$(1-\kappa n_{\mathrm{m}})\frac{\partial \eta}{\partial t} + \frac{1}{B}\frac{\partial Q}{\partial s} = \frac{i}{B}. \tag{1.9b}$$

1.3.2.2 Momentum Conservation

For the elemental control volume shown in Fig. 1.1 we now evaluate the four integrals in the momentum Eq. 1.3.

1. Unsteady term
   For the first term we obtain

$$\frac{\mathrm{d}}{\mathrm{d}t}\int_{\mathrm{CV}} \rho u \,\mathrm{d}\mathcal{V} = \rho \Delta s \frac{\partial}{\partial t}\int_{A} (1-\kappa n) u \,\mathrm{d}A.$$

Without the curvature term the integral is simply that of velocity u over the cross-section, which gives the discharge Q, exactly. In the curvature term, we approximate by assuming u constant. This is the most severe approximation we have yet made, however it is in a term that will be shown to be relatively unimportant. Again neglecting terms of magnitude $F^2\kappa A/B$ we obtain the contribution

$$\rho \Delta s (1-\kappa\tilde{n})\frac{\partial Q}{\partial t}. \tag{1.10}$$

2. Momentum flux term

The second term on the left of Eq. 1.3, $\overline{\int_{CS} \rho u \,\mathbf{u}_{\mathrm{r}} \cdot \hat{\mathbf{n}} \,\mathrm{d}S}$, has its most important contributions from the vertical transverse faces. On the upstream face, $\mathbf{u}_{\mathrm{r}} \cdot \hat{\mathbf{n}} = -u$, giving the contribution $-\rho \overline{\int_A u^2 \,\mathrm{d}A}$. The downstream face at $s + \Delta s$ has a contribution of a similar nature, but positive, all quantities having changed over the distance Δs. Again, taking a Taylor expansion gives the contribution:

$$\rho \Delta s \frac{\partial}{\partial s}\overline{\int_A u^2 \,\mathrm{d}A} = \rho \Delta s \frac{\partial}{\partial s}\int_A \left(\bar{u}^2 + \overline{u'^2}\right) \mathrm{d}A,$$

where to allow for turbulence, we assume that its fluctuations do not change the control surface, and express the time mean of the integral as the integral of the time mean of the integrand, separated into mean and fluctuating components such that $u = \bar{u} + u'$. Evaluating the integral requires a detailed knowledge of the flow distribution and its turbulent nature that is rarely available. Traditionally a Boussinesq momentum coefficient β has been introduced to allow for the non-uniformity of velocity distribution; we also use it to allow for the effects of turbulence (Fenton 2005), so that we simply write the contribution as

$$\rho \Delta s \frac{\partial}{\partial s}\left(\beta U^2 A\right) = \rho \Delta s \frac{\partial}{\partial s}\left(\beta \frac{Q^2}{A}\right), \tag{1.11}$$

where we have temporarily introduced the mean streamwise velocity $U = Q/A$. The contribution of this term is often small, which will be shown below.

The remaining contribution to momentum flux is from inflow such as tributary streams, or, less importantly, from flow seeping in or out of the ground, or from rainfall. In obtaining the mass conservation equation above, these were lumped together as an inflow i per unit length, such that the mass rate of inflow was $\rho i \Delta s$, (i.e. an outflow of $-\rho i \Delta s$). If this inflow has a mean streamwise velocity of u_i before it mixes with the water in the channel, the contribution is

$$-\rho \Delta s \beta_i i u_i, \tag{1.12}$$

where β_i is the Boussinesq momentum coefficient of the inflow. This contribution is unlikely to be known accurately or to be important, except where a significant fast-moving stream enters.

3. Pressure term

We assume that the vertical curvature of all streamlines is small, such that there are no centrifugal contributions. In passing, we note that there is a simple result which is known, but may easily be overlooked or forgotten, that in a real flowing fluid on a finite slope, the pressure p is actually *not* the widely-assumed hydrostatic value, as there is a transfer of momentum by shear throughout the flow, with a component of force directed upstream on each element of fluid, such that the free surface is inclined at an angle to the horizontal. It is a surface of constant pressure and other surfaces of constant pressure in the fluid will generally be approximately parallel to that of the surface. This leads to the expression for pressure in the fluid

$$p = \frac{\rho g(\eta - z)}{1 + (\partial\eta/\partial s)^2},$$

a known exact result for uniform flow (e.g. Chow 1959, #2–10). This cannot be termed a *hydrostatic* result, as the vertical component of resistance leading to the surfaces of constant pressure being tilted actually comes from the flow, which is not static, and that terminology is no longer available to us.

Here, for rivers, we make a small-slope approximation neglecting the term $(\partial\eta/\partial s)^2$, giving the *hydrostatic* pressure distribution, as used almost universally in open channel hydraulics, $p = \rho g(\eta - z)$. Differentiating, we find that the pressure gradient at any point is due only to the slope of the surface above, and the third contribution, the first on the right of Eq. 1.3 is

$$-\rho \Delta s g \int_A \frac{\partial\eta}{\partial s}\, \mathrm{d}A.$$

We have assumed that η is constant across the section, hence so is $\partial\eta/\partial s$ and the contribution to the momentum equation, becomes

$$-\rho \Delta s g A \frac{\partial \eta}{\partial s}. \tag{1.13}$$

Almost everywhere in a river this is a good approximation as variation is sufficiently long. It breaks down in flows over curved and steep structures.

4. Resistance forces
The forces of the boundary on the flow are incorporated using the Darcy-Weisbach formulation, which provides insights into the nature of the equations and some convenient quantifications of the effects of resistance. It is directly related to stress and force on the boundary, unlike the Gauckler-Manning formulation. Consider the Darcy-Weisbach expression for the shear force τ on a pipe wall (e.g. Sect. 6.3 of White 2009)

$$\tau = \frac{\lambda}{8} \rho V^2, \tag{1.14}$$

where λ is the Weisbach dimensionless resistance coefficient (for which the symbol f is often used, but here we follow the terminology of fundamental researchers in the field in the first half of the twentieth century), and V is the mean velocity in the pipe. Such an expression follows from a dimensional analysis of the pipe problem, suggesting its fundamental nature. We now consider the elemental channel slice shown in Fig. 1.1. An element of bed area is $(1 - \kappa n)\,\mathrm{d}P$. To evaluate the stress at any point we use Eq. 1.14 where V is now the local mean velocity in the vicinity of that part of the boundary, and where we attach a sign such that we write $-V|V|$ such that resistance is always opposite to the velocity. The fourth contribution to Eq. 1.3 becomes

$$\rho \Delta s \int_P \tau_s (1 - \kappa n)\,\mathrm{d}P = -\rho \Delta s \int_P \frac{\lambda}{8} V|V| (1 - \kappa n)\,\mathrm{d}P.$$

Our apparently-precise mathematical notation disguises somewhat the approximate physical nature of the expression. Evaluating this for a general section represents a difficult problem, as we usually do not know what the local velocity V is. As a model, we introduce a velocity distribution coefficient γ relating the local velocity to the mean velocity in the channel, such that $V|V| = \gamma Q|Q|/A^2$, where, of course, we know the variation of γ no better than that of V. This gives

$$-\rho \Delta s \frac{Q|Q|}{A^2} \int_P \frac{\lambda}{8} \gamma (1 - \kappa n)\,\mathrm{d}P \approx -\rho \Delta s \frac{Q|Q|}{A^2} \Lambda P, \tag{1.15}$$

where we have introduced the dimensionless resistance coefficient Λ

$$\Lambda = \frac{\tilde{\lambda}}{8} = \frac{1}{P}\int_P \frac{\lambda}{8}\gamma(1-\kappa n)\,\mathrm{d}P, \tag{1.16}$$

which we have shown expressed in terms of $\tilde{\lambda}$, the weighted mean of λ around the perimeter. We will be using the symbol Λ throughout this work, rather than $\tilde{\lambda}/8$. For steady uniform flow it will be shown to be related to the Chézy coefficient C as $\Lambda = g/C^2$ and it avoids the use of the number 8 in many equations. To evaluate the integral in Eq. 1.16 we usually have little idea of the variation of λ, γ, or indeed the details of the bottom topography around the section. Recognising that the mean values of both γ and the curvature term $(1-\kappa n)$ are roughly 1, under the uncertainties we will not include explicitly the curvature correction in Λ. More practically, we might write the integral as a finite sum of contributions from, say, sediment, vegetation, and different parts of the cross section, such as, say, for a glass-walled flume with a sand bed, as

$$\Lambda = \frac{g}{C^2} = \frac{\tilde{\lambda}}{8} = \frac{1}{P}\sum_i \frac{\lambda_i}{8}\gamma_i P_i. \tag{1.17}$$

The integral and sum of 1.16 and 1.17 have combined contributions to force linearly, which seems physically correct. Unfortunately in this field there has been much irrationality. Yen (2002, Table 8) lists 26 different compound or composite section formulae. Some of them some just weight contributions according to individual areas A_i, some just according to perimeters P_i. Most do not include allowance for the local velocity being different from the mean of the whole section. Some do not weight different contributions at all, but combine them imaginatively. For example, in 2003 in a major report on resistance one can find the possibilities suggested, using the overall Manning coefficient n in terms of individual contributions n_i:

$$n = \sum\nolimits_i n_i \quad \text{or} \quad n = \left(\sum\nolimits_i n_i^2\right)^{1/2} \quad \text{or} \quad \frac{1}{n} = \left(\sum\nolimits_i \frac{1}{n_i^2}\right)^{1/2}.$$

Recommendations were even given as to when each is to be preferred. Such nonsense seems not to have been publicly named as such.

Now we collect all the terms in Eq. 1.3, using contributions 1.10, 1.11 and 1.12 on the left, and terms 1.13 and 1.15 on the right. Dividing by $\rho\,\Delta s$, taking derivative terms to the left and others to the right, and expanding the fluid momentum term $\partial/\partial s(\beta Q^2/A)$, gives the balance equation for rate of change of momentum per unit mass of liquid and length of channel:

$$(1-\kappa\tilde{n})\frac{\partial Q}{\partial t}+2\beta\frac{Q}{A}\frac{\partial Q}{\partial s}-\beta\frac{Q^2}{A^2}\frac{\partial A}{\partial s}+gA\frac{\partial \eta}{\partial s}=-\Lambda P\frac{Q|Q|}{A^2}+\beta_i i u_i-\frac{Q^2}{A}\beta'(s). \tag{1.18}$$

This contains a mixture of three dependent variables, the discharge Q, the cross-sectional area A, and the surface elevation η. It is convenient to keep A as a symbol where it appears in a coefficient, but where it appears as a derivative $\partial A/\partial s$, we now express it in terms of the more practical quantity $\partial \eta/\partial s$. The calculation is surprisingly non-trivial, but it does reveal some interesting physical significance of terms in the equations.

1.3.3 Relating Area Derivative $\partial A/\partial s$ to Surface Slope $\partial \eta/\partial s$

Area A is the integral across the channel of the flow depth, from the bed $z = Z$ to the surface $z = \eta$:

$$A=\int_{n_R}^{n_L}(\eta - Z)\,\mathrm{d}n.$$

Differentiating with respect to s and using Leibniz' theorem for the derivative of an integral gives

$$\frac{\partial A}{\partial s}=\int_{n_R}^{n_L}\frac{\partial \eta}{\partial s}\,\mathrm{d}n-\int_{n_R}^{n_L}\frac{\partial Z}{\partial s}\,\mathrm{d}n+(\eta - Z)_L\frac{\partial n_L}{\partial s}-(\eta - Z)_R\frac{\partial n_R}{\partial s}. \tag{1.19}$$

We have assumed that free surface elevation η is constant across the channel, so that the first term becomes simply $(n_L - n_R)\,\partial\eta/\partial s = B\,\partial\eta/\partial s$. The second term is the integral across the channel of the downstream bed slope. We introduce the concept of $\tilde{S}$, the local mean downstream slope of the stream bed evaluated across the section:

$$\tilde{S}=-\frac{1}{B}\int_{n_R}^{n_L}\frac{\partial Z}{\partial s}\,\mathrm{d}n, \tag{1.20}$$

defined with a negative sign such that in the usual situation where the bed slopes downwards, Z decreasing with s, $\tilde{S}$ will be positive. It is probable that the bed topography is not known accurately and the bed slope would be estimated only approximately without using this expression. However, if the bottom geometry is known, the integral can be evaluated. It allows for the fact that $\partial Z/\partial s$ in a narrowing or widening stream is different from that of a prismatic one with of the same thalweg slope. Thus it incorporates what in other presentations is referred to as the non-prismatic contribution. This seems never to have been specifically evaluated and presented in any textbook or reference book that the author has read, but is usually written vaguely using something like $1/B\ \partial A/\partial x|_h$, suggesting it is the contribution if some mean thalweg depth is held constant. Its incorporation here as a contribution to mean bed slope seems simple. In any case, the mean bed slope is often poorly known, the term in the momentum Eq. 1.18 where it appears is multiplied by β, whose value is only approximately-known, and in any case it will be shown below that the contribution of the whole term is often small. All this shows that the bed slope term could be calculated precisely, but it is much more likely that it is able just to be estimated approximately, which is what is done in practice.

The remaining two terms in Eq. 1.19 are denoted by A_s^{V} which is the contribution to $\partial A/\partial s$ from vertical converging or diverging side walls:

$$A_s^{\mathrm{V}} = (\eta - Z)_{\mathrm{L}} \frac{\partial n_{\mathrm{L}}}{\partial s} - (\eta - Z)_{\mathrm{R}} \frac{\partial n_{\mathrm{R}}}{\partial s}, \tag{1.21}$$

where $(\eta - Z)_{\mathrm{L,R}}$ is the water depth at the left and right banks which are zero if the sides of the stream are not vertical, the usual case for natural streams and most canals, and so A_s^{V} is zero in this usual case. If, on the other hand, the sides are vertical, such as in a flume, they usually neither converge nor diverge ($\partial n_{\mathrm{L}}/\partial s = \partial n_{\mathrm{R}}/\partial s = 0$), when the term is also zero. Hence this term A_s^{V} contributes only for vertical-walled variable-width channels such as in Parshall flumes.

Combining the contributions to $\partial A/\partial s$ in Eq. 1.19 gives

$$\frac{\partial A}{\partial s} = B\frac{\partial \eta}{\partial s} + B\tilde{S} + A_s^{\mathrm{V}}. \tag{1.22}$$

1.3.4 The Long Wave Equations

The resulting one-dimensional long wave equations in terms of (η, Q), taking the mass conservation Eq. 1.9b and substituting Eq. 1.22 into Eq. 1.18, are

$$(1 - \kappa n_{\mathrm{m}})\frac{\partial \eta}{\partial t} + \frac{1}{B}\frac{\partial Q}{\partial s} = \frac{i}{B}, \tag{1.23a}$$

$$\begin{aligned}(1 - \kappa\tilde{n})\frac{\partial Q}{\partial t} &+ 2\beta\frac{Q}{A}\frac{\partial Q}{\partial s} + \left(gA - \beta\frac{Q^2 B}{A^2}\right)\frac{\partial \eta}{\partial s} \\ &= \beta\frac{Q^2}{A^2}\left(B\tilde{S} + A_s^{\mathrm{V}}\right) - \Lambda P\frac{Q|Q|}{A^2} + \beta_i i u_i - \frac{Q^2}{A}\beta'(s).\end{aligned} \tag{1.23b}$$

1.3.5 Discussion

- A complicated flow problem has been reduced to two quite simple first-order partial differential equations.
- There are few numerical parameters to approximate. The only really important one is the resistance coefficient Λ, which in fact is just the symbol for a complicated contribution as defined in Eqs. 1.16 and 1.17, and is usually poorly-known. The momentum coefficient β is usually known to within 5 %, say, but is in a term which is usually not important, as we shall see. Finally there is the rarely-required inflow momentum coefficient β_i, in a term that is also small if not non-existent.
- The resistance term has appeared as $-\Lambda PQ|Q|/A^2$. This use of the coefficient Λ here is non-standard, but the resistance term appears simply as the coefficient Λ multiplied by the square of the velocity to give the mean stress per unit mass on the bed, and multiplied by wetted perimeter P to give the force per unit mass and length of the stream. The significance seems clear and concise. Also, Λ has a simple relationship to Chézy's C, $\Lambda = g/C^2$, as will be shown. If one had used the Gauckler-Manning formulation, a complicated expression would be required. Alternatively, if one had used the actual Weisbach resistance coefficient, it would have appeared as $\tilde{\lambda}/8$, and the factor of 8 would then appear obtrusively in a number of places in the theoretical discussion below. Another alternative would have been to have used the symbol S_f, the ratio of resistance force to gravitational force, when the resistance term would have appeared as $-gAS_f$, introducing gravity into a term where it does not belong and where S_f does not reveal the nature of the term. The use of Λ seems to have advantages over these other formulations.
- The definition of local mean downstream slope $\tilde{S}$ allows evaluation for general geometries. However in most problems the details of the bed geometry are poorly known, and a gross approximation to the slope is used. There is no need to go to much trouble to evaluate any "non-prismatic" contribution.
- The bed geometry does appear importantly in the form of the cross-sectional area A, surface width B, and wetted perimeter P. Often these are not particularly well-known.
- The equations will enable us to find surface elevation and mean velocity in the stream, but they can do nothing to calculate primary or secondary velocity distributions, whether in a straight or curved channel.
- Here we have presented the equations in terms of the most practical quantities of (1) discharge Q, rather than mean velocity U, because inflows from floods and control structure formulae are all in terms of discharge, and (2) surface elevation η because the surface of water bodies, heads over control structures, and irrigation applications all use it. For straight channels, in Sect. 1.3.8 we also present a pair of equations in terms of Q and A, which are more concise, and useful for theoretical purposes. In Sect. 1.3.9 we present the differential equation for steady flow in terms of a depth-like quantity.

In all the presentations of long wave equations in books and research papers that the present author has read, none has actually presented the two equations in just two dependent variables, which is necessary for solution and for theory. Every presentation has used three such variables, either mean velocity U or discharge Q plus any one of the three related variables η, area A, and a variable which is called "depth", a local vertical co-ordinate which is the surface elevation relative to what is usually described as the bottom of the channel, even if that is a highly ambiguous quantity in a river. Similarly, nowhere in those many sources is there an explicit evaluation of the non-prismatic term as included in the generalised slope definition, Eq. 1.20. People who write hydraulic software, however, presumably have confronted these problems and developed the equations to a useable form.

1.3.6 Assumptions and Approximations

The equations are more accurate than often supposed. They have been obtained with very few really limiting approximations. These are, in approximate decreasing order of severity for applications to rivers:

- The resistance around the wetted perimeter, possibly including particles of varying size, bedforms, obstacles, and vegetation, is expressed in terms of a single coefficient. This is a severe approximation, however we have shown how the resistance of a compound section could be calculated, if one were in possession of all physical information. Usually that is not the case, and the approximation provides a convenient simple model, albeit a not well-known one. That is the only really important approximation. *Below this, the approximations are usually very good ones, or have little effect on results.*
- The free surface is horizontal across the section. This would not be so true in compound sections with flood plains. For a curved river it means that we have neglected terms of the order of the square of the Froude number multiplied by the ratio of mean depth to radius of curvature, which is almost always small.
- The momentum flux over the whole section is expressed in terms of a single coefficient β that encapsulates the effects of the velocity distribution and turbulence, which is known only to some 5 %. The momentum flux term will be shown to be, in any case, often unimportant.
- The slopes of the stream bed and surface are small.
- The longitudinal curvatures of the stream bed and surface are small, such that there are no centrifugal effects on pressure.
- The planform curvature effect in the time-derivative term in the momentum equation has been obtained by assuming velocity constant over the section. For a straight stream the term is exact.
- Fluid density is constant
- Turbulent fluctuations have no effect on the free surface.

1.3.7 Effects of Horizontal Curvature

- The equations with curvature in the horizontal plane are not much more complicated than the straight-channel equations. It is surprising how little they are affected by curvature. This compares with the rather more complicated equations obtained by Fenton and Nalder (1995), where cross-channel slope of the surface was included, and there were a number of extra terms. These were all of the order of $F^2 \times (A/B)/r$, the square of the Froude number multiplied by the ratio of mean depth to the radius of curvature. In meandering rivers, both are usually small, and the approximation that the free surface is horizontal across the channel is a good one. For curved high-speed flow in a flume it would be necessary to include the extra terms.
- In the present approximation, effects of curvature appear only in the time derivative terms and are simply expressed in terms of the dimensionless length ratios $\kappa\tilde{n} = \tilde{n}/r$, and $\kappa n_m = n_m/r$, which are expressions of the lateral displacement, relative to the longitudinal axis, of the centroid and the centre of the surface. As one might choose the longitudinal axis from a plan or aerial photograph, the obvious place for it is at the visual centre of the stream, such that $n_m = 0$ for a particular flow. For other flows, the centre might move laterally.
- If the stream cross-section is symmetric, when one would of course put the s axis at the centre of the section, then $\tilde{n} = 0$ also, and there are no effects of curvature in the governing equations here.
- On the other hand, in meandering rivers with moveable beds, cross-sections can be asymmetric and bends are deeper on the outside, whether turning to left or right. The centroid at $\tilde{n}$ is usually further from the centre of curvature than the middle of the surface n_m, as shown in Fig. 1.2. For a left-turning stream κ is positive, and if one places the longitudinal axis such that $n_m = 0$, $\tilde{n}$ will be negative, so that $\kappa\tilde{n}$ will be negative. For a right-turning stream, κ negative, $\tilde{n}$ will be positive, so that $\kappa\tilde{n}$ will again be negative. The effects do not cancel as a stream turns to left and right, they are consistently such that the coefficient $1 - \kappa\tilde{n}$ is greater than 1.
- However, that coefficient multiplies a time derivative in the momentum equation. We are about to consider the common approximation of a straight channel. It will be shown that for many streams, where disturbances are due to floods with a relatively large time scale, the time derivative term is indeed unimportant, and hence so is curvature. We will return to the effects of curved channels in Sect. 1.5.6 and show quantitatively that the effects of curvature on wave propagation can usually be avoided or ignored.

1.3.8 The Common Case—Straight Channel, No Inflow Momentum, No Vertical Converging Walls, Constant β

If we make the traditional approximation that the stream is straight, $\kappa = 0$, and with no vertical diverging/converging walls, no inflow momentum, and β constant, and now using x rather than s for the horizontal longitudinal co-ordinate, the long wave equations become:

$$\frac{\partial \eta}{\partial t} + \frac{1}{B}\frac{\partial Q}{\partial x} = \frac{i}{B}, \tag{1.24a}$$

$$\frac{\partial Q}{\partial t} + 2\beta\frac{Q}{A}\frac{\partial Q}{\partial x} + \left(gA - \beta\frac{Q^2 B}{A^2}\right)\frac{\partial \eta}{\partial x} = \beta\frac{Q^2}{A^2}B\tilde{S} - \Lambda P\frac{Q|Q|}{A^2}. \tag{1.24b}$$

For theoretical purposes we also obtain the equations in terms of A and Q as dependent variables, using Eq. 1.8: $\partial\eta/\partial t = 1/B \times \partial A/\partial t$ and Eq. 1.22: $\partial\eta/\partial x = 1/B \times \partial A/\partial x - \tilde{S}$. Regrouping terms, the equations can be written in a rather simple concise form:

$$\frac{\partial A}{\partial t} + \frac{\partial Q}{\partial x} = i, \tag{1.25a}$$

$$\frac{\partial Q}{\partial t} + \frac{\partial}{\partial x}\left(\beta\frac{Q^2}{A}\right) + \frac{gA}{B}\frac{\partial A}{\partial x} = gA\tilde{S} - \Lambda P\frac{Q|Q|}{A^2}. \tag{1.25b}$$

We mention a result for steady uniform flow that might be of interest. Firstly setting all derivatives in Eq. 1.25b to zero we obtain the Chézy-Weisbach equation:

$$U = \frac{Q}{A} = \sqrt{\frac{g}{\Lambda}\frac{A}{P}S}, \tag{1.26}$$

where we use $\Lambda = \tilde{\lambda}/8$, and where $\tilde{S} = S$ in this uniform slope case. That result is expected. What is interesting is to re-write this in terms of Froude number F as

$$F^2 = \frac{Q^2 B}{gA^3} = \frac{S}{\Lambda}\frac{B}{P}, \tag{1.27}$$

and as $B \approx P$ for wide channels, the square of the Froude number is approximately given by the ratio of slope S to resistance coefficient Λ, giving some significance and physical feeling for the resistance coefficient Λ. This means that for a particular reach of river, where slope S is effectively independent of flow, and where B/P also does not vary much with the flow, and neither does the resistance coefficient Λ, the Froude number F does not change much at a section, and is a characteristic number there for all flows. While a flood flow might look more dramatic than a more-common low flow, being faster and higher, the Froude number is roughly the same for both.

1.3.9 Steady Gradually-Varied Flow Equations

For the important case of steady flow, the equations become a pair of ordinary differential equations. In this case the mass conservation equation has the solution $Q = Q(x_0) + \int_{x_0}^{x} i(x')\mathrm{d}x'$. If there is no distributed inflow i, then the solution is $Q = Q(x_0) =$ constant. We can now use $Q|Q| = Q^2$ in the momentum equations as the flow is unidirectional, and here we consider only $A_x^{\mathrm{V}} = 0$ and $\beta'(x) = 0$ to give, from Eq. 1.24b, the gradually-varied flow equation for η:

$$\frac{\mathrm{d}\eta}{\mathrm{d}x} = \frac{\beta\tilde{S} - \Lambda\frac{P}{B}}{1/F^2 - \beta}. \tag{1.28}$$

This is usually presented in terms of a local depth-like quantity h, as section properties A, B, and P are usually so specified. However presentations do not define how h is determined in a natural channel, where the thalweg might not be known well. We introduce a reference axis $Z_0(x)$, which might be the real thalweg elevation if that were known, but it is more likely to be simply a straight line of constant slope, so that $\eta = h + Z_0$ and $\partial\eta/\partial x = \partial h/\partial x - S_0$, where $S_0 = -\partial Z_0/\partial x$ is the local slope of the axis (positive in the usual downward-sloping channel sense). Equation 1.29 becomes

$$\frac{\mathrm{d}h}{\mathrm{d}x} = \frac{S_0 + \beta(\tilde{S} - S_0)F^2 - \Lambda\frac{P}{B}F^2}{1 - \beta F^2}. \tag{1.29}$$

Both Eqs. 1.28 and 1.29 are valid for non-prismatic channels, using the generalised definition of slope $\tilde{S}$ in Eq. 1.20. Other presentations usually give an equation like 1.29 for prismatic channels, such that $\tilde{S} = S_0$, and where the symbol S_f is used for the resistance term. We prefer to keep the term explicit; it shows how simple it is and does not confuse us that we might be using energy. It is interesting that for wide channels, $P \approx B$, all variation with the dependent variable on the right of the differential Eqs. 1.28 and 1.29 is in the occurrences of $F^2 = Q^2B/gA^3$.

1.4 Non-dimensional Long Wave Equations and the Nature of Approximations

Here we non-dimensionalise the long wave equations to determine the relative magnitudes of terms in the equations. It will be seen that previous approaches have led to incorrect understanding of the nature of approximations to the equations and indeed to incorrect nomenclature.

Consider the long wave Eqs. 1.25a, b in terms of A and Q for a straight channel. We introduce dimensionless quantities denoted by asterisks, such that for the independent variables $t = t_* T$, and $x = x_* L$, where T and L are time and length scales, respectively. The dependent variables are scaled as $A = A_* A_0$ and $Q = Q_* U_0 A_0$, where A_0 is a cross-sectional area scale and U_0 is the scale of the mean velocity in a steady uniform reference flow, and all subsequent quantities with subscript 0 are those of the reference flow. For channel width we use the width scale B_0 such that $B = B_* B_0$, and for the perimeter we write $P = P_* P_0$, in terms of a perimeter scale P_0, which could well be the same as B_0 but it gives a simpler result if we use a different symbol. In addition we write the channel slope and resistance coefficient, even though they are already dimensionless, in terms of reference values with a 0 subscript as $\tilde{S} = S_0 S_*$, and $\Lambda = \Lambda_0 \Lambda_*$. The equations we obtain are

$$\frac{L}{U_0 T}\frac{\partial A_*}{\partial t_*} + \frac{\partial Q_*}{\partial x_*} = 0, \tag{1.30a}$$

$$\frac{U_0}{g S_0 T}\frac{\partial Q_*}{\partial t_*} + \frac{U_0^2}{g S_0 L}\frac{\partial}{\partial x_*}\left(\beta \frac{Q_*^2}{A_*}\right) + \frac{A_0/B_0}{S_0 L}\frac{A_*}{B_*}\frac{\partial A_*}{\partial x_*} + \Lambda_* P_* \frac{Q_*^2}{A_*^2} - A_* S_* = 0, \tag{1.30b}$$

where we have used the uniform flow Eq. 1.26 to write Λ_0 in terms of U_0. All starred (*) quantities have a scale of order 1.

In the traditional approach (e.g. Woolhiser and Liggett 1967, and subsequent researchers), the length and time scales of disturbances, and hence their velocity of propagation, are then assumed related by the mean fluid velocity, such that $T = L/U_0$, so that the equations become:

$$\frac{\partial A_*}{\partial t_*} + \frac{\partial Q_*}{\partial x_*} = 0,$$

$$F_0^2\left(\frac{\partial Q_*}{\partial t_*} + \frac{\partial}{\partial x_*}\left(\beta \frac{Q_*^2}{A_*}\right)\right) + \frac{A_*}{B_*}\frac{\partial A_*}{\partial x_*} + \frac{S_0 L}{A_0/B_0}\left(\Lambda_* P_* \frac{Q_*^2}{A_*^2} - A_* S_*\right) = 0,$$

where F_0 is the Froude number of the steady uniform flow used to relate the scales, $F_0^2 = U_0^2 B_0 / g A_0^3$.

Now we consider the well-known approximation to the momentum equation where both the time derivative term and the momentum flux term, identified by β, are neglected. This is always justified by stating that it is accurate for F_0^2 small, which is the coefficient multiplying those terms, and the approximation is identified as a small Froude number one. Hence, the equation with the remaining terms has been called the "kinematic" or "low inertia approximation".

However, the assumption that L and T are related by $T = L/U_0$ is, in general, not correct. For example, if waves were to travel at a speed of $\sqrt{gA_0/B_0}$, often presumed to be the case, then $T = L/\sqrt{gA_0/B_0}$ is more appropriate. In fact, as we will see below, L is a complicated function of T. There is no relationship like $T = L/U_0$ for us to use. We assert that T is imposed on the stream from outside in the form of a boundary condition, such as that of a flood hydrograph or a gate movement, and the channel responds with that value everywhere. Retaining T as a parameter, the equations become

$$\frac{\partial A_*}{\partial t_*} + \frac{U_0 T}{L}\frac{\partial Q_*}{\partial x_*} = 0, \tag{1.31a}$$

$$\frac{U_0}{gS_0T}\left(\frac{\partial Q_*}{\partial t_*} + \frac{U_0 T}{L}\frac{\partial}{\partial x_*}\left(\frac{\beta Q_*^2}{A_*}\right)\right) + \frac{A_0/B_0}{S_0 L}\frac{A_*}{B_*}\frac{\partial A_*}{\partial x_*} + \Lambda_* P_* \frac{Q_*^2}{A_*^2} - A_* S_* = 0 \tag{1.31b}$$

There is a factor U_0T/L in front of the x_* derivative grouped with the time derivative in each equation. We could have written the factor as U_0/c, where $c = L/T$ is the wave speed. We will see that it is a function of the wave period imposed; it always has a finite value.

More importantly, we see a quite different result in the momentum equation, where the time derivative and the fluid momentum term now have a coefficient $U_0/(gS_0T)$, such that neglecting those terms is not a "low inertia approximation" for F_0^2 small, but is actually valid when $U_0/(gS_0T)$ is small, that is, when the time scale T is large relative to $U_0/(gS_0)$. Hence, the well-known approximation of neglecting the time derivative and fluid inertia terms in the equation is actually a *very long-period wave approximation*, or, a *slow change approximation*, expressing the behaviour in time brought about by boundary conditions.

In view of this, the term "kinematic wave", originating with Lighthill and Whitham (1955) seems to be a misnomer. It would be better to use the terminology *very long-period wave.*

Equation 1.31b suggests, for sufficiently slow change in input conditions, and hence conditions in the stream, we could retain just the last three terms in the momentum Eq. 1.25b in (A, Q) or just those terms in Eq. 1.24b in (η, Q) which do not contain a time derivative or a coefficient β. We will examine this approximation in Sect. 1.5.5 and present the results of calculations that show clearly that the approximation is indeed one of slow variation, and possibly surprisingly, that it is valid for any Froude number—provided waves only propagate downstream, such as flood waves. For upstream propagation problems where a downstream boundary condition changes, such as tides in the sea, water levels in a reservoir, or downstream gate operations, then c is negative, and it and the wavelength can become small if the waves struggle against the current. This means that U_0T/L becomes large and the fluid momentum term can no longer be neglected. For the moment we note we have to change our newly-coined terminology and suggest the nomenclature for the use of the $\partial/\partial t = 0$ and $\beta = 0$ approximation to the momentum equation as the *Slow-Change-and-for-upstream-propagation-Slow-Flow* approximation, simplified to *Slow-change/Slow flow approximation.*

1.5 The Nature of Long Waves and Their Approximations

To obtain the actual behaviour of waves we linearise the long wave equations about a uniform steady flow and obtain mathematical solutions. This was done by Lighthill and Whitham (1955) and more recently by Ponce and Simons (1977). The important information of those works, giving the propagation behaviour of waves, and how wave speed and diminution are functions of wave length/period has apparently had not enough impact on the popular perception, and a common statement in textbooks is something like "*the* long wave speed $\sqrt{gA/B}$". Here we will see that there is no such thing as a long wave speed, as well as some other possible surprises.

1.5.1 Linearisation

We consider relatively small disturbances about a uniform flow with area A_0 and discharge Q_0 and write

$$A = A_0 + \varepsilon A_1 \quad \text{and} \quad Q = Q_0 + \varepsilon Q_1,$$

where ε is a small quantity. As A_0 and Q_0 are constant, all derivatives of A and Q in Eqs. 1.25a, b are of order ε so that the coefficients need only be written to zeroeth order, and the linearisation is simple. The only non-trivial operations are in the remaining resistance and slope terms on the right of Eq. 1.25b. We take them to the left of the equation and introduce the function ϕ:

$$\phi = \Lambda P \frac{Q^2}{A^2} - gAS_0, \tag{1.32}$$

where we have written $Q|Q| = Q^2$, as we consider small perturbations about a uniform flow that is unidirectional. The slope S_0 is assumed constant. The linearised equations are

$$\frac{\partial A_1}{\partial t} + \frac{\partial Q_1}{\partial x} = 0, \tag{1.33a}$$

$$\frac{\partial Q_1}{\partial t} + \left(C_0^2 - \beta^2 U_0^2\right)\frac{\partial A_1}{\partial x} + 2\beta U_0 \frac{\partial Q_1}{\partial x} + \left.\frac{\partial \phi}{\partial Q}\right|_0 Q_1 + \left.\frac{\partial \phi}{\partial A}\right|_0 A_1 = 0, \tag{1.33b}$$

in which the symbol for the mean fluid velocity $U_0 = Q_0/A_0$ has been used for simplicity, as well as the wave speed quantity $C_0 = \sqrt{gA_0/B_0 + \left(\beta^2 - \beta\right)U_0^2}$.

We will see that the term $\partial\phi/\partial Q|_0$ is a fundamental inverse time scale for long wave motion in channels. We introduce the symbol σ_0 for it, and evaluate it using

Eq. 1.32, neglecting any dependence of Λ on Q, which would appear if Λ were to depend on Reynolds number. Then we use the Chézy-Weisbach Eq. 1.26 for the reference flow: $U_0 = \sqrt{g/\Lambda_0 \times A_0/P_0 \times S_0}$ to eliminate Λ_0 in favour of U_0 to give

$$\sigma_0 = \left.\frac{\partial \phi}{\partial Q}\right|_0 = 2\frac{gS_0}{U_0}. \tag{1.34}$$

Expressing σ_0 in terms of the dependent quantity U_0 might seem strange, but this result is important for us—the right side appears in the leading coefficient of the dimensionless momentum Eq. 1.31b, so that the leading coefficient U_0/gS_0T of that equation, which we used to justify the terminology "Slow change approximation", can actually be written $2/(\sigma_0 T)$, where σ_0 has dimensions of T^{-1}, and we see that it is an important scale for wave behaviour. This will be made more specific below.

Now we use the Chézy-Weisbach Eq. 1.26 for the reference flow again, this time to eliminate U_0 and express the result in terms of the fundamental channel properties of resistance coefficient Λ_0, slope S_0, and depth scale A_0/P_0:

$$\sigma_0 = 2\sqrt{\frac{g\Lambda_0 S_0}{A_0/P_0}}. \tag{1.35}$$

We could argue by a rough electrical analogy that the resistance term in the momentum equation is equivalent to potential difference or voltage, while discharge Q is equivalent to current. As the derivative of voltage with respect to current gives electrical resistance, σ_0 can be thought of as a *resistance parameter* in our nonlinear case.

For the other derivative we obtain

$$\left.\frac{\partial \phi}{\partial A}\right|_0 = -3gS_0\left(1 - \frac{1}{3}\left(\frac{A_0}{P_0}\left.\frac{\partial P}{\partial A}\right|_0 + \frac{A_0}{\Lambda_0}\left.\frac{\partial \Lambda}{\partial A}\right|_0\right)\right). \tag{1.36}$$

In this case, both derivative terms in $\partial P/\partial A$ and $\partial \Lambda/\partial A$ in general may have finite contributions, the first expressing the effect of finite channel width, and the second because the resistance coefficient varies with depth and hence area, as we shall see.

It is mathematically and physically simpler if the equations are combined into a single equation by introducing a function $v(x,t)$ where $A_1 = \partial v/\partial x$ and $Q_1 = -\partial v/\partial t$, so that v satisfies the linear mass conservation Eq. 1.33a identically. The linearised momentum Eq. 1.33b becomes a Telegraph or Telegrapher's equation, a single equation in a single dependent variable:

$$\sigma_0\left(\frac{\partial v}{\partial t} + c_0\frac{\partial v}{\partial x}\right) + \frac{\partial^2 v}{\partial t^2} + 2\beta U_0\frac{\partial^2 v}{\partial t \partial x} - (C_0^2 - \beta^2 U_0^2)\frac{\partial^2 v}{\partial x^2} = 0. \tag{1.37}$$

We have introduced the symbol c_0, which is the ratio of the two derivatives

$$c_0 = -\frac{\partial\phi/\partial A|_0}{\partial\phi/\partial Q|_0},$$

and using Eqs. 1.34 and 1.36 gives:

$$c_0 = \frac{3}{2}U_0\left(1 - \frac{1}{3}\left(\frac{A_0}{P_0}\frac{\partial P}{\partial A}\bigg|_0 + \frac{A_0}{\Lambda_0}\frac{\partial \Lambda}{\partial A}\bigg|_0\right)\right) = \alpha U_0, \tag{1.38}$$

where for convenience, the symbol α has been introduced, which is roughly 3/2 as shown. The quantity c_0 is a wave velocity, which we are about to show is actually the velocity of propagation of very long waves.

1.5.2 Nature of Wave Propagation in Limiting Cases

The Telegraph equation is a partial differential equation of hyperbolic type, according to the usual classification of such equations using the coefficients of the second derivatives. The first order derivatives, multiplied by the resistance parameter σ_0, give it a dissipative nature, as we shall subsequently see. At this stage it is possible to anticipate results that will be established more fully below. The terms in the Telegraph equation 1.37 are in two groups: two first derivatives multiplied by σ_0, and three second derivative terms. We now examine the equation in two limits:

Very long waves: For disturbances that have a slow variation, such that $\partial v^2/\partial t^2 \ll \sigma_0 \partial v/\partial t$, which we will call "very long waves", with the same inequality for the other second derivatives and their coefficients, the last three terms in Eq. 1.37 can be neglected, and the equation becomes approximately

$$\frac{\partial v}{\partial t} + c_0\frac{\partial v}{\partial x} = 0.$$

This has a general solution $v = f_1(x - c_0 t)$, where $f_1(.)$ is an arbitrary function given by the initial and boundary conditions. This solution is a wave propagating downstream at speed $c_0 = \alpha U_0$. Such a wave is known as a "kinematic wave" and the equation is widely known as the "kinematic wave equation". However, no approximation has been made here by neglecting dynamical terms leaving supposedly kinematic or non-dynamic terms. We have shown instead that it is actually a very long wave approximation, as we foresaw in Sect. 1.4.

Not-so-long waves: In the other limit, for disturbances which are shorter, we use the term "not-so-long" waves, where $\partial^2/\partial t^2 \gg \sigma_0 \partial/\partial t$. Other second derivative contributions are of a similar dominant scale, and so Eq. 1.37 is approximated by

$$\frac{\partial^2 v}{\partial t^2} + 2\beta U_0 \frac{\partial^2 v}{\partial x \partial t} - (C_0^2 - \beta^2 U_0^2) \frac{\partial^2 v}{\partial x^2} = 0,$$

which is a second-order wave equation with solutions

$$v = f_{21}(x - (\beta U_0 + C_0)t) + f_{22}(x - (\beta U_0 - C_0)t)$$

where $f_{21}(.)$ and $f_{22}(.)$ are arbitrary functions determined by initial and boundary conditions. In this case the solutions are waves propagating upstream and downstream at velocities of $\beta U_0 \pm C_0$, such that in the usual terminology C_0 is "the long wave speed". We have shown here that it is actually the speed of waves that are not so long, apparently paradoxically: they are long enough that the pressure distribution in the fluid is still hydrostatic, but are in the short wave limit of such waves according to our criterion using the scale of time variation and the resistance parameter σ_0.

1.5.3 Solutions for Waves Periodic in Time

We now obtain a solution of the Telegraph equation valid for all time scales. We make the equation dimensionless by introducing dimensionless variables $t_* = \sigma_0 t$ and $x_* = \sigma_0 x / U_0$, and consider v to have been made dimensionless in some manner that does not matter, as the equation is homogeneous, and v occurs linearly in every term. It becomes

$$\frac{\partial v}{\partial t_*} + \alpha \frac{\partial v}{\partial x_*} + \frac{\partial^2 v}{\partial t_*^2} + 2\beta \frac{\partial^2 v}{\partial x_* \partial t_*} - (F_0^{-2} - \beta) \frac{\partial^2 v}{\partial x_*^2} = 0. \tag{1.39}$$

As the physical parameters α and β are constant for a particular channel, the equation contains a single parameter determining the nature of solutions, the Froude number in the form F_0^{-2}, such that it is actually a dimensionless gravity number.

We can represent any input to our system by a Fourier series in time, so we start by obtaining solutions for a single wave, periodic in time, of the form

$$v = \exp(\mathrm{i}(\kappa x_* - \omega t_*)),$$

where $\mathrm{i} = \sqrt{-1}$, and where the frequency ω is real but in general κ is a complex quantity whose nature determines the behaviour of the solutions in space. Substituting this solution into the dimensionless Telegraph Eq. 1.39 and dividing through by common terms gives the equation which is quadratic in κ:

$$\omega - \alpha\kappa - \mathrm{i}\omega^2 + 2\mathrm{i}\kappa\beta\omega + \mathrm{i}\kappa^2 (F_0^{-2} - \beta) = 0, \tag{1.40}$$

where the order of the terms corresponds to that in Eq. 1.39. The solution is

$$\kappa = \kappa_r + i\kappa_i = \frac{-(i\alpha/2+\beta\omega)+\varepsilon\sqrt{(i\alpha/2+\beta\omega)^2+(F_0^{-2}-\beta)(\omega^2+i\omega)}}{F_0^{-2}-\beta}, \tag{1.41}$$

where $\varepsilon = \pm 1$, which corresponds to downstream/upstream propagation. In terms of the physical independent variables the solution is

$$v = \exp\left(-\frac{\sigma_0\kappa_i}{U_0}x\right)\exp\left(i\frac{\sigma_0\kappa_r}{U_0}\left(x-\frac{\omega U_0}{\kappa_r}t\right)\right),$$

so that the physical propagation velocity is $\omega U_0/\kappa_r$, for which we will use the symbol c. From Eq. 1.41 κ_r depends on ω, and we have the important physical result that a wave periodic in time will propagate at a velocity dependent on the wave period/frequency, $c = \omega U_0/\kappa_r(\omega)$, and will decay in space due to resistance with a decay rate $\sigma_0\kappa_i(\omega)/U_0$, also dependent on wave period/frequency. We have seen in both "very long" and "not-so-long" limits that the behaviour becomes independent of period.

Now we present a figure showing the variation of wave velocity with period. To estimate over what range we should present solutions we consider Eq. 1.34 for σ_0, which gives $\sigma_0 T = 2gS_0T/U_0$ and take $g \approx 10\ \mathrm{ms}^{-2}$, and a representative value of $U_0 \approx 1\ \mathrm{ms}^{-1}$. To estimate the larger magnitudes of $\sigma_0 T$, we consider a flood that takes a day to rise ("time-to-rise") such that the period of the corresponding sine wave might be $T = 4$ days. On a stream of steeper gradient $S_0 = 10^{-3}$, this gives

$$\sigma_0 T = 2\frac{gS_0}{U_0}T \approx 2\times\frac{10\times 10^{-3}}{1}\times 4\times 24\times 3600 \approx 7000.$$

For the smallest likely value, we consider the fast movement of a gate on an irrigation canal on a gentle slope $S_0 = 10^{-4}$, with an opening time of 5 min such that $T = 20$ min, giving

$$\sigma_0 T = 2\frac{gS_0}{U_0}T \approx 2\times\frac{10\times 10^{-4}}{1}\times 20\times 60 \approx 2.4.$$

In view of these values we plot results over a range of dimensionless wave periods $1 \le \sigma_0 T \le 10^4$.

Figure 1.3 shows the dimensionless propagation velocity $c/\sqrt{gA_0/B_0}$ as a function of dimensionless period $\sigma_0 T$, with Froude number as parameter, as calculated from Eq. 1.41 for wide channels using $\alpha = 3/2$ and $\beta = 1$. The set of solid curves for positive c are for downstream propagation, setting $\varepsilon = +1$ in Eq. 1.41, those for negative c for upstream with $\varepsilon = -1$. The dashed lines will be explained further below.

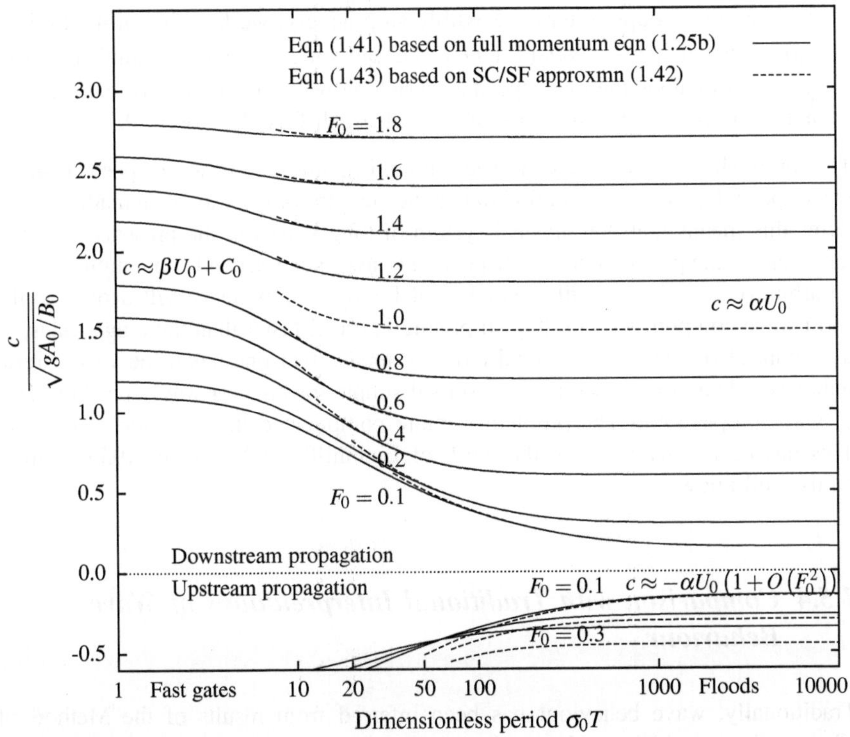

Fig. 1.3 Dependence of wave velocity on wave period and Froude number, with the slow-change/slow-flow approximation added for comparison

Not-so-long waves: for $\sigma_0 T < 2$, say, near the left of the figure, corresponding to rapid changes such as those due to gate movements, the curves are almost horizontal and as shown in Sect. 1.5.2, $c \approx \beta U_0 \pm C_0$ is a good approximation to the wave velocity for both downstream and upstream propagation, as widely believed in the hydraulics literature.

Intermediate length, $2 < \sigma_0 T < 200$ or 500: $\beta U_0 \pm C_0$ is no longer a good approximation and the propagation velocity continuously depends on wave period.

Very long waves, $\sigma_0 T > 200$ or 500: the curves are again almost horizontal, but for downstream propagation with a propagation velocity $c \approx c_0 = \alpha U_0$, as shown in Sect. 1.5.2, and more honoured in the hydrology literature. The conventional view is that this is from a low-inertia or small Froude number approximation. Figure 1.3 shows the surprising result that it is valid for all Froude numbers, as we foresaw in Sect. 1.5.2, and that the approximation is actually one of slow variation, for sufficiently long period waves.

For upstream propagation, when velocity of propagation is negative, in this very long wave limit our simple arguments in Sect. 1.5.2 as to when first and second derivative terms dominate, seem no longer to apply, and there are interactions between the groups of terms. Possibly surprisingly, we find that in the limit as $\sigma_0 T \to \infty$, that $c \approx -\alpha U_0 \left(1 + O\left(F_0^2\right)\right)$, and for small Froude number, waves travel upstream at the same speed as very long waves travel downstream. This is not obvious from the structure of the Telegraph Eqs. 1.37 or 1.39.

The figure shows that there is no such thing as a single long wave speed. Instead, wave speed depends on the period (and hence length) of the waves. In mathematical terms this means that the physical system of long waves in the presence of resistance shows the phenomenon of dispersion (wave speed depends on length), so that an arbitrary disturbance with a number of Fourier components will evolve as the short wave components, possibly surprisingly, travel faster than the longer ones, as the figure also shows. In practical terms, it means that one has to be very careful with any calculations based on an assumed simple formula for wave speed or ideas of wave propagation. The problem is more complicated than is widely believed. This has been obvious since the work of Lighthill and Whitham (1955), and is partly well-known.

1.5.4 Comparison with Traditional Interpretations of Wave Behaviour

Traditionally, wave behaviour has been inferred from results of the Method of Characteristics, which can be used to convert two partial differential equations to two pairs of ordinary differential equations. The first pair, from either Eqs. 1.24a, b or 1.25a, b, gives the paths of two characteristics in space:

$$\frac{\mathrm{d}x}{\mathrm{d}t} = \beta U \pm C, \quad \text{where } C = \sqrt{\frac{gA}{B} + U^2\left(\beta^2 - \beta\right)}.$$

Considering $\mathrm{d}x/\mathrm{d}t$ to be the propagation velocity we see that these are fully nonlinear equivalents of our earlier linearised results but which we showed applied *only in the not-so-long wave limit of those results*. The traditional belief using such results has been that all long waves travel at a unique speed C relative to the water. That this is not the case is shown by the remaining pair of ordinary differential equations for A and Q obtained from Eqs. 1.25a, b by the Method of Characteristics (using A or local water depth identifies a wave better than using surface elevation η):

$$\left(-\beta\frac{Q}{A} \pm C\right)\frac{\mathrm{d}A}{\mathrm{d}t} + \frac{\mathrm{d}Q}{\mathrm{d}t} = gA\tilde{S} - \Lambda P\frac{Q|Q|}{A^2}.$$

The right side is zero in steady uniform flow, but in all other general cases it is non-zero, such that $\mathrm{d}A/\mathrm{d}t$ and $\mathrm{d}Q/\mathrm{d}t$ on the left are also non-zero, so that area A and discharge Q are continually changing along a characteristic, as is well-known. Our point here is: if A is changing along a characteristic one cannot say that a definite wave, identified by a particular constant value of A (a measure of water depth), is travelling along that characteristic. Information of a certain type is doing that, but not what we would identify as a wave.

In contrast, in the previous section we showed rather greater complexity of behaviour by linearising the equations and obtaining solutions corresponding to actual waves. The false interpretation of wave behaviour using characteristics has hindered the understanding of wave motion in rivers.

1.5.5 A Common Approximation—Neglect of Time Derivative and Inertia Terms

Now we introduce the common approximation to the momentum equation where both the time derivative and inertia terms are neglected, which we showed in Sect. 1.4 was actually a slow change approximation, although the terms "low-inertia" or "kinematic" approximation have been used for this in the past. The momentum Eq. 1.25b becomes

$$\frac{gA}{B}\frac{\partial A}{\partial x} + \Lambda P \frac{Q^2}{A^2} - gA\tilde{S} = 0. \tag{1.42}$$

To examine the linear wave propagation properties we consider only the corresponding terms from the polynomial, Eq. 1.40, ignoring the second time derivative and the β terms, giving the quadratic equation

$$\omega - \alpha\kappa + \mathrm{i}\kappa^2 F_0^{-2} = 0,$$

with solution

$$\kappa = \mathrm{i}F_0^2\left(-\alpha/2 + \varepsilon\sqrt{\alpha^2/4 - \mathrm{i}\omega/F_0^2}\right). \tag{1.43}$$

This result is plotted as dashed lines on Fig. 1.3, and over at least half of the logarithmic scale of what we might call the spectrum, for longer downstream travelling waves the approximation shows remarkably good agreement with the full solution even for large super-critical Froude numbers.

An extra dashed line result has been plotted on Fig. 1.3 for critical Froude number $\beta F_0^2 = 1$ (with $\beta = 1$ here also), as the solution Eq. 1.43 is valid for that also.

We conclude that for downstream-propagating waves the simplified momentum Eq. 1.42 can be used for all practical Froude numbers for waves that are sufficiently long. It is not a low-inertia approximation.

A different situation arises when we consider upstream propagating waves, c negative, caused by downstream gate operations or tides. It can be seen on Fig. 1.3 that the results are now dependent on Froude number, even for the longest waves, and already for $F_0 = 0.3$ they are not accurate. The approximation seems to be not just one of *slow change* but also requires a *slow flow*. In view of these two different results for downstream and upstream propagation we will refer to Eq. 1.42 as the *Slow-Change/Slow-Flow momentum equation*, where the terminology is an abbreviation for the rather longer term *Slow-Change-and-for-upstream-propagation-Slow-Flow momentum equation*, as we foreshadowed in Sect. 1.4.

Now, provided those conditions are satisfied, the approximate momentum Eq. 1.42 written in terms of η instead of A becomes simply

$$gA\frac{\partial \eta}{\partial x} = -\Lambda P\frac{Q^2}{A^2}. \tag{1.44}$$

This gives an explicit equation for Q in terms of η, provided the resistance coefficient Λ does not depend on Reynolds number and hence on Q:

$$Q = A\sqrt{\frac{g}{\Lambda}\frac{A}{P}\left(-\frac{\partial \eta}{\partial x}\right)}, \tag{1.45}$$

which is a simple generalisation of the Chézy-Weisbach equation to non-uniform flow. If the flow were uniform, $\partial\eta/\partial x = -S$, for constant S, we would recover the original Eq. 1.26. Figure 1.3 shows that this simple equation, that one might have been tempted to write down in the belief that it was a rough approximation, is, in fact, a very good approximation to the full unsteady momentum equation where the primary interest is flood waves and disturbances with a slow rate of change, "slow" in the sense that $\sigma_0 T$ is greater than 10 or 20.

This level of approximation can be used to develop a fully nonlinear *Slow-Change/Slow-Flow routing equation* in a single variable, as done by Fenton and Keller (2001, p. 59) who believed incorrectly that it was a low-inertia approximation, without the insight obtained in this work that it is a very-long-wave approximation. We re-arrange the simplified momentum Eq. 1.42 in terms of conveyance $K = \sqrt{gA^3/\Lambda P}$ to give an explicit equation for Q in terms of A this time:

$$Q = K\sqrt{\tilde{S} - \frac{1}{B}\frac{\partial A}{\partial x}}, \tag{1.46}$$

Now consider the total volume of fluid upstream of a point x at a time t contained in a stream between the upstream boundary x_0 and the general point x:

$$V(x,t) = \int_{x_0}^{x} A(x',t)\,\mathrm{d}x'.$$

From simple calculus, the derivative of volume with respect to distance x gives the cross-sectional area: $\partial V/\partial x = A$. The time rate of change of V at a point is equal to the total rate at which the volume in the channel upstream is increasing, which is the inflow at the upstream end $Q(x_0, t)$ plus the distributed inflow $\int i\, dx$ minus Q, the volume rate that is passing the point. Writing these two results for A and Q in terms of derivatives of V:

$$A = \frac{\partial V}{\partial x} \quad \text{and} \quad Q = Q(x_0, t) + \int_{x_0}^{x} i(x')\, dx' - \frac{\partial V}{\partial t}, \tag{1.47}$$

and substituting both into the mass conservation Eq. 1.25a shows that it is identically satisfied, as we might have expected. Fatemeh Soroush (personal communication) suggested the incorporation of the $Q(x_0, t)$ term so that V is indeed just the volume upstream in the channel and not the total volume of fluid that will pass, including liquid not yet in the channel. Now we substitute A and Q as defined here in terms of V into the simplified momentum Eq. 1.46. As both breadth B and conveyance K can be written as functions of area $A = \partial V/\partial x = V_x$, we obtain the single equation in the single dependent variable V,

$$\frac{\partial V}{\partial t} + K(V_x)\sqrt{\bar{S} - \frac{1}{B(V_x)}\frac{\partial^2 V}{\partial x^2}} = \int_{x_0}^{x} i(x')\, dx', \tag{1.48}$$

in which the only approximation relative to the long wave equations has been that the variation with time is slow, $\sigma_0 T$ large. It is a fully nonlinear equation, and boundary conditions involving Q and η and their derivatives can be incorporated, using Eqs. 1.47 and the geometrical relationship between A and η at a point. The equation has been used by Soroush et al. (2013) to simulate flows in irrigation ditches, where it has some advantages in treating starting flows, where the long wave equations have problems. If Eq. 1.48 is linearised, the advection-diffusion approximation (e.g. Lighthill and Whitham 1955, Eq. 82) is obtained, revealing the nature of the equation and that computational limitations characteristic of diffusion equations might be encountered.

1.5.6 Streams with Curvature

Until now, it has been simpler to ignore the effects of curvature in discussing the nature of long waves and their approximations. We now consider the linearised equations with curvature corresponding to Eqs. 1.33a, b, where the time derivative in Eq. 1.33a is now modified by a curvature term and we make the slow change approximation of Sect. 1.5.5, in neglecting time derivative and fluid inertia terms in Eq. 1.33b:

$$(1-\kappa n_{\mathrm{m}0})\frac{\partial A_1}{\partial t}+\frac{\partial Q_1}{\partial s}=0,$$

$$C_0^2\frac{\partial A_1}{\partial s}+\sigma_0(Q_1-c_0A_1)=0.$$

Eliminating Q_1 gives

$$\frac{\partial A_1}{\partial t}+\frac{c_0}{1-\kappa n_{\mathrm{m}0}}\frac{\partial A_1}{\partial s}=\frac{C_0^2}{\sigma_0(1-\kappa n_{\mathrm{m}0})}\frac{\partial^2 A_1}{\partial s^2},$$

which is an advection-diffusion equation, showing that the advection velocity is

$$\frac{c_0}{1-\kappa n_{\mathrm{m}0}}=\frac{\alpha U_0}{1-\kappa n_{\mathrm{m}0}},$$

where $\alpha \approx 3/2$ as given by Eq. 1.38, and the diffusion coefficient is

$$\frac{C_0^2}{\sigma_0(1-\kappa n_{\mathrm{m}0})}=\frac{U_0A_0/B_0}{2S_0(1-\kappa n_{\mathrm{m}0})}.$$

Such results for the straight-channel case are well-known (e.g. Lighthill and Whitham 1955, Eq. 82). We might almost always choose a longitudinal axis coinciding with the centre of the surface such that $n_{\mathrm{m}} \approx 0$, and we recover the traditional result. We see that streamline curvature has little effect on flood-propagation behaviour with such a co-ordinate system.

We conclude that there is little point in considering stream curvature in flood propagation studies, and we will not consider it further.

1.6 Resistance to Flow

In Sect. 1.3.6 we judged the resistance term to be the least accurately-known component of the long wave equations. Here we examine a large data set from New Zealand and a smaller one from the United States of America to investigate the nature of resistance in rivers, as well as to provide a method of calculating it. We examine the variation with the relative size of the boundary particles and find that the results also depend on the state of the bed: whether or not the bed grains are coplanar, or whether individual grains are exposed to some degree, whether or not there is bed-load transport, whether or not there is suspended transport, and, whether there are bed-forms present. As the state of the bed can never really be known accurately, depending as it does on the recent history of the channel flows and their effects on bed grains, we assert that we can never really accurately know what the resistance will be. In a sense, this uncertainty frees us, allowing us just to make calculations using expected minimum and maximum resistance, frees us from the belief that any simulation is highly accurate, and allows us to use judgement as to what we can do with the results.

1.6.1 Traditional Formulae

The earliest theory for flow in an open channel with resistance was by Chézy in 1775, who obtained the formula for the mean velocity in a uniform flow

$$U = C\sqrt{\frac{A}{P}S}$$

where the coefficient C is a kind of conductance, and it has dimensions, for which Chézy gave a value which in our standard French-revolutionary (SI!) units was $31\ \mathrm{m}^{1/2}\,\mathrm{s}^{-1}$. The presence of a coefficient with units is a significant problem and hides somewhat the physical significance. If we introduce gravity and write $C = \sqrt{g/\Lambda} = \sqrt{8g/\bar{\lambda}}$, where $\bar{\lambda}$ is the mean value of the Weisbach dimensionless resistance coefficient around the perimeter, we obtain the Chézy-Weisbach Eq. 1.26

$$U = \frac{Q}{A} = \sqrt{\frac{g}{\Lambda}\frac{A}{P}S} = \sqrt{\frac{8g}{\bar{\lambda}}\frac{A}{P}S},$$

with the dimensionless resistance coefficient written in two different ways.

In the 1840s Gauckler proposed a modification to Chézy's law, where the mean velocity of flow in a waterway was assumed proportional to the 2/3 power of A/P. In the 1880s Manning also proposed such a formula, and it has usually been referred to as Manning's formula, although Gauckler-Manning would seem to be fairer:

$$U = \frac{1}{n}\left(\frac{A}{P}\right)^{2/3}\sqrt{S},$$

where n is Manning's coefficient, with dimensions $\mathrm{L}^{-1/3}\mathrm{T}$, and values are specified in SI units. In Europe it is often written in terms of Strickler's coefficient $k_{\mathrm{St}} = 1/n$. This equation has dominated the field of open channel hydraulics, but it has several disadvantages. It has little basis in fluid mechanics. The coefficient n has no physical significance other than its relation to other quantities obtained from its use. It is dimensional, which causes trouble if units other than SI are used. The most common method of obtaining n seems to be by looking at tables or pictures in books, where the important underlying roughness is not known or visible, or by contacting a friend to find out what he/she used for a similar stream.When introduced, the Gauckler-Manning equation was a simple approximation to the real variation of resistance with depth in a channel, but with computers, a more rational approach becomes feasible. It is certainly desirable.

Fifty years ago the ASCE Task Force on Friction Factors in Open Channels (1963) expressed its belief in the utility of using the Chézy-Weisbach formulation, noting that it was more fundamental than the Gauckler-Manning form, and was based on more research. The Task Force noted that the Gauckler-Manning equation could be used for fully rough flow conditions, however it presented a revealing figure for the variation of n with Reynolds number, which showed that with that equation there is continual decay of resistance with Reynolds number, even in the limit of large values, when there should be no variation, so that one could deduce that it is fundamentally flawed. In spite of all these problems, the Gauckler-Manning formulation has continued to dominate open channel hydraulics. The reason for the Chézy-Weisbach approach not being adopted may be that, even though the ASCE Task Force presented a number of experimental and analytical results, there was no simple general path to follow.

1.6.2 Field Results for the Weisbach Resistance Coefficient

Here we attempt to obtain understanding and a formula for the resistance coefficient using results from a number of field measurements. We first consider the results of Hicks and Mason (1991), a scholarly catalogue of 558 stream-gaugings from 78 river and canal reaches in New Zealand, of which 55 were sites with grading curves for boundary material, so that particle sizes were known. Neither vegetation nor bed-form resistance can be isolated. Hicks & Mason based their approach on Barnes (1967), who provided values of Manning's resistance coefficient n for a single flow at each of 50 separate river sites in the United States of America, of which boundary material details were given for 14. We also include those results here.

In both works cited, an energy approach was adopted to calculate Manning's n, using the Gauckler-Manning equation. The slope was assumed to be that of the energy grade line. If the cross-section expanded, an arbitrary allowance of half the change of kinetic head over the reach was deducted. The Coriolis energy coefficient was assumed to be 1, and an arbitrary weighted compounding of geometry at each cross- section was adopted, rather than solving a gradually-varied flow equation through the reach, and such mean values were combined. From the values of Manning's n so obtained, Hicks & Mason computed values of Chézy's coefficient $C = (A/P)^{1/6}/n$.

In this work, to avoid the assumptions of the energy approach, we took the steady form of the momentum Eq. 1.18 for no inflow and constant β:

$$-\beta \frac{Q^2}{A^2}\frac{dA}{ds} + gA\frac{d\eta}{ds} = -\Lambda P \frac{Q^2}{A^2},$$

and discretising over the whole length of the reach measured, between an initial point 1 and final point 2, a distance L downstream gave the formula for Λ in terms of information we had:

$$\Lambda = \frac{1}{P}\left(\beta \frac{A_2 - A_1}{L} + g\frac{A^3}{Q^2}S\right),$$

where S is the free surface slope (positive) as given by Hicks & Mason, while A_1 and A_2 were calculated from expansion data they provided. Our ignoring of all intermediate geometry, where there were more than two cross-sections, is not as good as their treatment, although both they and Barnes used an unnecessary geometrical mean for compound conveyances. Our use of $\beta = 1.1$ is probably better than the value of an energy coefficient with a value of 1 that they used, but it is still highly approximate. We used a constant value of $g = 9.8\ \mathrm{m\,s^{-2}}$, even though $g = 9.81\ \mathrm{m\,s^{-2}}$ is better for the bottom 2/3 of the South Island of New Zealand. We do not expect any results to be accurate enough for this to matter. In the case of the Barnes results, we calculated the resistance coefficient using $\Lambda = gn^2/(A/P)^{1/3}$.

From both catalogues we took the values of D_{84}, the boundary particle size for which 84 % of the material was finer, and from the values of A/P, calculated the relative roughness $\varepsilon_{84} = D_{84}/(A/P)$. Some of the New Zealand sites contained sufficient in-stream debris, obvious from photographs, or had such an irregular channel, that we neglected them. We were left with a total of 388 flows where we had enough stream, bed, and flow data to be able to compute the resistance coefficient Λ. It is a tribute to the scholarly nature of Hicks & Mason's work that we were able to do all this.

Our results, almost everywhere, were very close to those of Hicks & Mason. The only important differences were for the smallest values of resistance coefficient Λ, where some of their results were anomalously small. We took this to be a confirmation of our approach of not having to use the energy approach and not having to make assumptions as to expansion/contraction losses, so that we then used the momentum results throughout.

Results for Λ are plotted against relative grain size ε_{84} in Fig. 1.4, thin dashed lines connecting the points for different flows at a particular station. Where such a line is discontinuous but a circled point is not shown, it was ignored for the reasons described above. The points are plotted with a circle type corresponding to whether the bed particles were likely to be mobile or not, according to our calculations. If they were mobile, we expect that the force required to move them over each other and/or to support their suspension will lead to larger resistance than for a stable bed

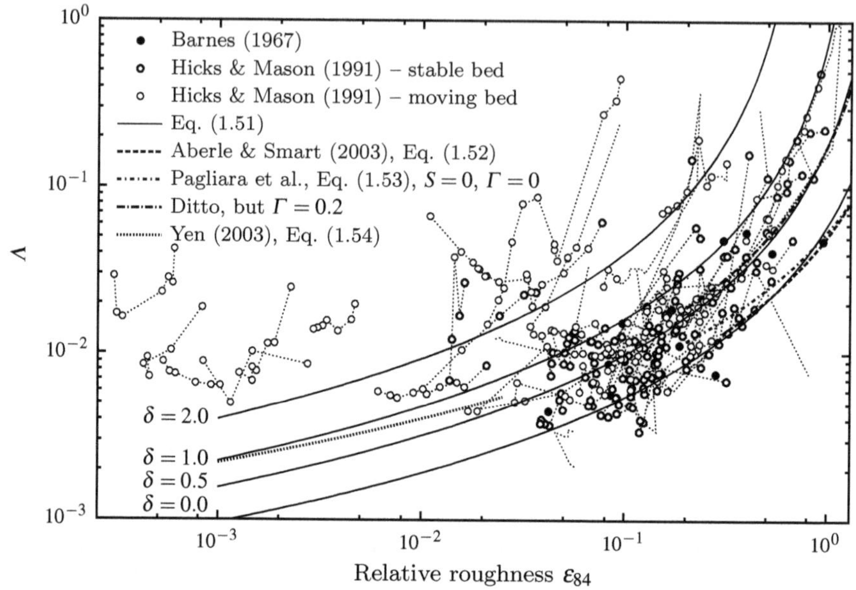

Fig. 1.4 Values of resistance coefficient $\Lambda = g/C^2$ obtained from field and laboratory measurements

composed of the same particles. The criterion for mobility we used was a modification of Shields' criterion for large grain Reynolds numbers, for which motion occurs. R.A. Bagnold (personal communication) suggested that the Shields diagram has been widely misinterpreted. He noted that for experiments with large particles (gravel, boulders, etc.) the bed was made flat by levelling the tops of the large particles, giving a limiting threshold dimensional shear stress of $\Theta \approx 0.045$,

$$\Theta = \frac{\tau}{\rho(G-1)gD}, \tag{1.49}$$

where G is specific gravity of the bed material and D is particle size. In nature, however, even large bed particles are free to project into the flow, and are likely to move at a smaller stress. Fenton and Abbott (1977) followed Bagnold's suggestion and examined the effect of protrusion of particles into the stream. They suggested that for large particles the value of Θ_{cr} was more like 0.01 for randomly-placed particles in nature rather than 0.045 for levelled beds. The numerical value of 0.01 was established only very approximately. Nevertheless we use it here, in the absence of any other result.

The dimensionless shear stress Θ can be expressed simply in terms of stream parameters. Substituting Eqs. 1.14 and 1.16 into Eq. 1.49, using D_{84} and $G = 2.65$, we used the criterion that motion, and hence greater resistance, is likely if

$$\Theta = \frac{1}{(G-1)}\frac{\Lambda U^2}{gD_{84}} = \frac{S}{(G-1)\,\varepsilon_{84}} > \Theta_{cr} \approx 0.01. \tag{1.50}$$

For smaller particles, when levelling of laboratory particles is less feasible and the Shields criterion should be more applicable, the threshold Θ_{cr} will be larger than 0.01, but as the general relationship passing from the Shields criterion to that for large randomly-placed particles has not been established, we used this constant value. On the figure we have plotted points for which $\Theta \le 0.01$ using circles with a thick line, and labelled them "stable". Points for which $\Theta > 0.01$ have been drawn as circles with a thin line and labelled "moving". It can be seen that all the results for small ε_{84} satisfy the latter criterion. Of interest, some of the points at the left of the figure, shown as mobile, are from the Waikato River, a major river with fine particles and well-known to have a highly mobile bed with large bed-forms.

We now describe Fig. 1.4 in detail, and use the information in it to build a model.

- The results of Barnes are quite compatible with those of Hicks & Mason. It can be seen that many of the results from each study are for large bed material $\varepsilon_{84} > 0.1$, possibly a reflection of the hilly and mountainous nature of New Zealand and of the country that Barnes used, often in the Rocky Mountains and the Pacific North-West of the United States of America.
- There is a wide scatter of results. We will now attempt to explain that. The points, we believe, have a tendency to group around three of the curves shown and the rest to be bounded below by the fourth (upper) curve shown. The curves have been drawn using the expression, found by trial and error:

$$\Lambda = \frac{0.06 + 0.06\,\delta}{(1.0 - 0.6\delta - \ln \varepsilon_{84})^2}, \tag{1.51}$$

  with values of $\delta = 0$, 0.5, 1 and 2. The parameter δ is an arbitrary one that we use to identify the state of the particles making up the bed. This will now be explained.
- The first grouping of points comprises those around the bottom curve. We hypothesise that these points, having the lowest resistance, are those forming beds where the particles are relatively co-planar such that the bed is armoured. We assigned $\delta = 0$ to this state, and used that in Eq. 1.51 to plot the curve. For large bed roughness this closely agrees with the equation obtained by Aberle and Smart (2003), re-written in terms of $\Lambda = \lambda/8$,

$$\Lambda = (4.41 - 3.54\,\ln \varepsilon_{84})^{-2}, \tag{1.52}$$

  for $0.3 < \varepsilon_{84} < 3$. Pagliara et al. (2008) obtained results that depended on slope S, and the fractional area Γ of large grains sitting on top of the bed. They obtained:

$$\Lambda = \left((3.04 - 7.82S)\left(1.4e^{-2.98\Gamma} - \ln \varepsilon_{84}\right)\right)^{-2}, \tag{1.53}$$

also plotted on Fig. 1.4 for $0.2 < \varepsilon_{84} < 2$ for $\Gamma = 0$, for a co-planar bed, and plotting using $S = 0$, although this was outside the recommended limit of $S \geq 0.024$. It can be seen that the curve for $\Gamma = 0$ agrees with Eq. 1.52 for very large roughness. For smaller, but still large, roughness, it is tending to our next curve and band of experimental points, which would correspond to those smaller roughnesses being less likely to be levelled, as we now suggest.

- The next grouping of points on Fig. 1.4 is around the second curve from the bottom, which can be seen to substantially coincide with the curve obtained from Eq. 1.53 using $\Gamma = 0.2$, corresponding to exposed boulders on top of the bed occupying 20 % of the surface area. Of course, with a number of these grains thus exposed, the resistance is greater. We assigned a value of $\delta = 0.5$ to this intermediate state, and plotted the second curve using this value in Eq. 1.51.
- Substituting $\delta = 1$ in Eq. 1.51 gives the third curve on the figure, passing through what we believe is the third grouping of particles. This is probably the state for the maximum resistance for a stable bed corresponding to exposed grains occupying something like 50 % of the surface area. Any more such grains will cause shielding of particles, the bed will start to resemble the co-planar case, and resistance will actually be reduced. We found that the dimensionless shear stress Θ for points near this curve was roughly $\Theta = \Theta_{\mathrm{cr}} = 0.01$, so motion was indeed incipient.
- Further evidence supporting our assertions is obtained from the expression proposed by Yen (2002, Eq. 19), who considered results from a number of experimental studies using fixed impermeable beds. We used his formula, converted to $\Lambda = \lambda/8$, used an infinite Reynolds number, and converted his equivalent sand roughness $\varepsilon_s = 2\varepsilon_{84}$, the coefficient of 2 being taken from the values given in Yen (2002, Table 8) to give the equation

$$\Lambda = 0.166\left(1.8 - \ln \varepsilon_{84}\right)^{-2}, \tag{1.54}$$

where we plot it for $\varepsilon_{84} \leq 0.025$, corresponding to Yen's recommended limit. Although its vertical position is slightly uncertain, because of the need to assume a ratio of ε_s to ε_{84}, it can be seen that the curve passes from our curve $\delta = 1$ for small particles, which are unlikely to have the tops levelled so that particles are exposed, to the second curve for larger particles, more likely to be levelled in the laboratory experiments.
- For points above the third curve almost all experimental points had shear stresses greater than critical, $\Theta > 0.01$. If particles move, not only do many particles protrude above others, increasing the stress, but there is the additional force required to maintain the sliding and rolling and jostling of all the particles. Hence, the resistance is greater. And, if there is a need to maintain particles in suspension, that will contribute also to resistance. We have shown the fourth curve as drawn for $\delta = 2$. There are relatively few results between the previous

curve and this one, so suggesting δ going from 1 to 2 implies some sort of transition to a new state, of finite bed-load motion at least. Possibly suspended load occurs for values of δ greater than this.

Aberle and Smart (2003), investigating the influence of roughness structure on flow resistance for large roughness found that they obtained less scatter of results if they used, instead of grain size, the standard deviation of the bed elevation as a characteristic roughness. This seems to be a good way of characterising the state of the bed rationally. However, it does not offer a rational extension to mobile beds and suspended transport. Pagliara et al. (2008), on the other hand, used the fractional area Γ of exposed grains as a parameter. This worked for small fractions, but it could not describe the probable reduction of resistance as more grains sit on top, forming another, higher, plane bed.

Both measures, standard deviation of bed elevation and fraction of superimposed grains, are more rational than the somewhat ad hoc parameter δ introduced here, and they may be used in the future. However, we did not have enough information to apply them here. We have found that δ could be used to calculate the resistance of beds in various states, from co-planar through to mobile and possibly even with a suspended load, and provided a simple means of predicting minimum and maximum values of resistance.

1.6.3 Conclusions and Recommendations

Figure 1.4 shows that the state of the bed is a more important determinant of resistance than the grain size. A hundred-fold change in ε_{84}, from 0.001 to 0.1, gives only a five-fold increase in resistance coefficient Λ, which is about the change corresponding to passing from $\delta = 0$ for an armoured bed of a particular grain size, to $\delta = 2$ for a mobile bed of the same material.

To use a plausible value for δ in practice, we need to know the answer to the question: what is the actual state of the bed? The answer, unfortunately, requires a detailed knowledge of the recent past history of flow rates, as well as the actual effect of any particular flow on bed particles of a particular size, let alone a mixture. A long period of constant flow would tend to remove exposed particles and to armour the bed, leaving the particles co-planar. Unless of course, there were a substantial fraction of larger grains in the sediment that might be unable to be removed by that flow. If the flow were to increase, maybe they would be removed, and the resistance decrease. Or, maybe the armoured particles would also start to be removed, when the resistance would increase. Or, maybe if the flow decreases, then particles start to be deposited, sitting above the bed and contributing to increased resistance. What happens then in the next change of flow? And what is the particle distribution along the channel upstream? These questions cannot be simply answered in any reach of a river at any particular time. Maybe this is actually a benefit—the uncertainty frees us and allows us to put calculations in a better perspective. We could calculate the *minimum* likely resistance, corresponding to the curve for $\delta = 0$, using Eq. 1.51. We could also calculate the *maximum* likely resistance, first by using Eq. 1.50 to see if Θ were less than $\Theta_{cr} \approx 0.01$. If it were, and the bed were stable, we could use $\delta = 1$ in Eq. 1.51, which gives a value of Λ typically about 250 % of the minimum. If the bed were unstable, we might choose $\delta = 2$ or more. In the mobile case it is possible that a relationship between δ and Θ exists, as all particles can be exposed, but that has not been found at the time of writing.

We can also make some rather simpler and possibly more useful inferences from Fig. 1.4, that for all sediment sizes $\varepsilon_{84} < 0.1$, typical of most streams, Λ is always in the relatively narrow range between $\Lambda \approx 0.004$ and $\Lambda \approx 0.010$, unless the bed grains are mobile.

The uncertainty, of being only able to calculate Λ in either of these approximate ways might, paradoxically, be of benefit, to remind us of our lack of precision.

In all of this discussion we have not mentioned bed-forms and the increased resistance caused by them. This has been addressed in considerable depth by Simons and Richardson (1962). There was no visual recording of such information in our data sources, so we are unable here to add any information, but in a sense, our criteria include them too: a certain relative grain size and hence a certain resistance would include bed-form effects for that condition.

1.7 Obstacles in a Stream

When water in a river or canal flows past an obstacle or obstacles such as bridge piers, baffle blocks, or woody debris, each obstacle exerts a resistance force that reduces the momentum of the flow. In the usual sub-critical flow this means that the water level upstream is higher than downstream. Most momentum loss to the flow occurs immediately at the upstream face of the obstacle. Initially the loss is confined to the fluid along the line of the obstacle; and as the turbulent flow proceeds downstream, the deficit is dispersed laterally throughout the flow.

In comparison, there are almost no energy losses at the front face of an obstacle, but because of the sudden local diversion of the flow, as the water flows downstream, interacting flows and turbulent flow processes cause energy losses for some distance. They are more difficult to measure or calculate, as they take place in boundary layers, shear layers, separation zones, vortices, and subsequent turbulent decay in the wake. For this reason the problem can be quantified more easily using momentum rather than energy as the underlying principle. Some standard software, however, uses neither discrete momentum nor discrete energy losses, but treats bridges as merely constrictions to the flow that change the channel geometry. The US Hydrologic Engineering Center River Analysis System (HEC-RAS) computer program, thoroughly documented in HEC-RAS (2010), in its section on bridges presents four alternative methods available for computing losses through a bridge, one of which is a momentum method. However, when the HEC evaluated the performance of three of the one-dimensional bridge modelling programs for a number of bridge sites (HEC 1995) losses at the bridges were included with the river roughness and distributed over the computational reach. Near bridges there were obvious differences from the actual measured water levels, which showed discrete drops at the bridges.

In principle it does not seem satisfactory to distribute the bridge losses in with boundary resistance. Using the momentum approach, with a discrete surface elevation difference between upstream and downstream means that finite obstacles can and should be considered to be hydraulic controls, where the surface elevation difference across the obstacle depends on the flow.

1.7.1 Momentum Formulation

The conservation of momentum principle for an obstacle in a prismatic channel, in terms of an upstream Sect. 1.1 and a downstream Sect. 1.2, as shown in Fig. 1.5 is $P = M_1 - M_2$, where P is the magnitude of the drag force and $M = \rho(gA\bar{h} + \beta Q^2/A)$ is the momentum flux at a section, where $\bar{h}$ is the depth of the centroid below the surface, and other symbols are as used previously. The drag is given by the conventional expression in terms of drag coefficient

$$P = \frac{1}{2}\rho C_{\mathrm{D}} v^2 a_1,$$

where a_1 is the blockage area of the object measured transverse to the flow with water level that at Sect. 1.1, C_{D} is the drag coefficient, and v, the fluid speed impinging on the object. We take this as being proportional to the upstream velocity, such that we write

$$v^2 = \gamma \left(\frac{Q}{A_1}\right)^2, \tag{1.55}$$

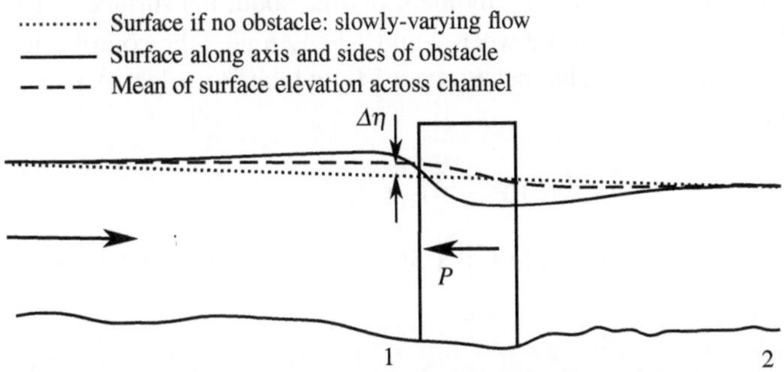

Fig. 1.5 Flow past a bridge pier showing real surface at the centreline and mean surface elevation

where γ is a coefficient that recognises that the velocity which impinges on the object is not necessarily equal to the mean velocity in the flow. For a small object near the bed, γ could be quite small; for an object near the surface such as a bridge deck $\gamma > 1$; for objects of a vertical scale that of the whole depth such as a bridge pier $\gamma \approx 1$. The momentum equation is then

$$\frac{1}{2}\gamma C_{\mathrm{D}}\frac{Q^2}{A_1^2}a_1 = \left(gA\bar{h} + \beta\frac{Q^2}{A}\right)_1 - \left(gA\bar{h} + \beta\frac{Q^2}{A}\right)_2. \tag{1.56}$$

A typical problem is in sub-critical flow, where the downstream water level is given, where we want to know how much the water level will be raised upstream if a bridge is built. As both $A_{1,2}$ and $\bar{h}_{1,2}$ are functions of the surface elevations $\eta_{1,2}$, conditions at 2 can be evaluated, while the equation becomes a nonlinear equation for the unknown η_1, which can be solved numerically using any method for such problems.

If the depth change is small, then an explicit approximate solution can be obtained, which reveals the nature of the problem and how the important quantities affect results (Fenton 2003). This will now be presented.

1.7.2 An Approximate Solution

We consider a small change of surface elevation $\Delta\eta$, presumed positive, between 2 and 1, as shown in Fig. 1.5. We write the series for the upstream area of the stream

$$A_1 \approx A_2 + B_2\Delta\eta + \cdots,$$

and similarly we write for the blockage area with the water level at 1 and 2 as

$$a_1 \approx a_2 + b_2\Delta\eta,$$

where b_2 is the surface width of the obstacle, which for submerged obstacles is zero. We now need to express the first moment of area about the surface at 1 in terms of that at 2. The increase in $A\bar{h}$ between 2 and 1 is $A\Delta\eta$ plus the contribution from the elemental strip across the channel $B_2\Delta\eta \times \frac{1}{2}\Delta\eta$, which is neglected as it is of higher order in $\Delta\eta$:

$$(A\bar{h})_1 \approx (A\bar{h})_2 + A_2\Delta\eta + \cdots.$$

Substituting into the momentum Eq. 1.56 for the quantities at 1 gives

$$\frac{1}{2}\gamma C_{\mathrm{D}}\frac{Q^2(a_2+b_2\Delta\eta)}{(A_2+B_2\Delta\eta)^2}=g(A_2\Delta\eta+\cdots)+\beta\frac{Q^2}{A_2+B_2\Delta\eta+\cdots}-\beta\frac{Q^2}{A_2}. \quad (1.57)$$

Recognising that $\Delta\eta$ is small we expand each side as a power series in $\Delta\eta$, neglecting terms like $(\Delta\eta)^2$. This gives a linear equation which can be solved to give an explicit approximation for the dimensionless drop across the obstacle $\Delta\eta/(A_2/B_2)$, where A_2/B_2 is the mean downstream depth:

$$\frac{\Delta\eta}{A_2/B_2}=\frac{\frac{1}{2}\gamma C_{\mathrm{D}}F_2^2}{1-\beta F_2^2}\frac{a_2}{A_2}. \quad (1.58)$$

This explicit approximate solution has revealed the important quantities of the problem to us and how they affect the result: velocity ratio parameter γ defined in Eq. 1.55, drag coefficient C_{D}, the Froude number $F_2^2=Q^2B_2/gA_2^3$, and the relative blockage area a_2/A_2. Equation (1.58) has been compared by Fenton (2003) with experimental results and found to be accurate.

For subcritical flow the denominator in Eq. (1.58) is positive, and so is $\Delta\eta$, so that the surface drops from 1 to 2, as we expect. If the flow is supercritical, $\beta F_2^2>1$, we find $\Delta\eta$ negative, and the surface rises between 1 and 2. If the flow is near critical ($\beta F_2^2\approx 1$) the change in depth will be large and the theory will not be valid, which is made explicit by the theory.

Another benefit of the approximate analytical solution of Eq. 1.58 is that it shows explicitly that an obstacle forms a control in the channel, such that $\Delta\eta$ is a function of Q^2, or Q a function of $\sqrt{\Delta\eta}$, in a manner analogous to a broad-crested weir. In numerical river models it should ideally be included as an internal boundary condition.

For multiple bodies, such as all the piers of a bridge, it seems quite reasonable to add the forces and so to lump the contributions together, given that the dimensions of a bridge are much smaller than the length scales in the channel, and so, in Eq. 1.58 to use

$$\gamma C_{\mathrm{D}}a_2=\sum_j \gamma_j C_{Dj}a_{2j},$$

summed over all local resistance elements j of a structure.

1.8 River Junctions

Almost all hydraulic research on river junctions has consisted of studies for simplified geometry such as flow in two rectangular channels with co-planar bottoms, where one enters the other, whose width remains unchanged thereafter. All studies have been for finite Froude numbers, as for the small Froude numbers generally observed in rivers there are few dynamic effects to measure. Much of this idealisation has been due to the formal application of the conservation of momentum, which is difficult for anything other than such idealised geometries, and which has limited extension to more complicated geometries.

Here we show using the slow-change approximation that any complexity of geometry loses importance, and it is possible to use the same sort of approximations for junctions that we have used for flow in a channel.

Consider the integral momentum theorem Eq. 1.1b. Substituting Eq. 1.2 from Gauss' divergence theorem gives the vector expression for a general control surface and volume encompassing all of a river junction:

$$\frac{\mathrm{d}}{\mathrm{d}t}\int_{\mathrm{CV}} \rho\mathbf{u}\,\mathrm{d}\mathscr{V} + \int_{\mathrm{CS}} \rho\mathbf{u}\,\mathbf{u}_{\mathrm{r}}\cdot\hat{\mathbf{n}}\,\mathrm{d}S = -\int_{\mathrm{CV}} \nabla p\,\mathrm{d}\mathscr{V} + \rho\mathbf{g}\mathscr{V} + \mathbf{T},$$

where $\mathbf{g}$ is the body force per unit mass, $\mathscr{V}$ is the volume, and $\mathbf{T}$ is the total resistance force on the boundary. We have shown above that the time derivative term and the fluid momentum term can be neglected for slow change of boundary conditions such as for the passage of floods, giving the slow-change approximation

$$\int_{\mathrm{CV}} \nabla p\,\mathrm{d}\mathscr{V} = \rho\mathbf{g}\mathscr{V} + \mathbf{T}.$$

As $\mathbf{g}$ has a single vertical z component, we take just the two horizontal components of the vector equation, and use the hydrostatic approximation $(\partial p/\partial x, \partial p/\partial y) = \rho g(\partial\eta/\partial x, \partial\eta/\partial y)$:

$$\rho g \int_{\mathrm{CV}} \left(\eta_x, \eta_y\right)\mathrm{d}\mathscr{V} = \left(T_x, T_y\right).$$

In view of the fact that resistance occurs mostly in a direction opposite to that of the main flow, we can just take a single streamwise component of the vector equation. Probably it is enough to take a slice of length Δs across the whole flow, and write for each arm, where η is now a function of just s and t:

$$gA\frac{\partial\eta}{\partial s} = -\Lambda P\frac{Q^2}{A^2},$$

just like Eq. 1.44 leading to the generalised Chézy-Weisbach Eq. 1.45. It seems that even in the complicated geometry of a river junction, the slow-change approximation to the momentum equation can be used, and we can treat the arms of a junction simply as part of each channel. For sufficiently slow change such as for flood routing, there is no need to go to any trouble to include fluid inertia contributions, even if one were using a model with all the terms included.

Fig. 1.6 Confluence of three rivers—from the top, Inn, Danube, and Ilz at Passau in Germany

We leave, as a partial counter-example, a photograph in Fig. 1.6 from Passau in Germany, 3 km from the border with Austria. It shows the confluence of three rivers: the River Danube coming from the middle of the right edge of the picture; the River Inn also coming from the right above that, with lighter-coloured, silt-laden, and possibly-colder Alpine water; and the smaller River Ilz coming from the bottom of the photograph with clearer, apparently darker water. Engineers needing to solve problems such as these confluences could in general use the simple approach that has been suggested above. However, in this particular problem, what has not been included in the above theory using homogeneous fluids, is the fact that the Inn water seems to be heavier than that of the Danube, and plunges underneath it at the centre top of the photograph, as well, of course, as starting to mix laterally right from the point of confluence so that at the interface, with a smaller density difference, the plunge occurs further downstream. Our section-averaged one-dimensional model could describe the overall problem approximately, but not the details of the inhomogeneous flow phenomena, which is where we end our presentation.

References

Aberle J, Smart GM (2003) The influence of roughness structure on flow resistance on steep slopes. J Hydraul Res 41:259–269

ASCE Task Force on Friction Factors in Open Channels (1963) Friction factors in open channels. J Hydraulics Div ASCE 89(HY2):97–143

Barnes HH (1967) Roughness characteristics of natural channels. Water-Supply Paper 1849, U.S. Geological Survey

Batchelor GK (2000) An introduction to fluid dynamics. Cambridge University Press, Cambridge
Birta LG, Arbez G (2013) Modelling and simulation, 2nd edn. Springer, London
Chow VT (1959) Open-channel hydraulics. McGraw-Hill, New York
Fenton JD (2003) The effects of obstacles on surface levels and boundary resistance in open channels. In: Proceedings of the 30th IAHR congress, vol C2, 25–29 August, Thessaloniki, pp 9–16
Fenton JD (2005) On the energy and momentum principles in hydraulics. In: Proceedings of the 31st congress IAHR, 11–16 September. IAHR, Seoul, pp 625–636
Fenton JD (2010) The long wave equations. Technical report, Alternative Hydraulics Paper 1. http://johndfenton.com/Papers/01-The-long-wave-equations.pdf
Fenton JD, Abbott JE (1977) Initial movement of grains on a stream bed: the effect of relative protrusion. Proc R Soc Lond Ser A 352:523–537
Fenton JD, Nalder GV (1995) Long wave equations for waterways curved in plan. In: Proceedings of the 26th congress IAHR, London, vol 1, pp 573–578
Fenton JD, Keller RJ (2001) The calculation of streamflow from measurements of stage. Technical report 01/6, Cooperative Research Centre for Catchment Hydrology, Melbourne. http://johndfenton.com/Papers/Calculation-of-streamflow-from-measurements-of-stage.pdf
Hamming RW (1973) Numerical methods for scientists and engineers, 2nd edn. McGraw-Hill, New York
HEC (1995) A comparison of the one-dimensional bridge hydraulic routines from: HEC-RAS, HEC-2 and WSPRO. Report RD-41, US Army Corps of Engineers, Institute for Water Resources, Hydrologic Engineering Center
HEC-RAS (2010) HEC-RAS river analysis system hydraulic reference manual. Report CPD-69, Version 4.1, US Army Corps of Engineers, Institute for Water Resources, Hydrologic Engineering Center
Hicks DM, Mason PD (1991) Roughness Characteristics of New Zealand Rivers. DSIR Marine and Freshwater, Wellington
Liggett JA (1993) Critical depth, velocity profiles, and averaging. J Irriga Drainage Eng 119 (2):416–422
Lighthill MJ, Whitham GB (1955) On kinematic waves. I: flood movement in long rivers. Proc R Soc Lond Ser A 229:281–316
Milne-Thomson LM (1968) Theoretical hydrodynamics, 5th edn. Macmillan, London
Pagliara S, Das R, Carnacina I (2008) Flow resistance in large-scale roughness condition. Can J Civ Eng 35(11):1285–1293
Ponce VM, Simons DB (1977) Shallow wave propagation in open channel flow. J Hydraul Div ASCE 103(12):1461–1476
Simons DB, Richardson EV (1962) The effect of bed roughness on depth-discharge relations in alluvial channels. Water-Supply Paper 1498-E, U.S. Geological Survey
Soroush F, Fenton JD, Mostafazadeh-Fard B, Mousavi S, Abbasi F (2013) Simulation of furrow irrigation using the slow-change/slow-flow equation. Agric Water Manag 116:160–174
White FM (2009) Fluid mechanics, 7th edn. McGraw-Hill, New York
Woolhiser DA, Liggett JA (1967) Unsteady, one-dimensional flow over a plane—the rising hydrograph. Water Resour Res 3(3):753–771
Yen BC (2002) Open channel flow resistance. J Hydraul Eng 128(1):20–39

Chapter 2
Turbulence in Rivers

Mário J. Franca and Maurizio Brocchini

Abstract The study of turbulence has always been a challenge for scientists working on geophysical flows. Turbulent flows are common in nature and have an important role in geophysical disciplines such as river morphology, landscape modeling, atmospheric dynamics and ocean currents. At present, new measurement and observation techniques suitable for fieldwork can be combined with laboratory and theoretical work to advance the understanding of river processes. Nevertheless, despite more than a century of attempts to correctly formalize turbulent flows, much still remains to be done by researchers and engineers working in hydraulics and fluid mechanics. In this contribution we introduce a general framework for the analysis of river turbulence. We revisit some findings and theoretical frameworks and provide a critical analysis of where the study of turbulence is important and how to include detailed information of this in the analysis of fluvial processes. We also provide a perspective of some general aspects that are essential for researchers/practitioners addressing the subject for the first time. Furthermore, we show some results of interest to scientists and engineers working on river flows.

Keywords Turbulence · Scales · Space-frame · Time-frame · River flow

2.1 Introduction

Similar to most flows of natural fluids, riverine flows are typically turbulent: turbulence is ubiquitous and represents a fundamental engine of transport, spreading and mixing. In particular, turbulence is the main sink of riverine flow total energy (E). Large turbulent eddies are responsible for the conversion of total flow energy

---

M.J. Franca (✉) · M. Brocchini
Laboratoire de Constructions Hydraulique, École Polytechnique Fédérale de Lausanne, Lausanne, Switzerland
e-mail: mario.franca@epfl.ch

M. Brocchini
DICEA, Università Politecnica delle Marche, Via Brecce Bianche 12, 60131 Ancona, Italy

P. Rowiński and A. Radecki-Pawlik (eds.), *Rivers—Physical, Fluvial and Environmental Processes*, GeoPlanet: Earth and Planetary Sciences,
DOI 10.1007/978-3-319-17719-9_2

into turbulent energy; the former break down into mid-sized eddies, which in turn become increasingly smaller until they are destroyed by viscosity, wherein the flow energy is finally lost to heat. Furthermore, turbulence is responsible for physical processes in rivers such as the transport and mixing of diluted or undiluted substances, temperature transfer, sediment motion and suspension, and geomorphological evolution. Hence, the consideration of turbulence is fundamental to riverine environmental applications and requires the use of an adequate theoretical analysis [see, for instance, the amount of research dedicated to turbulence in fluvial hydraulics recently published in Schleiss et al. (2014)].

Although more than 120 years have passed since the first Reynolds experiments and after over 80 years since Taylor's first attempts to build a mathematical framework for turbulent flows, much still remains to be done and river turbulence is still a challenge for researchers and engineers working in hydraulics and fluid mechanics [paradigmatically, Enzo Levi names a section of his book *That Annoying Turbulence* (Levi 1995)]. However, a proper description of river turbulence is fundamental to the evolution of recently emerged ecologically oriented research areas pertaining to river flows, which include eco-geomorphology, bio-geomorphology, eco-hydrology, eco-hydraulics and environmental hydraulics (Nikora 2010).

The rather generic definition of a turbulent flow as a highly unpredictable flow characterized by many scales suggests some quantification for such a definition. Hence, when a variable of a fluid flow fluctuates in time or space, with non-zero second-order or higher statistical moments calculated along these dimensions, we are in the presence of a turbulent flow. Flow variables typically include velocity and pressure, although density, concentration of solids and temperature are also imprinted by the characteristics of turbulence. Typical of turbulence is the influence of many different interacting time/space scales in the dynamics of the flow.

The Reynolds decomposition, which was introduced more than 100 years ago (Levi 1995), divides fluid variables into a mean and a fluctuating field (Monin and Yaglom 1971; Frisch 1995; Pope 2000). This decomposition is applied to the hydrodynamic equations and it is the first step to account for the fluctuation of the variables due to turbulence (see Fig. 2.1 and, later, Sect. 2.2.2). The effect of turbulence in the flows involves the following processes, which may be properly formalized mathematically: ensemble advection by the mean flow; dispersion, which is due to the diversity of movements at different scales; turbulent advection or transport by eddies associated with the fluctuating field; and viscous or molecular diffusion due to agitation at very small scales (Chassaing 2000). These processes influence the environmental properties of the flows and represent the main engines of transport, mixing, entrainment, detrainment and dissipation.

Other structural characteristics associated with the time and space heterogeneity of the flows, which are thus associated with fluctuations in the fluid properties, include secondary currents and large-scale vortices either with horizontal or vertical axis, which may be considered *quasi*-steady structures, and waves (surface or internal) traveling within the river, with time scales that are large and with space scales that are comparable with the water depth and the river width. They may be conditioned or even generated by turbulent instability at smaller scales and, in turn,

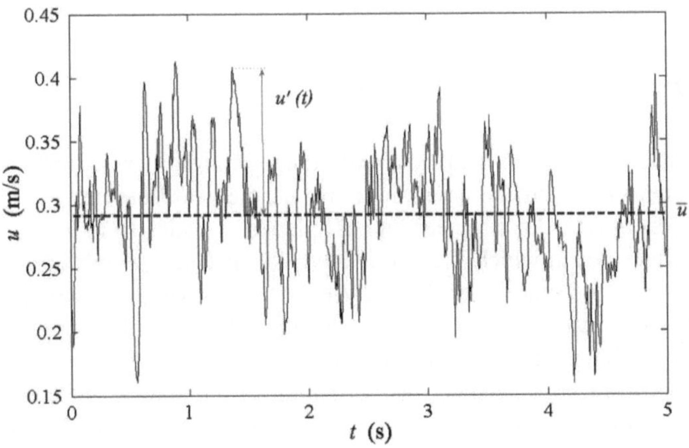

Fig. 2.1 Turbulent velocity signal and illustration of the application of Reynolds decomposition. t is the time in seconds, u is velocity, the *overbar* denotes turbulence-averaged components and ′ stands for the fluctuating component; see Sect. 2.2.2

may be responsible for changes in the overall turbulence of the flow, and include secondary currents in river bends (Blanckaert and De Vriend 2005; Termini and Piraino 2011; Nikora and Roy 2012; Abad et al. 2013), horizontal coherent vortices at the boundary between the main channel and floodplains (Knight and Shiono 1990; Shiono and Muto 1998; van Prooijen et al. 2005; Proust et al. 2013), and low-frequency large-scale flow structures traveling in low-submergence flows (Kirkbride and Ferguson 1995; Roy et al. 2004; Franca and Lemmin 2014).

The present paper is organized with a structure that focuses on the main methodological approaches used to address turbulent fluvial flows (e.g. time/space analysis, decomposition, etc.), rather than in terms of phenomenological aspects (e.g. coherent structures, mixing, etc.). The theoretical aspects discussed herein pertain to fluvial flows and concern the most common problems encountered by river engineers and scientists, namely, how to define time and spatial frameworks for turbulence analysis, the relation between turbulence and the transport of substances, diluted and undiluted (typical sediments), and simplifications that may be made in the analysis of river flows.

In the following, we first introduce basic definitions related to turbulence in rivers, including a discussion on scales characterizing fluvial flows; a presentation of the basic hydrodynamic equations used to address momentum, mass and species in the fluid; a description of energetic processes; and a discussion on several possible frameworks that can be used to study turbulent flows. Turbulent flows are characterized by fluctuations of the flow variables both in time and space; in view of the above-mentioned differences in time and space averaging, two analysis frameworks are discussed by means of practical examples. Concluding remarks are given at the end of the present contribution.

2.2 Background

2.2.1 Typical Scales of River Flows

River dynamics is characterized by a wide range of scales, which may vary from seconds to years (or even centuries), in terms of time, and from mm to km in terms of spatial dimensions. An engineer typically looks at scales from cm to km when addressing river flows, which may be related, for instance, to infrastructures constructed in the fluvial space or to the analyses of floods, whereas a biologist may be concerned with phenomena occurring at lenghts of mm or less. In view of our analysis, in which temporal and spatial analyses complement each other, we inspect typical scales from both perspectives pertaining to the analysis of environmental physical processes in rivers. Figure 2.2, which is adapted from Nikora (2007), represents an interpretation of how the flow energy is distributed through temporal and spatial scales present in fluvial systems.

Tiny time scales, which are related to viscous diffusion and dissipation processes, are orders of magnitude smaller than seconds and are ubiquitous within the flow body. Floods with return periods of years, or even centuries, although less frequent, are responsible for important geomorphic processes, and drastically change the shape of the rivers. Regular floods in rivers (with small return periods, i.e., one to tens of years) destroy riverbed and bank armoring, feed the sediment continuum along the rivers and promote the regular biota renovation of the riverbed and floodplains. Although being less energetic events, secondary currents and flow cells typically generated locally, continuously shape the river morphology by their persistence in time. Turbulent coherent structures have time scales of seconds but they may transmit sufficient momentum to promote sediment entrainment and suspension.

The river basin scale (on the order of km) regulates the amount of water, sediments and organic matter arriving at a given river section. Furthermore, rivers can be regarded as open-channel flows with highly heterogeneous beds and irregular boundaries. These different geometric scales influence the flow structure and are thus related to the scales of the turbulence. Spatial scales simultaneously present in the fluvial milieu are due to grain roughness (grain-scale), bedforms (river width-scale), protuberant elements (grain scale to river width scale; this is especially important in low relative submergence flows), and channel configuration (valley scale, e.g., bends and braiding). Natural obstacles (e.g., riffles, pools, tree trunks, and root wads) and man-made structures (e.g., bridge foundations, groynes, and stream restoration structures) introduce further complexity by adding locally induced scales to the flow. Scales of orders of magnitude smaller than mm exist and are linked to micro-organisms and to molecular and viscous processes such as diffusion and energy dissipation. Mean advection is found at the larger scales of the rivers and is thus conditioned by the channel configuration. Turbulence generation is typically influenced by flow boundaries where the grain roughness, bedforms and

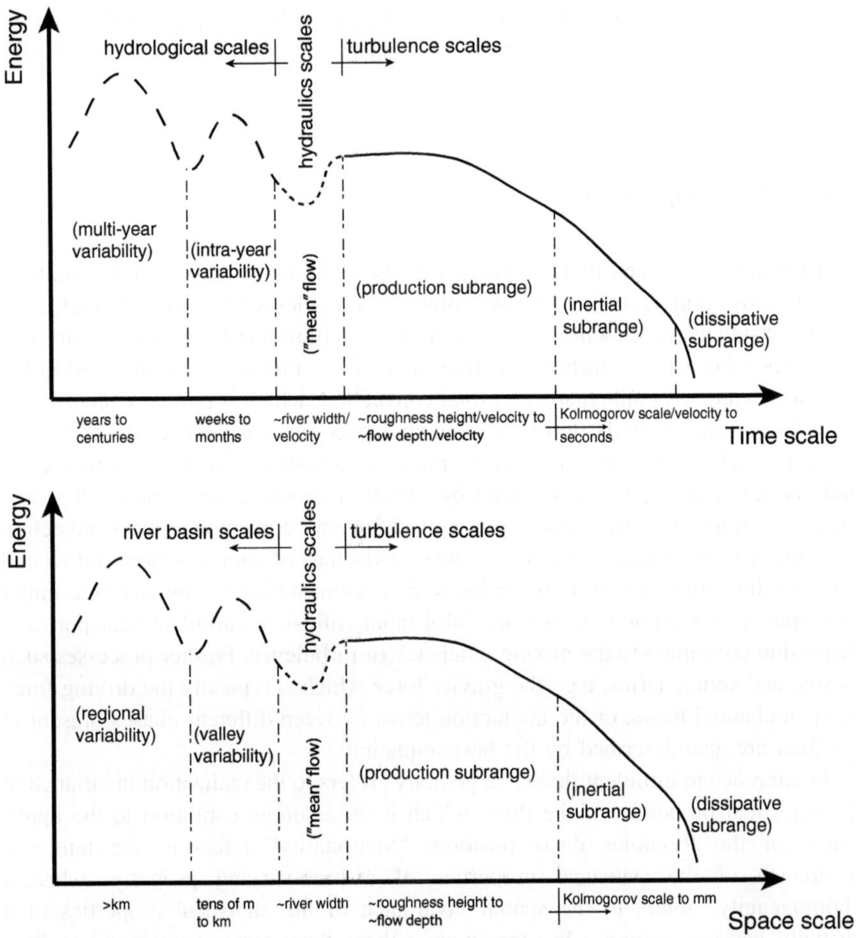

Fig. 2.2 Distribution of the flow energy through time and spatial scales in river systems (adapted from Nikora 2007)

channel protuberances may play a role. Turbulent structures are, in turn, responsible for the continuous shaping of rivers and modify the roughness and geometry of the river boundaries.

The development of a theoretical model capable of solving such a variety of scales is nearly impossible. Restrictions, whereby a selective scaling that is limited to the necessary and sufficient detail of the problem under study, must be imposed. Examples of practical applications of the "restricted scale treatment" include Large Eddy Simulation (LES) models, where turbulence at large scales is resolved, whereas the smaller scales are modeled, and the Double-Averaging Methodology (DAM), which is a conceptual framework whereby up-scaling is performed in the spatial and temporal sense, the flow is described at the largest spatial scales and the effects of spatial variability are included in the basic equations by means of dispersive-type terms (see Sect. 2.4.1).

2.2.2 Basic Equations

Conservation equations of flow properties, in addition to mass and momentum, include those corresponding to environmental variables such as diluted and undiluted substances, sediments and temperature. Within these equations, physical processes related to the turbulent nature of the flow affecting the conserved quantities are expressed mathematically: total convection, which is typically represented by a local (time partial) derivative and an advection term related to the mean advective field; diffusion, which corresponds to processes occurring at molecular and viscous scales and is represented by spatial gradients at very small scales; and turbulent transport (also called turbulent diffusion) corresponding to advection resulting from turbulent fluctuations. The co-existence of various scales of flow and flow pathlines in turbulent flows induces dispersion, which can be analyzed either in a spatial or temporal framework. Molecular diffusion, turbulent transport and dispersion contribute to the mixing efficiency of turbulence. Further processes such as sink and source terms, e.g., the gravity force which is typically the driving force of open-channel flows, or the interaction terms between different phases present in the fluid are also described by the basic equations.

In reference to turbulent flows, “ergodicity” refers to the realization invariance of the statistical properties of the flow, which is an essential condition to the application of the Reynolds decomposition. “Stationarity” refers to the temporal invariance of the statistical properties of a time-varying property, whereas “homogeneity” refers to the spatial invariance of the statistical properties of a spatially varying property. In other words, these three terms correspond to flow fields whereby the statistical properties of the turbulence are independent of the realization, time and position, respectively. Isotropic turbulence exists when the statistical properties are invariant, regardless of the direction considered for their analysis. Homogeneity and isotropy of turbulence are rarely found in nature but can be considered under given hypotheses and limited to several scales, hence enabling important simplifications in the analysis of the flows (Chassaing 2000; Pope 2000).

The Navier-Stokes equations, which correspond to the momentum equations of the flow, are likely adequate to describe the entire turbulence within a flow (Frisch 1995). Averaging the Navier-Stokes equations over the realization/time/space after application of Reynolds decomposition ($\theta_j = \overline{\theta}_j + \theta_j'$, where θ is any generic flow variable susceptible to turbulence, the overbar stands for the turbulence-averaged component, and $'$ stands for the fluctuating component) produces the so-called Reynolds-Averaged Navier-Stokes equations (RANS), which, together with the mass conservation equation, for an incompressible although not necessarily homogeneous fluid, become

$$\underbrace{\frac{\partial(\bar{\rho}\bar{u}_j)}{\partial t}}_{I}+\underbrace{\frac{\partial\left(\overline{\rho' u'_j}\right)}{\partial t}}_{II}+\underbrace{\frac{\partial(\bar{u}_k\bar{\rho}\bar{u}_j)}{\partial x_k}}_{III}+\underbrace{\frac{\partial(\bar{u}_k\overline{\rho' u'_j})}{\partial x_k}}_{IV}+\underbrace{\frac{\partial(\bar{u}_j\overline{\rho' u'_k})}{\partial x_k}}_{V}$$
$$=-\underbrace{\frac{\partial\bar{p}}{\partial x_j}}_{VI}+\underbrace{\mu\frac{\partial^2\bar{u}_j}{\partial x_k\partial x_k}}_{VII}-\underbrace{\frac{\partial\bar{\rho}\overline{u'_k u'_j}}{\partial x_k}}_{VIII}-\underbrace{\frac{\partial\overline{\rho' u'_k u'_j}}{\partial x_k}}_{IX}+\underbrace{\bar{\rho}g_j}_{X} \tag{2.1}$$

$$\frac{\partial\bar{u}_k}{\partial x_k}=0 \tag{2.2}$$

where k and j are indices running from 1 to 3 and represent the three Cartesian spatial directions (1 ≡ streamwise; 2 ≡ spanwise; 3 ≡ vertical); u stands for the velocity; x stands for the spatial coordinate; ρ stands for the density (time and space variant); t stands for time; p stands for pressure; μ stands for dynamic fluid viscosity; and g stands for the gravitational acceleration. After the application of the Reynolds decomposition followed by realization/time/space averaging, and after comparing Eq. (2.1) to the Navier-Stokes equations written for instantaneous quantities, new terms appear in the momentum equation and introduce non-linearity; these new terms are related to the fluctuations in velocity and density. The terms in the momentum Eq. (2.1), which is also known as Reynolds equation, denoted using roman capital letters are as follows: *I*, local derivative of averaged momentum; *II*, local derivative of momentum fluctuation; *III*, mean advection of averaged momentum; *IV*, mean advection of momentum fluctuation; *V*, mean advection in the j direction of the momentum fluctuation; *VI*, averaged pressure term; *VII*, viscous diffusion term; *VIII*, Reynolds stress term; *IX*, turbulent transport of momentum fluctuation (turbulent diffusion); and *X*, average of the body forces term or, in this case, simply the gravity term.

The quantity $\overline{u'_k u'_j}$ in RANS equations is the so-called Reynolds stress tensor, which is normalized by the density, and represents an additional stress in the flow compared to the Navier-Stokes equations. This non-linear term results from the Reynolds-averaging procedure of the Navier-Stokes equations. The elements in this tensor represent second-order moments of the fluctuating velocity field, normal moments or variances for the main diagonal elements and cross moments or co-variances for the non-diagonal elements. The main diagonal elements ($\overline{u'_k u'_j}$, where $k = j$) represent normal stresses along the three Cartesian directions ($k = 1, 2, 3$) and shear elements from this tensor ($\overline{u'_k u'_j}$, where $k \neq j$) correspond to shear stresses. The half trace of the Reynolds stress tensor per unit mass corresponds to the average turbulent kinetic energy (*tke*) of the flow (also per unit mass): $\frac{1}{2}\overline{u'_k u'_k}$. The anisotropy tensor is defined using the Reynolds stress tensor and is the basis for the analysis of turbulence states based on the Lumley triangle (Lumley and Newman 1977; Chassaing 2000; Jovanovic 2004; Dey 2014) (see Sect. 2.2.4).

The *tke* is extracted from the flow total energy (E), which, for an incompressible flow, is essentially represented by the kinetic energy $E \approx K = \frac{1}{2} u_k u_k$. The mean kinetic energy (*mke*) of the flow is composed of a contribution corresponding to the kinetic energy of the mean flow $\frac{1}{2}\overline{u_k}\,\overline{u_k}$ added to the *tke* ($\frac{1}{2}\overline{u_k' u_k'}$). The Reynolds stress tensor and mean and turbulent kinetic energy (*mke* and *tke*) have well-defined transport equations, as detailed in Chassaing (2000), Pope (2000), among others. A dissipation term is present in all these equations and is dependent on the (mean or fluctuating) strain rate.

Empirical expressions describing the distributions of the Reynolds stress tensor components and of the *tke* are abundant in the literature and provide good results for uniform flows in hydraulically smooth beds (Cardoso et al. 1989; Nezu and Nakagawa 1993; Kironoto and Graf 1994). However, when the flow bed is hydraulically rough, which is typical of fluvial flows, the vertical distribution on the turbulence quantities is locally dependent of the bed forms below the height where the influence of the bed is felt (see Nikora and Smart 1997; Smart 1999; Nicholas 2001; Franca 2005b; Franca and Lemmin 2006b, among others). This inner region of the flow corresponds to the "roughness layer" (Nikora and Smart 1997).

Turbulent transport (or diffusion), as shown in term *IX* of Eq. (2.1), is difficult to determine experimentally; usually, it is either taken as negligible or modeled by a Fickian-type law. For homogeneous fluids, such as clear-water flow, the density is constant, and thus, $\rho = \overline{\rho}$ and $\rho' = 0$, which considerably simplifies the transport equations. For a non-homogeneous system where the density differences in the fluid continuum are relatively small, the Boussinesq approximation may be used to simplify the hydrodynamic equations: the relative density variation is taken to be negligible when multiplying inertial terms and non-negligible when multiplying gravitational terms (Tennekes and Lumley 1972). Typically, the latter represents the buoyancy, which is the driving force of fluid movement due to density differences such as density currents (Simpson 1997).

For environmental applications, and in addition to momentum and mass conservation equations, other transport laws may be introduced for other environmental variables, such as the mass of substances, diluted or undiluted in the water and temperature (all these variables are turbulent variables and are susceptible to the Reynolds decomposition). Furthermore, the Reynolds-averaged transport equations for passive diluted and undiluted scalars (e.g., salt and sediments, respectively) may be written in terms of the volume fraction of the diluted species (Φ_s) or particle concentration of the undiluted species, respectively (c).

When the species is diluted (Φ_s), the transport velocity corresponds to the fluid velocity. When the species is undiluted (c), however, particles transported by the flow have their own momentum; thus, the advection field is given by the particle velocity (u^p) rather than by the fluid velocity, which is generally different. The behavior of species particles can be assessed as a function of the Stokes number, which is defined as the ratio between the time scale of particles to react and the Kolmogorov time scale (e.g., Soldati and Marchioli 2009). "Massive particles", or "inertial particles", are characterized by a Stokes number of order one or larger, whereas "water particles" are characterized by a vanishingly small Stokes number. The Stokes number significantly influences particle suspension and deposition; for example, large Stokes number particles that possess sufficient momentum may either coast through some accretion region and deposit by impacting directly at the flow bed or go into suspension. Otherwise, those particles that have not received sufficient momentum are forced to deposit on the bed due to gravity and inertia.

Species diffusion is typically modeled as a Fickian process, in practice by $\Gamma_m \frac{\partial^2}{\partial x_k^2}$, where Γ_m is a molecular diffusivity. Turbulent transport or diffusion may also be modeled as a Fickian process with Γ_t, which is the so-called turbulent diffusivity. In Fickian processes the property flux travels from regions of higher concentration to regions of lower concentration, proportionally to the concentration gradient. For a non-conservative substance (for instance suspended sediments that can deposit or erode from the riverbed), sink or source terms must be added when the integration of the species conservation equations is made.

Further transport equations pertinent to the analysis of environmental turbulent flows, such as those in terms of pressure, fluctuating momentum, Reynolds stress tensor, kinetic energy (mean, turbulent and instantaneous), species fluctuation and variance, and energy dissipation (cf. Monin and Yaglom 1971; Frisch 1995; Chassaing 2000; Pope 2000, among others), although not shown here, may be derived from these basic equations.

2.2.3 Energy-Based Description of Processes

As stated in Sect. 2.1, the energy transfer for which turbulence is responsible generally evolves through a so-called cascading process from the larger scales, which extract energy from the total flow, thus generating turbulent kinetic energy, to very small scales, where viscous dissipation takes place; the kinetic energy of the flow is eventually lost as heat. At the intermediate scales, an eddy fragmentation process occurs with minimal energy transfer; only break-up of the large scales into smaller ones is observed without substantially influencing the energy content of the flow. The scales of the intermediate eddies corresponding to this transfer belong to the so-called inertial sub-range (in contrast to the productive and dissipative sub-ranges for large and small eddies, respectively). The first description of the energy cascade process of turbulent flows was provided by Richardson in the 1920s and was later formalized and included in the papers on the theory of turbulence published by Kolmogorov in 1941 (Frisch 1995). Since then, much work has also been performed on nonlocal turbulence processes, in which eddies of sizes within the available ranges significantly interact (e.g. Nazarenko and Laval 2000). However, the authors are not aware of specific studies dedicated to riverine flows.

At the flow solid frontiers of open-channels, due to the non-slip boundary condition at the river bed and banks which are fixed (non-moving), the velocities are zero. Due to viscous effects, a thin layer exists near these frontiers where velocities are very small. Here, the flow is consequently laminar and this region of the flow is called the laminar or viscous layer (cf. Chassaing 2000). Important velocity gradients exist and generate unstable flow conditions, where the generation of turbulent eddies is promoted. Therefore, bounded flows have one main source of turbulence generation in their solid boundaries. Furthermore, the presence of solid boundaries protruding in the flow typically induces flow separation and, consequently, recirculation cells in shaded areas. Again, strong gradients between the lee of these protuberances and the external flow produce new turbulent structures that may be advected downstream, thus conditioning the flow structure. The size of the eddies generated by individual boundary protuberances scales, typically, with the dimension of the protruding elements. Examples of boundary singularities and irregularities in rivers causing eddy generation include sediment clusters or boulders protruding from the riverbed [submerged (Buffin-Bélanger and Roy 1998; Buffin-Bélanger et al. 2000; Franca 2005b) or emerging (Tritico and Hotchkiss 2005)], high relative roughness riverbeds (Baiamonte et al. 1995; Katul et al. 2008), vegetation patches (Tanino and Nepf 2008; Siniscalchi et al. 2012; Sukhodolova and Sukhodolov 2012; Ricardo 2014), wood debris or remains of organic elements (Shields et al. 2004; Blanckaert et al. 2014).

The wide range of the above-mentioned turbulent eddies are macroscale coherent structures belonging to the productive range. Coherent structures are formed, typically, in shear zones such as the bottom boundary layer and depth transition layers, and their generation is due to the interaction between regions with different momenta. They have a recognized role in the mechanisms of sediment entrainment in river flows (Séchet and Le Guennec 1999; Cellino and Lemmin 2004). At rough boundaries, which in rivers are typically rough sand or gravel beds and banks, dune and ripple fields, subaquatic plant canopies, or vegetation patches, a continuous shear zone is imposed on the flow, and periodical generation of coherent structures is observed, corresponding to a bursting process.

Chassaing (2000) introduced the concept of physical coherence together with statistical coherence. The physical coherence concept is related to the fact that turbulent structure properties (for example, geometric, kinematics and dynamics) change relatively slowly with respect to their representative scale domain; turbulent structure properties are, consequently, conveyed by the flow field. Coherent structures have a short life cycle; they cannot be identified with a time-averaged analysis and require an investigation based on time and space correlations or on flow visualization techniques. In laminar and transitional flows, coherent structures occur periodically, whereas in turbulent flows they occur chaotically in space and time. Hence, in addition to flow visualization techniques, conditional sampling and statistical techniques have to be used in the detection and characterization of coherent structures. Large- and small-scale coherent motion within the coherent structure range can still be distinguished. Large-scale motion typically scales with the boundary layer thickness (eddies scale with the flow depth in open-channel flows or with the depth-transition region in compound channels), whereas small-scale motion scales with parcels of the boundary layer (eddies scale with dimensions comparable to the roughness sizes). Each of these groups acts differently but possibly jointly in the flow energy balance.

Large-scale flow structures and secondary currents are characterized by their quasi-steadiness in time; however, they introduce 3D complexity to river flows. They are associated with mesoscale features (river depth and width) in the river geometry such as bends, meanders, floodplains and section changes. Large-scale flow structures can also be defined in the time frame and include waves and pulses traveling in the flow and interacting with the turbulent field.

2.2.4 Frameworks for Turbulence Analysis

The Eulerian and Lagrangian descriptions of the flows are equally valid for the derivation of the hydrodynamic equations and are easily related through the velocity field. However, the Eulerian approach is the most common in practice (Currie 1993). Both descriptions are associated with different techniques of flow observation, the Eulerian corresponding to an approach wherein the fluid variables (velocity, pressure, etc.) are measured continuously in time at one or more fixed points and the Lagrangian corresponding to an approach where the fluid variables are followed along the flow trajectories (i.e., the positions of particles or of diluted substances as a function of time). While Lagrangian techniques naturally provide a measure of dispersion and diffusion in turbulent flows, from a practical point of view the Eulerian techniques are simpler and widely used (cf. Romano et al. 2007; Ghilardi et al. 2014 for reviews of measurement techniques of turbulent flows and rivers).

Time and space scales of turbulence are often linked through the Taylor frozen-turbulence hypothesis by means of an advection velocity. This permits the linking, for instance, of the partition of the turbulent energy through scales pertaining to both dimensions. However, to apply Taylor's hypothesis and thus translate Eulerian observations in time into the space domain, the turbulent eddies are taken to be undeformable by the transport imposed by the mean and turbulent velocity fields (Chassaing 2000). Thus, generally, two analytical frameworks for the study of turbulent river flows pertaining to the two dimensions (time and space) are needed. Because time and space averaging do not commute for nonlinear processes, both are generally used independently as functions of the process of interest.

The technique based on anisotropy invariants proposed by Lumley and Newman (1977) (see, also, Chassaing 2000; Jovanovic 2004; Dey 2014 for further details) provides a methodology of classifying the anisotropy degree and nature of turbulence, which is irrespective of the reference system. The so-called Lumley triangle technique (or Lumley plots), which contains any turbulence state within its limits, enables the identification of the state of turbulence being studied in the flow regions in an analysis: 3D isotropic turbulence, 2D isotropic turbulence, and 1D turbulence. In the transition between these limiting states of turbulence, two characteristic types of turbulent structures are found: pancake-shaped turbulence, which corresponds to a situation where two of the fluctuation components are equal and considerably larger than the third, and cigar-shaped structures where two of the turbulence components, and hence the normal stresses values, are equal and the third has a substantially higher value. The analysis based on the Lumley plots, combined with other observational techniques, such as the quadrant analysis (Lu and Willmarth 1973; Nakagawa and Nezu 1977), permits the establishment of a proper framework for the analysis of turbulence and forms the basis of useful simplifications for the analysis of complex 3D flows (Mera et al. 2014). The knowledge of the nature of turbulent eddies in river flows allows the establishment of restricted spatial frameworks for the analysis of these flows, as will be seen in Sect. 2.4.

2.3 Time-Frame Analysis

2.3.1 Steady Flows

As mentioned in Sect. 2.2.2, stationarity or steadiness implies temporal invariance of the flow statistical properties; however, this enables spatial variations. In formal terms, and considering the property "mean momentum" as an example, this means that the local derivative in Eq. (2.1) (term *I*) does not exist, thus considerably simplifying the practical resolution of hydrodynamic equations.

The concept of stationarity is related to the time scale of the problem and to the statistical property under analysis. Considering a given time-scale and a statistical property, we consider the flow stationary when the value of this statistical property is stable, i.e., it does not change with the increase of the time-scale. This stationarity is conditioned to the statistical property under analysis; in the study of first-order moments (means), the time scale where these are stable is smaller than when considering higher order moments. Common statistical properties used in the analysis of turbulent river flows are the Reynolds stress tensor and the *tke* (based on second-order statistical moments). For these properties, the statistical invariance (defining an appropriate stationarity) is less demanding in terms of the time scale compared to, for instance, analyzing turbulent transport (third-order moment), as in the term *IX* of Eq. (2.1).

In the practical analysis of natural floods in a limited river reach, engineers commonly consider a constant peak flood discharge for flow modeling, which may last for several hours. This assumption of constant discharge is valid if the time response of the reach under analysis to flow changes is less than the duration of the peak discharge, i.e., if the travel time of the flood wave along the reach is smaller than the latter and if there is no considerable spatial variation in the discharge. However, in the analysis of an entire river basin, the time response of the reach, which may be comparable to the concentration time of the catchment, is not compatible with a typical flood duration; thus, stationarity can no longer be used.

The theoretical advancements in the understanding of complex 3D flows by physical modeling are typically made considering steady-state conditions. The stationary approach is valid and, sometimes, is the best method to study fundamental aspects of turbulent flows in an isolated fashion. The spatial characterization of turbulent quantities obtained under steady-state flow conditions allows identifying, quantifying and understanding the mechanisms of the transport of undiluted or diluted species (including dispersive and diffusive processes) and of the processes of erosion and deposition which are related to the shaping of the river morphology.

Under steady-state conditions, the determination of turbulence quantities, such as the Reynolds stresses [term *VIII* of Eq. (2.1)], is straightforward (Chassaing 2000). The turbulent structure of the flow in complex geometries and the interpretation of the related fluvial processes are determined through the analysis of the spatial distribution of the mean velocity and vorticity fields and of the second- and third-order moments present in the conservation equations, provided that the data is sufficient to obtain stable statistical moments. Leite Ribeiro et al. (2002), Proust et al. (2013), Ricardo et al. (2014) are examples of recent theoretical analyses of fluvial turbulent processes with highly complex boundary geometries performed by assuming steady-state flow conditions. When performing field studies, which are typically based on measurements of instantaneous velocities, researchers pay special attention to the steadiness of the flow to assure that measurements made in multiple spatial positions are representative and can be combined in a time-averaged description of the flow (Franca 2005a; Saggiori et al. 2012).

2.3.2 Unsteady Flows

When unsteadiness can no longer be ignored, i.e., when the time scale under analysis is larger than that of the mean flow variations, full Reynolds equations have to be considered. In unsteady flows, the inference of statistical moments may be made by means of different types of average operators (for their definition, cf., for instance, Chassaing 2000).

If in a turbulent flow the processes are ergodic, i.e., the flow statistical properties are invariant with their realization, the ensemble average allows proper estimates of turbulent properties of the flows. This consists of taking the average of all the possible realizations of one turbulent flow (Pokrajac and Kikkert 2011) and it is applied to transient processes in river flows such as wave passage, bore passage, dam-break flows, breaking waves, and ship waves (cf., Fig. 2.3).

In transient phenomena, a time interval may exist wherein the flow properties remain stationary, i.e., where statistical properties are invariant for some duration. The level of invariance of the statistical properties depends on the flow turbulence level and on the time interval: for instance, we may be able to obtain stable time-averaged quantities and stable quantities derived from second-order moments (Reynolds stresses, *tke*, etc.) but not stable higher statistical moments, hence the designation of *quasi*-stationary flows. For these cases, we may apply a zone or conditioned average along the *quasi*-stationary period to infer turbulent properties: recent examples of the application of zone averages to riverine flows include (Aleixo 2013), who used this approach for the statistical analysis of dam-break

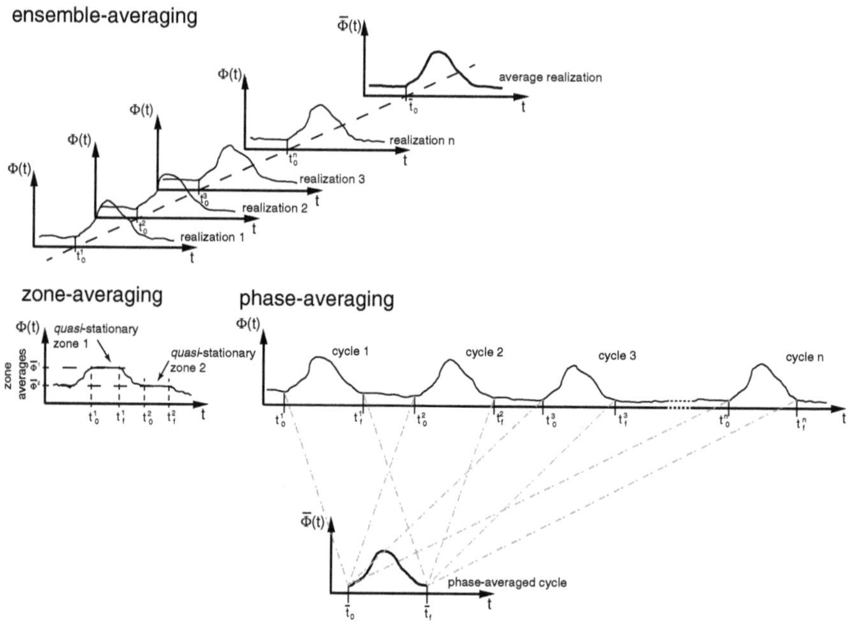

Fig. 2.3 Illustration of the type of averages that can be applied to turbulent signals: ensemble average, zone average and phase average

flows based on PIV high-frequency measurements, and (Nogueira 2014), who applied zone averages to obtain mean velocity profiles and Reynolds stresses in a lock-exchange density flow based on PIV measurements.

In the case of a periodic phenomenon, an average may be obtained over different cycles, with each being comparable to a realization of the phenomenon, to estimate turbulent properties; this is called phase (or also composite) averaging. Such an average is adequate for the analysis of time/space flow features which have a limited life cycle, and pseudo-periodicity is verified (pseudo-periodic cycles do not necessarily have a constant period and are not necessarily consecutive). Conditional sampling and statistical techniques have to be used in the detection and characterization of the cyclic processes. Techniques used for the detection of periodic events in open-channel flows include quadrant threshold analysis (Willmarth and Lu 1972; Nakagawa and Nezu 1977, among others), wavelet decomposition (Foufoula-Georgiou and Kumar 1994; Yoshida and Nezu 2004; Franca and Lemmin 2006a, among others), proper orthogonal decomposition (Berkooz et al. 1993; Holmes et al. 1998) and empirical mode decomposition (Huang et al. 1998; Franca and Lemmin 2014). Examples of the application of phase averages for obtaining statistical properties of turbulent flows include the following: Franca and Lemmin (2006a), where the identification and reconstruction of coherent structures was performed based on wavelet multi-resolution analysis for field measurements; Nogueira et al. (2014), where a cyclic pattern of growth and a decrease in the head of density currents was described and quantified in terms of kinematics and dynamics; and Franca and Lemmin (2014), where large-scale coherent structures were detected using Huang`s empirical mode decomposition (Huang et al. 1998) and reconstructed with phase-averaging techniques based on a Hilbert transform of the velocity signal (Huang et al. 1999).

Recently, several authors have developed methods to address upscaling techniques for flows with time-varying boundaries: Pokrajac and Kikkert (2011) presented the Reynolds-Averaged, Depth-Integrated Navier-Stokes equations applied to an upsloping transient bore where the water surface was not constant and contained air bubbles, and Nikora et al. (2013) presented double-averaged conservation equations for mobile-boundary conditions, namely, mobile rough beds. The range of application of these techniques is wide in the field of fluvial flows with time intermittent boundaries such as the mixing layer of density currents reproduced in laboratory (Lopes et al. 2013).

2.4 Space-Frame Analysis

2.4.1 Homogeneous and Heterogeneous Flows

Further simplifications of the hydrodynamic equations can be made by assuming the uniformity of the flow, implying that a spatial variation of flow velocities does not exist, $\frac{\partial}{\partial x_k} = 0$. A practical result of this simplification that is most frequently used by engineers is the relation between the bed shear stress and the slope of an open-channel, which, for a one-dimensional flow in a wide channel with rectangular cross-section, is $\tau_b = \rho g h S$ (h is the water depth, and S is the longitudinal bed slope of the channel). This relation is easily demonstrated when considering the one-dimensional integral form of the momentum conservation equation (Currie 1993) over a control volume (V_c) with surface S_c. For uniform flows, all terms become zero except the streamwise component of the gravitational force ($\int_{V_c} \rho g_1 dV$) and the bed shear force term ($\int_{S_c} \tau_b dS$), which must equilibrate, thereby resulting in the above expression for the determination of τ_b. In other words, the force corresponding to the streamwise component of the weight of the fluid inside the control volume (gravity-driven force) is balanced by the bed shear stress resistance force. This assumption thus requires the use of strong arguments, and it is commonly misused in highly varied flows such as mountain rivers or watercourses with complex geometries.

Although hard to find in natural rivers, uniform flows are commonly used to experimentally or numerically study turbulent flows. The empirical closures for the vertical distribution of the Reynolds stress tensor and *tke* in open-channel flows mentioned in Sect. 2.2.2 and that have been established and accepted for some decades are an example of this assumption.

Typically, natural rivers have boundaries which are very spatially varied and heterogeneous. For instance, for low relative submergence flows, i.e., for small values of h/D, where D is a geometric parameter that is representative of the bed roughness, there is no self similarity within the lower regions of the flow, especially within the troughs and crests of the riverbed. Here, the flow is highly 3D and heterogeneous, and accounting for spatial variability effects with the Reynolds equation becomes complex. To smoothen flow irregularities, a spatial averaging operation may be applied to the RANS equations, thus resulting in the so-called Double-Averaged (in both time and space) Navier-Stokes (DANS) equations (Nikora et al. 2007a). If averaging is performed at sufficiently large scales to smoothen the flow spatial heterogeneity, then uniformity may be assumed simplifying the hydrodynamic equations. The scale over which upscaling is performed should be sufficiently large to ensure that the statistical properties do not depend on the averaging volume anymore; thus, the flow may be treated as homogeneous. The variability of the flow at scales that are smaller than the upscaling domain is considered by means of additional (dispersive or form-induced) components that are incorporated into the DANS equations when these are derived. These additional terms characterize momentum and scalar fluxes in the flow regions under the influence of the heterogeneous boundaries and where flow similarity is not possible. In river engineering, it is often desirable to estimate the vertical distribution of streamwise flow velocity and momentum balances (including the determination of stresses and drag forces) based on simple assumptions made on scales that are sufficiently large to incorporate the smaller scale phenomena.

Given the high spatial variability of the flow characteristics in gravel-bed rivers with low relative submergence, the application of the double-averaged methods (DAM) using a minimum scale of one wavelength of the bedforms (Raupach and Shaw 1982) seems appropriate. The analysis and modeling of heterogeneous and irregular-bounded open-channel flows by means of double-averaging (DA) methods is presented by several authors for different problems in fluvial hydraulics (Aberle and Koll 2004; Manes et al. 2007; Franca et al. 2008; Stoesser and Nikora 2008; Ferreira et al. 2010, among others). Franca et al. (2010) and Mignot et al. (2009) present results of the application of double-averaging methods to the streamwise velocity and Reynolds stresses of gravel-bed open-channel flows.

Recently, a journal's special issue on the application of the double-averaging approach to rough-bed flows was published (cf. Nikora and Rowinski 2008), and an overview of its application to geophysical, environmental and engineering physics was given. A comprehensive overview of the methodology and its applications to environmental hydraulics was also recently provided in Nikora et al. (2007a, b). Spatial averaging in the context of the DA technique is commonly performed along horizontal surfaces parallel to the riverbed, and the main formalisms described in the literature refer mainly to this type of application. Similar upscaling methods may be applied to any heterogeneous boundaries of the flow, such as river banks or vegetation canopy, the latter constituting an open boundary between different regions of the flow.

2.4.2 Turbulence Evolving on the Vertical Plane

In wide rivers, the boundary layers produced by the banks are usually neglected because they occupy a small region of the flow; thus, if sufficiently wide, open-channel flows and the physical interpretation of energetic turbulent processes are treated as 2D (in the vertical-longitudinal plane, $x_1 - x_3$) along the water column. This applies to existing closures for the distribution of mean flow quantities and of turbulence-related quantities such as Reynolds stresses, *tke*, turbulent scales and related production, diffusion and dissipation terms (Nezu and Nakagawa 1993; Chassaing 2000; Pope 2000). Hence, these distributions depend only on the lower and upper boundaries of the flows.

In line with the consideration of a 2D plane is the analysis of energetic events that contribute to the production of Reynolds shear stress in the flow. The bursting phenomenon results from a class of coherent structures occurring mainly in the near-wall region. This results from a quasi-cyclic process that is composed of interactions in the four quadrants of a 2D Cartesian longitudinal framework, hereafter called shear events: outward interaction, ejection, inward interaction and sweep (Nezu and Nakagawa 1993). Shear events cannot be identified with time averaged analysis and, in addition to time and space correlation measurements and flow visualization techniques, require conditional sampling and statistic techniques for their detection and characterization. Important works on the analysis of turbulence events contributing to Reynolds stress production include Antonia and Atkinson (1973), Corino and Brodkey (1969), Grass (1971), Kim et al. (1971), Kline et al. (1967), Nakagawa and Nezu (1977) and, more recently, Hurther and Lemmin (2000) and Adrian et al. (2000). Nakagawa and Nezu (1977) made a prediction of the contribution of all types of shear events to the production of Reynolds stresses. To quantify their results, they used the quadrant threshold method developed by Corino and Brodkey (1969), Lu and Willmarth (1973), Willmarth and Lu (1972).

The role of the so-called vertical turbulence is the most important in terms of sediment mobilization and transport (Séchet and Le Guennec 1999; Cellino and Lemmin 2004). In particular, knowledge of the turbulent pressure fluctuations leading to drag and lift forces at the water-sediment interface is fundamental. Recent measurements performed with miniaturized piezoresistive pressure sensors, which are small and include high sensitivity and accuracy, have provided significant information on turbulent forces on single grains at river beds (Detert et al. 2010). One aspect that attracts considerable attention is the rate at which bed sediment is entrained by turbulent shear flow. The recent contribution of Zhong et al. (2011) attempted to model such an entrainment in terms of kinetic theory for multiphase flows. This has advantages over other theories in terms of (a) accounting for the influences exerted on the sediment by external forces and (b) providing a statistical description of the random motion of sediments due to turbulence. These advantages enable the kinetic theory to act as a bridge between microscopic and macroscopic scales of moving sediment particles.

Experimental studies were also performed to quantify the near-bed turbulence characteristics for fluvial flows under sediment entrainment limit conditions and in the presence of mobile beds. Gyr and Schmid (1997), using Laser Doppler Anemometry (LDA) measurements over a smooth sand bed, observed that the presence of intense intermittent sediment transport increases the extreme values of shear stress while the flow becomes more organized in the second and fourth quadrants, mainly increasing the importance of sweep events to turbulence production. The period between events in the second and fourth quadrants decreases considerably in the presence of sediment transport, thereby producing more frequent ejection and sweep events. More recently, Dey et al. (2011) found that the observed time-averaged streamwise velocity is further from the logarithmic for immobile beds than for entrainment-threshold beds. They also found, by means of a quadrant analysis, that in the near-bed flow zone, ejections and sweeps in immobile beds cancel each other, thereby giving rise to the outward interactions, whereas sweeps are the dominant mechanism causing sediment entrainment. Finally, the bursting duration for entrainment-threshold beds is smaller than that for immobile beds; in contrast, the bursting frequency for entrainment-threshold beds is larger than that for immobile beds. Santos et al. (2014) showed that, generally, the sediment transport of sand decreases the transported momentum and maximum shear stress values but increases their frequency of occurrence in time. The analysis of the probability distribution function of both ejections and sweeps shows an effect of sediment transport in terms of the reduction in the frequency of large events and in the increase of the frequency of small events. This may be due to the breaking of eddy coherence by sediment motion and is especially observed in the so-called pythmenic region (Ferreira et al. 2012).

Prandtl's first and second types of secondary flows which are perpendicular to the streamwise direction of the flow, have thus expression in vertical planes, transverse to the main flow direction (cf. Nezu and Nakagawa 1993; Nikora and Roy 2012). These secondary flows are divided into two classes: the first type of Prandtl's secondary flow is observed when the streamwise mean vorticity is enhanced by vortex stretching, occurring typically in river bends and meanders. The second type of secondary flows occurs due to turbulence heterogeneity, and no curvature of the principal flow direction is required to exist. Secondary current cells may have a signature in the river bed morphology and may be influenced in turn by the channel roughness (Tsujimoto 1989).

The vertical distribution of turbulence has also a significant interaction with the instream vegetation which in turn conditions the river morphology (e.g., Neary et al. 2012). Compared to unvegetated flows with the same discharge and slope, vegetation causes flow blockage and increases channel resistance. If the vegetation is submerged, the consideration of a 2D vertical-longitudinal plane on the analysis of the flow is still valid as a canopy-like flow occurs (Nepf 2012). In vegetated flow where the vegetation is emergent, three regions of the flow may generally be observed, whereby the flow is controlled by the riverbed and the vegetation (lower region), solely by the vegetation or by the vegetation and the free surface (upper region). The middle region is generally self-similar as concerns turbulent quantities.

2.4.3 Turbulence Evolving on the Horizontal Plane

Natural geomorphological features and man-made obstacles opposing the flow cause geometrically-induced gradients in river flows, thereby inducing coherent vortical structures with vertical axes. These structures may range from large-scale, slowly evolving eddies in regions of recirculation and flow stagnation to energetic shear layers, whirlpool-type vortices, and curvature-induced streamwise vortices. The ability to reproduce the influence of such vortices is fundamental for a number of theoretical and practical issues. From a theoretical point of view, the understanding of lateral transfer of momentum and scalar (for instance, in compound channel flows) are current topics of research. From a practical perspective, large-scale vortical structures with vertical axes may influence the design of bridge foundations resistant to scouring and the stabilization of stream banks.

Two main, complementary, approaches can be used to describe the generation of large-scale vortices in a river characterized by topographic bed changes. On the one hand, vortical structures can be regarded as the manifestation of the shear instability at the junction of two different streams (van Prooijen and Uijttewaal 2002; van Prooijen et al. 2005); on the other hand, they can be seen as the outcome of differential energy dissipation of shallow-water currents interacting with submerged obstacles (Brocchini et al. 2004; Soldini et al. 2004; Kennedy et al. 2006).

To discuss the role of large-scale eddies in association with topographic bed changes we refer to the simple case of the compound channel of Fig. 2.4, which shows macrovortices in the transition regions and a mean flow velocity distribution along the transversal direction (see Stocchino et al. 2011). Recent experimental investigations (Stocchino and Brocchini 2010), which were based on use of the PIV technique, revealed that the population and properties of macrovortices largely depend on the typical depth gradients, i.e., on the ratio r_h between the largest and smallest flow depths. Shallow flows ($r_h > 3$) are dominated by strong shearing and large macrovortices populate the transition region between the main channel and the floodplains. The mean streamwise velocity induced by intermediate flows ($2 < r_h < 3$) is characterized by a dip in the transition region, while it closely resembles that occurring in a rectangular channel in the case of deep flows ($r_h < 2$).

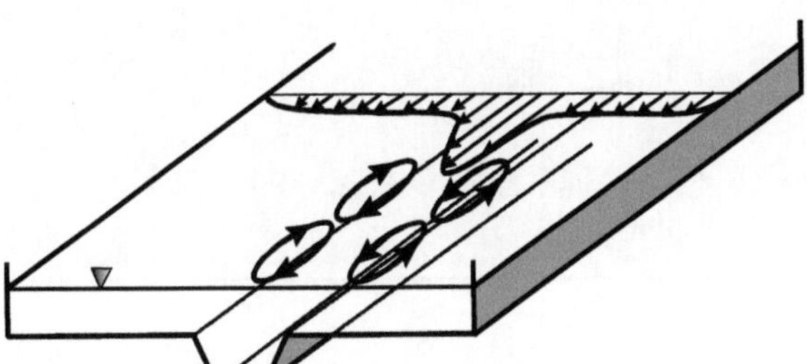

Fig. 2.4 Sketch of a typical compound-channel geometry with mean flow profile and large-scale vortices at depth transition

For both the latter cases, the shear in the transition region decreases, and the macrovortices are also generated in the wall boundary layer of the floodplains.

The visualization of macrovortices can be performed using typical indicators such as the Hua and Kline eigenvalue λ_+ of the local acceleration tensor (Hua and Kline 1998). An illustration of such a population is reported in Fig. 2.5 for the cases of shallow (top) and deep (bottom) flows. Shallow flows are characterized by large quasi-2D vortical structures with vertical axes, which behave as organized domains with distinct dynamical roles and dominate turbulence production; these flows are generally resolved considering 2D depth-averaged models. Under deep flow conditions, fewer macrovortices can be recognized, horizontal turbulence production becomes less prevalent, and 2D streamwise-spanwise approaches are no longer valid.

The mixing induced by these dynamical domains can be described in terms of both absolute (single-particle) and relative (e.g., particle pairs) statistics (e.g., Provenzale 1999; LaCasce 2008; Stocchino et al. 2011). Typically, under shallow flow conditions, macrovortices strongly influence the growth in time of the total absolute dispersion after an initial ballistic regime after their formation, leading to a

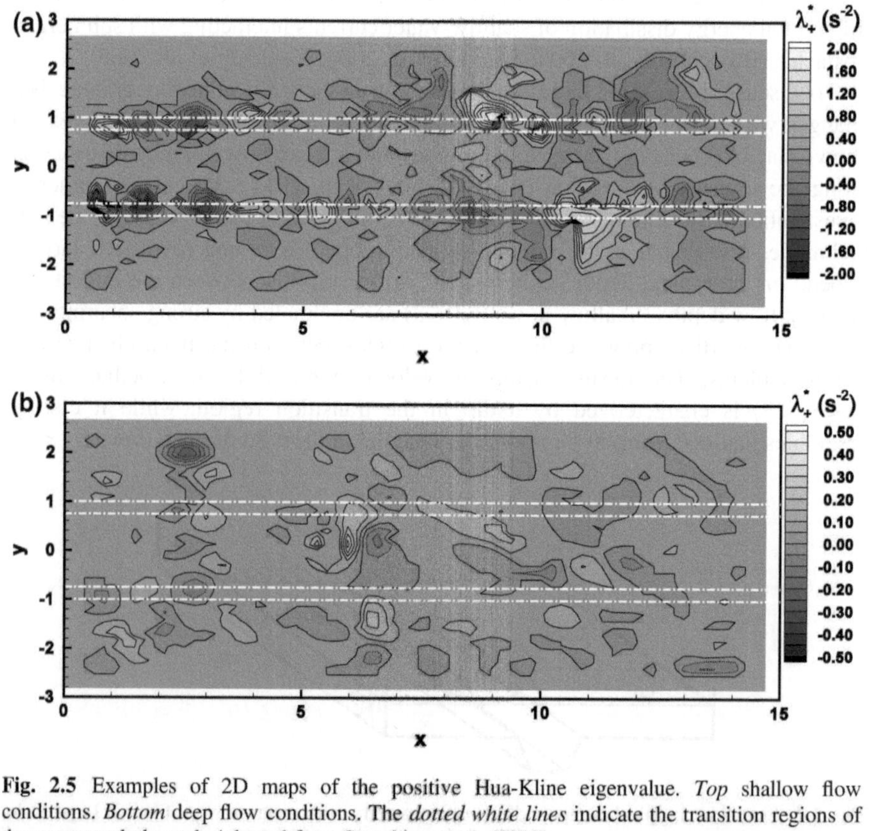

Fig. 2.5 Examples of 2D maps of the positive Hua-Kline eigenvalue. *Top* shallow flow conditions. *Bottom* deep flow conditions. The *dotted white lines* indicate the transition regions of the compound channel. Adapted from Stocchino et al. (2011)

non-monotonic behavior. Under deep flow conditions, on the contrary, the absolute dispersion exhibits a monotonic growth because the generation of transitional macrovortices does not occur. In all cases, an asymptotic diffusive regime is obtained. Multiple-particle dynamics are controlled by the ratio between the largest and smallest flow depths, r_h, and by the Froude number. Different growth regimes of the relative diffusivity exist as a function of the flow conditions. This is associated with different energy transfer processes, which show a different asymptotical shape as a function of the separation scales and the Froude number. An equilibrium regime is observed by analyzing the decay of the finite-scale Lyapunov exponents with the particle separations.

Further investigation on compound channels, namely, on the distribution of turbulent quantities and on energetic processes related to 2D macrovortices present at the interface between the main-channel and floodplain flows, include Bousmar and Zech (1999), Kara et al. (2012), Knight and Shiono (1990), Proust et al. (2013), Tominaga and Nezu (1991), van Prooijen et al. (2005).

As introduced earlier in Sect. 2.4.2, when the vegetation is emergent, a middle region of the flow exists and is generally controlled only by the vegetation, eventually becoming self-similar in terms of turbulent quantities. Here, the vegetation conditions local velocities, turbulence intensities, turbulent Reynolds stresses and their vertical and horizontal distributions (e.g., Nepf 1999). Ricardo et al. (2014) estimated the terms in the *tke* transport equation in a flow with emergent arrays of cylinders by considering velocity measurements in horizontal planes. With specific reference to the role of vegetation in turbulent flows evolving in compound channels, Koziol (2013) found, on the basis of dedicated experimental investigations, that trees placed on the floodplains do not significantly change the values of the relative turbulence intensity in the entire compound channel, but they do change the vertical distributions of the relative turbulence intensities in the three components in the floodplains and over the bottom of the main channel.

2.4.4 Turbulence and Vorticity Evolving from Vertical to Horizontal

A substantial amount of literature is devoted to the generation of horseshoe or hairpin types of vortices, in turbulent flows over plane boundaries (Chassaing 2000). The phenomenology associated with these vortices is somehow related to the experimental and field investigations of sand waves, which have documented turbulent events called "kolks" and "boils" (Matthes 1947; Coleman 1969); these events have a 3D signature in the flow, evolving from a 2D vertical-horizontal structure in a first stage. The so-called "kolk-boil" mechanism is one of predominant turbulent events occurring over sand waves in fluvial, estuarine, and marine coastal environments (Ha and Chough 2003). A kolk is a slowly rotating, upward-tilting vortex on the stoss face of a sub-aqueous bedform. A strong kolk may reach the water surface, create a cloudy columnar sediment-fluid mixture, and form a

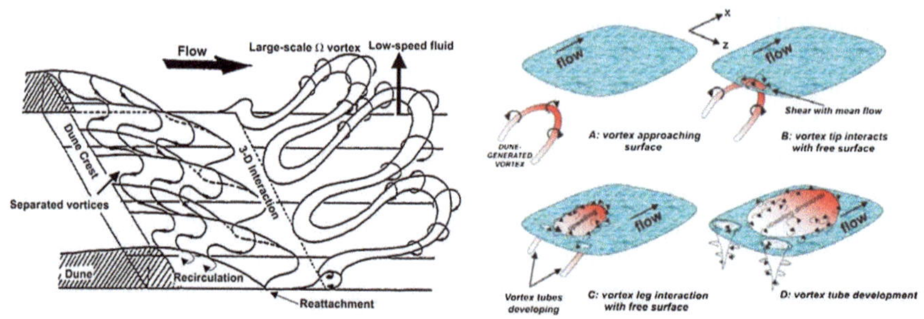

Fig. 2.6 *Left* model of vortex topology associated with dunes proposed by Nezu and Nakagawa (1993). *Right* model of flow illustrating the interaction of dune-related macro-turbulence with the water surface proposed by Best (2005). Adapted from Best (2005)

raised circular or oval patch at the air-water interface, which is referred to as a boil (Matthes 1947; Kostaschuk and Church 1993). These boils are created on a scale that is comparable to the flow depth as a first approximation.

These vortical structures are generated by a complex mechanism that is initiated at the reattachment point at the lee side of an obstacle, e.g., sand ripples and sand dunes (Nezu and Nakagawa 1993). Best (2005) proposed a model for the stages of interaction of a vortex loop with the surface and for the manner by which the phenomenon is manifested as different upwelling motions as the boil evolves and erupts on the surface (see Fig. 2.6). This model shows how the initial transverse vorticity is accompanied by vertical vorticity as the boil evolves and as the vortex legs of the vortex loop attach to the surface; this pattern is common in many natural rivers. It is important to note that Best (2005) also argues that this upwelling and flow surface interaction must induce subsequent downwelling toward the bed to satisfy flow continuity.

Flow in compound meandering channels is an example of the nature of tridimensional flows, combining both complex flows in a compound channel (with 2D vortical structures with a vertical axis) and in a meandered channel (with secondary flow cells with a streamwise axis, developing in a plane transverse to the flow). Experimental studies that focus on compound meandering morphologies are scarce (Shiono and Muto 1998; Shiono et al. 2008). Mera et al. (2014) characterized the hydrodynamics and turbulence patterns in a real compound meandering channel using the anisotropy invariants combined with quadrant analysis techniques.

2.5 Conclusions

Many textbooks are available on the basic theoretical aspects of turbulent flows and on their inclusion in fluid mechanics studies. Some examples of useful books are referenced in the present contribution, which focuses on specific aspects related to turbulent riverine flows. Although not exhaustive, the theories discussed herein were chosen based on their relevance to the study of fluvial processes, which means that the present paper is organized in terms of methodological approaches used to address turbulent fluvial flows, rather than in terms of phenomenological aspects. Furthermore, themes that are usually subject to misconceptions and erroneous interpretation by engineers and researchers studying river flows are discussed.

One of the main issues explored here is the notion of scales and their selective use. River flows are characterized by a large variety of scales (time or space scales). Using both temporal and spatial frameworks, we show that the concepts of ergodicity, stationarity and homogeneity depend on the relation between the typical time or space ranges of the phenomenon being studied and on the scale of turbulence evolving within this.

In addition, the focus of the analysis strictly depends on the scales of the phenomenon, which is illustrated in terms of the demands of the higher order statistical moments being studied. For example, the analysis of the mean flow requires a time range that is smaller than that needed by second-order analyses, where Reynolds stresses and *tke* may be included.

Only a partial overview of the basic equations for the study of turbulent flows is given. However, references of exhaustive formal analyses, as well as a discussion of sources of information on other transport equations of variables important for river flows, are provided. Topics related to mass, momentum and species conservation are also proposed. The latter topics are particularly important in terms of the environmental analysis of rivers, where the transport of species, diluted or not, is important. Problems of pollutants or salt mixing, as well as sediment transport, are analyzed.

Several frameworks for the analysis of turbulent flows are illustrated along with examples of results obtained within such frameworks. The variability of scales and phenomena typical of river flows generate a diversity of methods for solving problems. The simplifications needed for practical applications, such as the simplifications mentioned above, which include considering stationarity and homogeneity, as well as the eventual reduction of spatial dimensions for the analysis (i.e., from 3D to 2D), are also proposed, and examples are given.

Acknowledgements Mário J. Franca acknowledges the financial support by the European Fund for Economic and Regional Development (FEDER) through the Program Operational Factors of Competitiveness (COMPETE) and National Funds through the Portuguese Foundation of Science and Technology (RECI/ECM-HID/0371/2012 and PTDC/ECM/099752/2008). Maurizio Brocchini acknowledges the financial support from the EsCoSed Project, which is financed by the US-ONR through the NICOP Research Grant (N62909-13-1-N020). The PhD students Elena Battisacco, Reyhaneh S. Ghazanfari, Sebastián Guillén-Ludeña, Sebastian Schwindt and Jessica Zordan are acknowledged for their final check of possible inconsistencies in the text. The present work was completed while Maurizio Brocchini was Visiting Professor at the Laboratoire d'Hydraulique Environnementale, ÉCOLE POLYTECHNIQUE FÉDÉRALE DE LAUSANNE.

References

Abad JD, Frias CE, Buscaglia GC, Garcia MH (2013) Modulation of the flow structure by progressive bedforms in the Kinoshita meandering channel. Earth Surf Proc Land 38 (13):1612–1622

Aberle J, Koll K (2004) Double-averaged flow field over static armour layers. In: Greco M, Carravetta A, Della Morte R (eds) River flow 2004. Taylor & Francis Group, London

Adrian RJ, Christensen KT, Liu ZC (2000) Analysis and interpretation of instantaneous turbulent velocity fields. Exp Fluids 29:275–290

Aleixo RJF (2013) Experimental study of the early stages of a dam-break flow over fixed and mobile beds. PhD thesis, UCL, Louvain-la-Neuve

Antonia RA, Atkinson JD (1973) High-order moments of Reynolds shear stress fluctuations in a turbulent boundary layer. J Fluid Mech 58(3):581–593

Baiamonte G, Giordano G, Ferro V (1995) Advances on velocity profile and flow resistance law in gravel bed rivers. Excerpta 9:41–89

Berkooz G, Holmes P, Lumley JL (1993) The proper orthogonal decomposition in the analysis of turbulent flows. Annu Rev Fluid Mech 25(1):539–575

Best J (2005) The fluid dynamics of river dunes: a review and some future research directions. J Geophys Res Oceans. doi:10.1029/2004JF000218

Blanckaert K, De Vriend HJ (2005) Turbulence characteristics in sharp open-channel bends. Phys Fluids 17(5):055102

Blanckaert K, Han R, Pilotto F, Pusch M (2014) Effects of large wood on morphology, flow and turbulence in a Lowland River. In: Schleiss AJ, De Cesare G, Franca MJ, Pfister M (eds) River flow 2014. Taylor & Francis, Leiden

Bousmar D, Zech Y (1999) Momentum transfer for practical flow computation in compound channels. J Hydraul Eng 125(7):696–706

Brocchini M, Kennedy AB, Soldini L, Mancinelli A (2004) Topographically controlled, breaking-wave-induced macrovortices. Part 1. Widely separated breakwaters. J Fluid Mech 507:289–307

Buffin-Bélanger T, Roy AG (1998) Effects of a pebble cluster on the turbulent structure of a depth-limited flow in a gravel-bed river. Geomorphology 25(3):249–267

Buffin-Bélanger T, Roy AG, Kirkbride AD (2000) On large-scale flow structures in a gravel-bed river. Geomorphology 32(3):417–435

Cardoso AH, Graf WH, Gust G (1989) Uniform flow in a smooth open channel. J Hydraul Res 5:603–616

Cellino M, Lemmin U (2004) Influence of coherent flow structures on the dynamics of suspended sediment transport in open-channel flow. J Hydraul Eng 130(11):1077–1088

Chassaing P (2000) Turbulence en Mcanique des Fluides. Cépaduès-Éditions, Toulouse

Coleman JM (1969) Brahmaputra River: channel processes and sedimentation. Sediment Geol 3:129–329

Corino ER, Brodkey RS (1969) A visual investigation of the wall region in turbulent flow. J Fluid Mech 37(1):1–30

Currie IG (1993) Fundamental mechanics of fluids. McGraw Hill, Toronto

Detert M, Weitbrecht V, Jirka GH (2010) Laboratory measurements on turbulent pressure fluctuations in and above gravel beds. J Hydraul Eng 136:779–789

Dey S (2014) Fluvial hydrodynamics: hydrodynamic and sediment transport phenomena. Springer, Berlin

Dey S, Sarkar S, Solari L (2011) Near-bed turbulence characteristics at the entrainment threshold of sediment beds. J Hydraul Eng 137:945–958

Ferreira RML, Ferreira L, Ricardo AM, Franca MJ (2010) Impacts of sand transport on flow variables and dissolved oxygen in gravel-bed streams suitable salmonid spawning. River Research and Applications 26(10):414–438

Ferreira RML, Franca MJ, Leal JGAB, Cardoso AH (2012) Flow over rough mobile beds: Friction factor and vertical distribution of the longitudinal mean velocity. Water Resour Res 48(5): W05529

Foufoula-Georgiou E, Kumar P (1994) Wavelets in geophysics. Academic Press, San Diego

Franca MJ (2005a) A field study of turbulent flows in shallow gravel-bed rivers. PhD thesis, EPFL, Lausanne

Franca MJ (2005b) Flow dynamics over a gravel riverbed. In: Proceedings of the XXXI IAHR Congress, Seoul

Franca MJ, Lemmin U (2006a) Detection and reconstruction of coherent structures based on wavelet multiresolution analysis. In: Ferreira RML, Alves ECTL, Leal JGAB, Cardoso AH (eds) River flow 2006. Taylor & Francis Group, London

Franca MJ, Lemmin U (2006b) Turbulence measurements in shallow flows in gravel-bed rivers. In: Piasecki M (ed) 7th international conference on hydroscience and engineering. Session Wed1_S2: MINI-SYMPOSIUM fluvial hydraulics and river morphodynamics, College of Engineering, Drexel University, Philadelphia

Franca MJ, Lemmin U (2014) Detection and reconstruction of large-scale coherent flow structures in gravel-bed rivers. Earth Surf Proc Land 40(1):93–104

Franca MJ, Ferreira RML, Lemmin U (2008) Parameterization of the logarithmic layer of double-averaged streamwise velocity profiles in gravel-bed river flows. Adv Water Resour 31(6):915–925

Franca MJ, Ferreira RML, Cardoso AH, Lemmin U (2010) Double-average methodology applied to turbulent gravel-bed river flows. In: Dittrich A, Koll K, Aberle J Geisenhainer (eds) River flow 2010. Bundesanstalt fr Wasserbau, Braunschweig

Frisch U (1995) Turbulence. The legacy of A. N. Kolmogorov. Cambridge University Press, New York

Ghilardi T, Franca MJ, Schleiss AJ (2014) Bulk velocity measurements by video analysis of dye tracer in a macro-rough channel. Meas Sci Technol 25(3):035003

Grass AJ (1971) Structural features of turbulent flow over smooth and rough boundaries. J Fluid Mech 50(02):233–255

Gyr A, Schmid A (1997) Turbulent flows over smooth erodible sand beds in flumes. J Hydraul Res 35(4):525–544

Ha HK, Chough SK (2003) Intermittent turbulent events over sandy current ripples: a motion-picture analysis of flume experiments. Sed Geol 161:295–308

Holmes P, Lumley JL, Berkooz G (1998) Turbulence, coherent structures, dynamical systems and symmetry. Cambridge University Press, New York

Hua BL, Kline P (1998) An exact criterion for the stirring properties of nearly two-dimensional turbulence. Physica D 113(1):98–110

Huang NE, Shen Z, Long SR, Wu MC, Shih HH, Zheng Q, Yen N-C, Tung CC, Liu HH (1998) The empirical mode decomposition and the Hilbert spectrum for nonlinear and non-stationary time series analysis. Proc R Soc Lond A 454(1971):903–995

Huang NE, Shen Z, Long SR (1999) A new view of nonlinear water waves: the Hilbert Spectrum 1. Annu Rev Fluid Mech 31(1):417–457

Hurther D, Lemmin U (2000) Shear stress statistics and wall similarity analysis in turbulent boundary layers using a high-resolution 3-D ADVP. IEEE J Oceanic Eng 25(4):446–457

Jovanovic J (2004) The statistical dynamics of turbulence. Springer, Berlin

Kara S, Stoesser T, Sturm TW (2012) Turbulence statistics in compound channels with deep and shallow overbank flows. J Hydraul Res 50(5):482–493

Katul G, Wiberg P, Albertson J, Hornberger G (2008) A mixing layer theory for flow resistance in shallow streams. Water Resour Res 38(11):1250

Kennedy AB, Brocchini M, Soldini L, Gutierrez E (2006) Topographically controlled, breaking-wave-induced macrovortices. Part 2. Rip current topographies. J Fluid Mech 559:57–80

Kim HT, Kline SJ, Reynolds WC (1971) The production of turbulence near a smooth wall in a turbulent boundary layer. J Fluid Mech 50(01):133–160

Kirkbride AD, Ferguson R (1995) Turbulent flow structure in a gravel-bed river: Markov chain analysis of the fluctuating velocity profile. Earth Surf Proc Land 20(8):721–733

Kironoto BA, Graf WH (1994) Turbulence characteristics in rough uniform open-channel flow. Proc ICE-Water Marit Energy 106(4):333–344
Kline SJ, Reynolds WC, Schraub FA, Runstadler PW (1967) The structure of turbulent boundary layers. J Fluid Mech 30(04):741–773
Knight DW, Shiono K (1990) Turbulence measurements in a shear layer region of a compound channel. J Hydraul Res 28(2):175–196
Kostaschuk RA, Church MA (1993) Macroturbulence generated by dunes: Fraser River, Canada. Sed Geol 85:25–37
Koziol AP (2013) Three-dimensional turbulence intensity in a compound channel. J Hydraul Eng 139:852–864
LaCasce JH (2008) Statistics from Lagrangian observations. Prog Oceanogr 77:1–29
Leite Ribeiro M, Blanckaert K, Roy AG, Schleiss AJ (2002) Flow and sediment dynamics in channel confluences. J Geophys Res Earth Surf. doi:10.1029/2011JF002171
Levi E (1995) The science of water: the foundation of modern hydraulics. ASCE Press, New York
Lopes AF, Nogueira HIS, Ferreira RML, Franca MJ (2013) Laboratorial study of continuously fed low-submergence gravity currents over smooth and rough beds. In: EGU general assembly conference abstracts, Vienna
Lu SS, Willmarth WW (1973) Measurements of the structure of the Reynolds stress in a turbulent boundary layer. J Fluid Mech 60(3):481–511
Lumley JL, Newman GR (1977) The return to isotropy of homogeneous turbulence. J Fluid Mech 82(1):161–178
Manes C, Pokrajac D, McEwan I (2007) Double-averaged open-channel flows with small relative submergence. J Hydraul Eng 138(8):896–904
Matthes GH (1947) Macroturbulence in natural stream flow. Trans Am Geophys Union 28:255–265
Mera I, Franca MJ, Anta J, Peña E (2014) Turbulence anisotropy in a compound meandering channel with different submergence conditions. Adv Water Resour. doi:10.1016/j.advwatres.2014.10.012
Mignot E, Hurther D, Bartelhemy E (2009) On the structure of shear stress and turbulent kinetic energy flux across the roughness layer of a gravel-bed channel flow. J Fluid Mech 638:423–452
Monin AS, Yaglom AM (1971) Statistical fluid mechanics: mechanics of turbulence, vol I. MIT Press, Boston
Nakagawa H, Nezu I (1977) Prediction of the contributions to the Reynolds stress from bursting events in open-channel flows. J Fluid Mech 80(1):99–128
Nazarenko S, Laval JP (2000) Non-local two-dimensional turbulence and Batchelor's regime for passive scalars. J Fluid Mech 408:301–321
Neary VS, Constantinescu SG, Bennett SJ, Diplas P (2012) Effects of vegetation on turbulence, sediment transport, and stream morphology. J Hydraul Eng 138:765–776
Nepf HM (1999) Drag, turbulence and diffusion in flow through emergent vegetation. Water Resour Res 35(2):479–489
Nepf HM (2012) Hydrodynamics of vegetated channels. J Hydraul Res 50(3):262–279
Nezu I, Nakagawa H (1993) Turbulence in open-channel flows. IAHR monograph series, A.A. Balkema, Rotterdam, Netherlands
Nicholas AP (2001) Computational fluid dynamics modelling of boundary roughness in gravel-bed rivers: an investigation of the effects of random variability in bed elevation. Earth Surf Proc Land 26(4):345–362
Nikora V (2007) Hydrodynamics of gravel-bed rivers: scale issues. Dev Earth Surf Process 11:61–81
Nikora V (2010) Hydrodynamics of aquatic ecosystems: an interface between ecology, biomechanics and environmental fluid mechanics. River Res Appl 26(4):367–384
Nikora V, Rowinski PM (2008) Rough-bed flows in geophysical, environmental, and engineering systems: double-averaging approach and its applications. Acta Geophys 56(3):529–533

Nikora V, Roy AG (2012) Secondary flows in rivers: theoretical framework, recent advances, and current challenges. In Church M, Biron PM, Roy AG (eds) Gravel bed rivers: processes, tools, environments. John Wiley & Sons, New York

Nikora V, Smart GM (1997) Turbulence characteristics of New Zealand gravel-bed rivers. J Hydraul Eng 123(9):764–773

Nikora V, McEwan I, McLean S, Coleman S, Pokrajac D, Walters R (2007a) Double-averaging concept for rough-bed open-channel and overland flows: theoretical background. J Hydraul Eng 133(8):873–883

Nikora V, McLean S, Coleman S, Pokrajac D, McEwan I, Campbell I, Aberle J, Clunie D, Koll K (2007b) Double-averaging concept for rough-bed open-channel and overland flows: applications. J Hydraul Eng 133(8):884–985

Nikora V, Ballio F, Coleman S, Pokrajac D (2013) Spatially averaged flows over mobile rough beds: definitions, averaging theorems, and conservation equations. J Hydraul Eng 139(8):803–811

Nogueira HIS (2014) Experimental characterization of unsteady gravity currents developing over smooth and rough beds. PhD thesis, UC, Coimbra

Nogueira HIS, Adduce C, Alves E, Franca MJ (2014) Dynamics of the head of gravity currents. Environ Fluid Mech 14(2):519–540

Pokrajac D, Kikkert GA (2011) RADINS equations for aerated shallow water flows over rough beds. J Hydraul Res 49(5):630–638

Pope SB (2000) Turbulent flows. Cambridge University Press, New York

Proust S, Fernandes JN, Peltier Y, Leal JB, Riviere N, Cardoso AH (2013) Turbulent non-uniform flows in straight compound open-channels. J Hydraul Res 51(6):656–667

Provenzale A (1999) Transport by coherent barotropic vortices. Annu Rev Fluid Mech 31:55–93

Raupach MR, Shaw RH (1982) Averaging procedures for flow within vegetation canopies. Bound-Layer Meteorol 22:79–90

Ricardo AM (2014) Hydrodynamics of turbulent flows within arrays of circular cylinders. PhD thesis, EPFL, Lausanne

Ricardo AM, Koll K, Franca MJ, Schleiss AJ, Ferreira RML (2014) The terms of turbulent kinetic energy budget within random arrays of emergent cylinders. Water Resour Res 50(5):4131–4148

Romano GP, Ouellette NZ, Xu H, Bodenschatz E, Steinberg V, Meneveau C, Katz J (2007) Measurements of turbulent flows. In: Tropea C, Yarin AL, Foss JF (eds) Springer handbook of experimental fluid mechanics. Springer, Berlin

Roy AG, Buffin-Bélanger T, Lamarre H, Kirkbride AD (2004) Size, shape and dynamics of large-scale turbulent flow structures in a gravel-bed river. J Fluid Mech 500:1–247

Saggiori S, Rita S, Ferreira RML, Franca MJ (2012) Analysis of 3rd order moments on a natural vegetated flow. In: Proceedings of 2nd IAHR Europe congress, Munich

Santos BO, Franca MJ, Ferreira RML (2014) Coherent structures in open channel flows with bed load transport over an hydraulically rough bed. In: Schleiss AJ, De Cesare G, Franca MJ, Pfister M. (eds) River flow 2014. Taylor & Francis, Leiden

Schleiss AJ, De Cesare G, Franca MJ, Pfister M (eds) River flow 2014., Taylor & Francis, Leiden

Séchet P, Le Guennec B (1999) The role of near wall turbulent structures on sediment transport. Water Res 33(17):3646–3656

Shields FD Jr, Morin N, Cooper CM (2004) Large woody debris structures for sand-bed channels. J Hydraul Eng 130(3):208–217

Shiono K, Muto Y (1998) Complex flow mechanisms in compound meandering channels with overbank flow. J Fluid Mech 376:221–261

Shiono K, Spooner J, Chan TL, Rameshwaran P, Chandler JH (2008) Flow characteristics in meandering channels with non-mobile and mobile beds for overbank flows. J Hydraul Res 46 (1):113–132

Simpson JE (1997) Gravity currents: in the environment and the laboratory. Cambridge University Press, New York

Siniscalchi F, Nikora VI, Aberle J (2012) Plant patch hydrodynamics in streams: Mean flow, turbulence, and drag forces. Water Resour Res 48(1):W01513

Smart GM (1999) Turbulent velocity profiles and boundary shear in gravel bed rivers. J Hydraul Eng 125(2):106–116
Soldati A, Marchioli C (2009) Physics and modelling of turbulent particle deposition and entrainment: review of systematic study. Int J Multip Flows 35:827–839
Soldini L, Piattella A, Brocchini M, Mancinelli A, Bernetti R (2004) Macrovortices-induced horizontal mixing in compound channels. Ocean Dyn 54:333–339
Stocchino A, Brocchini M (2010) Horizontal mixing of quasi-uniform, straight, compound channel flows. J Fluid Mech 643:425–435
Stocchino A, Besio G, Angiolani S, Brocchini M (2011) Lagrangian mixing in straight compound channels. J Fluid Mech 675:168–198
Stoesser T, Nikora V (2008) Flow structure over square bars at intermediate submergence: large eddy simulation study of bar spacing effect. Acta Geophys 56(3):876–893
Sukhodolova TA, Sukhodolov AN (2012) Vegetated mixing layer around a finite-size patch of submerged plants: 1. Theory and field experiments. Water Resour Res 48(10):WR011804
Tanino Y, Nepf HM (2008) Laboratory investigation of mean drag in a random array of rigid, emergent cylinders. J Hydraul Eng 134(1):34–41
Tennekes H, Lumley JL (1972) A first course in turbulence. The MIT press, Cambridge
Termini D, Piraino M (2011) Experimental analysis of cross-sectional flow motion in a large amplitude meandering bend. Earth Surf Proc Land 36(2):244–256
Tominaga A, Nezu I (1991) Turbulent structure in compound open-channel flows. J Hydraul Eng 117(1):21–41
Tritico HM, Hotchkiss RH (2005) Unobstructed and obstructed turbulent flow in gravel bed rivers. J Hydraul Eng 131(8):635–645
Tsujimoto T (1989) Longitudinal stripes of alternate lateral sorting due to cellular secondary currents. In: Proceedings of XXX IAHR Cong, Ottawa
van Prooijen BC, Uijttewaal WSJ (2002) A linear approach for the evolution of coherent structures in shallow mixing layers. Phys Fluids 14(12):4105–4114
van Prooijen BC, Battjes JA, Uijttewaal WSJ (2005) Momentum exchange in straight uniform compound channel flow. J Hydraul Eng 131(3):175–183
Willmarth WW, Lu SS (1972) Structure of the Reynolds stress near the wall. J Fluid Mech 5 (1):65–92
Yoshida K, Nezu I (2004) Experimental study on air-water interfacial turbulent hydrodynamics and gas transfer in wind-induced open channel-flows. In: Proceedings of the 4th IAHR international symposium on environmental hydraulics, Hong Kong
Zhong D, Wang G, Ding Y (2011) Bed sediment entrainment function based on kinetic theory. J Hydraul Eng 137:222–233

Chapter 3
Principles of Mechanics of Bedforms

Domenico Ferraro and Subhasish Dey

Abstract Natural streambed does not exhibit a plane bed surface, but takes various geometrical forms known as bedforms. In this book chapter, the studies dealing with the formation mechanism of bedforms and their stability are discussed. The important feature of this chapter is the presentation of mathematical models proposed by various researchers.

Keywords Bedforms · Bed geometry · Bed instability · River bed

3.1 Introduction

In natural streams, the sediment bed conditions are far from a stable bed. In fact, considering non-cohesive sediments, one can find several bed configurations in each flow condition (subcritical or critical or supercritical), where the bed shear stress τ_0 induced by the flow exceeds threshold value τ_{0c} for the bed sediment motion. Engelund and Fredsøe (1982) showed how bedforms evolve with the Froude number defined as $Fr = U/(gh_d)^{0.5}$, where U is the depth-averaged velocity, g is the acceleration due to gravity, and h_d is the hydraulic depth defined as the ratio of flow area A to top width of flow T. With an increase in Froude number, a plane bed changes to various bedforms with a sequence: Plane bed → ripples → ripples on dunes → dunes → transition or washed out dunes → plane bed → antidune standing waves → antidune breaking waves → chutes and pools.

D. Ferraro (✉)
Dipartimento di Ingegneria Civile, Università della Calabria,
87036 Rende, CS, Italy
e-mail: domenico.ferraro@unical.it

S. Dey
Department of Civil Engineering, Indian Institute of Technology,
Kharagpur 721302, West Bengal, India
e-mail: sdey@iitkgp.ac.in

P. Rowiński and A. Radecki-Pawlik (eds.), *Rivers—Physical, Fluvial and Environmental Processes*, GeoPlanet: Earth and Planetary Sciences,
DOI 10.1007/978-3-319-17719-9_3

Fig. 3.1 Ripples

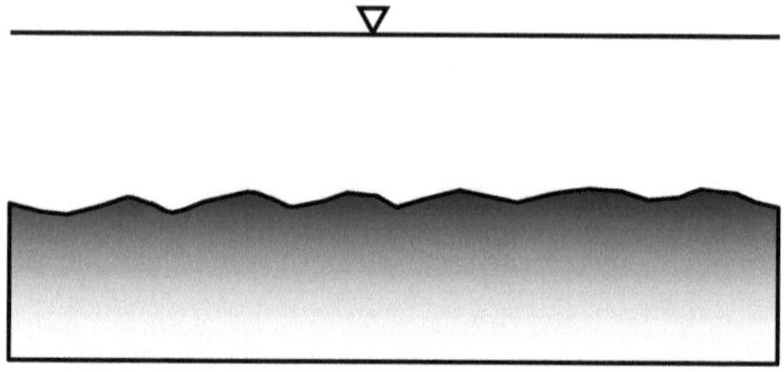

Ripples, as shown in Fig. 3.1, begin to form in subcritical flow condition ($Fr < 1$), when τ_0 marginally exceeds τ_{0c}, in presence of the viscous sublayer and are not able to affect the main flow zone (wall shear layer). The dimension of ripples (height and length) depends upon particle size (Yalin 1985; Baas 1993; Raudkivi 1997; Dey 2014). The arrangement of sediment particles forming ripples is such that a mild upstream slope (≈6°) is created by the induced shear stress; the downstream slope is relatively steep, equaling the angle of repose of sediment being the consequence of the fall of sediment particles from the crest of the ripples (≈32°). However, the mechanism of ripple formation is not so clear, although one can think about sediment bed as viscous fluid and ripples to occur because of Kelvin-Helmholtz instability between two different viscous fluids (Liu 1957), local intermittent disturbance (Raudkivi 1963, 1966), or small bed disturbance from near-bed turbulent burst (Williams and Kemp 1971).

With an increase in Froude number or in turn flow velocity, ripples evolve in ripples on dune with an increased height by a merger of adjacent ripples, as shown in Fig. 3.2 (Raudkivi 1997; Baas 1999; Valance 2005). Ripples on dune are characterized by individual deformed ripples on undeveloped dunes. Such a condition occurs for a transition from the hydraulically smooth flow to rough flow condition.

Dunes are formed in subcritical flow condition ($Fr < 1$) at relatively higher flow velocity and are able to affect the main flow zone (wall shear layer). In fact, the profile of dunes is out of phase with the free surface profile (Fig. 3.3). Such a behavior indicates how the velocity increases along the stoss-side (upstream) of a

Fig. 3.2 Ripples on dunes

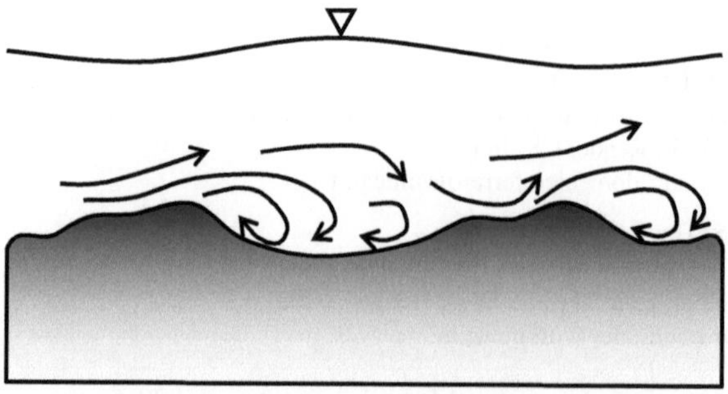

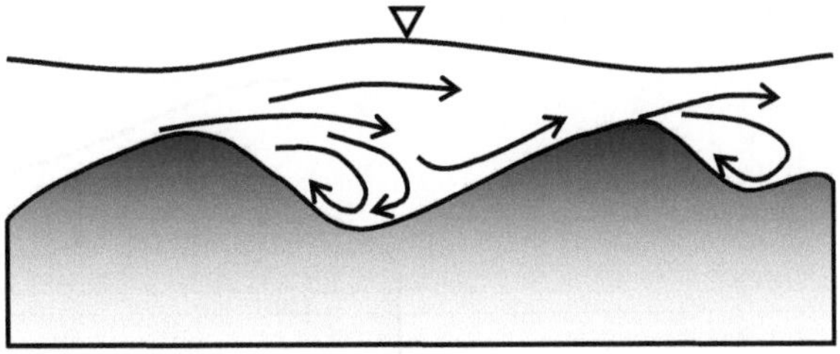

Fig. 3.3 Dunes

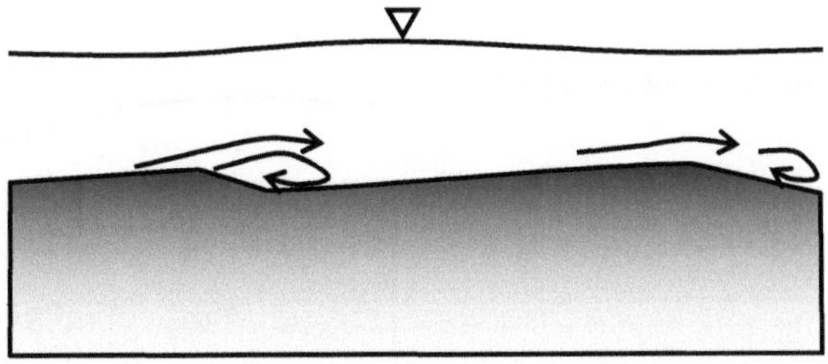

Fig. 3.4 Transition or washed out dunes

dune creating a mild-convexly curved profile of the dune. The sediment particles are transported over the crest of dunes falling down in the leeside (downstream). The leeside slope is thus equal to the angle of repose of sediments. The process of stoss-side erosion and leeside deposition causes to have a downstream migration of dunes. However, the formation of dunes was explained by Yalin (1977) as a consequence of quasi-periodic low-frequency large eddies creating local erosion and deposition of sediments.

At Froude number near to unity, one can observe transition from washed out dunes to reach plane bed, thus reducing the flow resistance and the flow depth (Fig. 3.4).

In supercritical flow condition ($Fr < 1$), antidunes are formed as a sinusoidal bed profile which is in phase with the free surface profile. Sediments in leeside are set in motion to develop the slope according to the flow profile. The turbulent flow helps the sediments to move up on the stoss-side, thus creating the upstream migration of antidunes (Fig. 3.5). Antidunes are not stable, however. They can grow (length and height-wise) until the free surface breaks down over the crest of antidunes making

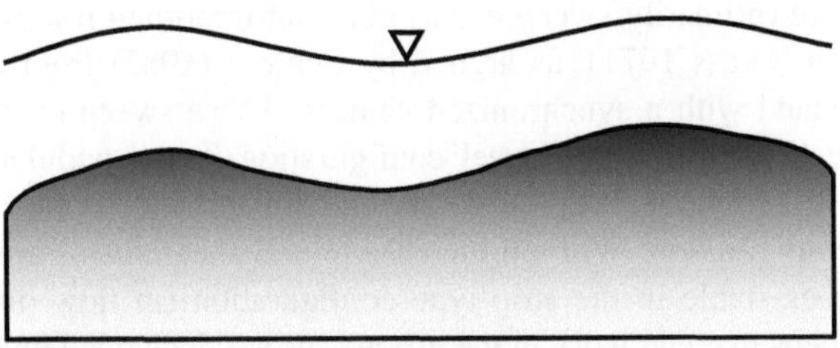

Fig. 3.5 Antidune standing waves

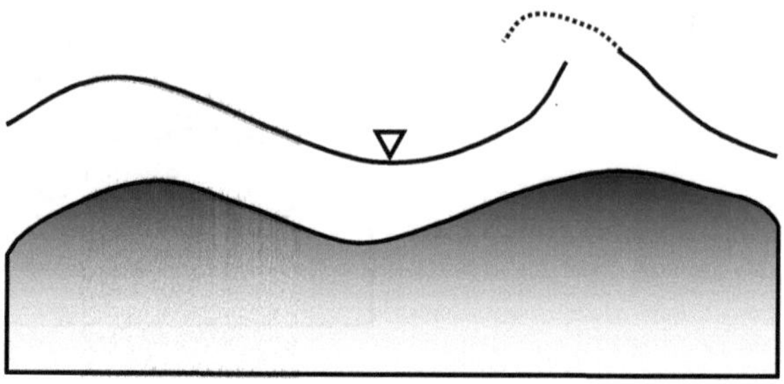

Fig. 3.6 Antidune breaking wave

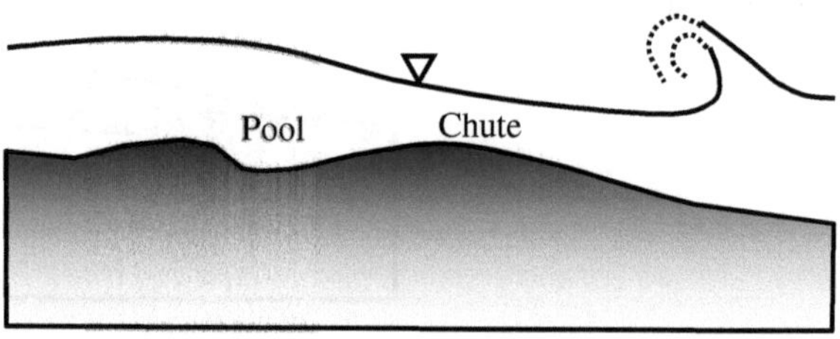

Fig. 3.7 Chutes and pools

whirling flow structures, called antidune breaking wave (Fig. 3.6). Then, the growing process of antidune is restarted and so on.

With further increase in Froude number, antidunes turn to chute-and-pool combinations (Fig. 3.7). These bedforms consist of sediment heaps followed by chutes (sliding channels). Two successive heaps and chutes make a pool. As a result, a succession of subcritical flow in the pools and a supercritical flow in the chutes are prevalent. However, formation of chutes and pools rarely occurs in an alluvial river.

3.2 Physical Processes

3.2.1 *Ripples*

The mechanism of ripple formation is not very well understood. Various theories based on different physical processes are reported in literature. It is believed that the coherent structure (primarily sweep event) plays an important role in initiating the sediment motion (Grass 1971), as argued by Schmid (1985) from an LDA measurement associated with a synchronized camera. These sweep events are able to modify a plane bed to an orange-peel configuration (small undulations) with an increase in the bed frictional roughness. The flow disturbance is also caused by the modified bed configuration. With an increase in bed shear stress, unstable orange-peel bed becomes stable in the strip-type configuration (in flow direction) channelizing the sweeps that can work in the grooves as are developed on the plane bed, as shown in Fig. 3.8.

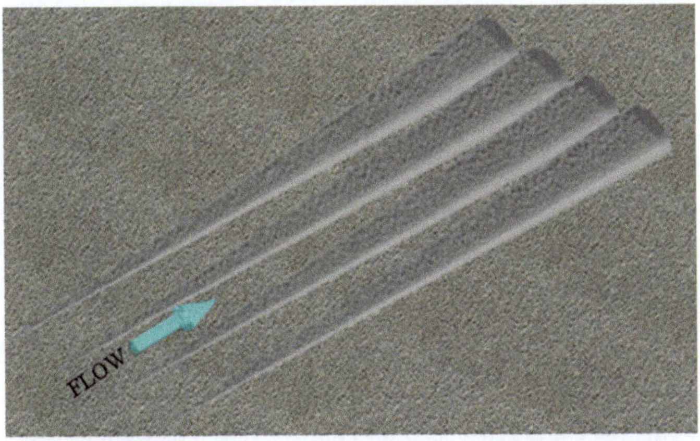

Fig. 3.8 Bed strips configuration

With an increase in bed shear stress, Gyr and Müller (1975) argued that the strip-like configuration changes to arrowhead-like bedforms with an apex angle of 25°. Such a value of the apex angle was also theoretically obtained by Landau (1944), Benney and Lin (1960) and Benney (1961). With a further increase in bed shear stress, the ripples begin to develop. However, ripple can also be formed, according to Schmid (1985), without following the above processes.

The near-bed coherent flow structure could result in the low pressure zone between sweep and ejection events (bubble vortex) as obtained from direct numerical simulations of turbulent boundary layer flows (Robinson 1991). Periodic property and two-dimensional spanwise structure, and also the geometry analogous to dunes, suggest that bedforms can be predicted by an adaption of the instability theory of a sand bed.

3.2.2 Dunes

Müller and Gyr (1983, 1986) suggested another scale for the dunes. In fact, these bedforms intrude into the main flow (shear layer) exposing them to the turbulence processes. They argued, however, that the flow separation at the crest of dunes is not stable. This phenomenon was also confirmed by Nezu and Nakagawa (1993), Bennett and Bridge (1995) and Baas and Best (2002). Kelvin-Helmholtz instability generates a spanwise vortex tube that twists on it creating bubbles in turbulence process. The dunes (due to their dimensions) can influence the flow depth making a quasi-stationary state.

The idea on feedback mechanisms between the mobile bed and the free surface of water may be helpful to understand the mechanism of dune formation. Turbulence structure and their statistical occurrence contribute to the formation of

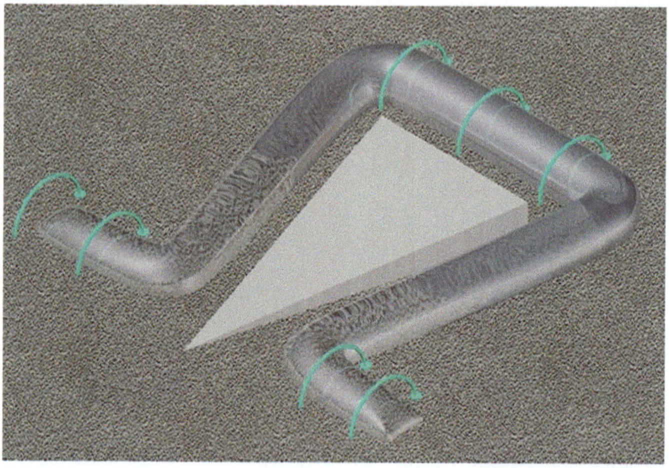

Fig. 3.9 Horseshoe shaped vortex

undular sediment bed which in turn produces a flow separation and an eddy system that can influence the main flow even up to the free surface. This process triggers a continuous interaction between the bed and the flow.

Taking into account the vortex involved in the process and with a modified definition of the mixing layer, the sediment transport and the deposition can be simulated in some discrete mathematical model (Martin and Meiburg 1994; Meiburg et al. 2000; Meiburg 2003).

It is interesting to note the interaction between dunes and free surface. Turbulent eruptions and boils are generated by dunes in a large-scale, as observed in laboratory experiments (Müller and Gyr 1983, 1986; Nezu and Nakagawa 1993; Gyr and Kinzelbach 2004; Best 2005) and fields as well (Matthes 1947; Korchokha 1968; Jackson 1976; Babakaiff 1993; Best et al. 2001). Another important behavior is the boils periodicity that can be linked with the dune height and in particular with the relative submergence of dunes with respect the flow depth (Babakaiff 1993). Babakaiff (1993) also noted further chaotic eruption boils and Lapointe (1992) related it with relative the roughness; whilst Gabel (1993) found that the flow structure is the cause of the curvature of the dune crest line. Some of the researches tried to link these vortexes to the shape similar to horseshoe shaped vortex as in Fig. 3.9 (Müller and Gyr 1983, 1986; Le Couturier et al. 2000).

3.2.3 Antidunes

In accordance with the Froude number ($Fr > 1$) in supercritical flow, the flow causes stationary sinusoidal waves altering the local scour process by making an erosion zone and a deposition zone, respectively. Such bedforms grow until they reach an equilibrium phase in sediments transport sense. The most important behavior of sediment transport process in antidunes is the upstream motion of the sand wave. However, antidunes can be stationary or can move downstream, as observed in some experiments (Kennedy 1961a; Simons et al. 1961).

3.3 Mathematical Models

3.3.1 *Exner's Model*

Exner's model is based on sediment continuity concept capable to predict bedform migration velocity (Exner 1925). Referring to Fig. 3.10, the continuity equation of sediment transport resulting in bed degradation is given by

$$(1-\rho_0)\frac{\partial \eta}{\partial t}+\frac{\partial q_T}{\partial x}=0 \Rightarrow \frac{\partial \eta}{\partial t}+\alpha_E\frac{\partial U}{\partial x}=0 \quad \wedge \quad q_T=(1-\rho_0)\alpha_E U \tag{3.1}$$

where η is the bed elevation with respect to a horizontal reference, t is the time, q_T is the bed-load transport rate (volume flux per unit width), x is the streamwise direction from a reference point, ρ_0 is the porosity of sediment, U is the depth-average flow velocity, and α_E is the erosion coefficient.

In Exner's model, a flat free surface is considered over bedforms (see Fig. 3.10), so that the depth-average flow velocity U can be obtained as

$$U=q/(h-\eta) \tag{3.2}$$

where h is the free surface elevation with respect the horizontal reference used for η and q is the discharge per unit width. Inserting Eq. (3.2) into Eq. (3.1) and integrating, one can obtain the solution at initial time $t = 0$ as a cosine function as follows:

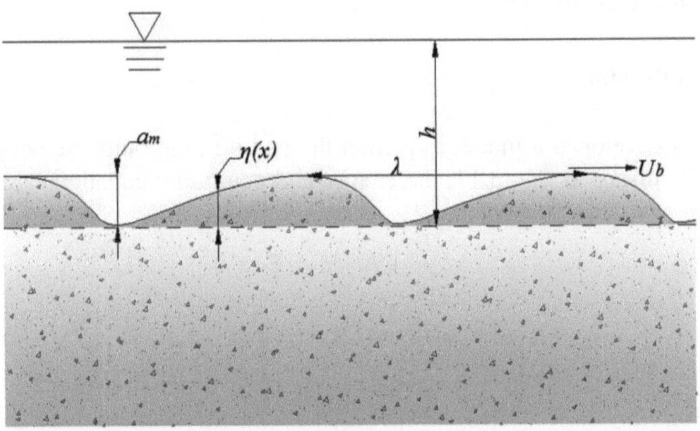

Fig. 3.10 Definition sketch of Exner's model

$$[\eta]_{t=0} = a_0 + a_m \cos(k_w x) \tag{3.3}$$

where a_m is the amplitude, $k_w = 2\pi/\lambda$, and λ is the wavelength of bedforms. Bedforms are assumed as a sand wave. The general solution for bedform amplitude is

$$\eta = a_0 + a_m \cos[k_w(x - U_b t)] \quad \wedge \quad U_b = \alpha_E q/(h - \eta)^2 \tag{3.4}$$

where U_b is the bedform migration velocity.

Exner's model was originally developed without considering bed friction. Next, the bed friction was introduced in the form of energy slope S_f. The dynamic equation for gradually varied unsteady flow can be written as

$$S_f = S_0 - \frac{\partial h}{\partial x} - \frac{U}{g} \cdot \frac{\partial U}{\partial x} - \frac{1}{g} \cdot \frac{\partial U}{\partial t} \Rightarrow \frac{\partial U}{\partial t} = -gS_f + gS_0 - g\frac{\partial h}{\partial x} - U\frac{\partial U}{\partial x} \tag{3.5}$$

For small bed slope ($S_0 \approx 0$), the fictional effect can be given by the relation $gS_f \approx k_f U$, and that is used in Eq. (3.1) to get

$$\frac{\partial^2 \eta}{\partial t^2} - m\frac{\partial^2 \eta}{\partial x \partial t} + k_f \frac{\partial \eta}{\partial t} - \alpha_E g \frac{\partial^2 \eta}{\partial x^2} = 0 \quad \wedge \quad m = \frac{gq}{U^2} - U \tag{3.6}$$

where k_f is the friction parameter. Integrating Eq. (3.6), one can obtain the following solution where initial condition is given by Eq. (3.3):

$$\eta = a_0 + a_m \exp\left[-\left(\frac{k_f}{2} - p\right)t\right] \cos\left\{k_w\left[x - \frac{m}{2p}\left(\frac{k_f}{2} - p\right)t\right]\right\} \tag{3.7}$$

where p is a function of k_f, m, λ and α_E.

3.3.2 *Kinematic Model*

3.3.2.1 Song's Model

Song (1983) developed a model to predict the bedform migration velocity and the direction. Referring to Fig. 3.11, the energy and continuity equations are

$$\frac{U^2}{2g} + h + \zeta = E \tag{3.8}$$

$$U(h + \zeta - \eta) = q \tag{3.9}$$

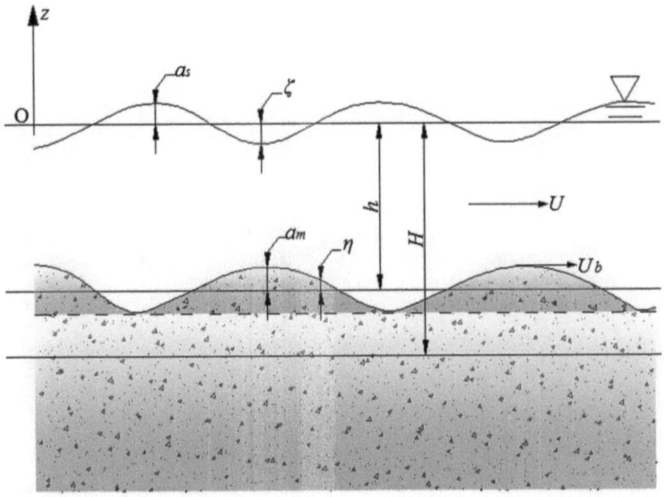

Fig. 3.11 Definition sketch of Song's model

Table 3.1 Bedform migration with respect to flow condition

Flow condition	Froude number	Bedform migration
Subcritical flow	$Fr < 1 \Rightarrow U_b > 0$	Downstream
Supercritical flow	$Fr > 1 \Rightarrow U_b < 0$	Upstream

where ζ is the free surface elevation with respect to mean flow level, η is the fluctuations of bed level with respect to mean bed level, and E is the total energy head.

In bedform phenomenon, the sediment continuity is a function of depth-averaged flow velocity. Thus, considering E and q constant as in Eqs. (3.8) and (3.9), respectively, the bedform migration velocity can be given by

$$U_b = \frac{\partial q_T}{\partial \eta} = \frac{\partial q_T}{\partial U} \cdot \frac{\partial U}{\partial \eta} = \frac{\partial q_T}{\partial U} \cdot \frac{U^2}{q(1 - Fr^2)} \tag{3.10}$$

In the above, the $\partial q_T / \partial U$ is a positive quantity, so the direction of bedform migration is governed by the quantity $1 - Fr^2$ according to the flow condition (subcritical or supercritical) as given in Table 3.1.

3.3.2.2 Núñez-González and Martín-Vide's Model

Núñez-González and Martín-Vide (2011) developed a model for antidune migration. The key issue in this model is to check if the free surface wave and the

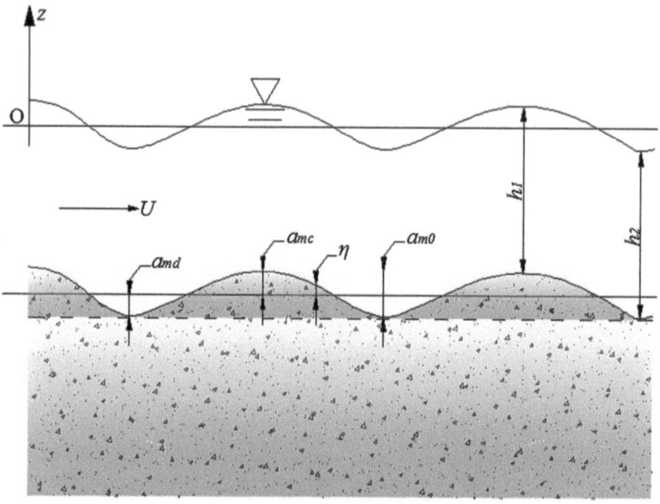

Fig. 3.12 Definition sketch of Núñez-González and Martín-Vide's model

Table 3.2 Direction of bedform migration with respect to phase parameters

Phase parameter	Phase condition	Bedform migration
h_1/h_2	<1	Downstream
	=1	Stationary
	>1	Upstream

antidune wave are in or out of phase. Referring to Fig. 3.12, Table 3.2 summarizes the direction of antidune migration with respect to the phase parameter, that is h_1/h_2, where h_1 and h_2 are the flow depths above the crest and the trough, respectively.

Following Fig. 3.12, energy balance at the crest and the trough (neglecting the loses) yields

$$\frac{U_1^2}{2g} + h_1 + a_{mc} = \frac{U_2^2}{2g} + h_2 + a_{md} \tag{3.11}$$

where a_{mc} and a_{md} are the average elevations of the crest and the trough of the bedforms with respect to mean bed level. For antidunes, flow depth is comparable with the sand wave fluctuations. So the pressure distribution is non-hydrostatic; and it is adjusted by the centrifugal effect. Equation (3.11) can then be rewritten as

$$\frac{U_1^2}{2g} + h_1\left(1 - \frac{1}{g}\cdot\frac{U_1^2}{r_1}\right) + a_{mc} = \frac{U_2^2}{2g} + h_2\left(1 + \frac{1}{g}\cdot\frac{U_2^2}{r_2}\right) - a_{md} \tag{3.12}$$

Table 3.3 Antidune migration with antidune mobility number

Antidune mobility number		Antidune migration
F_a	<1	Downstream
	=1	Stationary
	>1	Upstream

The symmetrically sinusoidal development of antidunes allows to make the following simplifications:

$$a_{m0} = 2a_{mc} = 2a_{md} = a_{mc} + a_{md}, \quad r = r_1 = r_2 \tag{3.13}$$

and then using the flow continuity, $q = U_1h_1 = U_2h_2$, the mean flow depth $h = (h_1h_2)^{0.5}$ and Eq. (3.13) into Eq. (3.12), one gets

$$a_{m0} + h_1 - h_2 = Fr^2h^3\left(\frac{1}{h_1} + \frac{1}{h_2}\right)\left[\frac{1}{h_1} + \frac{1}{2}\left(\frac{1}{h_2} - \frac{1}{h_1}\right)\right] \quad \wedge \quad Fr = \frac{q}{(gh^3)^{0.5}} \tag{3.14}$$

From Eq. (3.14), an *antidune mobility number* F_a can be defined as a function of Froude number Fr as

$$F_a = Fr \cdot h\left(\frac{2}{ra_{m0}}\right)^{0.5} = 1 \tag{3.15}$$

where

$$r = \frac{1}{2a_{m0}}\left(\frac{\lambda}{\pi}\right)^2 = \frac{1}{2a_{m0}}\left(\frac{2}{k_w}\right)^2 \quad \wedge \quad k_w = 2\frac{\pi}{\lambda} \tag{3.16}$$

where λ is the wavelength of antidunes. Thus,

$$F_a = Fr \cdot h \cdot k_w \tag{3.17}$$

Table 3.3 furnishes the antidune migration as a function of antidune mobility number F_a.

3.3.3 Stability Model

3.3.3.1 Kennedy's Model

Kennedy (1963) developed an analytical model considering the potential flow theory to examine the stability of the fluid-bed interface and the bedform

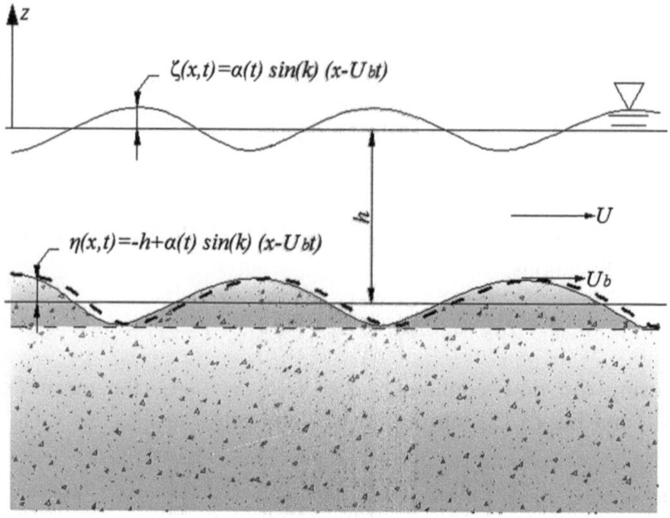

Fig. 3.13 Definition sketch of Kennedy's model

characteristics. Figure 3.13 shows the coordinate system and the sinusoidal bed and free surface profiles, which are represented by appropriate equations. The potential flow is governed by the Laplace equation and its two-dimensional form is

$$\frac{\partial^2 \phi}{\partial x^2} + \frac{\partial^2 \phi}{\partial z^2} = 0 \tag{3.18}$$

The boundary conditions are taken at the free surface and the bed elevations as

$$z = 0: \ \frac{\partial \zeta}{\partial t} + U\frac{\partial \zeta}{\partial x} = \frac{\partial \phi}{\partial z} \tag{3.19}$$

$$z = 0: \ U\frac{\partial \phi}{\partial x} + \frac{\partial \phi}{\partial t} = -g\zeta \tag{3.20}$$

$$z = -h: \ \frac{\partial \eta}{\partial t} + U\frac{\partial \eta}{\partial x} = \frac{\partial \phi}{\partial z} \tag{3.21}$$

The velocity function ϕ that satisfies the Laplace equation is

$$\phi(x, z, t) = U\alpha\left[\frac{1}{Fr^2 k_w h}\cos\mathrm{h}(k_w z) + \sin\mathrm{h}(k_w x)\right] \cos k_w (x - U_b t) \tag{3.22}$$

where α is the amplitude of the water-wave. The main assumption is to consider the quasi-steady flow $\partial\alpha/\partial t \ll U k_w \alpha$, and then the smaller growth rate cannot produce phase displacement between free surface and sand wave. Such assumption allows having only one coordinate system given by

$$\alpha_m(t) = \alpha(t)\left[1 - \frac{1}{Fr^2 k_w h}\tan\mathrm{h}(k_w h)\right]\cos\mathrm{h}(k_w h) \tag{3.23}$$

Equation (3.23) reveals the phase shift between bed and free surface waves as given by the following conditions:

$$\theta = 0:\ U^2 > \left(\frac{g}{k_w}\right)\tan\mathrm{h}(k_w h), \quad \theta = \pi:\ U^2 < \left(\frac{g}{k_w}\right)\tan\mathrm{h}(k_w h) \tag{3.24}$$

Dune migration is due to the sediment transport over the sand wave surface. The sediment continuity equation over a wavy bed is

$$\frac{\partial q_T}{\partial x} + \frac{\partial \eta}{\partial t} = 0 \quad \wedge \quad q_T = q_s b \tag{3.25}$$

where b is the channel width and q_s the sediment transport rate per unit width. Equation (3.25), however, does not take into account the flow separation in the leeside of dunes, while flow separation does exist in reality. Then it is possible to define U_b by Eq. (3.10). Furthermore, potential flow theory does not consider bed shear stress, which is required to have sediment motion. To estimate sediment transport rate, Kennedy chose the follow equation suggested by Alam et al. (1966):

$$q_T(x,t) = m\left[U - U_c + \frac{\partial\phi(x-\delta,h,t)}{\partial x}\right]^n \tag{3.26}$$

where m is a dimensional coefficient, n is a dimensionless exponent, U_c is the threshold velocity for sediment motion, and δ is the phase shift between free surface wave and sand wave. In addition, Kennedy (1969) considered *transport relaxation time* and *transport relaxation distance* in order to define the time needed to have sediment transport due to an increasing velocity and the distance needed to deposit sediment due to a decrease in velocity, respectively. Of course, such physical considerations are related to the bed shear stress and the flow velocity.

Expanding Eq. (3.26), it is possible to describe the phase of the free surface wave with respect to sand wave in accordance with Froude number

$$Fr^2 = \frac{1 + k_w h\tan\mathrm{h}(k_w h) + jk_w h\cot(jk_w h)}{(k_w h)^2 + [2 + jk_w h\cot(jk_w h)]k_w h\tan\mathrm{h}(k_w h)} \tag{3.27}$$

$$Fr^2 = \frac{j}{k_w h}\tan\mathrm{h}(k_w h) \tag{3.28}$$

Table 3.4 Bedform migration criteria

Bedforms	Equation	Phase relations	Bedform migration
Ripples	(27)	$jk_w h = 3\pi/2$	Downstream
Dunes	(27)	$jk_w h = 3\pi/2$	Downstream
Transitive forms	(28)	$jk_w h = 3\pi/2$	Upward
Plane bed	(28)	$jk_w h = \pi$	–
Antidunes	(27)	$jk_w h = \pi/2$	Downstream
Antidunes	(27)	$jk_w h < \pi/2$	Upstream

where $j = \delta/h$. Gyr and Hoyer (2006) gave the bedform migration criteria based on the above equation, as furnished in Table 3.4.

3.3.3.2 Hayashi's Model

Hayashi (1970) considered the boundary conditions similar to those of Kennedy (1969), except the modified version of Eq. (3.20) as follows:

$$z = 0 : \frac{1}{2}\left[\left(\frac{\partial \phi}{\partial x}\right)^2 + \left(\frac{\partial \phi}{\partial z}\right)^2\right] + \frac{\partial \phi}{\partial t} + g\zeta = \text{constant} \tag{3.29}$$

Using Laplace equation, Eq. (3.18), and the boundary conditions, Eqs. (3.19), (3.21) and (3.29), the velocity potential function is obtained as

$$\phi(x, z, t) = U\left[x - \alpha_m \frac{\cos\text{h}(k_w z) + Fr^2 k_w h \sin\text{h}(k_w x)}{\sin\text{h}(k_w h) - Fr^2 k_w h \cos\text{h}(k_w h)} \cos k_w (x - U_b t)\right] \tag{3.30}$$

The slowly varying amplitude $\alpha(t)$ brings back to Kennedy's assumption of quasi-steady flow ($\partial\alpha/\partial t \ll Uk_w\alpha$) and then $U_b \ll U$. The relationship between amplitudes of free surface wave and sand wave becomes

$$\alpha_s(t) = \alpha(t) \frac{Fr^2 k_w h}{[\tan\text{h}(k_w h) - Fr^2 k_w h] \cos\text{h}(k_w h)} \tag{3.31}$$

The continuity equation is

$$\frac{\partial q_T}{\partial x} + \frac{\partial \eta}{\partial t} = 0 \quad \wedge \quad q_T = \frac{q_s}{(1 - \rho_0)} \tag{3.32}$$

and again

$$q_T(x,t) = m\left[1 + c\frac{\partial}{\partial x}\eta(x-\delta,t)\right]\left[\frac{\partial}{\partial x}\phi(x-\delta,-h,t)\right]^4 \tag{3.33}$$

where m is a dimensional coefficient and c is a non-dimensionless constant.

Using Eqs. (3.30)–(3.33), expressing $\partial\phi/\partial x = U + u(x,z,t)$ and assuming $k_w\delta \ll 1$, yield

$$\alpha(t) = \alpha(0)\exp\left(\frac{mg^2c}{C}Fr^4k_w^2h^2\vartheta t\right) \tag{3.34}$$

where

$$C = \frac{c}{\delta}\left(\frac{U^2}{2g}\right)^{-1}, \quad \vartheta = C - 2Fr^2k_wh\left[\frac{1 - Fr^2k_wh\tan\mathrm{h}(k_wh)}{\tan\mathrm{h}(k_wh) - Fr^2k_wh}\right] \tag{3.35}$$

Equation (3.34) predicts the criteria for the formation of sand waves:

1. $\vartheta > 0$ is the criterion for unstable plane bed;
2. $\vartheta < 0$ is the criterion for stable plane bed; and
3. $\vartheta = 0$ and $\tanh(k_wh) - Fr^2k_wh = 0$ are the limits of sand wave formation.

Then, the limiting values of Fr^2 are obtained as follows:

$$Fr^2 = \left|\begin{matrix} Fr_2^2 \\ Fr_1^2 \end{matrix}\right| = \frac{1}{4k_wh\tan\mathrm{h}(k_wh)}\{C + 2 \pm [(C+2)^2 - 8C\tan\mathrm{h}^2(k_wh)]^{0.5}\} \tag{3.36}$$

$$Fr^2 = Fr_a^2 = \frac{\tan\mathrm{h}(k_wh)}{k_wh} \tag{3.37}$$

Hayashi defined the dunes formation criterion as $\alpha_s < \alpha_m$ and the antidunes criterion as $\alpha_s > \alpha_m$. Here, α_s and α_m are the amplitudes of free surface and bed waves, respectively.

The bedform formation criteria in terms of Froude number are

1. $Fr < Fr_1$ is the criterion for dune formation;
2. $Fr_a < Fr < Fr_2$ is the criterion for antidune formation; and
3. $Fr_1 < Fr < Fr_a$ and $Fr_2 < Fr$ are the criteria for plane bed to occur.

3.3.3.3 Song's Model

Song's (1983) model is based on Milne-Thompson (1960) theory of potential flow over a sinusoidal bed. Satisfying the complex velocity potential with a constant pressure at the free surface, he obtained the depth-averaged flow velocity as

$$U^2 = \left(\frac{g}{k_w}\right) \tan\mathrm{h}(k_w h) \tag{3.38}$$

The amplitudes of free surface wave and sand wave are related as

$$\alpha_m = \alpha_s \left[1 - \frac{g}{k_w U^2} \tan\mathrm{h}(k_w h)\right] \cos\mathrm{h}(k_w h) \tag{3.39}$$

The celerity generated by gravity is given by $C = [g \tanh(k_w h)/k_w]^{0.5}$. Equation (3.39) is also useful to understand if surface wave and bed wave are in or out of phase according to the sign of the square bracket quantity in Eq. (3.39). Then it can be determined if dunes or antidunes are prevalent.

Song argued that the near-bed flow velocity is a function of bed-load transport rate. This assumption helps to determine the bedform migration velocity. Such assumption is in accordance with potential flow motion; thus, errors are expected due to the flow separation that appears in the leeside of a dune. Near-bed flow velocity in streamwise direction is obtained as

$$u_0 = U_m(1 + k_w U^* \eta) \quad \wedge \quad U^* = \frac{1 - Fr^2 k_w h \tan\mathrm{h}(k_w h)}{\tan\mathrm{h}(k_w h) - Fr^2 k_w h} \tag{3.40}$$

where U^* is a dimensionless parameter representing bedform migration velocity.

A further assumption can be made on near-bed velocity to be approximately equal to the depth-averaged velocity, that is, $u_0 \approx U$. Then, the bedform migration velocity can be obtained from Eq. (3.1) as

$$U_b = \frac{\partial q_T}{\partial U} \cdot \frac{\partial U}{\partial \eta} = \frac{\partial q_T}{\partial U} k_w U U^* \Rightarrow U^* = U_b \left(\frac{\partial q_T}{\partial U} k_w U\right)^{-1} \tag{3.41}$$

Bedform migration can be identified as following:

1. $U < C$ and $U^* > 1$ are the conditions to occur dunes;
2. $U \geq C$ and $U^* < 0$ are the conditions to occur antidunes; and
3. $U \geq C$ and $U^* > 0$ are the conditions for the migration of antidunes downstream, as observed by Engelund (1970) and Fredsøe (1974).

3.4 Bose-Dey Instability Theory

A new theory of turbulent shear flow over a wavy bed was developed by Bose and Dey (2009) by using the Reynolds averaged Navier-Stokes and the time-averaged continuity equations to address the instability criterion of erodible beds leading to the formation of sand waves.

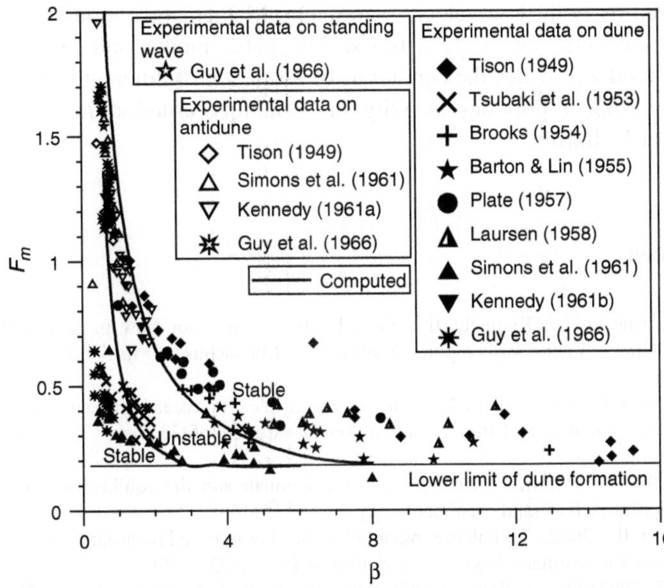

Fig. 3.14 Diagram of stability of sand waves given by Bose and Dey (2009) and its comparison with the experimental data (Tison 1949; Tsubaki et al. 1953; Brooks 1954; Barton and Lin 1955; Plate 1957; Laursen 1958; Simons et al. 1961; Kennedy 1961a, b; Guy et al. 1966)

Figure 3.14 shows the results obtained by Bose and Dey (2009). The curves of the Froude number F_m [$=U/(gh)^{0.5}$] versus nondimensional wave number β ($=k_w h$) decide upon a stability zone. For $F_m < 0.8$, the bed remains stable with the formation of dunes, while for $F_m \geq 0.8$, the bed remains unstable with the formation of standing waves and antidunes. [Note: As the derivation of Bose-Dey equations is quite lengthy, it is not included here. However, one can read it from Bose and Dey (2009) or Dey (2014).]

Importantly, a comparison of results obtained from various models is available in Dey (2014).

3.5 Conclusions

In lower flow regime (Froude number less than unity), ripples and dunes are formed; while in upper flow regime (Froude number greater than unity), antidunes are the governing bedforms. Ripples are formed in the presence of viscous sublayer in hydraulically smooth flow, where the bed shear stress exceeds the threshold value for the sediment motion, while dunes are formed in hydraulically rough flow. The dimension of the ripples depends on the sediment size and the flow velocity, but is independent of the flow depth. However, dunes and antidunes are characterized by both flow velocity and depth. Mathematical models proposed by various researchers are of two types: kinematic model and stability model. Kinematic model provides the conditions of formation of different bedforms; while the stability model provides stability or instability criterion in formation and migration of bedforms.

References

Alam AMZ, Cheyer TF, Kennedy JF (1966) Friction factors for flow in sand bed channels. Hydrodynamics Laboratory report number 78. Massachusetts Institute of Technology, Cambridge

Baas JH (1993) Dimensional analysis of current ripples in recent and ancient depositional environments. Geological Ultraiectina, vol 106. Department of Geology, University of Utrecht, Utrecht

Baas JH (1999) An empirical model for the development and the equilibrium morphology of current ripples in fine sand. Sedimentology 46(1):123–138

Baas JH, Best JL (2002) Turbulence modulation in clay-rich sediment-laden flows and some implications for sediment deposition. J Sediment Res 72:336–340

Babakaiff CS (1993) Flow hydraulics, bedforms and macroturbulence of Squamish River estuary, British Columbia. MSc thesis, Simon Fraser University, Burnaby, BC, Canada

Barton JR, Lin PN (1955) A study of sediment transport in alluvial channels. Report number 55JRB2, Department of Civil Engineering, Colorado A and M College, Fort Collins, Colorado

Bennett SJ, Bridge JS (1995) The geometry and dynamics of low-relief bed forms in heterogeneous sediment in a laboratory channel, and their relationship to water flow and sediment transport. J Sediment Res 65A(1):29–39

Benney DJ (1961) A non-linear theory for oscillations in a parallel flow. J Fluid Mech 10:209–236

Benney DJ, Lin C (1960) On the secondary motion induced by oscillations in a shear flow. Phys Fluids 3:656–657

Best J (2005) The fluid dynamics of river dunes: a review and some future research directions. J Geophys Res F: Earth Surf 110:F04S02

Best JL, Kostaschuk RA, Villard PV (2001) Quantitative visualization of flow fields associated with alluvial sand dunes: Results from the laboratory and field using ultrasonic and acoustic Doppler anemometry. J Visualiz 4:373–381

Bose SK, Dey S (2009) Reynolds averaged theory of turbulent shear flow over undulating beds and formation of sand waves. Phys Rev E 80:036304

Brooks NH (1954) Laboratory studies of the mechanics of streams flowing over a movable bed of fine sand. Doctoral thesis, California Institute of Technology, Pasadena

Dey S (2014) Fluvial hydrodynamics: hydrodynamic and sediment transport phenomena. Springer, Berlin

Engelund F (1970) Instability of erodible beds. J Fluid Mech 42:225–244

Engelund F, Fredsøe J (1982) Sediment ripples and dunes. Annu Rev Fluid Mech 14:13–37

Exner FM (1925) Über die wechselwirkung zwischen wasser und geschiebe in flüssen. Sitzungsberichte der Akademie der Wissenschaften 134(2a):165–203

Fredsøe J (1974) On the development of dunes in erodible channels. J Fluid Mech 64:1–16

Gabel SL (1993) Geometry and kinematics of dunes during steady and unsteady flows in the Calamus River, Nebraska, USA. Sedimentology 40:237–269

Grass AJ (1971) Structural features of turbulent flow over smooth and rough boundaries. J Fluid Mech 50:233–255

Guy HP, Simons DB, Richardson EV (1966) Summary of alluvial channel data from flume experiments, 1956–1961. United States geological survey water supply paper number, 462-1, Washington, DC

Gyr A, Hoyer K (2006) Sediment transport: a geophysical phenomenon. Fluid mechanics and its application, vol 82. Springer, The Netherlands

Gyr A, Kinzelbach W (2004) Bed forms in turbulent channel flows. Appl Mech Rev 57:77–93

Gyr A, Müller A (1975) Alteration of structures of sublayer flow in dilute polymer solutions. Nature 253:185–187

Hayashi T (1970) Formation of dunes and antidunes in open channels. J Hydraul Div 96(3):357–366

Jackson RG (1976) Sedimentological and fluid-dynamic implications of the turbulence bursting phenomenon in geophysical flows. J Fluid Mech 77:531–560

Kennedy JF (1961a) Stationary waves and antidunes in alluvial channels. Report number KH-R-2, WM Keck Laboratory of Hydraulics and Water Resources, California Institute of Technology, Pasadena

Kennedy JF (1961b) Further laboratory studies of the roughness and suspended load on alluvial streams. Report number KH-R-3, WM Keck Laboratory of Hydraulics and Water Resources, California Institute of Technology, Pasadena

Kennedy JF (1963) The mechanics of dunes and antidunes in erodible bed channels. J Fluid Mech 16(4):521–544

Kennedy JF (1969) The formation of sediment ripples, dunes and antidunes. Ann Rev Fluid Mech 1:147–168

Korchokha YM (1968) Investigation of the dune movement of sediments on the Polomet' River. Sov Hydrol Sel Pap 6:541–559

Landau LD (1944) On the problem of turbulence. Akad Nauk 44:339–342

Lapointe M (1992) Burst-like sediment suspension events in a sand bed river. Earth Surf Processes Landforms 17:253–270

Laursen EM (1958) The total sediment load of streams. J Hydraul Div 84(1):1–36

Le Couturier MN, Grochowski NT, Heathershaw A, Oikonomou E, Collins MB (2000) Turbulent and macro-turbulent structures developed in the benthic boundary layer downstream of topographic features. Estuarine Coastal Shelf Sci 50:817–833

Liu HK (1957) Mechanics of sediment-ripple formation. J Hydraul Div 83(2):1–23

Martin JE, Meiburg E (1994) The accumulation and dispersion of heavy particles in forced twodimensional mixing layers. I The fundamental and subharmonic case. Phys Fluids 6:1116–1132

Matthes GM (1947) Macroturbulence in natural stream flow. EOS Trans AGU 28:255–265

Meiburg E (2003) Numerical investigation of two-way coupling mechanisms in dilute, particle laden flows. In: Gyr A, Kinzelbach K (eds) Sedimentation and sediment transport, pp 149–154

Meiburg E, Wallner E, Pagella A, Riaz A, Haertel C, Necker F (2000) Vorticity dynamics of dilute two-way-coupled particle-laden mixing layers. J Fluid Mech 421:185–227

Milne-Thompson LM (1960) Theoretical hydrodynamics. Macmillan, New York

Müller A, Gyr A (1983) Visualisation of the mixing layer behind dunes. In: Sumer BM, Müller A (eds) Mechanics of sediment transport. Balkema, Rotterdam, pp 41–45

Müller A, Gyr A (1986) On the vortex formation in the mixing layer behind dunes. J Hydrul Res 24:359–375

Nezu I, Nakagawa H (1993) Turbulence in open-channel flows. Balkema, Rotterdam

Núñez-González F, Martín-Vide JP (2011) Analysis of antidune migration direction. J Geophys Res 116:F02004. doi:10.1029/2010JF001761

Plate EJOF (1957) Laboratory studies on the beginning of sediment ripple formation in an alluvial channel. Master thesis, Colorado State University, Fort Collins, Colorado

Raudkivi AJ (1963) Study of sediment ripple formation. J Hydraul Div 89(6):15–33

Raudkivi AJ (1966) Bed forms in alluvial channels. J Fluid Mech 26:507–514

Raudkivi AJ (1997) Ripples on streambed. J Hydraul Eng 123(1):58–64

Robinson SK (1991) Coherent motions in the turbulent boundary layers. Annl Rev Fluid Mech 23:601–639

Schmid A (1985) Wandnahe turbulente bewegungsabläufe und ihre bedeutung für die riffelbildung. Ph.D. thesis and report R22-85, Institute for Hydromechanics and Water Resources Management, ETH Zürich, Switzerland

Simons DB, Richardson EV, Albertson ML (1961) Flume studies using medium sand (0.45 mm). United States geological survey water supply paper number 1498-A, United States Government Printing Office, Washington, DC

Song CCS (1983) Modified kinematic model: application to bed forms. J Hydraul Eng 109 (8):1133–1151

Tison LH (1949) Origine des ondes de sable et des bancs de sable sous l'action des courants. Transactions of the International Association for hydraulic structures research, third meeting, report II-13, Grenoble

Tsubaki T, Kawasumi T, Yasutomi T (1953) On the influence of sand ripples upon the sediment transport in open channels. Rep Res Inst Appl Mech 2:241–256

Valance A (2005) Formation of ripples over a sand bed submitted to a turbulent shear flow. Eur Phys J B 45(3):433–442

Williams PB, Kemp PH (1971) Initiation of ripples on flat sediment beds. J Hydraul Div 97 (4):505–522

Yalin MS (1977) Mechanics of sediment transport. Pergamon, Oxford

Yalin MS (1985) On the determination of ripple geometry. J Hydraul Eng 111(8):1148–1155

Chapter 4
Scour Problems Downstream of Low-Head Hydraulic Structures

Stefano Pagliara and Michele Palermo

Abstract Low-head structures are widely used in river restoration. Their function is to regulate the sediment transport and at the same time they can assure optimal habitat for fish species in the river. They strongly affect river morphology because of the erosive processes occurring downstream of them. Thus, a correct design has to take into account several aspects, i.e. technical, economic, environmental, etc. The present chapter aims to present the most recent achievements in scientific literature regarding the design criteria for low-head structures. In particular, low-environmental impact structures will be analyzed and illustrated. In the last decades, the environmental sensibility has increasingly forced hydraulic engineers to propose design solutions which can conjugate both hydraulic functioning and environmental impact minimization. This chapter proposes a synthesis of criteria to predict scour characteristics of the stilling basin downstream of several low-head structure typologies. Namely, the scour process downstream of block ramps will be discussed and the effect of both stilling basin geometry and ramp configuration will be analyzed, for both clear water and live bed conditions. Furthermore, the erosive process downstream of rock grade control structures and stepped gabion weirs will be illustrated along with relationships to predict the characteristic lengths of the scour hole and dune.

Keywords Clear water · Live-bed · Low-head structures · Basin morphology · Scour process

---

S. Pagliara (✉) · M. Palermo
DESTEC—Department of Energy Engineering, Systems, Land and Construction, University of Pisa, Via Gabba 22, 56122 Pisa, Italy
e-mail: s.pagliara@ing.unipi

M. Palermo
e-mail: michele.palermo@ing.unipi.it

P. Rowiński and A. Radecki-Pawlik (eds.), *Rivers—Physical, Fluvial and Environmental Processes*, GeoPlanet: Earth and Planetary Sciences,
DOI 10.1007/978-3-319-17719-9_4

4.1 Introduction

Scour problems downstream and in correspondence with hydraulic structures are one of the most important topics in hydraulic engineering. The study of erosive phenomena and hydrodynamic mechanisms related to river restoration have been extensively analyzed in the last decades. Nevertheless, the approach to solve or optimize the design criteria has been deeply modified. In fact, river restoration structures have been substantially developed to propose solutions which can both assure a correct functioning of the hydraulic structure and, at the same time, minimize their impact in the natural contests in which they are located. Based on this last assumption, low-environmental impact hydraulic structures have become more and more popular. In particular, nowadays rock-made structures (characterized by different geometries) are widely used. They can efficiently control sediment transport and are much more integrated in the river systems. In other words, river restoration structures are currently designed to be both hydraulically efficient and to constitute a non-intrusive element in the natural contest. In this perspective, the present chapter aims to synthesize some of the most recent design criteria relative to several hydraulic structures. In particular, the analysis will be focused on block ramps, stepped gabion weirs and rock-grade control structures. Among these, block ramps are surely the most diffused eco-friendly hydraulic structures. They have been extensively studied, especially in the last decades (among others Veronese 1937; Bathurst 1978; Bathurst et al. 1981; Platzer 1983; Whittaker and Jaggi 1986; Bormann and Julien 1991; Breusers and Raudkivi 1991; Rice et al. 1998; Robinson et al. 1997; D'Agostino and Ferro 2004; Pagliara and Palermo 2008a, b, 2010, 2011a, b, 2012; Pagliara et al. 2008, 2009a, b, 2011, 2012; Oertel and Schlenkhoff 2012; Radecki-Pawlik 2013). They are characterized by a sloping bed on which blocks are disposed either in regular or irregular arrangement. The blocks constituting the ramp can be either loose or fixed on the bed. Generally, this structure typology is used instead of classical concrete check dams in order to regulate sediment transport. A correct design has to take into consideration several aspects. In particular, block ramp stability has to be assured in order to warranty its functioning. Nevertheless, both the dissipative mechanism occurring either in stilling basin or on the ramp itself have to be analyzed in order to optimize the ramp geometry. Finally, the erosive processes occurring downstream of them are a fundamental step of the design process. This chapter will be mainly focused on this last aspect, including both the effect of ramp and stilling basin geometry on the main scour lengths either in clear-water or in live-bed conditions. Successively, the analysis will be focused on both stepped gabion weirs and rock-grade control structures. Also these two hydraulic structures are characterized by a low-environmental impact. In particular, experimental tests were recently conducted by several researchers (among these Peyras et al. 1992; Essery and Horner 1978; Mossa 1998; Chinnararsi et al. 2008; Pagliara and Palermo 2013). It was proven that they are efficient structures in order to dissipate flow energy and regulate sediment transport. They are characterized by relatively small height (generally less than 3 m) and mainly they are used in torrential streams. In particular, rock grade-control structures are often used in series, such as concrete sills, in order to stabilize river bed. Downstream of them, water pools form and they constitute a suitable environment for fish species.

4.2 Block Ramps

Block ramps are river restoration structures which were initially used for torrential stream. Their height is generally not prominent and are mainly used instead of traditional concrete check dams in order to dissipate a greater amount of flow energy. In particular, the first studies regarding these structures were focusing on both blocks stability and arrangements, such as on the energy dissipation processes (Platzer 1983; Whittaker and Jaggi 1986; Rice et al. 1998; Robinson et al. 1997; Pagliara and Chiavaccini 2006a, b). Nevertheless, in the last decades, the design approaches have been substantially re-considered and enriched by several researches, who aimed to analyze both the hydraulic processes and scour mechanisms. In other words, the new tendency is to consider block ramps not just an isolated element in the natural contest, but part of it, i.e. the design criteria have to take into consideration both the structure itself and the river in situ conditions. Based on this perspective, the scour processes occurring downstream of block ramps assume a particular importance. In fact, both the erosive mechanism and the stilling basin configuration play a fundamental role in both dissipative process and ecological impact of structure (see for example Bormann and Julien 1991; D'Agostino and Ferro 2004; Pagliara and Palermo 2008a, b, 2010, 2011a, b; Pagliara et al. 2009a). Furthermore, both erosive and dissipative mechanisms are influenced by the river sediment transport dynamics. In particular, the erosive mechanism downstream of the structure depends on the approaching sediment coming from the upstream river branch (i.e., live-bed conditions or clear water conditions) as proved by Pagliara et al. (2011) and (2012). In the following a brief description of the effects on the erosive mechanism occurring downstream of a block ramp due to both stilling basin geometry and approaching sediment transport conditions will be furnished, along with design criteria to foresee the characteristics lengths of the scour geometry downstream of the structure. In the following Fig. 4.1, a picture of a block ramp is reported.

4.2.1 Erosive Processes in Clear-Water Conditions

Block ramps are eco-friendly hydraulic structures which have been widely used in the last decades. Their peculiarity is to be highly dissipative structures and at the same time they can easily adapt to the in situ conditions. The dissipative process occurring on a block ramp are mainly due to the ramp surface roughness and localized variation of stream slope. The dissipative process on the block ramp

Fig. 4.1 Picture of a block ramp built in a river in the Carpathian mountains (Poland)

depends on the relative roughness and on the location of the hydraulic jump. In the case in which the hydraulic jump is entirely occurring in the stilling basin, the global dissipative process, i.e. the energy dissipation between the ramp entrance and the downstream conjugate depth of the hydraulic jump, is more efficient. Nevertheless, the presence of a movable bed in the stilling basin determines the formation of erosive processes, which have to be taken into consideration in order to prevent structural problems. In the following the analysis of the scour process occurring in the stilling basin, in the case of a stable block ramp, clear-water conditions and hydraulic jump occurring in the stilling basin, will be analysed and discussed. In addition, the analysis will also focus on the effect of both stilling basin material gradation and geometry on the scour formation and morphology. In Fig. 4.2a–d, several sketches of block ramps are reported including the main hydraulic and geometric parameters. Namely, h_1 is the approaching flow depth at the ramp toe, h_2 is the tailwater depth (depth of water downstream of the hydraulic jump), z_{max} is the maximum scour depth in the stilling basin, l_0 is the maximum length of the scour hole, z_M is the ridge height, b is the ramp width and the stilling basin width in the case of prismatic channel without lateral expansions, B is the stilling basin width in the case of symmetrically enlarged configuration, A is the longitudinal expansion of the stilling basin, h_{max} is the maximum water depth in the expanded stilling basin, l_s is the longitudinal distance of the maximum scour depth from the ramp toe, l_d is the longitudinal distance of the maximum ridge height from

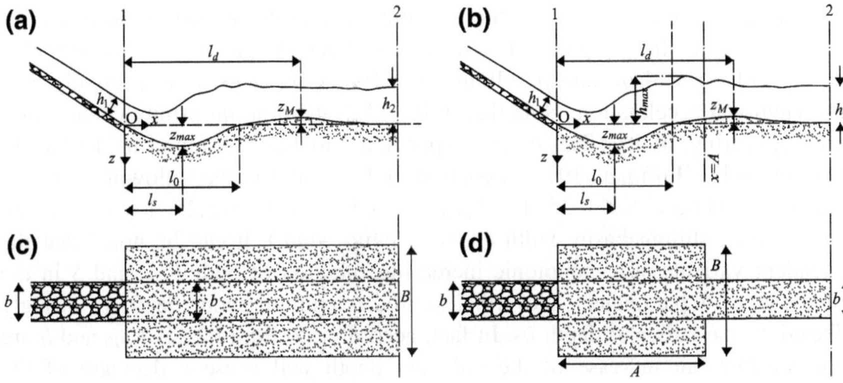

Fig. 4.2 Diagram sketches of block ramps: side view for **a** symmetrically enlarged stilling basin and for **b** longitudinally contracted stilling basin; plan view of **c** symmetrically enlarged stilling basin configuration and **d** longitudinally contracted stilling basin

the ramp toe, O, x and z are the origin, the longitudinal and vertical coordinates, respectively, of the adopted reference system. S is the ramp slope.

The analysis of the scour morphology downstream of block ramps in the case of a stilling basin having the same width of the ramp (i.e., $B/b = 1$) and under clear-water conditions was conducted by Pagliara and Palermo (2010). Authors analyzed the scour mechanism and proposed useful practical relationships to foresee the main geometric characteristics of the scour hole and ridge in the case in which a hydraulic jump (F_{MB} type) was entirely located in the stilling basin. According to Palermo et al. (2008), F_{MB} hydraulic jump is characterized by a peculiar flow recirculation in the stilling basin: the flow does not submerge the ramp toe and the sediment are transported both downstream of the scour hole and towards the ramp toe. Pagliara and Palermo (2010) noted that tailwater level h_2 (Fig. 4.2a) is playing a fundamental role in the scour process, as it is affecting the exiting jet diffusion length. In fact, the scour process dynamics appears quite similar to that occurring in the case of impinging jets for low vertical impingement angles. In this last case, Hoffmans (1998) proved that $z_{max} + h_2$ is depending on both hydraulic conditions and sediment characteristics. In the case of a block ramp, Pagliara and Palermo (2010) proposed the following functional relationship, where $\sigma = (d_{84}/d_{16})^{0.5}$, in which d_{xx} is the diameter for which $xx\%$ of the stilling basin sediment is finer:

$$(z_{max} + h_2)/h_1 = f(F_{dxx}, S, \sigma) \tag{4.1}$$

In addition, they developed a methodology and practical relationships to predict the main scour hole features. Namely, for the maximum scour depth, they found:

$$(z_{max} + h_2)/h_1 = (11.64 \cdot S + 0.70) \cdot \exp[(-0.64 \cdot S + 0.17) \cdot F_{d90}] \tag{4.2}$$

where $F_{d90} = v_1/(g' \cdot d_{90})^{0.5}$ is the densimetric Froude number in which v_1 is the approaching flow depth at the ramp toe (Sect. 4.1 in Fig. 4.2a) and $g' = [(\rho_s - \rho)/\rho] \cdot g$ with ρ_s and ρ stilling basin and water density, respectively, and g gravitational acceleration. Note that in Eq. (4.2) the non-uniformity coefficient σ is not appearing, as its effect on the dependent variable was found negligible (as proved also by Hoffmans 1998). Equation (4.2) is valid in the following range of parameters: $0.083 \leq S \leq 0.25$, $1 < F_{d90} < 4$, $1.15 < h_2/h_1 < 3.25$, $1 < \sigma < 2.8$ and $B/b = 1$ (i.e., stilling basin width equal to ramp width). It can be noted that the dependent variable is a monotonic increasing function of both F_{d90} and S in the tested range of parameters. Nevertheless, the maximum scour depth is strongly affected by the tailwater depth h_2. In fact, according to Eq. (4.2), if F_{d90} and S are kept constant, an increase of the tailwater depth will cause a decrease of the maximum scour depth.

For scour hole length, Pagliara and Palermo (2010) proposed a methodology in order to evaluate the parameter $L_0 = l_0/h_1$. According to Breusers and Raudkivi (1991), there is a close relation between the scour hole characteristics, i.e., the scour hole length can be expressed as only function of the maximum scour depth. Based on this assumption, Pagliara and Palermo (2010) found that:

$$L_0 = 7.42 \cdot \exp(0.37 \cdot Z_{maxcalc}) \tag{4.3}$$

where

$$Z_{\max calc} = [(z_{\max} + h_2)/h_1]_{\text{Eq. (4.2)}} - h_{2meas}/h_{1calc} \tag{4.4}$$

in which $[(z_{\max} + h_2)/h_1]_{\text{Eq.(4.2)}}$ is calculated using Eq. (4.2) and h_{2meas} is the measured value of h_2, whereas h_{1calc} (which is unknown in practical applications) is estimated assuming that uniform flow conditions are reached on the ramp. This last assumption is not strictly verified, especially for short ramps, but authors proved that the estimated h_{1calc} values (derived with the following procedure) are in very good agreement with those measured in experimental runs. h_{1calc} values can be derived using Manning's equation, assuming the hydraulic radius $R \approx h_1$. The coefficient n can be estimated adopting the relationships proposed by Pagliara and Chiavaccini (2006a), which was derived for block ramps as follows:

$$n = 0.064 \cdot (D_{50} \cdot S)^{0.110} \tag{4.5}$$

in which D_{50} is the mean diameter of the ramp material.

From Eq. (4.3) it can be easily noted that the scour length is a monotonic increasing function of the scour depth, as it is also occurring for plunging jets scour process.

Previous deductions were further developed, including stilling basin geometry effect on the scour hole. Successive tests (Pagliara et al. 2009a) proved that a symmetrically enlarged stilling basin is characterized by a different scour dynamic. In Fig. 4.2c a diagram sketch of a symmetrically enlarged stilling basin downstream of a block ramp is represented. Namely, in an enlarged stilling basin both hydraulic jump and scour morphology is 3D. The flow exiting from the ramp toe is concentrated axially because of the presence of lateral flow re-circulating vortexes. It was experimentally observed that the flow re-circulation in lateral zones of the enlarged channel increases for higher expansion ratios B/b. Pagliara et al. (2009a) tested three different enlargement ratios B/b (i.e., B/b = 1, 1.8 and 2.8) and ramp slopes ranging between 0.167 and 0.25. Authors introduced a modified densimetric Froude number $\tilde{F}_{d90}$ in order to take into consideration the effect of flow concentration at the ramp toe and to furnish relationships which are analytically identical to the proposed ones for the case B/b = 1 (Eqs. 4.2 and 4.3). Namely, the modified densimetric Froude number is defined as follows:

$$\tilde{F}_{d90} = F_{d90} \cdot (B/b)^{\left(150.5 \cdot S^2 - 43.8 \cdot S + 3.8\right)} \tag{4.6}$$

It can be easily noted that $\tilde{F}_{d90} = F_{d90}$ for B/b = 1. In addition, the modified densimetric Froude number is a monotonic increasing function of S in the tested range of ramp slope. This is mainly due to the fact that an increase of ramp slope contributes to concentrate axially the exiting flow, causing an increase of scour depth. Thus, the previous Eq. (4.2) can be re-written as follows and it is valid for $1 \leq B/b \leq 2.8$:

$$(z_{\max} + h_2)/h_1 = (11.64 \cdot S + 0.70) \cdot \exp\left[(-0.64 \cdot S + 0.17) \cdot \tilde{F}_{d90}\right] \tag{4.7}$$

Similarly, the non-dimensional scour hole length L_0 can be evaluated following the same procedure proposed by Pagliara and Palermo (2010) for B/b = 1. In the case of abrupt symmetrically enlarged stilling basin, Eq. (4.3) can be re-written as follows:

$$L_0 = 7.42 \cdot \exp\left(0.37 \cdot \tilde{Z}_{maxcalc}\right) \tag{4.8}$$

where

$$\tilde{Z}_{maxcalc} = \left[\left((z_{\max} + h_2)/h_1\right)_{\text{Eq. (4.7)}} - h_{2meas}/h_{1calc}\right] \cdot (B/b)^{-0.25} \tag{4.9}$$

with $\left((z_{\max} + h_2)/h_1\right)_{\text{Eq.(4.7)}}$ calculated using Eq. (4.7) and h_{2meas} and h_{1calc} derived as specified above. Note that both Eqs. (4.8) and (4.9) coincide with Eqs. (4.3) and (4.4) for B/b = 1. In particular, comparing Eqs. (4.9) and (4.4) it is evident that for $B/b > 1$, the scour hole morphology, which is essentially 3D, contributes to modify the relationship between maximum non dimensional scour depth and length. This is due to the fact that increasing B/b, the scour hole depth increases faster than the scour hole length, resulting to be confined by a more prominent surrounding ridge.

A further generalization of the effect of stilling basin geometry on the scour features was proposed by Pagliara and Palermo (2011a). Authors tested several geometric configurations of the stilling basin, including both transversal and longitudinal expansions (see Fig. 4.2b, d). Namely, the transversal expansion ratios tested B/b varied between 1 and 2.8, whereas the longitudinal expansion ratios $E = A/B$ tested ranged between 1 and 2. It was observed that also for this configuration, a lateral flow re-circulation takes place. Nevertheless, the presence of the downstream stilling basin contraction strongly modifies both scour hole morphology and flow features, contributing to confine the hydraulic jump occurring inside the expanded stilling basin. In presence of both longitudinal and transversal contractions, two other flow re-circulating zones take place in correspondence with the downstream transversal wall (i.e., where the channel has the same width of the ramp). The downstream flow re-circulation is strongly depending on the ratio h_2/h_1, as the hydraulic jump can shift towards or backward according to different tailwater level in the downstream contracted channel, eventually impacting on the transversal walls. In addition, the hydraulic jump appears symmetric and it is not characterized by periodical asymmetrical oscillations as in the case in which downstream contraction is absent. In the tested range of parameters, the water level in the stilling basin is generally higher than that occurring in the contracted channel. Nevertheless, due to the complex 3D scour morphology, water depth in the stilling basin can be sensibly different. This observation led Pagliara and Palermo (2011a) to assume a virtual stilling basin water depth h_A, which was derived considering the simplifying hypothesis that no energy dissipation occurs between sections $x = A$ (upstream of the contraction) and 2-2 of Fig. 4.2a, d. In other words, authors derived a fictitious water level h_A in section A-A ($x = A$) of the enlarged stilling basin from the following assumption:

$$h_2 + \frac{Q^2}{(bh_2)^2 \cdot 2 \cdot g} = h_A + \frac{Q^2}{(Bh_A)^2 \cdot 2 \cdot g} \tag{4.10}$$

Note that h_A derived from Eq. (4.10) is depending only on h_2 and $h_2 = h_A$ when no downstream contraction is present, i.e., $E = 0$ or $E = \infty$, representing the cases illustrated in Fig. 4.2c, in which the stilling basin has either the same width of the ramp ($B/b = 1$) or it is indefinitely transversally enlarged ($1 < B/b \leq 2.8$), respectively. Based on this observation, in the general case (presence of both transversal and longitudinal expansions), the functional relationship expressed by Eq. (4.1) can be modified as follows:

$$(z_{\max} + h_A)/h_1 = f(F_{dxx}, S, B/b, E, h_A/h_1) \tag{4.11}$$

When no longitudinal contraction is present, i.e., $E = 0$ or $E = \infty$, the proposed Eq. (4.7) is valid. For both longitudinally and transversally expanded stilling basin ($0 < E \leq 2$ and $1 < B/b \leq 2.8$), Pagliara and Palermo (2011a) proposed a modified densimetric Froude number, which includes the effects due to both stilling basin geometry and the tailwater. The modified Froude number was derived in such a way to preserve the analytical form of the general predicting equations reported above (Eqs. 4.2 and 4.7). Namely, authors introduced the following modified densimetric Froude number:

$$F^{**}_{d90} = \tilde{F}_{d90} \cdot f(E, h_A/h_1) \tag{4.12}$$

in which

$$f(E, h_A/h_1) = \left(-5.85 \cdot S^2 + 4.24 \cdot S - 0.49\right) \cdot E \cdot (h_A/h_1) + 1 \tag{4.13}$$

for $0 \le E \le 2$ and $1 \le B/b \le 2.8$ and

$$f(E, h_A/h_1) = 1 \tag{4.14}$$

for $E = \infty$ and $1 < B/b \le 2.8$.

Thus, the general predicting equation becomes:

$$(z_{max} + h_A)/h_1 = (11.64 \cdot S + 0.70) \cdot \exp\left((-0.64 \cdot S + 0.17) \cdot F^{**}_{d90}\right) \tag{4.15}$$

From the previous Eqs. (4.10), (4.13), (4.14) and (4.15), it easy to observe that if $E = 0$ and $B/b = 1$ (no longitudinal and transversal expansion) $h_A = h_2$ and Eq. (4.15) coincides with Eq. (4.2), whereas for $E = \infty$ and $1 < B/b \le 2.8$, $h_A = h_2$ and Eq. (4.15) coincides with Eq. (4.7).

Regarding the non-dimensional longitudinal length L_0, the findings of Breusers and Raudkivi (1991), such as those of Pagliara et al. (2009b) and Pagliara and Palermo (2010), are not confirmed in presence of both longitudinal and transversal stilling basin expansions, i.e., the scour hole length cannot be expressed as only function of the maximum scour depth. This occurrence is mainly due to the fact that the scour region can be either confined in the expanded stilling basin or extend to the contracted downstream channel, resulting in a scour length which strongly depends on both geometric configuration and hydraulic conditions. Thus, for $1.8 \le B/b \le 2.8$ and $1 \le E \le 2$, Pagliara and Palermo (2011a) proposed the following relationship to estimate L_0:

$$\ln(L_0) \cdot S = 0.30 \cdot \exp\left((-24.10 \cdot S^2 + 11.13 \cdot S - 1.03) \cdot F_{d90}\right) \tag{4.16}$$

Finally, in order to design the height h_w of the expanded stilling basin walls, Pagliara and Palermo (2011a) proposed to assume $h_w = 2 \cdot h_A$, for which a safe coefficient equal to 1.4 was considered.

4.2.2 *Erosive Processes in Live-Bed Conditions*

The presence of an upstream sediment transport contributes to deeply modify the scour hole morphology. Namely, live-bed scour conditions take place when the stilling basin is continuously fed by the upstream stream. It implies that equilibrium conditions are reached when the rate of supplied sediment equals that of removed sediment from the stilling basin and no significant variation in scour hole geometry can be observed.

In the case of a prismatic channel in which the stilling basin has the same width of the ramp (B/b = 1, see Fig. 4.2a, c), Pagliara et al. (2012) analysed the scour phenomenon and described the effect of upstream sediment concentration on the main geometric parameters of the scour hole. Namely, authors built a block ramp model, whose ramp slope S varied between 0.071 and 0.25. They conducted experiments varying the inlet sediment load $Q_{s\text{-}in}$ which was supplied from the ramp entrance. The relative inlet sediment concentration $C_{‰} = (Q_{s\text{-}in}/Q) \cdot 1000$, where Q is the flow discharge, ranged between 0.029 and 0.66 for S = 0.071, between 0.026 and 0.746 for S = 0.125 and between 0.032 and 0.84 for S = 0.25. Equilibrium conditions were achieved when $Q_{s\text{-}in} \approx Q_{s\text{-}out}$, where $Q_{s\text{-}out}$ is the outlet sediment load collected by a trap located at the end of the channel. The densimetric Froude numbers tested varied in the following ranges: $3.46 \le F_{d90} \le 4.35$ for $S = 0.071$, $4.34 \le F_{d90} \le 5.30$ for $S = 0.125$ and $4.86 \le F_{d90} \le 5.86$ for $S = 0.25$. According to the classification proposed by Palermo et al. (2008), for the mentioned range of parameters, "transport conditions" are achieved, i.e., no ridge formation occurs downstream of the scour hole and no equilibrium condition is reached as the scour hole gradually increases becoming flatter and longer if no sediment are supplied from upstream. In the case of live-bed conditions, Pagliara et al. (2012) showed that four different equilibrium profiles can be distinguished. Namely, authors classified and termed the four profiles as D_{LB} (scour profile with downstream bed degradation), E_{LB} (scour profile with downstream bed level coincident with the original one), A_{LB} (scour profile with downstream bed aggradation), N_{SL} (absence of scour hole in the stilling basin). Thus, both the inlet sediment concentration and hydraulic conditions play a fundamental role in the resulting equilibrium morphology. It implies that, in the case of live-bed conditions and for B/b = 1, the functional relationship expressed by Eq. (4.1) can be modified as follows:

$$(z_{\max} + h_2)/h_1 = f(F_{dxx}, S, C_{‰}) \tag{4.17}$$

Experimental observations led authors to state that the non-dimensional dependent variable $(z_{\max} + h_2)/h_1$ is deeply affected by the densimentric Froude number. It was experimentally observed that if the ramp configuration does not change and $C_{‰}$ is constant, the maximum scour depth increases for higher F_{d90}. Furthermore, experimental tests showed that if $C_{‰}$ increases, being constant all the other variables, scour depth reduces and eventually no scour hole takes place. Pagliara et al. (2012) proposed the following relationship to evaluate the maximum scour depth, which is valid in the mentioned range of parameters:

$$\frac{(z_{\max} + h_2)}{h_1} = (2.60S + 1.47) \cdot \exp[(0.28S + 0.07) \cdot F_{d90} + (2.33S - 0.96) \cdot C_{‰}] \tag{4.18}$$

Note that the previous equation is in agreement with experimental observations (i.e., it monotonically increases with F_{d90} and decreases with $C_{‰}$) and predicts well the totality of model data (±10 %). In addition, morphology analysis was extended to the non-dimensional scour length. Also in this case, non-dimensional scour length is not depending only on scour depth. This occurrence is due to the fact that in live-bed conditions it was experimentally proved that both geometric configuration and hydraulic parameters deeply affect the non-dimensional profiles shape. In fact, no profile similitude can be established as in the case of clear-water conditions. Therefore, authors proposed the following non-dimensional relationship to derive L_0:

$$\frac{L_0 \cdot S}{h_2/h_1} = 1.266 \cdot Z_m \tag{4.19}$$

with $Z_m = z_m/h_1$. Nevertheless, it can be interfered that the non-dimensional scour length is strictly related to non-dimensional scour depth. Note that Z_m is depending on the variables specified in Eq. (4.17); it implies that also L_0 is depending on them. The analysis of longitudinal profiles revealed that there is a close correlation between the maximum scour hole length and the longitudinal distance l_s of the transversal cross section in which the maximum scour depth takes place. The non-dimensional longitudinal position $L_s = l_s/h_1$ is related to the non-dimensional scour length, hydraulic parameters and ramp configuration as follows:

$$L_s = 4.489 \cdot \left[\frac{L_0 \cdot S}{h_2/h_1}\right]^{0.589} \tag{4.20}$$

Finally, the morphology analysis includes also the characteristic lengths of the ridge. Namely, both non-dimensional ridge height $Z_M = z_M/h_1$ and the longitudinal distance of the maximum ridge height location ($L_d = l_d/h_1$) from the ramp toe can be foreseen. It is evident that the following relationships apply only in the case in which a A_{LB} profile occurs, as, in other mentioned cases, no ridge takes place (i.e., for D_{LB} and E_{LB}). Just for example, Fig. 4.3a, b illustrates the existence fields of the classified profiles typologies in terms of F_{d90} and $C_{‰}$ for $S = 0.125$ and for different ranges of relative tailwaters h_2/h_1. Namely, the transition between the different profile typologies is sketched, along with the hydraulic conditions for which they are valid. For other tested slopes ($S = 0.071$ and S = 0.25), Pagliara et al. (2012) proposed similar graphs.

The previous figure shows that for lower inlet sediment concentration and relative tailwater conditions, mainly E_{LB} and A_{LB} profiles take place (Fig. 4.3a), whereas an increase of h_2/h_1 shifts the transition between the profile typologies toward lower $C_{‰}$ values (Fig. 4.3b). This occurrence is due to the fact that an increase of h_2 determines an increase of the diffusion length of the exiting flow in the stilling basin, resulting in a reduction of the erosive forces. Based on these observations and classification, Pagliara et al. (2012) proposed the following empirical relationships to evaluate both Z_M and L_d:

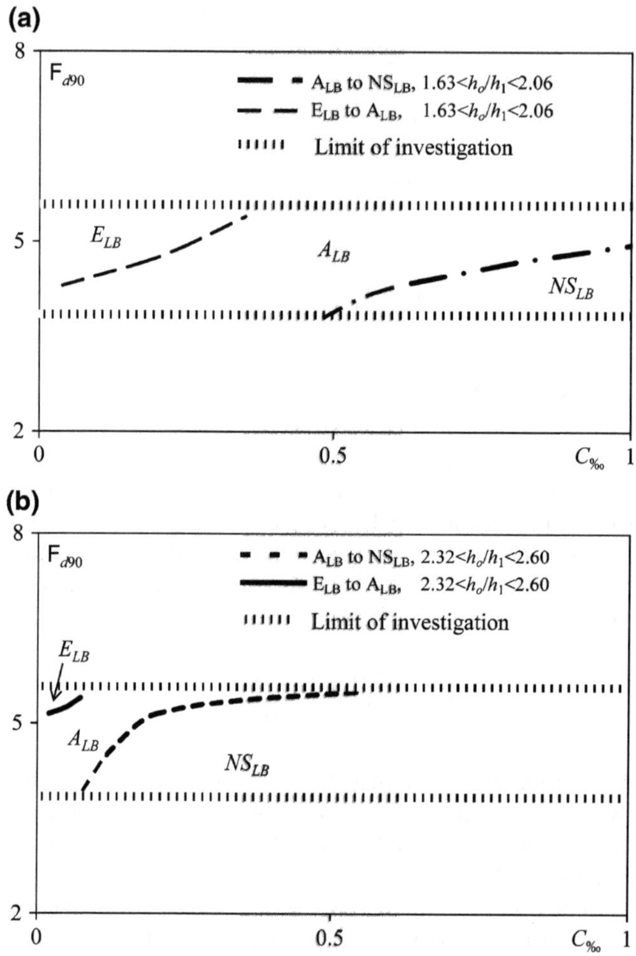

Fig. 4.3 Scour profile classification for $S = 0.125$ and **a** $1.63 < h_2/h_1 < 2.06$ and **b** $2.32 < h_2/h_1 < 2.60$

$$Z_M = \left(171.47 \cdot S^{2.45}\right) \cdot \exp[(9.04 \cdot S - 3.90) \cdot Z_m] \tag{4.21}$$

$$L_d = 8.030 \cdot \exp(0.068 \cdot L_0) \tag{4.22}$$

As shown by previous equations, there is a close correlation between scour hole geometric characteristics and ridge dimensions.

The analysis of scour morphology in live-bed conditions was also conducted in the case of a stilling basin which is both longitudinally and transversally enlarged by (Pagliara et al. 2011). In particular, authors analysed the configuration reported in Fig. 4.2d in the following ranges of parameters: $0.083 < S < 0.25$, $0.143 < C_{‰} < 0.574$, $1 < E = A/B < 2$, $1.8 < B/b < 2.8$. Furthermore, for $S = 0.083$,

F_{d90} ranged between 4.81 and 6.24 and h_2/h_1 between 1.29 and 1.69; for $S = 0.125$, F_{d90} ranged between 5.65 and 6.54 and h_2/h_1 between 1.53 and 1.72; and, for $S = 0.250$, F_{d90} ranged between 6.63 and 7.16 and h_2/h_1 between 1.65 and 2.06. Several preliminary tests were also conducted for $E = \infty$, in order to investigate the similarities and differences between the symmetrically enlarged stilling basin in clear-water and live-bed conditions. According to hydraulic jump classification proposed by Pagliara et al. (2009a) for symmetrically enlarged stilling basin, several similitudes can be pointed out in the case of $E = \infty$ for live-bed conditions. In this last case, only a Repelled Oscillatory Jump takes place (termed $R\text{-}O_{E=\infty}$). It is characterized by symmetrically re-circulating flow zones and by periodical deflections of flow towards the stilling basin lateral walls. The hydraulic jump is not forced to occur just downstream of the ramp toe and no prominent ridge forms downstream of the scour hole. This is mainly due to high F_{d90} values and to the fact that flow tends to flatten the downstream ridge. In the case of both longitudinal and transversal contraction (Fig. 4.2d), the flow pattern inside the stilling basin in live-bed conditions becomes more complex, and, according to the expansion ratio E, different hydraulic jump typologies and scour morphologies can take place, namely, Repelled Oscillatory Impact Jump ($R\text{-}O_{I\text{-}J}$) and Repelled Symmetric Free jump ($R\text{-}S_{F\text{-}J}$). $R\text{-}O_{I\text{-}J}$ hydraulic jump generally occurs for $E = 1$. It is characterized by a strong re-circulation in correspondence with the downstream transversal walls, on which the hydraulic jump directly impacts. Maximum scour depths generally occur close to the stilling basin walls. $R\text{-}S_{F\text{-}J}$ hydraulic jump is entirely located in the stilling basin and mainly occurs for $E = 2$. Maximum scour depth takes place in the center of the pool and the flow re-circulation is similar to that occurring for $E = \infty$. The stilling basin configuration determines substantial different erosive mechanisms inside the stilling basin and in the downstream contracted channel, resulting in the absence of (both transversal and axial) non-dimensional profiles similitude. Based on experimental tests results, Pagliara et al. (2011) derived useful empirical relationships to predict the non-dimensional parameter $(z_{max} + h_2)/h_1$ and the longitudinal distance l_s from the ramp toe of the transversal section in which the maximum scour depth occurs. Namely, they proposed the following equations:

$$\begin{aligned}\frac{\ln[(z_{\max}+h_2)/h_1]}{C_{\%o}} = {} & [(-3.00\cdot C_{\%o}^2 + 2.70\cdot C_{\%o} - 0.45)\cdot F_{d90} \\ & + (63.60\cdot C_{\%o}^2 - 61.50\cdot C_{\%o} + 15.00)] \\ & \times (33.60\cdot S^2 + 15.60\cdot S - 0.03)\end{aligned} \tag{4.23}$$

$$\ln(L_s) = 0.286 Z_{\max} + 1.450 \tag{4.24}$$

Note that both the variables B/b and E do not exhibit a substantial effect on the adopted dependent variable in Eq. (4.23). Thus, they do not appear in previous equations. This occurrence allowed to simplify the proposed empirical relationship and to furnish simple analytical tools to estimate the maximum scour depth.

4.3 Stepped Gabion Weirs

Stepped gabion weirs are flexible grade-control structures which are widely used, especially in mountain streams restoration. Their peculiarity is to dissipate a considerable amount of energy and to be easily built and rearranged in order to satisfy modified hydraulic conditions. Their hydraulic behavior has been studied by several researchers, among these Peyras et al. (1992), Chinnararsi et al. (2008) and Mohamed (2010). They pointed out the similarities of hydraulic behavior in the dissipative mechanism between this structure typology and stepped chutes. In particular, Pagliara and Palermo (2013) analyzed the onset conditions of different flow regime on the stepped gabion weirs in clear-water conditions. Based on the studies of Ohtsu et al. (2001), Pagliara and Palermo (2013) proposed a graph by which it is possible to foresee which flow regime is occurring on stepped gabion weirs knowing both geometrical characteristics and hydraulic parameters. They stated that three flow regimes may occur: Nappe Flow, Transition Flow and Skimming flow, respectively. Different flow regimes deeply influence the scour process downstream of the hydraulic structures. In addition, the erosive mechanism is also depending on which type of hydraulic jump is occurring in the stilling basin. Two jump typologies were distinguished, according to the classification proposed by Palermo et al. (2008): F_{MB} and S_{MB} hydraulic jump. This last hydraulic jump typology mainly occurs for high downstream tailwater level h_2. Furthermore, authors tested different boundary configurations of the structure. In particular, they tested four configurations termed as GW_0, GW_{imp}, GW_f and $GW_{f\text{-}imp}$. In Fig. 4.4, a comprehensive diagram sketch of the different configurations tested is reported, along with the main geometric and hydraulic parameters. H is the height of the structure and eventual filtering layer from the original stilling basin level, h is the upstream water depth, h_2 is the downstream water depth, $h_s = w_s$ are the height and width of the steps, respectively, l_s is the maximum scour hole length, z_{max} is the maximum scour depth, z_M is the ridge height, x and z are the longitudinal and vertical coordinates of the selected reference system. Figure 4.5 shows a picture of a stepped gabion weir built in Ripalimosani, Molise (Italy).

Authors conducted several tests in clear-water conditions in the presence of a filtering layer located upstream of the structure and other tests involving the presence of an impermeable covering, which did not allow the flow passage either inside the structure or in both filtering layer and structure. Therefore, in the absence of an upstream filtering layer, two configurations were tested: GW_0, characterized by the absence of any impermeable covering on the upstream part of the structure (base configuration); and GW_{imp}, for which an impermeable steel covering was located on the upstream part of the structure, in order to not allow an upstream flow infiltration in it. In the presence of an upstream filtering layer, the two tested configuration were: GW_f (base configuration with filtering layer) and $GW_{f\text{-}imp}$, characterized by an impermeable steel covering located on both the filtering layer and upstream part of the structure. It has to be specified that the filtering layer contains the same material constituting the stilling basin, whereas, the material

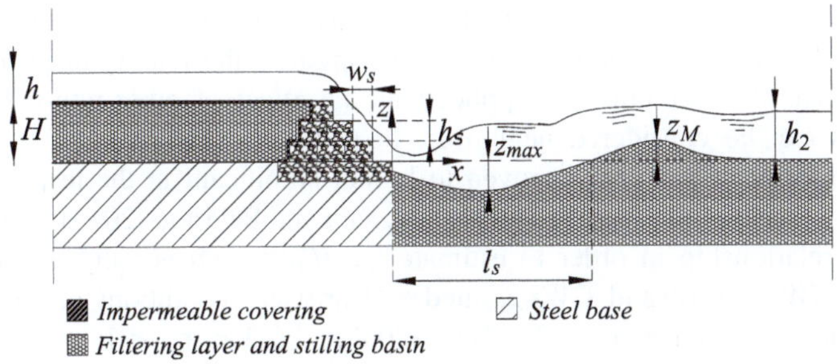

Fig. 4.4 Diagram sketch of a stepped gabion weirs along with the geometric and hydraulic parameters

Fig. 4.5 Picture of a stepped gabion weir built in Ripalimosani, Molise (Italy)

constituting the stepped gabion weirs is consistently bigger in size as it is common in practical applications. Pagliara and Palermo (2013) analysed the scour processes in clear-water conditions for all the mentioned structure configurations. They experimentally verified that, for each boundary structure condition, the maximum non dimensional scour depth z_{max}/E_0 can be expressed as only function of A_{50}:

$$\frac{z_{max}}{E_0} = f(A_{50}) \tag{4.25}$$

in which $A_{50} = q/[H \cdot [g \cdot d_{50} \cdot (\Delta\rho/\rho)]^{0.5}]$ is the parameter introduced by D'Agostino and Ferro (2004) for grade-control structures, where q is the unit discharge, d_{50} is the mean grain size of the stilling basin material, $\Delta\rho = \rho_s - \rho$ with ρ_s and ρ the stilling basin material and water densities, respectively, g is the gravitational acceleration and E_0 the total energy head upstream of the structure. Note, that the previous equation satisfactorily predicts the totality of experimental data, it means that for practical applications the effect of other non-dimensional parameters can be considered negligible. In particular, the effect of the relative tailwater h_2/H on z_{max}/E_0 was proven to be not significant in the tested range of parameters. Based on these observations, Pagliara and Palermo (2013) proposed the following relationship in order to estimate z_{max}/E_0, which is valid for the tested structures GW_{imp}, GW_f and $GW_{f\text{-}imp}$, under clear-water conditions and in the following ranges of parameters: $0.2 < h_2/H < 0.8$ and $0.3 < A_{50} < 1$.

$$\frac{z_{max}}{E_0} = 7.53 \cdot A_{50}^3 - 11.53 \cdot A_{50}^2 + 6.66 \cdot A_{50} - 1.16 \tag{4.26}$$

The following Fig. 4.6 shows the plot of the previous Eq. (4.26) along with the plots of the Eqs. (4.31)–(4.33) valid for rock grade-control structures which will be discussed in the next paragraph.

In this figure, the different scour evolution behaviour occurring between Nappe Flow and Skimming Flow regimes is evident. A clear transition takes place for $0.5 < A_{50} < 0.7$. The scour depth increases significantly faster with A_{50} in the case of Skimming Flow regime ($A_{50} > 0.7$) occurring on the structures.

In general, the shape of the transversal cross-section in which the maximum scour depth takes place is not 2D. A certain three-dimensionality always occurs, mainly due to the wall effects especially for low A_{50} values. Pagliara and Palermo (2013) found that there is a close relationship between the maximum scour depth z_{max} and the averaged value of the scour depth z_m in the section in which z_{max} occurs. Namely, they derived the following equation:

$$\frac{z_{max}}{z_m} = 1.00 + 0.15 \cdot A_{50}^{-2.00} \tag{4.27}$$

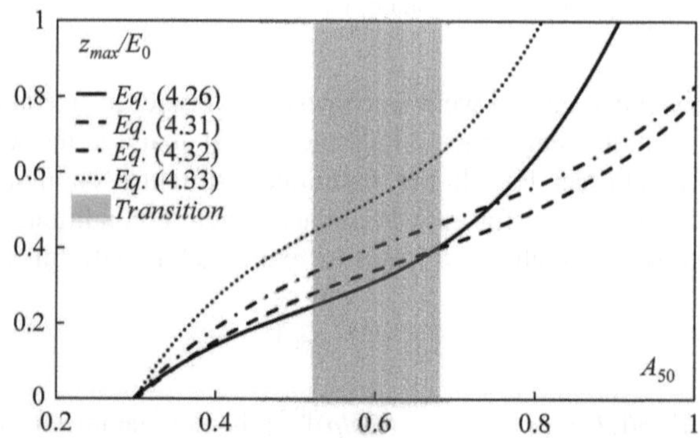

Fig. 4.6 Plots of Eqs. (4.26) and (4.31)–(4.33) valid for stepped gabion weirs, $B1$-$B2$, B_f and $B_{f\text{-}imp}$, respectively, along with the delimitation of the transition zone between Nappe Flow and Skimming Flow regimes

As it was observed also for block ramps, the ratio z_{max}/z_m is a monotonic decreasing function of the parameter A_{50}. It means that for higher discharges or densimetric Froude numbers the scour shape tends to become more 2D. Furthermore, authors furnished also a set of three useful empirical relationships to foresee the maximum non dimensional scour length $L_s = l_s/E_0$:

$$\frac{l_s}{E_0} = 4.2\frac{z_{max}}{E_0} \tag{4.28}$$

valid for GW_{imp} structures,

$$\frac{l_s}{E_0} = 3.6\frac{z_{max}}{E_0} \tag{4.29}$$

valid for GW_f structures and

$$\frac{l_s}{E_0} = 5.1\frac{z_{max}}{E_0} \tag{4.30}$$

valid for $GW_{f\text{-}imp}$ structures.

4.4 Rock Grade Control Structures

Rock grade control structures are also used for river restoration. As stepped gabion weirs and block ramps, they are flexible hydraulic structures which can be suitable for a large variety of natural contests. They can be used instead of concrete check dams and eventually to reconvert them. The main difference between rock grade control structures and block ramp is the downstream surface slope. Generally, block ramp is characterized by mild bed slopes, whereas rock grade control structures exhibit a considerable downstream and upstream surface slope. Figure 4.7 shows a picture of a rock grade control structure built in Cardoso river (Tuscany, Italy).

According to the rock sizes constituting the structure and to the hydraulic conditions, three flow regimes can occur on them. Pagliara and Palermo (2013) analyzed the hydraulics of rock grade control structures and found that similarities with stepped chutes and stepped gabion weirs can be pointed out. Authors conducted their studies using different experimental models, i.e., as for stepped gabion weirs, they analyzed both the hydraulic behavior and the scour mechanism in the stilling basin varying hydraulic and structure boundary conditions. Namely, they tested three different structure configurations, under clear-water conditions: $B1$-$B2$, B_f and $B_{f\text{-}imp}$, respectively. Rocks constituting the structures were linked together using a silicone glue (i.e., the structure is impermeable). Tests with base configuration $B1$-$B2$ were conducted without any upstream filtering layer. In addition, other experimental tests were conducted both in presence and absence of

Fig. 4.7 Picture of a rock grade control structure built in Cardoso river (Tuscany, Italy)

an impermeable covering located on an upstream filtering layer. The tested configuration with upstream permeable layer was termed B_f, whereas the configuration with a filtering layer made impermeable using a steel covering was termed $B_{f\text{-}imp}$. The filtering layer had the same height H of the structure from the original stilling basin bed level.

Also in the presence of rock grade control structures, two hydraulic jump typologies occur in the stilling basin, namely F_{MB} and S_{MB}. The transition between the two hydraulic jump typologies takes place for $h_2/H \approx 0.5$. The following figure is illustrating a comprehensive diagram sketch of the various tested configurations, along with hydraulic and geometric parameters (Fig. 4.8).

Following the approach exposed for stepped gabion weirs, Pagliara and Palermo (2013) derived three empirical equations to foresee the maximum non dimensional scour hole depth z_{max}/E_0, valid in clear-water conditions and in the same range of parameters specified for stepped gabion weirs:

$$\frac{z_{max}}{E_0} = 3.28 \cdot A_{50}^3 - 6.28 \cdot A_{50}^2 + 4.74 \cdot A_{50} - 0.95 \tag{4.31}$$

valid for structure configuration $B1$-$B2$,

$$\frac{z_{max}}{E_0} = 3.46 \cdot A_{50}^3 - 6.96 \cdot A_{50}^2 + 5.42 \cdot A_{50} - 1.09 \tag{4.32}$$

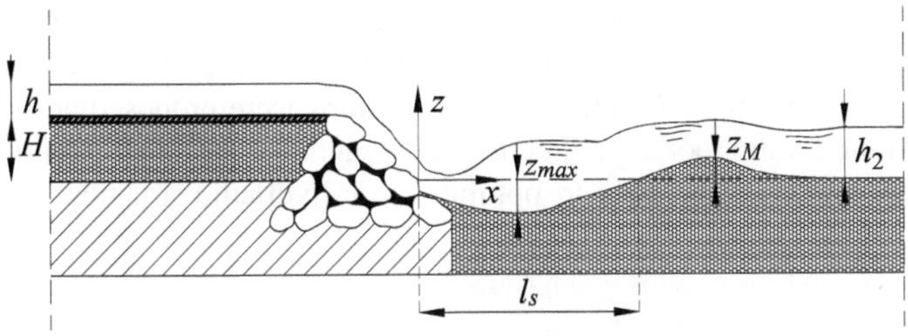

Fig. 4.8 Diagram sketch of a rock grade control structure along with the geometric and hydraulic parameters

valid for structure configuration B_f and

$$\frac{z_{\max}}{E_0} = 13.46 \cdot A_{50}^3 - 22.02 \cdot A_{50}^2 + 13.12 \cdot A_{50} - 2.32 \qquad (4.33)$$

valid for structure configuration $B_{f\text{-}imp}$.

As specified, the previous equations are plotted in Fig. 4.6. It can be easily inferred that also for rock grade control structures there is a different behaviour in terms of scour mechanism according to the flow regime occurring on the structure itself. In addition, the different boundary configurations determine a variation in erosive process. In fact, for stepped gabion weirs, it was found that one unique equation could satisfactorily predict the totality of experimental data. In this case, this finding is not confirmed. The maximum non dimensional scour depth occurs for configuration $B_{f\text{-}imp}$.

Pagliara and Palermo (2013) also proposed an empirical relationship to estimate the maximum scour length. Namely, they experimentally proved that the maximum scour length l_s/E_0 for all the rock grade control structures can be expressed by the following function:

$$\frac{l_s}{E_0} = 3.16 \frac{z_{\max}}{E_0} \qquad (4.34)$$

It is evident that there is a close relationship between the scour depth and length.

4.5 Conclusions

The present chapter synthesized some of the most relevant literature contributions for the design of low-head hydraulic structures. The analysis focused on the following low-environment impact structures: block ramps, stepped gabion weirs and rock grade control structures. The erosive process occurring in the downstream stilling basin was discussed for each one of the mentioned structure. It was shown that it is influenced by both structure configuration and stilling basin geometric characteristics. Several useful practical relationships were proposed and discussed. They constitute a valid tool to predict the main lengths of the scour morphology and can give designers a valid help to optimize both hydraulic structures and stilling basin geometry.

References

Bathurst JC (1978) Flow resistance of large-scale roughness. J Hydraulic Div ASCE 104 (12):1587–1603

Bathurst JC, Li RM, Simons DB (1981) Resistance equation for large-scale roughness. J Hydraulic Div ASCE 107(12):1593–1613

Bormann E, Julien PY (1991) Scour downstream of grade control structures. J Hydraulic Eng 117 (5):579–594

Breusers HNC, Raudkivi AJ (1991) Scouring IAHR Hydraulic structures manual. Balkema, Rotterdam, The Netherlands

Chinnarasri C, Donjadee S, Israngkura U (2008) Hydraulic characteristics of gabion-stepped weirs. J Hydraulic Eng 134(8):1147–1152

D'Agostino V, Ferro V (2004) Scour on alluvial bed downstream of grade-control structures. J Hydraulic Eng 130(1):24–37

Essery ITS, Horner MW (1978) The hydraulic design of stepped spillways. CIRIA, Report no. 33, London, UK

Hoffmans GJCM (1998) Jet scour in equilibrium phase. J Hydraulic Eng 124(4):430–437

Mohamed HI (2010) Flow over gabion weirs. J Irrig Drainage Eng 136(8):573–577

Mossa M (1998) Experimental study on the scour downstream of grade-control structures. In: Proceedings of the 26th Convegno di Idraulica e Costruzioni idrauliche. Catania, Italy, 9–12 Sept 1998

Oertel M, Schlenkhoff A (2012) Crossbar block ramps: flow regimes energy dissipation, friction factors, and drag forces. J Hydraulic Eng 138(5):440–448

Ohtsu I, Yasuda Y, Takahashi M (2001) Discussion of "onset of skimming flow on stepped spillways" by MR Chamani and N Rajaratnam. J Hydraulic Eng 127(6):522–524

Pagliara S, Chiavaccini P (2006a) Energy dissipation on block ramps. J Hydraulic Eng 132(1):41–48

Pagliara S, Chiavaccini P (2006b) Energy dissipation on reinforced block ramps. J Irrig Drainage Eng 132(3):293–297

Pagliara S, Palermo M (2008a) Scour control downstream of block ramps. J Hydraulic Eng 134 (9):1376–1382

Pagliara S, Palermo M (2008b) Scour control and surface sediment distribution downstream of block ramps. J Hydraulic Res 46(3):334–343

Pagliara S, Das R, Palermo M (2008) Energy dissipation on submerged block ramps. J Irrig Drainage Eng 134(4):527–532

Pagliara S, Palermo M, Carnacina I (2009a) Scour and hydraulic jump downstream of block ramps in expanding stilling basins. J Hydraulic Res 47(4):503–511

Pagliara S, Palermo M, Lotti I (2009b) Sediment transport on block ramp: filling and energy recovery. KSCE J Civ Eng 13(2):129–136

Pagliara S, Palermo M (2010) Influence of tailwater depth and pile position on scour downstream of block ramps. J Irrig Drainage Eng 136(2):120–130

Pagliara S, Palermo M (2011a) Effect of stilling basin geometry on clear water scour morphology downstream of a block ramp. J Irrig Drainage Eng 137(9):593–601

Pagliara S, Palermo M (2011b) Block ramp failure mechanisms: critical discharge estimation. Proc Inst Civ Eng Water Manag 164(6):303–309

Pagliara S, Palermo M (2012). Effect of stilling basin geometry on the dissipative process in the presence of block ramps. J Irrig Drainage Eng 138(11):1027–1031

Pagliara S, Palermo M (2013) Rock grade control structures and stepped gabion weirs: scour analysis and flow features. Acta Geophysica 61(1):126–150

Pagliara S, Palermo M, Carnacina I (2011) Expanding pools morphology in live-bed conditions. Acta Geophysica 59(2):296–316

Pagliara S, Palermo M, Carnacina I (2012) Live-bed scour downstream of block ramps for low densimetric Froude numbers. Int J Sedim Res 27(3):337–350

Palermo M, Das R, Pagliara S (2008) Hydraulic jump classification downstream of block ramps for non-uniform channel bed material. In: Proceedings of the 2nd international junior researcher and engineer workshop on hydraulic structures. Pisa, Italy, 30 July–1 Aug 2008, pp 129–134

Peyras L, Royet P, Degoutte G (1992) Flow and energy dissipation over stepped gabion weirs. J Hydraulic Eng 118(5):707–717

Platzer G (1983) Die Hydraulik der breiten Blocksteinrampe, rampenbeigung 1:10. Bundensanstalt, Wien, Austria

Radecki-Pawlik A (2013) On using artificial Rapid Hydraulic Structures (RHS) within mountain stream channels: some exploitation and hydraulic problems. In: Rowinski P (ed) Experimental and computational solutions of hydraulic problems, GeoPlanet: Earth and Planetary Sciences, Springer, Netherlands, pp 101–115

Rice CE, Kadavy KC, Robinson KM (1998) Roughness of loose rock riprap on steep slopes. J Hydraulic Eng 124(2):179–185

Robinson KM, Rice CE, Kadavy KC (1997) Design of rock chutes. ASAE Paper No. 972062, St. Joseph, Michigan

Veronese A (1937) Erosioni di fondo a valle di uno scarico. Annali Lavori Pubblici 75(9):717–726 [in Italian]

Whittaker J, Jaggi M (1986) Blockschwellen. Mitteilungen 91, Versuchsanstalt fur Wasserbau Hydrologie und Gloziologie, ETH, Zurich, Switzerland

Chapter 5
Current Research and Challenges Related to Shallow Flows

Wim Uijttewaal

Abstract Many environmental flows can be considered as shallow. The effects of shallowness are reflected in a strong influence of bed friction and in horizontal velocity gradients as a result of variations in bathymetry and roughness. The horizontal length scales, being generally larger than the vertical length scales, dominate the flow pattern through horizontal momentum exchange. However, three-dimensional effects do play an important role as non-uniformity of bed roughness gives rise to secondary circulation driven by turbulence anisotropy. This not only results in vertical motion, but also hampers the development of eddies formed in the horizontal shear layers. This chapter addresses a few examples where the flow structure is affected by groynes, and examples where roughness variations are the dominant mechanisms creating horizontal shear layers. In the latter cases, the contribution of secondary circulation to horizontal momentum transfer cannot be neglected, demonstrating the importance of three-dimensional effects in shallow flows.

Keywords Shallow flows · Flow structures · Mixing · Turbulence · Roughness transitions · Secondary circulation

5.1 Introduction

Environmental flows can often be considered as shallow. Particularly in lowland rivers, estuaries and intertidal areas, the water depth is much smaller than the horizontal dimensions of the flow domain. This puts restrictions on the degrees of freedom the flows and the turbulence structures therein have. By definition the disparate horizontal and vertical length scales are reflected in the flow properties and

W. Uijttewaal (✉)
Department of Hydraulic Engineering, Faculty of Civil Engineering and Geosciences, Delft University of Technology, Delft, The Netherlands
e-mail: W.S.J.Uijttewaal@tudelft.nl

P. Rowiński and A. Radecki-Pawlik (eds.), *Rivers—Physical, Fluvial and Environmental Processes*, GeoPlanet: Earth and Planetary Sciences,
DOI 10.1007/978-3-319-17719-9_5

associated transport processes. Induced by the flow geometry or inflow conditions, gyres and other large-scale coherent structures can extend over horizontal distances that are much larger than the water depth and have the capacity to transport mass and momentum over such distances. Examples can be found with river confluences, groyne field flows, shallow wakes, compound channels and tidal flats (see also Uijttewaal and Booij (2000), Jirka and Uijttewaal 2004; Uijttewaal 2014).

5.2 Simplifying a Three-Dimensional Flow

In most modelling applications the shallowness of the flow suggests the use of a depth-averaged approach, despite the fact that the velocities are far from uniform over the vertical. The no-slip boundary conditions at the bed link the properties of the bed and its variations to the flow. It is therefore important to notice that in contrast with deep water the details of the near-bed processes have a large effect on the flow as a whole. Integrating the ensemble averaged Navier-Stokes equations over the water depth results in the shallow water equations in which only a few parameters representing bed friction and horizontal mixing, are remaining.

Reducing all 3D processes to a 2D description allows for a simpler interpretation of the flows but at the same time prohibits the representation of the many essentially 3D-effects present in a turbulent environmental flow. A velocity component u can be interpreted as being composed of three parts: the depth and ensemble averaged mean velocity $\bar{u}$, the deviation of the mean velocity from the depth averaged one $\tilde{u}$ and the fluctuation with respect to the local mean u'; Thus $u = \bar{u} + \tilde{u} + u'$.

By making this distinction it is possible to separate the 2D and 3D mean motion from the fluctuating component. The latter, u', can be further subdivided into horizontal large-scale ($>h$) quasi-2D fluctuations and 3D small-scale fluctuations ($<h$).

The depth averaged continuity and momentum equations for the streamwise x direction with velocity component $\bar{u}$ and transverse y direction with component $\bar{v}$, include bed friction, horizontal turbulence viscosity, pressure p and body forces k, and read for a constant water depth D:

$$\frac{\partial \bar{u}}{\partial x} + \frac{\partial \bar{v}}{\partial y} = 0 \tag{5.1}$$

$$\frac{\partial \bar{u}}{\partial t} + \bar{u}\frac{\partial \bar{u}}{\partial x} + \bar{v}\frac{\partial \bar{u}}{\partial y} = -\frac{1}{\rho}\frac{\partial \bar{p}}{\partial x} - \frac{c_f}{D}\bar{u}\sqrt{\bar{u}^2 + \bar{v}^2} + \nu_t \nabla^2 \bar{u} + k_x \tag{5.2}$$

$$\frac{\partial \bar{v}}{\partial t} + \bar{u}\frac{\partial \bar{v}}{\partial x} + \bar{v}\frac{\partial \bar{v}}{\partial y} = -\frac{1}{\rho}\frac{\partial \bar{p}}{\partial y} - \frac{c_f}{D}\bar{v}\sqrt{\bar{u}^2 + \bar{v}^2} + \nu_t \nabla^2 \bar{v} + k_y \tag{5.3}$$

in which ρ is the mass density, c_f is the bed friction coefficient, and ν_t the horizontal eddy viscosity. All aspects relevant for the particular flow field need to be expressed in terms of the latter two parameters. The contributions that hinder the flow such as bed roughness, vegetation, and permeable obstacles need to be parameterized in an effective friction coefficient, and all processes that contribute to horizontal mixing are absorbed in the turbulent viscosity.

The turbulent viscosity in these equations results from two sources i.e. the production of turbulence in the boundary layer characterized by the friction velocity $u*$ and the water depth D:$\nu_{t3D} \propto u^{*}D$, and the contribution resulting from the horizontal mixing $\nu_{t2D} \propto U_c\delta$. The latter one is important in locations where the horizontal velocity differences characterized by U_c are substantial and the horizontal mixing length scale δ is much larger than the water depth D. Note that for convenience the velocities in this paragraph are indicated by the symbol u, but are not necessarily restricted to the streamwise component. Considering the relative importance of bed friction in comparison with horizontal mixing we obtain the so called bed friction number S, which for a shallow mixing layer reads (Chu and Babarutsi 1988):

$$S = \frac{\frac{c_f}{D}U_c^2}{\nu_{t2D}\frac{d^2\bar{u}}{dy^2}} = \frac{\frac{c_f}{D}U_c^2}{U_c\delta\frac{\Delta U}{\delta^2}} = \frac{c_f\delta U_c}{D\Delta U} \tag{5.4}$$

Here U_c is a characteristic advection velocity, ΔU a characteristic velocity difference. For large values of S friction stabilizes the flow whereas for small values shear instabilities give rise to the formation of large eddy structures. A similar parameter was defined for shallow wakes (Chen and Jirka 1995). Based on experimental data, critical values for S can be found that demarcate stable from unstable conditions. It does however not provide any information on the length scales or intensities of the large eddy structures.

For many engineering applications a 2D description suffices and solutions can be found using limited resources. The simplest solutions are found in cases where the influence of bed friction and pressure gradients are small and self-similar solutions can be found from the notion that the advection of mean momentum is balanced by the shear stresses. Archetypical examples are free jets, wakes and mixing layers in which the transverse length scale is directly linked to the downstream distance on the basis conservation principles and dimensional analysis (Townsend 1956). The self-similarity is destroyed as the water depth introduces another length scale that remains rather constant in contrast with the varying length scales of the evolving free-shear flows. However by including the bed friction length scale in the analysis of a shallow mixing layer, van Prooijen and Uijttewaal (2002) found a self-similar solution for the conditions where the velocity difference is small in comparison with the mean streamwise velocity, given by:

$$\delta(x) = \alpha\frac{\Delta U}{U_c}\frac{D}{2c_f}\left(1 - \exp\left(-\frac{2c_f}{D}x\right)\right) \tag{5.5}$$

in which α is the growth rate as found for unbounded shear layers ($\alpha = 0.085$) and the velocity difference decreasing over a characteristic distance $D/2c_f$, With the decreasing velocity difference the effects of friction become more important leading to an increase of the friction number (see Eq. 5.4).

The depth-average description of the flow basically contains two parameters that in addition to the boundary conditions determine the flow. These are the bed friction coefficient c_f and the effective horizontal turbulent viscosity ν_t. Both are parameterizations of the complex processes associated with the flow losing momentum against bed friction and momentum exchange in the horizontal plane. For shallow wakes and jets similar approaches can be taken in which the friction number plays the same role as above, discriminating between bed friction dominated or large-eddy dominated momentum transfer.

5.3 Eddy Structures

In cases where eddy structures have the chance to develop, the dynamics are governed by the flow geometry and the generation mechanism. Three mechanisms can be discerned (Jirka 2001): I Topographical forcing, II Internal transverse shear instabilities, and III Secondary instabilities of the base flow

I. The strongest forcing is due to the geometry of the flow domain. Obstacles and irregularities of the boundaries give rise to flow separation in the form of detachment of the wall boundary layer. This transverse shear tends to roll up and form spatially growing structures. Examples are found in shallow wakes, sudden expansions etc. As with the Von Kármán vortex street, the formation of vortices requires a minimum Reynolds number, but due to the shallowness an additional requirement prescribes that the bed friction should be weak enough (small value of S) to allow the formation of vortices (Chen and Jirka 1995). As the horizontal dimensions of the obstacle introduce a length scale for shed vortices they can be generated at a scale that is much larger than the water depth.
II. Shear layers are formed as a result of upstream disturbances, confluences of rivers, jets etc. Such shear layers are intrinsically unstable (Brown and Roshko 1974) and result in Kelvin-Helmholz instabilities provided these are not damped by bed friction. This also requires a small value of S (Ingram and Chu 1987; van Prooijen and Uijttewaal 2002).
III. Spontaneous formation of large coherent structures can occur as equally signed vortices have the tendency to merge into a large vortex. By such mechanism 'random' vorticity fluctuations can organize into a structured large scale vortex pattern (Sommeria 1986). This process is promoted by two-dimensional forcing in terms of stratification, background rotation or other vertical confinements that aligns the vorticity directions (Uijttewaal and Jirka 2003).

The evolution of the mean flow properties described above is not directly addressing the emergence and evolution of the eddy structures that are formed in the unstable shear layer. That information is found from a stability analysis of the shear flow. The simplest form, a linear stability analysis, with the incorporation of bed friction, (Chen and Jirka 1998), and the effects of 3D-boundary layer turbulence in the form an eddy viscosity (van Prooijen and Uijttewaal 2002) results in estimates of the length scales that develop in the unstable shear layer. By including the stabilizing effects of a turbulent viscosity and bed friction also provides information about the negative growth, and thus the evolution of the spectral distribution of energy in the large scales. Although a linear stability analysis is only valid for small-amplitude disturbances, van Prooijen and Uijttewaal (2002) demonstrated that for the simple conditions of their laboratory study the spectral evolution of shear layer vortices could be reproduced quite well by accumulating the growth and reduction rate for each length scale along the downstream direction of the mixing layer.

A more complete analysis of the mixing layer evolution and the structures therein can be found using more advanced non-linear analyses or eddy resolving 3D numerical simulations (Rodi et al. 2013).

5.4 Applications to Groyne Fields in Rivers

As mentioned in the introduction, environmental flows such as rivers can generally be considered as shallow, thus giving rise to the formation of large eddy structures, particularly when the flow is disturbed by structures that are built in the main channel, in the flood plains or at the banks.

In many lowland rivers, groynes are used for bank protection. Slender structures oriented perpendicular to the river axis keep the high flow velocities away from the erodible banks. In this way the flow is blocked in the near-bank region and the confined cross-sectional area leads to higher velocities in the centre of the river with a consequent deepening of the main channel, which is beneficial for navigation purposes. For normal flow conditions the equilibrium bed level can be 'adjusted' by choosing the proper length for the groynes fitting with the desired width of the main channel.

The common flow field found in groyne fields with an aspect ratio close to unity consists of a single gyre that fills up the whole groyne field (Fig. 5.1). The circulation is driven by the momentum exchange with the main stream through the interfacial mixing layer that can be considered as shallow.

In the corners near the bank small counter rotating gyres are found. With this geometry a stable circulation is obtained which flows rather smoothly at about 30 % of the main stream velocity. When the distance between the groynes increases, resulting in an aspect ratio of about $w/l = 0.5$, the circulation cell becomes elongated and separates from the bank (see also Chen and Ikeda 1997; Kimura and Hosoda 1997; Uijttewaal et al. 2001; Weitbrecht and Jirka 2001). This provides room for a

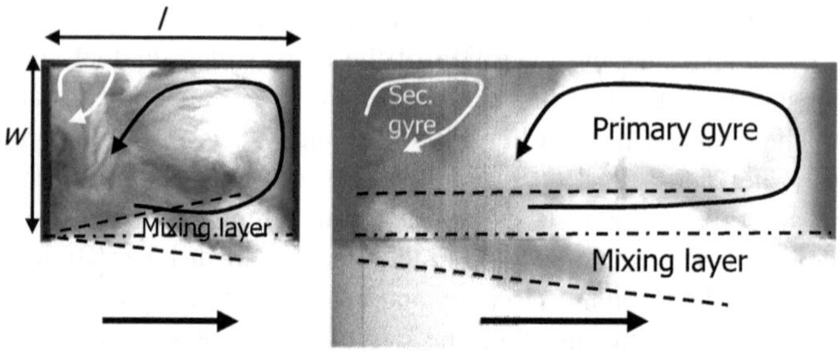

Fig. 5.1 Flow patterns, indicated by *arrows*, as observed with dye exchange experiments for two different aspect ratios w/l = 0.7 *left*, w/l = 0.3 *right*. The *thick arrows* in the lower parts of the figures indicate the direction of the main stream

much larger secondary gyre rotating in the opposite direction. The secondary gyre gets its momentum from the primary gyre via an intermediate mixing layer resulting in a velocity of approximately 30 % of the speed of the primary gyre. There appears to be no direct contact between the secondary gyre and the main stream. Its flow velocity and the exchange with the main stream are therefore very small. The mixing layer has more time and space to develop and is on average wider than in the square groyne field. It starts already with a significant width and grows further downstream over the length of the groyne field. This is due to the vortex shedding that occurs at the groyne tip and the velocity gradient sustaining and enhancing the vortical motion.

Due to the dynamics in the flow in the mixing layer the instantaneous velocity field differs considerably from the time averaged flow pattern. This is nicely visualised in the comparison in Fig. 5.2. The primary and secondary gyre look smooth and well defined in the time averaged picture. This pattern can hardly be recognised from the instantaneous picture where a vortex detaches from the groyne tip and merges with the primary gyre. The strong deformation of the gyre pattern and the large contribution of the dynamic eddies to the exchange of mass and momentum gives an impression of the difficulties associated with the forecasting of mixing and transport. The dominant length scale of horizontal mixing is intermittently governed

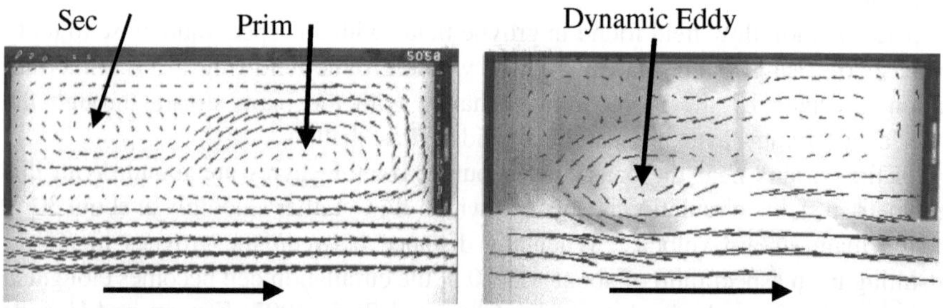

Fig. 5.2 Difference between time averaged and instantaneous flow field (Uijttewaal et al. 2001)

by the dynamic eddies that do not show up in the mean flow field. Moreover, in most applications in rivers on the downstream side of the groyne tip a scour hole is formed which is often associated with the dynamics of the flow in that area. The formation of these scour holes can be reproduced in downscaled laboratory experiments (Garde et al. 1961; Gill 1972).

In the dye-exchange experiment of Fig. 5.2 (right panel), the background of the vector plot shows a dye visualisation of the associated exchange of mass. Starting with a groyne field homogeneously coloured the exchange of mass gradually washes out the dye, first from the primary gyre, later on also from the slower exchanging secondary gyre. The figure also shows the important role played by the dynamic eddy in this process. Setting up a simple mass balance with the assumption of a uniform exchange velocity over the length of the interface between the groyne field and the main stream demonstrates that the characteristic time of exchange is proportional to the width of the groyne field w and inversely proportional to the velocity in the main stream U_r. This time scale can be found from a time series of the decreasing mean concentration averaged over the groyne field.

Figure 5.3 shows the decrease in time of the integrated dye content as measured in the two different groyne fields of Fig. 5.1. The logarithmic scale for the concentration makes the exponential decay visible as a straight line. This holds for more than 90 % of the rapid exchange of the large groyne field while the smaller groyne field shows a gradual deviation from this line starting beyond 50 %.

The exponential decay suggests a first order process where the decrease in concentration is proportional to the concentration itself and is governed by a single time constant. However, the experiments of Fig. 5.3 where performed using the same groyne field width and main stream velocity which should have resulted in the same exchange time for both cases. The visualization of this process (Fig. 5.1) shows that the dynamics of the exchange process are affected by the groyne field length. The largest groyne field in this example allows for a further development of the mixing layer such that larger eddies are formed contributing to a faster exchange of mass per unit of interfacial area between the mainstream and the groyne field.

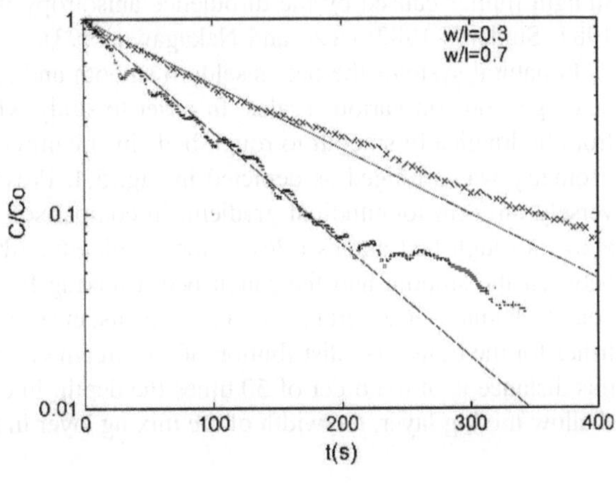

Fig. 5.3 Decay of the integral dye content of an initial homogeneous concentration distribution for groyne fields of different aspect ratios (Uijttewaal et al. 2001)

It appears correct to assume a single retention or relaxation time in modeling the exchange but care should be taken of the specific geometrical aspects (Altai and Chu 1997; Uijttewaal et al. 2001; Weitbrecht and Jirka 2001; Weitbrecht et al. 2008).

With the exchange process characterized by a single parameter the effect of groyne fields on the dispersion in rivers can easily be represented in a one-dimensional model (Mazijk 1996).

5.5 Three-Dimensional Effects

The above considerations are valid in cases where the flow is predominantly horizontal and dominated by bed friction. Identifying shear layers and incorporating their contribution to horizontal mixing could justify a 2D approach (depth averaged) in the modelling of such flows.

However, by definition no flow is two-dimensional and plenty of examples of shallow flows can be found that are, in addition to the omnipresent 3D-turbulence, affected by 3D-phenomena as part of the mean flow. These can be due to abrupt variations in the water depth in transverse (Tominaga and Nezu 1991) or streamwise direction (Ali and Uijttewaal 2014). Another effect that relates to transverse variations is found in curved flows as in river bends (see Van Balen et al. 2010). In those cases vertical motion is induced by flow separation and associated secondary circulation.

5.5.1 Roughness Induced Shear Layer

More subtle and less well known 3D effects are found for shallow flows in simple uniform geometries. Well known examples are the secondary circulations in straight flumes caused by the turbulence anisotropy in the corners (Naot and Rodi 1982; Studerus 1982; Nezu and Nakagawa 1993).

In natural systems the bed is seldom smooth and roughness distributions can be heterogeneous on various scales. In order to study what happens at the transitions from hydraulically smooth to rough beds in streamwise and transverse directions a geometry was arranged as depicted in Fig. 5.4. Provided the presence of a transversely uniform longitudinal gradient, in comparison with a smooth bed, the flow above a rough bed attains a lower mean velocity. Above the transverse transition between the smooth and the rough bed a mixing layer develops. As with a compound channel, at a certain downstream distance an equilibrium situation establishes for the transverse distribution of the streamwise velocity. For the case shown, this distance is of the order of 50 times the depth. In contrast with the classical free shallow mixing layer, the width of the mixing layer in this case remains of the order

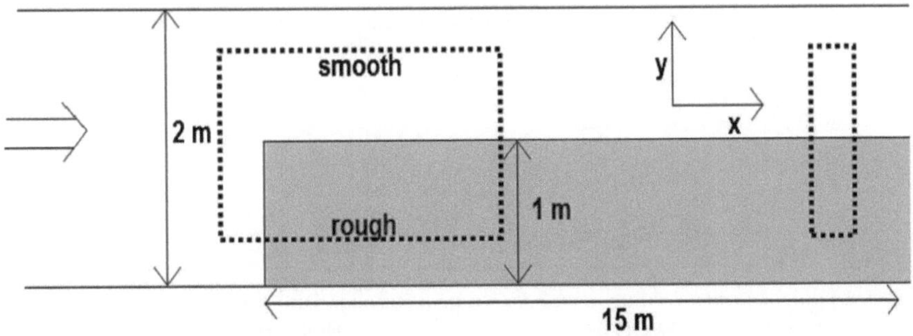

Fig. 5.4 Sketch of an experiment on mixing layer formation due to roughness difference. *Top view* of the experimental configuration

of the water depth indicating that another mechanism is governing the momentum transfer (Vermaas et al. 2011).

Surprisingly, no large eddy structures are found in this case. This explains why the mixing layer in Fig. 5.5 remains narrow despite the large downstream distance. Apparently here the water depth determines the dominant length scale of mixing. It turns out that for this configuration the transverse roughness change gives rise to a circulation cell in the plane perpendicular to the main stream as shown in Fig. 5.6. Though the magnitude of the transverse velocity is small, it is large enough to prevent mixing layer eddies to be formed. It is clear from Fig. 5.6 that near the bed close to the transition the flow is pushed away from the rough towards the smooth side. The formation of streamwise vorticity is known to occur where the anisotropy in the turbulence is strong (Perkins 1970). The corner eddies in a straight open channel are examples of the same phenomenon. The strength of the cell is influenced by the abruptness of the lateral change in bed roughness.

As the circulation cell is bounded by the vertical dimensions of the flow, the mixing layer width is also restricted to this size. Any large-scale structure that

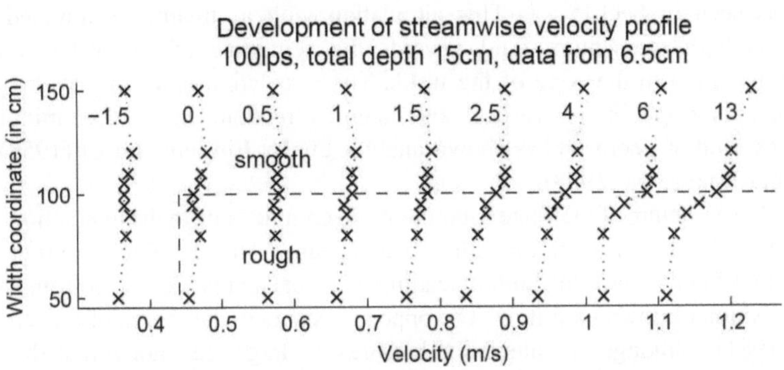

Fig. 5.5 Measured transverse profiles of streamwise velocity are labeled with downstream distance (m) and the velocity scale is shifted 0.1 m/s for each curve (Vermaas et al. 2011)

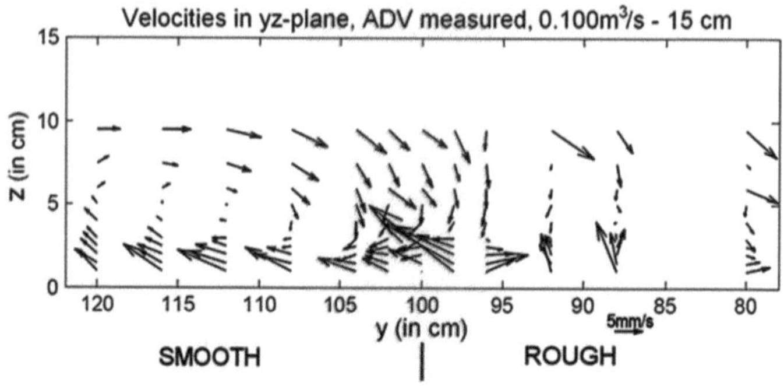

Fig. 5.6 Time averaged velocities in a cross-sectional plane in a developed flow over parallel lanes that differ in roughness. ADV measurements (Vermaas et al. 2011)

would develop in the mixing layer is advected and deformed by the circulation cell before it can attain a significant amplitude. These phenomena can also be encountered in compound channel flows when the transition between the main channel and flood plain is rather abrupt (Tominaga and Nezu 1991).

5.5.2 Roughness Induced Wake

Unlike solid obstructions, a porous obstruction in shallow flows allows mass to flow through it, which decreases the streamwise velocity gradient implying an increasing stability of the wake as shown by Zong and Nepf (2012) for a patch of simulated vegetation. However, depending on the dimensions of the obstruction and its porosity, interaction between the separate mixing layers inside the obstruction can occur, suggesting a possible decrease in stability in the wake.

A shallow wake induced by roughness heterogeneity includes an additional mechanism responsible for exchanging momentum, namely the secondary circulation as seen in Sect. 5.5.1. This circulation with a streamwise directed axis, transports high momentum fluid towards the centerline of the wake and low momentum fluid to the edge of the wake. The circulation is again driven by turbulence anisotropy in the vertical and lateral direction, e.g. a discontinuity in roughness and/or geometry, as shown analytically by Einstein and Li (1958), and Nezu and Nakagawa (1993).

The flow structure of the wake becomes more complex due to the interaction of the secondary circulation with the other mechanisms. The circulation brings high momentum fluid towards the bottom leading to a local increase of the bottom friction at the position of downward flow. The opposite occurs in the region of upward flow (Ikeda 1981). Although the interaction between the large scale horizontal shear and the secondary circulation has never been described in the wake of a roughness patch, its interaction has been examined by for instance Vermaas et al. (2011).

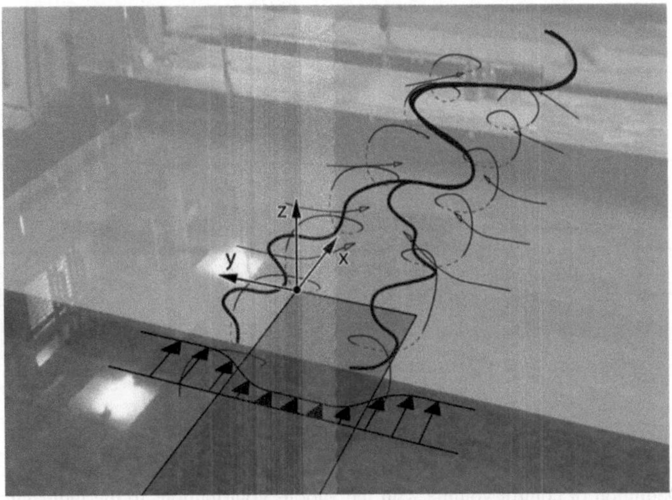

Fig. 5.7 Photograph of experimental set-up including a schematization of the mechanisms of momentum transfer, i.e. lateral advection, large scale horizontal shear and the secondary circulation. Note the definition of the coordinate system. The relatively *dark rectangular* area is the downstream end of the 6 × 0.5 m^2 roughness patch

The secondary circulation is suggested to dampen the growth of the large scale horizontal structures due to wake inward velocity component at the surface. The circulation results in a deformation these horizontal structures, which indicates the difficulty of proper representing the flow field by means of a quasi two-dimensional modeling approach. In Fig. 5.7 a conceptual picture is shown of the flow structure in the wake of the roughness patch against the background of the experimental set-up of a 3 m wide 20 m long shallow flume (Voermans and Uijttewaal 2013).

The roughness patch is 0.5 m wide and 6 m long consisting of irregularly shaped stones with a nominal diameter of $d_n = 21$ mm. Velocities were measured using an Acoustic Doppler Velocimeter (Vectrino).

The mechanisms of mass and momentum exchange contributing to the recovery of the wake can be recognized qualitatively, their relative importance however needs to be quantified from experimental data. The quantification is established by a momentum balance. Figure 5.8 shows the control volume of this balance that covers the range downstream from the patch. In transverse direction it ranges from the line of symmetry towards the left hand side. The balance is based on the shallow water equations; see for instance Vreugdenhil (1994) which for the streamwise depth averaged momentum results in Eq. (5.6). Note that in case of a constant water depth, this is the equivalent of Eq. (5.2). The momentum exchange caused by the flow structures is denoted by T_{ij} (Eq. 5.7), where i and j take the values of the coordinate directions x and y.

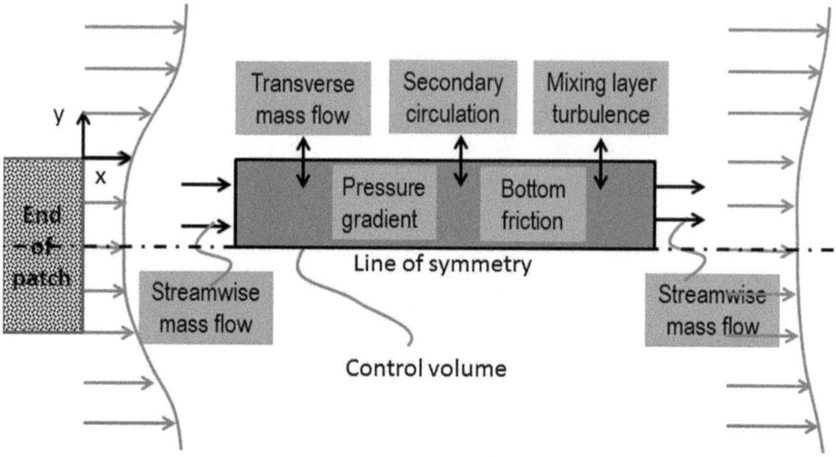

Fig. 5.8 Definition of control volume with the various contributions to the momentum balance, seen from above. On the *left* the end of the roughness patch is indicated. The control volume has lateral coordinates of $y = -0.25$ m (axis of symmetry) and $y = 0$ m (edge of the patch). The longitudinal coordinates as used with the experimental data, are $x = 5$ m and $x = 6$ m

Integration over the section at which the force is acting leads to the balance equation for the control volume and the formulation of the individual terms (Eq. 5.8).

$$\underbrace{\frac{\partial(D\bar{u})}{\partial t}}_{I}+\underbrace{\frac{\partial(D\bar{u}^2)}{\partial x}}_{II}+\underbrace{\frac{\partial(D\bar{u}\bar{v})}{\partial y}}_{III}+\underbrace{gD\frac{\partial D}{\partial x}}_{IV}-\underbrace{\frac{1}{\rho}\tau_b}_{V}-\underbrace{\frac{\partial(D\overline{T_{xx}})}{\partial x}}_{VI}-\underbrace{\frac{\partial(D\overline{T_{xy}})}{\partial y}}_{VII}=\underbrace{F_x}_{VIII} \tag{5.6}$$

$$T_{ij}=\frac{1}{D}\int_{z_b}^{D}\left[\underbrace{\nu\left(\frac{\partial u_i}{\partial x_j}+\frac{\partial u_j}{\partial x_i}\right)}_{IX}-\underbrace{u_i'u_j'}_{X}+\underbrace{\tilde{u}_i\tilde{u}_j}_{XI}\right]dz \tag{5.7}$$

Term *I* is zero since with this statistically stationary flow, the time averaged values of the quantities have been taken and not the ensemble averaged ones. Term *II* is indicated in Fig. 5.8 and represents the difference between the streamwise momentum at the beginning and the end of the control volume. The transverse mass flow, the hydrostatic pressure and the bottom friction correspond to terms *III, IV* and *V,* respectively. The normal stresses of term *VI* is considered small in comparison with term *VII* that represents a shear stress, frequently modeled by means of an eddy viscosity, i.e. the viscous stresses, turbulent stresses and the differential advection as indicated by *IX*, *X* and *XI,* respectively. Term *IX* is considered negligible in comparison with the turbulent stresses of term *X*, for sufficiently high Reynolds numbers and at some distance from the wall. Term *XI* represents the secondary circulation and is an additional term originating from the depth-averaging of the Reynolds-equations (Tennekes and Lumley 1972).

Term *VIII* has been set to zero since no external forces are taken into account.

Integration of Eq. (5.6) over the depth and the perimeter of the domain leads to the momentum balance Eq. (5.8) for the control volume and the formulation of the remaining individual force terms associated with the above terms *II*, *V*, *III*, *X*, *XI* and *IV*, respectively.

$$M + T_b = MF + Mix + SC + Pr, \text{ with} \tag{5.8}$$

$$M = D\rho \int_{y_1}^{y_2} \left(\overline{u_{out}^2} - \overline{u_{in}^2} \right) dy \tag{II}$$

$$T_b = \rho \int_{x_1}^{x_2} \int_{y_1}^{y_2} -u_*^2 \, dy \, dx \tag{V}$$

$$MF = D\rho \int_{x_1}^{x_2} -\bar{u}\bar{v} \, dx \tag{III}$$

$$Mix = D\rho \int_{x_1}^{x_2} -\overline{u'v'} \, dx \tag{X}$$

$$SC = D\rho \int_{x_1}^{x_2} -\overline{\tilde{u}\tilde{v}} \, dx \tag{XI}$$

$$Pr = \rho g D \int_{x_1}^{x_2} \int_{y_1}^{y_2} -\frac{dD}{dx} dy dx \tag{IV}$$

Figure 5.9 shows the contributions to the momentum balance as a function of the normalized longitudinal distance. The main mechanisms contributing to the recovery of the wake show decreasing exchange with downstream distance, since the driving mechanisms are absent or decreasing as well, except for the constant contribution of the hydrostatic pressure force (*Pr*) and the dissipation of momentum by bottom shear (*T*) which tends to increase after a certain distance. Although the local absolute velocity of the secondary circulation (*SC*) is larger than the lateral advection (*MF*) (Fig. 5.3), the total transport of momentum at the given boundary is less for the secondary circulation since the upper and lower part of the water

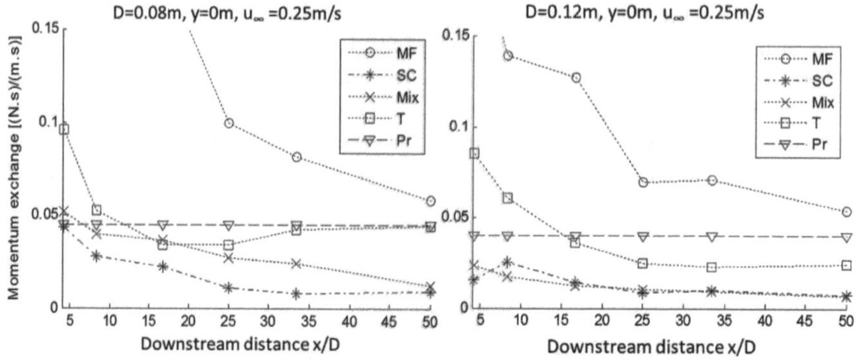

Fig. 5.9 Momentum exchange at the boundary of the control volume as a function of the streamwise position, see Fig. 5.8. For a water depth D = 8 cm (*left*), and 12 cm (*right*)

column have an opposite directed lateral velocity component. The contribution of the large scale horizontal structures (*Mix*) to the recovery of the wake is similar to the contribution of the secondary circulation for the 0.12 m water depth case whereas the contributions by the large scale horizontal structures are larger for the 0.08 m water depth case. All mechanisms do play a significant role in the structure of the wake behind a roughness patch and associated momentum transport. This implies that an accurate description of the flow requires a three-dimensional modeling approach or a smart parameterization of the observed effects.

The critical stability parameter *S* was found to be of the order of 0.1 using theoretical (Chu et al. 1991) as well as experimental considerations (Chen and Jirka 1995). It should be noted that this is based on a quasi-two-dimensional assumption which suggests that the effects of the roughness are uniformly affecting the full water column. Based on the experimental data of our experiment, the stability parameter happens to be rather low i.e. of the order of 0.001, which is the result of the high shear velocity caused by the lateral roughness heterogeneity. This would suggest for the large horizontal scale, according to the definition of the parameter, a negligible dissipation of turbulent kinetic energy (TKE) by bottom friction in comparison with the production of TKE by lateral shear, resulting in a continuous increase of the large horizontal structures i.e. an unstable wake. However, visual observations show the opposite with a stable wake for the 12 cm water depth case and an 'unsteady bubble' for the 8 cm water depth case. Likewise, Fig. 5.9 shows that bottom friction cannot be neglected.

The discrepancy between the stability parameter from the experiment and the critical parameter found by others is suggested to originate within its definition, since the flow field of a wake behind a roughness patch is not adequately described by a quasi-two-dimensional modeling approach. The secondary circulation has a dampening effect on the growth of the large horizontal structures, which is not taken into account in the parameter.

5.6 Challenges

Considering shallow flows in laboratory conditions helps understanding the physical processes under controlled conditions. From the above addressed flow phenomena it is seen that even with rather simple geometries complex three-dimensional processes govern the evolution of the large-scale eddy structures.

However, the context of shallow flow research is found in natural free-surface flows as in rivers, these real flows contain far more complexity and are far more variable than what can be reproduced in the laboratory. These complexities relate for example to vegetation, variations in flow geometry, curvature effects, free surface effects, erosion and transport processes of sediment and other suspended, floating or dissolved matter. Large-scale modelling with inclusion of small-scale features such as vegetation, porous beds and (porous) structures is therefore still a challenge.

It should be mentioned that experience teaches that the multitude of features affecting the flow in natural rivers may have little effect on the predictability of reach-scale properties. Predicting mean flow velocities and water levels is for an important part governed by the geometry and roughness of the bed and uncertainties therein. The quality of most predictions depends therefore on the information present on bathymetry, bed material and vegetation. Nevertheless local flow features will be important for local interferences as with choosing the locations for water intakes and out-falls, identifying locations that are prone to siltation or scour, specification of biotopes suitable for certain species etc..

References

Ali S, Uijttewaal WSJ (2014) Flow resistance of vegetated oblique weir-like obstacles during high water stages. Hydrol Earth Syst Sci 18:1–14

AltaiW, Chu VH (1997) Retention time in a recirculating flow. In: Proceedings of XXVII IAHR congress. San Francisco, pp 9–14

van Balen W, Uijttewaal WSJ, Blanckaert K (2010) Large-eddy simulation of a curved open-channel flow over topography. Phys Fluids 22:075108

Brown GL, Roshko A (1974) On density effects and large structure in turbulent mixing layers. J Fluid Mech 64:775

Chen D, Ikeda S (1997) Horizontal separation flows in shallow open channels with spur dikes. J Hydrosci Hydraulic Eng 15(2):15–30

Chen D, Jirka GH (1995) Experimental study of plane turbulent wakes in a shallow water layer. Fluid Dyn Res 16:11–41

Chen D, Jirka GH (1998) Linear stability analysis of turbulent mixing layers and jets in shallow water layers. J Hydraulic Res 36(5):815–830

Chu VH, Babarutsi S (1988) Confinement and bed-friction effects in shallow turbulent mixing layers. J Hydraulic Eng 114:1257–1274

Chu V, Wu J-H, Khayat R (1991) Stability of transverse shear flows in shallow open channels. J Hydraulic Eng 117:1370–1388

Einstein H, Li H (1958) Secondary currents in straight channels. Trans Am Geoph Union 39:1085–1088

Garde RJ, Subramanya K, Nambudripad KD (1961) Study of scour around spur-dikes. J Hydraulic Div Am Soc Civ Eng 87(6):23–38
Gill MA (1972) Erosion of sand beds around spur dikes. J Hydraulic Div Am Soc Civ Eng 98 (9):1587–1602
Ikeda S (1981) Self-formed straight channels in sandy beds. J Hydraulic Div 107(4):389–406
Ingram RG, Chu VH (1987) Flow around islands in rupert bay: an investigation of the bottom friction effect. J Geophys Res 92(C13):14521–14533
Jirka GH (2001) Large scale flow structures and mixing processes in shallowflows. J Hydraulic Res 39(6):567–573
Jirka GH, Uijttewaal WSJ (2004) Shallow flows: a definition. In: Jirka HG, Uijttewaal WSJ (eds) Proceedings of the international conference on shallow flows. Balkema, Delft, The Netherland, pp 3–11
Kimura I, Hosoda T (1997) Fundamental properties of flows in open channels with dead zone. J Hydraulic Eng 123:98–107
van Mazijk A (1996) One-dimensional approach of transport phenomena of dissolved matter in rivers. PhD-thesis Delft University of Technology, Delft, The Netherlands
Naot D, Rodi W (1982) Calculation of secondary currents in channel flow. J Hydraulic Div ASCE 108(8):948–968
Nezu I, Nakagawa H (1993) Turbulence in open-channel flows, IAHR monograph series. Balkema, The Netherlands
Perkins HJ (1970) The formation of streamwise vorticity in turbulent flow. J Fluid Mech 44 (4):721–740
van Prooijen BC, Uijttewaal WSJ (2002) A linearized model for the evolution of large-scale turbulence structures in shallow mixing layers. Phys Fluids 14:4105–4114
Rodi W, Constantinescu G, Stoesser T (2013) Large-eddy simulation in hydraulics. In: Davies PA (ed) IAHR-monograph. CRC Press/Balkema, The Netherlands
Sommeria J (1986) Experimental study of the two-dimensional inverse energy cascade in a square box. J Fluid Mech 170:139–168
Studerus F (1982) Sekundärströmingen im offenen Gerinne über rauhen Längsstreifen. PhD thesis Eidgenössischen Technischen Hochschule Zürich, Switzerland
Tennekes H, Lumley J (1972) A first course in turbulence. MIT Press, Cambridge, MA
Tominaga A, Nezu I (1991) Turbulent structure in compound open-channel flows. J Hydraulic Eng 117:21–41
Townsend AA (1956) The structure of turbulent shear flow. Cambridge University Press, Cambridge, UK
Uijttewaal WSJ, Booij R (2000) Effects of shallowness on the development of free-surface mixing layers. Phys Fluids 12:392–402
Uijttewaal WSJ, Lehmann D, van Mazijk A (2001) Exchange processes between a river and its groyne fields: model experiments. J Hydraulic Eng 127:928–936
Uijttewaal WSJ, Jirka GH (2003) Grid turbulence in shallow flows. J Fluid Mech 489:325–344
Uijttewaal WSJ (2014) Hydrodynamics of shallow flows, application to rivers. J Hydraulic Res 52 (2):157–172
Vermaas DA, Uijttewaal WSJ, Hoitink AJF (2011) Lateral transfer of streamwise momentum caused by a roughness transition across a shallow channel. Water Resour Res 47:W02530
Voermans J, Uijttewaal WSJ (2013) Turbulence in the wake of a roughness patch. In: Proceedings of 35th IAHR congress. Chengdu, China
Vreugdenhil C (1994) Numerical methods for shallow-water flow. Kluwer, Dordrecht, The Netherlands
Weitbrecht V, Jirka GH (2001) Flow patterns and exchange processes in dead zones of rivers. In: Proceedings of 29th IAHR congress, theme B. Beijing, pp 439–445
Weitbrecht V, Socolofsky SA, Jirka GH (2008) Experiments on mass exchange between groin fields and main stream in rivers. J Hydraulic Eng 134:173–183
Zong L, Nepf H (2012) Vortex development behind a finite porous obstruction in a channel. J Fluid Mech 691:368–391

Chapter 6
One-Dimensional Modeling of Flows in Open Channels

Dariusz Gąsiorowski, Jarosław J. Napiórkowski and Romuald Szymkiewicz

Abstract In this chapter, modeling of the unsteady open channel flow using one-dimensional approach is considered. As this question belongs to the well-known and standard problems of open channel hydraulic engineering, comprehensively presented and described in many books and publications, our attention is focused on some selected aspects only. As far as the numerical solution of the governing equations is considered, one can find out that essentially there are no problems with provided accuracy. Usually, the implementation of the Saint Venant equations (i.e. the full dynamic wave model) for any case study is successful as long as the basic assumptions introduced during their derivation are fulfilled. Otherwise, some computational difficulties can occur. For this reason we would like to draw the readers' attention only to such situations when special computational tricks or simplification of the governing equations should be applied.

Keywords 1D open channel flow · Numerical solutions · Simplified equations · Linear approximation · Lumped parameter simulators

D. Gąsiorowski · R. Szymkiewicz
Faculty of Civil and Environmental Engineering, Gdańsk University of Technology, Narutowicza 11/12, 80-233 Gdańsk, Poland
e-mail: gadar@pg.gda.pl

R. Szymkiewicz
e-mail: rszym@pg.gda.pl

J.J. Napiórkowski (✉)
Institute of Geophysics, Polish Academy of Sciences, Ks. Janusza 64, 01-452 Warsaw, Poland
e-mail: jnn@igf.edu.pl

P. Rowiński and A. Radecki-Pawlik (eds.), *Rivers—Physical, Fluvial and Environmental Processes*, GeoPlanet: Earth and Planetary Sciences, DOI 10.1007/978-3-319-17719-9_6

6.1 One-Dimensional Unsteady Open Channel Flow Equations

Open-channel unsteady flow may be considered as a process of propagation of long waves with small amplitudes in shallow water. The accelerations and velocities in vertical direction are negligibly small in relation to the accelerations and velocities in horizontal directions. Domination of one component of the velocity vector is a typical attribute of the open channel flow. However, this dominated flow velocity vector component varies over wetted cross-section of a channel, so it is a function of 3 spatial co-ordinates and time. Usually, the flow velocity vector component related the longitudinal channel axis is averaged over wetted cross-section. Consequently, 2 spatial independent variables are eliminated. This feature allows us to consider the flow process as spatially 1D phenomenon. Similarly to the other types of water flows, the governing equations are derived starting from two principles of conservation:

- mass conservation law,
- momentum conservation law.

For the first time it was carried out in 1871 by Barré de Saint Venant to describe the process of propagation of the flood waves (Chanson 2004). Assuming that:

- motion of water is gradually varied,
- channel bed slope is small,
- flow is considered as a one-dimensional process,
- velocity distribution over cross-section is uniform,
- pressure is governed by the hydrostatic law,
- friction slope is estimated as for the steady flow,
- lateral inflow does not influence the momentum balance.

The following system of the partial differential equations has been derived

$$\frac{\partial A}{\partial t} + \frac{\partial Q}{\partial x} = 0. \tag{6.1}$$

$$\frac{\partial U}{\partial t} + U\frac{\partial U}{\partial x} + g\frac{\partial H}{\partial x} - g \cdot s + g \cdot S = 0. \tag{6.2}$$

where:

t time,
x space co-ordinate,
A wetted cross-sectional area of the channel,
Q discharge,
U flow velocity averaged over wetted cross-section,
H flow depth,
g acceleration due to gravity,

s channel bottom slope,
S friction slope.

In the literature, these equations are called the system of Saint Venant equations.

Equation (6.1) is the continuity equation derived from the mass conservation principle, while Eq. (6.2) is a dynamic equation derived from the momentum conservation principle. It should be added that the system of Saint Venant equations can be expressed in many forms depending on the assumed dependent variables, geometry of the channel and other factors taken into account. All of these forms are systematically reviewed; see, for instance, Singh (1996). Moreover, theirs derivation can be performed in various ways. It seems that the most consistent approach is to derive the unsteady open channel flow equations from general equations of hydrodynamics, i.e. from the Reynolds averaged Navier-Stokes equations. In such approach, the effects of introduced simplifications and assumptions can be easily followed. However, derivation of the governing equations for unsteady flow can be carried out in a simplified manner, by balancing the fluxes and forces acting on the considered control volume with a priori accepted assumption of uniform velocity flow distribution over a channel cross-section. Next, when the governing equations are derived, an additional factor is introduced to correct the momentum. Moreover, in particular circumstances, apart from the factors taken by Saint Venant into consideration, some additional ones can be included into the continuity and dynamic equations. Consequently, more developed forms of both equations occur:

$$\frac{\partial A}{\partial t} + \frac{\partial Q}{\partial x} = q. \tag{6.3}$$

$$\frac{\partial U}{\partial t} + \beta \cdot U \frac{\partial U}{\partial x} + g \frac{\partial H}{\partial x} - g \cdot s + g \cdot S = \frac{a}{\rho \cdot H} |W| W - \frac{1}{\rho} \frac{\partial P_a}{\partial x} + \frac{\mu^D}{\rho} \frac{\partial^2 U}{\partial x^2} \tag{6.4}$$

where:

q lateral inflow,
β corrective parameter,
ρ water density,
W component of the wind velocity acting along channel axis,
a empirical coefficient characterizing friction between water surface and atmosphere,
P_a atmospheric pressure,
μ^D coefficient of longitudinal dispersion of the momentum.

The factor β which corrects the momentum (Abbott 1979):

$$\beta = \frac{1}{U^2 \cdot A} \int\int_A u^2 \cdot dA \tag{6.5}$$

is called the Boussinesq coefficient (Chanson 2004). It represents the ratio of the actual momentum and the momentum calculated with the averaged velocity U.

Equation (6.4) is a more general form of the dynamic equation for 1D unsteady open-channel, gradually varied flow. It contains all significant components which can determine the flow process. Namely, at the right hand side of Eq. (6.4) the succeeding terms represent the following processes: wind action on the water surface, influence of the spatial variation of the atmospheric pressure and diffusive transport of the momentum. The only process omitted in the derivation is the lateral water inflow distributed along the channel.

In practical applications it is more convenient to use the flow rate and the water stage instead of flow depth and velocity as the dependent variables. The equations with new variables can be obtained using the following relations:

$$H = h - Z, \tag{6.6}$$

$$s = -\partial Z/\partial x, \tag{6.7}$$

$$Q = U \cdot A \tag{6.8}$$

where:

h water stage with regard to assumed datum,
Z bed elevation with regard to assumed datum.

If all the terms at the right hand side are neglected and, additionally, if we take into consideration the fact that the wetted cross-sectional area is a function of the water stage, i.e. $A = A(h)$ while the water stage is a function of time $h = h(t)$, then the following well-known version of the unsteady flow equations is obtained:

$$\frac{\partial h}{\partial t} + \frac{1}{B}\frac{\partial Q}{\partial x} = \frac{q}{B} \tag{6.9}$$

$$\frac{\partial Q}{\partial t} + \frac{\partial}{\partial x}\left(\frac{\beta \cdot Q^2}{A}\right) + g \cdot A\frac{\partial h}{\partial x} + g \cdot A \cdot S = 0 \tag{6.10}$$

where B is channel width at the water surface level whereas the friction slope S is estimated as for the steady flow usually using the Manning formula, i.e.

$$S = \frac{n^2}{R^{4/3}}\frac{|Q|Q}{A^2} \tag{6.11}$$

where R is hydraulic radius of a channel cross-section while n is the Manning roughness coefficient. Equations (6.9) and (6.10) constitute the most popular mathematical model of the unsteady open channel flow, applied in many professional codes widely disseminated among the engineers.

6.2 Initial and Boundary Conditions

Both continuity and dynamic equations constitute the system of first order quasi-linear partial differential equations. Its solution is searched in the following domain:

$$0 \leq x \leq L \quad \text{and} \quad t \geq 0$$

where L denotes the length of considered channel reach. To this order, the following initial-boundary problem is formulated: determine the functions $h(x, t)$ and $Q(x, t)$ which satisfy the Eqs. (6.9) and (6.10) in the solution region and which simultaneously satisfy the auxiliary conditions imposed over its limits. The system of Saint Venant equations has two families of the characteristics. As both families are real, its type is classified as hyperbolic. The run of characteristics is related to the flow regime. For supercritical flow both positive and negative characteristics have a similar slope, whereas for the subcritical flow their slopes have opposite signs. Assume subcritical (tranquil) flow over the considered channel reach. For such kind of flow the run of characteristics is displayed in Fig. 6.1. Remember that for the hyperbolic equations the following rule is valid: at any limit the number of imposed additional conditions must be equal to the number of the characteristics which enter the solution domain from this limit. Therefore, for the assumed subcritical flow the following additional information must be prescribed (Fig. 6.1):

- initial conditions

$$h(x, t = 0) = h_i(x) \quad \text{and} \quad Q(x, t = 0) = Q_i(x) \text{ for } 0 \leq x \leq L,$$

- boundary conditions

$$h(x = 0, t) = h_o(t) \quad \text{or} \quad Q(x = 0, t) = Q_o(t) \text{ for } t \geq 0$$

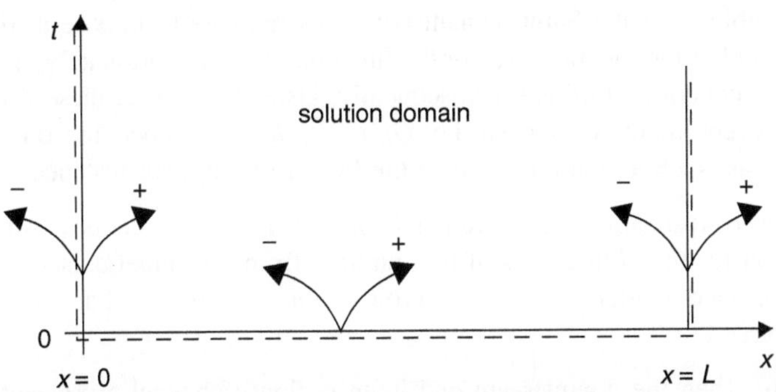

Fig. 6.1 Two families of the characteristics of the Saint Venant equations for the subcritical flow

and

$$h(x = L, t) = h_L(t) \quad \text{or} \quad Q(x = L, t) = Q_L(t) \text{ for } t \geq 0.$$

where the functions $h_i(t), Q_i(t), h_o(t)$ or $Q_o(t)$ and $h_L(t)$ or $Q_L(t)$ are known.

As far as practical implementation of the system of unsteady flow equations is considered, the initial conditions, i.e. the functions $h_i(x)$ and $Q_i(x)$ for $0 \leq x \leq L$ must be computed. To this order, the steady gradually varied flow over considered open channel reach is assumed at $t = 0$. This kind of flow is governed by the equations resulting from Eqs. (6.9) and (6.10). When the time derivatives are neglected, these equations can be reduced to the following ones:

$$\frac{dQ}{dx} = q, \tag{6.12}$$

$$\frac{d}{dx}\left(h + \frac{\alpha \cdot Q^2}{2g \cdot A^2}\right) = -\frac{n^2 \cdot Q^2}{R^{4/3} \cdot A^2}. \tag{6.13}$$

Since Eq. (6.10) has been reformed to the mechanical energy equation then, instead of the momentum correction using the corrective factor β, the correction of energy using the factor α should be done. To obtain the water surface profile $h_i(x)$ and the flow rate distribution $Q_i(x)$ along considered channel reach of length L, the ordinary differential Eqs. (6.12) and (6.13) are solved numerically with the following initial conditions: $Q_i(x = L) = Q_0$ and $h_i(x = L) = h_0$, where Q_0 and h_0 are initial values of the flow rate and the water stage.

Note that numerical integration of Eqs. (6.12) and (6.13) using the implicit trapezoidal rule leads to the well-known step method (Chow 1973), commonly applied in hydraulic engineering for the steady, gradually varied flow. Usually the step method is derived directly from the principle of energy conservation applied for two neighboring cross-sections lying along channel axis, i.e. for a discrete system.

Another important problem is dealing with possibility of determination of the required boundary condition at the downstream end x = L. Proper formulation of solution problem for the Saint Venant equations requires to impose at the downstream end either the function $h_L(t)$ or the function $Q_L(t)$. Unfortunately, usually this requirement cannot be fulfilled. Consequently, instead of one of these functions, a relation between them is imposed, i.e. $Q_L(t) = f(h_L(t))$. Depending on the actual circumstances, such a relation can take the form resulting, for instance, from:

- formula for discharge over a weir if such a weir exists at the downstream end,
- condition for the critical flow if free outflow from a channel exists,
- mass balance written for the reservoir if the considered channel enters the reservoir.

Very often, when the downstream end is an ordinary channel cross-section, as a relation between both functions the rating curve can be imposed. However, this should be realized carefully since the rating curve is an appropriate relation between the water stage and the discharge for the steady uniform flow only. Consequently, in the case of unsteady flow such a condition affects the computed results in the vicinity of the downstream end. To reduce influence of such inappropriate condition, the considered channel reach should be artificially extended so that the improper condition acts far from the controlled channel cross-section.

6.3 Numerical Solution of the 1D Unsteady Open Channel Flow Equations

The open channel unsteady flow equations can be solved using the numerical methods only. This means that we expect to obtain the approximated solution. At the very beginning to this order, the method of characteristics has been applied. Its detailed description for flow with free surface was given by Abbott (1979). Some inconveniences related to the non-linearity of the Saint Venant equations and non-uniform grid points, very troublesome in the natural channels caused that this method has been given up. It was replaced by various schemes of the finite difference method, which appeared to be the most suitable for solving the 1D partial differential equations. During more than last 40 years, many algorithms have been proposed. Their review is given by Cunge et al. (1980) and Singh (1996). Among them one can find the explicit and implicit schemes. Some of them use the staggered grid while others non-staggered one. The available schemes represent different approximation accuracy. Many years of applications provided valuable experiences which allowed to distinguish the most interesting methods for solution. It seems that the most willingly applied is the finite difference implicit four point scheme known as the box scheme (Liggett and Cunge 1975: Cunge et al. 1980). One can find out that this scheme is a basic numerical tool for solving the unsteady flow equations. It is highly appreciated by the users for the following features (Cunge et al. 1980):

- scheme uses non staggered grid which allows computing of both unknown functions at the same node,
- scheme relates the variables coming from both neighboring nodes, x_j and x_{j+1}, which allows to use non-uniform grid point, i.e. variable space interval;
- scheme represents variable order of accuracy controlled by two involved weighting parameters;
- scheme ensures exact solution of the system of linear wave equations for appropriately selected values of the space and time step, i.e. Δx and Δt,
- - scheme is implicit and absolutely stable so that for the reason of numerical stability no restriction for the value of time step exists,
- scheme allows to introduce the imposed boundary conditions in a very simple manner.

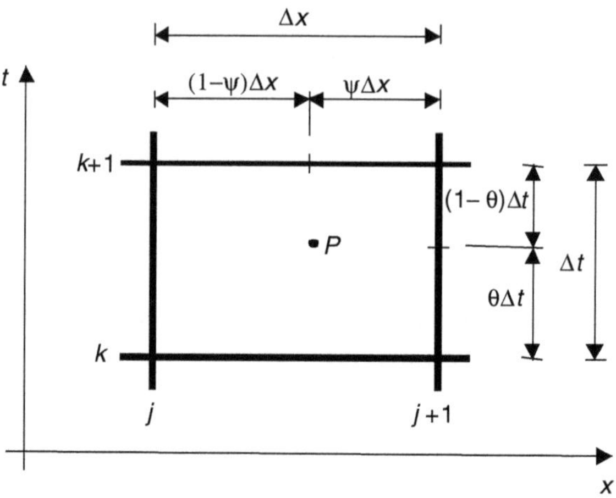

Fig. 6.2 Mesh applied by the finite difference implicit box scheme

Let us remember briefly the idea of box scheme and its most important numerical properties. For numerical solution of the Saint Venant equations using the finite difference box scheme, the solution domain is covered with a grid of dimensions $\Delta x \times \Delta t$ as presented in Fig. 6.2. In a single mesh, containing 4 nodes (j, k), $(j + 1, k)$, $(j, k + 1)$ and $(j + 1, k + 1)$, approximation of all derivatives and functions is carried out at point P using the linear interpolation between the mentioned nodes. Its position is determined by two weighting parameters, ψ and θ, both ranging from 0 to 1.

Approximation of the partial differential equations in each mesh provides the system of non-linear algebraic equations, which in each time step is completed using the imposed boundary conditions and next it is solved using the iterations.

Numerical stability analysis carried out using the Neumann method for the following system of linear wave equations

$$\frac{\partial H}{\partial t} + H_0 \frac{\partial U}{\partial x} = 0, \tag{6.14}$$

$$\frac{\partial U}{\partial t} + g \frac{\partial H}{\partial x} = 0, \tag{6.15}$$

where:

H flow depth,
U flow velocity,
H_0 averaged flow depth considered as constant,

obtained by simplification of Eqs. (6.1) and (6.2), shows that the box scheme is absolutely stable on the condition that:

$$\psi \leq 1/2 \tag{6.16}$$

$$\theta \geq 1/2 \tag{6.17}$$

Very interesting conclusions result from accuracy analysis performed for the same system using the modified equation approach (Fletcher 1991). Replacing of all nodal values of both functions H and Q in algebraic equations approximating the partial differential ones provides the following system of modified linear wave equations:

$$\frac{\partial H}{\partial t} + H_0 \frac{\partial U}{\partial x} = \nu_1 \frac{\partial^2 H}{\partial x^2} + \nu_2 \frac{\partial^2 U}{\partial x^2} + \varepsilon_1 \frac{\partial^3 H}{\partial x^3} + \varepsilon_2 \frac{\partial^3 U}{\partial x^3} + \ldots, \tag{6.18}$$

$$\frac{\partial U}{\partial t} + g \frac{\partial H}{\partial x} = \nu_1 \frac{\partial^2 U}{\partial x^2} + \nu_3 \frac{\partial^2 H}{\partial x^2} + \varepsilon_1 \frac{\partial^3 U}{\partial x^3} + \varepsilon_3 \frac{\partial^3 H}{\partial x^3} + \ldots, \tag{6.19}$$

The coefficients ν_1, ν_2, ν_3 and ε_1, ε_2, ε_3 associated with the higher order derivatives occurring at the right hand side of the modified Eqs. (6.18) and (6.19) have numerical roots. They are given as:

$$\nu_1 = g \cdot H_0 \cdot \Delta t \left(\theta - \frac{1}{2} \right), \tag{6.20}$$

$$\nu_2 = H_0 \cdot \Delta x \left(\frac{1}{2} - \psi \right), \tag{6.21}$$

$$\nu_3 = g \cdot \Delta x \left(\frac{1}{2} - \psi \right), \tag{6.22}$$

$$\varepsilon_1 = \frac{g \cdot H_0 \cdot \Delta x \cdot \Delta t}{2} (1 - \psi - \theta), \tag{6.23}$$

$$\varepsilon_2 = \frac{H_0 \cdot \Delta x^2}{6} \left((3\theta - 2) C_r^2 + (3\psi - 1) \right), \tag{6.24}$$

$$\varepsilon_3 = \frac{g \cdot \Delta x^2}{6} \left((3\theta - 2) C_r^2 + (3\psi - 1) \right). \tag{6.25}$$

where C_r is the Courant number

$$C_r = \frac{\sqrt{g \cdot H_0} \Delta t}{\Delta x}. \tag{6.26}$$

Note that all terms with derivatives of higher order occurring at the right hand side of Eqs. (6.18) and (6.19) were omitted while approximating the linear wave equations using the box scheme. The coefficients ν_1, ν_2, ν_3 are associated with the dissipative properties of the box scheme while the coefficients ε_1, ε_2, ε_3 are associated with the dispersive properties.

The derived system of modified equations allows concluding on all numerical properties of the box scheme. Note that for $\theta = 1/2$, $\psi = 1/2$ and $C_r = 1$ the box scheme provides exact solution of the linear wave equations. In such a case, all terms at the right hand sides of Eqs. (6.18) and (6.19) vanish. Note as well as that assuming $\psi = 1/2$ the numerical diffusion generated by the scheme is strongly limited. In such a case the modified equations become

$$\frac{\partial H}{\partial t} + H_0 \frac{\partial U}{\partial x} = \nu_1 \frac{\partial^2 H}{\partial x^2} + \varepsilon_1 \frac{\partial^3 H}{\partial x^3} + \varepsilon_2 \frac{\partial^3 U}{\partial x^3} + \ldots, \tag{6.27}$$

$$\frac{\partial U}{\partial t} + g \frac{\partial H}{\partial x} = \nu_1 \frac{\partial^2 U}{\partial x^2} + \varepsilon_1 \frac{\partial^3 U}{\partial x^3} + \varepsilon_3 \frac{\partial^3 H}{\partial x^3} + \ldots, \tag{6.28}$$

with

$$\nu_1 = g \cdot H_0 \cdot \Delta t \left(\theta - \frac{1}{2} \right), \tag{6.29}$$

$$\varepsilon_1 = \frac{g \cdot H_0 \cdot \Delta x \cdot \Delta t}{2} \left(\frac{1}{2} - \theta \right) \tag{6.30}$$

$$\varepsilon_2 = \frac{H_0 \cdot \Delta x^2}{12} \left(1 - C_r^2 \right), \tag{6.31}$$

$$\varepsilon_3 = \frac{g \cdot \Delta x^2}{12} \left(1 - C_r^2 \right). \tag{6.32}$$

This version of box scheme (with $\psi = 1/2$) is known as the Preissmann scheme, usually applied for numerical integration of the Saint Venant equations. Note that in practical application the value of coefficient of numerical diffusion can reach the order of a couple thousand m^2/s. Fortunately, its influence becomes remarkable only when the steep waves occur. Typical role of this term is to suppress spurious oscillations occurring in the numerical solution. To this order $\theta > 1/2$ should be applied. Cunge et al. (1980) recommend $\theta = 2/3$.

More detailed information on the numerical solution of the Saint Venant equations using both the finite difference method and the finite element method is given, for instance, by Cunge et al. (1980), Singh (1996), or Szymkiewicz (2010).

6.4 Extension of Application of the 1D Unsteady Open Channel Flow Equations

The system of Saint Venant equations has been derived with the following fundamental assumptions: the open channel flow is a gradually varied one, the channel has compact cross-sections, the free surface of water exists and the flow depth is still greater than zero. On condition that these assumptions are valid, the results provided by the system usually are very satisfying. However, in engineering practice it sometimes happens that some of the above-mentioned assumptions locally and temporarily are not preserved. To overcome the difficulties occurring while applying the Saint Venant equations in such circumstances, some particular approaches are used. As a matter of fact, in such situations we try to apply the Saint Venant equations beyond the region of their validity. Let us recall some of them.

Locally the flow profile in an open channel can vary so rapidly that in fact a discontinuity occurs. A very steep wave can occur in many circumstances. For instance, the steep waves can be generated by the operation of hydro-power plant, when the turbines are started or stopped suddenly or caused by sudden opening or closing of a gate. The extreme shock wave occurs in an open channel, when a dam separating different water levels is suddenly removed (break dam problem).

For those types of steep waves, the system of Saint Venant equations is formally not valid, since in its derivation the vertical acceleration has been omitted. On the other hand, in hydraulic engineering, when the main subject of interest is flow considered in large scale, the local untypical but complicated phenomena can be less important and their internal structure may be neglected. For instance, length of the hydraulic jump is small comparing with the spatial dimension of practically applied grid. Consequently, it is spatially reduced to a point in which the water surface is discontinuous. The Saint Venant equations are capable to reproduce such a case of flow on condition that they are expressed in particular, so-called conservative form. This means that the forms of equations have to satisfy the integral form of conservation principles from which they were derived. Note that the system of Saint Venant equations can be written in integral form as well as in differential form (Chanson 2004; Cunge et al. 1980). Both representations are equivalent as long as smooth solutions are considered. The difference becomes noticeable when discontinuities occur. The second required condition concerns the numerical method applied. It is assumed that the standard schemes of the finite difference method based on the Taylor series expansion can break down near discontinuity arising in the solution (LeVeque 2002). An appropriate method should ensure required solution accuracy. First of all, it is expected that the considered conservation principles will be satisfied on the numerical grid. This requirement is satisfied particularly exactly by the finite volume method based on the equations written in integral form. However, to solve successfully some flow problems of shock wave type one can use the finite difference methods as well. The flow of shock wave type means that we consider flow with large gradients rather than containing pure discontinuity. In such a case the computation should be carried out very carefully. First of all, as it was stated previously, the Saint Venant equations must be expressed in the conservative form. Secondly, their solution needs a particular numerical approach. There are known three possible techniques (Cunge et al. 1980).

In the first approach, called the shock fitting method, the problem of discontinuity propagation and the unsteady flow between the discontinuities is split and considered separately. Traveling discontinuities are analyzed using the method of characteristics, whereas for solution of the Saint Venant equations describing flow in the regions limited by discontinuities, the standard methods are used.

The second approach, called pseudoviscosity method, requires introduction of an artificial diffusive term into the dynamic equation. This term, having smoothing properties, allows us to control the solution near discontinuity. Its particular form relating intensity of generated diffusion with the wave steepness ensures that this extra term acts strongly only locally, being insignificant far away from the discontinuity. This approach is applicable for non-dissipative schemes (Potter 1973; Cunge et al. 1980).

The last possible approach bases on the solution using the dissipative methods. In this approach it is assumed that the process of smoothing can be ensured by the numerical diffusion generated by the applied scheme. Controlling artificial diffusion one can obtain an acceptable solution. To this order, one can apply for instance the Preissmann scheme which is able to generate numerical diffusion.

The disadvantage of the presented approaches is that they base on the differential equations. This can be avoided by using the finite volume method. The finite volume method has very important feature—it ensures very good conservation of the transported quantity since it uses the flow equations in the integral form. The finite volume method appears to be an effective tool for solving the propagation of shock waves, particularly for extreme auxiliary conditions. However, it deals with rather idealized situation, when the homogenous equations are solved. Unfortunately, the method cannot be simply implemented for more realistic cases. The presence of source terms in the form of bed and grade line energy slope requires applying non-trivial approaches to overcome occurring difficulties. Detailed description of the method and its implementation for solution of the 1D shallow water equations is given by LeVeque (2002), among others.

Very often the Saint Venant equations are solved for a storm sewer network. Such a system is constituted by closed conduits where the flow is usually free-surface (Fig. 6.3a).

However, every now and then, after heavy rain, the conduits can be locally filled to the top and pressurized flow occurs. Since it happens only temporarily one can apply the so-called "Preissmann slot" (Cunge et al. 1980) shown in Fig. 6.3b. Owing to this concept it is possible to continue the computation using the same mathematical model, i.e. the system of Saint Venant equations. When the water level in conduit reaches its top, then the further increasing pressure does not cause an increase of the cross-section parameters. They are still constant, corresponding to the entirely filled conduit. The assumed width of slot should be relatively small (after Cunge et al. (1980)—of the order of 0.01 m) so that the mass balance remains not affected.

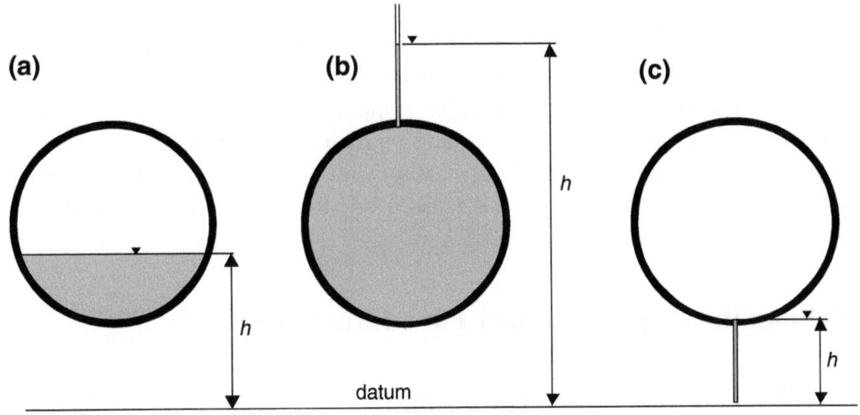

Fig. 6.3 Flow in closed conduit: **a** with free surface; **b** pressurized; **c** with dry bottom

A similar idea can be also applied in open channels with a temporarily dry bed. This can arise similarly, in a storm sever network or in the channels of watering systems. We have to remember that the equations of unsteady flow are valid for positive values of depth. In such a case one can apply a concept opposite to the "Preissmann slot", called "Abbott slot" (Abbott and Basco 1989). The slot goes down from the level of bottom, as presented in Fig. 6.3c. This technique allows us to continue the computations even if a channel bed becomes temporarily dry.

In hydraulic practice, compound channel cross-sections are encountered, for example, if the river has flood plains as presented in Fig. 6.4. When the water stage exceeds the level of flood plain, the flow process becomes more complicated, since it takes place in parts of the channel having different hydraulic properties. If 1D Saint Venant equations are used to model the unsteady flow, they should be modified and special treatment of the cross-section is needed. Usually in channel cross-section two parts playing different roles in flow process are distinguished (see for instance Abbott and Ionescu 1967). The first one, called active, is connected with the main channel and it is involved in the dynamic equation. The second part of section is called inactive, since it serves as a reservoir, which stocks the water only. This part which together with the active part constitutes the whole area of section is involved in the continuity equation.

Such an approach requires modification of the original Saint Venant equations. Although many attempts are still undertaken for modeling flood wave propagation in

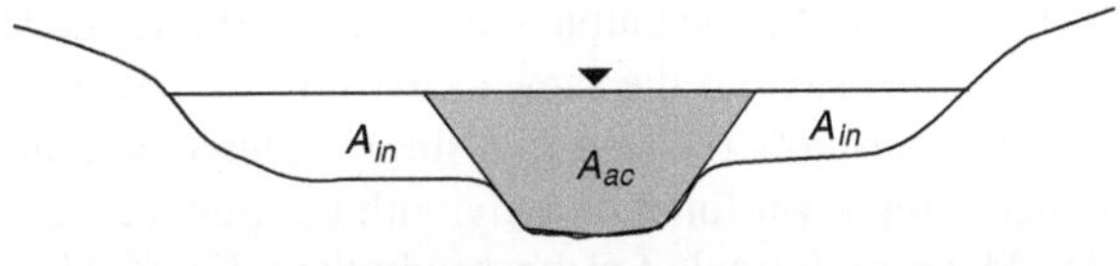

Fig. 6.4 Channel of compound cross-section

the rivers with adjacent large flood plains it seems that nowadays such approaches became less interesting. Actually, such a case of flow should be considered rather as a 2D process with the river bed being partially and temporarily dry. It can be modeled with one of the widely available computer codes for 2D unsteady flow. Expectations that 1D model will be able to reproduce accurately being in fact 2D process, is not founded. The problems dealing with propagation of the flood wave over flooded area next to a river are discussed by Moussa and Cheviron (2015).

6.5 Simplified Open Channel Flow Equations

The system of Saint Venant equations, referred to as the dynamic wave model, requires relatively large number of data representing channel and initial-boundary conditions. The difficulties in acquiring such data cause that instead of the full Saint-Venant equations, simplified models are often used for modeling of the flood wave propagation in open channels. Such models require less input data and the numerical algorithms applied to their solution are simpler and more effective than in the case of the dynamic wave model. The simplified models are obtained by omitting less important terms in the dynamic equation of Saint Venant system. Analysis performed by Cunge et al. (1980) and Henderson (1996) shows that there are physically founded reasons to neglect the terms of less significance in the dynamic equation. There are known two simplified models: the kinematic wave model and the diffusive wave model. Both models are based on the original continuity equation and on the appropriately simplified dynamic equation.

The kinematic wave model is derived by elimination of the inertial and pressure terms in dynamic equation and simultaneously the continuity equation is taken in the unchanged form (6.1). According to this simplification, the kinematic wave model is constituted by the following system of equations:

$$\frac{\partial A}{\partial t} + \frac{\partial Q}{\partial x} = 0 \qquad (6.33)$$

$$s = S \qquad (6.34)$$

where:

s slope of bottom,
S slope of the energy line (friction slope).

Equation (6.34) is a simplified dynamic equation describing the steady uniform flow where the channel bed slope is the same as the slope of energy grade line. The kinematic wave model is usually reduced to a single differential transport equation with regard to a single unknown function only with the flow rate as the dependent variable. Using the Manning formula for the steady flow, Eq. (6.34) is rewritten as:

$$A = a \cdot Q^m \tag{6.35}$$

with

$$a = \frac{1}{\left(\frac{s^{1/2}}{n \cdot p^{2/3}}\right)^{3/5}}, \tag{6.36a}$$

$$m = \frac{3}{5}. \tag{6.36b}$$

If we assume a wide and shallow channel with constant value of wetted perimeter p, then Eq. (6.35) can be differentiated with respect to time. Substituting the result of differentiation into the continuity Eq. (6.33) yields the kinematic wave model in the form of advection equation:

$$\frac{\partial Q}{\partial t} + C(Q)\frac{\partial Q}{\partial x} = 0. \tag{6.37}$$

where $C(Q)$ is the kinematic wave celerity given as:

$$C(Q) = \frac{1}{a \cdot m \cdot Q^{m-1}} \tag{6.38}$$

The second simplified form of Saint-Venant equations—the diffusive wave model—is obtained by omitting the inertial force in the dynamic Eq. (6.2). Thus, this model is constituted by the continuity Eq. (6.1) and the simplified dynamic Eq. (6.2), which can be rewritten as:

$$\frac{\partial H}{\partial x} + \frac{|Q|Q}{k^2} - s = 0 \tag{6.39}$$

where the coefficient k, called conveyance, is a function of the cross-sectional parameters only:

$$k = \frac{1}{n}R^{2/3} \cdot A \tag{6.40}$$

Similarly to the previously obtained kinematic wave model, the system of Eqs. (6.33) and (6.39) can be reduced to a single differential equation. The derivation of the diffusive wave equation with the flow rate as dependent is performed by differentiation of both above-mentioned equations in order to eliminate the derivatives of the water depth (function $H(x,t)$). Rearranging of obtained relations with additionally assumed constant width of channel gives the diffusive wave equation in the following final form:

$$\frac{\partial Q}{\partial t} + C(Q)\frac{\partial Q}{\partial x} - D(Q)\frac{\partial^2 Q}{\partial x^2} = 0 \tag{6.41}$$

where C is the kinematic wave celerity and D is the coefficient of hydraulic diffusivity given, respectively, as:

$$C(Q) = \frac{Q}{k \cdot B}\frac{\partial k}{\partial H}, \tag{6.42a}$$

$$D(Q) = \frac{k^2}{2B|Q|} \tag{6.42b}$$

For wide and shallow channel with constant value of the wetted perimeter and for the slope of energy line replaced by the bottom slope, the kinematic wave celerity C given by Eq. (6.42a) coincides with Eq. (6.38), whereas the hydraulic diffusivity (6.42b) becomes (Eagelson 1970):

$$D(Q) = \frac{Q}{2B \cdot s} \tag{6.43}$$

The diffusive wave model (6.41) is an advection-diffusion transport equation.

The theory of the kinematic wave was given by Lighthill and Whitham (1955) whereas the diffusive wave theory was presented by Hayami (1951). The properties and applicability of both above-mentioned models have been discussed by many authors, e.g. Miller and Cunge (1975), Woolhiser and Liggett (1967), Ponce, Li and Simmons (1978) and Ponce (1990). It is worth to add that the kinematic and diffusive wave models are frequently formulated with respect to the discharge and to the water surface elevation as the dependent variables. However, other forms of both models are also possible. A comprehensive review focusing on the different forms of these models as well as their applicability for open channel flow and floodplain problems can be found in Singh (1996).

The dynamic wave model as well as its simplified forms—kinematic wave model and diffusive wave model—constitute the distributed models where all dependent variables occurring in equations are the functions of time and space. 1D flood routing can be also carried out using the Muskingum model where the variability in space is completely eliminated. This model is derived from the storage equation

$$\frac{dV}{dt} = Q_j - Q_{j-1} \tag{6.44}$$

where:

V volume of water stored by channel reach of length Δx,
Q_j inflow entering a channel reach through the upstream cross-section x_j,
Q_{j+1} outflow entering a channel reach through the downstream cross-section x_{j+1},

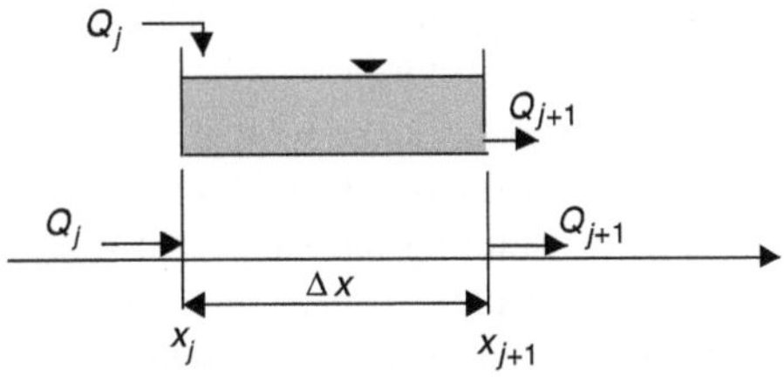

Fig. 6.5 Interpretation of the lumped models—the volume of water V stored by a channel reach of length Δx between the cross-sections x_j and x_{j+1}

obtained by spatial integration of the continuity Eq. (6.33) over a channel reach of length $\Delta x = x_{j+1} - x_j$ (Fig. 6.5).

Equation (6.44) requires an additional relationship between the storage V, inflow Q_j and outflow Q_{j+1}, which for the Muskingum model has the following form (Chow, Maidment and Mays 1988):

$$V(t) = K\left(X \cdot Q_j + (1 - X)Q_{j+1}\right) \tag{6.45}$$

where the parameter X is dimensionless, while the parameter K has the dimension of time and represents the travel time of wave propagating between two cross-sections of channel reach. Substitution of Eq. (6.45) into storage equation yields the well-known Muskingum model:

$$X\frac{dQ_j}{dt} + (1 - X)\frac{dQ_{j+1}}{dt} = \frac{1}{K}\left(Q_j - Q_{j+1}\right) \tag{6.46}$$

While using the Muskingum model, the channel reach is commonly represented by a cascade of $M - 1$ reservoirs rather than by a single reservoir. In this case, the transformation of flood wave by each reservoir is described by Eq. (6.46), where the index of cross-section takes the values $j = 1, 2, 3, \ldots, M$.

It appears that Eq. (6.46) constitutes a separate family of the simplified models. As a matter of fact, the Muskingum model should be considered as a semi-discrete form of the kinematic wave equation (Szymkiewicz 2002). For the first time the kinematic character was noticed by Cunge (1969). Using an accuracy analysis, he proved that an attenuation in solution is caused by numerical diffusion which is controlled by the parameter X and a space interval Δx. Cunge (1969) proposed such a value of the parameter X which allows to generate the numerical diffusion equal to the hydraulic diffusivity present in the linear diffusive wave Eq. (6.41). Consequently, an equivalence of both models with regard to the solution is satisfied for the following condition:

$$X = \frac{1}{2} - \frac{Q}{2B \cdot s \cdot C \cdot \Delta x} \tag{6.47}$$

Usually this version of the Muskingum method is called the Muskingum-Cunge model (Chow, Maidment and Mays 1988).

The simplified flood routing models can be written as a system of equations or as a single transport equation. In the first case, the problem of physical interpretation is rather simple. The equations constituting the simplified model represent the mass and momentum conservation principles, respectively. The troubles arise when the simplified model is reformed to a single transport equation. As it is obtained from two equations representing different principles of conservation then the following question seems to be relevant: which principle of conservation is represented by the final single transport equation? It appears that no problem exists in the case when linear version of the simplified equations is considered. In such a case, both mass and momentum conservation principles are satisfied perfectly. However, very often the non-linear simplified transport Eqs. (6.37) and (6.41) with $C(Q)$ and $D(Q)$ are used for flood routing. The idea of such approach is to increase the reproduction accuracy of flood wave propagation. Unfortunately, in such a case new challenging problems usually occur.

The non-linear partial differential equations can be used in a conservative (divergent) or non-conservative (non-divergent) form. Application of adequate conservative form of differential equation causes that the corresponding equation satisfies the integral form of conservation principle from which it was derived. Consequently, the conservation form allows to avoid the balance errors. Unfortunately, the standard way of derivation of the kinematic wave Eq. (6.37) and the diffusive wave Eq. (6.41) provides to the non-conservative forms, which does not guarantee the mass conservation. The numerical calculations performed by Gąsiorowski and Szymkiewicz (2007) indicate that the non-conservative forms generate in numerical solution of the diffusive wave equation mass balance error which can achieve a significant value.

Let us begin with consideration of the kinematic wave Eq. (6.37). It appears that it is possible to derive another non-linear but conservative form. To this order, the steady flow Eq. (6.35) is directly substituted into the continuity Eq. (6.33). Assuming that parameter a is constant, one obtains (Gąsiorowski and Szymkiewicz 2007):

$$a\frac{\partial Q^m}{\partial t}+\frac{\partial Q}{\partial x}=0 \tag{6.48}$$

In order to show the conservative and non-conservative features of non-linear equations, the kinematic wave Eq. (6.37) written in the non-conservative form is integrated with regard to x over considered channel reach of length L bounded by upstream ($x = 0$) and downstream ($x = L$) ends:

$$\frac{\partial}{\partial t}\int_0^L Q dx = C\cdot Q|_{x=0} - C\cdot Q|_{x=L} + R \tag{6.49}$$

$$R = \int_0^L Q\frac{\partial C}{\partial x}dx = \int_0^L \left(\frac{1}{a \cdot m}Q^{1-m}\frac{\partial Q}{\partial x} + \frac{1}{a \cdot m}Q^2\frac{\partial Q^{-m}}{\partial x}\right)dx \tag{6.50}$$

Similar integration over considered reach performed for the conservative form of the kinematic wave Eq. (6.48) yields:

$$a\frac{\partial}{\partial t}\int_0^L Q^m dx = Q|_{x=0} - Q|_{x=L} \tag{6.51}$$

Comparing both integral equations, (6.49) and (6.51), a significant difference is apparent. From Eq. (6.51) it results that the time variation of the total quantity Q^m stored over considered channel reach is caused only by the net advective flux through upstream and downstream ends of channel. Thus, the form of Eq. (6.48) is conservative and represents the mass conservation principle. However, in the case of Eq. (6.37) besides the net flux through the channel endpoints there is also an additional unbalanced term R. This term results from non-conservative form of the non-linear equation and causes that the integral Eq. (6.49) does not represent a global conservation law. As a consequence, the kinematic wave equation written in the non-conservation form (6.37) cannot ensure the mass conservation in numerical solution. The form of relation (6.50) indicates that the mass balance errors depend on the spatial derivative of the flow rate $\partial Q/\partial x$ Thus, this error achieves a significant value for rapidly varied waves with a strong gradient. Moreover, it is worth to add that the balance errors resulting from improper form of non-linear kinematic wave equation can be amplified by the numerical diffusion generated in the solution (Gąsiorowski 2013). Consequently, these errors depend also on the numerical parameters of the numerical algorithm applied to the solution of non-linear equation.

As far as the diffusive wave equation is considered, it is impossible to obtain its adequate conservative form with the flow rate satisfying the mass conservation principle (Gąsiorowski and Szymkiewicz 2007). This is caused by the way of standard derivation of the diffusive wave equation, which leads to the non-divergent form of diffusive term containing the second derivative:

$$D(Q)\frac{\partial^2 Q}{\partial x^2} \tag{6.52}$$

The resulting form of this term cannot be transformed to its divergent form, and consequently the mass balance errors cannot be eliminated from numerical solution.

Similar difficulties with respect to the conservative properties, as for the kinematic wave equation, can be observed in the case of the non-linear Muskingum model as well. When instead of constant parameters K and X they are related to the solution, then a variable-parameter Muskingum model is obtained. Such approach has been presented, for example, by Ponce and Yevjevich (1978), Tung (1984), Ponce and Chaganti (1994) and Mohan (1997). According to the studies performed by Tung (1984), Mohan (1997), Tang, Knight and Samuels (1999a, b), Perumal and Sahoo (2008), Perumal and Price (2013), if the non-linear storage relation or the variable parameters are applied in the various forms of the Muskingum model, then the numerical solution is frequently affected by mass balance errors. It seems that the reason of this fact is associated with an improper non-conservative form of differential equations. Therefore, the derivation and application of the variable Muskingum model without an analysis of its conservative forms can lead to invalid results. The conservative aspects with respect to the form of the non-linear equations of Muskingum model were reported by Gąsiorowski (2009) as well as by Reggiani, Todini and Meißner (2014).

More information on the conservative aspects of the non-linear advection-diffusion equations is comprehensively discussed for instance by Gresho and Sani (1998).

6.6 Linear Approximation of the One-Dimensional Open Channel Flow Equations

In estuarine hydraulics, the aim is to predict levels or velocities at various points in the channel, given the variation of water level at the downstream end. In problems of flood routing, the aim is to predict the level and/or the flow at the downstream end of the channel when given the level or the flow at the upstream end. In hydrologic flood routing, only the upstream boundary condition is properly specified and the downstream boundary is either ignored or crudely approximated. By studying the linearized St Venant equations for a finite channel reach with a properly defined boundary condition at each end, a basis for the analysis of the errors involved in the solution due to inadequate specification of one of the boundary conditions can be provided.

The problem of flood routing involves the prediction of the hydrograph of flow $Q(t)$, or flow depth $H(t)$ (or velocity $U(t)$, or area $A(t)$) at the selected channel cross-sections. The problem involves the solution of the St. Venant Eqs. (6.1, 6.2) subject to a given initial condition and two appropriate boundary conditions.

Since the system of equations is a hyperbolic one, two boundary conditions are required and for the case of tranquil flow (i.e. the Froude number less than 1) one of these boundary conditions must be prescribed at each end of the reach. No analytical solution is available and the problem must be solved either by simplification of the equations or by same method of numerical approximation as that described in Sects. 6.3 and 6.5.

The complete non-linear set of Eqs. (6.1, 6.2) can be solved analytically by considering variations from a steady state trajectory (Dooge and Napiórkowski 1987, Napiórkowski and Dooge 1988).

The second Eq. (6.2) can be written in terms of the same dependent variables as the first (Q and A) by substituting $Q(x,t)/A(x,t)$ for the velocity $U(x,t)$ in Eq. (6.2) and grouping the terms to obtain the set of equations (Dooge et al., 1982):

$$\frac{\partial A}{\partial t} + \frac{\partial Q}{\partial x} = 0 \tag{6.33}$$

$$(1 - F^2) g \frac{A}{B} \frac{\partial A}{\partial x} + \frac{2Q}{A} \frac{\partial Q}{\partial x} + \frac{\partial Q}{\partial t} = gA(s - S) \tag{6.53}$$

where F is the Froude number defined by

$$F^2 = \frac{Q^2 B}{gA^3} \tag{6.54}$$

To compute the linearized second-order equation we make use of expansion of nonlinear terms in Eq. (6.53) in a Taylor series around the uniform steady state (Q_o, A_o)

$$\begin{aligned} Q(x,t) &= Q_o + Q'(x,t) + e_Q(x,t) \\ A(x,t) &= A_o + A'(x,t) + e_A(x,t) \end{aligned} \tag{6.55}$$

and limitation of this expansion to the first order increments $Q'(x, t)$, $A'(x, t)$; $e_Q(x, t)$, $e_A(x, t)$ represent the higher-order terms (that is, the error of the linear approximation)

$$\begin{aligned} &\frac{\partial A'}{\partial t} + \frac{\partial Q'}{\partial x} = 0 \\ &(1 - F_0^2) g h_o \frac{\partial Q'}{\partial x} + 2u_o \frac{\partial Q}{\partial x} - \frac{\partial Q}{\partial t} = gA_o\left(-\frac{\partial S}{\partial Q} Q' - \frac{\partial S}{\partial A} A'\right) \end{aligned} \tag{6.56}$$

with (see Dooge and Napiórkowski 1987)

$$\frac{\partial S}{\partial Q} = 2\frac{s}{Q_o}; \quad \frac{\partial S}{\partial A} = -2c_k \frac{s}{Q_o}; \quad c_k = -\frac{\partial S}{\partial A} \Big/ \frac{\partial S}{\partial Q} = \frac{dQ}{dA}; \quad u_o = \frac{Q_o}{A_o}; \quad h_o = \frac{A_o}{B_o} \tag{6.57}$$

In Eq. 6.57 c_k is the kinematic wave speed as given by Lighthill and Whitham (1955). For convenience, we may define a parameter m as the ratio of the kinematic wave speed to the average velocity of flow, $m = c_k/u_o$. The parameter m is a function of the shape of channel and of the area of flow. For a wide rectangular channel with Manning friction m is always equal to 5/3. For shapes of channel other than wide rectangular, m will take on different values.

Making the necessary substitution to eliminate the dependent variable *A*'(*x*, *t*) in Eq. 6.56 and leaving a single dependent variable *Q*'(*x*, *t*), one gets a second order partial differential equation for the perturbation *Q*'(*x*, *t*) from the steady uniform reference area Q_o.

$$(1 - F_0^2)gh_o\frac{\partial^2 Q'}{\partial x^2} - 2u_o\frac{\partial^2 Q}{\partial x \partial t} - \frac{\partial^2 Q}{\partial t} = gA_o(-\frac{\partial S}{\partial A}\frac{\partial Q'}{\partial x} + \frac{\partial S}{\partial Q}\frac{\partial Q'}{\partial t}) \qquad (6.58)$$

For any given channel, the relative error due to linearization varies with the inflow hydrograph being routed and with the choice of reference conditions for the linearization. Equation (6.58) can be generalized to a large number of choices for the dependent variable. A perturbation flow potential can be defined as the function $\phi'(x,t)$ whose partial derivative with respect to *x* gives the minus perturbation from the reference area and partial derivative with respect to time *t* gives the perturbation from the reference discharge (see Deymie 1935; Dooge et al. 1987a), $\partial\phi'/\partial x = -A'(x,t)$; $\partial\phi'/\partial t = Q'(x,t)$. It can be shown that perturbation flow potential $\phi'(x,t)$ and any linear function of the perturbation flow potential $\phi'(x,t)$ represent a solution of Eq. (6.58), e.g. area of flow *A*'(*x*,*t*), average water depth *H*'(*x*, *t*), average velocity *U*'(*x*,*t*), the Froude number *F*(*x*,*t*), etc.

The solution of (6.58) for an upstream boundary condition and a semi-infinite wide rectangular channel with Chezy friction was discussed in Lighthill and Whitham (1955), and Dooge and Harley (1967). The two-point boundary problem in which both an upstream and a downstream boundary condition are taken into account has been reported in the hydrologic literature, e.g., by Dooge and Napiórkowski (1987), Napiórkowski and Dooge (1988) and Moramarco et al. (1999). In this section the basic case of tranquil flow is considered (Froude number less than one), in which *Q*'(*x*,*t*) is prescribed both at the upstream boundary *x* = 0 and at the downstream boundary *x* = *L*. The problem is to solve Eq. (6.58) subject to zero initial condition and subject to boundary conditions:

$$Q'(0,t) = Q_u(t); \quad Q'(L,t) = Q_u(t) \qquad (6.59)$$

The solution can be sought in terms of the Laplace transform. Equation (6.55) when transformed to the Laplace transform domain becomes:

$$(1 - F_0^2)gh_o\frac{d^2\bar{Q}}{dx^2} + \left(-2u_o p + gA_o\frac{\partial S}{\partial A}\right)\frac{d\bar{Q}}{dx} - \left(p^2 + gA_o\frac{\partial S}{\partial Q}p\right)\bar{Q} = 0 \qquad (6.60)$$

where parameter *p* is the complex number frequency, $\bar{Q}(x,p)$ is the Laplace transform of *Q'*(*x*, *t*). Equation (6.60) is a second-order homogeneous ordinary equation, so the solution can be written in the general form:

$$\bar{Q}(x,p) = C_1\exp[\lambda_1(p)x] + C_2\exp[\lambda_2(p)x] \qquad (6.61)$$

where λ_1 and λ_2 are the roots of the characteristic equation for Eq. (6.60) and are given by:

$$\lambda_{1,2} = ep + f \pm \sqrt{ap^2 + bp + c} \tag{6.62}$$

The parameters a, b, c, e, f are given in terms of hydraulic variables by the following relationships:

$$a = \frac{1}{gh_o\left(1 - F_o^2\right)^2}; \quad b = \frac{2s}{u_o h_o}\frac{1 + (m-1)F_o^2}{\left(1 - F_o^2\right)^2}; \quad c = \left(\frac{ms}{h_o}\right)^2 \frac{1}{\left(1 - F_o^2\right)^2}$$
$$e = \frac{u_o}{gh_o}\frac{1}{1 - F_o^2}; \quad f = m\frac{s}{h_o\left(1 - F_o^2\right)} \tag{6.63}$$

Having determined the parameters $C_1(p)$ and $C_2(p)$ from the boundary conditions, one can write $\bar{Q}(x, p)$ in terms of the $h_u(x,s)$ and $h_d(x,s)$ which may be defined as the Laplace transforms of the responses of the channel

$$\bar{Q}(x, p) = h_u(x, p)Q_u(p) + h_d(x, p)Q_d(p) \tag{6.64}$$

where the downstream and upstream linear channel responses are given by

$$h_u(x, p) = \exp[(ep + f)x]\frac{\sinh\left[\sqrt{ap^2 + bp + c}(L - x)\right]}{\sinh\left[\sqrt{ap^2 + bp + c}L\right]} \tag{6.65}$$

$$h_d(x, p) = \exp[(ep + f)(L - x)]\frac{\sinh\left[\sqrt{ap^2 + bp + c}x\right]}{\sinh\left[\sqrt{ap^2 + bp + c}L\right]} \tag{6.66}$$

After the inversion of Eq. (6.64) to the time domain, the original function $Q'(x, t)$ in the time domain is determined from the corresponding boundary conditions through the relationship (Dooge and Napiórkowski 1987):

$$Q(x, t) = h_u(x, t)Q_u(t) + h_d(x, t)Q_d(t) \tag{6.67}$$

The transfer functions $h_u(x, t)$ and $h_d(x, t)$ have two distinct parts (Dooge and Napiórkowski 1987):

$$h_u(x, t) = h_u^1(x, t) + h_u^2(x, t); \quad h_d(x, t) = h_d^1(x, t) + h_d^2(x, t) \tag{6.68}$$

The first part of the transfer function due to upstream input, which may be termed as the head of the wave, is given by

$$h_u^1(x,t) = \sum_{n=0}^{\infty} \exp(-2nL\alpha_1 - \alpha_2 x)\delta(t - nt_o - x/c_1) - \sum_{n=1}^{\infty} \exp(-2nL\alpha_1 + \alpha_3 x)\delta(t - nt_o - x/c_2) \tag{6.69}$$

where

$$\alpha_1 = b/2\sqrt{a}; \quad \alpha_2 = \alpha_1 - f; \quad \alpha_3 = \alpha_1 + f; \qquad c_{1,2} = u_{1,2} = u_o \pm \sqrt{gh_o}; \quad t_o = L/c_1 - L/c_2 \tag{6.70}$$

The behaviour of the first part of the solution is shown in Fig. 6.6. It can be seen that the head of the wave moves downstream at the dynamic speed c_1 (along the first characteristic, see Fig. 6.1) in the form of a delta function of exponentially declining volume proportional to $\exp(-\alpha_2 x)$. At $x = L$ the delta function is reflected with inversion of sign and then is propagated upstream at the speed c_2 (along the second characteristic, see Fig. 6.1) and with a heavier damping factor $\exp[-\alpha_3(L - x)]$, then is reflected again at $x = 0$ to move in a downstream direction and so on until the volume of the head of the wave becomes negligible.

The second part of the solution, which may be termed the body of the wave, is given by:

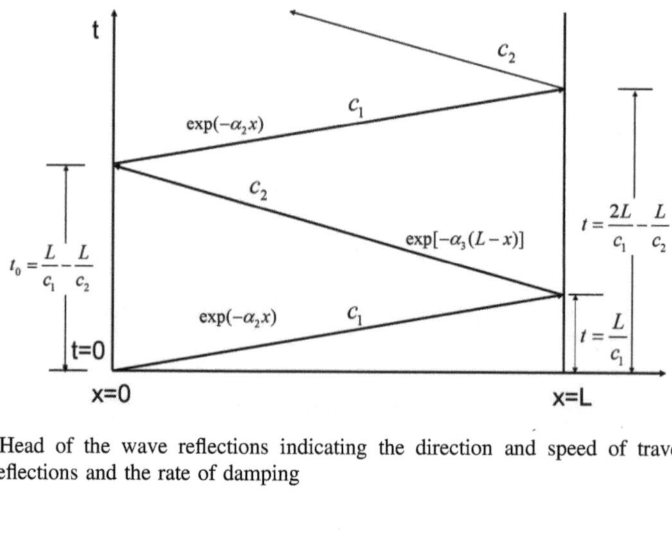

Fig. 6.6 Head of the wave reflections indicating the direction and speed of travel, interval between reflections and the rate of damping

$$
\begin{aligned}
h_u^2(x,t) = & \sum_{n=0}^{\infty} \exp(-\beta_1 t + \beta_2 x) h (1/c_1 - 1/c_2)(2nL + x) \\
& \times \frac{I_1\left[2h\sqrt{(t - nt_o - x/c_1)(t + nt_o - x/c_2)}\right]}{\sqrt{(t - nt_o - x/c_1)(t + nt_o - x/c_2)}} 1(t - nt_o - x/c_1) \\
& - \sum_{n=0}^{\infty} \exp(-\beta_1 t + \beta_2 x) h (1/c_1 - 1/c_2)[2(n+1)L + x) \\
& \times \frac{I_1\left\{2h\sqrt{[t - (n+1)t_o - x/c_2][t + (n+1)t_o - x/c_1]}\right\}}{\sqrt{[t - (n+1)t_o - x/c_2][t + (n+1)t_o - x/c_1]}} 1[t - (n+1)t_o - x/c_2]
\end{aligned}
\tag{6.71}
$$

where I_1 is a modified Bessel function of the first kind, 1[t] is a unit step function, and the remaining parameters are given by:

$$\beta_1 = b/2a \quad \beta_2 = f - be/2a; \quad h = \sqrt{b^2/4 - ac)}/2a \tag{6.72}$$

The solution is in the form of an infinite series which seems too complicated for practical application in river flow forecasting. However, due to heavy damping, only the first few terms of the two transfer functions would normally be required, and the polynomial approximation of the first order modified Bessel function (Abramowitz and Stegun 1965) is sufficiently accurate and can be easily calculated.

As in the case of the head of the wave, the body of the wave is subject to successive reflection at both downstream and upstream boundaries but moves and dissipates more slowly than the head of the wave.

The flow at any intermediate point in the reach is determined by the upstream boundary condition $Q_u(t)$ and the downstream boundary condition $Q_d(t)$ in accordance with Eq. (6.67). It is clear that if the value of $Q_d(t)$ is very much larger than $Q_u(t)$ then this boundary condition will have a dominant influence on conditions throughout most of the reach. In many cases, however, the two boundary conditions are of the same order of magnitude. Accordingly, it is instructive to check how the position of the downstream boundary conditions affects the shape of the impulse responses $h_u(x,t)$ and $h_d(x,t)$. For the purpose of illustration, a broad rectangular channel with Manning friction ($m = 5/2$) is considered. The results are represented in terms of dimensionless independent variables defined with the help of the bottom slope s, the depth h_o, and the velocity u_o, for the steady uniform reference conditions:

$$x' = x\frac{s}{h_o}; \quad t' = tu_o\frac{s}{h_o} \tag{6.73}$$

Figures 6.7 and 6.8 illustrate the effect of the position of the downstream control on the shape of the body of the wave for $F = 0.2$ and $F = 0.8$, respectively, for two values of the length factor: $x' = 1$ (a short channel), and $x' = 20$ (a long channel) and the selected locations of the downstream boundary $L' = x' + \Delta x$, namely $\Delta x \in [0.1, 0.3, 0.5, 1.0, \infty]$.

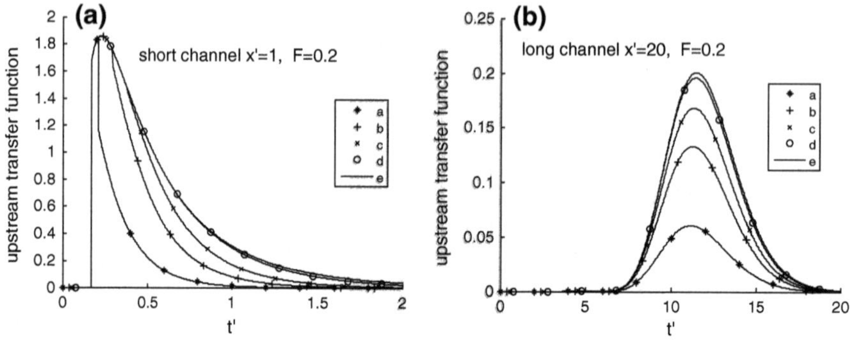

Fig. 6.7 Upstream transfer function for a reference Froude number $F_o = 0.2$, and selected locations of the downstream boundary $L' = x' + \Delta x$, **a** $\Delta x = 0.1$; **b** $\Delta x = 0.3$; **c** $\Delta x = 0.5$; **d** $\Delta x = 1.0$; **e** $\Delta x \to \infty$

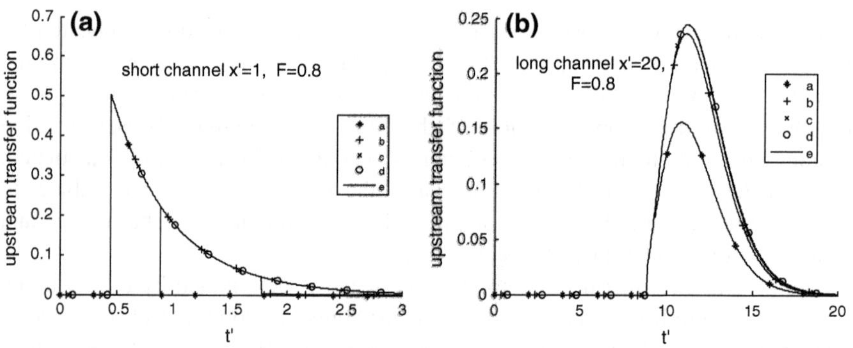

Fig. 6.8 Upstream transfer function for a reference Froude number $F_o = 0.8$, and selected locations of the downstream boundary $L' = x' + \Delta x$, **a** $\Delta x = 0.1$; **b** $\Delta x = 0.3$; **c** $\Delta x = 0.5$, d) $\Delta x = 1.0$, **e** $\Delta x \to \infty$]

The behaviour shown in Fig. 6.7 can be explained in terms of successive reflections shown in Fig. 6.6. For values of t' less than $[L'/c_1' - (L' - x')/c_2']$ only the first term in the first sum in Eq. 6.66 and the first term in Eq. 6.68 differ from zero. For values of t' greater than $[L'/c_1' - (L' - x')/c_2']$ the first term in the second sums in both equations comes into play because of the first reflection of the wave by the zero downstream boundary condition at $t' = L'/c_1'$. For values of t' greater than $t' = L'/c_1' - L'/c_2^1$ the second term in the first sums in both equations, (6.66) and (6.68), becomes effective because of the reflection by the upstream boundary condition. Each reflection brings a new term into effect at a time appropriate to the position in the channel.

Even Fig. 6.7 for a short channel shows only one reflection and thus confirms that only the leading terms in the infinite series are significant.

The solution for the special case of the downstream movement of the flood waves in a semi-infinite channel is well known (Lighthill and Whitham 1955; Dooge and Harley 1967). The general solution for a semi-infinite channel was published by Dooge et al. (1987a). This special case of the semi-infinite channel has the solution

$$h_u(x,t) = \delta(t - x/c_1)\exp(-\alpha_2 x)$$
$$\exp(-\beta_1 t + \beta_2 x)h(x/c_1 - x/c_2)\frac{I_1\left[2h\sqrt{(t-x/c_1)(t+x/c_2)}\right]}{\sqrt{(t-x/c_1)(t+x/c_2)}} \tag{6.74}$$

The properties of the impulse response for a linearized channel of any shape and any friction law were studied by Dooge et al. (1987b) using the cumulants and shape factors of the general response and the amplitude and phase spectra. It was confirmed that even for this very general case the average downstream movement is given exactly by the kinematic approximation. It was shown that for very long waves the attenuation approaches zero, whereas for very short waves the amplitude decreases exponentially with distance.

For the downstream transfer function, the head of the wave is given by

$$h_d^1(x,t) = \sum_{n=0}^{\infty} \exp[-2nL\alpha_1 - \alpha_3(L-x)]\delta[t - nt_o + (L-x)/c_2]$$
$$-\sum_{n=0}^{\infty} \exp[-(2n+1)L\alpha_1 - fL - \alpha_2 x]\delta(t - nt_o + L/c_2 - x/c_1) \tag{6.75}$$

and is subject to reflection at the two ends of the reach as in the case of $h_u^1(x, t)$. The body of the wave in the case of downstream transfer function is given by

$$h_d^2(x,t) = \sum_{n=0}^{\infty} \exp[-\beta_1 t - \beta_2(L-x)]h(1/c_1 - 1/c_2)[2nL + (L-x)]$$
$$\times \frac{I_1\{2h\sqrt{[t - nt_o + (L-x)/c_2][t + nt_o + (L-x)/c_1]}\}}{\sqrt{[t - nt_o + (L-x)/c_2][t + nt_o + (L-x)/c_1]}} 1[t - nt_o + (L-x)/c_2]$$
$$-\sum_{n=0}^{\infty} \exp[-\beta_1 t - \beta_2(L-x)]h(1/c_1 - 1/c_2)[2(n+1)L + x]$$
$$\times \frac{I_1\{2h\sqrt{(t + nt_o - x/c_2 + L/c_1)(t - nt_o + L/c_2 - x/c_1)}\}}{\sqrt{(t + nt_o - x/c_2 + L/c_1)(t - nt_o + L/c_2 - x/c_1)}} 1(t - nt_o + L/c_2 - x/c_1) \tag{6.76}$$

The solution is in the form of an infinite series which seems too complicated for practical application in river flow forecasting. However, due to heavy damping, only the first few terms of the two transfer functions would normally be required, and the polynomial approximation of the first order modified Bessel function (Abramowitz and Stegun 1965) is sufficiently accurate and can be easily calculated.

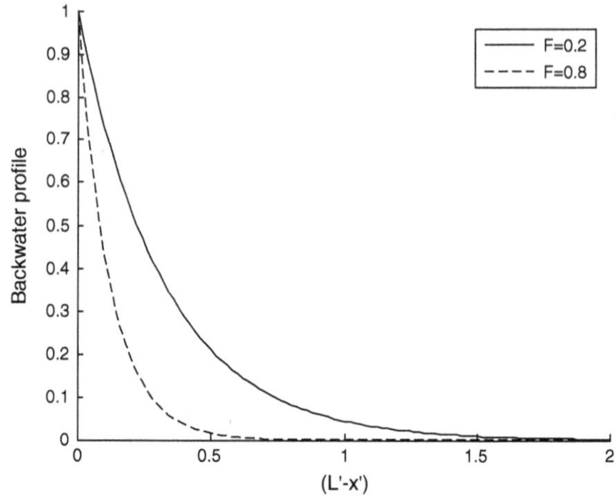

Fig. 6.9 Backwater profile due to a constant unit excess of downstream flow

As an illustration of the effect of the transmission a change in the value of $Q_d(t)$ at the point $(L' - x')$, the change in flow due to a constant downstream boundary $Q_d(t) = 1$ is calculated. The backwater curve is shown in Fig. 6.9 for $F_o = 0.2$ and $F_o = 0.8$. It is clear from Fig. 6.9 that the backwater effect is effective only for $(L' - x') < 1.2$ for $F_o = 0.2$ and for $(L' - x') < 0.5$ for $F_o = 0.8$.

6.7 Modeling of Unsteady Open Channel Flow by Means of Stochastic Transfer Function

Solution of the flood operating problem in the system of reservoirs requires repeated solving of unsteady flow equations for successively generated operation scenarios. Thus the solution algorithms applied in such a case should be maximally efficient, not only in respect of the computer capabilities requirements, but—particularly important in this case—time of computations required to obtain the solution as well.

In order to facilitate the computations, lumped parameter simulators of a distributed flow routing (e.g. St. Venant equations) are usually used in the multiobjective optimization of a water reservoir management system. The simulator can advantageously apply the Stochastic Transfer Function (STF) approach together with a nonlinear transformation of variables. Experience gained by Romanowicz et al. (2004, 2006) indicated that STF models are compatible with distributed model predictions at cross-sections where observations are available. The STM model is calibrated on historical data and on distributed model realizations for the parts of the river where the observations are not available. In Romanowicz and Beven (1998), a lumped model based on a Stochastic Transfer Function (STF) approach was used to update on-line flood inundation forecasts for the River Culm, UK.

The STS model is stochastic, enabling derivation of prediction uncertainty in a straightforward manner; therefore, it is suitable for scenario analysis of the water management system under uncertain climatic conditions. Estimated probability of water levels at the cross-sections along the river enables the derivation of probability maps of inundation at different times of the year.

At the reach scale, the discrete-time STF describes the open channel dynamics and can be presented as:

$$x_k = \frac{B(z^{-1})}{A(z^{-1})} u_{k-d}; \quad y_k = x_k + \zeta_k \tag{6.77}$$

where k is a discrete time period, u_{k-d} denotes STF model input (flow or water level), x_k is the underlying "true" flow or water level, is the noisy observation of this variable, d is a time delay, while $A(z^{-1})$ and $B(z^{-1})$ are polynomials of the following form:

$$\begin{aligned} A\left(z^{-1}\right) &= 1 + a_1 z^{-1} + a_2 z^{-2} + \cdots + a_p z^{-p} \\ B\left(z^{-1}\right) &= 1 + b_1 z^{-1} + b_2 z^{-2} + \cdots + b_m z^{-m} \end{aligned} \tag{6.78}$$

in which z^{-1} is the backward shift operator. The additive error ζ_k in (6.77) is usually both heteroscedastic and autocorrelated in time. It is assumed to account for all the uncertainty at the output of the system that is associated with the inputs affecting the model, including measurement error, unmeasured inputs, and uncertainties associated with the model structure. The orders of the polynomials, p and m, are identified from the data during the data-based identification process. When flow is used as a routing variable, water levels are derived from the rating curve equation.

Stochastic transfer function model and a nonlinear transformation of model variables was recently successfully applied to combined reservoir management and flow routing on the Upper Narew River, northeast Poland (Romanowicz et al. 2010).

References

Abbott MB (1979) Computational hydraulics-elements of the theory of free surface flow. Pitman, London

Abbott MB, Basco DR (1989) Computational fluid dynamics. Longman Scientific and Technical, Harlow

Abbott MB, Ionescu F (1967) On the numerical computation of nearly-horizontal flows. J Hydr Res 5:97–117

Abramowitz M, Stegun IA (1965) Handbook of mathematical functions. Dover Publications Inc., New York

Chanson H (2004) The hydraulics of open channel flow: an introduction, 2nd edn. Elsevier, Amsterdam

Chow VT (1973) Open channel hydraulics. Mc Graw-Hill, New York
Chow VT, Maidment DR, Mays LW (1988) Applied hydrology. McGraw-Hill International Editors, New York
Cunge JA (1969) On the subject of a flood propagation computation method (Muskingum method). J Hydraul Res 7(2):205–230
Cunge J, Holly FM, Verwey A (1980) Practical aspects of computational river hydraulics. Pitman Publishing, London
Deymie P (1935) Propogation d'une intumescence allongée. Revue Générale de l'Hydraulique 3:112–135
Dooge JCI, Harley BM (1967) Proceedings of International Hydrology Symposium, Fort Collins, Colorado, paper no. 8, 1, 57–63
Dooge JCI, Strupczewski WG, Napiórkowski JJ (1982) Hydrodynamic derivation of storage parameters of the Muskingum model. J Hydrol 54:371–387
Dooge JCI, Napiórkowski JJ (1987) The effect of the downstream boundary conditions in the linearized St. Venant equations. Quart J Mech Appl Math 40, part 2, pp 245–256
Dooge JCI, Napiórkowski JJ, Strupczewski WG (1987a) The linear downstream response of a generalized uniform channel. Acta Geophysica Polonica 35(3):279–293
Dooge JCI, Napiórkowski JJ, Strupczewski WG (1987b) Properties of the generalized linear downstream channel response. Acta Geophysica Polonica 35(4):405–416
Eagleson PS (1970) Dynamic hydrology. McGraw-Hill, New York
Fletcher CAJ (1991) Computational techniques for fluid dynamics, vol. I. Springer, Berlin
Gąsiorowski D (2009) Flood routing by the non-linear Muskingum model: conservation of mass and momentum. Arch Hydro-Eng Environ Mech 56(3–4):3–19
Gąsiorowski D (2013) Balance errors generated by numerical diffusion in the solution of non–linear open channel flow equations. J Hydrol 476:384–394
Gąsiorowski D, Szymkiewicz R (2007) Mass and momentum conservation in the simplified flood routing models. J Hydrol 346(1–2):51–58
Gresho PM, Sani RL (1998) Incompressible flow and the finite-element method, vol .1. Advection-diffusion. John Wiley, Chichester
Hayami S (1951) On the propagation of flood waves. Bulletin 1, Disaster Prevention Research Institute, Kyoto University, Kyoto, Japan
Henderson FM (1996) Open channel flow. Macmillan Company, New York
LeVeque RJ (2002) Finite volume methods for hyperbolic problems. Cambridge University Press, Cambridge
Liggett JA, Cunge JA (1975) Numerical methods of solution of the unsteady flow equations. In: Mahmood K, Yevjewich V (eds) Unsteady flow in open channels. Water Resources Publishing, Fort Collins
Lighthill MJ, Whitham GB (1955) On kinematic waves, I: flood movement in long rivers. Proc Roy Soc London Ser A 229:281–316
Moramarco T, Fan Y, Bras RL (1999) Analytical solution for channel routing with uniform lateral inflow. J Hydr Eng ASCE 125:707–713
Moussa R, Cheviron B (2015) Modeling of floods—State of the art and research challenges. This issue
Miller WA, Cunge JA (1975) Simplified equations of unsteady flow. In: Miller WA, Yevjewich V (Eds) Unsteady flow in open channels. Water Resources Publishing, Fort Collins
Mohan S (1997) Parameter estimation of non-linear Muskingum models using genetic algorithm. J Hydraul Eng ASCE 123(2):137–142
Napiórkowski JJ, Dooge JCI (1988) Analytical solution of channel flow model with downstream control. Hydrol Sci J 33(3), part 6, pp 269–287
Perumal M, Price RK (2013) A fully mass conservative variable parameter McCarthy-Muskingum method: theory and verification. J Hydrol 502:89–102
Perumal M, Sahoo B (2008) Volume conservation controversy of the variable parameter Muskingum-Cunge method. J Hydraul Eng ASCE 134(4):475–485

Ponce VM (1990) Generalized diffusion wave equation with inertial effects. Water Resour Res 26 (5)
Ponce VM, Chaganti PV (1994) Muskingum-Cunge method revised. J Hydrol 163:433–439
Ponce VM, Li RM, Simmons DB (1978) Applicability of kinematic and diffusion models. J Hydraul Divis ASCE 104(3):353–360
Ponce VM, Yevjevich V (1978) Muskingum-Cunge methods with variable parameters. J Hydraul Divis ASCE 104(12):1663–1667
Potter D (1973) Computational physics. Wiley, London
Reggiani P, Todini E, Meißner D (2014) A conservative flow routing formulation: Déjà vu and the variable-parameter Muskingum method revisited. J Hydrol 519:1506–1515
Romanowicz R, Beven K (1998) Dynamic real-time prediction of flood inundation probabilities. Hydrol Sci J 43:181–196
Romanowicz RJ, Young PC, Beven KJ (2004) Data assimilation in the identification of flood inundation models: derivation of on-line multi-step ahead predictions of flows. In: Webb B, Arnell N, Onf C, MacIntyre N, Gurney R, Kirby C (eds) Proceedings of the BHS international conference: hydrology, science and practice for the 21st century, London, pp 348–53
Romanowicz RJ, Young PC, Beven KJ (2006) Data assimilation and adaptive forecasting of water levels in the river Severn catchment, United Kingdom. Water Resour Res 42:W06407. doi:10.1029/2005WR004373
Romanowicz RJ, Kiczko A, Napiórkowski JJ (2010) Stochastic transfer function model applied to combined reservoir management and flow routing. Hydrol Sci J 55(1):27–40
Singh VP (1996) Kinematic wave modelling in water resources: surface water hydrology. Wiley, New York
Szymkiewicz R (2002) An alternative IUH for hydrological lumped models. J Hydrol 259:246–253
Szymkiewicz R (2010) Numerical modeling in open channel hydraulics. Springer, Berlin
Tang X, Knight DW, Samuels PG (1999a) Volume conservation characteristics of the variable parameter Muskingum-Cunge method for flood routing. J Hydraul Eng ASCE 125(6):610–620
Tang X, Knight DW, Samuels PG (1999b) Variable parameter Muskingum-Cunge method for flood routing in a compound channel. J Hydraul Res 37(5):591–614
Tung YK (1984) River flood routing by non-linear Muskingum method. J Hydraul Divis ASCE 111(12):1447–1460
Woolhiser DA, Ligget JA (1967) Unsteady one-dimensional flow over a plane: the rising hydrograph. Water Resour Res 3(3)

Chapter 7
Modeling of Floods—State of the Art and Research Challenges

Roger Moussa and Bruno Cheviron

Abstract This chapter presents a state of the art review and research challenges in modeling flood propagation and floodplain inundation. The challenges for flood inundation models are directly linked to the representation of flow processes, to the formulation of theoretical physical laws and to practical considerations. First, we review the various structures of coupled spatially distributed hydrological-hydraulic models and the corresponding spatial representation of flow processes. Second, we present the theoretical basis of 1-D and 2-D Saint-Venant "shallow water" equations with overbank flow, the approximation of Saint-Venant models such as the Diffusive Wave and the Kinematic Wave models and then discuss the domains and limits of applications of each type of models. Practical considerations linked to numerical solution schemes, boundary conditions and model parameterization, calibration, validation and uncertainty analysis were also considered. Finally, the discussion addresses the research challenges for guiding the modeler, according to the principle of parsimony, in seeking the simplest modeling strategy capable of (i) a realistic representation of the physical processes, (ii) matching the performances of more complex models and (iii) providing the right answers for the right reasons.

Keywords Floods · Modelling · Hydraulic · Saint-Venant equations · Diffusive wave · Numerical solutions

R. Moussa (✉)
INRA, UMR LISAH "Laboratoire d'étude des Interactions entre Sol - Agrosystème - Hydrosystème", 2 Place Pierre Viala, 34060 Montpellier Cedex 1, France
e-mail: moussa@supagro.inra.fr

B. Cheviron
IRSTEA, UMR G-EAU "Gestion de L'Eau, Acteurs et Usages", 361 Rue Jean-François Breton, BP 5095, 34196 Montpellier Cedex 5, France
e-mail: bruno.cheviron@irstea.fr

P. Rowiński and A. Radecki-Pawlik (eds.), *Rivers—Physical, Fluvial and Environmental Processes*, GeoPlanet: Earth and Planetary Sciences,
DOI 10.1007/978-3-319-17719-9_7

7.1 Introduction

Floods are the most impacting natural disasters and are a major threat to human life and infrastructures (see for example Fig. 7.1 and the compilation on European flash floods in Gaume et al. 2009). They have a large impact on the socio-economic context, emphasizing the importance of being able to predict them: the demands for predictions of flooding scenarios have markedly increased in recent years, from stakeholders and from the authorities. Many spatially distributed hydrological and hydraulic models turned towards flood predictions have been developed in the

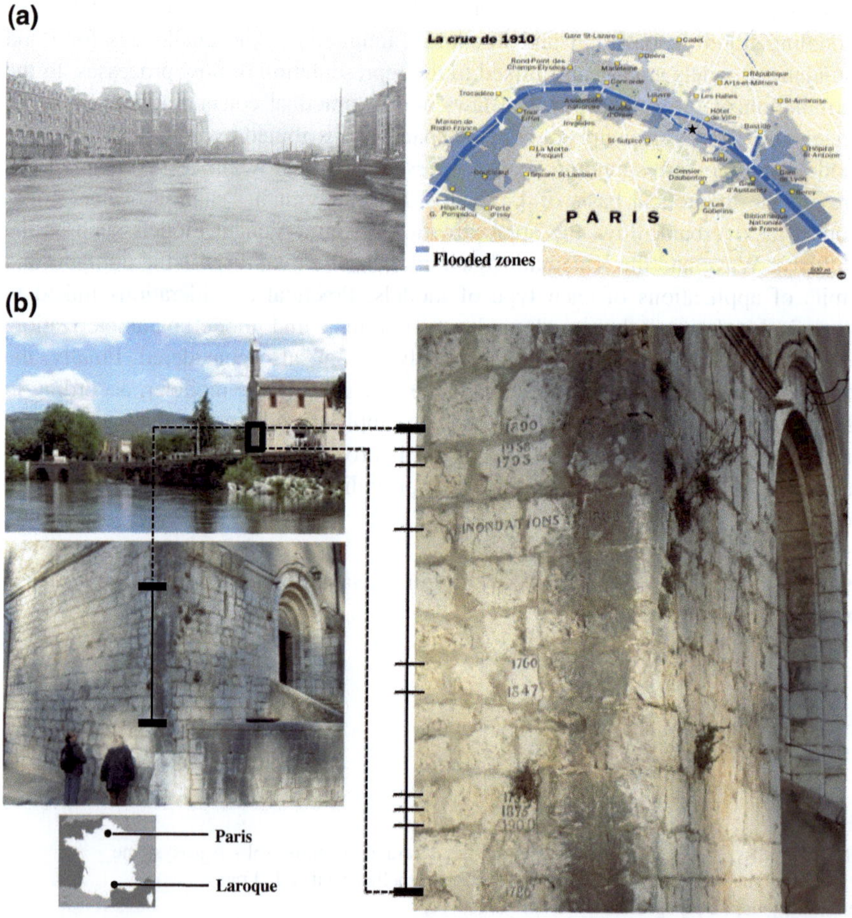

Fig. 7.1 Examples of historical floods in France. **a** The 1910 great flood of the Seine river in Paris: cathedral of Notre-Dame (*left*) and extension of the flooded area (*right*; the location of the cathedral is indicated by *). **b** Flood level mark on Laroque's church (catchment area 750 km²) in the Hérault Mediterranean region (southern France) indicating the maximum water level of successive Mediterranean flash floods since 1735 (*Photos* R. Moussa)

literature (see a synthesis in Stewart et al. 1999; Bates and de Roo 2000; Hunter et al. 2007). However, one may discuss the adequacy of the chosen model structure (existing or to be developed) and equations (classical or new formulation) from the angle of the announced objectives of the study and the required data (available or to be collected). For example, advances in applied research may either be directly linked to new modeling concepts and progresses in theoretical developments or to the increase of computational power, to the enlarged access to high quality elevation data from LiDAR and more generally to the ability of remote sensing techniques to identify the spatial patterns of the landscape properties.

Flood inundation models can be classified according to the number of dimensions in which they represent the spatial domain and flow processes therein, one- (1-D), two- (2-D) or three-dimensional (3-D) models (Cunge 1975). However, whilst the real shallow flow processes are fully 3-D, the numerical resolution of the partial differential equations representing this process is still one of the main challenges for hydraulics and mathematics. Limiting factors include the need to accurately represent the spatial domain, numerical instabilities, uncertainties and equifinality in model parameterization, difficult calibration, and consequent computational time (e.g. Booker et al. 2001; Morvan et al. 2002; Nicholas and McLelland 2004). Hence, the 1-D "shallow water" equations of Saint-Venant (1871) or 2-D "shallow water" equations obtained by depth averaging the Navier-Stokes equations (Navier 1822; Stokes 1845) or approximations have proven adequate, especially for dynamically varying flows in compound channels, given the type and quality of data typically available for model construction, calibration and uncertainty analysis (Bates and De Roo 2000; Hunter et al. 2005a, b; Werner et al. 2005).

For practical purposes, flood inundation models should also generate the required hydraulic information depending on the operational objectives by coping with the input data (Smith et al. 2004). However, the quantities predicted by these models, such as water depth and discharge in the channel and in the floodplain, should be recognized as uncertain (Romanowicz et al. 1994, 1996; Aronica et al. 1998, 2002; Romanowicz and Beven 2003; Hunter et al. 2005a; Pappenberger et al. 2005). Therefore, the choice by the modeler of the appropriate flood inundation model (e.g. Saint-Venant, Diffusive Wave, Kinematic Wave or Uniform flow laws, by order of decreasing complexity) and the corresponding spatial representation (1-D or 2-D), is guided by the terms that can be neglected in the Saint-Venant equations, but also by the availability and the accuracy of input data (intrinsic uncertainty and sampling steps in space and time: topography, hydraulic properties of river reaches and the inundation zone, extension of the inundation zone, and the flood hydrographs). For all these reasons, recent researches in hydraulic modeling examine reduced complexity approaches (Hunter et al. 2007). In particular, it is often hypothesized that uncertainties over the representation of topography and roughness coefficients, rather than those incurred through mathematical simplification of equations, will dominate and thus influence model results (Cunge et al. 1980; Bates and De Roo 2000; Hunter et al. 2007).

This chapter presents a state of the art review and research challenges in modeling flood propagation and floodplain inundation, and is structured into four parts:

(i) the representation of flow processes in spatially distributed hydrological-hydraulic models; (ii) the progress in the formulation of theoretical physical laws; (iii) practical considerations such as numerical solutions, parameterization, calibration, validation and uncertainty analysis; (iv) research challenges.

7.2 Representation of Flow Processes and Model Structure

Figure 7.2 shows a representation of a channel and the floodplain. The channel has one main section and two side sections representing the floodplain on the right and left banks.

A flood consists of a wave propagating downstream, with two phases: the hydrograph rise (Fig. 7.3a–d) and the hydrograph recession (Fig. 7.3d–g). When the bankful flow depth (noted *h* in Fig. 7.3) is reached, water ceases to be contained in the main river channel and water spills onto adjacent floodplains. For flow depth below bankful depth (Fig. 7.3a, g), the flow processes can be represented by a simple 1-D approach. For the overbank flow cases (Fig. 7.3b–f), the situation is more complex. Generally, the floodplain flow is represented by a 2-D approach, whilst at the channel–floodplain interface the development of intense shear layers leads to a strongly turbulent and 3-D flow field which is a limit-case of application of the 1-D and 2-D approaches (Tominaga and Nezu 1991; Babaeyan-Koopaei et al. 2002).

The representation of flow processes (Knight and Shiono 1996) involving coupled hydrological and hydraulic models is presented in Fig. 7.4. Hydrological models can be classified by increasing order of complexity from lumped (Fig. 7.4a left) to fully spatially distributed (Fig. 7.4a right), and enable to predict the input hydrograph and the lateral flow in the floodplain. Figure 7.4b focuses on the downstream floodplain part and exemplifies the choice of a hydraulic model which also can be classified by increasing order of complexity from 1-D storage cell models (Fig. 7.4b left) to 2-D hydraulic models (Fig. 7.4b right). The river channel and the floodplain are represented as a series of cross-sections often perpendicular to the flow direction. The spatial representation consists either in 1-D flow routing in the main channel (Fig. 7.4b left), coupled to the exchanges between the main channel and the floodplain (R_1 and R_2), between floodplain cells (F_1, F_2, F_3 and F_4), and exchanges with surrounding zones (E_1 and E_2), or a complete 2-D flow routing in both the channel and the floodplain (Fig. 7.4b right).

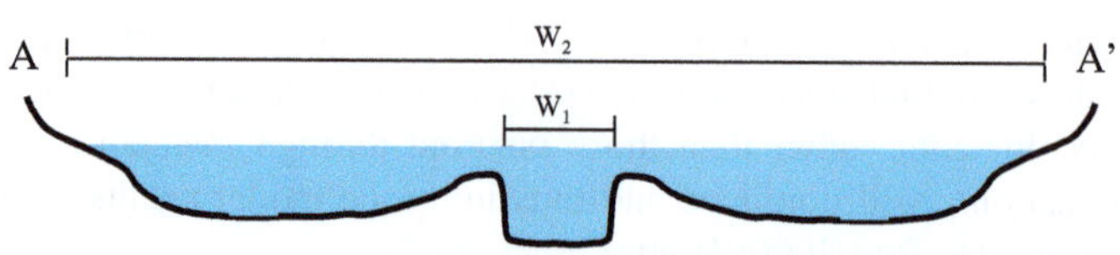

Fig. 7.2 A channel consisting of one main section (width W_1) and the floodplain (width W_2) (section A–A' on Fig. 7.4)

Fig. 7.3 Main overbank flow processes during the hydrograph rise (successively **a**, **b**, and **c**), the equilibrium (**d**) and the hydrograph recession (successively **e**, **f**, and **g**) where: h is the bankful flow depth, y_1 the water level in the main channel and y_2 the water level in the floodplain

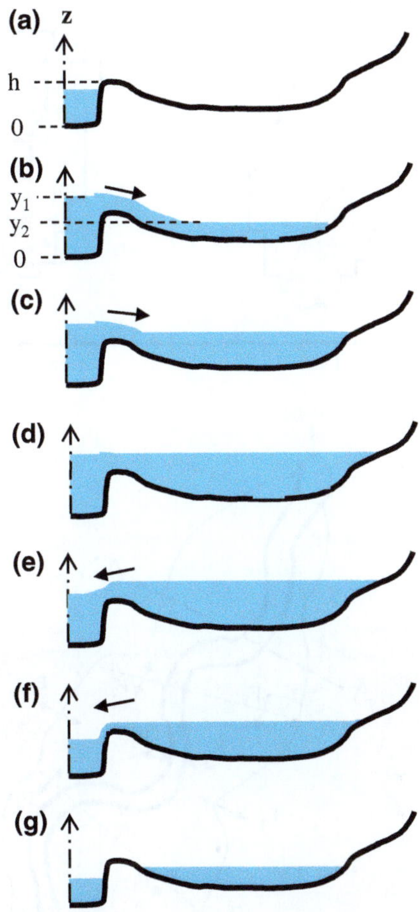

Because of the complexity of the processes of flow exchanges (R_i, F_i and E_i in Fig. 7.4b left), the probable absence of sufficient time records on the flows and the uncertain data on the spatial extension of the inundated zone, modelers are prone to choose conceptual representations that respect the principle of parsimony. Modeling approaches therefore tend to rely on the minimum complexity process representation, provided it is still capable of realistic dynamic simulations.

7.3 Theoretical Considerations

Free-surface flow models may roughly be sorted into levels of decreasing refinement, judging from the richness of their physical content and basis, by order: the Navier-Stokes equations (noted NS: Navier 1822; Stokes 1845), their average in

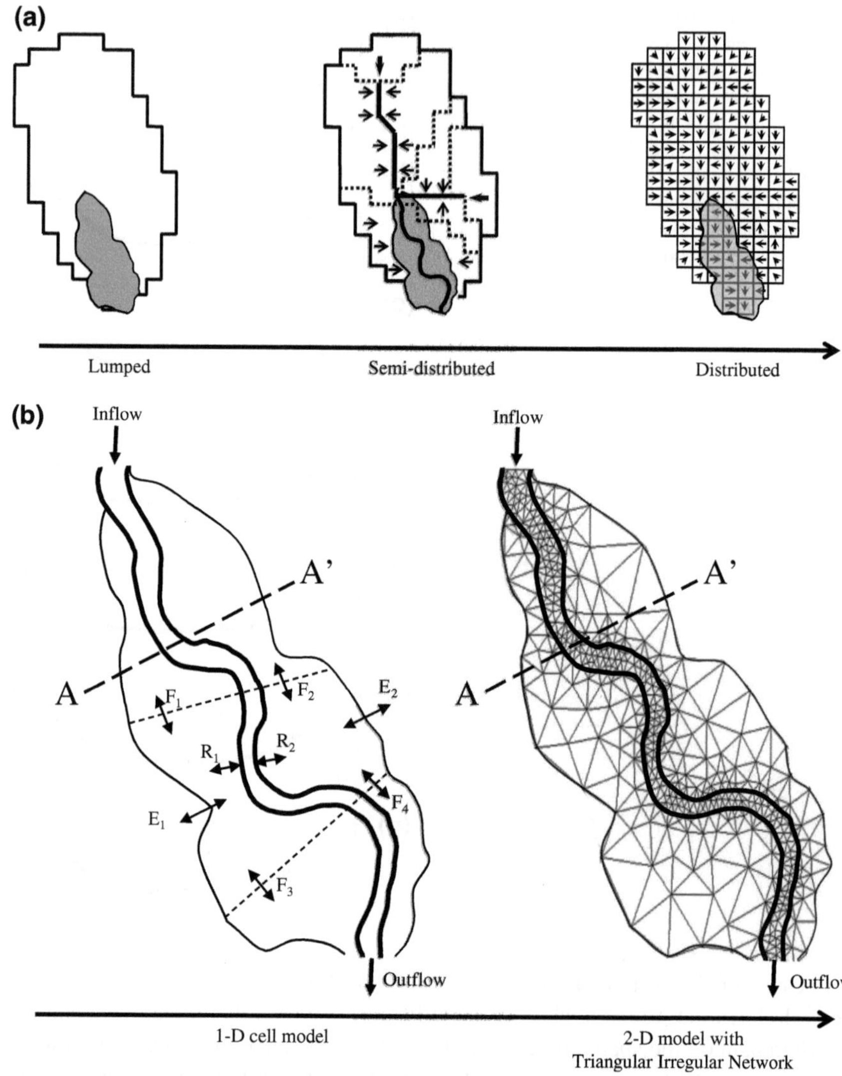

Fig. 7.4 Coupling hydrological (**a**) and hydraulic (**b**) models. Sketch (**b**) focuses on the downstream floodplain part in *grey* on sketch (**a**). **a** Hydrological model spatial representation (the floodplain location is represented in *grey*). **b** Hydraulic model spatial representation on the floodplain (zone in *grey* in **a**).

time termed Reynolds-Averaged Navier-Stokes equations (RANS: Reynolds 1895), the depth-averaged Saint-Venant equations (SV: Saint-Venant 1871) and further approximations (referred to as ASV), among which the Diffusive Wave (DW: Hayami 1951) and Kinematic Wave (KW: Lighthill and Whitham 1955).

The most popular approaches to modeling fluvial hydraulics, and thus implicitly overbank flow and flood inundation, have been 1-D solutions of the Saint-Venant equations which underlie many engineering packages (see a synthesis in Bates and De Roo 2000; Hunter et al. 2007). This section presents the 1-D Saint-Venant equations with overbank flow, then their approximations, the exchange flow between the channel, the floodplain and cells, and finally the 2-D Saint-Venant equations.

7.3.1 The 1-D Saint-Venant Equations with Overbank Flow

The two equations describing mass and momentum in the Saint-Venant system are (Sivapalan et al. 1997)

$$\text{Continuity equation} \qquad \frac{\partial A}{\partial t} + \frac{\partial Q}{\partial x} = q_a \tag{7.1}$$

$$\text{Momentum equation} \qquad \frac{1}{gA}\frac{\partial Q}{\partial t} + \frac{1}{gA}\frac{\partial}{\partial x}\left(\beta\frac{Q^2}{A}\right) + \frac{\partial y}{\partial x} + S_f - S = 0 \tag{7.2}$$

where x is the longitudinal distance [L], t is time [T], A is the cross-sectional area [L^2], Q is the discharge [L^3T^{-1}], q_a is the lateral flow per unit channel length [L^2T^{-1}], y is the flow depth [L], g is the acceleration due to gravity [LT^{-2}], S is the river bed slope [-], S_f the slope of energy line [-] and β is the Boussinesq's momentum correction coefficient [-]. The magnitudes of the various terms in Eqs. (7.1) and (7.2) are given in the literature (e.g. Henderson 1966; Kuchment 1972).

The slope of energy line S_f is usually calculated using a friction law, as for example the Manning formula:

$$S_f = n_c V^2 R_h^{-m} \tag{7.3}$$

where V is the flow velocity [LT^{-1}], R_h is the hydraulic radius [L], n_c the coefficient of roughness in the channel [$L^{1/3}T^{-1}$], and m a constant ($m \approx 4/3$).

The momentum coefficient β represents the non-uniformity of fluid velocities across A and ranges from 1 for prismatic channels to 1.3 for river valleys with floodplains (Henderson 1966).

The term $q_a(x,t)$ represents the lateral flow (inflow or overbank flow). Let $R(t)$ [L^3T^{-1}] be the total lateral inflow hydrograph (Moussa 1996)

$$R(t) = \int_0^L q_a(x, t)\,dx \Rightarrow q_a = \frac{dR}{dx} \tag{7.4}$$

L being the length of the river reach. For this problem, the boundary conditions are $Q(x, 0)$, the upstream inflow $Q(0, t)$ and the downstream outflow $Q(L, t)$.

7.3.2 The 1-D Approximation of Saint-Venant Equations with Overbank Flow

The full Saint-Venant equations are most often not needed and modelers have therefore simplified the equations by neglecting terms in the momentum equation whenever possible (Ponce and Simon 1977; Katopodes 1982; Daluz Vieira 1983; Ferrick 1985; Romanowicz et al. 1988; Moussa and Bocquillon 1996b, 2000). Various terms of the momentum Eq. (7.2) may be sufficiently small to be neglected, leading to the following simplifications: the Kinematic Wave KW (terms IV and V), the Diffusive Wave DW (terms III, IV and V), the steady Dynamic Wave (terms II, III, IV and V), and the Gravity Wave GW (terms I, II, III and V). Figure 7.5 shows the approximation zones of the Saint-Venant equations with overbank flow in a non-dimensional diagram in function of the Froude number Fr and the non-dimensional period $T_+ = TVS_f/y$ (where T is the wave length), for two different cases: without a floodplain ($W_2/W_1 = 1$ where W_1 and W_2 are respectively the width of the main channel and the flooded area; Fig. 7.5 left), and with a floodplain (example of $W_2/W_1 = 20$ in Fig. 7.5 right, among other cases tested by Moussa and Bocquillon 2000). We observe that the domain of application of models in the (T_+, Fr^2) plane changes with the W_2/W_1 ratio that characterizes the extension of the inundation zone.

In most practical applications, the acceleration terms in the momentum balance Eq. (7.2) can be neglected since they are small in comparison to the remaining terms. By combining the two Eqs. (7.1) and (7.2), we obtain the Diffusive Wave equation (Moussa 1996)

$$\frac{\partial Q}{\partial t} + C\left(\frac{\partial Q}{\partial x} - q_a\right) - D\left(\frac{\partial^2 Q}{\partial x^2} - \frac{\partial q_a}{\partial x}\right) = 0 \tag{7.5}$$

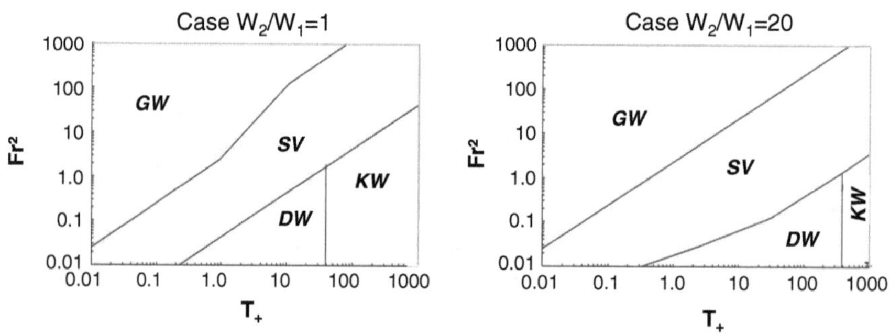

Fig. 7.5 Approximation zones of the Saint-Venant equations (Saint-Venant; *DW* Diffusive Wave; *KW* Kinematic Wave; *GW* Gravity Wave) in a non-dimensional diagram function of the Froude number $Fr = V/(gy)^{0.5}$ and a non-dimensional period $T_+ = TVS_f/y$ (V velocity; y water depth; g acceleration due to gravity; S_f slope of the energy line; T wave length) for two different cases: without overbank flow $W_2/W_1 = 1$ and with an example of overbank flow $W_2/W_1 = 20$ (from Moussa and Bocquillon 2000)

where *C* and *D* are non-linear functions of the discharge *Q* (and consequently the flow depth *y*) generally known as the celerity [LT^{-1}] and diffusivity [L^2T^{-1}], respectively. The *C*(*y*) and *D*(*y*) relations for rectangular sections are

$$Q = V\,B\,y \qquad C = \frac{dQ}{dA} = \left[\frac{5}{3} - \frac{4}{3}\frac{y}{B + 2.y}\right]V \qquad D = \frac{Q}{2\,S\,B} \tag{7.6}$$

where *A* is the cross-sectional area [L^2] of the flow. Each reach of the channel network, therefore, has three characteristics: *S*, *B* and n_c. For each value of *y*, the *Q*, *C* and *D* terms can be calculated, then each reach is also characterized by two relations: *C*(*Q*) and *D*(*Q*).

In the particular case when the pressure-gradient term could also be neglected, the Diffusive Wave becomes the Kinematic Wave model ($D = 0$). The Diffusive Wave can thus be considered a higher order approximation than the Kinematic Wave approximation (Katopodes 1982; Daluz Vieira 1983; Ferrick 1985; Ponce 1990).

7.3.3 Exchange Flow Between the 1-D Main Channel, the Floodplain and Cells

For modeling the flow exchange between the main channel and the floodplain (R_1 and R_2 in Fig. 7.4b left), weir type equations are generally used and two cases are distinguished depending on whether the weir is submerged or not: (i) in the case of non-submerged weir, the equation of broad-crested weir for clear overflow weir is used to calculate the flow as a function of the water level in the main channel y_1 and the bankful flow depth *h* (Fig. 7.3b, f); (ii) in the case of submerged weir by the tail water, the equation of submerged crested weir is used to calculate the flow as a function of the water level in the main channel y_1 and the water level in the floodplain y_2 (Fig. 7.3c, e). Note that the intermediate case where $y_1 = y_2$ (Fig. 7.3d) corresponds to the time of occurrence of peak flow when the equilibrium is reached between water levels in the main channel and in the floodplain.

In the case of Fig. 7.3b, d, *R*(*t*) can be calculated using the equation of broad-crested weir for clear overflow weir

$$R(t) = KL(y_1 - h)\sqrt{2g(y_1 - h)} \tag{7.7}$$

where *K* [-] is a constant which depends on the shape of the lateral weir (generally $0.3 < K < 0.6$). Note that *R*(*t*) is negative for overbank flow (flow from the main channel towards the floodplain) during the hydrograph rise (Fig. 7.3b), and positive for flows from the floodplain to the main channel during the hydrograph recession (Fig. 7.3f). If the weir is submerged by the tail water (e.g. Fig. 7.3c, e), the constant K used in Eq. (7.7) is multiplied by a corrective term function of y_1, y_2 and

h (Chow 1959; Carlier 1980). The same reasoning applies for flow exchanges between adjacent cells (F_1, F_2, F_3 and F_4 in Fig. 7.4b left), and when moving from 1-D to 2-D models.

Hence, each floodplain cell is treated as a storage volume. The change in cell volume over time is therefore described by the in- and out- fluxes during a time step, including the flows exchanged between the main channel and the floodplain, and the flows exchanged between adjacent cells in the floodplain (Fig. 7.3). Digital Elevation Models (DEMs) are used to construct the "Water level—Volume—Area of the inundated zone" relationships which characterize a given cell. The model enables to simulate water depths in the main channel and in the floodplain, and hence the volume and the area of the inundation extent.

7.3.4 The 2-D Saint-Venant Equations

A depth-averaging procedure of the Navier-Stokes equations leads to the 2-D Saint-Venant equations (Benqué et al. 1982) which can also be derived from mass and momentum conservation in the plane of motion (e.g. Saint-Venant 1871; Cunge et al. 1980) under the following assumptions (Fig. 7.4b right):

- Vertical velocities are neglected.
- The pressure field is assumed hydrostatic.
- The bottom slope is assumed small.
- A uniform horizontal velocity field is assumed across the water layer.
- Turbulence effects are not explicitly accounted for.
- Friction formulae are usually taken from uniform flow conditions.

We obtain

$$\frac{\partial \mathrm{U}}{\partial \mathrm{t}}+\frac{\partial}{\partial \mathrm{x}}(\mathrm{F}+\mathrm{F_d})+\frac{\partial}{\partial \mathrm{y}}(\mathrm{G}+\mathrm{G_d})=\mathrm{H}+\mathrm{I} \tag{7.8}$$

where U is the variables vector, F and F_d and G and G_d are the convective and diffusive fluxes vectors in the x- and y-direction, respectively, H is the friction and slope source term vector and I the infiltration source vector with

$$U=\begin{bmatrix} y \\ yu \\ yv \end{bmatrix}, F=\begin{bmatrix} yu \\ yu^2+gy^2/2 \\ yuv \end{bmatrix}, G=\begin{bmatrix} yv \\ yuv \\ yv^2+gy^2/2 \end{bmatrix}, F_d=\begin{bmatrix} 0 \\ -\varepsilon y\frac{\partial u}{\partial x} \\ -\varepsilon y\frac{\partial v}{\partial x} \end{bmatrix},$$
$$G_d=\begin{bmatrix} 0 \\ -\varepsilon y\frac{\partial u}{\partial y} \\ -\varepsilon y\frac{\partial v}{\partial y} \end{bmatrix}, H=\begin{bmatrix} 0 \\ gy(S_{ox}-S_{fx}) \\ gy(S_{oy}-S_{fy}) \end{bmatrix} \text{ and } I=\begin{bmatrix} -i_r \\ -ui_r/2 \\ -vi_r/2 \end{bmatrix} \tag{7.9}$$

where y is water depth [L], u and v the velocity components [LT^{-1}], g is the acceleration due to gravity [LT^{-2}], ε a kinematic viscosity and S_{ox} and S_{oy} are the bed slopes [-] in the x- and y-direction, respectively, which are assumed small. The term i_r represents the infiltration rate into the ground that can be estimated using empirical formulae (Mahmood and Yevjevich 1975). As for 1-D equations, the 2-D Saint-Venant equations can be resolved using finite-difference or finite element numerical schemes which are time consuming and can cause numerical instabilities. Recent research focuses on the derivation of simplified 2-D equations designed to be solved explicitly at very low computational cost as suggested by Bates et al. (2010) or to the parallelization of model runs (Neal et al. 2010). Therefore, future theoretical researches on flood inundation modeling needs to address the development of approaches to guide the modeler in the choice of the adequate simplest parsimonious model structure (e.g. 1-D, 2-D, cells) and equations (e.g. NS, RANS, SV or ASV such as DW or KW), the adequate friction law, and the corresponding model analysis or control as synthesized in Table 7.1.

7.4 Practical Considerations

This section presents practical considerations for applications on study cases, proposing guidelines for the choice of models in relation to their numerical solution schemes, model parameters and boundary conditions, calibration, validation, uncertainty analysis.

7.4.1 Numerical Solution Schemes

The full Saint-Venant equations have only a few exact solutions, in particular cases (see a synthesis in Delestre 2010). Finite difference and finite element schemes have thus been developed in order to provide widely-applicable numerical solutions the full Saint-Venant, the Diffusive and the Kinematic Wave equations (Liggett and Woolhiser 1967; Cunge et al. 1980; Galland et al. 1991; Bates et al. 1992, 1995; Horritt 2000). Compared to 1-D models, the 2-D models enable better representation of the flow paths but the modeler may encounter serious issues in the construction of finite-difference (Crank and Nicholson 1947; Richtmeyer and Morton 1967; Remson et al. 1971; Smith 1978; Moussa and Bocquillon 1996a), or finite-element systems (Cooly and Moin Cooley and Moin 1976; Szymkiewicz 1991; Blandford and Ormsbee 1993; Marks and Bates 2000; Horritt 2002; Horritt et al. 2006) as well as in the choice of methods for solving them (Cunge et al. 1980; Horritt and Bates, 2001a, b, 2002). However, both the 1-D and 2-D approaches need an adequate discretization of time and space in order to obtain stable and reliable numerical resolutions.

The time derivative term may be discretized in several ways, either using explicit or implicit schemes (Smith 1978). Explicit solutions are often favored as they allow

Table 7.1 Advantages (+) and drawbacks (−) of the four complexity levels (NS, RANS, SV, ASV) regarding free-surface flows, friction laws, bed erosion and model analysis or control

		Navier-Stokes (NS)	Reynolds-averaged Navier-Stokes (RANS)	Saint-Venant (SV)	Approximations to Saint-Venant (ASV)
Free-surface flows	+	Allow the finest scales and strong 3D effects without periodicity in spatial and temporal patterns. Provide "exact" numerical solutions: improve or question our conceptual understanding. Excellent for laminar flows	Intended for time-averaged turbulence modelling. Relevant for velocity distributions and profiles in very shallow flows. Allows grain-scale as well as space-averaged descriptions	The tilting point in complexity: cover a wide variety of contexts and scales, also admit contextual ASV simplifications or upgrades towards NS and RANS performances	Stable analytical solutions in fluvial or channel flows over simple slowly-varied topographies. Allow wide-scale flow routing in networks
	−	Highly resource-demanding: limited to small spatio-temporal domains or to a few control simulations, unless used with degraded resolution. Require specific programming skills	Hypothesize relevancy of the information contained in the time-average of turbulent velocity fluctuations: often best with sufficiently stationary flows	A priori not relevant for stratified flows, flows significantly diverted by obstacles or flows with strong (variations in) vertical velocity components. Strictly speaking, suitable within the gradually-varied flow hypothesis	Require simple flow structures associated with small relative roughness: often used for simplicity outside their strict domain of validity
Friction laws	+	Allow adaptable definitions of friction in wall-laws for high inundation ratios. By contrast, permit the specification of any boundary topography for macro-roughness elements	Allow calculation of viscous and/or turbulent stresses, support the useful quadrant analysis of Reynolds stresses and give rise to equations identifying form-induced stresses	Bottom friction appears in the momentum equation (not as a boundary condition) allowing simpler empirical adjustments. Implicit friction-velocity relations (Darcy-Weisbach) give more realistic simulations	The simplicity of the often-used Manning expression where *n* values are associated to bottom topography via tables
	−	Problematic transition from negligible to non-negligible roughness: how to use wall laws or explicit topographies? What is the threshold in relative roughness and/or Reynolds number?		The conceptual and numerical connection between friction in the depth-averaged momentum equation and the real bottom boundary condition is not straightforward	The lack of analytical or empirical depth-dependent formulations of friction makes it sometimes simplistic or inaccurate, e.g. for overland flows

(continued)

Table 7.1 (continued)

		Navier-Stokes (NS)	Reynolds-averaged Navier-Stokes (RANS)	Saint-Venant (SV)	Approximations to Saint-Venant (ASV)
Analysis and/or control	+	Statistical methods indicate the range of the solution. Then adjoint sensitivity analysis and variational data assimilation allow its optimal control even with time-dependent spatially-distributed boundary conditions (friction and erodibility parameters)		Compatible either with traditional statistical methods but also prone to "simple" deterministic variational methods (forward sensitivity analysis, sensitivity equations)	Scarcity or prior reduction of the parameterisation before analysis is favourable to deterministic methods
	–	Simple deterministic methods do not fit complexity in the parameterisation. Global statistics may fail to captivate the complex spatio-temporal patterns of the solution		Due to the structure of the SV and ASV equations, friction terms may be easily tested but not their connection to the real bottom boundary conditions, except in "upgraded" SV formulations	

the straightforward integration of flow models within a dynamic Geographic Information Systems (GIS) environment (Burrough 1998), but the time step must be chosen small enough to satisfy the Courant–Friedrichs–Lewy conditions (Courant et al. 1928) and prevent instabilities in the numerical solution. This often leads to a time step which is very small compared to that of the physical phenomena (Bradbrook et al. 2004). In some cases though, for example these of complex topographies, there is no way to guarantee stable conditions. By contrast, the often more time-consuming implicit schemes allow much larger time steps, more compatible with the slow evolution of flood events, which also ensure the unconditional stability of the numerical solution.

For the discretization of space, in the early storage cell models (Cunge et al. 1980), the studied area was subdivided into a series of cells (1–100 km^2) that correspond to distinct flood compartments (Fig. 7.4b left). The recent developments of GIS software and the availability of DEMs now permit the fusion of this storage cell concept with the raster data format typically used in GIS (Bates and De Roo 2000; Bradbrook et al. 2004). The inevitable drawback of such enabled ad hoc spatial discretization is the increase in computational time as highly refined, irregular and adaptive grids often exceed 10^6 of cells (e.g. Hunter et al. 2005a, b).

7.4.2 Model Parameters and Boundary Conditions

A flood inundation model requires the specification of three key data items: topographic data to construct the model grid or cells, an estimation of the roughness coefficient for each reach and cell, and the definition of boundary conditions for the flow.

Until the late 1980s, topographic data were mostly acquired by ground survey and photogrammetry. These methods are expensive and time consuming. Later on, topographic data available on national survey maps were largely used but these data tend to be of low accuracy with poor spatial resolution in the floodplain (Bates et al. 1992). Recently, novel remote sensing techniques such as airborne laser altimetry (e.g. Marks and Bates 2000) have largely overcome the previous limitations. High-quality elevation data derived from airborne systems (e.g. LiDAR) can be collected at $\sim$1-m horizontal resolutions with a vertical accuracy of 0.15 m RMS error (Bates 2004).

Hydraulic models also require the specification of roughness parameters in the channel and on each cell (e.g. the "tables of roughness" in Carlier 1980, p. 540–542; see a synthesis in Vidal 2005). It has proven very uncertain that such roughness coefficients are sufficient to enable accurate predictions, especially due to the discrepancies between the measurements scale and the model grid scale. Calibration is therefore usually undertaken in order to identify the optimized empirical values for roughness coefficients (Beven 2000).

Flow boundary conditions may be assigned using the quantitative hydrometric data arising from stage and discharge measurements. However, despite the wide coverage ensured by monitoring networks and detailed surveys allowed by gauging stations in most countries, it should be noted that the quantification of flood flows is subject to important errors. These typically affect the estimation of stream depth and flow velocity, even more when the highest discharges trigger bed instabilities and instant channel adjustments (e.g. flash floods and formative discharges; Wainwright 1996). Moreover, discharge records are often derived from rating curves with substantial expected uncertainties for large overbank flows. Another source of uncertainty arises from the difficulties in estimating the exchanges with the surrounding catchments which are simply not considered in most models.

7.4.3 Model Calibration, Validation and Uncertainty Analysis

Regardless of their internal complexity, flood inundation models require verification and independent calibration–validation to establish both the quality of and the confidence in the hydraulic information generated (Klemeš 1986). The assessment of model performance requires a structured sequence of numerical experiments that are simple enough to isolate the effect of the conceptual algorithm being studied. Analytical solutions of the governing equations have been shown to provide useful tests for a variety of hydrodynamic schemes (e.g. Hunter et al. 2005b). These approaches typically involve the use of a reduced form of the governing equations, hence the formulation of a hopefully equivalent though simplified problem, in order to work with analytical expressions.

Traditionally, flood inundation models have been calibrated and validated using either the water levels or discharge values available from networks of hydrometric gauging stations. Even when benefiting from narrow temporal sampling intervals, any too sparse spatial data do not allow testing the performances of distributed models. By contrast, a move towards finer spatial resolutions in the parameterization of topography gives increased confidence in the model's distributed predictions. Numerous authors have therefore tested distributed hydraulic models using maps of inundation, whose extent was derived from such diverse sources as post-event trash line surveys (Romanowicz and Beven 2003), aerial photos (Yu and Lane 2006), airborne and satellite Synthetic Aperture Radar (SAR) data (Horritt 2000; Bates et al. 2006; Stephens et al. 2012; Yamazaki et al. 2012; Léauthaud et al. 2013) and post-event LiDAR survey of flood deposits (Lane et al. 2003). Validation has also been attempted with discharge data internal to a reach (Romanowicz et al. 2004; Hunter et al. 2005a, b) or point measurements of maximum water elevation (Hervouet 2000).

7.4.4 Model Choice

In the literature various flood inundation models were developed on the basis of the resolution of the full 1-D or 2-D Saint-Venant equations or their approximations:

e.g. TELEMAC-2D and 3D (Galland et al. 1991; Hervouet and Van Haren 1996), LISFLOOD (Bates and De Roo 2000), HEC-RAS (2002), MIKE11 (2003) or MASCARET (2012).

Figure 7.6 shows a broader synthesis of 125 articles indicating what category of equations was chosen by authors for various practical applications of flood inundation and free-surface flow equations, across multiple spatiotemporal scales (*L, T*) defined as the domain length (5 cm < *L* < 500 km) and the time duration of the process (0.1 s < *T* < 1 yr). The plots indicate how increasing spatiotemporal scales tend to be associated with decreasing complexity in the choice of flow models, sorted here into four levels of refinement: Navier-Stokes (NS), Reynolds-Averaged Navier-Stokes (RANS), Saint-Venant (SV) and Approximations to Saint-Venant (ASV: e.g. Diffusive Wave, Kinematic Wave, Uniform flow formulae). A transverse analysis involves forming *L/T* ratios, searching for clues to model selection according to these "system evolution velocities" (or similar concepts used in geomorphology; Paola et al. 1992; Allen 2008; Paola et al. 2009) or governed by flow typologies that would exhibit specific *L/T* ratios. In most of the literature sources, the choice of a given model has not been thoroughly justified nor was the result of a

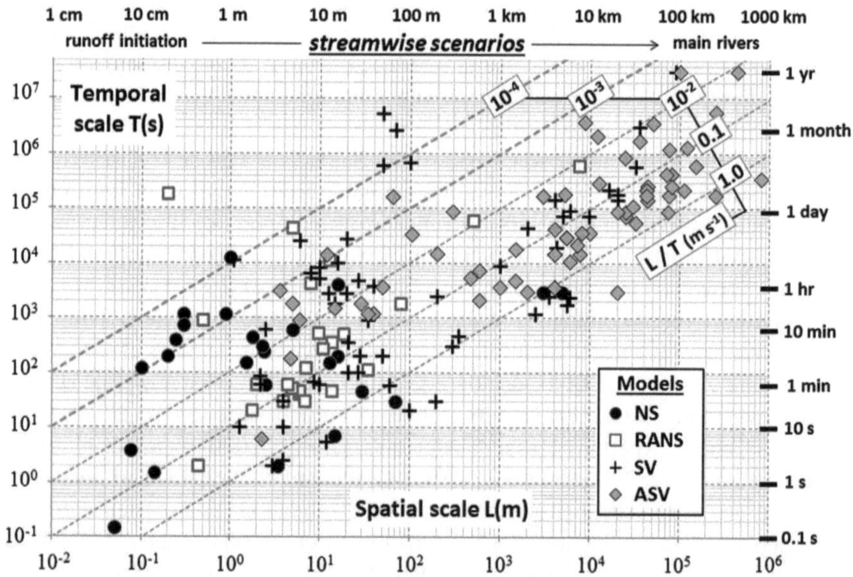

Fig. 7.6 A broader synthesis of 125 articles (each point) indicating what category of equations was chosen by authors for various practical applications of flood inundation and free-surface flow equations, across multiple spatiotemporal scales of the flow domain (*L* domain length; *T* temporal duration). It shows how increasing spatiotemporal scales (L, T) tend to be associated with decreasing complexity in the choice of flow models, sorted here into four levels of refinement: Navier-Stokes (NS), Reynolds-Averaged Navier-Stokes (RANS), Saint-Venant (SV) and Approximations to Saint-Venant (ASV such as the Diffusive Wave or the Kinematic Wave) (from Cheviron and Moussa, article in preparation)

formal, explicit contextual analysis. It nevertheless appeared to depend on the main flow processes and typologies (overland flows, high-gradient flows, flows over significant bedforms or fluvial flows; Montgomery and Buffington 1997; Church 2002), the spatiotemporal (L, T, y) scales where y is the water depth, the non-dimensional flow characteristics (e.g. Reynolds number, Froude number, river bed slope, inundation ratio, non-dimensional period) but also on the habits, experience and personal choice of the modeler.

7.5 Research Challenges

For practical applications, the starting point is the scope of the application and the endpoint is the evaluation of the objective function for deciding upon the success or the failure of the modeling strategy. Hence, a hydrological-hydraulic application can be represented by a symbiotic (interdependence) link between the model type, site data and objective function triplet. A question arises on how to guide the modeler in the choice of an adequate triplet. According to the principle of parsimony, modelers should seek the simplest modeling strategy capable of (i) a realistic representation of the physical processes, (ii) matching the performances of more complex models and (iii) providing the right answers for the right reasons. Figure 7.7 depicts ways to address the (i, ii, iii) problem by sorting models, data and objective functions along an “axis of complexity”. For each application, the modeler has to find the optimal (model, data, objective function) triplet, seen as a global optimization process.

7.5.1 Model Type

The elected hydrological-hydraulic model should be sought between existing models or be developed on purpose.

Traditionally, hydrological models have been classified as lumped, semi-distributed or distributed (Figs. 7.4a and 7.7a). In terms of spatial discretization and resolution, these hydrological model categories can be placed on an axis of complexity, from lumped to spatially distributed, covering various possible combinations between empirical to physically based equations and fine to coarse spatiotemporal steps.

The spatial representations in hydraulic models can be classified from simple 1-D to complex 2-D and 3-D approaches, also taking place on an axis of complexity (Fig. 7.7a). Although 1-D codes are computationally very efficient, they suffer from a number of drawbacks when applied to floodplain flows. These include the inability to simulate lateral diffusion of the flood wave, the discretization of topography as cross-sections rather than as surfaces (Samuels 1990), and the difficulty to represent connectivity (Trigg et al. 2013). All of these fundamental constraints can be

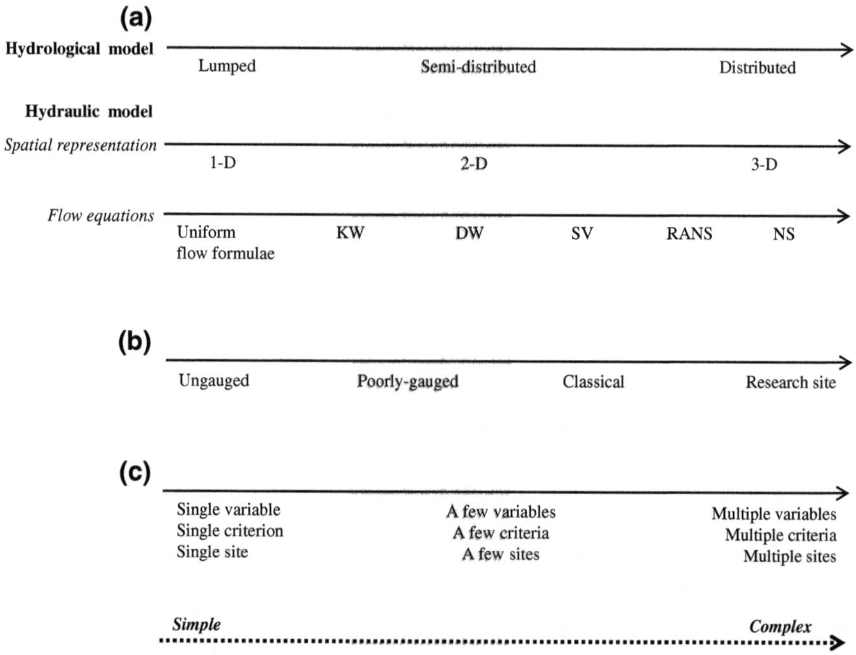

Fig. 7.7 Conceptual representation by sorting the model type, the site data and the objective function along an "axis of complexity"

overcome with 2-D codes (Fig. 7.4b) and numerous classes of 2-D schemes have been developed in response. These can be broadly distinguished as either full solutions of the Saint-Venant equations (Katapodes and Strelkoff 1979; Zhang and Cundy 1989) or approximations. However, solving the full Saint-Venant equations over complex topographies using finite difference (Smith 1978), finite element (Blandford and Ormsbee 1993) or finite volume (Hirsch 1988) algorithms may cause instability and convergence issues because of the highly nonlinear, hyperbolic nature of the governing equations.

Finally, flood inundation and more generally free-surface flow equations can be placed on an axis of complexity: Uniform flow formulae, Kinematic Wave, Diffusive Wave, Saint-Venant, Reynolds-Averaged Navier-Stokes and Navier-Stokes (Figs. 7.5, 7.6 and 7.7a; Table 7.1). The Saint-Venant equations are the reference a tilting point in complexity, as the development of numerical resolutions and simplifications remains a major challenge for hydraulic and mathematics. In order to simplify the Saint-Venant model, Uniform flow formulae (i.e. Manning) and weir type equations (e.g. Cunge et al. 1980; Estrela 1994; Romanowicz et al. 1996) or the Kinematic Wave equation (Woolhiser and Liggett 1967; Singh 1994; Bates and De Roo 2000) were largely used for practical applications. The Diffusive Wave model was also used to model flood routing in channels when overbank flow occurs (e.g. Todini 1996; Moussa et al. 2007; Moussa and Bocquillon 2009; Trigg et al. 2009).

While the full Saint-Venant equations require very detailed representations of the topography, simplified models such as the Diffusive Wave, the Kinematic Wave and the Uniform flow formulae do not enable the representation of all hydraulic processes at play during flood events.

7.5.2 Site Data

The elected site data should either be chosen among existing data or collected on purpose. These site data can also be placed on an axis of complexity depending on their consistency and accuracy (Fig. 7.7b) from (i) the ungauged sites where no data are available, (ii) the poorly-gauged sites where scarce data or data with high uncertainty are available, (iii) the long term classical hydrological-hydraulic data such as daily or hourly rainfall-runoff-meteorological data to (iv) the experimental research basins and/or observatories with long-term fine-time-step records.

Site data may be sorted into spatial and temporal data. Spatial data include maps (e.g. DEM, LiDAR, roughness, soil, landuse, radar, remote sensing), pixel size, and the GIS tools used for their analysis. Temporal data include temporal series corresponding to various quantities (e.g. rainfall, water depth, discharge, runoff, water-table) characterized by the sampling step, duration and total number of elements in the records.

7.5.3 Objective Function

Throughout the last decades, an important change of the scope of hydrological-hydraulic applications has taken place, with subsequent changes in the objective functions resorted to. As previously mentioned, the development of hydrological and hydraulic sciences has been directly linked to the progresses in understanding processes, in theoretical model development (e.g. computational facilities: numerical techniques, data assimilation, thorough model exploration, inverse calculus), and in data acquisition (new devices, remote sensing, LiDAR). However, there remains an important need for research on classical hydrological-hydraulic modeling for engineering applications in predicting floods, designing water supply infrastructures and for water resources management, from the headwater catchment to the regional scale. More recently, hydrological-hydraulic modeling has become an indispensable tool for many interdisciplinary projects, such as predicting pollution and/or erosion incidents, the impact of anthropogenic and climate change on environmental variables such as water, soil, biology, ecology, or socio-economy. The direct consequence is a significant increase of the complexity of the objective function, from simple mono-site (e.g. one-point), mono-variable (e.g. the water depth) and mono-criterion (e.g. the error on peakflow) to complex multi-site (e.g. large number of points within a catchment), multi-variable (e.g. water depth, hydrograph, water table, concentrations, ecological indicators, economic impact) and multi-criteria (e.g. errors on peakflow, volume, RMSE) objective functions (Fig. 7.7c; see a synthesis in Madsen 2000; Dawson et al. 2007).

7.5.4 Seeking the Optimum Balance Between Modeling, Data and Objective Functions

There is often a mismatch between model types, site data and objective functions. First, models were developed independently from the specificities of the study site and available data, prior to the definition of any objective function. In using hydrological-hydraulic models, the context of their original purpose and development is often lost, so that they may be applied to situations beyond their validity or capabilities. Second, site data are often collected independently of the objectives of the study. Third, the objective function must be specific to the application but also meet standard practices in evaluating model performance, in order to compare modeling results between sites and to communicate the results to other scientists or stakeholders.

The question of choosing the right (model type, site data and objective function) triplet for a given project is crucial. The choice of the optimal data-model couple to reach a predefined objective is not straightforward. We need a framework to seek the optimum balance between the model, data and the objective function as a solution for a hydrological-hydraulic problem, on the basis of the principle of parsimony as "*Simplicity is the ultimate sophistication*" (Leonardo da Vinci).

References

Allen PA (2008) Time scales of tectonic landscapes and their sediment routing systems. Geol Soc London Spec Publ 296:7–28

Aronica G, Hankin BG, Beven KJ (1998) Uncertainty and equifinality in calibrating distributed roughness coefficients in a flood propagation model with limited data. Adv Water Resour 22(4):349–365

Aronica G, Bates PD, Horritt MS (2002) Assessing the uncertainty in distributed model predictions using observed binary pattern information within GLUE. Hydrol Process 16(10): 2001–2016

Babaeyan-Koopaei K, Ervine DA, Carling PA, Cao Z (2002) Velocity and turbulence measurements for two overbank flow events in river Severn. J Hydraulic Eng:891–899

Bates PD (2004) Remote sensing and flood inundation modelling. Hydrol Process 18:2593–2597

Bates PD, De Roo APJ (2000) A simple raster-based model for flood inundation simulation. J Hydrol 236:54–77

Bates PD, Anderson MG, Baird L, Walling DE, Simm D (1992) Modelling floodplain flow with a two-dimensional finite element scheme. Earth Surf Proc Land 17:575–588

Bates PD, Anderson MG, Hervouet J-M (1995) Initial comparison of two two-dimensional finite element codes for river flood simulation. Proc Institution Civil Eng Water Maritime Energy 112:238–248

Bates PD, Wilson MD, Horritt MS, Mason DC, Holden N, Currie A (2006) Reach scale floodplain inundation dynamics observed using airborne synthetic aperture radar imagery: data analysis and modelling. J Hydrol 328(1–2):306–318
Bates PD, Horritt MS, Fextrell TJ (2010) A simple inertial formulation of the shallow water equations for efficient two-dimensional flood inundation model. J Hydrol 387:33–45
Benqué JP, Hauguel A, Viollet PL (1982) Engineering applications of computational hydraulics, vol 2. Pitman Publications, England
Beven KJ (2000) Uniqueness of place and process representations in hydrological modeling. Hydrol Earth Syst Sci 4(2):203–213
Blandford GE, Ormsbee LE (1993) A diffusion wave finite element model for channel networks. J Hydrol 142:99–120
Booker DJ, Sear DA, Payne AJ (2001) Modelling three-dimensional flow structures and patterns of boundary shear stress in a natural pool-riffle sequence. Earth Surf Proc Land 26(5):553–576
Bradbrook KF, Lane SN, Waller SG, Bates PD (2004) Two dimensional diffusion wave modelling of flood inundation using a simplified channel representation. Int J River Basin Manage 2(3): 211–223
Burrough PA (1998) Dynamic modelling and geocomputation. In: Longley PA, Brooks SM, McDonnell RM, Macmillan B (eds) Geocomputation: a Primer. Wiley, Chichester, pp 165–191
Carlier M (1980) Hydraulique générale et appliquée. Eyrolles, Paris 565 pp
Chow VT (1959) Open-channel hydraulics. McGraw-Hill, New York, p 680
Church M (2002) Geomorphic thresholds in riverine landscapes. Freshwater Biol, pp 541–557
Cooley RL, Moin SA (1976) Finite element solution of Saint-Venant equations. J Hydraulic Division Am Society Civil Eng 102(HY6):759–775
Courant R, Friedrichs K, Lewy H (1928) Über die partiellen Differenzengleichungen der mathematischen Physik. Mathematische Annalen (in German) 100(1):32–74
Crank J, Nicholson P (1947) A practical method for numerical integration of solutions of partial differential equations of heat-conduction type. Proc. Cambridge Philos. Soc, 43:50
Cunge J (1975) Two-dimensional floodplain flow. In: Mahmood K, Yevjevich V (eds) Unsteady flow in Open Channels. Water Resources Publications, Fort Collins, Colorado, pp 705–762
Cunge J, Holly FM, Verwey A (1980) Practical aspects of computational river hydraulics. Pitman Advanced Publishing Program, London, p 420
Daluz Vieira JH (1983) Conditions governing the use of approximations for the Saint-Venant equations for shallow water flow. J Hydrol 60:43–58
Dawson CW, Abrahart RJ, See LM (2007) Hydrotest: a web-based toolbox of evaluation metrics for the standardised assessment of hydrological forecasts. Environ Model Softw 22:1034–1052
Delestre O (2010) Simulation du ruissellement d'eau de pluie sur des surfaces agricoles. Ph.D. thesis, University of Orléans, France
Estrela T (1994 Use of a GIS in the modelling of flows on floodplains. In: White WR, Watts J (eds) Proceedings of the 2nd international conference on river flood hydraulics, Wiley, pp 177–189
Ferrick MG (1985) Analysis of river wave types. Water Resour Res 21:209–212
Galland JC, Goutal N, Hervouet J-M (1991) TELEMAC: a new numerical model for solving shallow water equation. Adv Water Resour 14(3):143–148
Gaume E, Bain V, Bernardara P, Newinger O, Barbuc M, Bateman A, Blaškovicová L, Blöschl G, Borga M, Dumitrescu A, Daliakopoulos I, Garcia J, Irimescu A, Kohnova S, Koutroulis A, Marchi L, Matreata S, Medina V, Preciso E, Sempere-Torres D, Stancalie G, Szolgay J, Tsanis I, Velascom D, Viglione A (2009) A compilation of data on European flash floods. J Hydrol 367:70–78
Hayami S (1951) On the propagation of flood waves. Disaster Prev Res Inst Bull 1:1–16
HEC-RAS (2002) HEC-RAS, river analysis system, user' manual. US Army Corps of Engineers, Hydrological Engineering Center, Davis, CA, report N° CPD-68
Henderson FM (1966) Open channel hydraulics. MacMillan and Co., New York
Hervouet J-M (2000) A high resolution 2-D dam-break model using parallelization. Hydrol Process 14(13):2211–2230

Hervouet J-M, Van Haren L (1996) Recent advances in numerical methods for fluid flows. In: Anderson MG, Walling DE, Bates PD (eds) Floodplain Processes. Wiley, Chichester, pp 183–214
Hirsch C (1988) Numerical computation of internal and external flows. Fundamentals of numerical discretization, vol 1. Wiley, Chichester. 515 pp
Horritt MS (2000) Calibration of a 2-dimensional finite element flood flow model using satellite radar imagery. Water Resour Res 36(11):3279–3291
Horritt MS (2002) Evaluating wetting and drying algorithms for finite element models of shallow water flow. Int J Numerical Methods Fluids 55(7):835–851
Horritt MS, Bates PD (2001a) Effects of spatial resolution on a raster based model of flood flow. J Hydrol 253:239–249
Horritt MS, Bates PD (2001b) Predicting floodplain inundation: raster-based modelling versus the finite-element approach. Hydrol Process 25(5):825–842
Horritt MS, Bates PD (2002) Evaluation of 1D and 2D numerical models for predicting river flood inundation. J Hydrol 268:87–99
Horritt MS, Bates PD, Mattinson MJ (2006) Effects of mesh resolution and topographic representation in 2D finite volume models of shallow water fluvial flow. J Hydrol 329:306–314
Hunter NM, Bates PD, Horritt MS, De Roo APJ, Werner MGF (2005a) Utility of different data types for calibrating flood inundation models within a GLUE framework. Hydrol Earth Syst Sci 9(4):412–430
Hunter NM, Horritt MS, Bates PD, Wilson MD, Werner MGF (2005b) An adaptive time step solution for raster-based storage cell modelling of floodplain inundation. Adv Water Resour 28(9):975–991
Hunter NM, Bates PD, Horritt MS, Wilson MD (2007) Simple spatially-distributed models for predicting flood inundation: a review. Geomorphology 90:208–225
Katopodes ND (1982) On zero-inertia and kinematic waves. J Hydraulic Eng Am Society Civil Eng 108(HY11):1380–1385
Katapodes ND, Strelkoff T (1979) Two-dimensional shallow waterwave model. J Eng Mech Div ASCE 105(EM2):317–334
Klemeš V (1986) Operational testing of hydrologic simulationmodels. Hydrol Sci J 31(1):13–24
Knight DW, Shiono K (1996) River channel and floodplain hydraulics. In: Anderson MG, Walling DE, Bates PD (eds) Floodplain processes. Wiley, Chichester, pp 139–182
Kuchment LS (1972) Matematicheskoye modelirovanye rechnogo stoka (Mathematical Models of River Flow). Gidrometeoizdat, Leningrad, 190 pp. (in Russian)
Lane SN, James TD, Pritchard H, Saunders M (2003) Photogrammetric and laser altimetric reconstruction of water levels for extreme flood event analysis. Photogramm Rec 18(104): 293–307
Léauthaud C, Belaud G, Duvail S, Moussa R, Grünberger O, Albergel J (2013) Characterizing floods in the poorly gauged wetlands of the Tana River Delta, Kenya, using a water balance model and satellite data). Hydrol Earth Syst Sci 17:3059–3075
Liggett JA, Woolhiser DA (1967) Difference solutions of the shallow-water equations. J Eng Mech Div Am Soc Civil Eng 93(EM2):39–71
Lighthill MJ, Whitham GB (1955) On kinematic waves. I. Flood movement in long rivers. II. A theory of traffic flow on long crowded roads. Proc R Soc London A229:281–345
Madsen H (2000) Automatic calibration of a conceptual rainfall-runoff model using multiple objectives. J Hydrol 235:276–288
Mahmood K, Yevjevich V (1975) Unsteady flow in open channels vols I and II, Water Resources Publications
Marks K, Bates PD (2000) Integration of high-resolution topographic data with floodplain flow models. Hydrol Process 14:2109–2122
MASCARET (2012) MASCARET: a 1-D open-source software for flow hydrodynamic and water quality in open channel networks. In: Goutal N, Lacombe J-M, Zaoui F, El-Kadi-Abderrezzak K (ed) River flow 2012—Murillo, pp 1169–1174
MIKE 11 (2003) A modelling system for river and channels. Short introduction tutorial, DHI Water and Environment, 88 pp

Montgomery DR, Buffington JM (1997) Channel-reach morphology in mountain drainage basins. Geol Soc Am Bull 109(5):596–611

Morvan H, Pender G, Wright NG, Ervine DA (2002) Three dimensional hydrodynamics of meandering compound channels. J Hydraul Eng—ASCE 128(7):674–682

Moussa R (1996) Analytical Hayami solution for the diffusive wave flood routing problem with lateral inflow. Hydrol Process 10(9):1209–1227

Moussa R, Bocquillon C (1996a) Algorithms for solving the diffusive wave flood routing equation. Hydrol Process 10(1):105–124

Moussa R, Bocquillon C (1996b) Criteria for the choice of flood-routing methods in natural channels. J Hydrol 186(1–4):1–30

Moussa R, Bocquillon C (2000) Approximation zones of the Saint-Venant equations for flood routing with overbank flow. Hydrol Earth Syst Sci 4(2):251–261

Moussa R, Bocquillon C (2009) On the use of the diffusive wave for modelling extreme flood events with overbank flow in the floodplain. J Hydrol 374:116–135

Moussa R, Chahinian N, Bocquillon C (2007) Distributed hydrological modelling of a Mediterranean mountainous catchment—model construction and multi-site validation. J Hydrol 337:35–51

Navier LMHN (1822) Mémoire sur les lois du movement des fluids. Mém. Acad. Sci. Inst. France 6:389–440

Neal JC, Fewtrell TJ, Bates PD, Wright NG (2010) Comparison of three parallelization methods for 2D flood inundation models. Environ Model Softw 25:398–411

Nicholas AP, McLelland SJ (2004) Computational fluid dynamics modelling of three-dimensional processes on natural river floodplains. J Hydraul Res 42(2):131–143

Paola C, Heller PL, Angevine CL (1992) The large-scale dynamics of grain-size variation in alluvial basins. I: Theory. Basin Res 4:73–90

Paola C, Straub K, Mohrig D, Reinhardt L (2009) The "unreasonable effectiveness" of stratigraphic and geomorphic experiments. Earth Sci Rev 97:1–43

Pappenberger F, Beven KJ, Horritt MS, Blazkova S (2005) Uncertainty in the calibration of effective roughness parameters in HEC-RAS using inundation and downstream level observations. J Hydrol 302:46–69

Ponce VM (1990) Generalized diffusive wave equation with inertial effects. Water Resour Res 26(5):1099–1101

Ponce VM, Simons DB (1977) Shallow wave propagation in open channel flow. J Hydraulic Division Am Soc Civil Eng 103(HY12):1461–1476

Remson RD, Hornberger GM, Molz FJ (1971) Numerical methods in subsurface hydrology. Willey Interscience, New York 399 pp

Reynolds O (1895) On the dynamical theory of incompressible viscous fluids and the determination of the criterion. Phil Trans R Soc A 186:123–164

Richtmeyer RD, Morton KW (1967) Difference methods for initial value problems. Interscience Publishers, New York 405 pp

Romanowicz RJ, Beven KJ (2003) Estimation of flood inundation probabilities as conditioned on event inundation maps. Water Resour Res 39(3):1061–1073

Romanowicz RJ, Dooge JCI, Kundzewicz ZW (1988) Moments and cumulants of linearized St. Venant equation. Adv Water Resour 11:92–100

Romanowicz RJ, Beven KJ, Tawn JA (1994) Evaluation of predictive uncertainty in nonlinear hydrological models using a Bayesian approach. In: Barnett V, Turkman F (eds) Statistics for the environment: water related issues, vol 2. Wiley, Chichester, pp 297–318

Romanowicz R, Beven KJ, Tawn J (1996) Bayesian calibration of flood inundation models. In: Anderson MG, Walling DE, Bates PD (eds) Floodplain processes. Wiley, Chichester, pp 333–360

Romanowicz RJ, Beven KJ, Young PC (2004) Assessing uncertainty in assessing flood risk. In: Reeve D (ed) Flood risk assessment. Institute of Mathematics and its Applications, Southend-on-Sea, pp 127–136

Saint-Venant B (1871) Théorie du mouvement non permanent des eaux, avec application aux crues des rivières et à l'introduction des marées dans leurs lits. Comptes Rendus des Séances de l'Académie des Sciences 73(147–154):237–240
Samuels PG (1990) Cross section location in one-dimensional models. In: White WR (ed) International conference on river flood hydraulics. Wiley, Chichester, pp 339–350
Singh VP (1994) Accuracy of kinematic wave and diffusion wave approximations for space-independent flows. Hydrol Process 8:45–62
Sivapalan M, Bates BC, Larsen JE (1997) A generalized, non-linear, diffusion wave equation: theoretical development and application. J Hydrol 192:1–16
Smith GD (1978) Numerical solution of partial differential equations. Clarendon Press, Oxford 304 pp
Smith MB, Seo DJ, Koren VI, Reed SM, Zhang Z, Duan Q, Moreda F, Cong S (2004) The distributed model intercomparison project (DMIP): motivation and experiment design. J Hydrol 298:4–26
Stephens EM, Bates PD, Freer JE, Mason DC (2012) The impact of uncertainty in satellite data on the assessment of flood inundation models. J Hydrol 414–415:162–173
Stewart MD, Bates PD, Anderson MG, Price DA, Burt TP (1999) Modelling floods in hydrologically complex lowland river reaches. J Hydrol 223:85–106
Stokes GG (1845) On the theories of the internal friction of fluids in motion. Trans Cambridge Philos, Soc 8
Szymkiewicz R (1991) Finite-element method for the solution of the Saint Venant equations in an open channel network. J Hydrol 122:275–287
Todini E (1996) The Arno rainfall-runoff model. J Hydrol 175:339–382
Tominaga A, Nezu I (1991) Turbulent flow structure in compound channel flows. J Hydraul Eng Am Soc Civ Eng 117:21–41
Trigg MA, Wilson MD, Bates PD, Horritt MS, Alsdorf DE, Forsberg BR, Vega MC (2009) Amazon flood wave hydraulics. J Hydrol 374(1):92–105
Trigg MA, Michaelides K, Neal JC, Bates PD (2013) Surface water connectivity dynamics of a large scale extreme flood. J Hydrol 505:138–149
Vidal J-P (2005) Assistance au calage de modèles numériques en hydraulique fluviale—Apports de l'intelligence artificielle. Ph.D., Institut National Polytechnique de Toulouse, France, 282 pp
Wainwright J (1996) Infiltration, runoff and erosion characteristics of agricultural land in extreme storm events, SE France. Catena 26:27–47
Werner M, Blazkova S, Petr J (2005) Spatially distributed observations in constraining inundation modelling uncertainties. Hydrol Process 19(16):3081–3096
Woolhiser DA, Liggett JA (1967) Unsteady one-dimensional flow over a plane: the rising hydrograph. Water Resour Res 3(3):753–771
Yamazaki D, Baugh CA, Bates PD, Kanae S, Alsdorf DE, Oki T (2012) Adjustment of a spaceborne DEM for use in floodplain hydrodynamic modeling. J Hydrol 436–437:81–91
Yu D, Lane SN (2006) Urban fluvial flood modelling using a twodimensional diffusion-wave treatment. Part 1: mesh resolution effects. Hydrol process 20(7):1541–1565
Zhang WH, Cundy TW (1989) Modeling of two-dimensional overland-flow. Water Resour Res 25(9):2019–2035

Chapter 8
Numerical Modeling of Free-Surface Flows in Practical Applications

Mario Oertel

Abstract Nowadays, numerical 1-D and 2-D flow simulations represent state-of-the-art solutions in hydraulic engineering. Hence, 1-D and 2-D as well as coupled simulations are enrooted in consulting companies and used as basic tools to answer various questions from flood protection to flow optimization. An exception are 3-D flow simulations. Due to complex boundary conditions and high calculation requirements, 3-D CFD simulations are expert tools for universities and research institutions. Here, detailed simulation models, e.g. for hydraulic structure flow optimization, are built up under massive time consumption. The present chapter deals with the basics of numerical flow simulations in practical applications. Additionally, various example flow simulations are mentioned to give an impression about the development of numerical modeling in practical application for the past decade.

Keywords Free-surface flow · Numerical simulation · CFD · Turbulence model

8.1 Introduction

Numerical flow simulations represent a modern tool to determine flow depths and flow velocities in rivers and hydraulic structures. But also flood durations of floodplains can be calculated via numerical simulations. Basically, three various hydrodynamic simulation models are distinguished: (1) one-dimensional, (2) two-dimensional, and (3) three-dimensional models. Thereby, the choice depends on the particular problem and boundary conditions. Nowadays, 1-D and 2-D flow simulation models represent the state-of-the-art in hydraulic engineering. Progressively a coupling of various models can be observed to increase calculation speed without loosing important information. Problems from flood protection up to

M. Oertel (✉)
Hydraulic Engineering Section, Civil Engineering Department, Lübeck University of Applied Sciences, Mönkhofer Weg 239, 23562 Lübeck, Germany
e-mail: mario.oertel@fh-luebeck.de

P. Rowiński and A. Radecki-Pawlik (eds.), *Rivers—Physical, Fluvial and Environmental Processes*, GeoPlanet: Earth and Planetary Sciences,
DOI 10.1007/978-3-319-17719-9_8

flow optimization can be answered with 1-D and 2-D models. Three-dimensional CFD[1] simulations are still expert systems and usually will be set up by universities and research institutions, because complex boundary conditions and high computer resources are necessary. Additionally, 3-D models are limited to relatively small investigation areas due to disproportionate increase of calculation cells with increased model size.

During the last 15 years, numerical simulations became more and more famous due to increased computer capacities and resulting model size as well as calculation speed. Where theory exists since decades, numerical models boom nowadays, while they can be used cost- and time-effective. Keywords like cluster computer, main storage, hard drive capacities and 64 bit systems can be mentioned. What is established as standard today has been unimaginable some decades before. Nowadays, a workstation PC can have the following specifications: 32 GB RAM, up to 8 cores with 3.5 GHz each, and storage capacities up to 3 TB. Considering a numerical 3-D simulation with some million cells can generate result file sizes with several ten GB, it can be clearly shown that a management of these data became feasible not before the last three to five years. Additionally, powerful processors and larger main storage capacities decrease calculation times and allow practicable simulations.

8.2 Basics of Numerical Flow Simulations

8.2.1 Dimensionality

Before creating numerical simulation models, the dimensionality of the problem must be analyzed. An over-dimensionalized model will lead to large calculation times and cost-expensive simulations without increasing model quality aspects. An under-dimensionalized model can produce major errors and uncertainties, e.g. due to not respected velocity components. Therefore, it is important to chose the correct model dimensionality for the specific hydraulic problem by analyzing boundary conditions and general flow situations within the model domain.

8.2.2 Basic Equations

Generally, numerical simulations follow two basic equations: (1) continuity equation, and (2) momentum equations. Based on the Navier-Stokes (N-S) equations, which cannot be solved analytically, e.g. the Reynolds-Averaged-Navier-Stokes-Equations (RANS) or the Shallow-Water-Equations (SWE) are developed, down to the Bernoulli equation or the flow equations. Details can be found in the literature

[1]CFD = Computational Fluid Dynamics.

(e.g. Abbott 1979; ASCE Task Commitee 1988; Ferziger and Perić 2002; Laurien and Oertel 2013; Martin 2011; Pironneau 1989; Tan 1992, and many more).

Basically, an ideal flow will be presumed—frictionless and incompressible. Therefore, Euler developed flow equations of ideal fluids, combined of continuity and momentum equations. The continuity equation states that the sum of in- and outflow corresponds to the temporal alteration of mass per volume element. According to Euler, an incompressible fluid is presumed, which means that the substantial conduction of density is equal to zero. Following, the continuity equation for ideal flows in form of cartesian coordinates is given as:

$$\frac{\partial u}{\partial x}+\frac{\partial v}{\partial y}+\frac{\partial w}{\partial z}=0 \tag{8.1}$$

where: x, y, z = coordinates, u, v, w, = velocity components.

The momentum equation, developed by Euler, describes the temporal alteration of impulse corresponding to the difference between the sum of incoming and outgoing pulse currents plus the sum of external forces. Momentum equations can be written as:

$$\begin{aligned}
&\frac{\partial u}{\partial t}+u\frac{\partial u}{\partial x}+v\frac{\partial u}{\partial y}+w\frac{\partial u}{\partial z}=-\frac{1}{\rho}\frac{\partial p}{dx}+f_x\\
&\frac{\partial v}{\partial t}+u\frac{\partial v}{\partial x}+v\frac{\partial v}{\partial y}+w\frac{\partial v}{\partial z}=-\frac{1}{\rho}\frac{\partial p}{dy}+f_y\\
&\frac{\partial w}{\partial t}+u\frac{\partial w}{\partial x}+v\frac{\partial w}{\partial y}+w\frac{\partial w}{\partial z}=-\frac{1}{\rho}\frac{\partial p}{dz}+f_z
\end{aligned} \tag{8.2}$$

where: t = time, f_i = force density (force per mass), outer forces, ρ = density, p = pressure.

For real flow simulations Navier and Poisson (1827 and 1831) developed the N-S-equations, taking into account the relation between tension and deflection rate and the Stokes' hypothesis. For incompressible flows these can be written as:

$$\begin{aligned}
&\rho\left(\frac{\partial u}{\partial t}+u\frac{\partial u}{\partial x}+v\frac{\partial u}{\partial y}+w\frac{\partial u}{\partial z}\right)=f_x-\frac{\partial p}{\partial x}+\mu\left(\frac{\partial^2 u}{\partial x^2}+\frac{\partial^2 u}{\partial y^2}+\frac{\partial^2 u}{\partial z^2}\right)\\
&\rho\left(\frac{\partial v}{\partial t}+u\frac{\partial v}{\partial x}+v\frac{\partial v}{\partial y}+w\frac{\partial v}{\partial z}\right)=f_y-\frac{\partial p}{\partial y}+\mu\left(\frac{\partial^2 v}{\partial x^2}+\frac{\partial^2 v}{\partial y^2}+\frac{\partial^2 v}{\partial z^2}\right)\\
&\rho\left(\frac{\partial w}{\partial t}+u\frac{\partial w}{\partial x}+v\frac{\partial w}{\partial y}+w\frac{\partial w}{\partial z}\right)=f_z-\frac{\partial p}{\partial z}+\mu\left(\frac{\partial^2 w}{\partial x^2}+\frac{\partial^2 w}{\partial y^2}+\frac{\partial^2 w}{\partial z^2}\right)
\end{aligned} \tag{8.3}$$

Herein, the left term represents advection and the right term forces, pressure, and viscosity. Volume forces f_i are gravitational force, tidal force, and Coriolis force.

To solve the N-S-equations, several methods are used, as Reynolds-Averaged-Navier-Stokes (RANS), Large Eddy Simulation (LES), Direct Numerical Simulation (DNS). The RANS equations are famous for CFD free-surface simulations. These equations are i.a. averaged over time to neglect small and fast fluctuating eddies. Thereby, new unknowns are formed, which can only be solved by approximations. These approximations will be defined by chosen turbulence models. The RANS equation can be written as:

$$\rho\frac{\partial \bar{u}_i}{\partial t}+\rho\left(\frac{\partial \bar{u}_i \bar{u}_j}{\partial x_j}\right)=\bar{f}_i-\frac{\partial \bar{p}}{\partial x_i}+\mu\frac{\partial}{\partial x_j}\left(\frac{\partial \bar{u}_i}{\partial x_j}+\frac{\partial \bar{u}_j}{\partial x_i}-\rho\overline{u_i' u_j'}\right) \tag{8.4}$$

where: $\overline{u_i'}$ = time averaged mean velocity, i,j = index of cell notation, $\overline{u_i' u_j'}$ = Reynolds stress tensor from turbulence model. Additional details can be found in the literature (e.g. Jakirlic 1997).

For 2-D hydraulic flow problems, the Shallow Water Equations (SWE)—St.-Venant equations—were developed, which allow a good approximation in free-surface flows like channels and rivers (Wesseling 2001). The SWE are another simplification of the Reynolds equations by averaging the flow velocity over flow depth. These equations can be used, where the vertical velocity component can be neglected. The SWE can be written as (Nujic 2006; Tan 1992):

$$\frac{\partial c_1}{\partial t}+\frac{\partial c_2}{\partial x}+\frac{\partial c_3}{\partial y}+c_4=0 \tag{8.5}$$

with:

$$c_1=\begin{bmatrix} H \\ uh \\ vh \end{bmatrix} \quad c_2=\begin{bmatrix} uh \\ u^2h+0.5gh^2-vh\frac{\partial u}{\partial x} \\ uvh-vh\frac{\partial v}{\partial x} \end{bmatrix}$$

$$c_3=\begin{bmatrix} vh \\ uvh-vh\frac{\partial u}{\partial y} \\ v^2h+0.5gh^2-vh\frac{\partial v}{\partial y} \end{bmatrix} \quad c_4=\begin{bmatrix} 0 \\ gh(I_{Rx}-I_{Sx}) \\ gh(I_{Ry}-I_{Sy}) \end{bmatrix}$$

where: $H = h + z$ = water surface elevation, h = flow depth, z = bottom elevation, u, v = velocity components in x-, y-direction, g = acceleration due to gravity, v = viscosity, $I_{R,i}$ = friction slope, $I_{S,i}$ = bottom slope.

Additionally to the mentioned equations there is a plurality of approximations of the N-S-equation—shown in Fig. 8.1.

8.2.3 Discretization

This section gives a raw overview of numerical discretization methods that are used for flow simulations. Details can be found in the special literature (e.g. Ferziger and Perić 2002; Martin 2011).

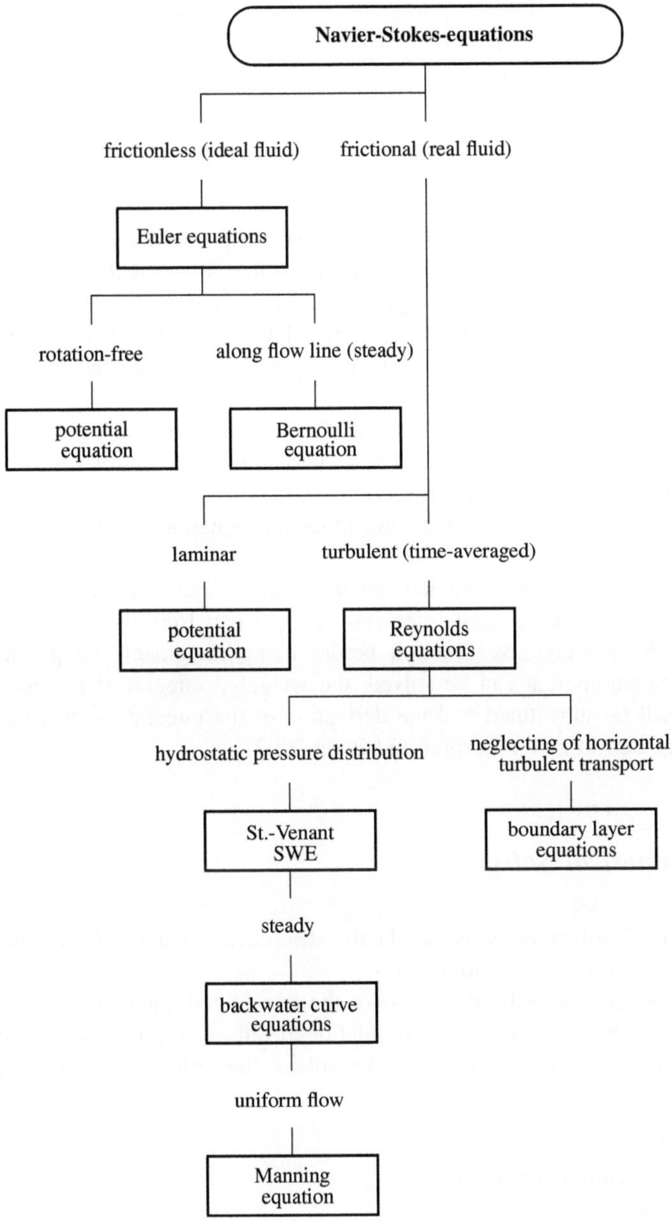

Fig. 8.1 Overview of N-S-equations approximations (Rutschmann 2014)

Because basic equations of flow conditions are described by partial differential equations and hence cannot be solved analytically; simplifications, based on approximations and dimensional analysis, must be used. To solve those basic equations a numeric discretization method, where approximations of time or area are applied, is used commonly. There are three main discretization methods (Ferziger and Perić 2002):

- Finite Difference Method,
- Finite Volume Method,
- Finite Element Method.

The Finite Difference Method is the oldest method to solve partial differential equations numerically and was developed by Euler. The calculation is carried out by covering the flow field within a grid and approximating the differential equation on every grid cell corner. Therefore, partial derivatives are replaced by variables based on corner values, resulting in an algebraic equation that can be solved (Ferziger and Perić 2002).

The Finite Volume Method is based on the integral of the conservation equation. The conservation equation is applied to a finite number of control-volumes. For the main focus of the volume, variables are solved and transferred by interpolating the surface of the volume. Thus, the built algebraic equation can be solved (Martin 2011).

For the Finite Element Method the mean flow field is divided into a set of discrete volumes (3-D) or finite elements (2-D). Apart from that, all equations are multiplied by a weighting function before being integrated. To get non-linear algebraic equations that can be solved, the weighted integral of the conservation equation will be substituted and the derivative of the integral on every grid point value is equated to zero (Ferziger and Perić 2002).

8.2.4 Numerical Solver

The numerical solver consists of (1) the discretization method, (2) the solving method, and (3) the error estimation.

There are many possibilities to solve the considered equations for flow simulations. Thereby the accuracy of simulation results is influenced by the chosen solver. Hence, to choose a satisfying solver the following aspects must be considered:

- accuracy,
- size of the computational mesh,
- time steps,
- symmetric or not symmetric equation,
- convergence and efficacy.

This section gives a short overview of solvers which are frequently used for fluid simulations. Due to the plurality of possibilities, only relevant methods (discussed in Ferziger and Perić 2002; Martin 2011) are shown in Fig. 8.2. All solvers apply matrix notation and result in approximations.

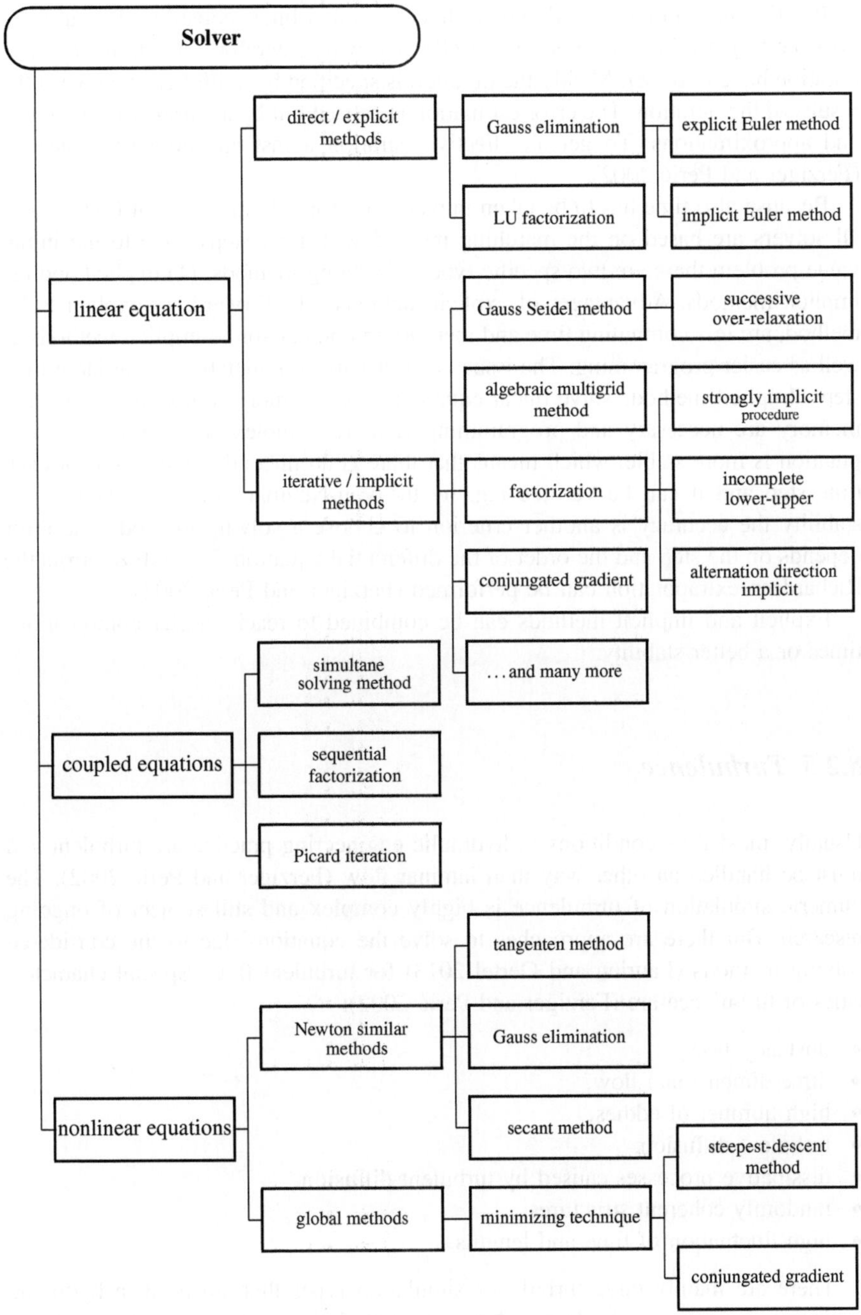

Fig. 8.2 Overview of solving methods

Iteration is commonly used to solve linear and non-linear equations, because it is possible to get a high accuracy in an effective way. Therefore, iteration steps and duration have to be set. Mostly the duration is specified by a difference between the results of the iteration. The error estimation is carried out by a plurality of iterations and approximations. To get an effective iteration a fast convergence is needed (Ferziger and Perić 2002).

Because also time has to be taken into account for solving transient flow, nearly all solvers are based on the marching method with time steps. Due to the initial value problem there are two specific types of solving methods: (1) explicit and (2) implicit methods. Advantages of explicit methods, like forwards or explicit Euler method, are less computing time and memory in comparison to implicit methods as well as easier programming. The issue is an unstable computation using large time steps. Implicit methods solve linear equations. Hence, more computation time and memory are necessary and programming is more complex. Therefore, the computation is more stable, which means that there is an unconditional result for each time step and it can be verified e.g. by the von-Neumann-method. Next to the stability the accuracy is another criterion to choose a solving method. The error depends on the step and the order of the differential equation. To analyze errors the Richardsen extrapolation can be performed (Ferziger and Perić 2002).

Explicit and implicit methods can be combined to reach shorter computations times or a better stability.

8.2.5 Turbulence

Usually, most flow conditions in hydraulic engineering practice are turbulent and must be handled an other way than laminar flow (Ferziger and Perić 2002). The numeric simulation of turbulence is highly complex and still subject of ongoing research. But there are approaches to solve the equations due to the considered solving methods (Laurien and Oertel 2013) for turbulent flow. Special characteristics of turbulence are (Ferziger and Perić 2002):

- unsteady flow,
- three-dimensional flow,
- high number of eddies,
- turbulent diffusion,
- dissipative processes caused by turbulent diffusion,
- randomly coherent structures,
- high fluctuation of time and lengths.

There are mainly three turbulence simulation types that are used in hydraulic engineering practice: (1) Direct Numeric Simulation (DNS), (2) Large Eddy Simulation (LES), and (3) Reynolds-Average-Navier-Stokes (RANS). For RANS, turbulence models are necessary to solve the Reynolds stress tensors. Consecutively, simulation methods for turbulent flows are explained briefly.

Further information is given in the literature (e.g. Ferziger and Perić 2002; Laurien and Oertel 2013; Martin 2011).

DNS is the most precise method to simulate turbulence because no approximations must be made. Only numeric discretization is necessary to solve the Navier-Stokes-equation (Ferziger and Perić 2002). Therefore, a high mesh resolution is required to allow the description of all dynamic flow in time and space (Laurien and Oertel 2013). The size of the mesh is based on the physical size of the investigation area but has to be as large as the largest turbulence. Hence, the mesh resolution has to be smaller than the Kolmogoroff-scale, which is influenced by the viscosity to reflect the total dissipation of kinetic energy. DNS results are very detailed, but long calculation times are necessary. Consequently, DNS is mostly used to qualitatively and quantitatively describe physical phenomena (Ferziger and Perić 2002).

LES is a large-scale simulation, compared to DNS. Small eddies are neglected and the course of velocity due to time is averaged. This method is appropriate because large-scale eddies have more influence on turbulence. The calculation time is still long and a precise definition of variables have to be made (Ferziger and Perić 2002). To solve the equations, approximations are needed—the so-called turbulence models, e.g. eddy viscosity, subgrid-scale model, dynamic model or dynamic Lagrange-model which can be found in the literature (e.g. Ferziger and Perić 2002). The solving methods in Sect. 8.2.4 can be used in this case.

RANS is usually used in hydraulic engineering practice. The equations (shown in Eq. 8.4) can be solved by using turbulence models, which include approximations. There are two kinds of turbulence models: (1) eddy viscosity models, and (2) Reynolds stress models (RSM); also categorized as the so-called algebraic models or differential equations (one- or two-point-closure, transport models) (Laurien and Oertel 2013). In algebraic turbulence models, further equations, which describe e.g. the eddy viscosity, are added to the RANS-equations. Examples are the Prandtl's mixing-length concept, the Baldwin-Lomax model (eddy viscosity models) or algebraic Reynolds stress models. Two-point-closure is applied in the eddy viscosity model, e.g. in the k-ε-model or k-ω-model. The RSM is approximated by e.g. τ-ε-model or τ-ω-model (Laurien and Oertel 2013). The k-ε-model is the mostly chosen turbulence model for 3-D practical hydraulic problems. Generally, eddy viscosity models are used for a huge amount of numerical flow simulations in hydraulic engineering.

8.3 Advantages and Disadvantages of Free-Surface Numerical Simulations

With increasing quantity of numerical simulation products and achieved simulation projects also their advantages and disadvantages must be discussed. For free-surface problems, following aspects can be mentioned as advantages:

- easy variation of boundary conditions,
- fast results for varied boundary conditions,
- case-studies can be arranged cost- and time-effective, compared to experimental models,
- real-time calculations and worst-case scenarios,
- and more.

Contrary, disadvantages can be, e.g.:

- reliability,
- errors and uncertainties,
- approximative solutions, e.g. turbulence models,
- and more.

Discussing advantages and disadvantages, the question of model validation and reliability of numerical simulation results can be addressed. Therefore, it must be noticed that a good data background is essential for numerical simulations. Next to geometrical boundary conditions, e.g. cross sections and digital terrain models (DTM), also hydraulic boundary conditions, e.g. rating curves and water surface levels, must be given. These data can be used to calibrate the numerical model and to validate numerical simulation results. For complex scenarios, sensitive analysis can help to reduce errors and to identify influences of various variables.

By knowing advantages and disadvantages of numerical flow simulations and including a detailed validation and calibration study, also critics can be satisfied when discussing benefits and cost-effectiveness in accurate detail.

8.4 Application Area of Free-Surface Numerical Flow Simulations

The application area of numerical flow simulations range from flood protection to flow optimization. During the last years the range of application is huge and also small engineering offices use numerical simulation products as standard programs like CAD and GIS and a couple of these. Subsequently, some areas of application will be mentioned as examples, being separated in those before, after and during the event.

- Before the event, e.g.:
  - floodplain area flow velocity determination for flood maps and flood management plans,
  - preparation of rating curves
  - dimensioning of polder areas,
  - flow optimization (e.g. for dike influences or complex 3-D hydraulic structure flows),
  - sediment transport and water quality models.

- After the event, e.g.:
  - verification of flood discharge development and improvement of forecast models,
  - verification of available rating curves,
  - reports concerning damage courses.
- During the event, e.g.:
  - real-time simulations to forecast floodplains and to determine evacuation areas, e.g. after dike breaching),
  - assimilation of flow processes and forecast of flood peaks,
  - calculation of flood durations, e.g. to evaluate dike stability.

8.5 Free-Surface Numerical Simulation Software

8.5.1 One-Dimensional Flow Models

Nowadays, one-dimensional numerical flow simulation models are used as standard tools and often are available at no charge with complex model characteristics and features. Software products like e.g. HEC-RAS, MIKE11, JABRON, WSPWIN, etc., can be used for steady as well as unsteady calculation of water surface levels and flow velocities in main channel direction, based on one-dimensional St.-Venant equations or energy conservation. To use 1-D models adequately, no major flow velocity components in *y*- and *z*-direction (width and depth) should exist due to averaging these values. Hence, no velocity distribution in width and depth can be reported by using 1-D numerical codes, which limits their applicability. Typical hydraulic structures, like e.g. weirs and bridges, are included with simple analytical approaches (e.g. Poleni formula) within the 1-D model. Using 1-D numerical simulations river reaches more than 100 km can be simulated and analyzed.

8.5.2 Two-Dimensional Flow Models

Numerical simulation models with more than one dimension usually neglect the vertical velocity components. Hence, these products are used as two-dimensional models, based on the depth averaged shallow water equations. Herein, the velocity component on the *z*-axis (velocity *w*) is integrated over flow depth and velocity components on the *y*-axis (velocity *v*) can be described in detail. Typically, 2-D numerical simulations are arranged to identify unsteady flooding processes, e.g. after dike breaching, or to analyze complex flow patterns at discontinuities in river reaches. Available software products are amongst others HYDRO_AS-2D,

MIKE21, and TELEMAC2D. If the vertical velocity component is of major interest (e.g. modeling a cooling water inlet) also (pseudo) 3-D models are available, like Delft3D or TELEMAC3D. Additionally, sediment transport models are available, which will not be discussed in detail in this chapter. Often numerical software codes allow a linking to external systems. In this context, rivers can be developed as 1-D models with available profile data and floodplain areas will be attached as 2-D model from digital terrain data. Due to climate change discussions and adaptation of rainfall-runoff scenarios, also flash flood events are simulated e.g. using a combined analytical model (sewage system) and 2-D model (surface area). These simulations allow the determination of flow paths through sealed residential and industry areas after reaching discharge capacities of sewage system pipes. Nowadays, 2-D models often include urban canyons, formed by buildings and flow guiding structures. These have an influence due to circulation processes and flow resistance. Automatic processes can be used to include surface roughness by land use data or breaking edges for flow guiding elements or forced nodes. Depending on the automatic level and time capacities to build up the model, river reaches of 20–50 km can be achieved without difficulty.

8.5.3 Three-Dimensional CFD Flow Models

Computational Fluid Dynamics (CFD) flow models usually are three-dimensional models and involve a more detailed resolution of velocity components in all spatial directions, allowing a detailed analysis of highly complex flow situations. CFD models are still time-consuming expert systems with complex boundary conditions. The model domain usually is limited to some 100 m—for complex models with huge amounts of calculation cells only some 10 m can be achieved. Software products are e.g. FLOW-3D, OpenFoam, ANSYS CFX, and more. Focusing on free-surface flow problems the Volume-Of-Fluid (VOF) method (Hirt and Nichols 1981; Nichols and Hirt 1975) is used to determine the sharp water surface. Therefore, an additional differential equation is adopted and solved, giving information on the filling level of each cell in percent and the declination of the water surface within this cell. Hence, the free-surface can be determined as sharp line. Application areas of CFD simulation in hydraulic engineering are extensive, e.g. weir flow, filling and emptying lock systems, block ramp flow (e.g. Oertel 2012; Oertel and Schlenkhoff 2012) and many more. Limitations can be found concerning cell numbers (directly linked with PC's main storage capacity), boundary conditions, general model sizes, and calculation capacities. Additionally, it is rarely possible to simulate complex two-phase flow problems (e.g. Bung 2009) for hydraulic structures like spillways. Contrary, non-aerated free-surface flows and friction factors can be determined adequately using 3-D CFD models. This can be confirmed by hybrid modeling progressively.

8.6 Examples of Free-Surface Numerical Flow Simulations

In this section example free-surface 2-D and 3-D flow simulations are presented and compared due to the development of calculation time during the last years and the contrary benefit of 3-D simulations compared with 2-D. Therefore, the following numerical models are used: (1) 2-D model Overath, (2) 2-D model Hattingen, (3) 2-D model Wetter to Hattingen, (4) 3-D CFD model gauge Hattingen. Models (1)–(3) are used to give a benchmark on calculation times and model (4) is compared with a selective area of model (2).

For 2-D flow simulations the software package Surface-water Modeling System (SMS) was used in combination with the calculation module HYDRO_AS-2D, which is based on the SWE. 3-D flow simulations were performed in FLOW-3D, which implements the RANS equations, k-ε turbulence model and VOF method.

Re. (1): Overath is located along the river Agger and is prone to flooding. In a 2003 study (see Oertel 2003) flood plains for several flood discharges and unforeseeable dike failure events were carried out. Table 8.1 gives detailed model specifications. In Fig. 8.3 an example result for a dike breach is given. It can be noticed that the 2-D model is applicable to determine detailed flow traces in plan view to identify main flow paths and, e.g., evacuation areas. Hence, a 2-D model can provide more information than simple 1-D simulation models.

Re. (2) and (3): After extreme flood events in August 2007 at the Ruhr catchment area in North Rhine-Westphalia, Germany, a study was assigned to analyze discharge curves at the gauges Wetter and Hattingen at the river Ruhr. Therefore, hydrodynamic numerical 2-D flow simulations were carried out in 2008 (see Oertel et al. 2009). Flow measurements during the flood event indicated significant deviations from the discharge curves at both gauges for extreme water levels. This also could be confirmed by numerical results from the developed 2-D models. The focus of 2-D modeling was to analyze vegetational influence on gauge's rating curves. In 2013 both models, Hattingen and Wetter, were combined to only one 2-D numerical model at the Hydraulic Engineering Section at Lübeck University of Applied Sciences (see Fig. 8.4). The main aim was to perform a benchmark regarding the calculation times. Table 8.1 gives model specifications for these models.

Table 8.1 2-D and 3-D model specifications

Model	Grid points	Elements	River reach length	Year
1. Overath (2-D)	22,650	30,470	4.0 km	2003
2. Hattingen (2-D)	32,500	57,000	5.4 km	2008
3. Wetter–Hattingen (2-D)	271,500	480,500	ca. 30.0 km	2013
4. Hattingen (3-D)	–	13 Mio.	ca. 300 m	2014

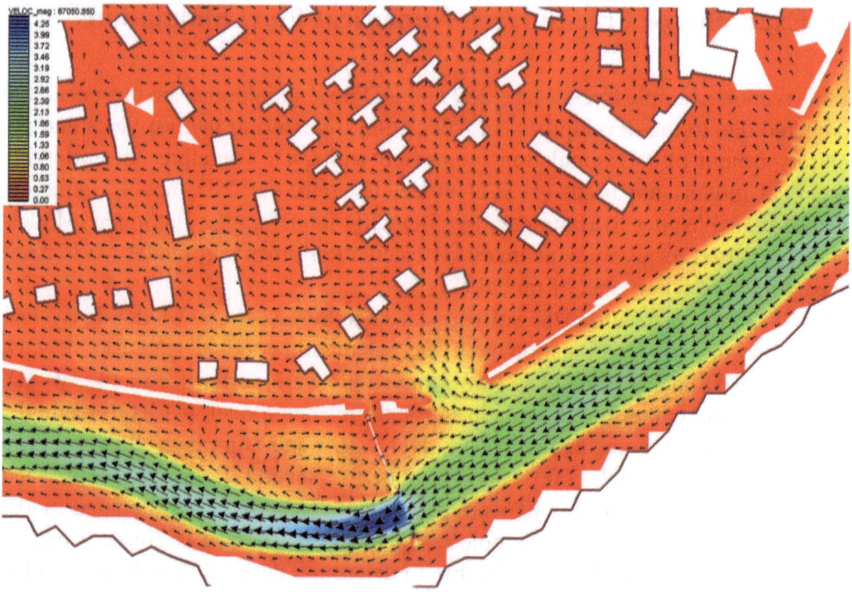

Fig. 8.3 Example result of 2-D flow through polder area after a fictive dike breach

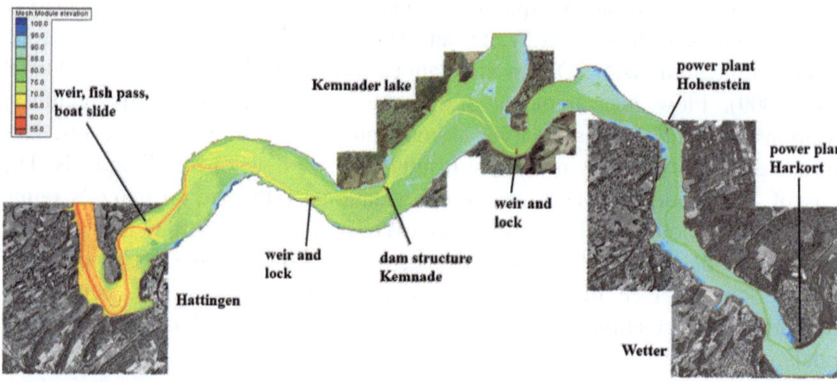

Fig. 8.4 Model domain of 2-D model from Wetter to Hattingen, aerial photos: Ruhrverband

To compare calculation times, selective model runs from 2003 to 2008 were performed on a modern workstation in 2014. Table 8.2 gives results for calculation times for all mentioned 2-D models. Individual example simulations show an identifiable improvement of calculation time during the past 10 years. To reach a steady final result of a mean discharge flow event, a time benefit up to 97 % can be observed e.g. for the Overath 2-D model. Even when comparing a simulation of 2008 in respect to 2010 with a simulation of 2014 there is 80 % time benefit. It can be assumed that this trend will be resumed during the next years.

Table 8.2 Comparison of example calculation times of various 2-D simulations

Model overath: simulation time steady flow event t_{sim} = 6000–25,000 s			
	MQ = 20 m^3s^{-1}	HQ100 = 289 m^3s^{-1}	HQ1000 = 462 m^3s^{-1}
2003 1 core, 1.0 GHz, Pentium III, 512 MB RAM	102 min	289 min	700 min
2014 8 cores, 3.5 GHz i7, 32 GB RAM	3 min (97 %)	76 min (73 %)	87 min (87 %)
Model Hattingen: simulation time for steady flow event t_{sim} = 36,000 s, t_{out} = 3600 s			
	$Q = 75$ m^3s^{-1}	$Q = 360$ m^3s^{-1}	$Q = 916$ m^3s^{-1}
2008 4 cores, 3.0 GHz, 4 GB RAM	51 min	60 min	78 min
2014 8 cores, 3.5 GHz i7, 32 GB RAM	7 min (86 %)	11 min (81 %)	17 min (78 %)
Model Hattingen: simulation time for transient flow event $t_{sim} = 2,700,000$ s, $t_{out} = 21,600$ s			
2010 4 cores, 3.3 GHz, 8 GB RAM	1087 min		
2014 8 cores, 3.5 GHz i7, 32 GB	206 min (81 %)		
Model Wetter to Hattingen: simulation time for steady flow event $Q = 100$ m^3s^{-1},, $t_{sim} = 720,000$ s, $t_{out} = 7200$ s			
2014 8 cores, 3.5 GHz i7, 32 GB	787 min		

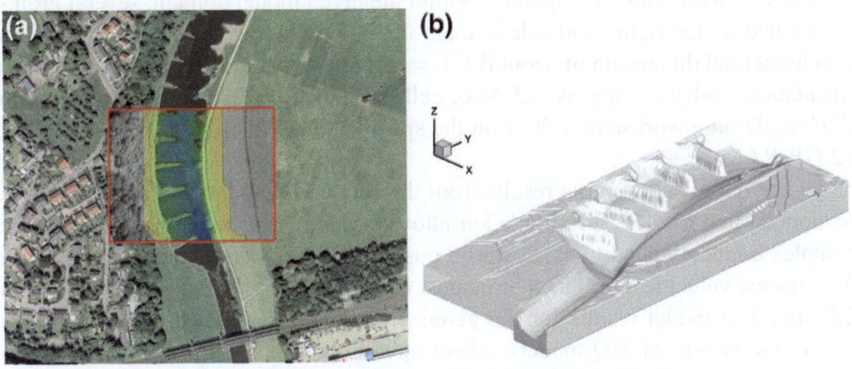

Fig. 8.5 Model domain of 3-D CFD model upstream gauge Hattingen, **a** plan view, aerial photo: Ruhrverband, **b** 3D detail view

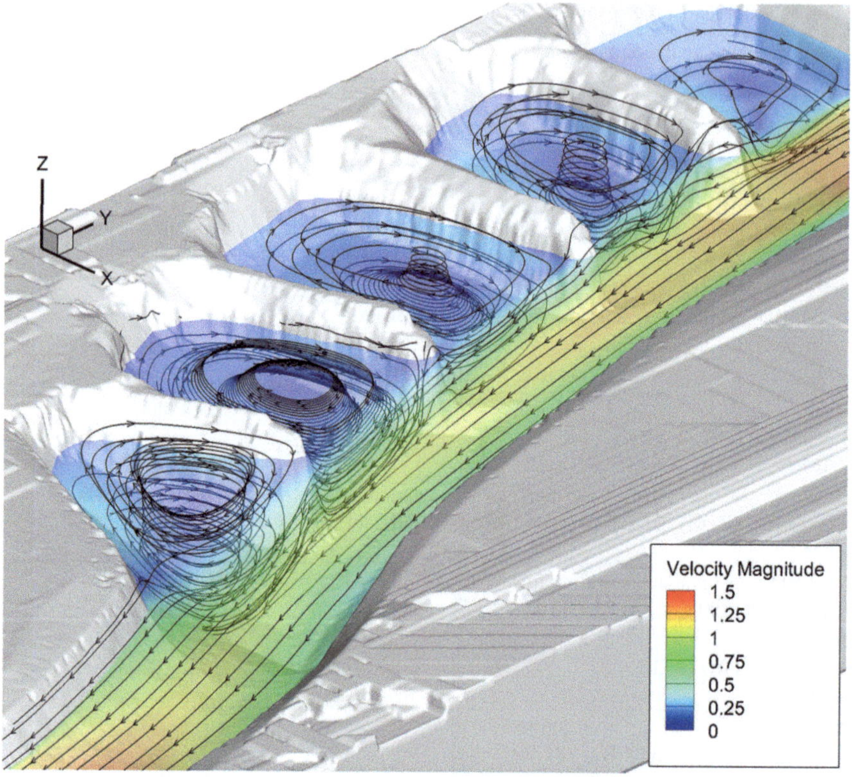

Fig. 8.6 Example results from 3-D CFD model for groin field flow

Re. (4): As an example, a 3-D CFD model is build up for the near field of the inflow area at gauge Hattingen to give an idea about today's simulation possibilities on high end workstation computers. Within the given model domain, several groins are located on the right hand side in main flow direction (see Fig. 8.5). The shown area has a total dimension of about B × L = 130 × 290 m and is represented within a calculation grid with approx. 13 Mio. cells. Simulations take up to 2 days with FLOW-3D on a workstation PC with the specifications: Intel i7, 8 cores, 3.5 GHz, 32 GB RAM.

Figure 8.6 gives example results from the 3-D CFD model (FLOW-3D). It can be noticed that a fully 3-D simulation allows a deep view into flow processes for complex boundary conditions. Powerful post-processor tools can be used to analyze flow traces, velocities, turbulent structures etc. graphically. A comparison between 2-D and 3-D model results for the given investigation area (see Fig. 8.7) shows clearly the benefit of 3-D models, where three-dimensional flow effect are significant. Where a 2-D numerical simulation provides only depth-averaged velocity components, a 3-D simulation includes detailed velocity vectors for all directions.

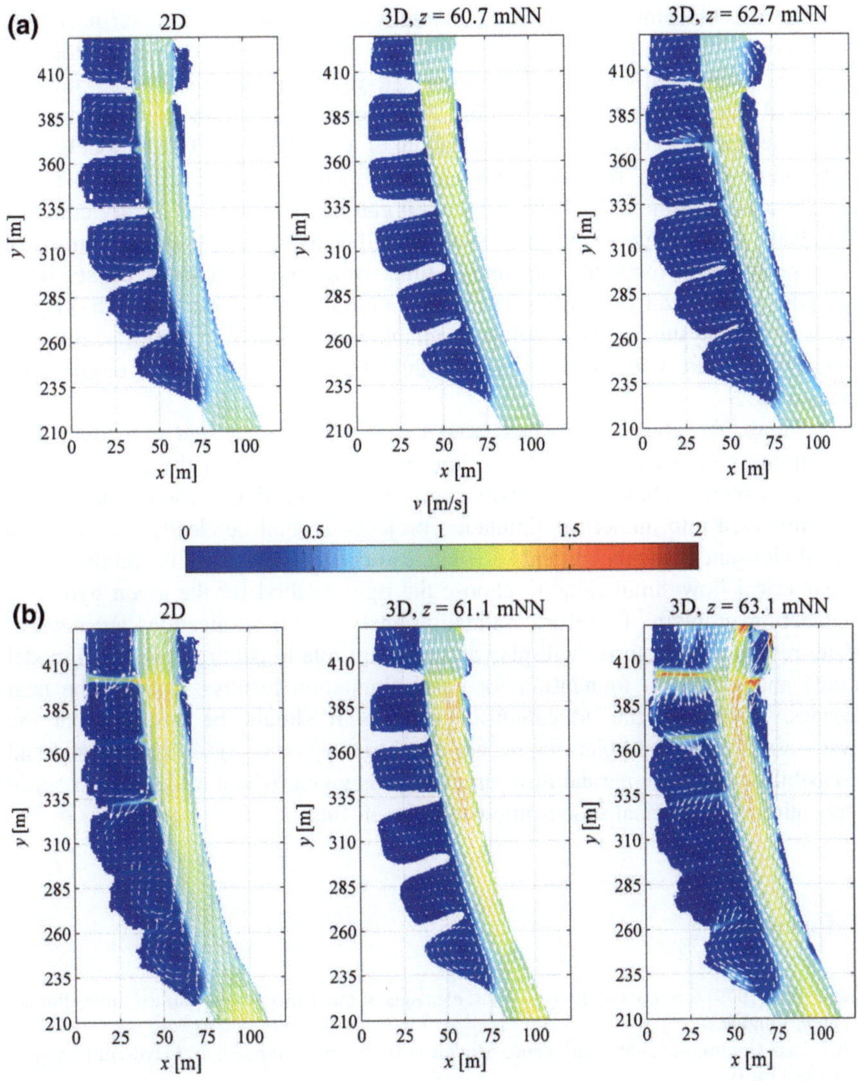

Fig. 8.7 Result comparison from 2-D depth-averaged and 3-D CFD model

8.7 Conclusions and Outlook

Numerical simulations of free-surface flows represent a modern tool to solve a large number of problems. With accurate data selection a calibrated model can produce good results e.g. for decision makers in a cost-effective way. 1-D and 2-D models became state-of-the-art and are deep-rooted in engineering companies. Contrary, 3-D CFD numerical models are mostly applied by universities and research institutes due to complex boundary conditions and high class computer necessity. With increasing computer capacities a boom can be observed during the last 15 years, which led to several projects and research activities in the field of numerical free-surface flow simulations. The rapid development of computer and storage technology enables the calculation of complex 3-D CFD models at a workstation PC with several million cells, which appeared inconceivable two decades ago. Within the next 10–15 years it can be expected that highly complex two-phase flows (e.g. air-water-mixture) can be solved by standard numerical software codes in moderate calculation time. Additionally, a focus on sediment transport models can be assumed to answer complex questions on transport processes in river systems. These will be available in 2-D software products. But also 3-D CFD codes will become state-of-the-art tools in hydraulic engineering companies.

In future, the above-mentioned steep trend will continue and numerical flow simulations will become more and more essential tools, with the need of good educated experts. These experts will come from universities, where students get more involved into numerical simulation projects. It shall be clearly stated, that a critical view and basic theoretical knowledge are necessary to understand the nature of numerical flow simulations to choose the right method for the given hydraulic engineering problem. Therefore, validation analysis and verifications are needed. Water research laboratories will play an important role to confirm numerical model results and to give a foundation for exact simulation results—also for the next decades. Regarding the education of students it should be ensured that the awareness that "colored pictures are not equal to real results" is communicated and a sensibility due to numerical flow simulations is given, which lead to a successful application of numerical model improvements in future.

References

Abbott MB (1979) Computational hydraulics, elements of the theory of free-surface flows. Pitman Publ, London

ASCE Task Commitee (1988) Turbulence Models in Hydraulic Computation. J Hydraul Eng Spec Issue 114(9)

Bung DB (2009) Zur selbstbelüfteten Gerinneströmung auf Kaskaden gemäßigter Neigung. Dissertation Thesis, University of Wuppertal, Shaker, Aachen

Ferziger JH, Perić M (2002) Computational methods for fluid dynamics, 3rd rev. edn. Springer, Berlin

Hirt CW, Nichols BD (1981) Volume of fluid (VOF) method for the dynamics of free boundaries. J Computat Phys 39(1):201–225

Jakirlic S (1997) Reynolds-Spannungs-Modellierung komplexer turbulenter Strömung, Herbert Utz- Verlag

Laurien E, Oertel H Jr (2013) Numerische Strömungssimulation—Grundlagen und Modelle—Lösungsmethoden—Qualität und Genauigkeit, 5th rev. edn. Springer, Berlin

Martin H (2011) Numerische Strömungssimulation der Hydromechanik—Grundlagen und Methoden. Springer, Berlin Heidelberg

Nichols BD, Hirt CW (1975) Methods for calculating multi-dimensional, transient free surface flows past bodies. In: Proceedings first international conference numerical methods ship hydrodynamics, Gaithersburg, ML, 20–23 Oct 1975

Nujic M (2006) HYDRO_AS-2D User Manual

Oertel M (2003) Numerische zweidimensionale Strömungssimulation für Poldergebiete hinter Flussdeichen unter besonderer Berücksichtigung von Deichbrüchen. Diplma Thesis. Hydraulic Engineering Section, University of Wuppertal

Oertel M (2012) Cross-bar block ramps: flow regimes—flow resistance—energy dissipation—stability. Habilitation Thesis, University of Wuppertal, Shaker, Aachen

Oertel M, Schlenkhoff A, Morgenschweis G (2009) Überprüfung von Abflusskurven für die Pegel Wetter und Hattingen an der Ruhr nach den Hochwasserereignissen im August 2007 mit Hilfe von numerischen 2-D-Modellen. In: Hydrologie und Wasserbewirtschaftung (HyWa), 53(6). BAW, Koblenz

Oertel M, Schlenkhoff A (2012) Cross-bar block ramps: flow regimes, energy dissipation, friction factors, drag forces. J Hydr Eng 138(5):440–448

Pironneau O (1989) Finite element methods for fluids. Masson, Paris

Rutschmann P (2014) CFD Grundlagen. lecture script, TU Munich

Tan W (1992) Shallow water hydrodynamics. Elsevier, Amsterdam

Wesseling P (2001) Principles of computational fluid dynamic, Vol 29. Springer Series in Computational Methematics, Berlin

Chapter 9
Laboratory Models of Free-Surface Flows

Daniel B. Bung

Abstract Hydraulic modeling is the classical approach to investigate and describe complex fluid motion. Many empirical formulas in the literature used for the hydraulic design of river training measures and structures have been developed using experimental data from the laboratory. Although computer capacities have increased to a high level which allows to run complex numerical simulations on standard workstation nowadays, non-standard design of structures may still raise the need to perform physical model investigations. These investigations deliver insight into details of flow patterns and the effect of varying boundary conditions. Data from hydraulic model tests may be used for calibration of numerical models as well. As the field of hydraulic modeling is very complex, this chapter intends to give a short overview on capacities and limits of hydraulic modeling in regard to river flows and hydraulic structures only. The reader shall get a first idea of modeling principles and basic considerations. More detailed information can be found in the references.

Keywords Physical modeling · Similitude · Open channels · Hydraulic structures

9.1 Introduction

For more than 500 years, people have tried to better understand the nature of complex flows. For instance, several famous records and sketches on flow phenomena were published by Leonardo Da Vinci in the late 15th century. Sir Isaac Newton (1642–1727) was the first to develop basic approaches on the similitude between a prototype and its scaled reproduction, i.e. the model. These approaches were advanced by other researchers developing modeling laws which are well recognized nowadays. These model laws were first applied to scaled models at the

D.B. Bung (✉)
FH Aachen University of Applied Sciences, Bayernallee 9, 52066 Aachen, Germany
e-mail: bung@fh-aachen.de

P. Rowiński and A. Radecki-Pawlik (eds.), *Rivers—Physical, Fluvial and Environmental Processes*, GeoPlanet: Earth and Planetary Sciences,
DOI 10.1007/978-3-319-17719-9_9

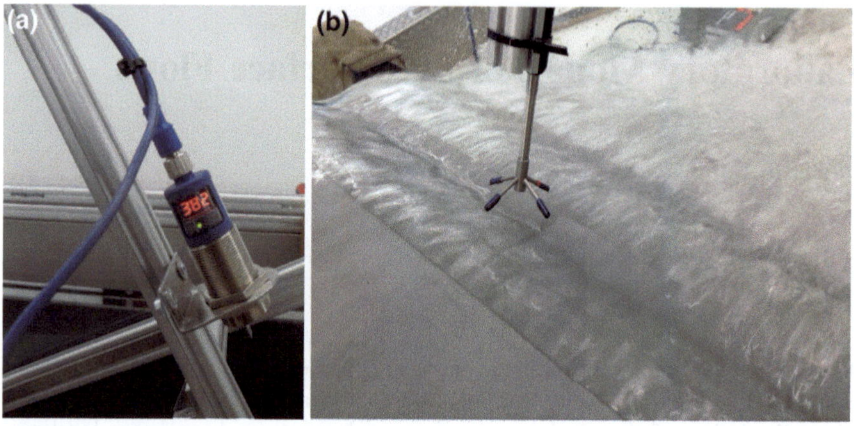

Fig. 9.1 Measuring devices for laboratory studies. **a** Ultrasonic displacement sensor for accurate measurement of water levels. **b** Acoustic Doppler Velocimeter (ADV) for measuring of local three-dimensional velocity components (*here* installed in the non-aerated part of stepped weir before submerging into water)

end of the 19th century. In particular, William Froude (1810–1879) and Osborn Reynolds (1842–1912) shall be mentioned in the context as representatives of many important individuals of fluid mechanics in these days.

Hughes (1993) stated that physical modeling is still an important technique to solve hydraulic engineering problems even in a time that computers become more and more powerful and numerical models more reliable. Even two decades later, this statement is still valid.

Hydraulic modeling allows to directly face the effects on flow features induced by changes of boundary conditions. Physical hydraulic models may also be operated to verify theoretical and numerical models or to draw some conclusion in regard to flow phenomena that are still not fully understood to date. Further development of measurement techniques allows for new investigation methods and gives high-precision results. Some up-to-date equipment are illustrated in Fig. 9.1.

On the other hand, the setup of a physical model is often relatively expensive and time-intensive. A change of boundary conditions is oftentimes not trivial and needs to be considered in the early stages of planning. Later adaption to other conditions may be difficult.

9.2 Similitude

In order to extend the findings from a scaled physical model to prototype scale, hydromechanical similitude between both scales must be guaranteed. This means that a geometric, kinematic and dynamic similitude are needed to ensure a similar motion and, by consequence, similar forces acting in the fluid. In a flowing water body, the following forces must be considered (forces due to elasticity are neglected for water as an incompressible fluid):

- a force due to acceleration of the fluid (inertia) F_i,
- a force due to gravity F_g,
- a force due to pressure in the flow field F_p,
- a force due to inner friction (viscosity) F_v,
- a force due to surface tension at the air-water interface F_s.

It yields the following balance of forces:

$$F_i = F_g + F_p + F_v + F_s \tag{9.1}$$

Considering a general length L and time T and neglecting F_p because it is a dependent force, Eq. 9.1 becomes:

$$\rho \times L^3 \times L/T^2 = \rho \times L^3 \times g + \mu \times \frac{L/T}{L} \times L^2 + \sigma \times L \tag{9.2}$$

where ρ is the density of water in [t/m^3], g is the gravity acceleration in [m/s^2], μ is the dynamic viscosity in [kg/(m × s)] and σ is the surface tension in [kN/m]. A dynamic similitude now claims that all forces have the same scale factor:

$$\frac{F_{i,N}}{F_{i,M}} = \frac{F_{g,N}}{F_{g,M}} = \frac{F_{v,N}}{F_{v,M}} = \frac{F_{s,N}}{F_{s,M}} \tag{9.3}$$

and by consequence

$$\frac{\rho_N \times L_N^4/T_N^2}{\rho_M \times L_M^4/T_M^2} = \frac{\rho_N \times L_N^3 \times g_N}{\rho_M \times L_M^3 \times g_M} = \frac{\mu_N \times L_N^2/T_N}{\mu_M \times L_M^2/T_M} = \frac{\sigma_N \times L_N}{\sigma_M \times L_M} \tag{9.4}$$

with index N referring to prototype scale and index M to model scale, respectively. Full hydromechanical similitude could be achieved if the model would be operated by a fluid with density, viscosity and surface tension that fulfills this constraint. Unfortunately, this fluid does not exist and models are commonly operated by the same fluid as in nature, i.e. water. With the length scale $\lambda_L = L_N/L_M$, the time scale $\lambda_T = T_N/T_M$ as well as $\rho_N/\rho_M = 1$, $\mu_N/\mu_M = 1$, $\sigma_N/\sigma_M = 1$ and $g_N/g_M = 1$, Eq. 9.4 becomes:

$$\frac{\lambda_L^4}{\lambda_T^2} = \lambda_L^3 = \frac{\lambda_L^2}{\lambda_T} = \lambda_L \tag{9.5}$$

It must be noted that this equation is only valid for $\lambda_L = \lambda_T = 1$, i.e. the prototype scale. It is not possible to properly scale all acting prototype forces in a scaled hydraulic model. Every scaled model implements certain errors. However, this error can be minimized if the most relevant counter-acting force is considered together with the inertia force and properly scaled. For free-surface flows which is dealt with in this chapter, the predominant force is gravity.

9.3 Froude's Law

The relation between the inertia force and the gravity force in open channels is generally expressed by the Froude number:

$$\mathrm{F} = \sqrt{\frac{\text{inertia}}{\text{gravity}}} = \sqrt{\frac{\rho \times V \times v^2/L}{\rho \times V \times g}} = \frac{v}{\sqrt{g \times L}} \tag{9.6}$$

where V is the volume of the water body moving with the velocity v.[1]

The basic idea of Froude's law is that the dimensionless Froude number is the same in model and nature for a similar flow condition. This means that a sub-/supercritical flow in nature must be sub-/supercritical in the model as well. The flow may then be described by the same governing differential equations with transferrable boundary conditions upstream or downstream. Identical Froude numbers yield:

$$\frac{v_N}{\sqrt{g \times L_N}} = \frac{v_M}{\sqrt{g \times L_M}} \tag{9.7}$$

On this basis, the scaling factors for all physical properties may be derived, e.g.:

- Velocity factor: $\lambda_v = v_N/v_M = \sqrt{L_N/L_M} = \sqrt{\lambda_L}$
- Time factor: $\lambda_T = \lambda_L/\lambda_v = \lambda_L/\sqrt{\lambda_L} = \sqrt{\lambda_L}$
- Discharge: $\lambda_Q = \lambda_v \times \lambda_A = \sqrt{\lambda_L} \times \lambda_L^2 = \sqrt{\lambda_L^5}$
- etc.

9.3.1 Model Accuracy

Data from the laboratory needs to be scaled to prototype dimensions according to the appropriate model scale factors. However, the modeler must be aware of inaccuracies which may derive from different sources. In detail, model effects and scale effects may occur.

[1]It must be noted that similar considerations yield the Reynolds number (viscosity), Weber number (surface tension) and Mach number (elasticity) if the gravity force is replaced by the other forces.

Fig. 9.2 Model investigation on stilling basin performance downstream of a steep stepped spillway (Bung et al. 2012), channel width is 1 m to avoid model effects

9.3.1.1 Model Effects

Model effects may arise from an inappropriate idealization of the prototype conditions in the laboratory. For example, three-dimensional flow patterns in the stilling basin downstream of a weir model (see Fig. 9.2) may not fully develop when the model is installed in a narrow flume. Depending on the purpose of such a study, this effect may be more (e.g. stilling basin design) or less (e.g. determination of weir coefficient) severe. Reflections from walls may also lead to some undesired effect as well as incorrect water levels due to wall effects when water levels are visually investigated through a transparent sidewall. On the other hand, some effects being present in the prototype system are difficult to reproduce, e.g. raised water levels by wind. These exemplary problems emphasize that some experience is required by the modeler in order to avoid or at least to recognize such effects and in this case, to evaluate those in regard to not misinterpret the data.

9.3.1.2 Scale Effects

As discussed above, model laws, such as the Froude's law for open channel flows, are based on the idea to just account for the major force counter-acting the inertia force. This assumption is reasonable as long as the model scale does not fall below a critical value where other forces may become significant. A few simple examples shall illustrate this problem:

- If the Froude's law is considered, the scale factor for the dimensionless Reynolds number[2] in a model with identical fluid becomes $\lambda_R = \lambda_v \times \lambda_L = \lambda_L^{3/2}$. Thus, the Reynolds number is never properly scaled in a Froude model and viscosity effects may lead to disturbed results. Considering the precondition that the flow

[2]Note: the Reynolds number is defined as: $\mathrm{R} = \rho \times v \times D/\mu$.

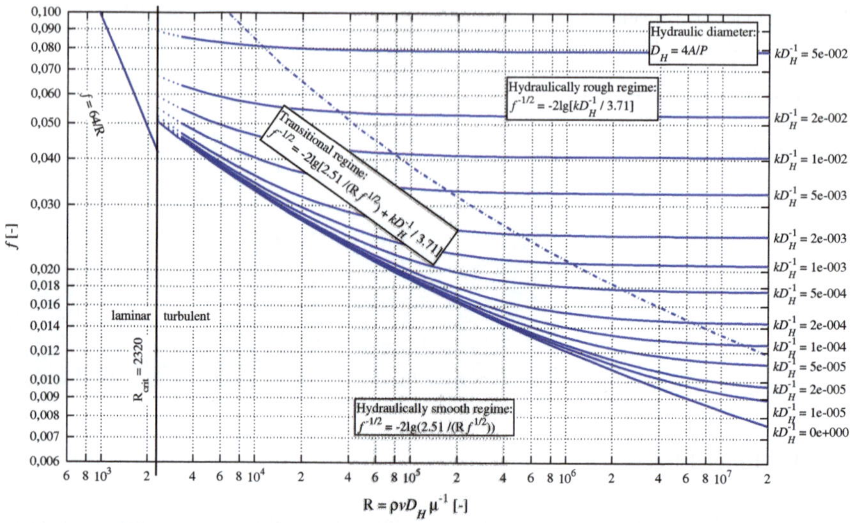

Fig. 9.3 Moody diagram for determination of flow resistance factor f as a function of relative roughness k/D_H (with $D_H = 4 \times A/P$ the hydraulic diameter, A the cross-sectional area and P the wetted perimeter) and Reynolds number R

resistance f in model and prototype must be equal, it has to be ensured that hydraulic rough conditions are met in the model (i.e. the flow resistance f is no more dependent on R, see Fig. 9.3) which in turn is only met in large scale models. Alternatively, the modeler may come up with the idea to chose a smaller surface roughness k in order to ensure a properly scaled f even for a small Reynolds number in the model. However, it must be recognized that scaling of bed roughness is generally difficult, particularly for a small roughness in prototype, e.g. for a concrete surface.

- A small scale weir overflow model becomes more and more affected by surface tension when the water level at the weir crest is low for a given flow velocity.
- Investigations on sediment transport may be affected by scale effects when particles need to be scaled to a smaller size, thus reaching undesired cohesive characteristics.
- Laboratory studies with focus on air-water flows, e.g. hydraulic jump downstream of a weir, must be modeled in large scale as air bubbles in prototype tend to break up when reaching a critical size, particularly in turbulent flows. By consequence, air bubbles have roughly the same size, in model and in prototype, and are thus not properly scalable. Air bubble lift forces may play a more significant role in model as in nature.

Numerous studies have been performed to determine the critical scale to prevent scale effects. Several publications on scale effects for models with different focus are included in Kobus (1984). Recently, a comprehensive review on critical model scales was presented by Heller (2011). If no information on scale effects and minimum model scales is found in the literature, the modeler should consider evaluating a series of models and differing size scale before setting up the final full model in order to check from which scale on the data lead to differing results when scaled up to prototype scale according to the relevant scale factors. Laminar flows however, where the flow depth is low and the velocity is small, should be modeled in prototype scale to reproduce the viscous effects correctly.

9.4 Open Channels

9.4.1 Models with Fixed Bed

Physical models with fixed beds are used with different focuses. In earlier times, the main interest was to determine water levels and flooded ares for different flood events. Other concerns could be be found in the effects of river training measures or the formation of 2D or 3D flow patterns (e.g. eddy formation in groin fields). It must be confessed that such studies are commonly performed numerically nowadays. Large laboratory capacities in terms of space and discharge are required to operate physical river models. Consequently, such studies are very cost-intensive while numerical 2D and 3D river models are available for free as open-source codes. However, physical river models are still a useful method to investigate more complex flow situations, which may be caused by e.g. jet formation and mixing procedures at an cooling water outfall structure (see Sect. 9.5.4).

9.4.2 Models with Movable Bed

River models with movable beds are used for investigation of bed load and sediment transport. Local erosion and deposition processes may be in the focus for waterways modeling and in models of inlet structures to hydropower plants. Such models need to fulfill the same hydraulic similitude as developed above, but further considerations on particle dynamics are needed. These particle dynamics can be described by the sediment Froude number

$$\mathrm{F}_s = \frac{\tau_0}{g \times (\rho_s - \rho) \times d_s} \tag{9.8}$$

and the sediment Reynolds number

$$\mathrm{R}_s = \frac{\sqrt{\tau_0 \times \rho} \times d_s}{\mu} \tag{9.9}$$

where τ_0 is the bed shear stress, ρ_s the particle density and d_s the particle diameter. Both dimensionless numbers are known from the famous Shields diagram for onset of particle movement and must be the same in model and prototype. With this precondition, relevant scale factors for sediment modeling may be derived, such as the particle size scale and the particle time scale, and a model sediment material (and grain size curve) can be chosen. For further information see (ASCE 2000) and (Kobus 1980). One resulting difficulty is caused by the particle time scale which differs from the flow time scale derived above. Sediment movement is faster in the model than in the nature. The modeler may address this issue by comparing the model data with sediment data from the field.

The total bed load or sediment load being transported through the model can be measured by collecting and weighing the material at the model outflow boundary. Depending on the test duration, the model size and the study purpose, it may be necessary to ensure that the same amount of material is added at the inflow boundary in order to provide enough material for the test.

9.4.3 Distorted Models

Model scales for large river models are often limited due to given space and/or discharge capacity in the laboratory. However, too small model scale may result in an improper scaling of Reynolds numbers and friction factors. One technique to overcome this problem is to set up a distorted model in which the length scale number in vertical direction $\lambda_{L,v}$ is smaller than the length scale number in horizontal direction $\lambda_{L,h}$. By distorting the model, the flow velocity and flow depth is increased compared to an undistorted model with $\lambda_{L,h}$ in all directions and a larger Reynolds number is obtained (the influence of surface tension is decreased as well). By consequence, a hydraulic rough condition can set in. In this case, different scale factors must be regarded for different hydraulic parameters which can be found in Kobus (1980). It must be noted that the scaling of surface roughness in a distorted model is even more difficult than in an undistorted model. Proper scaling of roughness needs to be achieved by calibration using prototype data. It must also be taken into account that three-dimensional flow patterns, such as vertical eddy formation around a bottom structure, become difficult to transfer to the prototype in a distorted model ASCE (2000). The applicability of distorted models must be carefully evaluated. Some model tests, e.g. the three-dimensional mixing procedure in the near field of an outfall structure, do not allow model distortion.

9.4.4 Vegetational Flow

In the recent past, several studies were conducted which focused on the effects of vegetation in natural channels on the river flow. One major feature which must be addressed is that vegetation may be deformed under hydrodynamic conditions and the absolute roughness becomes a function of discharge. In hydraulic models, the vegetation is often modeled by some artificial plants which must then have a similar geometry and stiffness. Proper scaling of the vegetation characteristics is not trivial. The modeler may also pay attention to river vegetation when a study on sediment transport is planned as some relevant interaction can occur. Vegetational flow features are presented in detail in Aberle (2015) (this issue). The flow resistance nature of the vegetation can also change seasonally. For further information on the topic, the reader is referred to that chapter which also includes some exemplary model studies.

9.4.5 Debris and Ice

Modeling of debris and ice are challenging (and rare) tasks which may come up in early design stages of weirs, locks or hydropower plants and which require correct reproduction of drift and accumulation of ice and debris. For some questions, special attention must be paid to the proper modeling of the material to be used in the laboratory to reproduce the rigidity and eventually the thermal properties in case of ice. Furthermore, wind may play an important role, particularly when the load on a structure is of interest. Debris and ice modeling are no standard problems, see ASCE (2000) for additional details regarding derivation of relevant scale factors.

9.5 Hydraulic Structures

9.5.1 Aerated Flow

9.5.1.1 Introduction

Aeration mainly occurs in high-turbulent flows. The most important causes for aeration in rivers are hydraulic jumps and mixing downstream of weir overflows (see Fig. 9.4). For hydraulic jumps, the amount of entrained air depends on the inflow Froude number. In case of a weir, a growing turbulent boundary layer develops at the weir crest. If the structure is high enough, self-aeration sets in at the point of intersection of the boundary layer and water surface. In both cases, the entrainment of air leads to a bulking of the flow depth, i.e. the air bubbles are transported along the whole water column which generates higher water levels than for clear water. Knowledge of the amount of entrained air is thus important for the hydraulic design, e.g. for sidewall heights. Consideration of entrained air is also relevant for energy dissipation calculations when the bulked flow depth is measured in the laboratory.

Some weirs are used specifically to force air entrainment in rivers in order to re-oxygenate the water. Oxygen is transferred across the air-water interface of the bubbles. However, it must be considered that air bubbles are roughly of the same

Fig. 9.4 Aeration at hydraulic structures. **a** Aeration at a weir downstream of Oker dam in Germany; note the strong aeration at the *bottom outlet* in the background. **b** Aeration on a *cross-bar* block ramp due to local hydraulic jumps (Oertel and Bung 2012)

size in model and in prototype. Scale effects are thus very likely when oxygenation tests are performed in scaled model.

9.5.1.2 Air-Water Mixture

Description of the air-water mixture body becomes difficult when the flow is fully aerated. Close to the bottom, the air-water mixture is characterized by air bubbles transported in the water, while in higher elevation larger air pockets being entrapped between surface waves may be found (Wilhelms and Gulliver 2005). Above this elevation, mainly ejected droplets are found (see Fig. 9.5). By consequence, void fraction C is non-uniformly distributed along the water column. A typical void fraction profile for a stepped weir is also presented in Fig. 9.5.

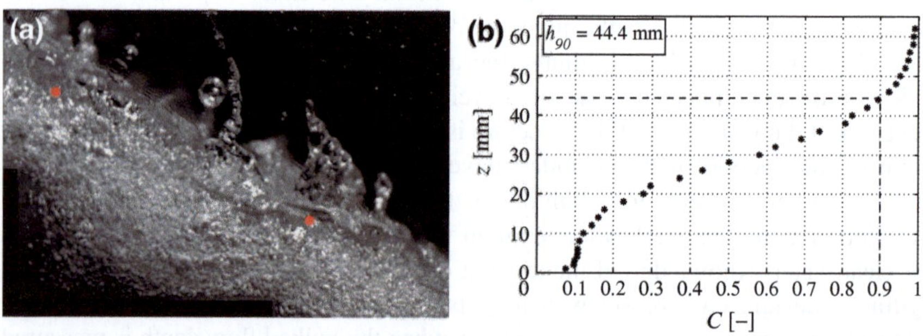

Fig. 9.5 Air-water mixture in self-aerated high-speed flows on a stepped weir model (width 50 cm, step height 6 cm, slope 1:2, discharge 35 l/s). **a** Photograph captured by a high-speed camera with 700 fps; the *markers* indicate the water level with 90 % time-averaged void fraction which is often taken into account as idealized water surface. **b** Void fraction profile measured at the stepped edges with a conductivity probe

In most studies, air-water flow properties were measured by means of intrusive needle probes which consist of fine tips that pierce the air bubbles. These probes work on basis of different conductivity characteristics between air and water (conductivity probe) or different optical refraction characteristics (fibre optical probes). The principle of these probes is discussed in Chanson (2002) amongst others. Recent studies show that modern non-intrusive methods may also be applied. Leandro et al. (2014) present a calibration method which can be used to obtain air concentration profiles and velocity fields from high-speed camera movies. Even the data from ultrasonic displacement sensors, normally known to be inaccurate in flows with irregular surface, can give some reasonable results as demonstrated by Chachereau and Chanson (2011) in hydraulic jumps and Bung (2013) in self-aerated chute flows.

9.5.1.3 Oxygenation Potential

Due to the large amount of air-water interface in aerated open channel flows, the oxygenation potential is high. This may be of interest to enrich water from a reservoir with oxygen when discharged to the downstream river. As mentioned above, scale effects cannot be avoided when investigations are performed on a scaled model due to similar air bubble sizes in the model and in prototype. However, in large-scale models the measured data give at least an idea of relative potentials when different situations are compared. Several recent studies presented direct oxygenation measurements where the water was de-oxygenated by addition of sodium sulfite before the test and the oxygen content recorded over time [(Avery and Novak 1978), (Bung 2011), (Essery et al. 1978), (Toombes and Chanson 2005) a.o.]. The references include more information on the technique, further information may also be found in ASCE (2007).

9.5.1.4 Scale Effects

Many publications address the very important topic of scale effects in aerated flows. Accordingly, model tests should be carried out on large-scale models. For self-aerated flows, a minimum scale is often given by 1:10 [e.g. (Chanson 1996)] and a minimum Reynolds number of 10^5 at the aeration point [e.g. (Kobus 1985)]. Newer publications identify a minimum Reynolds number of 2 to 3 $\times$ 10^5 to properly represent the turbulent properties [e.g. (Pfister and Chanson 2012)].

9.5.2 Scouring

For the modeling of scour processes, the basic considerations on particle dynamics from Sect. 9.4.2 dealing with sediment transport models are still valid. Scouring models are typically detailed models of flow at hydraulic structures, i.e. with larger scale than river models with movable beds. For the operation of a scour model, two important aspects are essential:

1. Scouring is a long-term process in nature which takes relatively less time in the model due to different time scales as discussed in Sect. 9.4.2. For the safe design of hydraulic structures the maximum values for the scour depth and size are important, i.e. the equilibrium state which sets in after a certain time under steady-state flow conditions. It must thus be ensured that this equilibrium state is obtained in the model. As scour dimensions are usually measured in dry conditions (after the model run), a sensitivity analysis in terms of a series of model runs is required for comparison of results from tests with different duration.
2. As mentioned above, the scour dimensions are measured in dry conditions. It must be ensured by the operator that the scour is not deformed during the run-off process. Seepage through the loose bed may lead to erosion in the scour hole and adulterate the results.

Usually, laser distance meters are used for measuring of scour depth. The measuring spot is tiny and the application in small scour becomes possible. Acoustic displacement meters are characterized by a larger measuring area due to spreading of the acoustic beam. These probes are less accurate for small scour holes where the acoustic beam size is in the range of the scour dimensions.

More information on scouring processes including some exemplary model studies can be found in Pagliara (2015) (this issue).

9.5.3 Structural Vibration

Knowledge about potential vibrations is essential in early design stages for any regulation structure. Severe damages may happen in case where excitation frequencies coincide with structural frequencies, i.e. the eigenfrequencies which depend on the structure's mass and stiffness. Besides a self-excited vibration which may be induced by a gate movement, flow-induced vibration are the most important vibration source. In this case, flow turbulence leads to fluctuating pressures around the structure. In general, two different ways of modeling are possible for vibration tests.

1. Proper modeling and scaling of all structural parameters, i.e. dimensions, mass and stiffness: Structural vibrations can then be directly inspected. However, scaling of the structure's stiffness is very difficult, particularly for complex structures.
2. Proper modeling of the hydraulic conditions and the structure's dimensions only (see Fig. 9.6): The structure's mass and stiffness are disregarded and a fully rigid model is built. In this case, only the external flow fluctuations can be measured in the model. Th excitation potential is then evaluated by comparison of the

Fig. 9.6 Rigid physical model (scale 1:35) of a high-head radial gate for vibration test purposes (Schlurmann and Bung 2012), note the pressure transducers installed in *bottom* and at the skin plate of the gate used to measure pressure fluctuations

major frequencies with mathematically derived eigenfrequencies of the structure. The modeler must then decide which eigenfrequencies are relevant for a global vibration. Generally, eigen modes higher than the 6–7th modes may be considered to lead to local, less severe vibration modes only.

The phenomenon of structural vibration in hydraulic engineering is addressed in ICOLD (1996). General information on its modelling is presented by Kobus (1980) while exemplary studies may be found in Chowdhury et al. (1997) and Schlurmann and Bung (2012).

9.5.4 Outfalls

Effluents from outfall structures often have a different density than the ambient water body. Thus, upward (positive) or downward (negative) directed lift forces due to the different specific weight affect the turbulent diffusion. In this case, a densimetric Froude number[3] becomes relevant (Turner 1966):

$$\mathrm{F}_d = \frac{v}{\sqrt{g \times (\Delta\rho/\rho) \times L}} \tag{9.10}$$

where $\Delta\rho$ is the difference in density between effluent and ambient water.

Generally, the near field around the outfall must be distinguished from the far field. In the near field, the mixing is mainly caused by small-scale turbulence between the jet and the surrounding water body. In the far field, large-scale

[3]For temperature-driven density differences an alternative formulation is given in Riester et al. (1980).

turbulent diffusion maintains the mixing process. Thus, near field investigations need a detailed reproduction of hydraulic and geometrical conditions at the structure, while for far field investigations proper modeling of the river characteristics is required. However, the transition between near field and field is hard to define. The modeler should therefore ensure correct modeling for all regions.

As the turbulence is the decisive parameter for the mixing, the modeler must pay attention to possible scale effects. As mentioned earlier, the Reynolds number can never be scaled correctly in a Froude model. According to Kobus (1980), a minimum Reynolds numbers of 2000 for the jet and 3000 for the river, respectively, must be achieved to avoid those scale effects. It must be noted that degradation time scales of pollution due to biological and chemical processes cannot be scaled.

Models are mostly operated by adding some tracers which show a comparable mixing characteristic as the real effluent. For example, the tracer could have a certain salinity leading to a different density. The mixing could then be observed by measuring the decreasing salinity at numerous downstream cross sections. In laboratories with closed water cycles, the modeler must consider the travel time of the tracer to re-arrive at the model to avoid an undesired increase of density at the model inlet.

References

Aberle J (2015) Hydrodynamics of vegetated channels. In: Rowinski PM, Radecki-Pawlik A (eds) Rivers—physical, fluvial and environmental processes, Springer (this issue), Berlin

Avery ST, Novak P (1978) Oxygen transfer at hydraulic structures. J Hydr Div HY11:1521–1540

ASCE (2000) Hydraulic modeling: concepts and practice. Bulletin 7, ASCE manuals and reports on engineering practice. American Society of Cicvil Engineers, Reston

ASCE (2007) Measurement of oxygen transfer in clean water (ASCE standard 2-06). American Society of Cicvil Engineers, Reston

Bung DB (2011) Fließcharakteristik und Sauerstoffeintrag bei selbstbelüfteten Gerinneströmungen auf Kaskaden mit gemäßigter Neigung (Flow characteristics and oxygenation in self-aerated skimming flows on embankment cascades, in German). ÖWAW 3–4(2011):76–81

Bung DB, Sun Q, Meireles I, Viseu T, Matos J (2012) USBR type III stilling basin performance for steep stepped spillways. In: 4th IAHR symposium on hydraulic structures, Porto

Bung DB (2013) Non-intrusive detection of air–water surface roughness in self-aerated chute flows. J Hydr Res 51(3):322–329

Chachereau Y, Chanson H (2011) Free-surface fluctuations and turbulence in hydraulic jumps. Exp Thermal Fluid Sci 35(6):896–909

Chanson H (1996) Air bubble entrainment in free-surface turbulent shear flows. Academic Press, San Diego

Chanson H (2002) Air-water measurements with intrusive phase-detection probes: can we improve their interpretation? J Hydr Eng 128(3):252–255

Chowdhury MR, Hall RL, Pesantes E (1997) Flow-induced vibration experiments for a 1:25-scale-model flat wicket gate. Technical report SL-97-4, U.S. Army Corps of Engineers. Waterways Experiment Station, Louisville

Essery ITS, Tebutt THY, Rasaratnam SK (1978) Design of spillways for re-aeration of polluted waters. Technical report 72. Construction Industry Research and Information Association (CIRIA), Birmingham

Heller V (2011) Scale effects in physical hydraulic engineering models. J Hydr Res 49(3):293–306
Hughes SA (1993) Advanced series on ocean engineering. In: Liu PL-F (ed) Physical models and laboratory techniques in coastal engineering, vol 7. World Scientific Publishing, Singapore
ICOLD (1996) Vibrations of hydraulic equipment for dams: review and recommendations. Bulletin 102, Commission International des Grands Barrages/Committee on hydraulics for dams, Paris
Kobus H (1980) Hydraulic modelling. Bulletin 7. German Association for Water Resources and Land Improvement. Parey, Hamburg
Kobus H (1984) Symposium on scale effects in modelling hydraulic structures. University of Stuttgart, Stuttgart, Hydraulic Engineering Institute
Kobus H (1985) An introduction to air-water flows in hydraulics. University of Stuttgart, Hydraulic Engineering Institute, Stuttgart
Schlurmann T, Bung DB (2012) Experimental investigation of flow-induced radial gate vibrations at Lower Subansiri dam. In: 6th Chinese–German joint symposium on hydraulic and ocean engineering, Keelung
Leandro J, Bung DB, Carvalho R (2014) Measuring void fraction and velocity fields of a stepped spillway for skimming flow using non-intrusive methods. Exp Fluids. doi:10.1007/s00348-014-1732-6
Oertel M, Bung DB (2012) Characeristics of cross-bar block ramp flows. In: 4th IAHR symposium on hydraulic structures, Porto
Pagliara S (2015) Energy dissipation and scouring problems in rivers. In: Rowinski PM, Radecki-Pawlik A (eds) Rivers—physical, fluvial and environmental processes, Springer (this issue), Berlin
Pfister M, Chanson H (2012) Discussion of scale effects in physical hydraulic engineering models. J Hydr Res 50(2):244–246
Riester JB, Bajura RA, Schwarz SH (1980) Effects of water temperature and salt concentration on the characteristics of horizontal buoyant submerged jets. J Heat Transfer 102:557–562
Toombes L, Chanson H (2005) Air–water mass transfer on a stepped waterway. J Environ Eng 131 (10):1377–1386
Turner JS (1966) Jets and plumes with negative or reversing buoyancy. J Fluid Mech 26:779–792
Wilhelms SC, Gulliver JS (2005) Bubbles and waves description of self-aerated spillway flow. J Hydr Res 43(5):522–531

Chapter 10
Measurements of Turbulence Structure in a Compound Channel

Adam Paweł Kozioł and Janusz Kubrak

Abstract Experimental research was undertaken to investigate the changes in spatial turbulence intensity, water turbulent kinetic energy, the time and spatial macro-scale, scales of turbulent eddies (macro- and microeddies) in a compound channel. Three tests for two various roughness values were realized. The surface of the main channel bed was smooth and made of concrete, whereas the floodplains and sloping banks were covered by cement mortar composed of terrazzo. Instantaneous velocities were measured with the use of a three-component Acoustic Doppler Velocimeter (ADV). The distributions of relative turbulence intensity (u'/U, v'/U, w'/U) in the main channel and on the floodplains were presented. It was found that the longitudinal (u'/U) and transverse (v'/U) turbulence values decreased from the bottom upwards to the floodplain elevation (z/h = 0.56) in the main channel, but remained constant above the floodplain level. Vertical relative turbulence intensity (w'/U) increased going up from the bottom until z/h = 0.15, decreased until about z/h = 0.7, and then increased again upwards to the water surface. The distributions of relative turbulence intensities were described with regression equations. The distributions of turbulent kinetic energy at different water depths were described by regression equations. Vertical distributions of turbulent kinetic energy on the floodplains and over the banks of the main channel were divided into three zones. Over the bottom of the main channel, four zones were determined, containing the middle zone of the flow field divided into two zones of different trends. Measurements of instantaneous velocities are used to investigations of the longitudinal sizes of the smallest eddies (Kolmogorov's microscale). Presented analyses were based on the already published results.

Keywords Compound channel · Turbulence intensity · Turbulent kinetic energy · Sizes of the smallest eddies · Floodplains

---

A.P. Kozioł (✉) · J. Kubrak
Department of Hydraulic Engineering, Warsaw University of Life Sciences, Nowoursynowska Str. 159, 02-787 Warszawa, Poland
e-mail: adam_koziol@sggw.pl

P. Rowiński and A. Radecki-Pawlik (eds.), *Rivers—Physical, Fluvial and Environmental Processes*, GeoPlanet: Earth and Planetary Sciences,
DOI 10.1007/978-3-319-17719-9_10

10.1 Introduction

Lowland rivers during flood are characterized by a compound cross-section and unsteady flow. The characteristics of turbulence are not well recognized in such a situation. The studies on flow conditions in channels of the compound cross-section confirmed that between the main channel and the floodplain, a transitional zone with strongly developed eddy structures exists. The intensive turbulence zone results from discontinuity of the velocity field. As such discontinuity cannot be held in the viscous fluid, the high velocity gradient leads to an intensive exchange of the water mass and momentum, causing continuous variation of the velocity profile. In the transitional zone, the rapid change of the flow velocity occurs, maintaining the main flow direction. The transitional zone has a finite width and is characterized with the non-zero velocity gradient. It causes a rotation of the stream in the main channel and its circulation, the appearance of eddy currents and waving of the water surface on the interface of the channel and the floodplain.

One may suspect that an additional flow resistance occurs due to the interaction process, i.e. an intensive momentum exchange and a strong mixing of water between the deep main channel and the adjacent shallow floodplains. To make the considerations simpler, it is convenient to study the geometrical and temporal complexities separately, i.e. to study the turbulence structure in a compound cross-section at steady state conditions and to study turbulence under unsteady state condition. For a long time, research on turbulence characteristics has been performed for both small- and large-scale cross-sectional models of single and compound channels (Czernuszenko and Lebiecki 1980; McQuivey et al. 1971; Nezu and Rodi 1986; Knight and Shiono 1990; Rowiński et al. 1998, 2002; Mazurczyk 2007; Yang et al. 2007; Sanjou et al. 2010; Terrier et al. 2010; Mera et al. 2014). In addition, more and more research has been performed on turbulence structure in natural rivers (Czernuszenko and Lebiecki 1989; Nikora et al. 1994; Nikora and Smart 1997; Babaeyan-Koopaei et al. 2002; Stone and Hotchkiss 2007; Sulaiman et al. 2012; Xie and Pan 2013). In the present study, experimental investigations performed in a laboratory compound channel are considered. The main aim of this paper is the analysis of turbulence variation and relative turbulence intensity distributions, turbulent kinetic energy of water and sizes of eddies in a channel of a compound cross-section. Presented analyses were based on the already published results.

10.2 Description of the Problems

Shallow floodplains occurring in nature are usually rough and a lot of attention has been paid recently to the investigations of the influence of various roughness conditions on the flow structure. Open channel flow is, by nature, three-dimensional. Even in a straight laboratory channel, turbulence can cause instantaneous velocities v_i and w_i. The turbulence intensity is one of the most important characteristics of turbulent water flow, being defined in all three directions from the velocity data as follows:

$$u' = \sigma_x = \sqrt{\overline{u_i'^2}}, \quad v' = \sigma_y = \sqrt{\overline{v_i'^2}}, \quad w' = \sigma_z = \sqrt{\overline{w_i'^2}} \qquad (10.1)$$

where x, y, and z, are the streamwise, lateral, and vertical directions, respectively. However, many researchers prefer the relative turbulence intensity given by u'/U, v'/U, and w'/U, where U = time-averaged point velocity in the x direction.

Nezu and Nakagawa proposed the equations describing relative turbulence intensity distributions in single channels for steady two-dimensional flow (Nezu and Nakagawa 1993). The conditions of water flow in a compound channel are significantly different from conditions in channels of a single cross-section. In the case of the former channel type, the shape of the cross-section varies, and there are different roughness values in the main channel and on the floodplains. Due to this, the flow structures that occur in rivers of a compound cross-section are very complex due to at least three mechanisms: boundary-generated turbulence, free shear layer turbulence and velocity fluctuations associated with perturbations in the longitudinal secondary flow cells (Shiono and Knight 1991; Tominaga and Nezu 1991). Knight and Shiono (1990) determined the constants for the equations of turbulence intensity in a compound channel at its centerline, which differ from constants given by Nezu and Nakagawa (1993). They concluded also that the three-dimensional nature of the flow in both corner and shear layer regions is significant, for the overbank flow cases, and, therefore, that the intensities deviate very strongly from their nominally two-dimensional values, in particular in the side slope area.

Babaeyan-Koopaei et al. (2002) study, based on research results for a natural river, show that near the main channel and in the midpoint between the floodplain and the main channel, measured data deviate from the Nezu and Nakagawa's (1993) equations. The flow behaviour is certainly not two dimensional at these locations and the three-dimensional nature of the flow in both the main channel and the main channel/floodplain interface is evident. The validation of the Nezu and Nakagawa's equation (1993) was also performed on the basis of the model results, presented in this paper, and also described by Czernuszenko et al. (2007). Modelled coefficients found here are very similar to the values by Nezu and Nakagawa at the channel centreline, but other coefficients are considerably different from those achieved for two-dimensional steady flow.

Water velocity pulsations are caused by turbulent vortex structures. Dimensions vortices change from largest to smallest, which occurs within the turbulent kinetic energy dissipation is due to the viscosity of water (Yokosi 1967). Large-scale vortices are unstable and quickly disintegrate. The final step is to change the distribution of turbulent vortices of the kinetic energy into kinetic energy of molecular motion or heat (Yokosi 1967). The analysis of the transport of kinetic energy allow autocorrelation function and spectral.

Turbulent kinetic energy (TKE) is the mean kinetic energy per unit mass of fluid in relation with turbulent flow eddies. Physically, it is characterized by the measured, mean squared (RMS) velocity fluctuation. On the basis of three-dimensional turbulent velocity values—along 3 axes: {x, y, z}, the longitudinal turbulent intensity u', the transverse turbulent intensity v' and the vertical turbulent intensity w' were calculated, and then the turbulent kinetic energy (TKE), k, was calculated (Nezu and Nakagawa 1993) according to the following formula:

$$k = 0.5\left(u'^2 + v'^2 + w'^2\right) \tag{10.2}$$

The turbulent kinetic energy is produced by shear and wakes. In principle, shear production converts mean kinetic energy to turbulence and at the same time wakes are created as an effect of work against form drag (Rowiński and Mazurczyk 2006).

According to Grinvald and Nikora (1988), the share of longitudinal turbulence intensity in kinetic energy is dominant and ranges from 55 to 80 %. In the analysed channel, the share of longitudinal turbulence intensity in the kinetic energy was about 50 %. The lower percentage share of longitudinal turbulence intensity, in the analysed case, results from a complex cross-sectional area and relatively small main channel roughness, which contributes to the process of water mixing between the main channel and the floodplains for the whole analysed channel width (Mazurczyk 2007). This process significantly influences the increase of turbulence intensity in the transverse direction. This is why the determination of turbulence structure, especially in the case of channels with a complex cross-section, requires the measurements of velocities with proper probes at least in two directions—the longitudinal and the transverse one. Usually, the vertical turbulence intensity has the lowest share in the kinetic energy, not exceeding several percents. Grinvald and Nikora (1988), basing on their research results, proved that the share of vertical turbulence intensity in the kinetic energy decreases downwards to the bottom. The measurements in the field proved their observations, however only for main channels.

Nezu and Nakagawa (1993) proposed the description of the distribution of vertical turbulent kinetic energy in the form of exponential regression equation, valid for the intermediate region of single channels for steady two-dimensional flow. Knight and Shiono (1990) found that in a channel of a complex cross-section, there is a similarity of turbulent kinetic energy distribution, and the best fit to the exponential regression equation was obtained for the channel centreline. They found also that the data revealed the three dimensional nature of the flow in regions where strong lateral shear and secondary flow exit.

The external scale of turbulence is determined by the sizes of macroeddies. Evaluation of sizes of macroeddies must be preceded by the determination of time-scales of macroeddies. To this end, autocorrelation functions $R(t)$ were used for this evaluation. The functions exhibit very similar forms of decaying curves with an alteration of the domains of the positive and negative values. Basing on autocorrelation functions, Euler time-macroscales were derived T_E:

$$T_E = \int_0^\infty R(t)dt \tag{10.3}$$

which are the measure of the slowest changes in the turbulent flow caused by macroeddies. For a steady and uniform turbulent flow, when mean velocity in a given point significantly exceeds the velocity of fluctuations, there exists, according to Taylor's hypothesis, a direct relationship of temporal and spatial Eulerian autocorrelation functions. Basing on Taylor's relationships between the spatial L and temporal T_E turbulence macroscale, following formula for mean, longitudinal sizes of macroeddies L in a channel, the following can be written:

$$L = UT_E \tag{10.4}$$

where U is the mean-time longitudinal velocity in a point. In this paper, sizes of macroeddies are related to local values of a depth for easier comparison.

A size of microeddies stands for the smallest size of eddies which are present in the turbulent flow of water. Kolmogoroff and Taylor proposed the spatial scale of microeddies, where Taylor's microscale depends both on the macroscopic motion by means of the fluctuating velocity and on dissipative characteristics, whereas Kolmogoroff's microscale depends only on dissipative and viscous characteristics. This paper presents the Kolmogoroff's microscale η. In order to determine the size of microeddies η, the spectrum density function $S(\omega)$ was calculated for instantaneous, longitudinal velocities and next the proper frequency subrange of velocities was determined, and for that subrange the rate of energy dissipation ε was analysed (Kozioł 2012). The spectrum density function $S(\omega)$ expresses kinetic energy of eddies for the frequency range (ω, $\omega + d\omega$) and this function is not uniform versus frequency. Exemplary spectrum density functions for the courses of instantaneous longitudinal velocities are given in the section with results. These functions take the highest values for the frequencies of the averaged motion, while the lowest values are reached for the highest frequencies. The analysis of the spectrum density function involves, among others, the determination of the frequency subrange (inertial subrange), whose existence is assumed by Kolmogoroff's hypothesis, and where the local anisotropy hypothesis is also valid as well as the "−5/3" power law (Nikora 1999). Nikora (1999) has proposed an explanation for the existence of the regions in which an inverse "−1" power law is fulfilled. Such a subrange is called the inertial subrange of a constant energy stream, coming from all scales of eddies: from the largest eddies to the smallest ones. The determination of the inertial subranges enables to find turbulence scales and energy dissipation rates. The aforementioned values can be calculated by applying the following formulas (Nikora et al. 1994):

$$S(\omega) = C_1 \varepsilon^{2/3} U^{2/3} (\omega)^{-5/3} \tag{10.5}$$

$$\eta = \left(\frac{\nu^3}{\varepsilon} \right)^{1/4} \tag{10.6}$$

where: C_1 = 0.48—constant (Monin and Yaglom 1975), η—size of microeddies (Kolmogoroff's microscale), ν—molecular (kinematic) viscosity. Equations (10.5) and (10.6) are valid on the condition that we apply the Taylor hypothesis of "frozen turbulence". Equation (10.4), known as the Kolmogoroff's power law "−5/3", is used to determine energy dissipation rates in an inertial subrange. The rates of energy dissipation were calculated as arithmetical means, derived from the course of a spectrum function for a particular subrange. Knowing the energy dissipation rates, the sizes of Kolmogoroff's microeddies η were calculated by applying Eq. (10.6).

In the trapezoidal, compound channel model used here, one- and three-dimensional velocity measurements were performed in order to estimate basic turbulence parameters in a channel cross-section and to determine turbulence structure on floodplains (Kozioł et al. 1998; Rowiński et al. 1998; Kozioł 2000, 2002, 2008, 2011, 2013; Rowiński et al. 2002; Czernuszenko et al. 2007; Mazurczyk 2007).

10.3 Experimental Setup and Procedure

The experiments considered herein were carried out in the Hydraulic Laboratory of the Department of Hydraulic Engineering, Faculty of Civil and Environmental Engineering at the Warsaw University of Life Sciences—SGGW. A straight open channel (16 m long and 2.10 m wide) with a symmetrically trapezoidal cross section was used for the laboratory tests (Figs. 10.1 and 10.2). The main channel width was 30 cm and the floodplain width was 60 cm. The banks were inclined at a slope of 1:1. The channel bed slope of the channel was 0.5 ‰. Water discharge values were recorded with the use of a 540 mm diameter circular weir. A row of PCV pipes was installed in the upstream approach channel reach to straighten the flow. A uniform and steady flow was used in every case. The water surface was

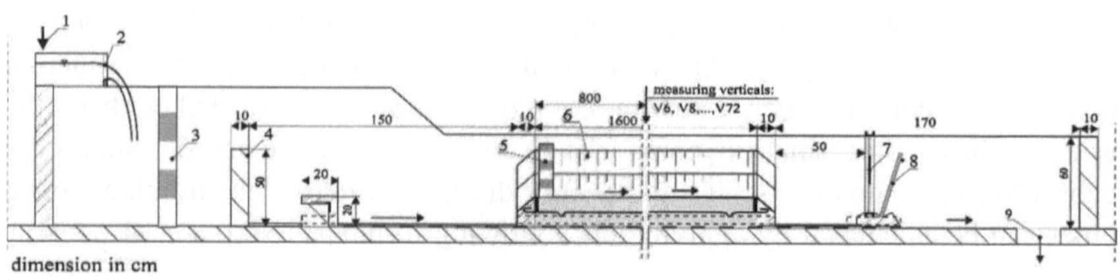

Fig. 10.1 Scheme of the laboratory flume. *1* Inflow, *2* circular weir, *3* honeycombed wall, *4* overfall, *5* row of PCV pipes, *6* model of the channel, *7* the open-work-weir, *8* adjustable overflow weir, *9* outflow

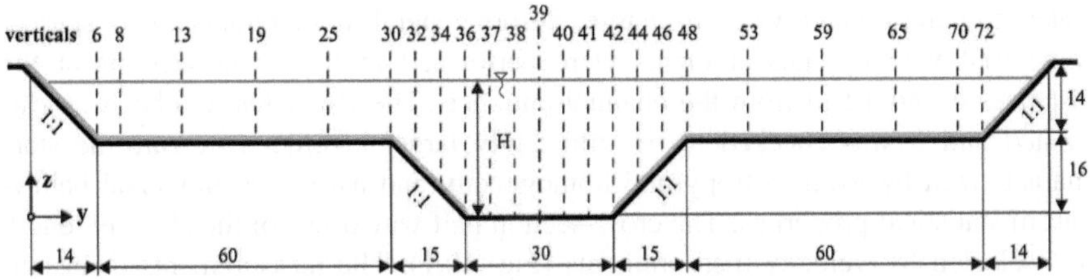

Fig. 10.2 Scheme of a laboratory cross-section for test 1, 2 and 3 in a channel with the smooth bed of the main channel and rough floodplains (dimensions in cm)

kept parallel to the bed during the experiments. The water surface slope was measured by recording the pressure differences between readings of piezometers located along the centreline of the channel bed at the distances 4 and 12 m from the channel entrance.

Three tests for two roughness values were conducted (Fig. 10.2). The surface of the main channel bed was smooth (Manning's roughness coefficient $n = 0.011\ \mathrm{m}^{-1/3}\ \mathrm{s}$) and made of concrete whereas the floodplains and all sloping banks were covered by cement mortar composed with terrazzo with grains of 0.5–1 cm in diameter (Fig. 10.2). However, the studies showed that the calculated Manning's roughness coefficient and the absolute roughness of both surfaces had been different due to slight variations in mortar application procedures. The Manning's roughness coefficient for the rough surfaces of the floodplain equalled about $n = 0.018\ \mathrm{m}^{-1/3}\ \mathrm{s}$ for the left and $0.025\ \mathrm{m}^{-1/3}\ \mathrm{s}$ for the right area. The values of average Manning's coefficient and absolute roughness of the channel surface were determined from the Manning equation and the Colebrook-White equation on the basis of the average velocity values of the flow measured in those parts of the channel. The obtained roughness amounted to $k_s = 0.00005$ m for the smooth surfaces, $k_s = 0.0074$ m for the rough surface of the left floodplain, and $k_s = 0.0124$ m for the rough surface of the right floodplain. The hydraulic parameters of each experimental case are given in Table 10.1.

During the experiments the following parameters were measured: water levels in the main channel and on the floodplains, all three components of point velocities,

Table 10.1 The hydraulic parameters of experiment tests

Parameter	Test		
	1	2	3
Discharge Q (l/s)	95.2	81.1	61.5
Depth in the main channel H (cm)	28.3	26.4	24.1
Depth on the floodplain h_f (cm)	12.3	10.4	8.1
Relative depth $D_r = h_f/H$	0.43	0.39	0.34
The sloping banks (–)	1:1		
The bed slope i (‰)	0.5		

water temperature and water discharge. Before general measurements were started, some trial velocity measurements were performed in few cross-sections at the distances 4 and 12 m from the channel entrance. The Reynolds number was calculated and it was checked to be sufficiently large, in order to create the state characterized by local isotropy and homogeneity and associated universal behaviour of statistical properties. The cross-section half way down of the channel length was selected for velocity measurements (Fig. 10.1). The measurements of instantaneous velocity were carried out at 250 points at 23 verticals—six on each floodplain and eleven in the main channel (Fig. 10.2).

Instantaneous velocities were measured with the use of a three-component Acoustic Doppler Velocimetry (ADV) manufactured by Sontek Inc. The acoustic sensor was mounted on a rigid stem attached to a specially designed trolley allowing for its detailed positioning. The ADV works on pulse-to-pulse coherent Doppler techniques in relatively high temporal resolution. The ADV proved to yield a good description of the turbulence characteristics when certain conditions related to the flow itself and the configuration of the instrument were satisfied. The measurements were conducted with a maximum frequency of 25 Hz in the velocity range of 0–1.0 m/s with an accuracy of 0.25 cm/s. A sampling duration of 6 min was determined from initial tests to be adequate to obtain stationary turbulence statistics and was, therefore, used. Buffin-Bélanger and Roy (2005) report that, for most turbulent statistics, a sufficient record length for the measurements is 60–90 s. In cases of our experiments, even longer time series were recorded to provide reliability of data and constancy of higher order velocity moments. The measured velocity field was obviously of a stochastic nature and the stationarity and ergodicity of the process was checked. It is important to note that sporadic abrupt spikes appearing in velocity time series were removed to avoid their significant influence on the variance and further statistical moments. Spikes were removed with the program WinADV, which applied the combinations of two filters, the phase-space threshold despiking filter Goring and Nikora with modified Wahl (2000, 2003) and the filter of the minimum SNR. Information on various recommended methods of despiking the data are to be found in the study of Goring and Nikora (2002).

10.4 Results of Relative Turbulence Intensity

In this section our analysis is based on the already published results, which were presented in Kozioł (2013) and we are using them to explain a change of the distributions of relative turbulence intensity in the channel of compound cross-section. The intensity of the longitudinal relative turbulence of the main channel on the floodplain was investigated on the basis of laboratory measurements of water velocities—given by the test 1 presented in Fig. 10.3. The results of two remaining tests, 2 and 3, provide similar distributions of the intensities and would lead to

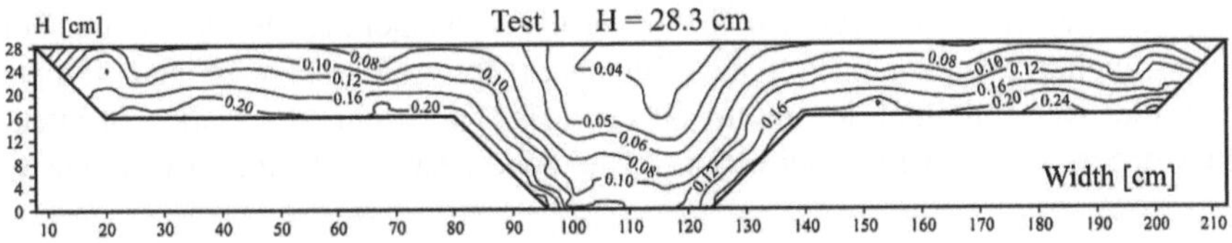

Fig. 10.3 Contour lines for longitudinal relative turbulent intensities (u'/U) in a cross-section channel in test 1 (Kozioł 2013)

similar observations. The maximum intensity of the longitudinal relative turbulence was reported on the floodplains and in the transitional area between the main channel and the floodplain (Fig. 10.3). Over the sloping banks of the main channel the intensity was significantly lower and the minimum was reached over the bottom of the main channel. On the contrary, values on the floodplain were approximately two times higher than in the main channel. The intensities are affected by the roughness of floodplains and sloping banks of the main channel and also by the interactions of zones of the different velocity. Previous studies with this model of the compound channel showed that the relative turbulence intensity increases significantly with the bed roughness and higher values are expected at the interface between shallow and deep areas (Rowiński et al. 2002). As it can be seen in Fig. 10.3, the distribution of the intensity is weakly asymmetric. The intensity of the relative turbulence is slightly stronger at the left interface and left parts of the main channel. The asymmetry is caused by the different roughness of the bottom: the highest roughness is on the right sloping banks and on the right floodplain. The bulges at the interface (Fig. 10.3) were not present, probably due to the strong influence of the rough bottom and the uniform value of the transverse velocity on the width of this area.

In Fig. 10.4 the vertical profiles of the intensity of the relative turbulence in three directions for test 1 are presented. The intensities were shown against the relative depth z/h, where z denotes distance from the bed, h depth in a vertical, z and h are relative to the floodplain bed and the main channel bed, depending on the location. For the depth z/h in the range of 0.04–0.92, intensities decreases significantly in the direction of the water surface. Near bottom maximum values of intensities result from the highest gradients of mean velocity (Fig. 10.4). The intensities in verticals over the bed of the main channel (V37-V41) do not vary considerably (Fig. 10.5), as it can be noticed for verticals over the sloping banks of the main channel (V32-V36, Fig. 10.4) and to verticals on the floodplains (V6-V30, Fig. 10.4). The most noticeable gradient of the relative turbulence intensity, between three directions occurs over the sloping banks of the main channel. For both floodplains it is less significant.

The study confirms the results achieved in natural rivers (Babaeyan-Koopaei et al. 2002) that longitudinal relative turbulence intensity is dominant over other directions (Fig. 10.4). As it can be seen in Fig. 10.4, the lowest intensities were achieved for vertical direction, both for the main channel and the floodplain. For depths higher than $z/h = 0.5$, the differences between transverse and vertical intensities were minimal, for almost all verticals. The strongest variation of intensities occurs for the vertical direction in the zone near the channel bed. Intensity

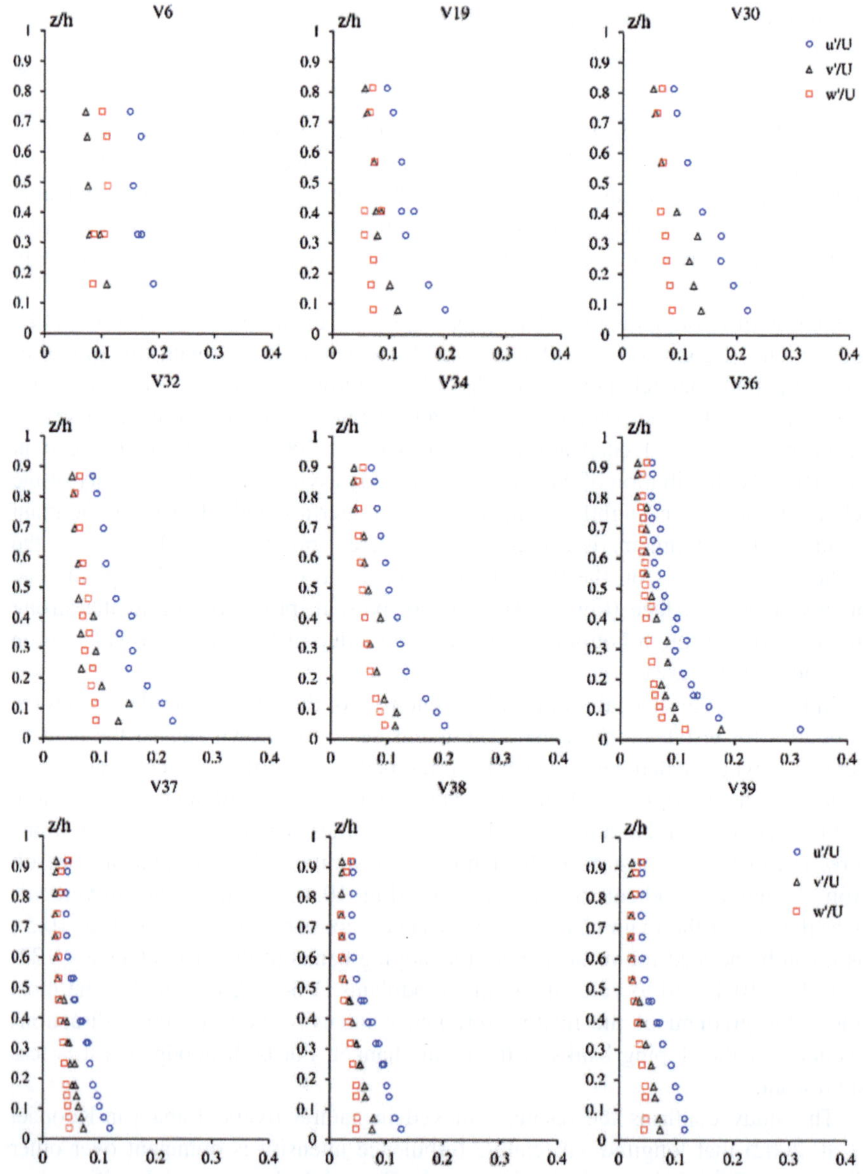

Fig. 10.4 Vertical distributions of relative turbulence intensities in three directions on the floodplains and in the main channel in test 1 (Kozioł 2013)

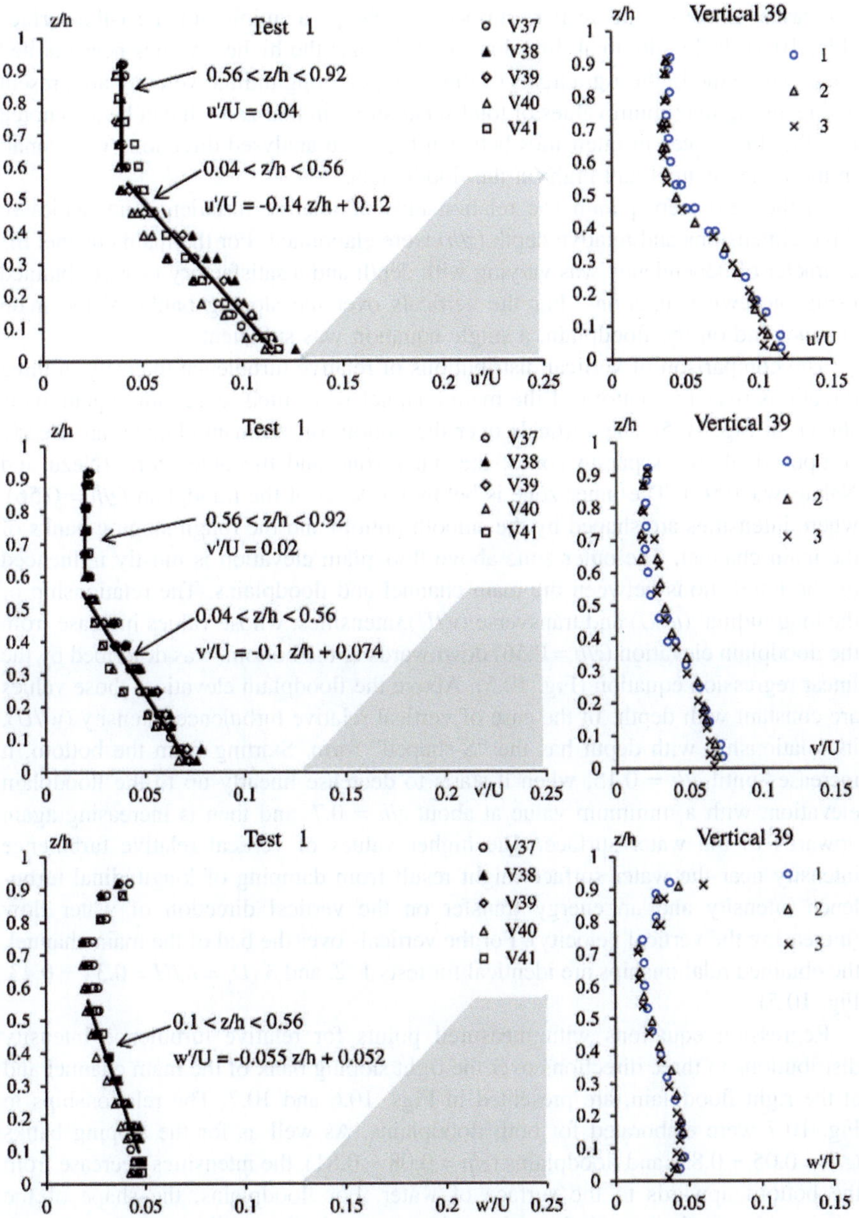

Fig. 10.5 Vertical distributions of relative turbulence intensities in three directions over the bed of the main channel (Kozioł 2013)

decreases towards the water surface, reaching a minimum near-the-surface (Fig. 10.4). In longitudinal direction, intensity has the highest values near the bed zone, where the highest gradients of the averaged, longitudinal velocity are present and resulting maximum values of total shear stress in that zone. It can be also noted that the differences in intensities between the three analysed directions were small in the main channel and high on the floodplains.

In the next step, parametric relationships of relative turbulence intensities in three dimensions and relative depth (z/h) were elaborated. For the main channel the character of dependency was varying with depth and a satisfactory fit was obtained using piecewise functions. For the verticals over the sloping banks of the main channel and on the floodplain, a single equation was sufficient.

The comparison of vertical distributions of relative turbulence intensity in three directions over the bottom of the main channel with fitted regression equations is shown in Fig. 10.5. The verticals over the bottom of the main channel are clearly composed of two separate zones: the inner zone and the outer zone (Nezu and Nakagawa 1993). The inner zone is below the level of the floodplain ($z/h = 0.56$), where intensities are shaped by the smooth bottom and the rough sloping banks of the main channel. The outer zone above floodplain elevation is mostly influenced by the interactions between the main channel and floodplains. The relationship of the longitudinal (u'/U) and transverse (v'/U) intensities, whose values increase from the floodplain elevation ($z/h = 0.56$) downwards to the bottom, was described by the linear regression equation (Fig. 10.5). Above the floodplain elevation, those values are constant with depth. In the case of vertical relative turbulence intensity (w'/U), its relationship with depth has the "S-shaped" form. Starting from the bottom, it increases until $z/h = 0.15$, when it starts to decrease linearly up to the floodplain elevation, with a minimum value at about $z/h = 0.7$, and then is increasing again upwards to the water surface. The higher values of vertical relative turbulence intensity near the water surface might result from damping of longitudinal turbulence intensity and an energy transfer on the vertical direction of water flow (increasing the vertical velocity). For the verticals over the bed of the main channel, the obtained relationships are identical for tests 1, 2, and 3 ($D_r = h_f/H = 0.34 - 0.43$, Fig. 10.5).

Regression equations with measured points for relative turbulence intensity distributions in three directions over the right sloping bank of the main channel and at the right floodplain, are presented in Figs. 10.6 and 10.7. The relationships in Fig. 10.7 were elaborated for both floodplains. As well as for the sloping banks ($z/h = 0.05 - 0.85$) and floodplains ($z/h = 0.08 - 0.81$), the intensities decrease from the bottom upwards to the surface of water. For floodplains, the shape of the relationship of relative turbulence intensity is very similar in all tests. In test 3, with the lowest water level, intensities (w'/U) do not change with depth, because of the strong influence of the bottom roughness. The complex nature of the distribution of vertical relative turbulence intensities in three directions confirms a three-dimensional nature of water flow in a channel of a compound cross-section.

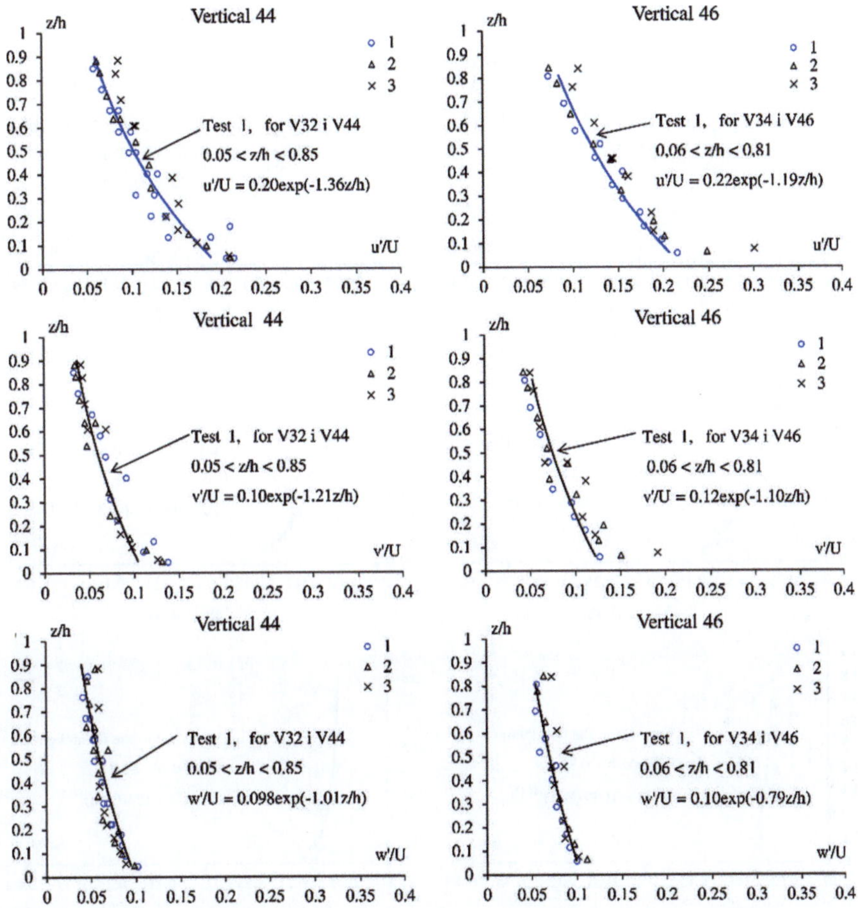

Fig. 10.6 Vertical distributions of relative turbulence intensities in three directions over the right sloping bank of the main channel (Kozioł 2013)

10.5 Turbulent Kinetic Energy in the Compound Channel

Figure 10.8 presents vertical turbulent kinetic energy distributions (TKE) in the main channel and on the floodplains, achieved in test 1 (smooth main channel bottom, rough surface of the main channel banks and floodplains). Turbulent kinetic energy is calculated on the basis of the turbulence intensity, and that is why vertical distributions of the TKE are almost identical to the distributions of the relative turbulence intensity in the longitudinal direction, in relation to the relative depth z/h (Kozioł 2013).

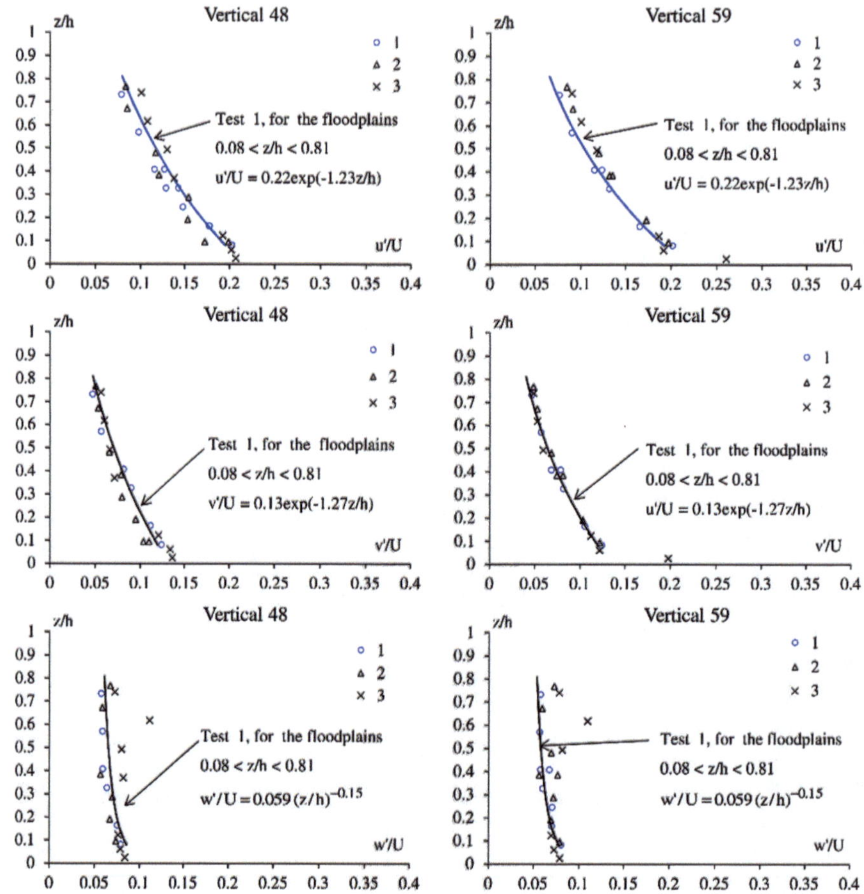

Fig. 10.7 Vertical distributions of relative turbulence intensities in three directions on the right floodplain (Kozioł 2013)

Figure 10.9 presents vertical distributions k/U_*^2 over the bottom and the linear regression equations, describing the energy distribution versus the water depth in the centreline of the main channel (vertical 39). Friction velocity U_* was determined from measured Reynolds stress distribution (Table 10.2; Nezu and Nakagawa 1993; Rowiński et al. 2005). In the channel of a compound cross- section, with rough floodplains and rough main channel banks, it is not possible to apply one equation for vertical turbulent energy k/U_*^2 distribution over a smooth main channel bottom, which is also proved by research of Knight and Shiono (1990). The work by Knight and Shiono (1990) focused on a channel with a smooth bottom surface and of a similar cross-section, and as they suggested, there was the similarity of energy k/U_*^2

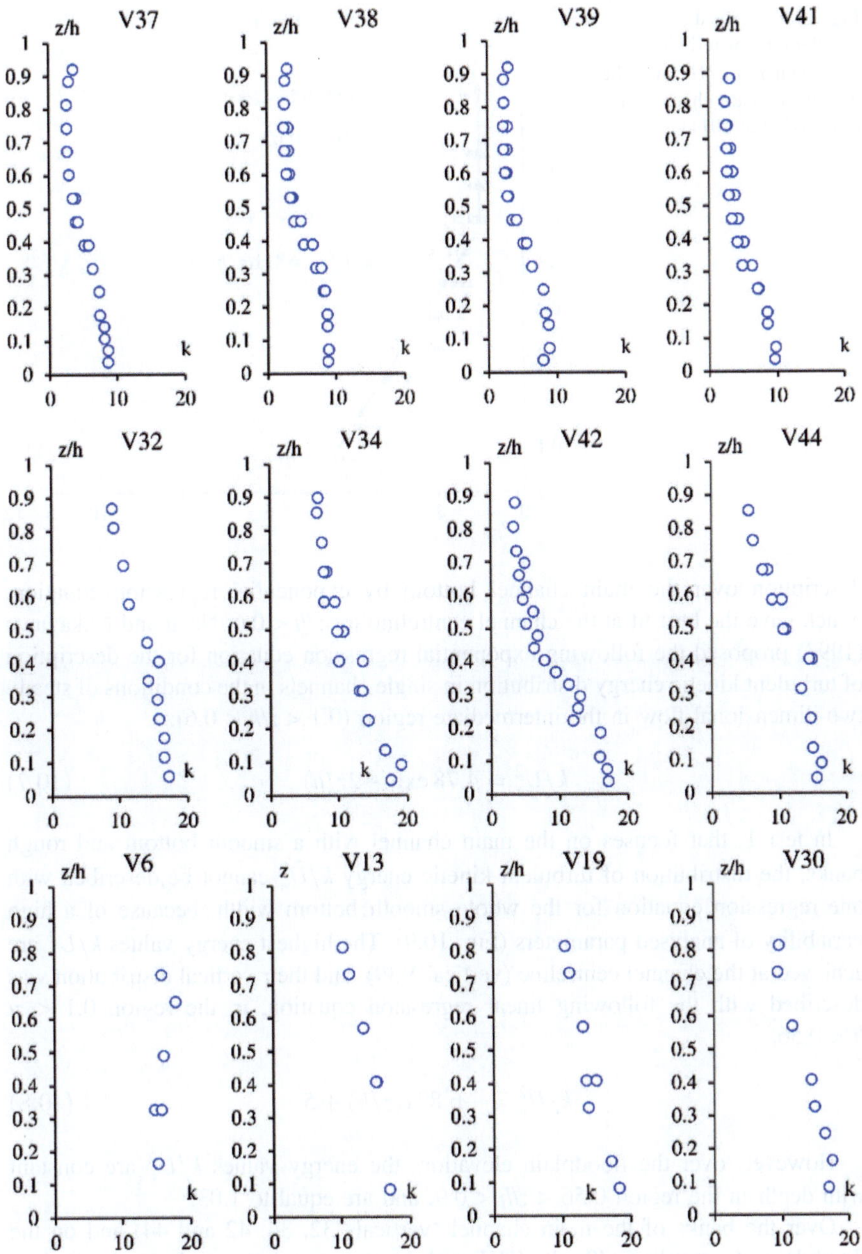

Fig. 10.8 Vertical distributions of turbulent kinetic energy k (cm^2/s^2) in test 1

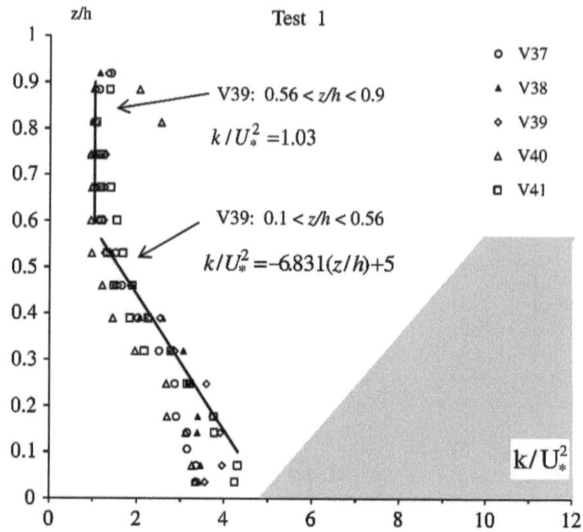

Fig. 10.9 Vertical distributions of turbulent kinetic energy k/U_*^2 over the bed of the main channel in test 1 (Kozioł 2011)

description over the main channel bottom by exponential regression equation, which gave the best fit at the channel centreline for $z/h < 0.6$. Nezu and Nakagawa (1993) proposed the following exponential regression equation for the description of turbulent kinetic energy distribution in single channels in the conditions of steady two-dimensional flow in the intermediate region ($0.1 < z/h < 0.6$):

$$k/U_*^2 = 4.78\exp(-2z/h) \quad (10.7)$$

In test 1, that focuses on the main channel with a smooth bottom and rough banks, the distribution of turbulent kinetic energy k/U_*^2 cannot be described with one regression equation for the whole smooth bottom width, because of a high variability of analysed parameters (Fig. 10.9). The highest energy values k/U_*^2 are achieved at the channel centreline (vertical V39), and their vertical distribution was described with the following linear regression equation, in the region $0.1 < z/h < 0.56$,

$$k/U_*^2 = -6.831(z/h) + 5 \quad (10.8)$$

However, over the floodplain elevation, the energy values k/U_*^2 are constant with depth in the region $0.56 < z/h < 0.9$, and are equal to 1.03.

Over the banks of the main channel (verticals 32, 34, 42 and 44) and on the floodplains (verticals 6–30) the TKE values increase downwards to the bottom (Fig. 10.8), where the best fit is achieved by linear regression equation, in the region $0.1 < z/h < 0.81$ (Fig. 10.10). Turbulent kinetic energy on floodplains is considerably higher than in the main channel.

Table 10.2 Friction velocity U_* (m/s)

Exp.	Verticals													
	6	8	13	19	30	32	34	37	38	39	40	41	42	44
1	0.014	0.022	0.021	0.023	0.022	0.016	0.019	0.014	0.015	0.016	0.015	0.015	0.018	0.018

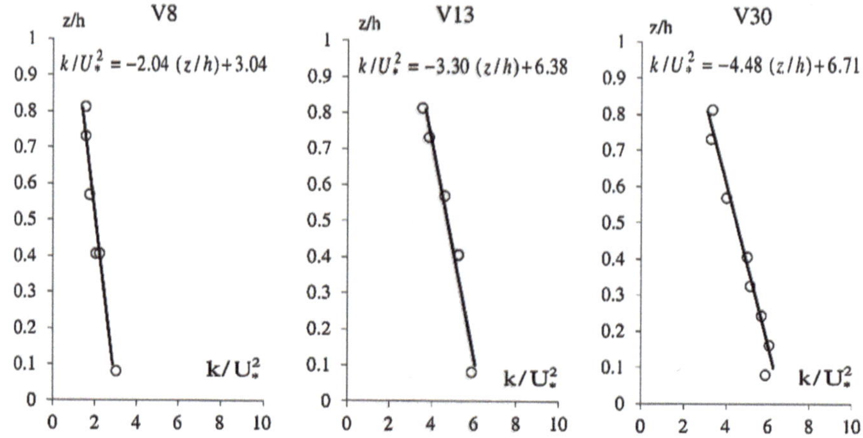

Fig. 10.10 Vertical distributions of turbulent kinetic energy k/U_*^2 on the floodplains in test 1 (Kozioł 2011)

10.6 Temporal Scale of Macroeddies in the Stream

The velocity components were measured with the ADV at the 0.04 s sampling interval. The exemplary autocorrelation time functions are shown in Fig. 10.11. They reveal a similar variability, regardless of the position of the measuring point. The values of the autocorrelation function quickly decrease over time, and then intermittently oscillate around zero. According to the performed research, the time required by the autocorrelation function to reach the zero point is in a range of 0.8–12 s. The computational results for channel sections and different variants are given in Table 10.3.

The calculated values of time macroscales on the floodplains and in the main channel are given for all analysed tests in Table 10.3. The achieved Euler time-

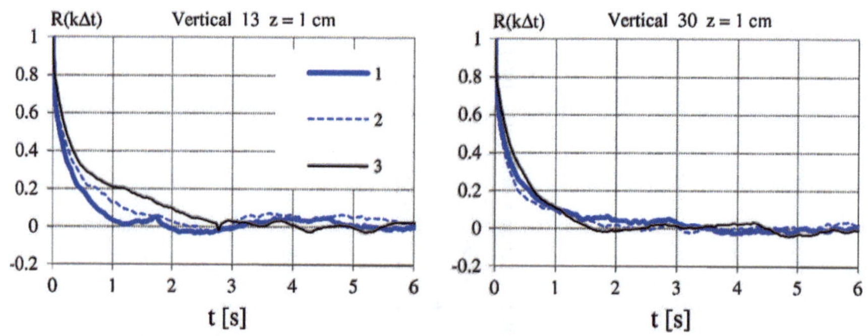

Fig. 10.11 Example time autocorrelation functions calculated for three tests, z—distance of measuring point from the bed, 1 to 3—numbers of test

Table 10.3 The times t of achieving the value of zero by autocorrelation function and the macro time-scales T_E for three tests

Area	Test 1		Test 2		Test 3	
	H = 0.283 m		H = 0.264 m		H = 0.241 m	
	t (s)	T_E (s)	t (s)	T_E (s)	t (s)	T_E (s)
The left flood plains	0.9–7.0	0.1–0.6	0.8–4.8	0.1–0.7	1.1–6.0	0.1–0.7
The main channel	1.0–12.0	0.1–1.2	0.9–6.1	0.1–0.6	1.0–4.7	0.1–0.6
The right flood plains	0.8–5.7	0.1–0.6	0.8–4.9	0.1–0.7	1.0–3.7	0.1–0.6

macroscale values ranged from 0.1 to 0.7 s on the floodplains, while in the main channel they changed from 0.1 to 1.2 s.

10.7 Spatial Longitudinal Scales of Macroeddies in the Stream

The calculated mean, longitudinal sizes of macroeddies and relative sizes of macroeddies (*L*/*h*), related to local values of depths *h* in an analysed measurement vertical, are presented in Table 10.4.

Figure 10.12 illustrates the contour lines for the longitudinal relative sizes of macroeddies (*L*/*h*). The results show that for all tests, the relative sizes of macroeddies are higher on the floodplains than in the main channel, the highest sizes appear on the floodplains in the interaction zone, close to the main channel, and the smallest ones occur over the main channel bed.

Figure 10.13 illustrates the changes of relative sizes of macroeddies (*L*/*h*) in relationship with the relative depth *z*/*h*, at selected measurement verticals on the floodplains and in the main channel. The distribution of longitudinal sizes of macroeddies in verticals does not reveal a constant tendency of changes for the whole transverse channel cross-section. In general, relative sizes of macroeddies in a vertical either decreases downwards the bed or fluctuates around a constant value (Figs. 10.12 and 10.13).

In the main channel centreline (V39, tests 1–3, Fig. 10.13) relative sizes of macroeddies in a vertical fluctuates around 0.5 of the stream depth. According to

Table 10.4 The calculated longitudinal sizes of macroeddies *L* (spatial macro-scale)

Area	Test 1		Test 2		Test 3	
	L (cm)	*L* (cm)	*L* (cm)	*L*/*h*	*L* (cm)	*L*/*h*
The left floodplains	4–19	4–19	2–24	0.2–2.3	4–20	0.5–2.5
The main channel	3–26	3–26	2–21	0.1–1.2	2–17	0.1–1.0
The right flood plains	1–23	1–23	2–24	0.1–2.3	2–13	0.2–1.6

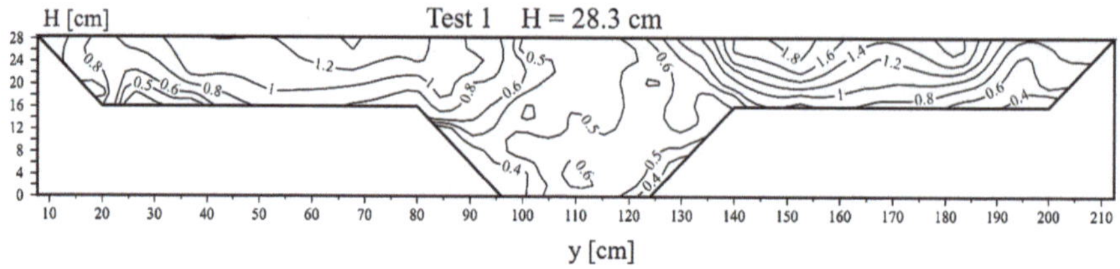

Fig. 10.12 Contour lines for longitudinal relative sizes of macroeddies (*L*/*h*)

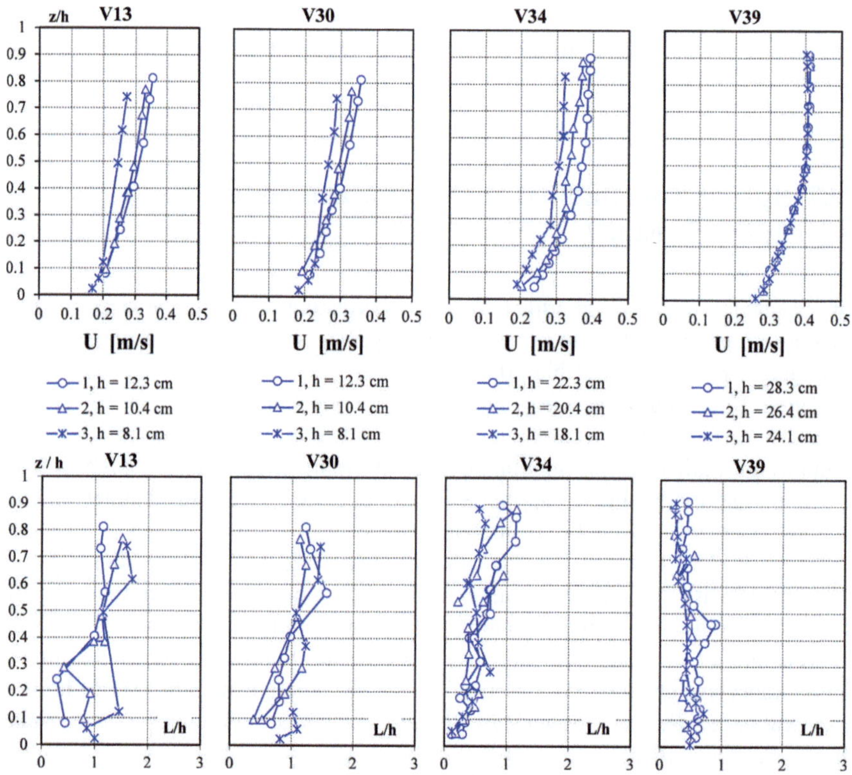

Fig. 10.13 Example vertical distributions of longitudinal velocities and relative sizes of macroeddies on floodplains

earlier investigations for the smooth channel (Kozioł 2000), the largest relative sizes of macroeddies were achieved for the main channel centreline (V39). The smallest sizes of macroeddies appeared in the channel with rough floodplains and rough sloping banks (tests 1–3). In the main channel in the smooth channel, when the relative depth reached $z/h = 0.5$, there were the largest relative sizes of macroeddies, equal to 4.4 times the stream depth; however, downwards the bed and upwards water surface the sizes decrease, reaching 0.8 of a stream depth at the bed, and 0.2 of a stream depth at the water surface. The increase of the roughness of the floodplains and main channel sloping banks (tests 1–3) resulted in a decrease of water velocity and also reduced the sizes of macroeddies. The increase of roughness only on the floodplains resulted in the decrease in sizes of macroeddies at the bed and at the water surface only, but in the middle depth zone it almost did not change and amounted to about one time stream depth. However, the increase of main channel sloping banks roughness in tests 1–3 brought about the decrease in sizes of macroeddies in the main channel to 0.5 of the stream depth (V39, Fig. 10.13).

Over the sloping banks of the main channel, sizes of macroeddies are the largest by the floodplain and the smallest in the vertical located between the sloping bank and the main channel bed (Fig. 10.12). In the vertical located between the sloping bank and the bed of the main channel, the largest sizes of macroeddies occur in the smooth channel and in the channel with rough floodplains, and the sizes are one time the stream depth (Kozioł 2000). The smallest sizes of macroeddies appeared in the channel with rough floodplains and rough sloping banks (tests 1–3), and it decreased downwards the bed from 1 to 0.1 times the stream depth. Over the sloping banks of the main channel close to the floodplains, where the most considerable interaction between the main channel and the floodplain exists, the largest sizes of macroeddies appear only in the smooth channel, and at the bed it reaches 2 times the stream depth. The smallest sizes of macroeddies occur in the case of rough floodplains and rough sloping banks of the main channel (tests 1–3), and here the size decreases downwards the bed from 1.2 to 0.1 times the stream depth (V34, Fig. 10.13). In the smooth channel, sizes of macroeddies increase downwards the bed, and in the channel with rough floodplains they remain constant with depth (L/h = 1.5 approximately).

On the floodplains, the largest relative sizes of macroeddies are achieved in the smooth channel (Kozioł 2000), and reach 4 times the stream depth. The increase of floodplain and the main channel sloping banks roughness resulted in the lowest sizes of macroeddies in tests 1–3 (V30, Fig. 10.13), which appear in a vertical near the bed and close to the sloping bank of the floodplain. The largest relative sizes of macroeddies on the floodplain are observed at the floodplain/main channel interface. On both floodplains, sizes of macroeddies decrease downwards the bed, from 1.7 to 0.1 times the stream depth (tests 1–3, V13-V30, Fig. 10.13).

10.8 Spatial Longitudinal Scales of Microeddies in the Stream

Figure 10.14 presents the frequency spectra S(f) of instantaneous longitudinal velocities in analysed tests. The analysis of the spectrum density function enabled to determine an inertial subrange, then the energy dissipation rates (Eq. 10.5) and the sizes of Kolmogorov's microeddies η (Eq. 10.6) were calculated.

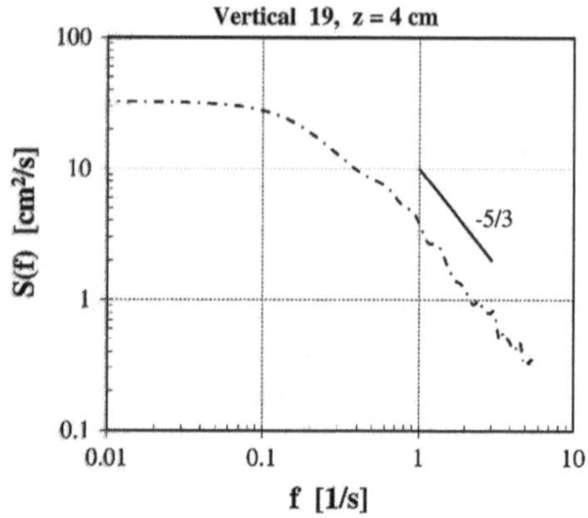

Fig. 10.14 Frequency spectra *S*(*f*) of instantaneous longitudinal velocities

Figure 10.15 presents vertical distribution of longitudinal velocities and longitudinal sizes of microeddies in tests 1, 2 and 3 for floodplains. The calculated values of microeddies are very small in comparison with the value of the velocity (Fig. 10.13) and with the sizes of macroeddies *L* which most often are presented in relation to the water depth (Fig. 10.15). All calculated sizes of microeddies are in the order of decimal parts of a millimetre, and they vary from 0.013 to 0.044 cm. Most of calculated values do not vary significantly in the channel, and they carry out approximately 0.02 cm, while the increase of sloping banks roughness in the main channel resulted in such a growth of microeddies size in the main channel that they reached from 0.024 cm to 0.044 cm in length (test 1, V39, Fig. 10.15). It was concluded that the increase of the floodplain roughness and main channel sloping banks roughness did not result in vital changes of the microeddies size on the floodplain. However, the increase of floodplain roughness and the increase of the main channel sloping banks roughness resulted in a growth of microeddies size in the main channel (Fig. 10.15).

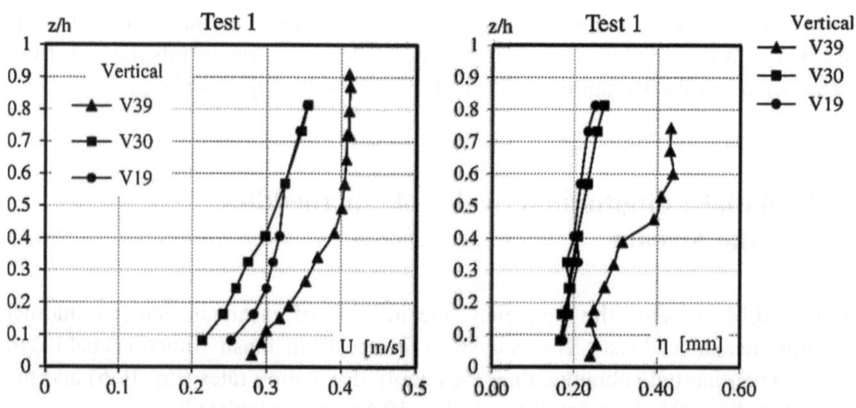

Fig. 10.15 Vertical distribution of longitudinal velocities and sizes of microeddies in test 1

The analysed microscale was within the range: 0.01–1 cm. This statement is consistent with the scientific results achieved for free streams and near the wall regions (Czernuszenko and Lebiecki 1989; Nikora et al. 1994; Nikora and Smart 1997; Mazurczyk 2007).

10.9 Conclusion

The description of water turbulence in compound channels is complicated, insufficient and still requires the explanation of many scientific problems, despite a considerable interest in this field of research. Contemporary 3D measurements enable a significant refinement of turbulence structure description. The consideration of the interaction between the main channel and floodplains in the case of compound channels requires repeated recognition of turbulence characteristics and more complete turbulence structure description. These issues justified the scope of the performed research. The presented results are based on instantaneous, 3-dimensional velocity measurements in a compound channel with rough floodplains and rough channel banks and with smooth main channel bottom.

The research was performed for the compound channel with rough floodplains and sloping banks of the main channel, with smooth bottom of the main channel, and the analysed water depth range was: $D_r = h_f/H = 0.34 - 0.43$. The highest values of relative turbulence intensity were measured on the floodplains and at the floodplain junction with the side slope. Lower values of relative turbulence intensity were reached over the sloping banks of the main channel, and the lowest were present over the bottom of the main channel.

In general, in the compound channel, vertical relative turbulence intensity distributions can be divided into three zones on the floodplains and over the sloping banks of the main channel, and into four zones of the flow field in the main channel: near the bottom zone, middle outer zone, middle inner zone, and the "free-surface region".

In the channel, the TKE values increase downwards to the bottom on floodplains and over main channel banks, where in general the best fit of the TKE distribution is reached by linear regression equations in the range: $0.1 < z/h < 0.81$.

Over the bottom of the main channel, from the floodplain elevation, the TKE values increase downwards to the bottom, and the best fit for the description of that increase is the linear regression equation (Eq. 10.8) in the range $0.1 < z/h < 0.56$. Over the floodplain elevation, the TKE values are constant with depth in the range: $0.56 < z/h < 0.9$.

The areas of the presence of increased turbulent kinetic energy are the channel banks, the zone of channel and floodplain interaction. The transverse turbulence intensity reaches high values in compound channels, especially in the interaction zones and on the floodplains, and they exert a high influence on the values of turbulent kinetic energy.

The performed research proved the vertical distribution of turbulent kinetic energy and the division into 3 zones of flow field on floodplains and over channel banks, and four zones over the bottom of the main channel: near the bed regions, the intermediate regions inner and outer, and at water level regions.

In case of rough floodplains and rough sloping banks of the main channel, sizes of macroeddies become larger on the floodplains than in the main channel, in verticals the size increases upwards the water surface on the floodplains and over the main channel sloping banks, but over the bed of the main channel it is almost constant with depth. The largest macroeddies appear on the floodplain/main channel interface, and the lowest are found over the main channel bed. The calculated values of sizes of microeddies varied within the ranges of 0.013–0.044 cm. The sizes of the microeddies are greater in the centreline of the main channel than on the floodplains; however, the smallest are in the main channel/floodplain interface. The calculated longitudinal sizes of microeddies are very small in comparison with the value of the velocity (0.05–0.45 m/s) and to the scale of macroeddies which most often are presented in relation to the water depth. The values of the microeddies are in the order of decimal parts of a millimetre, and that is why it is difficult to conclude significant changes of their values, as a result of bed roughness.

References

Babaeyan-Koopaei K, Ervine DA, Carling PA, Cao Z (2002) Velocity and turbulence measurements for two overbank flow events in River Severn. J Hydraul Eng 128(10):891–900

Buffin-Bélanger T, Roy AG (2005) 1 min in the life of a river: selecting the optimal record length for the measurement of turbulence in fluvial boundary layers. Geomorphology 68:77–94

Czernuszenko W, Lebiecki P (1980) Turbulent characteristics of stream in open channel. Archiwum Hydrotechniki (Poland) 27(1):19–38

Czernuszenko W, Lebiecki P (1989) The turbulence in river-flows (in Polish). Archiwum Hydrotechniki (Poland) 36(1–2):17–34 (Engl. Summ.)

Czernuszenko W, Kozioł A, Rowiński PM (2007) Measurements of 3D turbulence structure in a compound channel. Arch Hydro-Eng Environ Mech 54(1):3–21. ISSN 1231-3726 (Institute of Hydro-Engineering PAS)

Goring DG, Nikora VI (2002) Despiking acoustic Doppler velocimeter data. J Hydraul Eng 128 (1):117–126

Grinvald DI, Nikora VI (1988) River turbulence (in Russian). Hydrometeoizdat, Leningrad

Knight X, Shiono X (1990) Turbulence measurements in a shear layer region of a compound channel. J Hydraul Res 28(2):175–196

Kozioł A (2000) Longitudinal sizes of the largest eddies in the compound channel. Prz Nauk Wydz Inż Kształt Środow 18:151–159 (in Polish)

Kozioł A (2002) The Kolmogorov microscale in the compound channel (in Polish). Wiadomości Instytutu Meteorologii i Gospodarki Wodnej, Tom XXV (XLVI) (Poland), vol 1, pp 37–42 (Engl. Summ.)

Kozioł A (2008) Investigation of the time and spatial macro-scale of turbulence in a compound channel (in Polish). Acta Scientiarum Polonorum—Architectura 7(4):15–23 (Engl. Summ.)

Kozioł A (2011) Turbulent kinetic energy of water in a compound channel. Ann Warsaw Univ Life Sci (SGGW), Land Reclam 43(2):193–205

Kozioł A (2012) The Kolmogoroff's microscale eddies in a compound channel. Ann Warsaw Univ Life Sci (SGGW), Land Reclam 44(2):121–132. doi:10.2478/v10060-011-0068-7

Kozioł A (2013) Three-dimensional turbulence intensity in a compound channel. J Hydraul Eng 139(8):852–864. doi:10.1061/(ASCE)HY.1943-7900.0000739

Kozioł A, Kubrak J, Kuśmierczuk K (1998) Turbulence characteristics of a stream in a compound channel (in Polish). Zeszyty Problemowe Postępów Nauk Rolniczych Polskiej Akademii Nauk, Wydział Nauk Rolniczych i Leśnych (Poland), vol 458, pp 75–87

Mazurczyk A (2007) Scales of turbulence in compound channels with trees on floodplains. Publs Inst Geophys Pol Acad Sc E-6(390)

McQuivey RS, Keefer TN, Shirazi MA (1971) Basic data report on the turbulent spread of heat and matter. USA department of the interior geological survey, Open-file Report, Fort Collins, Colorado, p 166

Mera I, Franca MJ, Anta J, Peña E (2014) Turbulence anisotropy in a compound meandering channel with different submergence conditions. Adv Water Resour. doi:10.1016/j.advwatres.2014.10.012

Monin AS, Yaglom AM (1975) Statistical fluid mechanics: mechanics of turbulence, vol 2. MIT Press, Cambridge

Nezu I, Nakagawa H (1993) Turbulence in open channel flows. Balkema Rotterdam, The Netherlands

Nezu I, Rodi W (1986) Open-channel flow measurements with a laser Doppler anemometer. J Hydraul Eng 112(5):335–355

Nikora VI (1999) Origin of the "−1" spectral law in wall-bounded turbulence. Phys Rev Lett 83 (4):734–736

Nikora VI, Smart GM (1997) Turbulence characteristics of New Zealand gravel-bed rivers. J Hydraul Eng 123(9):764–773

Nikora VI, Rowiński P, Sukhodolov A, Krasuski D (1994) Structure of river turbulence behind warm-water discharge. J Hydraul Eng 120(2):191–208

Rowiński PM, Mazurczyk A (2006) Turbulent characteristics of flows through emergent vegetation. In: Ferreira RML, Alves ECTL, Leal JGAB, Cardoso AH (eds) River flow 2006. Taylor & Francis Group, London

Rowiński P, Czernuszenko W, Kozioł A, Kuśmierczuk K, Kubrak J (1998) Longitudinal turbulence characteristics in a compound channel under various roughness conditions. In: Proceedings of the 3rd international conference on hydro-science and -engineering. Cottbus, Berlin, Germany

Rowiński PM, Czernuszenko W, Kozioł AP, Kubrak J (2002) Properties of streamwise turbulent flow filed in an open two-stage channel. Arch Hydro-Eng Environ Mech 49(2):37–57

Rowiński PM, Aberle J, Mazurczyk A (2005) Shear velocity estimation in hydraulic research. Acta Geophys Pol 4:567–583

Sanjou M, Nezu I, Suzuki S, Itai I (2010) Turbulence structure of compound open-channel flows with one-line emergent vegetation. In: Proceedings of 9th international conference on hydrodynamics, Shanghai, China, pp 560–564

Shiono K, Knight DW (1991) Turbulent open-channel flows with variable depth across the channel. J Fluid Mech 222(7):617–646

Stone MC, Hotchkiss RH (2007) Turbulence descriptions in two cobble-bed river reaches. J Hydraul Eng 133(12):1367–1378. doi:10.1061/(ASCE)0733-9429(2007)133:12(1367)

Sulaiman MS, Sinnakaudan SK, Shukor MR (2012) Near bed turbulence measurement with acoustic Doppler velocimeter (ADV). KSCE J Civ Eng 17(6):1515–1528. doi:10.1007/s12205-013-0084-8

Terrier B, Robinson S, Shiono K, Paquier A, Ishigaki T (2010) Influence of vegetation to boundary shear stress in open channel for overbank flow. In: Dittrich A, Koll K, Aberle J, Geisenhainer P (eds) River flow 2010. Bundesanstalt für Wasserbau, 8 pp

Tominaga A, Nezu I (1991) Turbulent structure in compound open-channel flows. J Hydraul Eng 117(1):21–41

Wahl TL (2000) Analyzing ADV data using WinADV. In: ASCE joint conference on water resources engineering and water resources planning and management, Minneapolis, USA
Wahl TL (2003) Discussion of "despiking acoustic doppler velocimeter data" by Derek G. Goring and Vladimir I. Nikora. J Hydraul Eng, ASCE 128(1):117–126
Xie D, Pan C (2013) A preliminary study of the turbulence features of the tidal bore in the Qiantang River, China. J Hydrodyn Ser B. doi:10.1016/S1001-6058(13)60439-4
Yang K, Cao S, Knight DW (2007) Flow patterns in compound channels with vegetated floodplains. J Hydraul Eng 133(2):148–159
Yokosi S (1967) The structure of river turbulence. Bull Dis Prev Res Inst 17(2):1–29

Chapter 11
Uncertainty Analysis in River Modelling

Jord J. Warmink and Martijn J. Booij

Abstract Uncertainty analysis is an essential step in river modelling. Knowledge of the uncertainty is crucial for a meaningful interpretation of the model results. In this chapter we describe the whole process of an uncertainty analysis in four steps: identification, prioritization, quantification and propagation. In each step the rationale behind choosing a method is described and illustrated with an example of the design water level computation of the Dutch river Waal with a 2D hydrodynamic model. The sources of uncertainty related to the case study are identified and their (relative) importance is determined using expert opinions combined with a novel uncertainty identification method. Subsequently, the sources with the largest effect on the design water levels are individually quantified and propagated using Monte Carlo analysis to yield the quantified uncertainty in the design water levels. The uncertainty analysis provided information about the reliability of the model results and about further actions to possibly reduce the uncertainty and their benefits in terms of increased accuracy.

Keywords Uncertainty analysis · Hydrodynamic modeling · Expert elicitation · Monte carlo analysis · River rhine

11.1 Introduction

Uncertainty analysis is an essential step in river modelling. Knowledge of the uncertainty is crucial for a meaningful interpretation of the model results (Pappenberger and Beven 2006). In an uncertainty analysis, usually only the effects of statistical uncertainty of the input and the parameters on the model outcomes are quantified; however, the uncertainty is not limited to statistical uncertainty in input and parameters (see examples in Saltelli and Funtowicz 2014). Uncertainty is

J.J. Warmink (✉) · M.J. Booij
Twente Water Centre, University of Twente, Enschede, The Netherlands
e-mail: j.j.warmink@utwente.nl

P. Rowiński and A. Radecki-Pawlik (eds.), *Rivers—Physical, Fluvial and Environmental Processes*, GeoPlanet: Earth and Planetary Sciences,
DOI 10.1007/978-3-319-17719-9_11

defined by Walker et al. (2003) as any deviation of the unachievable ideal of complete deterministic knowledge of the relevant system. This definition implies that uncertainty is the absence of knowledge and therefore depends on the available amount of information. There are three reasons for the absence of knowledge that classify uncertainty. Firstly, the ontological uncertainties or natural variability, which is inherent to any natural system and cannot be reduced. For example, the weather forecast at a certain location in the future. Secondly, epistemic uncertainties or imperfect knowledge which are uncertainties that can be reduced by more and higher quality measurements or improved process understanding. Thirdly, ambiguities, which are defined by Dewulf et al. (2005) as multiple, equally valid frames of reference, or in other words, different views of a problem. Gathering more information does not reduce ambiguities, but rather discussions to convince other people might reduce this uncertainty source. In this chapter we will focus on these three types of uncertainties that arise in environmental modelling and their role in the uncertainty analysis.

Uncertainty analysis is often considered a burden in hydraulic and hydrological modelling (Pappenberger and Beven 2006). In their paper, Pappenberger and Beven (2006) argue that (1) uncertainty analysis helps in understanding physical processes, (2) uncertainty analysis results in better modelling practice by making assumptions explicit and (3) knowledge about the uncertainty results in better decisions. Many authors therefore argue that uncertainty analysis should be part of the modelling cycle (Jakeman et al. 2006; Pappenberger and Beven 2006; Refsgaard et al. 2007). However, an uncertainty analysis should not be executed at the end of the modelling cycle, after model development, calibration and validation, as suggested by Jakeman et al. (2006), but should be part of all phases in any model study (Refsgaard et al. 2007; Warmink et al. 2010). For example, the identification of uncertainties in the initial stage helps in the communication between experts, modelers and stakeholders about the uncertainties, thereby reducing uncertainty in the rest of the model study.

An uncertainty analysis consists of five steps (Van der Sluijs et al. 2005a): (1) identification, (2) importance assessment, (3) quantification of the sources of uncertainty, (4) propagation of the sources of uncertainty to the model outcomes, and (5) the communication of uncertainty (see Fig. 11.1). Leskens et al. (2014) argue that modelling and uncertainty analysis should be integrated in the decision making process, making this final step obsolete.

In this chapter we describe the first four steps of an uncertainty analysis. Each step will be illustrated with an example for a case study of the river Rhine in the Netherlands. In Sect. 11.2 the case study is described. Sections 11.3–11.6 each describe one of the steps. Section 11.7 presents a comparison of two methods used for the case study and the conclusions are given in Sect. 11.8.

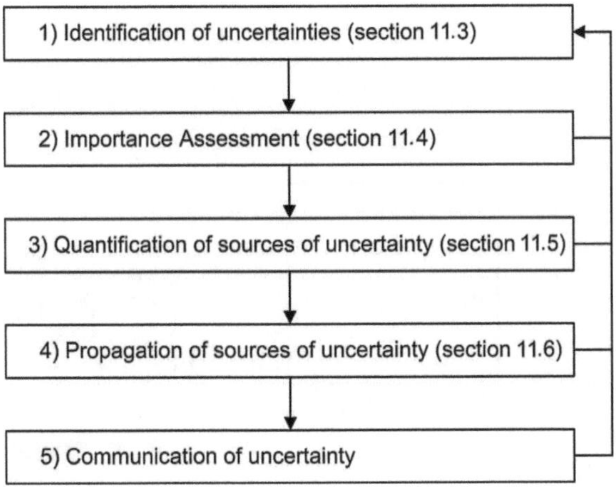

Fig. 11.1 The five steps in an uncertainty analysis and section in this chapter where they are described

11.2 Case Study Description

River flooding is a serious threat in the Netherlands. Strong dikes have been constructed to protect the land from flooding. After the 1993 and 1995 (near) flood events, the Dutch government laid down that every 5 years the safety of the primary dikes has to be evaluated against a design discharge (Van Stokkom et al. 2005). This design discharge is based on the statistical analysis of historical discharge series. The two-dimensional hydrodynamic river model, WAQUA, is the official standard for calculating design water levels for flood protection measures based on the design discharge (Rijkswaterstaat 2007). The design discharge for the upper part of the Dutch river Rhine corresponds to a return period of 1250 years and a magnitude of 16,000 m^3/s at Lobith, the station at the Dutch–German border where the river enters the Netherlands (Rijkswaterstaat 2007). The WAQUA model uses the depth-averaged shallow water equations to compute the design water levels along the river Rhine branches, using the design discharge as upstream input. The model schematization consists of the geometry of the river bed, and mapped characteristics of the flow channel (e.g. vegetation and structures, such as weirs and spill-ways). The model has been calibrated on the highest recorded discharge, which occurred in 1995 (Van den Brink et al. 2006), by adapting the hydraulic roughness of the main channel.

We modelled only the largest Rhine branch: the river Waal (Fig. 11.2), using the 2007 version of the WAQUA model. The model used a staggered curvilinear grid with 148,334 grid cells with a cell size of approximately 40 × 40 m. Water depths were simulated from river kilometer 867 to 960, along the river. The digital elevation model of the river Waal was based on multibeam echo-sounding data for the main channel, and laser altimetry and photogrammetry data for the flood plains. A constant discharge of 10,667 m^3/s was set as the upstream boundary condition, which is two third of the design discharge at station Lobith, as approximately two

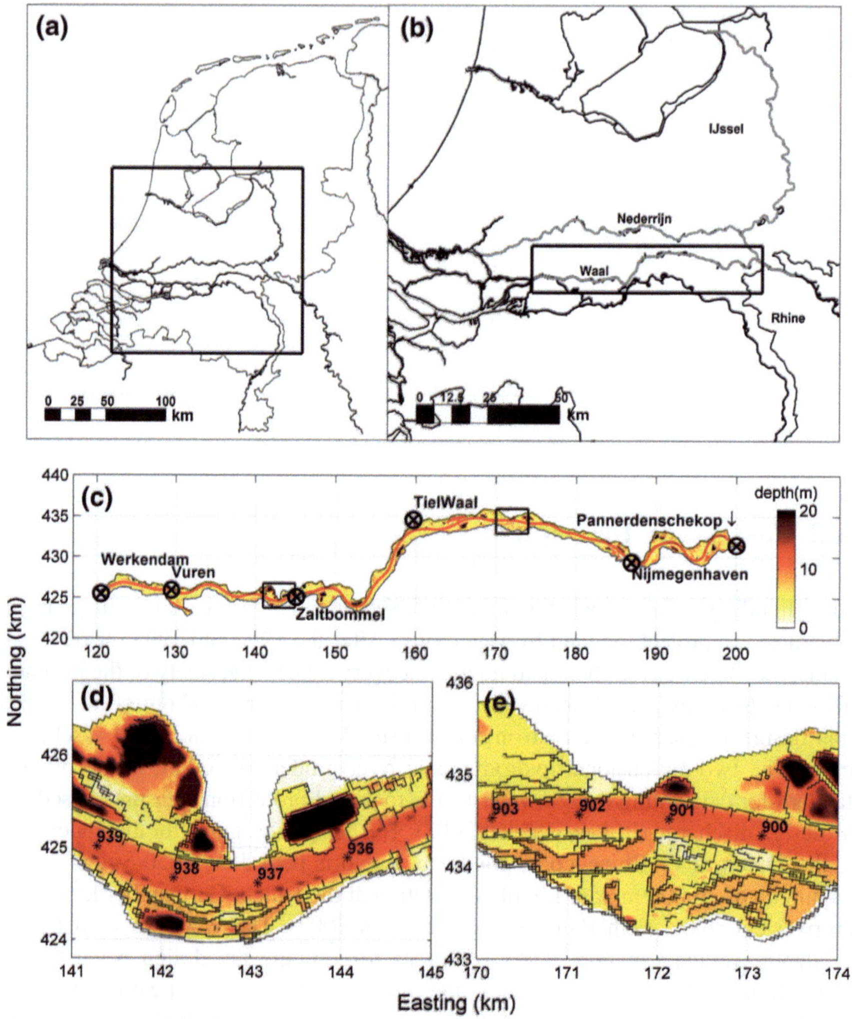

Fig. 11.2 Location of the Waal river in the Netherlands. **a** The Netherlands, **b** location of the Waal distributary, **c–e** water depths computed by the WAQUA model for the upper part of the river Waal (adapted after Warmink 2011)

third of the design discharge flows into the river Waal (Warmink et al. 2013a). The exact discharge fraction into the Waal is uncertain and depends on many factors (Rijkswaterstaat 2007). The downstream boundary condition near Werkendam (Fig. 11.2) was set to a constant water level of 4.8 m above Dutch ordnance datum (NAP), that corresponds to the design discharge (Rijkswaterstaat 2007). A simulation with a constant discharge of 72 h takes about 2.5 h on a 2.4 GHz computer with 4 GB of memory.

The computed design water levels are used as input for the five yearly dike safety assessment. Uncertainties in the water levels are acknowledged (Van den Brink et al. 2006) and accounted for using a freeboard of 0.5 m. This conservative approach might lead to over-dimensioning of the required dike heights and unnecessary costs that might be avoided by detailed knowledge of the uncertainty in the water levels.

In the new Dutch Delta Program (Delta Programme 2015) explicit attention is given to uncertainties, showing that the quantification of uncertainties becomes more important. In the EU Flood Directive (European Parliament 2007, Directive 2007/60/EC), it is obliged for each member state to have flood maps to identify the vulnerable areas. Furthermore, the UK Environmental Agency publishes flood risk maps for the public. These developments show that quantified knowledge of the uncertainties is increasingly requested by policy makers and water managers and novel methods for uncertainty analysis are required.

11.3 Step 1: Identification of Uncertainties

In recent uncertainty analysis studies about river modelling, often only the uncertainties that can easily be quantified are taken into account, such as uncertainties in model input and parameters (e.g. Refsgaard et al. 2006; Hall et al. 2005; Bates et al. 2004). The uncertainties in model context and model structure are often omitted in the analysis. In such cases, it is likely that the model outcome uncertainty is underestimated (Refsgaard et al. 2006). Refsgaard et al. (2006) showed the influence of the model structure on the vulnerability of the area around Copenhagen, Denmark, for groundwater pollution. They compared the results of five models developed by different renowned consultancy companies and showed that the assumptions for model development led to significantly different results. This example showed that not only the uncertainties in input and parameters are important, but model context and the chosen model structure can also have a very large influence.

Another problem is that the identification of uncertainties is often carried out in an unstructured manner. The conclusions of the uncertainty analysis are then a result of a suboptimal identification, which might result in an inaccurate uncertainty analysis (Warmink et al. 2010). For example, in expert opinion studies of Van der Sluijs et al. (2005b) and Krayer von Krauss et al. (2004) the uncertainties that were included in the analysis strongly depended on the expert.

The objective of the identification of uncertainties is to yield a list of unique and well-defined uncertainties as a starting point for an uncertainty analysis. Unique implies that the resulting uncertainties are complementary and do not overlap (Warmink et al. 2010). So, a single uncertainty should not be included twice or more in the final list. The second criterion: well-defined means that it can be analyzed using a single method to quantify its uncertainty.

11.3.1 Methodology to Identify Sources of Uncertainty

To describe and identify uncertainties in a structured and consistent manner, a classification scheme or uncertainty matrix is used (Warmink et al. 2010). Sources of uncertainty arise at numerous locations in a model. These locations are related to the steps in the modelling cycle, and the choices and simplifications made during the ongoing process of model development (Jakeman et al. 2006). Walker et al. (2003) presented an uncertainty classification matrix with three dimensions of uncertainty: (1) the *location*, which is where the uncertainty manifests itself in the model, (2) the *level*, which is where the uncertainty manifests itself along the (continuous) spectrum between deterministic knowledge and total ignorance, and (3) the *nature* of the uncertainty, which consist of natural variability (unreducible uncertainty), epistemic uncertainty (lack of knowledge) and ambiguities (Dewulf et al. 2005; Van den Hoek et al. 2013). Warmink et al. (2010) assumed that every uncertainty can be classified for each of these three dimensions.

The dimensions of uncertainty provide the starting point for the uncertainty identification method described in Warmink et al. (2010). The first step is a global identification, in which the initial uncertainties in the model environment are gathered. To acquire a comprehensive list of uncertainties, all uncertainties should be listed following the locations according to Walker et al. (2003). This can be done by the researcher or by a panel of experts. Besides statistical uncertainties also other sources of uncertainty need to be considered and the preliminary list should be as complete as possible. Secondly, one by one, the listed uncertainties are classified in the uncertainty matrix for all three dimensions. If an uncertainty cannot be classified in a single class, for example because an uncertainty is statistical, but also has a qualitative component, the uncertainty needs to be further specified by subdividing it into two different uncertainties. This process is repeated for all uncertainties until a list of unique and well-defined uncertainties is obtained. Warmink et al. (2010); Fig. 11.3 developed decision trees to help during the classification of uncertainties and to test if an uncertainty is well-defined using questions.

11.3.2 Application to the Case Study: Identification of Uncertainties

For the case study of the river Waal, Warmink et al. (2010) identified all uncertainties in the model. In the first step they assessed the uncertainties in all model

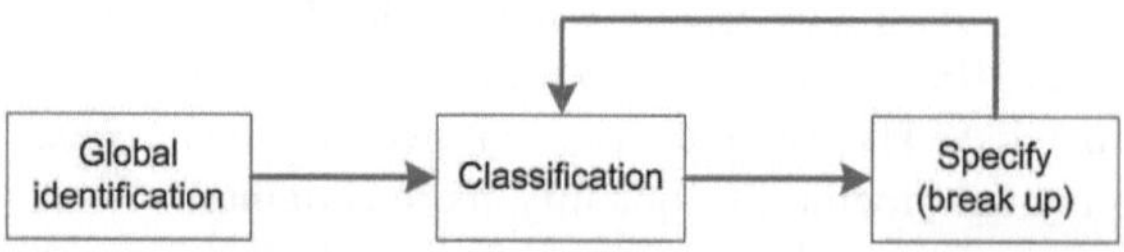

Fig. 11.3 Iterative method to identify uncertainties (adapted after Warmink et al. 2010)

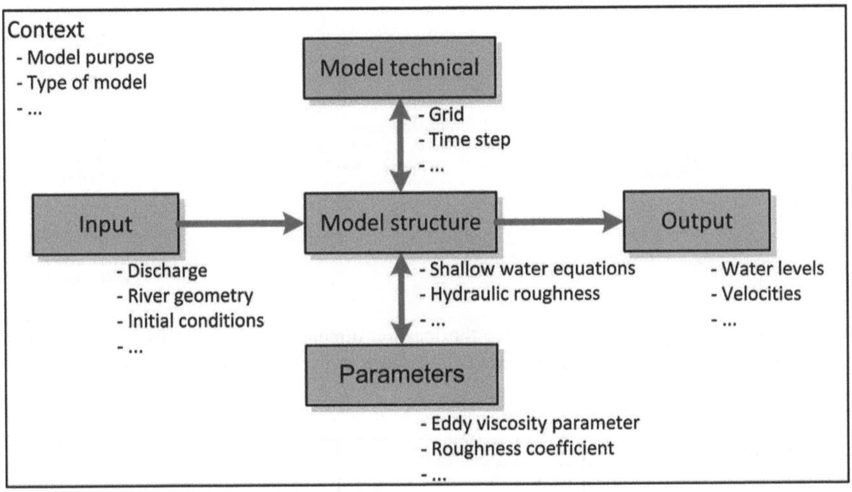

Fig. 11.4 Example of the global identification of uncertainties for the WAQUA model for the river Waal (adapted after Warmink et al. 2010)

locations (i.e. context, input, model structure, model technical, parameters). The initial uncertainties for the WAQUA model were obtained from the elicitation of expert opinions involved in the development and application of this model. Figure 11.4 shows some examples of the uncertainties for the WAQUA model for different locations in the model. The purpose of the model was to predict distributed water levels under design conditions for the river Waal and therefore the type of model and related model boundaries were chosen. In case of an event outside the model boundaries, context uncertainties arise. For example, it was assumed that design discharge events occur in the winter period. However, the 1988 and 2001 peak discharge events occurred in early spring, so the density of the vegetation and thus the hydraulic roughness was much larger than assumed in the model, which introduced a large uncertainty due to the chosen model context. Also, the assumption that the circumstances during calibration were similar to the circumstances during the design discharge introduced uncertainty in the model outcomes, which is shown in Warmink et al. (2007). Furthermore, the WAQUA model neglected (local) variations in the height of the river bed during peak discharges and 3D effects in the river bed morphology were also omitted or included in a constant parameter.

In the second step, the initially identified uncertainties were classified in the uncertainty matrix (Fig. 11.6). For example, the hydraulic roughness is often mentioned in literature as an uncertain parameter (e.g. Bates et al. 2004). Classifying this uncertainty using the decision trees in Warmink et al. (2010) yielded that the hydraulic roughness was an input, a parameter and part of the model structure (see Fig. 11.5 for an example). Therefore, the uncertainty needed to be further specified. In the Waal case, the hydraulic roughness consisted of, e.g., roughness of the main channel, vegetation roughness, and energy losses over weirs.

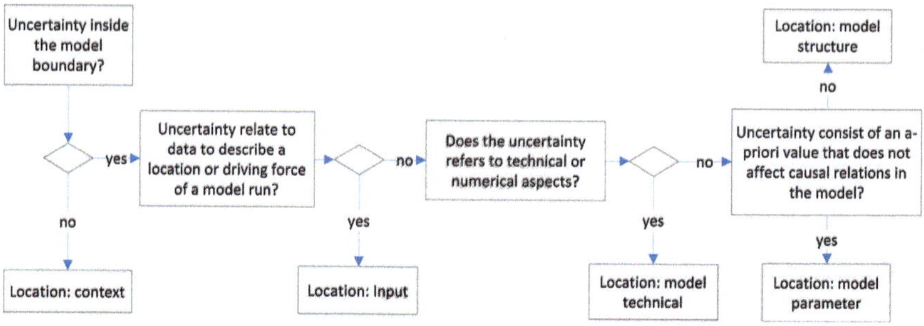

Fig. 11.5 Example of a decision tree for the location dimension only (adapted after Warmink et al. 2010). The decision trees for the other dimensions are presented in Warmink et al. (2010)

Description	Location					Level				Nature		
	Context	Input	Model structure	Model technical	Parameter	Statistical	Scenario	Qualitative	Recognized ignorance	Natural variability	Epistemic	Ambiguity
Roughness	x	x	x	x	x	x	x		x	x	x	
↘ MC roughness			x		x	x	x				x	
↘ MC roughn. equation			x				x				x	
↘ Parameters in equation					x	x					x	
↘ etc…												

Fig. 11.6 Uncertainty matrix modified after Walker et al. (2003) with an example to identify sources of uncertainty in the roughness in the WAQUA model. Uncertainty sources depicted in bold imply that further specification is required, because they belonged to more than one class for one of the dimensions of uncertainty (*MC* main channel)

Now, the decision trees of Warmink et al. (2010) needed to be followed again for these new uncertainties. The roughness of the main channel did not relate to data (within this model context) and did not relate to technical or numerical aspects of the model. However, the main channel roughness was not necessarily a constant. Therefore, again a further specification was required: the roughness of the main channel consisted of an empirical equation and of the parameters within this equation. This distinction enabled the classification of the empirical equation in the model structure location and the parameters in the model parameters location (Fig. 11.6). Their level was classified as scenario and statistical, respectively, and their nature was epistemic, because their uncertainty can be reduced by better understanding of the natural processes.

This example showed that, after proper specification, the hydraulic roughness consisted of, amongst others, the parameters in the roughness equation for the main channel and the structure of the equation itself. This specification enables the selection of methods for further analysis using, for example, the table of Refsgaard et al. (2007), which prescribes that the methodologies that are suitable for the main channel roughness equation are: expert elicitation, multiple model simulation and scenario analysis. For the parameters the methods are inverse modelling or quality assurance. However, more methods are available. The final choice for the method to use depends on the available resources (time and effort) and the required accuracy (Refsgaard et al. 2007).

The thorough analysis of possible uncertainties can significantly improve the accuracy of the subsequent steps of the uncertainty analysis. The resulting list of uncertainties may become very long and not all sources of uncertainty are equally important. Therefore, in the second step the uncertainties need to be prioritized.

11.4 Step 2: Prioritizing Uncertainties (Importance Assessment)

The uncertainties that are most important to include in an uncertainty analysis are the ones that have the largest effect on the model outcomes. These uncertainties are selected based on a sensitivity analysis or importance assessment. Traditionally, sensitivity analysis was defined as a local measure of the effect of a given input on a given model output (Saltelli et al. 2004). However, the effect on model outcomes depends both on the sensitivity of the model and the magnitude of the uncertainty. Therefore, statistical methods that are used to determine the importance of a source of uncertainty go hand in hand with uncertainty analysis (Saltelli et al. 2008), leading to the problem that all uncertainties need to be quantified before they can be prioritized, which is unfeasible due to their large number.

11.4.1 Methodology to Prioritize Sources of Uncertainty

To determine the parameters that have the greatest influence on the model performance, quantitative methods exist to filter out sensitive parameters. These methods are usually employed in combination with calibration and focus on parameters that are non-observable (Gan et al. 2014). To evaluate the influence of each parameter on the model performance, sensitivity analysis methods have been used by many people (e.g. Brugnach 2005; Hall et al. 2009). There are many different sensitivity analysis (SA) approaches, which can be distinguished in two groups: local SA and global SA. Local SA methods explores the change of the model outcomes by varying one parameter while keeping the others constant. Global SA approaches examine the changes in model outcomes by all parameters at the same time. Gan et al. (2014) present an overview of many local and global SA methods and developed a piece of software to perform SA using different approaches and sampling techniques.

Another method to prioritize uncertainties is using expert opinion elicitation (e.g. Krayer von Krauss et al. 2004; Van der Sluijs et al. 2005b; Refsgaard et al. 2007).

The advantage of expert elicitation methods is that little data are required, no large run times of complex mathematical models are involved and experts can include qualitative and recognized ignorance uncertainties in their evaluation. The disadvantage is that experts always have a certain degree of subjectivity and different opinions are generally difficult to compare. To reduce the degree of subjectivity, the selection of experts is of high importance (Warmink et al. 2011) and the expert should be made aware of biases in subjective judgment (Refsgaard et al. 2007).

11.4.2 Application to the Case Study: Prioritization of Uncertainties

Warmink et al. (2011) used expert elicitation to prioritize the uncertainties in the WAQUA model of the Dutch river Waal for design water level computations. They used expert elicitation because the focus was not only on parameter uncertainty and the identification proved that also uncertainties in model structure and its context might be important. For the selection of experts they used a Pedigree analysis to determine the weight of each expert based on four criteria: expertise with model code development, expertise with WAQUA projects, experience with WAQUA modelling and number of publications about WAQUA. 42 experts were asked to indicate their expertise for these criteria and the 11 experts with a score above 75 % were interviewed face-to-face. They were asked to identify the sources of uncertainty following the method described in step 1. Additionally, they were asked to quantify these uncertainties and their effect on the design water levels. The experts had difficulties in quantifying the uncertainty itself and were more comfortable with quantifying their effect on the water levels.

The results indicated that the upstream discharge (uncertainty 1 in Fig. 11.7) has the largest effect on the design water levels with an average effect of ±53 cm. The final

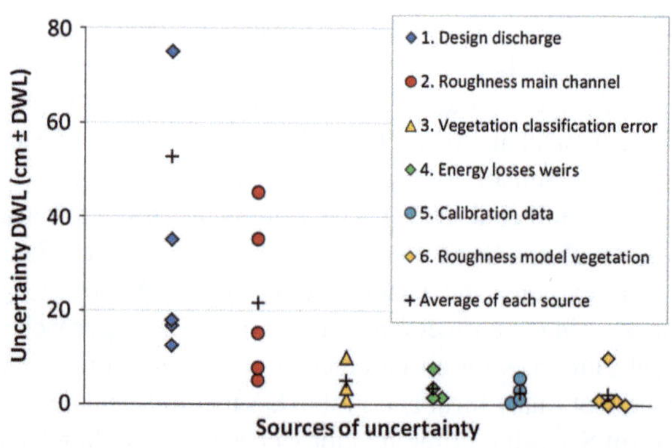

Fig. 11.7 Prioritization (and quantification) of the six largest uncertainties by the experts. Each *dot* represents the opinion of an expert, the '+' sign denotes the average of each source. Only the top 6 ranked uncertainties are shown (after Warmink et al. 2011)

range of possible water levels is therefore 106 cm wide between the lowest and highest estimate, according to the average expert opinions. The next two important sources of uncertainty are the equation used to compute the roughness of the main channel (ranked 2) and the schematization of the vegetation type in the floodplains (e.g. if there is a forest patch in the model, it should also be there in the field; ranked 3).

Based on these results, Warmink et al. (2011) selected the uncertainties to include in the further analysis. The quantification of the uncertainties by the experts is only a simple method and is not very accurate, but it is thought to be sufficient for prioritization. However, it should be noted that the differences between uncertainty sources 3–6 were very small and, therefore, they should be considered equally important. The prioritization based on expert elicitation clearly resulted in five groups of uncertainty: (1) most experts agreed that is was very important (unc. 1–2 in Fig. 11.7), (2) most experts agreed they were probably important (unc. 3–6 in Fig. 11.7), (3) some experts thought they were not important, but one expert thought it was, (4) most experts thought it was not important and (5) only a single expert mentioned it, but thought it was important.

The results themselves had some degree of uncertainty and the selection of which uncertainties to include in the further analysis depended on the available resources and the required accuracy. The prioritization could have been improved by organizing a workshop with the same experts and ask them to reach a consensus in a discussion on the ranking and the magnitude. However, in this case the simple method was sufficient to determine the ranking and no further action was taken.

The uncertainty in the design discharge is agreed to have a large contribution to the uncertainty in the design water levels (e.g. Silva et al. 2001; Van Gelder and Mai 2008). However, in Dutch river management practice, the design discharge can be considered a political decision even though it is computed from statistical extrapolation to the return period of 1250 years. In government protocols (e.g. Rijkswaterstaat 2007), the design discharge for the river Rhine is assumed fixed at 16,000 m^3/s, although in the future a new design discharge of 18,000 m^3/s is anticipated. In this study the design discharge was assumed deterministic even though it is subject to considerable uncertainty. For the further analysis, the uncertainty in the roughness predictor for the hydraulic roughness of the main channel (ranked 2nd) was selected. Additionally, the uncertainty due to the vegetation classification error (ranked 3rd) and the vegetation roughness predictors (ranked 6th) are included to show the effect of including more than 1 source of uncertainty (see Sects. 11.6 and 11.7).

11.5 Step 3: Quantification of Uncertainty Sources

The choice to quantify uncertainties limits the analysis to the uncertainties that are of statistical or scenario level. By using methods to quantify uncertainties, the qualitative uncertainties and recognized ignorance uncertainties are ignored. Van der Sluijs et al. (2005a) showed uncertainties from a non-technical point of view, to include those associated with problem framing, indeterminacies and value-ladenness. Van den Hoek et al. (2013) describe a methodology to deal with ambiguities between stakeholders in flood management projects. These methods are outside the scope of this chapter.

11.5.1 Methodology to Quantify Sources of Uncertainty

After the importance assessment, the selected uncertainties need to be quantified. The objective of quantifying uncertainties is to determine their distributions and correlations as input for the uncertainty propagation in the next step. The probability distributions and their correlations are generally poorly known (Refsgaard et al. 2007) and their quantification is usually the most difficult step in an uncertainty analysis. The quantification method depends on the level and the location of the uncertainties. Most model inputs represent a physical characteristic and can therefore be measured in the field. So data analysis techniques can be used to quantify their distribution. However, the accuracy of the quantification then depends on the quality and availability of the data. For uncertainties that cannot be described by a probability distribution, but only by means of alternatives, scenario methods can be used. It is essential to realize that the accuracy of the final estimated uncertainty is limited by the accuracy of the quantification of the sources. This third step is, therefore, one of the most crucial and difficult to carry out.

There are many methods to quantify the sources of uncertainty (see Refsgaard et al. 2007 for a list of examples). The requirements on the quantification of the sources of uncertainty depend strongly on the method used for propagation. Furthermore, the availability of data prescribes how (accurate) uncertainties can be quantified. Uncertainty quantification methods can be distinguished in direct or indirect methods. Direct methods quantify the uncertainty by means of analyzing the uncertainty source itself, without running the model. Indirect methods are only applicable if data of the model output are available, so inverse modelling is required to estimate the uncertainty of the model inputs and parameters. Expert elicitation is a direct method to quantify the uncertainty if only very few data are available, in which the subjective judgment is represented as a 'subjective' probability density function (PDF) reflecting the expert's degree of belief (Refsgaard et al. 2007). If more data are available, the PDF can be constructed using a statistical analysis of the data, however large data sets are required, especially if the interest lies in the tail of the distributions which is often the case for exceedance of threshold type of problems. The Data Uncertainty Engine (Van Loon and Refsgaard 2005) is an example of a tool to construct PDFs of uncertainties including temporal and spatial variability and correlations.

Indirect methods, such as inverse modelling, are used for combined calibration and uncertainty quantification. They only require data of the output of models and prior estimates of the possible range of the sources of uncertainty and a performance measure. In these methods the model is run multiple times using randomly selected input from the prior ranges. The resulting set of model outcomes is compared to the observations using a performance measure to weigh the effect of parameter variation on model performance. This results in posterior distributions of the model inputs and parameters, thereby quantifying their uncertainty (e.g. Bates et al. 2004). Some of these methods reject the parameter sets below a certain acceptance level of the objective function (e.g. GLUE; Beven and Binley 1992), to only include the "behavioural" parameter sets. These methods can be extended to account for model structure errors using Bayesian Model Averaging methods (Todini 2008).

The advantage of indirect methods is that they include the correlation between model parameters and they account for the remaining uncertainty over the parameters that were assumed deterministic. The latter is also a disadvantage, because these methods attribute all differences between the model and observations to the uncertain parameters, which leads to overestimation of the uncertainty of certain parameters. Direct methods on the other hand require explicit quantification of the correlation between parameters and omit the uncertainty in the deterministic parameters. The latter might lead to an underestimation of the total uncertainty. Quantification of the correlations is difficult and therefore parameters are often assumed independent. Furthermore, indirect methods are often very time consuming, because many model simulations are required to cover the parameter space (see Zhou et al. 2014; Gan et al. 2014). This was the reason to use direct statistical quantification of the sources of uncertainty in our case study.

11.5.2 Application to the Case Study: Uncertainty Quantification

To quantify the uncertainty in the main channel roughness for the river Waal, Warmink et al. (2013a, b) used the Monte Carlo method, because this method does not assume a certain shape of the probability distributions of the sources of uncertainty and data for quantification were available. The objective was to quantify the uncertainty in the hydraulic roughness of the main channel during design conditions. Identification using the method from Sect. 11.3 showed that this uncertainty consisted of the uncertainty in the hydraulic roughness equation, the uncertainty in the data used as input for the equation and the extrapolation to the design conditions.

The hydraulic roughness of the main channel is mainly determined by bed forms on the river bed. In many rivers, river dunes are the dominant bed forms. River dunes are rhythmic patterns caused by the interaction of the turbulent flow with the sandy bottom. The height of river dunes is in the order of 10–30 % of the water depth and their length (distance between two consecutive crests) is approximately ten times their height (e.g. Van Rijn 1984). Recirculating eddies develop at steep lee sides of dunes, resulting in a flow separation zone, which causes energy losses and therefore resistance. River beds are highly dynamic during floods. The lengths and heights of dunes increase and decrease during floods as a result of the changing flow conditions.

Warmink et al. (2013a) used field measurements of bed form and flow characteristics from the studies of Julien et al. (2002) and Wilbers and Ten Brinke (2003). The data were measured during the 1995, 1997 and 1998 flood waves in the river Rhine 200 m upstream of the first bifurcation in the Netherlands (Fig. 11.2). Data consisted of measured dune heights, dune lengths and flow characteristics (water levels, discharges and water surface slope) at various times during the flood wave. Warmink et al. (2013a) selected five roughness models that performed well for these field data. The roughness was expressed using the Nikuradse roughness height, k_N with unit meter.

To quantify the uncertainty in bed form roughness under design conditions, two sources of uncertainty were considered: firstly, the uncertainty due to the choice of the roughness model, that is, the variability between the k_N values under design conditions, and secondly, the uncertainty due to the extrapolation from the measured to the design conditions. The latter also accounted for the variability in the recorded discharge data. The uncertainty in the measurements of the dune and flow characteristics was not taken into account (Warmink et al. 2013a). Figure 11.8 shows that there was a large variation between the results of the roughness models and that the variation increases with increasing discharge.

Warmink et al. (2013a) extrapolated the predicted roughness to the design return period using the Generalized Extreme Value (GEV) distribution for each roughness model separately (Fig. 11.9). The uncertainty in the observations was expressed by a confidence interval around the extrapolated value. The width and shape of the distribution at the design return period described the uncertainty in the roughness

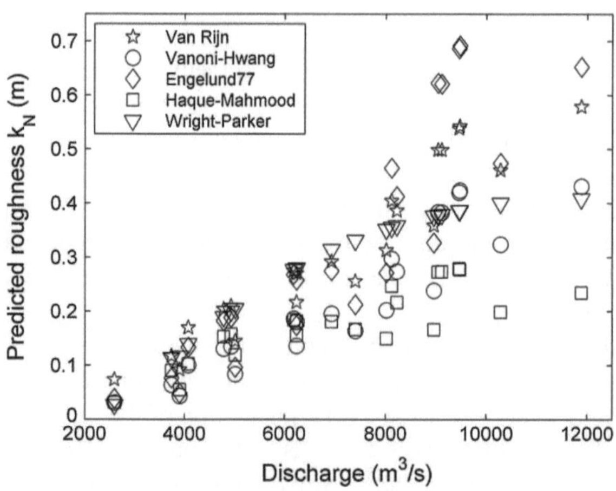

Fig. 11.8 Main channel roughness predicted by the five roughness models (Van Rijn 1984; Vanoni and Hwang 1967; Engelund 1977; Haque-Mahmood 1983 and Wright and Parker 2004) for the 1995, 1997 and 1998 discharge waves in the river Rhine (adapted after Warmink et al. 2013a)

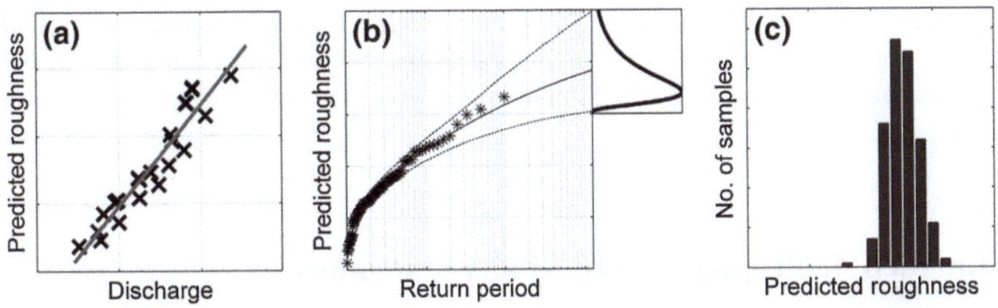

Fig. 11.9 Method for the quantification of the uncertainty due to parameters for 1 roughness model, **a** parameterization of roughness, **b** extrapolation to design return period, **c** resulting probability distribution at the design return period (after Warmink et al. 2013a)

for each roughness model. The confidence interval increases with increasing return period, thereby reflecting the uncertainty due to extrapolation including the uncertainty in the data. This heteroscedastic pattern is also shown in Fig. 11.8 showing that the magnitude of the uncertainty depends on the return period. Lowering of the return period will therefore lead to reduced uncertainty. Other ways to decrease the width of the confidence interval are to increase the number of data points and include observations at higher return periods.

To quantify the uncertainty due to the roughness models, the method described above was repeated for all five roughness models. Combining the final distributions resulted in the uncertainty in the hydraulic roughness of the main channel for the river Rhine just upstream of the bifurcation point (Fig. 11.10). The hydraulic

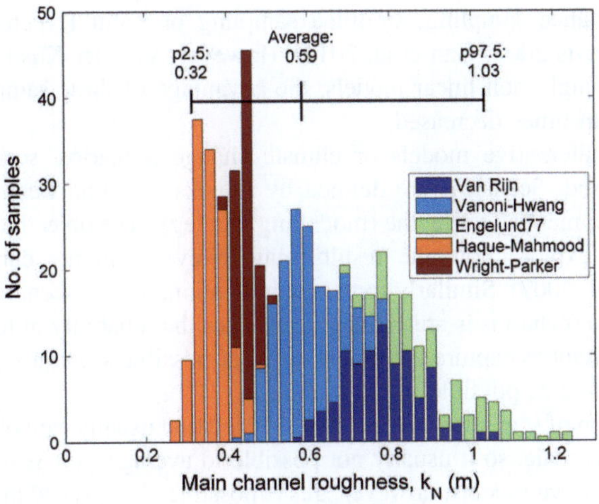

Fig. 11.10 Uncertainty in the main channel roughness under design conditions due to combined model and extrapolation uncertainty. Stacked histogram showing each roughness model separately and the final PDF of the main channel roughness with the average and 2.5 and 97.5 percentile. Total sample size was 1000 (adapted after Warmink 2011)

roughness of the main channel had a 95 % confidence interval between 0.32 and 1.03 m with an average of 0.59 m. The resulting samples were the input for the propagation using Monte Carlo Simulation to yield the uncertainty in the water levels.

11.6 Step 4: Propagation to Model Outcomes

The final step of a quantitative uncertainty analysis is to propagate the quantified sources of uncertainty to the model outcomes. For this purpose, Monte Carlo based methods are the most common ones. The method to propagate uncertainty is related to the level of the uncertainty and is already chosen before the quantification of the sources of uncertainty. We discuss only Monte Carlo Simulation and scenario analysis, but many more methods (e.g. inverse modelling) are available.

11.6.1 Methodology to Propagate Sources of Uncertainty

In Monte Carlo Simulation, the model is run multiple times for randomly selected input values. The advantage of Monte Carlo Simulation (MCS) is that it does not impose assumptions on the shape of the distribution and correlations can (if quantified) be taken into account. However, in case of a computationally time consuming model the run times may become significant. To reduce the computational demand, various methods exist to reduce the required number of samples, such as importance sampling, stratified sampling or Latin Hypercube sampling (Helton and Davis 2003; Gan et al. 2014). However, Van der Klis (2003) showed that for increasingly non-linear models, the advantage of these sampling methods for reducing run times decreased.

In case of alternative models or climate change scenarios, scenario analysis methods are used. Scenarios are defined by Loucks and Van Beek (2005) as an external development affecting the (modelling) strategy. The objective of a scenario analysis is to explore how the results may evolve from the current situation (Refsgaard et al. 2007). Similarly to the quantification, also in scenario analysis the definition of the scenarios is crucial and determines the reliability of the results. It is therefore important to capture the whole range of possible scenarios (or models) to describe the range of possible future realities.

The definition of scenarios implies that their relation usually cannot be quantified on a continuous scale, so it usually not possible to average over several scenarios. In case of alternative models, however, this is possible if they yield the same output variables. Bayesian model averaging (BMA) is the most often used method to combine the results of different model structures. It assumes a fixed probability distribution around individual model forecasts and uses a performance measure to assign a weight to the individual models. Recently, Parrish et al. (2012) developed a new BMA method that used dynamic pdfs and sequential BMA techniques during a flood wave. Their method continuously selected the best model, thereby significantly reducing the uncertainty in the discharge forecasts for a catchment in Mississippi, USA. Their research showed that under different conditions different model structures gave the best results.

11.6.2 Case Study Example for Uncertainty Propagation

After the prioritization, Warmink et al. (2013b) selected three uncertainties which were amongst the six largest sources of uncertainty for the design water level estimation for the river Waal. The three sources were: (1) the roughness of the main channel, (2) the vegetation classification error and (3) the vegetation roughness model. They used Monte Carlo Simulation for the propagation of uncertainty source 1 and 2 (for which the PDF was quantified) and scenario analysis (alternative model structures) for the third source. Warmink et al. (2013b) used a crude sampling method, because due to the non-linearity of the model an advanced sampling method was not expected to significantly reduce the required number of model runs.

The quantification of the uncertainty due to the main channel roughness has been shown in the previous section. The uncertainty due to the vegetation classification error was quantified by Straatsma and Huthoff (2011) and represented as a percentage of the vegetation types in the model that matched field observations. The uncertainty due to the vegetation roughness model was described by four alternative models that predicted the vegetation roughness as a function of water depth, where each of the four models was considered equally likely. Warmink et al. (2013b) carried out four sets of Monte Carlo simulations with 500 samples each. Three of these sets corresponded to the individual uncertainty sources, where one source was randomly sampled and the other fixed at the default value. In the last set of runs, all three sources of uncertainty were randomly varied in the Waal river section. Results were compared as water levels along the centre of the river and as pdfs at a single point (near river kilometer 893; see Fig. 11.2c).

The uncertainty due to the main channel roughness resulted in a wide range of possible water levels (Fig. 11.11). While the input distribution was skewed left, the skewness in the resulting water levels was significantly reduced. The uncertainty due to the vegetation classification resulted in a smaller range, but with a more skewed distribution with a heavy tail, indicating a small probability with a significant increase of water levels. The vegetation roughness model yielded a relatively small uncertainty range. Therefore, the assumption that all models were equally likely had a negligible effect on the final uncertainty range. In case this uncertainty would have a large effect on the water levels it is advisable to use more advanced methods (such as Bayesian Model Averaging) to account for this source of uncertainty. Combining the three sources of uncertainty resulted in a wide distribution with a heavy tail, similarly to a log-normal distribution with a 95 % confidence interval of 68 cm. Given an average water depth of approximately 12 m

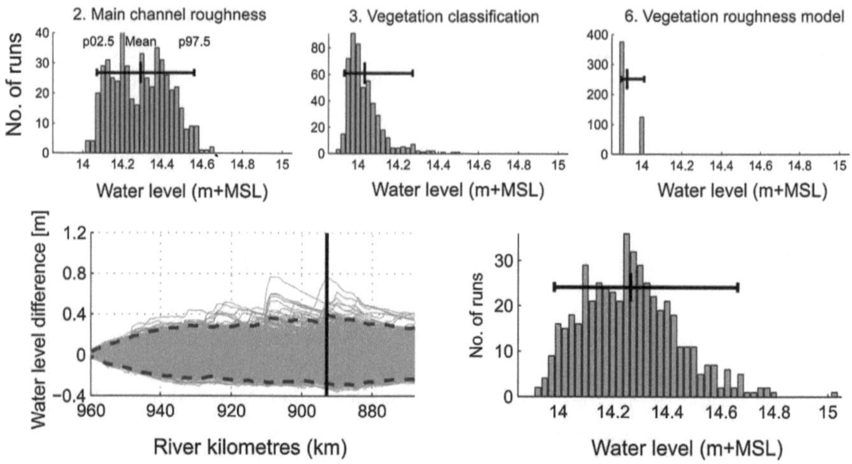

Fig. 11.11 Monte Carlo results. *Top* results of individual uncertainty sources. Ranked 2nd main channel (*left*), 3rd vegetation classification (*middle*) and 6th vegetation models (*right*). *Bottom* combined uncertainties showing water level difference from the mean along the river (*left*) and probability distribution of water levels at river kilometer 893. The mean and range between the percentiles 2.5 and 97.5 are depicted by the *black lines* (adapted after Warmink et al. 2013b)

(dependent on the highly variable bed level), this implies an uncertainty of 5.6 %, which seems quite low, but is highly significant in Dutch river management practice. An accuracy assessment of this confidence interval showed that 500 simulations yielded an error of 2 cm in the 97.5 percentile (Warmink et al. 2013b).

11.7 Comparison of Step 3 and 4 for the Case Study

The example for the Dutch river Rhine showed that it is possible to explicitly quantify the uncertainty starting from identification, prioritization and quantification of the sources and model outcomes. Many assumptions were taken during the process. Warmink et al. (2013b) quantified only three out of the six most important sources of uncertainty: the main channel roughness (ranked 2nd), the vegetation classification (ranked 3th) and the vegetation roughness model (ranked 6th). The propagation in the previous section showed that the 6th ranked uncertainty only had a negligible effect on the water level uncertainty. Therefore, the first step to improve this analysis would be to include the uncertainty sources ranked 4th and 5th. It is not expected that the uncertainty sources that were ranked lower by the experts will significantly affect the uncertainty range.

The uncertainties are quantified by two different methods: (1) expert opinions (Warmink et al. 2011) and (2) quantification of sources followed by Monte Carlo propagation (Warmink et al. 2013a, b). Table 11.1 shows that the experts estimated the same order of magnitude as the quantification study. However, they slightly

Table 11.1 Comparison of quantification of uncertainties by expert opinions and quantification based on Monte Carlo Simulation (MCS)

Rank	Uncertainty source	MCS (cm)	Experts (cm)	Experts (cm)
2	Main channel roughness	49	43	43
3	Vegetation classification	34	10	10
4	Energy losses due to weirs	–	–	7
5	Calibration data	–	–	5
6	Vegetation model	12	4.3	4.3
7	Bathymetry accuracy	–	–	5
	Sum [using eq. (11.1)]	61	44	45
	Sum (using Monte Carlo)	68	–	–

Values are maximum range (experts) and 95 % confidence intervals (MCS) for the situation of approximately 12 m water depth. The MCS column shows the results presented in Sects. 11.5 and 11.6. The last two columns show the uncertainty quantified by the experts: one column only showing the uncertainties that were also quantified by MCS and the last column showing only the top 7 ranked uncertainties. The bottom two rows show their combined effect using Eq. 11.1 and MCS

underestimated the uncertainty of the sources. The experts estimated the uncertainty due to the main channel roughness quite well (49 cm using MCS vs. 43 cm according to the experts), but underestimated the uncertainty due to the vegetation classification (34 vs. 10 cm). Overall, the experts in this case proved to be able to give a reliable estimate of the contribution of different uncertainty sources.

The uncertainty in the water levels due to the individual sources can be combined assuming a normal distribution and independence by Morgan and Henrion (1990):

$$var_{total} = \sum var_{source,i} \tag{11.1}$$

This equation states that the total variance in model output due to various sources (var_{total}) equals the sum of the variances of model output due to the individual sources ($var_{source,i}$). The bottom two rows in Table 11.1 show that the total uncertainty (expressed as the standard deviation) estimated by the experts for the three selected uncertainty sources is 44 cm using Eq. (11.1). The total uncertainty from the Monte Carlo Simulation was 68 cm, which is the same order of magnitude, but slightly underestimated by the experts. The difference between combining the individual contributions using Eq. (11.1) and the Monte Carlo Simulation is small, 61 versus 68 cm. This shows that the model responds relatively linearly to the uncertainties and that the assumption of independent and normally distributed uncertainties is reasonable. In the last column, Table 11.1 shows an estimate of combining the top 7 ranked uncertainties (omitting discharge uncertainty) using Eq. (11.1). Inclusion of the three sources (ranked 4th, 5th and 7th), indicated that adding more sources of uncertainty to the analysis will probably have a small effect on the final uncertainty range.

The final step in an uncertainty analysis is the communication of uncertainties. The case study examples showed that the water levels (based on a discharge with a return period of 1250 years) have a range of approximately 68 cm. This kind of information is useful for the policy makers (Pappenberger and Beven 2006) who, for example, decide how high the dikes should be. Based on this uncertainty information they can decide if the freeboard that accounts for the uncertainty is sufficient. Figure 11.11 shows that the uncertainty varies along the river, based on river geometry. A possibility is therefore to vary the freeboard along the river based on river geometry. However, it is the choice of the decision maker to define the acceptable level of uncertainty given the costs of heightening of the dikes.

Additionally, the uncertainty analysis provided information about further actions to reduce the uncertainty and their benefits in terms of increased accuracy. For the case study it might be beneficial to improve the mapping of vegetation types in the floodplain areas. This can reduce the uncertainty in the water levels from 68 to approximately 50 cm [using Eq. (11.1)]. Another approach is to improve the estimate of the roughness of the main channel. However, because this uncertainty is caused by limited knowledge of the roughness model, its reduction requires an extensive scientific study.

11.8 Conclusions

The objective of this chapter was to give an overview of the first four steps of an uncertainty analysis and show their application for the quantification of the uncertainty in the design water levels in the river Waal in the Netherlands. We have presented a methodology to explicitly quantify the uncertainty in a complex model. The uncertainties were unravelled in separate components, subsequently quantified and then combined to yield the total effect on the uncertainty in the design water levels.

1. The first step was to identify uncertainties and led to a coherent list of unique sources of uncertainty.
2. In the second step, the uncertainty due to the upstream discharge (an input uncertainty) and the main channel roughness equation (i.e. the model structure) proved to have the largest effect on the design water levels for the river Waal case according to expert opinions. The magnitude of the uncertainties quantified by the experts proved to be similar to the Monte Carlo results.
3. In the third step, we showed the explicit quantification of the roughness in the main channel due to both model uncertainty and extrapolation uncertainty to yield a probability distribution of the uncertainty source.
4. Finally, the combined effect of the most important sources resulted in a spatially variable uncertainty in the design water levels. Including more uncertainties did not significantly affect the final uncertainty range.

The application of an uncertainty analysis is useful for policy makers. It provides information about the reliability of the model results, but also information about how to reduce the uncertainty and how much can be gained. A thorough identification in the early stages of any (modelling) study contributes to a better communication between decision makers and can therefore improve the decisions.

Acknowledgments This research was supported by the Technology Foundation STW, applied science division of the Netherlands Organisation for Scientific Research (NWO), and the technology program of the Ministry of Economic Affairs. We thank the Dutch Centre for Water Management for providing the WAQUA model to do the analysis. Furthermore, we thank all the co-authors: Freek Huthoff, Hanneke Van der Klis, Menno Straatsma and Suzanne Hulscher for their assistance during the research. The research in Sect. 11.6 was also supported by the Flood Control 2015 program (www.floodcontrol2015.com).

References

Bates PD, Horritt MS, Aronica G, Beven KJ (2004) Bayesian updating of flood inundation likelihoods conditioned on flood extent data. Hydrol Process 18(17):3347–3370. doi:10.1002/hyp.1499

Beven K, Binley A (1992) The future of distributed models: model calibration and uncertainty prediction. Hydrol Process 6:279–298. doi:10.1002/hyp.3360060305

Brugnach M (2005) Process level sensitivity analysis for complex ecological models. Ecol Model 187(2–3):99–120. doi:10.1016/j.ecolmodel.2005.01.044

Dewulf A, Craps M, Bouwen R, Taillieu T, Pahl-Wostl C (2005) Integrated management of natural resources: dealing with ambiguous issues, multiple actors and diverging frames. Water Sci Technol 52 (6):115″U–124

Delta Programme (2015) Working on the delta. The decisions to keep the Netherlands safe and liveable. Ministry of Infrastructure and the Environment and Ministry of Economic Affairs. www.deltacommissaris.nl. Accessed 18 May 2015

Engelund F (1977) Hydraulic resistance for flow over dunes. Progress report of the Institute for Hydrodynamic and Hydraulic Engineering 44, Technical University Denmark

European Parliament (2007) Directive 2007/60/EC of the European Parliament and of the Council of 23 October 2007 on the assessment and management of flood risks. Official Journal of the European Union, L 288/27

Gan Y, Duan Q, Gong W, Tong C, Sun Y, Chu W, Ye A, Miao C, Di Z (2014) A comprehensive evaluation of various sensitivity analysis methods: a case study with a hydrological model. Environ Model Softw 51:269–285. doi:10.1016/j.envsoft.2013.09.031

Hall JW, Tarantola S, Bates PD, Horritt MS (2005) Distributed sensitivity analysis of flood Inundation model calibration. J Hydraul Eng131 (2):117–126. doi: 10.1061/(ASCE)0733-9429(2005)131:2(117)

Hall JW, Boyce SA, Wang Y, Dawson RJ, Tarantola S, Saltelli A (2009) Sensitivity analysis for hydraulic models. J Hydraul Eng 135(11):959–969. doi:10.1061/(ASCE)HY.1943-7900.0000098

Haque MI, Mahmood K (1983) Analytical determination of form friction factor. J Hydraul Eng 109(4):590–610

Helton JC, Davis FJ (2003) Latin hypercube sampling and the propagation of uncertainty in analyses of complex systems. Reliab Eng Syst Saf 81(1):23–69. doi:10.1016/S0951-8320(03)00058-9

Jakeman AJ, Letcher RA, Norton JP (2006) Ten iterative steps in development and evaluation of environmental models. Environ Model Softw 21(5):602–614. doi:10.1016/j.envsoft.2006.01.004
Julien PY, Klaassen GJ, Ten Brinke WBM, Wilbers AWE (2002) Case study: bed resistance of Rhine river during 1998 flood. J Hydraul Eng 128(12):1042–1050. doi: 10.1061/(ASCE)0733-9429(2002)128:12(1042)
Krayer von Krauss MP, Casman EA, Small MJ (2004) Elicitation of expert judgments of uncertainty in the risk assessment of herbicide-tolerant oilseed crops. Risk Anal 24(6):1515–1527. doi:10.1111/j.0272-4332.2004.00546.x
Leskens J, Brugnach M, Hoekstra AY, Schuurmans W (2014) Why are decisions in flood disaster management so poorly supported by information from flood models? Environ Model Softw 53:53–61. doi:10.1016/j.envsoft.2013.11.003
Loucks DP, Van Beek E (2005) Water resources systems planning and management, an introduction to methods, models and applications. Unesco Publishing, Paris and WL | Delft Hydraulics, the Netherlands. ISBN 92-3-103998-9
Morgan MG, Henrion M (1990) Uncertainty: a guide to dealing with uncertainty in quantitative risk and policy analysis. Cambridge University Press, Cambridge. ISBN 0-521-36542-2
Pappenberger F, Beven KJ (2006) Ignorance is bliss: or seven reasons not to use uncertainty analysis. Water Resour Res 42:W05302. doi:10.1029/2005WR004820
Parrish MA, Moradkhani H, DeChant CM (2012) Toward reduction of model uncertainty: integration of Bayesian model averaging and data assimilation. Water Resour Res 48(3): W03519. doi:10.1029/2011WR011116
Refsgaard JC, Van der Keur P, Nilsson B, Müller-Wohlfeil D, Brown J (2006) Uncertainties in river basin data at various support scales—example from Odense pilot river basin. Hydrol Earth Syst Sci Discuss 3(4):1943–1985. doi:10.5194/hessd-3-1943-2006
Refsgaard JC, Van der Sluijs JP, Lajer Hojberg A, Vanrolleghem PA (2007) Uncertainty in the environmental modelling process: a framework and guidance. Environ Model Softw 22 (11):1543–1556. doi:10.1016/j.envsoft.2007.02.004
Rijkswaterstaat (2007) Hydraulische randvoorwaarden primaire waterkeringen, voor de derde toetsronde 2006–2011 (HR 2006). Ministry of Transportation, Public Works and Water Management (in Dutch)
Saltelli A, Funtowicz S (2014) When all models are wrong. Issues Sci Technol winter 2014:79–85
Saltelli A, Tarantola S, Campologno F, Ratto M (2004) Sensitivity analysis in practice, a guide to assessing scientific models. Wiley, London. ISBN 0-470-87093-1
Saltelli A, Ratto M, Andres T, Campolongo F, Cariboni J, Gatelli D, Saisana M, Tarantola S (2008) Global sensitivity analysis, the primer. Wiley, West Sussex. ISBN 978-0-470-05997-5
Silva W, Klijn F, Dijkman J (2001) Room for the Rhine branches in the Netherlands: what the research has taught us. RIZA report 2001.031, Ministry of Public Works, Transportation and Water Management and WL|Delft Hydraulics, the Netherlands
Straatsma MW, Huthoff F (2011) Uncertainty in 2D hydrodynamic models from errors in roughness parameterization based aerial images. J Phys Chem Earth 36(7–8):324–334. doi:10.1016/j.pce.2011.02.009
Todini E (2008) A model conditional process to assess predictive uncertainty in flood forecasting. Int J River Basin Manage 6(2):123–137
Van den Brink NGM, Beyer D, Scholten MJM, van Velzen EH (2006) Onderbouwing hydraulische randvoorwaarden 2001 voor de Rijn en zijn takken. RIZA report 2002.015, RIZA, Lelystad, the Netherlands. ISBN 90-3695-322-7 (in Dutch)
Van den Hoek RE, Brugnach M, Mulder JPM, Hoekstra AY (2013) Analysing the cascades of uncertainty in flood defence projects: how "not knowing enough" is related to "knowing differently". Glob Environ Change. doi:10.1016/j.gloenv-cha.2013.11.008
Van der Klis H (2003) Uncertainty analysis applied to numerical models of bed morphology. Ph. D. thesis, Delft University of Technology, Delft, the Netherlands
Van der Sluijs JP, Craye M, Funtowicz S, Kloprogge P, Ravetz J, Risbey J (2005a) Combining quantitative and qualitative measures of uncertainty in model-based environmental assessment: the NUSAP system. Risk Anal 25(2):481–492. doi:10.1111/j.1539-6924.2005.00604.x

Van der Sluijs JP, Risbey JS, Ravetz J (2005b) Uncertainty assessment of VOC emissions from paint in the Netherlands using the NUSAP system. Environ Monit Assess 105(1–3):229–259. doi:10.1007/s10661-005-3697-7

Van Gelder PHAJM, Mai CV (2008) Distribution functions of extreme sea waves and river discharges. J Hydraul Res 46(Special Issue 2):280–291

Van Loon E, Refsgaard JC (eds) (2005) Guidelines for assessing data uncertainty in river basin management studies. Geological survey of Denmark and Greenland, Copenhagen, pp 182. Available on http://www.harmonirib.com

Van Rijn LC (1984) Sediment transport, part III: bed forms and alluvial roughness. J Hydraul Eng 110(12):1733–1754. doi: 10.1061/(ASCE)0733-9429(1984)110:12(1733)

Van Stokkom HTC, Smits AJM, Leuven, RSEW (2005) Flood defence in the Netherlands: a new era a new approach. Water International 30(1):76-87. doi:10.1080/02508060508691839

Vanoni VA, Hwang LS (1967) Relation between bed forms and friction in streams. J Hydraul Div 93(HY3):121–144

Walker WE, Harremoës P, Rotmans J, van der Sluijs JP, van Asselt MBA, Janssen P, Krayer von Kraus MP (2003) Defining uncertainty, a conceptual basis for uncertainty management in model–based decision support. Integr Assess 4(1):5–17

Warmink JJ (2011) Unravelling uncertainties. The effect of hydraulic roughness on design water levels in river models. Ph.D. thesis, University of Twente, pp 185

Warmink JJ, Booij MJ, Van der Klis H, Hulscher SJMH (2007) Uncertainty of water level predictions due to differences in the calibration discharge. In: Proceedings of the international conference on adaptive and integrated water management, CAIWA2007. Basel, Switserland, p 18

Warmink JJ, Janssen JAEB, Booij MJ, Krol M (2010) Identification and classification of uncertainties in the application of environmental models. Environ Model Softw 25(12):1518–1527. doi:10.1016/j.envsoft.2010.04.011

Warmink JJ, Booij MJ, Van der Klis H, Hulscher SJMH (2013a) Quantification of uncertainty in design water levels due to uncertain bed form roughness in the Dutch river Waal. Hydrol Process 27:1646–1663. doi:10.1002/hyp.9319

Warmink JJ, Straatsma MW, Huthoff F, Booij MJ, Hulscher SJMH (2013b) Uncertainty of design water levels due to combined bed form and vegetation roughness in the Dutch River Waal. J Flood Risk Manage 6:302–318

Warmink JJ, Van der Klis H, Booij MJ, Hulscher SJMH (2011) Identification and quantification of uncertainties in a hydrodynamic river model using expert opinions. Water Resour Manage 25 (2):601–622. doi:10.1007/s11269-010-9716-7

Wilbers AWE, Ten Brinke WBM (2003) The response of subaqueous dunes to floods in sand and gravel bed reaches of the Dutch Rhine. Sedimentology 50(6):1013–1034. doi:10.1046/j.1365-3091.2003.00585.x

Wright S, Parker G (2004) Flow resistance and suspended load in sand-bed rivers: simplified stratification model. J Hydraul Eng 130(8):796–805. doi:10.1061/(ASCE)0733-9429(2004) 130:8(796)

Zhou H, Gomez-Hernandez JJ, Li L (2014) Inverse methods in hydrogeology: evolution and recent trends. Adv Water Resour 63:22–37. doi:10.1016/j.advwatres.2013.10.014

Part II
Fluvial Processes

Chapter 12
Channel Stability: Morphodynamics and the Morphology of Rivers

Michael Church

Abstract River morphology and morphodynamics are subject to governing conditions that include the volume and timing of water flows, the volume and calibre of sediment introduced into the river, the nature of bed and bank materials and vegetation, and the geologic and topographic setting of the river, including landscape gradient, climate and human interference. Water flows set the scale of the channel and the sediment regime lends distinctive character to river morphology. Consequently, rivers can usefully be classified on the basis of scale, sediment calibre and gradient, leading to the distinction of steep, intermediate and low gradient channels that generally correspond to channels of high, intermediate and low boundary roughness. Channel changes associated with the downstream passage of sediment constitute the morphodynamics of rivers. Primary morphodynamical effects include channel deformation, channel division, and channel gradation. A fundamental lesson is that rivers transporting a significant charge of bed material sediment necessarily have a lateral style of instability: lateral displacement of the channel is a normal part of their equilibrium function. Hence the required channel zone is larger than the presently active channel. Implications of this circumstance are considered in the context of channel management and river restoration.

Keywords Channel management · Channel stability · Fluvial sedimentation · Morphodynamics · River morphology

---

M. Church (✉)
Department of Geography, The University of British Columbia,
Vancouver, BC V6T 1Z2, Canada
e-mail: mchurch@geog.ubc.ca

P. Rowiński and A. Radecki-Pawlik (eds.), *Rivers—Physical, Fluvial and Environmental Processes*, GeoPlanet: Earth and Planetary Sciences,
DOI 10.1007/978-3-319-17719-9_12

12.1 Governing Conditions

Stream channels exist to evacuate water from the surface of the land. The flow of water may also mobilize earth materials so that streams, to a greater or lesser degree, shape their channels. The consequent form of the channel and the history of channel changes depend upon four principal governing conditions:

1. the volume and time distribution of water that is supplied from upstream and the land surface;
2. the volume, timing and calibre of sediment that is introduced into the channel;
3. the character of the bed and bank materials, including vegetation, through which the river flows;
4. the geological history of the riverine landscape; in particular, the topographic gradient down which the water and sediment are transferred.

Secondary factors that may be important include climate, in particular, the occurrence of a freezing winter and a seasonal ice regime, or an extended dry season; land use in the contributing drainage basin; and direct human interference.

12.1.1 Volume and Timing of Water Flows

The volume of water that flows through the channel—in particular, the magnitude of flows that mobilize channel-bed sediments and thereby shape the channel—sets the scale of the channel (Fig. 12.1). Since runoff, including peak flows, varies with the size of the contributing drainage basin, river channel scale varies systematically through the drainage basin (Leopold 1994). It is widely supposed that flood flow of some relatively frequent recurrence—in the range 1.5–2.5 years (that is, approximately the mean annual flood: *cf.* Wolman and Miller 1960)—is the dominant or 'channel forming' flow in the sense that it is the flow that creates the greatest degree of morphodynamic change. Alternately, 'bankfull' is often invoked as the most effective flow because this represents the maximum flow that occurs within the confines of the channel, hence exerts the largest stresses on the channel boundaries, whilst rarer overbank flows have negligibly more severe effect within the channel. However, there is no consistent correlation between flow frequency and bankfull nor, in fact, between flood frequency and effectiveness in creating morphodynamic change. Large rivers flowing in relatively fine, easily mobilised sediments may indeed have their channels shaped mainly by frequently recurring flows, but headwater cobble or boulder channels may be subjected to channel reforming disturbance only once in decades, or even more rarely (Fig. 12.2). There may be scale-related consistency of channel disturbance on a regional scale but, if there is any universal character at all underlying channel disturbance, it is simply that larger rivers tend to undergo channel change more frequently but less severely than small channels. That is because they experience more powerful streamflows and generally are bounded by finer, more easily mobilized sediments.

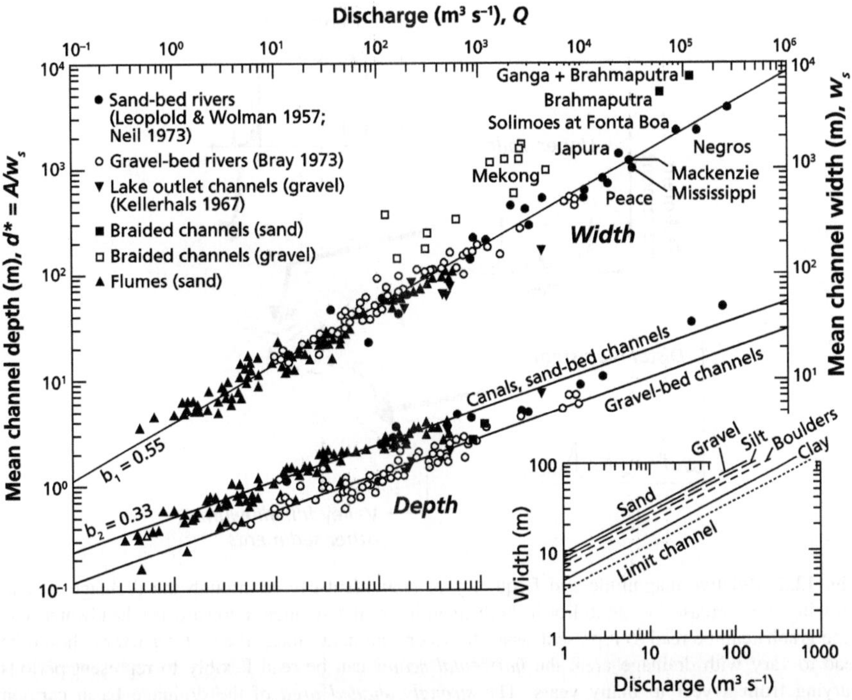

Fig. 12.1 Scaling relations for river channel geometry. Channels that are dynamic models of each other scale as $Q^{0.4}$ in both width and depth so that geometric similarity is preserved. Alluvial channels are distorted toward relatively increased width as Q increases (width exponent = 0.55; depth exponent = 0.33; both values exhibit some variation amongst individual data sets); that is, the width to depth ratio increases. *Inset* variation in channel width due to material properties, which covers the range of variability in the main diagram: the 'limit channel' is the narrowest mechanically stable channel for a given flow (Lane 1957). These scaling relations have been formalized for rivers as the 'hydraulic geometry' (Leopold and Maddock 1953). (Modified after Fig. 6.2 in Church 1992)

The duration of inundation and the season in which it occurs may significantly condition the channel edge and riparian zones. Significant correlations have been demonstrated between plant communities and elevation above normal water levels, hence usual duration of inundation (e.g., Woodyer 1968; Teversham and Slaymaker 1976) and between certain invertebrates and elevation (Radecki-Pawlik and Skalski 2008). The 'lower limit of continuous terrestrial vegetation' is a significant channel boundary that is more or less well-defined on most stream banks and is interpreted as the limit of the 'active channel' (Fig. 12.3). In many jurisdictions, this is also the legal limit of the river channel, though it may not correspond with the morphologically determined 'bankfull' stage.

While synoptic variations in flow change the area of the wetted channel with high frequency (time scale of days, or hours in small streams), the area of the active channel is naturally modified only on time scales of hydroclimatic change, which

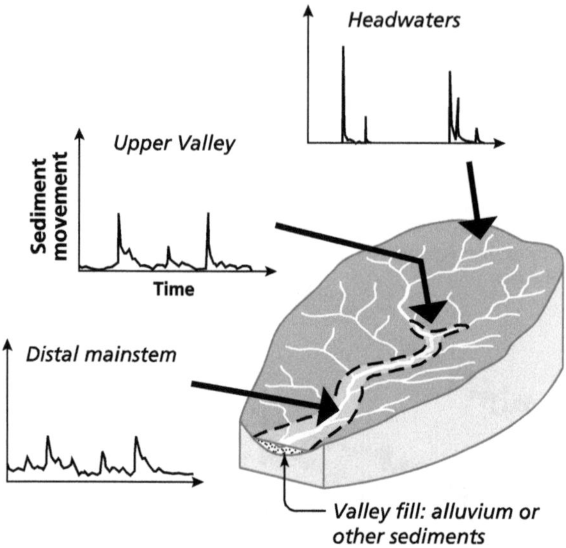

Fig. 12.2 Relative magnitude and frequency of significant events through the drainage system showing the increase in variability in both magnitude and frequency toward the headwater The time graphs can be read to represent water flows or sediment fluxes: the *vertical scales* should be read to vary with drainage area; the *horizontal scales* can be read flexibly to represent periods varying from a year to many years. The *strongly shaded area* of the drainage basin cartoon indicates the 'headwater' region of the drainage basin, where streams are directly coupled to adjacent hillslopes—that is, they receive increments of sediment directly overbank; the *lightly shaded sector* indicates the mainstream, where the channel is decoupled from adjacent slopes by the presence of alluvial deposits: sediments are received from upstream and by erosion of alluvial banks. Partially coupled reaches are common in mid-basin

Fig. 12.3 Cross-section of a river illustrating significant water level thresholds (modified after Osterkamp and Hedman 1982). The diagram is *vertically* exaggerated to emphasize the different levels

may be long. However, human regulation of streamflows—by dam building and water diversion—induces rapid channel change. An increase in channel width and area in response to a secular increase in high flows may occur quickly (over days to a few years, depending on the mobility of bed and bank sediments) because conditioned by the hydraulic force of the increased flows, whereas a reduction in high flows elicits a response measured in years (small streams) to decades (large rivers) because controlled by the rate at which riparian vegetation advances across the former channel margins. This latter transformation is climatically sensitive, being more rapid in the humid tropics than in temperate and boreal regions.

12.1.2 Volume and Calibre of Sediment

River channel morphology and channel morphodynamics are principally conditioned by the volume and character of sediment moving through the channel. When predominantly fine-grained material (fine sand, silt) is delivered to the channel it is largely carried in suspension in the water column and there is substantial overbank deposition during floods. This builds high, semi-cohesive banks that maintain a relatively narrow, deep channel that habitually meanders, or is anastomosed (that is, a low gradient channel divided about channel islands into two or more channels), and changes position slowly in a roughly predictable way. Coarser sediments (coarse sand, gravel, cobbles) are transported on the bed and deposited within the channel. Such deposits deflect the flow, creating a relatively wide, shallow channel with non-cohesive lower banks. Channels then exhibit less regular migration, dominated by lateral activity that sometimes undergoes rapid and unpredicted change. According to the sediment supply, such channels may meander irregularly, adopt an anabranched style (exhibiting division about channel islands or normally emergent bars into two or more branches), become braided, or wander (that is, exhibit characteristics of both braided and anabranched channels) (see Fig. 12.4).

Many rivers flow through sediments that they have previously deposited: their channels are termed *alluvial channels*. These rivers certainly are competent to reform their channel since they previously moved the sediment that makes up the bed and banks. Here we find consistent relations between streamflow and the channel dimensions of width and depth, hence also velocity. Such scaling relations are termed *hydraulic geometry* (Fig. 12.1). Since the actual form of the channel is conditioned by the material that forms the bed and banks, distinctive hydraulic geometries occur in different materials (Fig. 12.1, inset). However, the strength of granular materials has only a limited range, whilst the power of flows may vary over orders of magnitude through time at one place and downstream through the river system. The width to depth ratio (aspect ratio) of an alluvial channel is governed by bank material strength; as rivers become larger, they chiefly become relatively wider (Fig. 12.1) because neither bed nor bank material strength increases in proportion to the increase in the hydraulic forces that would be exerted in a self-similar channel. This effect distributes the force of the flow over a proportionally wider bed, which serves to increase channel stability. For these reasons, a small channel is not, in general, a model of a larger one.

Material calibre (grain size) and the force imposed by the flow on the granular boundary determine the intensity of sediment transport. A measure of the force

Fig. 12.4 Multi-thread channels: **a** Rivière Bléone, at Digne, France, a low-order braided channel, view downstream: note the rock-faced embankment on the left bank (photo by H. Piégay); **b** Kootenay River, British Columbia, Canada, a sandy anastomosed channel; view downstream; **c** wandering gravel-bed channel incorporating both braid bars and anabranches: Kemano River, British Columbia, Canada, view downstream

imposed on the boundary by the flow is given by $\tau = \rho g d S$, in which τ is shear stress (force per unit area), ρ is water density, g is the acceleration of gravity, d is flow depth and S is the energy gradient (usually assumed in regime studies to be equivalent to the water surface gradient over the length of a reach, which is then approximately equivalent to channel gradient). Resistance to the flow is distributed between a part applied directly to the solid boundary and a part absorbed by accelerations and deceleration of the flow induced by larger geometric elements of the channel (bedforms, variations in width, channel bends) and by turbulence.

The effective stress available to move sediment, then, is $\tau_{\text{eff}} = \alpha\tau$, $\alpha < 1.0$. The capacity of a flow to move material is indexed by the Shields relation, $\tau_* = \tau_{\text{eff}}/g(\rho_s - \rho)D$, in which τ_* is the Shields number, a non-dimensional expression of the stress applied directly to the grains on the boundary, ρ_s is the sediment density, and D is the grain size. For large materials (gravel and larger), $\tau_* < 0.1$ generally, and sediment movement is sporadic for most competent flows. When $0.1 < \tau_* < 1.0$, sand may be moved at substantial rates over the bed, with fine to medium sand and silt in suspension; for $\tau_* > 1.0$ general suspension of sand occurs. Church (2006) gave an extended discussion of sediment transport regimes in relation to channel morphology.

Fine materials are, then, moved over a wide range of flows and may be moved in high concentrations for a significant distance, whereas coarse materials are moved only in high flows and at relatively low concentrations for shorter distances. This means that bed reshaping is a relatively common occurrence in fine sediments: such channels may be termed *labile channels* (meaning, easily deformed); coarse materials form *threshold channels*; that is, sediment movement nearly always occurs near the threshold for movement and significant bed reshaping is a less common occurrence. In labile channels, bed sediments are easily mobilized. In threshold channels, however, the wide range of grain sizes present leads to size-selective transport, the development of a surface layer of relatively coarse material that moves less readily than finer fractions and 'hiding' of the finer material below the surface. Consequently, sediment flux is limited by the necessity to displace elements of the coarse surface before significant amounts of finer material can be entrained. A further consequence is that, if sediment supply to the channel is reduced, the bed surface is apt to coarsen further as finer grains are winnowed from the bed and transported away (Dietrich et al. 1989) so the characteristic difference between the size of sediments on the bed surface and transported sediment increases. Of course, sediment supply from the land surface mediates these conditions, hence human agency may influence channel processes and morphology by mobilizing or immobilizing sediment in the course of land use.

Variations in sediment supply to streams may occur in concert with hydroclimatic change as runoff from the land changes, or can occur relatively abruptly as the result of major landslides delivering large volumes of sediment to a channel, changes in land use that change sediment supply to the channel, or direct mining of riverbed sediments from the channel. The most dramatic short-term changes in river morphology are effected by abrupt changes in sediment supply. But despite its importance in establishing the morphological style of a river, sediment transport—in particular, the transport of the 'bed material' that constitutes the bed and lower banks of the channel—is difficult to measure and not commonly measured. River investigators must, then, often make qualitative inferences about the character of sediment transport by considering stream power, or by regarding the morphology of the channel and the character of the bed sediments. In the latter case, one in effect infers cause from effect (see Church 2006, for a discussion of the bases for inference).

12.1.3 The Character of Bed and Bank Materials

Not all channels have entirely alluvial boundaries. Rock and non-alluvial sediments may constitute one or both banks or the channel bed, and present more or less firm resistance to erosion. This may significantly affect the stability of the channel and also constrain channel orientation. Rivers with one or more non-alluvial boundary can at best be considered to be 'semi-alluvial'; rock-bound channels are non-alluvial. Non-alluvial boundaries may also be non-lithified sediments, including glacial sediments, streamside colluvium, loess, and peat. It may, further, include relict, often semi-lithified alluvium from an earlier era. Such materials may variously resist erosion strongly, or perform little differently than contemporary alluvium.

Riparian vegetation may have significant effects on bank strength, hence on the overall form of the river channel, provided the rooting depth is as great as the channel is deep. This generally limits vegetation effectiveness to rivers smaller than some limit size that will vary geographically according to the character of the native vegetation. For example, Beechie et al. (2006) report that, in Pacific Northwest North America, channels less than 15–20 m in width remain stable because of root reinforcement of bank strength. Vegetation may have further effects in channels with seasonally exposed bed areas by becoming established in-channel. Provided woody vegetation can withstand subsequent high flows, it then encourages sediment deposition and the eventual development of a channel island. Downed wood instream also has a significant impact on sedimentation, leading to island formation, and may direct the flow so as to increase or attenuate bank attack.

The composite strength of bank materials has a significant influence on the time scales for channel change, including normal lateral activity (see Sect. 12.2) and on the aspect ratio of the channel. Figure 12.5 presents a classification of the range of channel conditions, encompassing boundary conditions and sediment mobility, that may result. Each type exhibits a characteristic style of channel instability.

Summary classification of stream channels

Boundaries	Threshold *(large material)*	Labile *(finer material)*
Non-alluvial	Rock-bound or refractory sediments	Easily eroded sediments
Partly alluvial	Cobble/gravel bed; Rock or Quaternary sediment banks	Gravel/sand bed; Rock or Quaternary sediment banks
Fully alluvial	Cobble/gravel bed and banks	Gravel/sand bed, sand/silt banks
	Supply-limited, Upland type	*Transport-limited, Valley type*

Fig. 12.5 Channel classification. The more detailed description of boundary materials gives some emphasis to Quaternary materials that are encountered by many rivers, particularly within the limits of Quaternary glaciations. The effects of bankside and instream woody vegetation are not incorporated into this diagram

12.1.4 Geological History

The most important control exerted by the river setting is the topographic gradient down which the water and sediment load of the river must be transferred (that is valley or surface gradient, *not* the channel gradient). Most rivers flow in valleys established over a long geological history and valley gradient is accordingly imposed. River adjustment in the medium term is restricted to the possibility for the river to adopt a meandered habit so that its gradient declines below that of the valley. This is a particularly important constraint even in major rivers in regions that were affected by Pleistocene glaciation and is generally true of small headwater channels with limited erosive capacity. However, rivers flowing over purely depositional surfaces, such as alluvial fans or extensive floodplains, may indeed have conditioned their own topographic gradient as the result of a long period of aggradation.

Landscape history and physiographic setting impose additional constraints on river morphology and morphodynamics. Rivers flowing in valleys may be subject to lateral and/or vertical control over channel position. Vertical control is exercised in the form of resistant materials, usually rock, that resist erosion and so form a local grade control for the river. Rivers may be confined or entrenched in narrow valleys. Conversely, on surfaces of considerable lateral extent, in particular on river-constructed floodplains, deltas and alluvial fans, absence of lateral constraint permits avulsions—sudden lateral displacements of the channel when, due to excessive water volume or blockage by sediment or downed wood, the flow wholly or in part abruptly departs from its former channel to form a new channel (or reoccupy a former channel) on the adjacent surface.

12.1.5 Human Interference

Humans affect river channels by direct structural interference, as by the construction of dams, weirs and artificial channels, or by stabilizing bank position (that is, inhibiting lateral erosion). Rivers are also affected by land surface change on the contributing drainage basin that changes the water and/or sediment supply to the stream. Forestry, agricultural land conversion and urbanization each have characteristic effects on stream systems. Human activities in each of these categories occur with a characteristic intensity in the landscape, hence it is a reasonable generalization that impacts of land use are proportionally most severely visited upon smaller, headwater channels. Nevertheless, the cumulative impact of human activity may strongly affect even the largest rivers, many of which have been directly altered by structures (Church et al. 2009). In particular, most of the larger rivers of Europe have been modified by channel rectification or the construction of dams or weirs. Human impacts may be transient but more often have definitively changed the river morphology and morphodynamics.

12.2 River Regime and Channel Equilibrium

The governing conditions vary systematically along a river. Some changes are abrupt, such as the change in flow and sediment load at a tributary junction (Ferguson and Hoey 2008), or the change in channel morphology induced when a river crosses a geological boundary. Other changes may be subtle, such as when sediment grain size changes downstream under the influence of abrasion or selective transport, or sediment load changes in a long reach of slowly degrading channel. River channels must be analyzed reach by reach since the important forcing and boundary conditions are well-defined only on a reach basis. Because of the importance of sediment as a primary governing condition, a useful concept is that of the *sedimentary link* in a channel system (Rice and Church 1998), the distance between successive significant increments to the sediment load. Mostly, these would be tributary junctions that also designate hydrological links, but significant non-fluvial sediment sources do occur, especially in hillslope-coupled, headwater streams (Fig. 12.2) that flow directly along the base of adjacent, failing hillslopes.

Within sedimentary and hydrological links (reaches) forcing and boundary conditions are essentially constant (except in reaches in which the river approaches a grade control or end point that forces a continual change in slope) and alluvial rivers adjust their boundaries by erosion, transfer and deposition of sediment so that they approach a state of 'equilibrium'—that is, a state in which the channel is just able to pass the water and sediment supplied from upstream, hence remains stable in average dimensions. Another way to express this condition is to say that the channel is '*in regime*'. In this state, the channel dimensions are well-predicted by the equations of hydraulic geometry appropriate to the bounding materials (Fig. 12.1), a concept directly descended from ideas about the regime (stable) condition of unlined irrigation canals (Eaton 2013).

It is critically important to realize that an alluvial channel in equilibrium is not, as commonly supposed, absolutely stable. Although the channel dimensions remain stable, on average, any river that transports significant quantities of bed material—that is, sediments that constitute the bed and lower banks of the river—must move laterally to accommodate erosion and deposition of those materials as they are staged along the channel by varying flows. The channel is, at best, 'pattern stable', meaning that its geometry, including planform, retains the same *average* character. Failure to appreciate this condition has led to many unsuccessful attempts to stabilize river channels absolutely.

To achieve a regime condition, there must be an adjustment of the channel form and gradient so that all the energy of the river, except that required to pass the water and sediment supplied from upstream, is consumed in overcoming resistance to flow (Eaton et al. 2004). The river has four principal means to achieve this:

1. adjustment of grain size of the sediments exposed on the bed surface (by size-preferred erosion or deposition), or of the dimensions of labile primary bedforms (such as sand dunes);
2. adjustment of channel width by bank erosion or sediment deposition in bars;

3. adjustment of channel macroform elements, that is, the development of pools, riffles and bars, or assumption of a meandered habit;
4. adjustment of the mean gradient of the river by net degradation or aggradation.

The first, and most easily achieved, adjustment, because it entails only differential pickup and deposition of the various sizes of granular sediments present on the bed, alters the grain friction and micro- to mesoform resistance to flow of the channel. This is a significant means of adjustment as the strength of flow and magnitude of sediment transport vary with the waxing and waning of a flood: both grain size on the bed surface and sedimentary bedforms such as ripples and dunes may be changed as the flow changes. The second adjustment affects the aspect ratio of the channel which, in turn, affects the bed material transport because it affects the force imposed on the bed by the flow. The combination of these first two effects adjusts the transport capacity of the river to match the incoming sediment supply. Width may also change within the period of a flood but more commonly changes through a sequence of floods. The third and fourth adjustments change the rate of energy expenditure of the river by altering the macroform geometry and channel gradient (in the fourth case, within the limit set by valley gradient). Macroform geometry may change within the time scale of a flood by transient scour and fill of pools and riffles along the channel but larger scale adjustments such as meander growth require years to decades. These adjustments require at least local net displacement of sediment. Gradient adjustments require redistribution of sediment over long distances and may require decades to millennia (in the largest rivers) to complete (see Sect. 12.3.6 for a discussion of time scales).

Channel changes associated with the passage downstream of sediment, and changes described above that are associated with regime adjustments, constitute the morphodynamics of rivers. River channel morphology is the autogenerated result of the interaction amongst flow, sediment load and initial channel form working on a variety of scales from granular to channel system, leading to the occurrence of distinctive channel types (Kleinhans 2010). Since the calibre of the sediment and the energy required to move the sediment are principal controls on the transport and disposition of bed material, hence on channel morphology, river gradient and bed material supply and calibre are reasonable parameters upon which to base a classification of river channels. Volume of flow (in effect, channel scale) must also be considered since the relative size of channel and the sediments is a significant factor mediating sediment mobility, while the flow and gradient together determine the capacity to move sediment.

12.3 Channel Classification and Morphodynamics

12.3.1 Morphological Classification

River channels have customarily been classified principally according to channel planimetric form as straight, meandered or braided (e.g., Leopold and Wolman 1957), with anabranched channels introduced as an additional distinctive type

(Nanson and Knighton 1996). In fact, rivers exhibit much greater variety of form than is admitted by this spare classification (even though these major distinctions of morphological style remain fundamental) and different bases for classification are possible.

Channel and sediment scales are important classificatory criteria because the relative size of channel and sediment lends distinctiveness to channel morphology; the largest materials are found in the headward, smallest channels, and on the steepest gradients (Fig. 12.6). We may define channels with high granular roughness as ones in which flow depth, d, is typically comparable with grain size, D, so that relative roughness, $D/d \approx 1.0$: channel scale is comparable with the scale of individual boundary elements. Individual clasts accordingly constitute significant form elements of the channel that act to divide the flow into a series of jets. Conversely, in rivers with low granular roughness, one might arbitrarily say $D/d < 0.05$, grains on the bed contribute collectively to skin friction but do not individually affect overall channel geometry or function. A deep shear flow results. Between these limits lies a class of intermediate channels which experience wake flows dominated by eddies shed by submerged but relatively prominent grains. These divisions discriminate principal sources of flow resistance (see Ferguson 2007, Fig. 1a) but they remove any sense of absolute scale. Nonetheless, most high roughness channels are small, headwater channels, most low roughness channels are major rivers, while channels of intermediate roughness appear in the landscape between these end-members (Fig. 12.6).

Channels of high and intermediate granular roughness are typically threshold channels; ones of low granular roughness are typically transitional or labile. Furthermore, high roughness channels are typically hillslope-coupled, channels of intermediate roughness are often partially coupled, and ones with low roughness are characteristically decoupled. But while these correspondences define a characteristic organizational structure for drainage basins, they are not without exception. In regions of low relief, for example, headmost channels may be miniature low roughness channels enclosed in peat, turf or heath, or woodland streams of intermediate roughness.

Within this framework, both single and multi-thread channels are recognized. Figure 12.7 illustrates a range of channel morphologies that are defined by scale, sediment calibre and gradient, and notes the resultant characteristic (in)stability of the channel. An even more nuanced pattern of classification was presented by Kellerhals et al. (1976) which disaggregated channel properties into different categories for sinuosity (meander form), presence of islands (anabranched, of which anastomosed channels are a subcategory), style of in-channel bar forms, and style of instability (morphodynamics). That classification recognized that features that commonly are dichotomized in single-dimensioned classifications may in fact occur together (such as meandering and islands). In summary, rivers exhibit a continual range of variation in morphology.

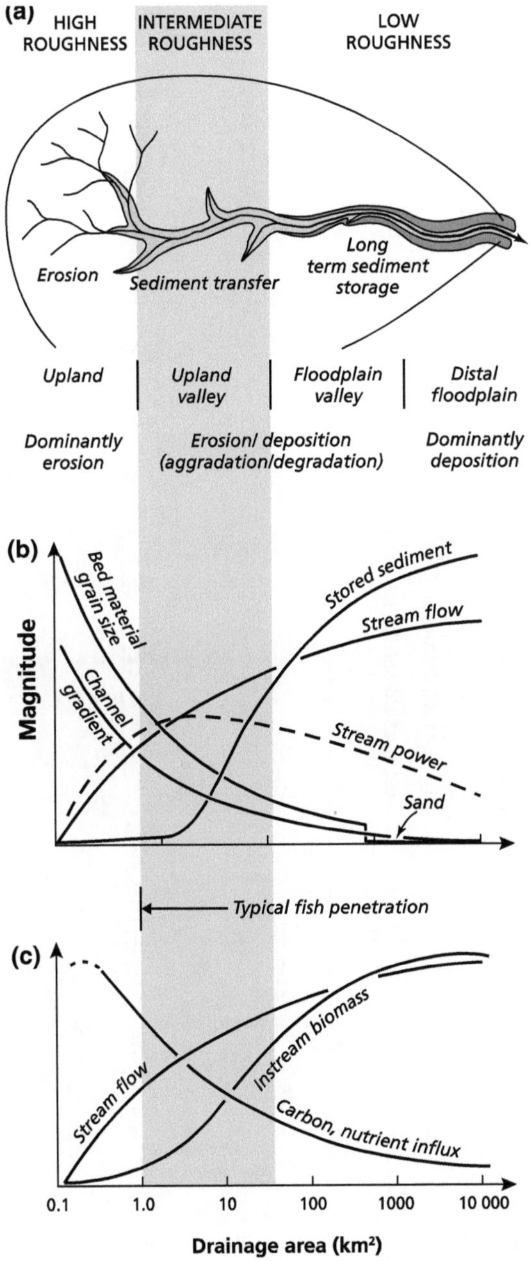

Fig. 12.6 Variation of flow and channel properties through the drainage system (partly after Schumm 1977). **a** drainage basin map showing zones of distinctive sediment behavior (only a fraction of the hillslope-coupled upland channels is shown); **b** pattern of variation of principal quantities related to flow, sediment flux and channel morphology; **c** pattern of variation of some ecological phenomena that are systematically related to drainage structure. The characteristic distribution of channel 'roughness types' (see text) is superimposed. (Modified from Church 2002, Fig. 1)

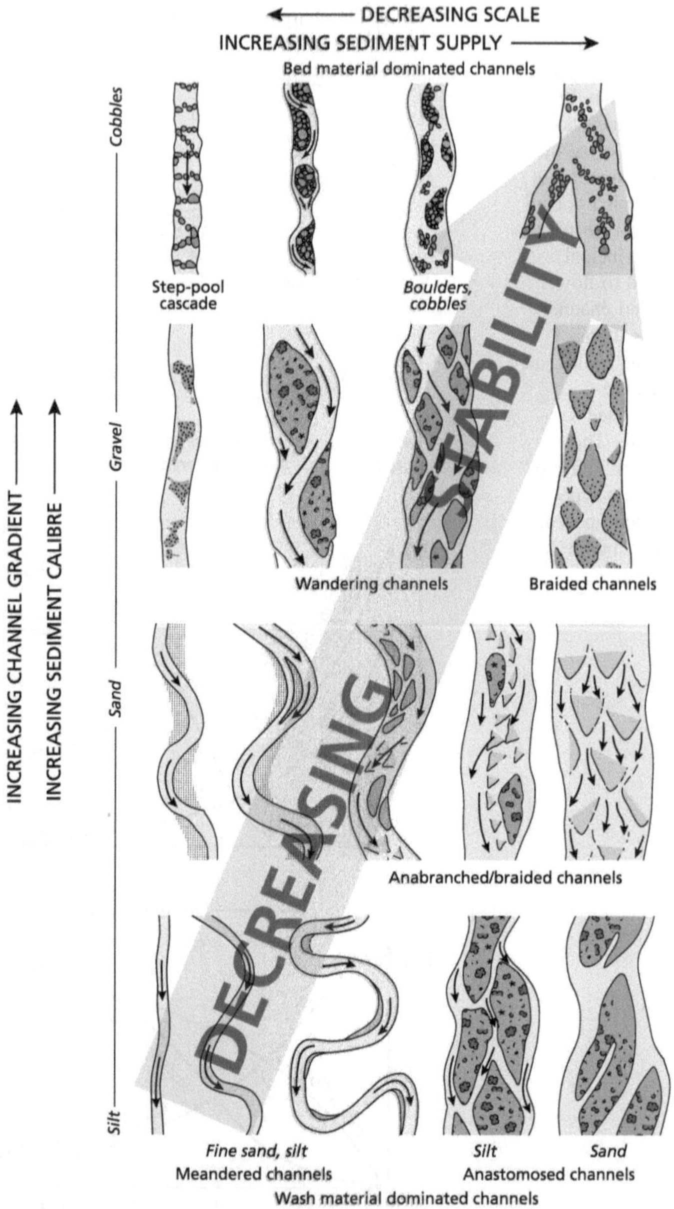

Fig. 12.7 Morphological channel types according to scale, sediment calibre and sediment flux. The diagram can be presented in two dimensions because of the characteristic correlations between sediment calibre and gradient and between sediment volume and stability. (Modified from Church 2006, Fig. 2)

12.3.2 Morphodynamics: Channels of High Roughness

High roughness occurs essentially exclusively in channels flowing on boulders, cobbles or coarse gravel. They are mainly small channels on relatively steep gradients (Fig. 12.8)—upland to montane headwaters (Church 2010)—though rivers on lower gradients but with a local source of boulders may have high roughness. In the smallest, steepest channels the larger individual clasts may have dimensions similar to those of the channel itself (Fig. 12.9). Hence instability of individual grains may affect the stability of the entire channel.

On gradients steeper than about 3° (5 %) individual grains as large as the channel is deep should not be stable (when the critical Shields number for mobilization, $\tau_{*c} \approx 0.03$, a widely accepted value for exposed grains). But they certainly are found in even steeper channels. It has been found that $\tau_{*c} > 0.03$ in steep channels and increases as slope increases (Mueller et al. 2005). Stones in steep channels form jammed chains that retain them in place (Church and Zimmermann 2007), and these form 'steps' so that a characteristic step-pool morphology results. Such 'force chains' are rarely more than 5*D* in length, which represents an approximate limit width for sediments to be stable in such channels. The source of the largest keystones in the channel usually is the immediately adjacent hillslope or, in glaciated terrain, erosion out of till lining the channel. They are not fluvial elements in the strict sense since the stream is rarely able to move them and does so mainly by undermining them, unless a debris flow—a mass movement of all mobilizable material in the channel—occurs. Most of the channel gradient is controlled by the steps. In forests, downed wood may also form steps. The intervening pools often

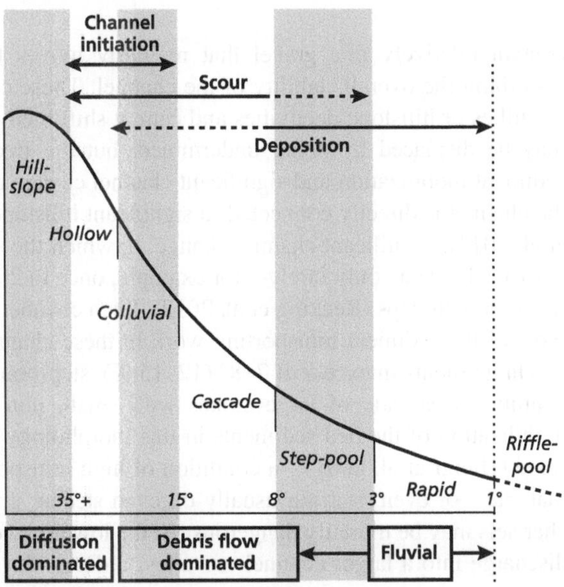

Fig. 12.8 Gradient-determined sequence of high roughness channels. (Modified from Montgomery and Buffington 1997, their Fig. 4)

Fig. 12.9 Vens stream, an active step-pool channel, gradient 0.15 (8.5°), French Alps (photo by A. Recking)

contain relatively fine gravel that regularly moves through the system without disturbing the overall stability of the channel. These channels are usually confined in gullies or hillslope declivities and cannot shift their course. Individual keystones may be displaced by being undermined, but the frequency of step-destabilizing sediment mobilization and significant channel change depends upon whether or not the channel is directly connected to significant hillslope sediment sources (Recking et al. 2012). Significant channel change, in which the step structure is substantially destroyed, occurs only rarely—for example, once in 20–50 years in active channels in the French Alps (Recking et al. 2012). Such channel changing events accomplish most of the sediment transporting work in these channels (see Fig. 12.2).

On gradients in excess of 7–8° (12–15 %), step-pools give way to a more or less continuous cascade of large stones with small, non-spanning pools. Significant mobilization of the bed sediments in this morphology may propagate into a debris flood (Hungr et al. 2001)—a condition of high transport of gravel- to cobble-sized material—or even, starting usually on even steeper gradients, a debris flow: these channels may be mortally dangerous. At the lower end of such channels, where they discharge into a larger channel or valley, a debris fan develops with characteristic angle of 3–5° (5–9 %)—lower for muddy sediments—and here the stream can easily avulse out of its channel during a major sediment transporting event. The entire fan is potentially an active zone (Kellerhals and Church 1990).

Maintaining the stability of these channels in the presence of engineering works (such as road crossings) critically depends upon maintaining the structure of locked keystones in the channel. Stabilizing steep, unstable gullies entails building sequences of check dams that act as artificial steps with much reduced gradients between them to discourage the movement of the coarser clasts.

12.3.3 *Morphodynamics: Channels of Intermediate Roughness*

'Intermediate' channels are ones in which individual bed material grains no longer dominate the morphology, but still strongly condition the flow. One commonly finds $0.1 < D/d \leq 1$ and wake-dominated flows. Many gravel-bed rivers fall into this class. Gravel derives from physical rock weathering in montane or upland regions and so gravel-bedded rivers are characteristic of montane valleys and forelands. Because of the wide grain size distribution and size specific differences in transport, a coarse surface layer develops that serves to reduce general transport. Further, structural reinforcement of bed sediments occurs on these stream beds in the form of stone clusters and stone networks composed of larger clasts anchored by the largest, least mobile clasts (Church et al. 1998). Imbrication, the stone-on-stone layering of bed grains—a virtually universal phenomenon in gravel-bedded streams—also acts to increase the stability of the bed as finer material is effectively hidden below the imbricate surface layer (Fig. 12.10).

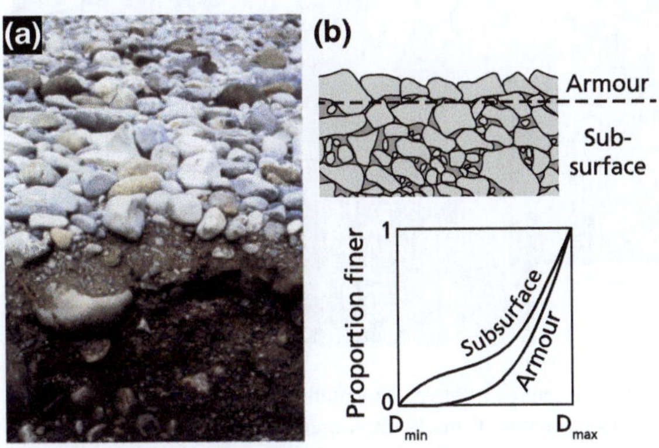

Fig. 12.10 **a** Bed sediment stratigraphy in a gravel-bed channel, showing armoured and imbricated surface; **b** cartoon of typical granular structure and cumulative size distributions. Photo by S.P.Rice

Sediment transport over these structurally constrained channel beds occurs in three stages. In stage 1, material introduced from upstream is passed over the bed without disturbance of the local bed material. Such mobile material would be mostly sand and substantially finer than local surficial bed material. In stage 2, the local bed material begins to be mobilized but entrainments are more or less sporadic: only some of the material is mobilized at any time (the condition known as 'partial transport') and the largest material may remain entirely immobile (resulting in 'size-selective transport'). The bed remains intact and deposition on bar faces remains an incremental process, so that lateral shifting of the channel is relatively slow. In stage 3, reached only in rare high flows, the bed is generally mobilized and rapid bank erosion may occur as the direct result of water attack. An individual flood of exceptional magnitude may, then, reform the entire channel (Fig. 12.11). An important outcome of the bed conditions found in channels of high and intermediate roughness is that bed material transport is normally orders of magnitude smaller than classical hydraulic formulae predict (Bathurst 2007; Yager et al. 2012) because of the constraints posed by size-segregated and structurally interlocked bed material.

Once relative roughness declines below about 0.1, grains may be stacked to form riffles and bars. Typical morphology is for a sequence of shallow pools and extended rapids on gradients of 1–2° (2–3 %), and for pool and riffle morphology to emerge on gradients lower than 1° (2 %) where a pool-riffle-bar triplet is the characteristic macroform feature of the channel. In these circumstances relative roughness may be much less than 1 locally (in the pools) and so at low to

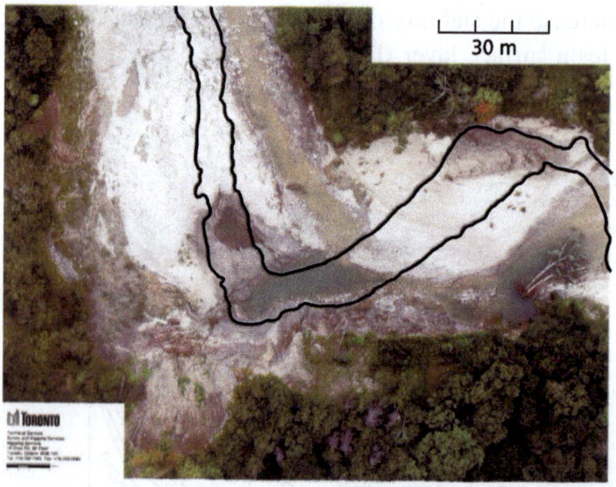

Fig. 12.11 Rapid enlargement of a gravel-bed channel of intermediate roughness by a 500 m^3 s^{-1} flood in Highland Creek, Toronto, Canada, in August, 2005. The Shields number for the flow was approximately 0.3 for a 100 mm bed grain. The *black outline* indicates the position of the well-vegetated pre-flood channel banks. (Church 2010, Fig. 9.6, reproduced by permission of John Wiley & Sons, Ltd.: original courtesy of the City of Toronto Engineering Department and Professor Joseph Desloges)

moderately high flows one has an alternation of shallow, wake-dominated flow over the riffle decaying to a deep shear flow in the pool (MacVicar et al. 2013). Channel morphodynamics remains dominated by the bed material load even if a substantial volume of fine material moves through the channel. Bed material is deposited as gravel sheets along bar edges, so bars grow laterally into the channel, forcing erosion on the opposite bank (Fig. 12.12). A floodplain is thus constructed by lateral deposition and consists primarily of bed material initially deposited as bar sediments, with a topset bed of sand deposited from suspension during overbank flooding. Bars form in alternate sequence on opposite sides of the channel, leading to a usually irregular meandering tendency. Much investigation has gone into attempts to understand the alternating regularity of bar formation, a process that leads to full-scale meander form in finer grained materials. The fundamental aspects of bar formation are reviewed by Rhoads and Welford (1991) in a paper that, for practical purposes, remains informative.

It is clear that this lateral style of instability is an intrinsic aspect of bed material staging downstream and so an aspect of equilibrium processes in rivers of this type. One expects the rate of lateral channel shifting and total channel zone width to depend upon the magnitude of bed material transport through the channel (Wickert et al. 2013). Further, the consequent age distribution of the floodplain area favors younger surfaces and the proportion of the channel zone not reoccupied by active channel becomes smaller with time (Fig. 12.13a) at a rate dictated by sediment throughput. It is important to recognize, then, that the floodplain is part of the channel zone: it consists of sediment that is stored for a greater or lesser length of time. Left unconstrained, the river will eventually reoccupy floodplain sites and

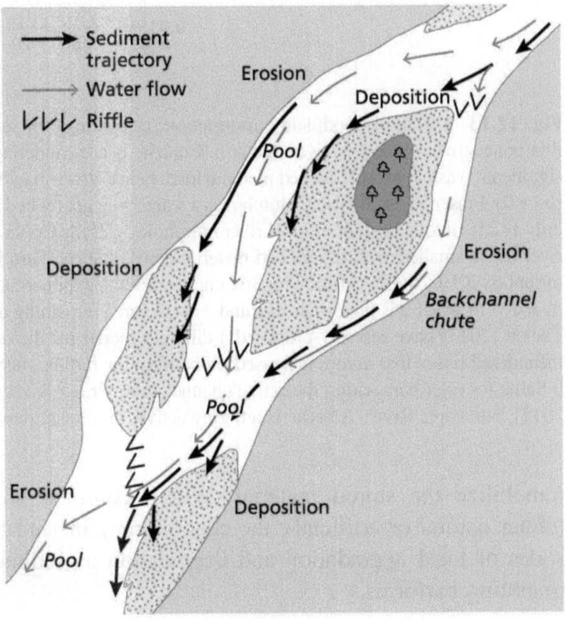

Fig. 12.12 Pattern of gravel movement in a pool-riffle system (*dashed arrows*): low-order braiding illustrated

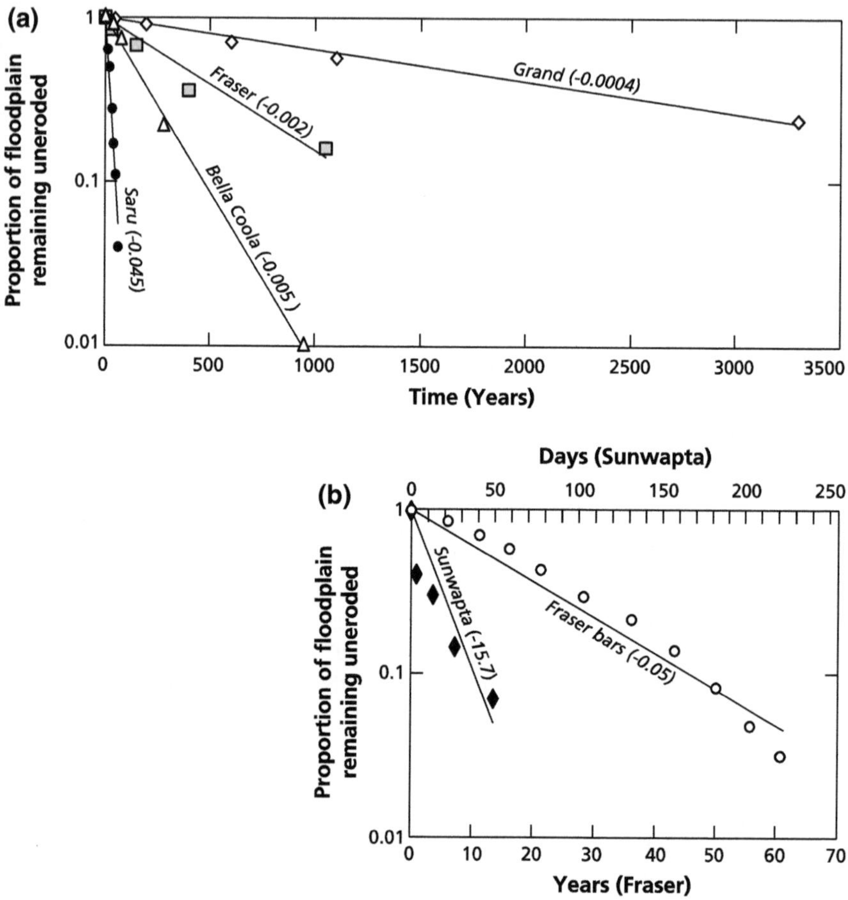

Fig. 12.13 **a** Rate of floodplain consumption (reworking) for several rivers with different sediment flux concentration: Grand River, Ontario, Canada, is a meandering sand-silt transporting river with a significant wash load and low bed material load, hence strong banks and a low rate of lateral movement ($b \rightarrow 0$); Fraser River, British Columbia, is a wandering gravel bed river in which the bed material load, only 1–2 % of total load, dominates river morphology; Bella Coola River, British Columbia, is a similar river but is smaller, with higher bed material concentration; Saru River, Japan, is an active mountain torrent ($b \ll 0$). Numbers following river names are the exponential decay constants, b, in the equation $A_r = e^{-bt}$ in which t is floodplain age and A_r is the area remaining unreworked at that age. Bradley and Tucker (2013) have recently proposed a different model for the distribution of age on floodplains of meandered rivers that favors increased probability for further survival of successively older surfaces. **b** Same for river bars within the active channel zone: Fraser River as above (data of Ham and Church 2012); Sunwapta River, Alberta, is a highly active, proglacial outwash river

remobilize the stored material. Where confined within erosion-resistant banks (either natural or artificial), the channel may instead be subjected to repeated episodes of local aggradation and degradation as sediment is staged downstream in migrating barforms.

Channels of intermediate roughness are common on the floors of montane valleys where topographic confinement is lost—often the stream is flowing on a valley fill of its own sediments. Major sediment transporting events can lead to channel-blocking deposits locally, forcing the channel to avulse into a new course. Such avulsions remain relatively rare but, on an otherwise little disturbed valley floor, serial avulsions over many decades, or even centuries, create a series of side channels and isolated pools that constitute unusually rich stream and wetland habitat.

In forest environments, drop-in wood, including trees toppled off undercut banks, may induce riffles or steps in the channel, promote scour pools, and even generate avulsions. The latter is particularly likely when many pieces drift into a channel-spanning jam. Conversely, along relatively shallow channels the root network of streamside trees commonly penetrates to the full depth of the channel and provides important bank reinforcement and channel stabilization.

Recking et al. (2013) have recently published an excellent monograph on the geomorphology and management of torrents and montane rivers—that is, channels of high and intermediate roughness.

12.3.4 Morphodynamics: Channels of Low Roughness

Channels of low roughness and deep shear flows are normally larger channels, though they may remain small in sand or silt on low gradients. They are bounded by materials varying from cobbles to fine sand and present a kaleidoscope of channel styles.

In gravel: processes are similar to those in channels of intermediate roughness. The gravel bedload—which may constitute as little as 1 or 2 % of the total sediment load in large gravel-bed rivers, but which nevertheless still dominates the morphology—moves from bar to bar; bars grow laterally and so a lateral style of instability characterises the channel (Fig. 12.12). Chute cutoffs are common on the

Fig. 12.14 Chute cutoff in the Ain River, France (photo by Hervé Piégay)

landward edge of the bars (Figs. 12.12 and 12.14). To maintain uninterrupted downstream sediment transfer, total channel zone width must be some multiple—depending on the magnitude of bed material transport—of current active channel width. Dunne et al. (2010) amplify the relations amongst flow, sediment provision and lateral migration rate.

Lateral erosion and deposition in heterogeneous gravel leads to channel patterns that may vary from irregularly sinuous or irregularly meandered, through wandering to braided form according to the total bed material concentration (that is transport rate in comparison with channel scale). Rate of lateral movement and the time required for the channel to develop and consume a bar also depends upon the volume of sediment moved through the channel, relative to the size of the channel (Fig. 12.13b). Accordingly, the time for the channel to laterally rework its floodplain depends on channel size, sediment volume and the valley width.

Where channels encounter resistant valley walls or non-alluvial sediments, the channel is forced to change direction and, here, bars are often 'forced' to develop as current slackens upstream of the bend or constriction. Such bars persist while sediment is staged through them. Where the channel is unconstrained, however, and not too tortuous, bars may be 'free' to migrate downstream and the movement of bedload may approximately equal the rate of bar shifting. But even in unconstrained circumstances, a channel may become sufficiently tortuous that bars become anchored on tight bends for extended periods. At such places the channel eventually may avulse.

In sand: bed material movement occurs over a much wider range of flows; channels are labile. Sands move on the bed or in intermittent suspension. Where sands are relatively coarse, they move principally on the bed and may build a sequence of alternating bars that grow by progressive accumulation of sand sheets on the downstream flank so that they progress downstream. Again, channel pattern may vary from single-thread meanders through anabranched channels to braided form. In low gradient channels carrying a fine sand and silt load, anabranching takes the special form of anastomosis, in which avulsions into the floodplain may abruptly create extensive channel islands. In larger single-thread channels transporting sand, alternate bars develop into progressing meanders (Fig. 12.15a). In more widely graded or finer sands, much of the material moves in suspension and the channel, more easily able to mould the boundary to the flow, develops more or less regular meanders (Fig. 12.15b). A combination of progression and loop extension occurs while, in still finer sediments, loop extension and cutoffs come to dominate. In this process it appears that outer bank erosion, locally widening the channel and thereby creating a slackwater zone on the convex bank opposite, encourages sediment deposition there and extension or progression of the meander bend (van de Lageweg et al. 2014). The resulting ratio $S_{valley}/S_{channel}$ = length of channel/length of valley defines the sinuosity of the channel. Channel sinuosity and meander belt width may vary according to how different is the valley gradient from the gradient that the river must maintain to pass its sediment load.

Because finer sediments can be suspended high in the water column, bars build to significant height and carry a topset bed of fine sand or silt that also appears in the

Fig. 12.15 Styles of meandered sand-bed channel deformation

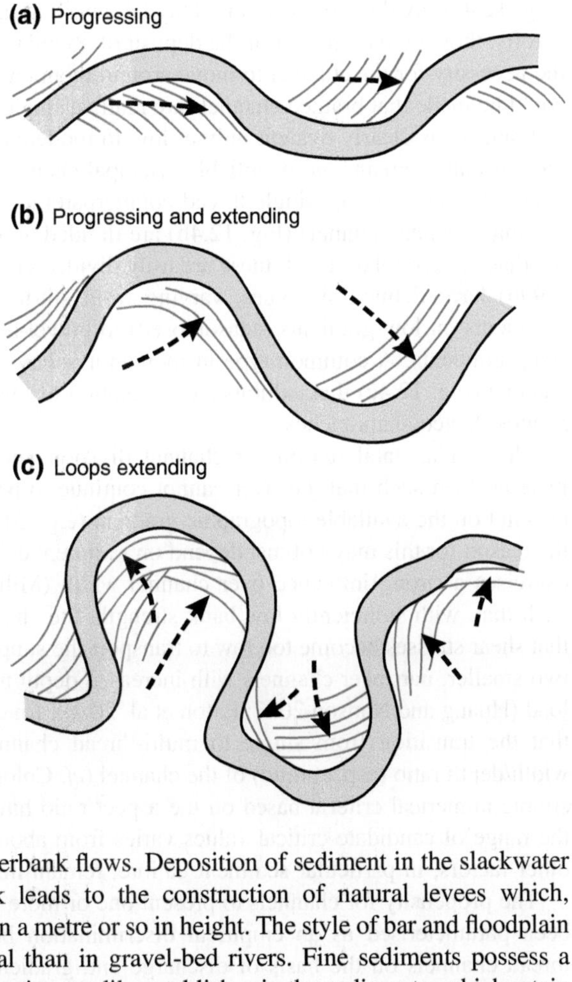

floodplain, the deposit of overbank flows. Deposition of sediment in the slackwater zone immediately overbank leads to the construction of natural levees which, however, are rarely more than a metre or so in height. The style of bar and floodplain development is more vertical than in gravel-bed rivers. Fine sediments possess a degree of cohesion and vegetation readily establishes in the sediments, which retain moisture more abundantly than gravels, so root reinforcement of channel banks is common. Hence, bank erosion is apt to be slower and more uniform than on gravel banks and may to some degree be predictable where a history of channel deformation (e.g. from air photographs or historical maps) is available. Rate of floodplain reoccupation still follows an exponential relation, but with a considerably greater time constant than in laterally active gravel-bed channels (Fig. 12.13a).

12.3.5 Morphodynamics: Channel Division

Channels that are divided—that is, having more than one branch—have been classified as 'braided' or 'anabranched' (see Fig. 12.7). Braided channels

(Fig. 12.4a) divide about bars that may be awash at high flow or result from chute cutoffs: they are the product of the deposit of abundant bed material in-channel and the necessity for the channel to move around them (Ashmore 1991a). An argument could be made that braided channels are really a single-channel form as the division of branches is clearly evident only at low to moderate flows. Despite the division, there usually remains an identifiable principal channel and braided channels retain many characteristics of single-thread counterparts (Ashmore 2001).

Anabranched channels (Fig. 12.4b) are divided about channel islands that build to adjacent floodplain level; these are truly divided channels. Nanson and Knighton (1996) have defined six types, varying from silt-transporting rivers in cohesive sediments on low gradients (anasomosed) to gravel-dominated rivers on relatively steep gradients. A common type in mountain valleys is the so-called 'wandering' channel (Fig. 12.4c) that exhibits a combination of low-order braiding and channel islands, hence anabranches.

The fundamental reason for channel division appears to be a surfeit of bed material load such that the river cannot continue to pass the load through a single channel on the available topographic gradient (e.g., Mueller and Pitlick 2013). But the reason for this may not just depend on sediment delivery; channel bank strength exercises a strong influence over channel width (Millar 2000; Eaton et al. 2004) such that, with sufficiently low bank strength, the channel may widen to the point that shear stresses become too low to transport the supplied load. Then division into two smaller, narrower channels with increased depth may more efficiently pass the load (Huang and Nanson 2007; Eaton et al. 2010). Linear stability analyses confirm that the transition from single to multi-thread channels depends largely on the width/depth ratio (aspect ratio) of the channel (*cf.* Colombini et al. 1987) so various simple numerical criteria based on the aspect ratio have been proposed. However, the range of candidate critical values varies from about 50 to 100, suggesting that other factors, in particular sediment calibre, remain important.

The propensity for channels to present one or more branches has conventionally been parameterised as an empirical discrimination between braided and single-thread channels on the basis of discharge and gradient (Lane 1957; Leopold and Wolman 1957) with the addition of sediment grain size (Henderson 1963; Ferguson 1987, amongst others). These empirical formulations encompass stream power and grain inertia (hence, presumably, the capacity for grains to be moved, or not, in a single talweg) but they do not incorporate the important bank strength parameter, nor did these analyses separately treat anabranched channels.

Basing their analysis on a rational formulation of river regime equations due to Millar (2005) for gravel-bed streams and approximate results from linear stability analysis, Eaton et al. (2010) arrived at a discriminant criterion, referred to channel gradient, that distinguishes single-thread channels from multi-thread ones:

$$S_c = 0.40\mu'^{1.41} Q*^{-0.43}$$

in which $Q* = Q/\{D_{50}^2[(s-1)gD_{50}]^{1/2}\}$ is a non-dimensional representation of streamflow (incorporating grain size as a direct variate in the analysis) and μ' is

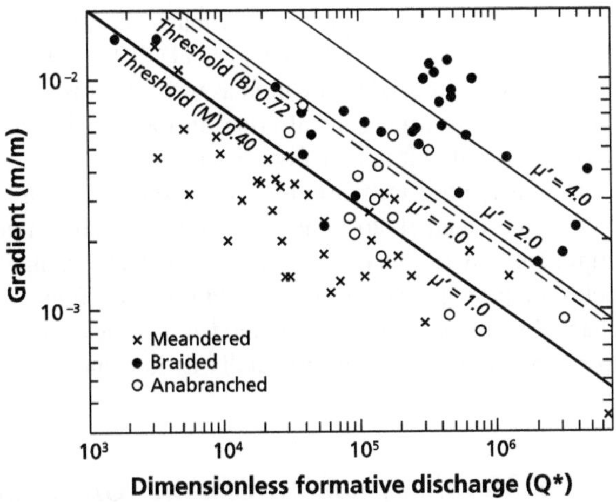

Fig. 12.16 Discrimination of river channel styles using the equation of Eaton et al. (2010) (figure provided by B. Eaton). The coefficient for the meandered-multithread threshold is 0.40; that for the upper limit of anabranched channels is 0.72

relative bank strength, that is bank strength as a proportion of bed strength.[1] For gravel channels in which the lower banks are composed of bed material, $\mu' \approx 1$ and the formula reduces in form to one proposed by Henderson (1963) (Fig. 12.16). To further discriminate braided channels from anabranched ones, Eaton et al. proposed the coefficient 0.72. Since the formulation is based on regime equations for a gravel bed, it presumably applies strictly to such channels, including only the steepest types of Nanson and Knighton's anabranched channels. It is also difficult to assign a value to μ', but Eaton et al. indicated that values may vary from 2 to 5 for increasingly heavily defended banks. Experimental evidence of the importance of vegetation in mediating bank strength has been supplied by Tal and Paola (2007) and by Braudrick et al. (2009). A similar discrimination for channels in fine sediments has not been worked out but it is likely that rivers transporting sand obey similar rules.

Mechanistically, while braided channels divide about medial bars that represent deposited bedload, and wandering channels may develop in cases where vegetation becomes established on high bar surfaces, anabranched channel systems may also develop by avulsion into the floodplain. Avulsion is promoted by aggradation in the case, described by Huang and Nanson (2007), that the channel can no longer pass the supplied load. A special case of anabranching—termed anastomosis—occurs in low gradient rivers carrying fine sand and silt in deep channels with high banks (Makaske 2001). Avulsion often occurs through a low point in the river bank, termed a crevasse, leading the river to form a new channel in the floodplain. Where low-gradient channels flow through a delta into a standing water body, or where the river enters a flood basin (a low, frequently or permanently flooded section of a floodplain), sediment is dropped near the channel mouth, creating a bifurcation and

[1]Note that there are errors in the printed equation in the original paper of Eaton et al. (2010). The paper of Huang and Nanson (2007) also contains typographical errors in displayed equations.

the establishment of two channels that may remain stable if water and sediment divide proportionally (Kleinhans et al. 2013). Hence, anastomosis develops by channel extension and bifurcation as well as avulsion. Slingerland and Smith (1998) have presented a mechanical criterion for avulsion in the general case of channel division in a floodplain or flood basin.

In considering the various channel types discussed above, from single-thread meandered channels to braided ones, it is finally important to realize that no discriminant criterion has been proven perfect: channel styles exhibit overlapping fields of principal governing conditions (see Figs. 12.7 and 12.16) and evident channel style may, in face of extended periods of drought or floods, appear to shift from one type to another, especially as in-channel vegetation comes or goes.

12.3.6 Morphodynamics: Channel Gradation

Key elements of channel adjustment are channel elevation and gradient. These may be altered through a change in sinuosity or in channel pattern, or by the raising or lowering of mean bed elevation via channel gradation, or via a change in downstream grade control. 'Gradation' is the systematic change in channel elevation due to net deposition or erosion of sediment. Degradation reduces slope as a counter-adjustment to reduced sediment supply or increased runoff. In contrast, aggradation occurs as a means of increasing slope and restoring depleted sediment transport capacity when flow is reduced or to accommodate increased sediment load. Changes in channel cross-sectional morphology usually accompany channel gradation (Brandt 2000). In bedload dominated systems (meaning, in general, gravel-bed rivers) sediment evacuation from a reach—the signal effect of degradation—may occur principally as bank erosion with only limited lowering of the more or less effectively armoured bed. However, severe degradation will lower the bed substantially (e.g., Liebault et al. 2013). On the other hand, aggradation in such rivers occurs as a vertical rise in bed level as sediments are deposited on the channel bed. In contrast, in suspended load dominated systems (in general, sand-bed or silty rivers) degradation is dominated by vertical lowering of the bed, whilst aggradation is dominated by lateral deposits that also narrow the channel or raise the floodplain. The difference in characteristic bank strength mediates these outcomes.

Downstream grade controls leading to aggradation or degradation include establishment or removal of weirs, establishment of a dam and a new base level in the reservoir, a change in sea level, meander cutoffs that increase river gradient and induce upstream progressing degradation, or successful erosional attack on former hard points. Galay (1983) gave a comprehensive review of the causes of river bed degradation that remains instructive today.

Slope adjustment of a river is most readily achieved by an adjustment of channel sinuosity, which entails bank erosion and local redistribution of sediment. In instances where the lateral activity of the channel is constrained (by non-erodible banks or by the fact that the river is already flowing on the limit topographic gradient) pulsed influx of sediment to the channel causes local aggradation (Nicholas et al. 1995). Such local accumulations may result from auto-generated variations in sediment flux in the channel (see Ashmore 1991b); they may also result from the entry of a landslide or debris flow into the channel. Subsequent dissipation of the aggraded material may occur by dispersion from the point of entry or by downstream translation, depending on flow conditions (Lisle 2008). A river with attendant floodplain that aggrades significantly is likely to avulse into the floodplain and may not return to an equilibrium state until the entire floodplain has aggraded.

Self-adjustment of the topographic gradient down which the river flows requires the net deposition or evacuation of sediment through the entire system and this may take a very long time. DeVries (1975) proposed an approximate formulation of the time scale for system-wide adjustment. In a slightly simplified form (that is, dropping $O[1]$ constants) it was presented by Castelltort and Van Den Driessche (2003) as

$$t_{50} = L^2/K$$

in which $K = bV_b/wS$ is a sediment diffusion coefficient, t_{50} is the time for channel bed adjustment to be 50 % complete, L is the length of the channel, b is the exponent of a bed sediment transport relation, V_b is the mean annual *bulk* volume of the bed material transported (about 1.6× the mineral volume), and w is the river channel width. Subsequent investigators (e.g. Dade and Friend 1998) have proposed variations on this formulation, while DiSilvio and Nones (2014) have recently abstracted similar approximations from simplified analyses of long profile adjustment to a sediment supply perturbation. Depending on the exact time scale selected (the diffusive effect is asymptotic, so no final time can be given), differences of $O[1]$ are observed amongst the formulations, but they do not materially affect the order of magnitude results, considering that they are all gross approximations. For observed values of sediment volume and channel geometry for streams of all sizes, the formula indicates that single reaches of length in the range 15–30 channel widths may compensate a gradient or load perturbation within one year to about 10 years; in detail, the flow regime will influence the result. However, for adjustments to propagate through a river system of order 100 km length requires of order 10^3–10^4 years, depending on the sediment charge, while a large system with length of order 10^3 km requires of order 10^4–10^6 years to adjust. Blum (2007, Table 30.1) gives estimates for some of the world's major rivers that are consistent with this range. (The foregoing estimates are based on b = 1.5, the common exponent for high rate sediment transport. In gravel-bed channels, for which most significant transport occurs in stage 2, b falls in the range 3–5 and adjustment times become commensurably longer.)

The predictions are by no means exact: notably, they do not—for the case of aggradation—include provision for floodplain aggradation. But, if they fall even within an order of magnitude, then no significant stream system anywhere is apt to be in equilibrium with current flow and sediment regimes, which vary on much shorter time scales. The implication is that channel equilibrium is a local phenomenon. This conclusion is consistent with results of Pizzuto (1992) who found, in a numerical model study of equilibration of an entire fluvial system, that hydraulic geometry (i.e., reach-scale effects) was quickly stabilised, but system-wide effects, commensurable with sediment storage time in the system, is of the order of millenia. Harvey (2002) gave a more detailed discussion of mechanisms affecting adjustment times. Of course gradient adjustments through an entire fluvial system are apt to be sufficiently subtle as not to be noticed over periods of decades or even longer, except in the case of extreme engineered changes.

12.3.7 Rockbound Channels

Channels confined within competent rock are non-alluvial (Fig. 12.5). If one or both banks consist of rock (Fig. 12.17), lateral movement is constrained on practical time scales. If the bed is rock, then the channel will not quickly degrade below the rock floor (Turowski et al. 2008), though erosion of the bed may occur by rock abrasion and rock quarrying (see Sklar and Dietrich 2004; Chatanantavet and Parker 2009). Conversely, the bed may aggrade by the deposition of incoming sediment load and be protected from rock incision. Channel beds erode to rock when the incoming sediment load is smaller than the capacity of the stream to move sediment by virtue of either a locally high gradient or restricted upstream sediment sources. Hence, rock exposure is a significant indicator of conditions governing the channel.

It is important to recognise that not all non-alluvial bounding sediments may be competent rock. Poorly lithified and unlithified material may form a channel boundary and may be eroded nearly as readily as alluvium. In formerly glaciated regions, contemporary rivers are frequently incised into and have banks comprising glacial till, outwash or lacustrine sediments. Over-consolidated sediments may resist ready erosion. If incision is relatively deep, so that banks are high, lateral mobility of the channel will be constrained simply by the large volume of material that must be entrained to allow the channel to move.

Fig. 12.17 Upper Drôme River near Luc-en-Diois, southern France, a rock-bound channel with partial sedimentary cover (photo by S.P. Rice)

The average geometry of rockbound channels is little different than alluvial channels (Montgomery and Gran 2001; Wohl and David 2008) though they are subject to abrupt changes in width and depth, especially in rock canyons where the stream course and morphology are apt to be controlled by rock structure.

12.4 Morphodynamical Considerations in River Restoration

12.4.1 The Problem

Since antiquity, rivers have been significantly engineered in one way or another to control water flows. The legacy of western engineering began in northern Europe in the Middle Ages with the installation of weirs for water power and fishery control (ca.1200CE) and channelization for navigation (ca.1400CE). Principal reasons to engineer river channels today (see Gregory 2006) include damming for water power, water supply and flow control, bank 'hardening' to prevent erosion, embankment (dyke) construction to inhibit out-of-channel flooding, channel realignment ('channelization') to facilitate land use or navigation, and diversion of water for human use. The 20th century saw massive gravel borrowing from many rivers (Rinaldi et al. 2005) but the destructive effects of this activity have been recognized and the practice largely ended. These activities more or less dramatically change the conditions governing channel morphology and morphodynamics:

- Dams change the magnitude and timing of water flows downstream and trap the sediment load of the river; upstream, river gradient is modified in the approach to the reservoir; downstream, morphodynamic tendencies may be radically altered—significant degradation is a common but not universal outcome;
- Bank hardening interferes with normal staging downstream of bed material sediments since it prevents the lateral movement associated with bed material deposition and reentrainment: it thereby interdicts floodplain renewal;
- Embankments isolate the river from its floodplain, eliminating transient storage of flood water and renewal of floodplain soils: they also raise the level of flood waters within the channel;
- Bank hardening and embankments may also directly promote bed aggradation or degradation if the chosen channel width is not appropriate for the passage of the bed material load of the river;
- Channelization, which almost always entails straightening of the channel, increases channel gradient, possibly leading to channel-bed scour and reduction of aquatic habitat diversity;
- Water diversion changes the flow magnitude and regime of the river, with effects on sediment transporting capability;
- All of the above measures may adversely affect aquatic and riparian ecology by altering flow regimes and reducing habitat diversity.

Each of these effects moves the channel out of the local equilibrium it may have gained by which it was able to pass its water and sediment load while remaining 'in regime' (pattern-stable), and usually deprives the river of the means to adjust successfully. But in the attempt to regain a stable geometry, the channel will attempt to do further erosional or depositional work—work that may undermine the purpose (and often the structures) of the original engineering work. Case studies are legion (e.g. Gilvear 1993; Steiger et al. 2000; Liebault and Piégay 2001; Surian and Rinaldi 2003; Wyżga 2008; Grant 2012) and often complex.

For example, Landon et al. (1998) identified gravel mining in the formerly braided gravel-bed Drôme River of southeastern France as the reason for metres of degradation in the latter half of the 20th century. But cessation of mining has not led to recovery of the river because, after centuries of upland erosion caused by land clearance and poor agricultural practices, rural depopulation in the 20th century and reafforestation have dramatically reduced sediment yield from upland tributaries. Riverside structures have been undermined and the braided river aquatic habitat persists in only some reaches of the river. Along most of the river, however, flood hazard has been substantially reduced.

Engineers have, of course, learned the effects of interference with river channels and use computational methods to design projects so that they compensate for changes in flow and sediment passage in the immediate reach. But computational methods for rivers remain inexact. Nor is it easy to find a fixed design that accounts for the dramatic variation in flows and sediment transport experienced in a river. Far field effects are rarely considered. Furthermore, the governing conditions themselves will change within the lifetime of the engineering works, while the river usually has been deprived of the freedom to adjust to those changes. Often, such changes are initiated by human activity, most notably when land use change alters the sediment supply to a river system (e.g. Lach and Wyżga 2002; Kondolf et al. 2002; Toone et al. 2014).

Beyond the primary effect on channel morphodynamics, the constraints placed on the river introduce constraints on aquatic, riparian and floodplain ecosystems, each of which is strongly driven by the hydrological and sediment regimes of the river. In recent years, the repair of ecological function has become the chief focus of efforts to 'restore' engineered channels to something more like their former state (see Beechie et al. 2010). 'Process-based river restoration' is defined by Beechie et al. as reestablishment of the 'normative rates and magnitudes of physical, chemical and biological processes that create and sustain river and floodplain ecosystems'. Such a goal is often not possible: major installations such as dams will not soon go away, while land use often prevents reincorporation of floodplains into the river system. Primary focus must be placed on the hydromorphological processes, which are an essential prerequisite to satisfactory ecological restoration. In such a perspective partial restoration of river processes may still be possible.

Within a process-based focus the most critical issue is the transport of sediment because sediment load determines the character of the river and, if the transporting capacity of a reach does not match the bed material influx, then aggradation or degradation will occur and usually undermine the intended outcome. The most difficult management problems arise in those rivers that carry significant bed material load, for these are ones that need to be laterally active in order to pass the load successfully. Deprived of this possibility they may experience significant aggradation or degradation in a confined reach. But deprivation of lateral freedom is the most common of all river engineering treatments. We must, therefore, consider bed material transport in relation to the morphodynamics of engineered channels. We must consider it even though bed material transport is so difficult to measure, and so variable in the short term (because it is a significantly non-linear correlate of flow) that we usually have only qualitative information—inferred from an appraisal of the river morphology—of its status.

12.4.2 Channel Responses to Engineered Constraints

Most high roughness channels are topographically confined, single thread, and of relatively high gradient. Significant morphodynamic change consists of local or general bed destabilization, including the mobilization of keystones that stabilize the bed. On sufficiently steep gradients (generally >25 % or 15°), a mass debris flow may result that invariably spreads outside the channel in the terminal zone—usually on a colluvial fan where the high gradient channel debouches into a valley. These sites are dangerous but are often occupied because mass flow events are rare and so the danger is disregarded (and in mountain terrain these are sometimes the only surfaces that can practically be occupied). Debris floods may also occur in such channels and may choke and escape the channel in its terminal zone. The rate of accumulation of deposits is less rapid and the deposition less damaging to structures, but still potentially destructive and dangerous. At many sites in the alpine regions of Europe where colluvial fans have been settled, flights of check dams have been constructed in the upper channel (Fig. 12.18a) to focus the drop at the dam overfalls and reduce intervening gradient and sediment mobilization, while elaborate and expensive channels are constructed to pass debris flows and floods through the settlement without major damage (Fig. 12.18b).

In some places high roughness channels are found on relatively low gradients (<3 % or 2°); these are almost always channels flowing through lag deposits of non-fluvial origin (e.g. glacial deposits). Such channels are usually stable for the large stones are immobile on the low gradient and the channel is non-alluvial or only partly alluvial.

Channels of intermediate roughness are characteristically cobble-gravel channels found in the upper valleys of hill country and in montane valleys (Fig. 12.19). They are apt to receive significant quantities of bed material derived from steep headward tributaries and hillslopes. In many cases they are naturally aggrading systems since the larger sediment is left behind as river gradient declines. Much of the material is moved as bedload and deposited in barforms that impose on the channel an irregular or raggedly meandered style of lateral instability. Channels become braided if the sediment load is sufficiently great. These are the most laterally active

Fig. 12.18 **a** Grade control check dam in the torrent Le Riou Bourdoux, near Saint Pons in the Ubaye River catchment, French Alps. The works were constructed in the 19th century (photo by S. P. Rice); **b** Channel construction on a colluvial (debris flow deposit) fan to protect concentrated settlement (Carinthia, Austria). Both views upstream

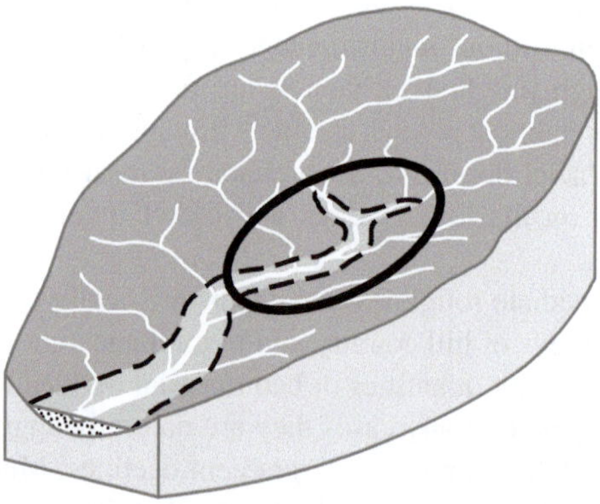

Fig. 12.19 Illustrating the zone of greatest lateral activity in most river systems: upper valleys receiving coarse sediment inputs from many upland tributaries. Compare Fig. 12.6; zone of intermediate roughness

of all channels. Efforts are often made to stabilize them or, at least, to confine their activity to a reduced channel zone by building strongly reinforced embankments, usually employing large stone (Fig. 12.4a). If the remaining channel zone width is sufficient to sustain hydraulic conditions that can transport the incoming sediment load through the reach in the manner that it did before confinement this strategy can be successful. But if the selected channel zone is too narrow, then aggradation may occur until the gradient has been increased sufficiently to force sediment through the reach or, if very high velocities are forced in order to pass sediment, degradation may occur locally as additional sediment is entrained. Either way, the river control works will eventually be threatened or fail (see Davies and McSaveney 2006) while the underlying problem of sediment passage remains unresolved.

Channels of low roughness present a range of levels of lateral activity, depending on the magnitude of bed material load. Many lowland rivers carry modest loads of relatively fine material (sand and silt) that can be accommodated within the active channel. Channel banks, often reinforced by woody vegetation, are sufficiently strong to withstand the normal range of shear stresses imposed by the river, hence the channel is stable (Fig. 12.20). In other cases, banks can be reinforced sufficiently to create stability. Many rivers in central and northwestern Europe are so conditioned, whether by virtue of natural bank strength or by the construction of rock or masonry walls. Stability is not guaranteed, however; a substantial increase in bed material load to the river or an exceptional flood may

Fig. 12.20 Stable lowland channels. **a** a stable reach of the River Seine, France, upstream from Paris (photo by H. Piégay); **b** Garonne at Toulouse, France, a constrained channel (photo by Wojsyl from Wikipedia)

Fig. 12.21 Meander loop extension on a lowland river: River Manifold near Hayesgate, U.K. Lack of bankside vegetation is a contributing factor to the level of lateral activity. (photo by S.P. Rice)

destabilize the channel or create significant problems of aggradation within the confined channel zone.

Given sufficient bed material load and insufficient bank strength, rivers of low roughness are laterally active, much like those of intermediate roughness. The style of instability is more likely, however, to consist of meander loop development (Fig. 12.21). Often, initial erosion augments bed material supply such that the problem propagates downstream. Similarly, the often-practiced remedy of local bank reinforcement to discourage erosion simply displaces the problem downstream. A true remedy must consider control of sediment production and bank stabilization over an extended reach.

There is evidence that many lowland floodplains in central and northwestern Europe (and elsewhere in the world) originally featured anabranched channel systems under heavy forest cover (Brown 1997; Hughes 2003). Some such systems survive, usually in simplified form, today. There is a close connection between the forest and the multiple channel formation, which derives from stream rerouting by downed wood allied with strong bank reinforcement by standing vegetation. These systems are unusually diverse biologically and effective in passage and sequestering of flood waters. Sustenance or reintroduction of anabranched systems, where feasible and ecologically appropriate, would be enlightened ecological management.

12.4.3 Design

The situations described above raise the question of design methods to determine a suitable channel zone width. Shields et al. (2003) and Millar (2005) have proposed computational methods based on regime equations to estimate active channel geometry. But in both cases an underlying assumption is that the channel will be stable. Further, a design flow must be selected, for which there is no clear methodology, nor is it clear how the resulting channel form should be varied to account for changes through pool and riffle sequences or channel bends (Wilcock 2012). In any case, laterally active channels require additional room. The term 'channel zone' is employed here to recognize the occurrence of a more or less substantial bar area or 'exposed' floodplain that will not be water-covered during most flows, but may be susceptible to erosional attack in the short to intermediate term (years to decades). The channel zone is equivalent to the 'erodible river corridor' of Piégay et al. (2005), who have made similar arguments.

Appropriate channel-zone space may be estimated by several means, depending on available morphological or historical evidence. Most simply, one may use purely empirical criteria based on observation of the morphological style of the river. For a braided channel a minimum channel zone would consist of the presently active channel plus an area on each side occupied by recent to sub-recent bars (Fig. 12.22a). The limit of mature riparian forest might be a useful indicator. Given

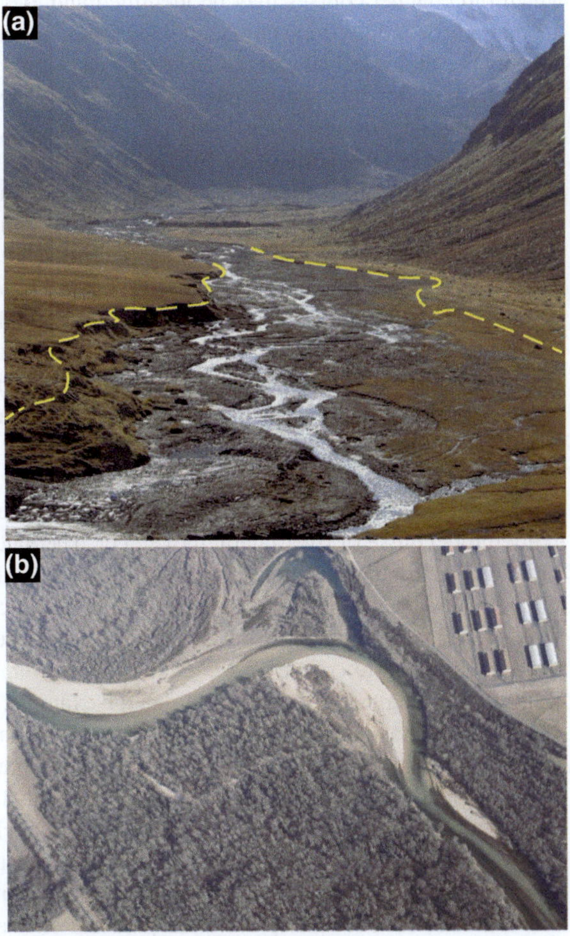

Fig. 12.22 Defining the 'channel zone' for laterally active channels: **a** low order braided channel at low flow, upper Őtztal, Őtztaler Alps, Austria: channel zone outlined; **b** meander bend on the Ain River at Mollon, France, showing a channel zone adapted to meander amplitude: recent cutoff in the upper portion of the photo (photo courtesy of H. Piégay)

such a crude estimation, however, a margin of safety varying from metres to tens of metres (depending on the size of the river and the evident degree of lateral activity) might prudently be added. As a general rule, the greater the bed material load, the wider must be the channel zone relative to the active channel width. For meandered channels, a channel zone may be defined by the maximum meander amplitude in the subject reach. For purely progressing meanders, this may be easily defined. However, for partially or principally loop extension styles of meandering, the possibility for loop lateral extension or compound loop development will make specification less clear.

Such subjective judgments can be made more objective by examining the history of river lateral displacement. For laterally active rivers, an air photo history that goes back at least fifty years is now available in most places, providing opportunity to map recent lateral activity. In many parts of Europe, reasonably precise mapping may extend back for as much as two centuries. It is important to confirm and possibly extend mappings by geomorphological investigations on the ground, by which old channel courses may be identified and even dated using methods based on the age of woody vegetation, radioisotopes, remanent luminescence, or other chemical or biological signatures. In some cases, it may be possible to construct the distribution of floodplain surface age (*cf.* Fig. 12.13) by amassing evidence from all of the foregoing means taken together. In that case, age and distance from the present active channel might be interpreted in terms of the probability for reoccupation by the channel. Then an objective (although arbitrary) basis can be established to define that part of the area best assigned to the channel zone. For floodplains with a high rate of sediment turnover (e.g. decades to reoccupy the entire surface: see Fig. 12.13, Saru River), the entire floodplain should be considered to lie within the channel zone; for those floodplains with slow turnover (e.g. centuries to millennia: Fig. 12.13, Fraser River and Grand River), it is reasonable to expect to be able to isolate some portions from the river for secure human occupance. Individual decisions will depend upon the nature of floodplain occupance, the level of investment in protective measures that is considered appropriate, and the residual risk factor that is considered bearable.

Morphodynamic modelling has also been proposed as a means to define the likely future course of river lateral movement. The method may give insight locally (around a particular meander bend, for example) but, in general, input data requirements are very onerous and the results will include small errors that inevitably propagate into major errors as the simulation is extended into the future. For projecting the future course of a laterally active river, numerical modelling is best utilised as a diagnostic tool (indicating, for example, likely degree of activity or sensitivity to changes in forcing conditions) rather than a predictive one.

It is finally necessary to emphasize that past and present levels of lateral activity (or of aggradation or degradation) in a river channel do not necessarily indicate future activity. One must also consider the effect that changes in runoff or sediment supply may have on the morphodynamic activity of the river. For this, one must consider land management in the contributing catchment and regional trends in hydroclimate, both for interpreting the past and envisaging the future.

Because of its importance in establishing the morphodynamical style of the river, a fundamental approach to channel stabilization consists of controlling the supply of bed material to the river. For upland channels draining limited areas, this matter is best tackled by managing land use to reduce sediment supply to the point that channels are just able to pass the load. In this endeavour, one must recognize that too little sediment supplied to a river may induce significant degradation (and thereby degrade fish habitat). Many European montane foreland rivers have experienced significant degradation during the twentieth century following reafforestation of abandoned upland farms with consequent reduction in sediment supply (Lach and Wyżga 2002; Kondolf et al. 2002). Hence there may be exceptional cases in which one wishes to encourage an increase in sediment yield to arrest undesirable river degradation (see Landon et al. 1998). For larger rivers, sediment control may be part of regional land planning criteria, but direct actions to constrain channel activity are more often the practical recourse. In such circumstances—and contrary to most past practice—the best management strategy always is to leave as much room as possible to accommodate the normal river lateral activity—activity associated with the downstream passage of the sediment load.

Acknowledgments Joseph Desloges furnished the data on floodplain longevity and Brett Eaton provided the data on discrimination of channel styles. Paul Carling, Rob Ferguson, Hervé Piégay, Alain Recking and Steve Rice reviewed the text and generously contributed many photographic examples of rivers. Eric Leinberger made the illustrations. Thanks to all of these colleagues whose contributions and comments have made this a better paper. The opinions expressed in the paper remain those of the author alone.

References

Ashmore PE (1991a) How do gravel-bed rivers braid? Can J Earth Sci 28:326–341

Ashmore PE (1991b) Channel morphology and bed load pulses in braided, gravel-bed streams. Geogr Ann 73A:37–52

Ashmore PE (2001) Braiding phenomena: statics and kinetics. In: Mosley MP (ed) Gravel-bed Rivers V (Proceedings volume of the 5th international gravel-bed rivers workshop, Christchurch and Franz Josef, New Zealand, 2000), New Zealand Hydrological Society, Wellington, New Zealand, pp 95–120

Bathurst JC (2007) Effect of coarse surface layer on bedload transport. J Hydraul Eng 111: 625–643

Beechie TJ, Liermann M, Pollock MM, Baker S, Davies J (2006) Channel pattern and river-floodplain dynamics in forested mountain river systems. Geomorphology 78:124–141. doi:10.1016/j.geomorph.2006.01.030

Beechie TJ, Sear DA, Olden JD, Pess GR, Buffington JM, Moir H, Roni P, Pollock MM (2010) Process-based principles for restoring river ecosystems. Bioscience 60:209–222

Blum MD (2007) Large river systems and climate change. In: Gupta A (ed) Large rivers: geomorphology and management. Wiley, Chichester, 627–659

Bradley ND, Tucker GE (2013) The storage time, age, and erosion hazard of laterally accreted sediment on the floodplain of a simulated meandering river. J Geophys Res Earth Surf 118: 1–12. doi:10.1002/jgrf.20083

Brandt SA (2000) Classification of geomorphological effects downstream of dams. Catena 40:375–401

Braudrick CA, Dietrich WE, Leverich GT, Sklar LS (2009) Experimental evidence for the conditions necessary to sustain meandering in coarse-bedded rivers. Proc Natl Acad Sci (USA) 106:16936–16941

Bray DI (1973) Regime relations for Alberta gravel bed rivers. In: 7th Canadian Hydrology Symposium Proceedings, National Research Council of Canada, pp 440–452

Brown AG (1997) Biogeomorphology and diversity in multiple-channel river systems. Glob Ecol Biogeogr Lett 6:179–185

Castelltort S & Van Den Driessche J (2003) How plaubible are high-frequency sediment supplydriven cycles in the stratigraphic record? Sed Geol 157:3–13

Chatanantavet P, Parker G (2009) Physically based modeling of bedrock incision by abrasion, plucking, and macroabrasion. J Geophys Res 114:F04018

Church M (1992) Channel morphology and typology. In: Calow P, Petts GE (eds) The rivers handbook (Sect. 2.6). Blackwell, Oxford, pp 126–143

Church M (2002) Geomorphic thresholds in riverine landscapes. Freshw Biol 47:541–557

Church M (2006) Bed material transport and the morphology of alluvial river channels. Ann Rev Earth & Planetary Sci 34:325–354

Church M (2010) Mountains and montane channels. In: Burt T, Allison R (eds) Sediment cascades. Wiley, Chichester, pp 19–53

Church M, Zimmermann A (2007) Form and stability of step-pool channels: research progress. Water Resour Res 43(21):W003415. doi:10.1029/2006WR005037

Church M, Hassan MA, Wolcott JF (1998) Stabilizing self-organized structures in gravel-bed stream channels. Water Resour Res 34:3169–3179

Church M, Burt TP, Galay VJ, Kondolf GM (2009) Rivers. In: Slaymaker O, Spencer T, Embleton-Hamann C (eds) Landscape change in the 21st century. Cambridge University Press, Cambridge, pp 98–129

Colombini M, Seminara G, Tubino M (1987) Finite-amplitude alternate bars. J Fluid Mech 181:213–232

Dade WB, Friend PF (1998) Grain-size, sediment-transport regime, and channel slope in alluvial rivers. J Geol 106:661–675

Davies TR, McSaveney MJ (2006) Geomorphic constraints on the management of bedload-dominated rivers. J Hydrol (NZ) 45:69–88

deVries M (1975) A morphological time scale for rivers. In: Proceedings of the 16th Congress of the International Association for Hydraulic Research, São Paulo, Brazil, vol 2, pp 17–23

Dietrich WE, Kirschner JW, Ikeda H, Iseya F (1989) Sediment supply and the development of a coarse surface layer in gravel-bedded rivers. Nature 340:215–217

DiSilvio G, Nones M (2014) Morphodynamic reaction of a schematic river to sediment input changes: analytical approaches. Geomorphology 215:74–82. doi:10.1016/j.geomorph.2013.05.021

Dunne T, Constantine JA, Singer MB (2010) The role of sediment transport and sediment supply in the evolution of river channel and floodplain complexity. Jpn Geomorphol Union Trans 31 (2):155–170

Eaton BC (2013) Hydraulic geometry: empirical investigations and theoretical approaches. In: Schroder J, Wohl EE (eds) Treatise on geomorphology, fluvial geomorphology, vol 9. Academic Press, San Diego, pp 313–329

Eaton BC, Church M, Millar RG (2004) Rational regime model of alluvial channel morphology and response. Earth Surf Proc Land 29:511–529. doi:10.1002/esp.1062

Eaton BC, Millar RG, Richardson S (2010) Channel patterns: braided, anabranching, and single-thread. Geomorphology 120:353–364. doi:10.1016/j.geomorph.2010.04.010

Ferguson RI (1987) Hydraulic and sedimentary controls of channel pattern. In: Richards K (ed) River channels: environment and processes. Blackwell, Oxford, pp 129–158

Ferguson RI (2007) Flow resistance equations for gravel- and boulder-bed streams. Water Resour Res 43: W05427, 12 pp. doi: 10.1029/2006WR005422

Ferguson R, Hoey T (2008) Effects of tributaries on main-channel geomorphology, Ch. 10. In: Rice S, Roy AG, Rhoads BL (eds) River confluences, tributaries and the fluvial network. Wiley, Chichester, 183–208

Galay VJ (1983) Causes of riverbed degradation. Water Resour Res 19:1057–1090

Gilvear DJ (1993) River management and conservation issues on formerly braided river systems; the case of the River Tay, Scotland. Geol Soc Lond Spec Publ 75:231–240

Grant GE (2012) The geomorphic response of gravel-bed rivers to dams: perspectives and prospects. In: Church M, Biron PM, Roy AG (eds) Gravel-bed rivers: processes, tools, environments. Wiley, Chichester, pp 165–181

Gregory KJ (2006) The human role in changing river channels. Geomorphology 79:172–191

Ham DG, Church M (2012) Morphodynamics of an extended bar complex, Fraser River, British Columbia. Earth Surf Proc Land 37:1074–1089. doi:10.1002/esp.3231

Harvey AM (2002) Effective time scales of coupling within fluvial systems. Geomorphology 44:175–201

Henderson FM (1963) Stability of alluvial channels. Am Soc Civ Eng Trans 128:657–686

Huang HQ, Nanson GC (2007) Why some alluvial rivers develop an anabranching pattern. Water Resour Res 43: W07441, 12 pp. doi:10.1029/2006WR005223

Hughes FMR (ed) (2003) The flooded forest: guidance for policy makers and river managers in Europe on the restoration of floodplain forests. FLOBAR 2. Department of Geography, University of Cambridge, Cambridge. http://www-flobar.geog.cam.ac.uk. Accessed 30 Mar 2014

Hungr O, Evans SG, Bovis MJ, Hutchinson JN (2001) A review of the classification of landslides of the flow type. Environ Eng Geosci 7:221–228

Kellerhals R (1967) Stable channels with gravel-paved bed. Proc J Waterways Harbours Div 93 (1):63–84

Kellerhals R, Church M (1990) Hazard management on alluvial fans. In: Rachocki AH, Church M (eds) Alluvial fans: a field approach. Wiley, Chichester, pp 335–354

Kellerhals R, Church M, Bray DI (1976) Classification and analysis of river processes. J Hydraulics Div 102:813–829

Kleinhans MG (2010) Sorting out river channel patterns. Prog Phys Geogr 34:287–326

Kleinhans MG, Ferguson RI, Lane SN, Hardy RJ (2013) Splitting rives at their seams: bifurcations and avulsion. Earth Surf Proc Land 38:47–61

Kondolf GM, Piégay H, Landon N (2002) Channel response to increased and decreased bedload supply from land use change: contrasts between two catchments. Geomorphology 45:35–51

Lach J, Wyżga B (2002) Channel incision and flow increase of the upper Wisłoka river, Southern Poland, subsequent to the reafforestation of its catchment. Earth Surf Proc Land 27:445–462

Landon N, Piégay H, Bravard J-P (1998) The Drôme River incision (France): from assessment to management. Landscape Urban Plann 43:119–131

Lane EW (1957) A study of the shape of channels formed by natural streams flowing in erodible material. United States Army Corps of Engineers, Missouri River Division, Technical Report, Omaha, Nebraska

Leopold LB (1994) A view of the river. Harvard U.P, Cambridge, MA 298p

Leopold LB, Maddock TJ Jr (1953) The hydraulic geometry of stream channels and some physiographic implications. United States Geological Survey, Professional paper 252

Leopold LB, Wolman MG (1957) River channel patterns: braided, meandering and straight. Unites States Geological Survey, Professional Paper 282B

Liebault F, Piégay H (2001) Assessment of channel changes due to long-term bedload supply decrease, Roubion River, France. Geomorphology 36:167–186

Liebault F, Lallias-Tacon S, Cassel M, Talaska N (2013) Long profile responses of alpine braided rivers in SE France. River Res Appl 29:253–266. doi:10.1002/rra.2615

Lisle TE (2008) The evolution of sediment waves influenced by varying transport capacity in heterogeneous rivers. In: Habersack H, Piegay H, Rinaldi M (eds) Gravel bed rivers VI: from process understanding to river restoration. Elsevier, Amsterdam, pp 443–472

MacVicar BJ, Obach L, Best JL (2013) Large scale coherent flow structures in alluvial pools. In: Venditti JG et al (eds) Coherent flow structures at earth's surface. Wiley, Chichester, pp 243–259

Makaske B (2001) Anastomosing rivers: a review of their classification, origin and sedimentary products. Earth Sci Rev 53:149–196

Millar RG (2000) Influence of bank vegetation on alluvial channel patterns. Water Resour Res 36:1109–1118

Millar RG (2005) Theoretical regime equations for mobile gravel-bed rivers with stable banks. Geomorphology 64:207–220

Montgomery DR, Buffington JM (1997) Channel-reach morphology in mountain drainage basins. Geol Soc Am Bull 109:596–611

Montgomery DR, Gran KB (2001) Downstream variations in the width of bedrock channels. Water Resour Res 37:1841–1846l

Mueller ER, Pitlick J (2013) Sediment supply and channel morphology in mountain river systems: 1, Relative importance of lithology, topography and climate. J Geophys Res Earth Surf 118: 1–18. doi:10.1002/2013JF002843

Mueller ER, Pitlick J, Nelson JM (2005) Variation in the reference Shields stress for bedload transport in gravel-bed streams and rivers. Water Resour Res 41:W04006. doi:10.1029/2004WR003692

Nanson GC, Knighton AD (1996) Anabranching rivers: their cause, character and classification. Earth Surf Proc Land 21:217–239

Neill CR (1973) Hydraulic geometry of sand rivers in Alberta. In: 7th Canadian Hydrology Symposium Proceedings, National Research Council of Canada, 453–461

Nicholas AP, Ashworth PJ, Kirkby MJ, Macklin MG, Murray T (1995) Sediment slugs: large-scale fluctuations in fluvial sediment transport rates and storage volumes. Prog Phys Geogr 19:500–519

Osterkamp WR, Hedman ER (1982) Perennial streamflow characteristics related to channel geometry and sediment in Missouri River basin. United States Geological Survey, Professional Paper 1242

Piégay H, Darby SE, Mosselman E, Surian N (2005) A review of techniques available for delimiting the erodible river corridor: a sustainable approach to managing bank erosion. River Res Appl 21:773–789

Pizzuto JE (1992) The morphology of graded gravel rivers: a network perspective. Geomorphology 5:457–474

Radecki-Pawlik A, Skalski T (2008) Bankfull discharge determination using the new invertebrate bankfull assessment method. J Water Land Dev 12:145–154

Recking AJ, Leduc P, Liebault F, Church M (2012) A field investigation of the influence of sediment supply on step-pool morphology and stability. Geomorphology 139–40:53–66. doi:10.1016/j.geomorph.2011.09.024

Recking A, Richard D, Degoutte G (eds) (2013) Torrents et rivières de montagne: dynamique et aménagement. Versailles, Editions Quæ, 334 pp

Rhoads BL, Welford MR (1991) Initiation of river meandering. Prog Phys Geogr 15:127–156

Rice SP, Church M (1998) Grain size along two gravel-bed rivers: statistical variation, spatial pattern and sedimentary links. Earth Surf Proc Land 23:345–363

Rinaldi M, Wyżga B, Surian N (2005) Sediment mining in alluvial channels: physical effects and management perspectives. River Res Appl 21:805–828

Schumm SA (1977) The fluvial system. NY, Wiley 338p

Shields FD, Copeland RR, Klingeman PC, Doyle MW, Simon A (2003) Design for stream restoration. J Hydraulic Eng 129:575–584

Sklar LS, Dietrich WE (2004) A mechanistic model for river incision into bedrock by saltating bed load. Water Resour Res 40:W06301

Slingerland R, Smith ND (1998) Necessary conditions for a meandering-river avulsion. Geology 26:435–438

Steiger J, Corenblit D, Vervier Ph (2000) Les ajustements morphologiques contemporains du lit mineur de la Garonne, France, et leurs effets sur l'hydrosystème fluvial. Zeitschrift für Geomorphologie, N.F. Supp.-Bd. **122**: 227–246

Surian N, Rinaldi M (2003) Morphological response to river engineering and management in alluvial channels in Italy. Geomorphology 50:307–326

Tal M, Paola C (2007) Dynamic single-thread channels maintained by the interaction of flow and vegetation. Geology 35:347–350

Teversham JM, Slaymaker O (1976) Vegetation composition in relation to flood frequency in Lillooet River valley, British Columbia. Catena 3:191–201

Toone J, Rice SP, Piégay H (2014) Spatial discontinuity and temporal evolution of channel morphology along a mixed bedrock-alluvial river, upper Drôme River, southeast France: contingent responses to external and internal controls. Geomorphology 205:5–16

Turowski JM, Hovius N, Wilson A, Horng M-J (2008) Hydraulic geometry, river sediment and the definition of bedrock channels. Geomorphology 99:26–38

van de Lageweg WI, van Dijk WM, Baar AW, Rutten J, Kleinhans MG (2014) Bank pull or bar push: what drives scroll-bar formation in meandering rivers? Geology 42:319–322

Wickert AD, Martin JM, Tal M, Kim W, Sheets B, Paola C (2013) River channel lateral mobility: metrics, time scales, and controls. J Geophys Res Earth Surf 118. doi:10.1029/2012JF002386, 17 pp

Wilcock PR (2012) Stream restoration in gravel-bed rivers. In: Church M, Biron PM, Roy AG (eds) Gravel-bed rivers: processes, tools, environments. Wiley, Chichester, pp 137–149

Wohl E, David GCL (2008) Consistency of scaling relations among bedrock and alluvial channels. J Geophys Res 113:F04013. doi:10.1029/2008JF000989

Wolman MG, Miller JP (1960) Magnitude and frequency of forces in geomorphic processes. J Geol 68:54–74

Woodyer KD (1968) Bankfull frequency in rivers. J Hydrol 6:114–142

Wyżga B (2008) A review on channel incision in the Polish Carpathian rivers during the 20th century. In: Habersack H, Piégay H, Rinaldi M (eds) Gravel-bed rivers VI: from process understanding to river restoration. Elsevier, Amsterdam, pp 525–555

Yager EM, Turowski JM, Rickenmann D, McArdell BW (2012) Sediment supply, grain protrusion, and bedload transport in mountain streams. Geophys Res Lett 39:L10402. doi:10.1029/2012GL051654

Chapter 13
Principles of Bedload Transport of Non-cohesive Sediment in Open-Channels

Rui M.L. Ferreira, Marwan A. Hassan and Carles Ferrer-Boix

Abstract This text addresses the particular case of motion and causes of motion of granular material as bedload in the fluvial domain. The aim is to perform and overview of key concepts, main achievements and recent advances on the description of the processes involved in erosion, deposition and transport of sediment in open-channels. The theoretical functional relations describing both the initiation of motion and the sediment transport are introduced. The classical problem of the initiation of motion of particles is treated at grain and at reach scales, accounting for the stochastic nature of flow. Concepts of granular kinematics and methods for quantifying the sediment transport rate in rivers are presented. The latter results from the interactions between the flow and the particles on the bed surface. The sediment transport rate, which has been shown to have a stochastic behaviour, is converted to a lumped statistic distribution. Finally, some field and laboratory techniques for measuring sediment transport, accounting for its inherent fluctuations, are introduced.

Keywords Bedload · Incipient motion · Bedload mechanics · Measuring techniques

R.M.L. Ferreira (✉)
CERIS, Instituto Superior Técnico, Universidade de Lisboa, Lisboa, Portugal
e-mail: ruif@civil.ist.utl.pt

M.A. Hassan · C. Ferrer-Boix
Department of Geography, The University of British Columbia, Vancouver, BC, Canada
e-mail: marwan.hassan@geog.ubc.ca

C. Ferrer-Boix
e-mail: cfboix@geog.ubc.ca

P. Rowiński and A. Radecki-Pawlik (eds.), *Rivers—Physical, Fluvial and Environmental Processes*, GeoPlanet: Earth and Planetary Sciences,
DOI 10.1007/978-3-319-17719-9_13

13.1 Introduction

Most interventions in the fluvial domain require understanding of the processes of entrainment, transport and deposition of the granular material that composes the channel banks and the bottom boundary. Classical problems include land erosion, flood management, design of stable channels, river restoration, prediction of morphologic evolution of natural streams, local scour or reservoir sedimentation, among others (Garde and Ranga Raju 1985). Hazard mitigation has expanded the field; fluvial engineering is often needed to solve problems involving debris and torrential flows, in particular dam-break flows.

The body of knowledge on the motion of the granular phase within a fluid flow remains, however, limited and fragmented. Unlike fluid flow, whose spatial (Eulerian) continuum description by the continuity and Navier-Stokes equations is widely accepted, the dynamics and kinematics of granular material set in motion by a continuous fluid phase or moving together with a fluid phase are not amenable to a universal description. If the solid fractions are high and gravity is a key source of momentum, the mixture of fluid and granular material may be described as a continuum albeit with a complex rheology, determined mostly by granular interactions (e.g., Iverson 1997). If, on the contrary, grains are mostly picked up and displaced by the action of fluid (low solid fractions), the contribution of granular momentum to the flow of both phases may be neglected and the motion of the granular phase is best described with stochastic tools (e.g., Ancey et al. 2008). In between these two extreme cases, for which, incidentally, there are no universally accepted descriptions, the granular phase may exhibit an infinite range of kinematic and dynamic behaviours.

Finding regular patterns in this vast domain of possibilities has constituted a research challenge that has traditionally been carried out independently over several geomorphological and engineering disciplines and determined by their ultimate applications, notably in coastal or river domains. Within each, it has been dealt with at several spatial scales, from the grain-scale, at which research has been mostly at fundamental level, to basin scale, including engineering scales appropriate to formulate problems of channel morphology. Within each discipline there have been outstanding achievements, notably that the structure of the laws describing the macroscopic aspects of the granular-fluid flows are, in most cases, relatively simple and can be devised through a judicious combination of dimensional analysis, laboratory work and field investigation. However, these progresses have seldom resulted from the integration of processes characterized at grain-scale. In other words, the vast amount of research work carried out at grain-scale has yet to be harmonized to match results obtained at macroscale into a consistent theoretical body.

In this respect, it may be argued that the first true modern research program concerned with the physics of granular motion was that of Einstein (1937) (see also Vanoni 1984; Ettema and Mutel 2004). Unlike most previous works (reviews in Mavies et al. 1935 or Williams et al. 1937), Einstein derived a law to quantify the flux of granular material through a channel cross-section (a “bedload function”) from fundamental kinematic and dynamic considerations on grain motion and grain equilibrium and not from particular measurements of weight of granular material displaced over time intervals. In this sense, his approach was universal, independent of the particular application for which the law may find application and of the channel in which the granular motion was observed. Furthermore, he fully acknowledged the stochastic nature of grain motion at low and moderate rates of transport and he incorporated the best knowledge of turbulence and boundary layer hydrodynamics existing at the time.

Einstein (1950) formula is known to have formal shortcomings and to incorporate some formulations from fluid mechanics that are now outdated but it is safe to name Einstein’s probabilistic approach the first paradigm for the study of the motion of granular material and its morphological consequences. Further developments in fluid mechanics, namely in what concerns turbulent flows, and in mechanics of granular flows, have provided the study of granular-fluid flows with the formal apparatus to expand Einstein’s research program and to create others, notably that of Bagnold (1954, 1956).

Currently, the description of the motion and causes of motion of the granular discrete phase encounters still difficulties at quite fundamental levels. These include the identification of the scales at which an Eulerian-continuum approach is justified, the articulation between the probabilistic definition of sediment discharge, the stochastic nature of the (turbulent) near-bed flow and the development of bed forms or the relation between Lagrangian and Eulerian quantities describing sediment fluxes. These, among other fundamental issues, have been addressed in recent research works (Furbish et al. 2012a; Ballio et al. 2014; Ma et al. 2014). However, the premises and formal apparatuses employed in such works are often different, which is typical of research fields yet to reach maturity. This hinders the effective dissemination of key advances to the broad research communities dealing with granular-fluid mixtures.

This text, addresses the particular case of the motion and causes of motion of granular material in the fluvial domain. Attention is restricted to non-cohesive natural granular material susceptible to be transported as bedload by water or by the action of gravity. Such granular material will be henceforth designated as sediment. The aim of this text is to perform and overview of key concepts, main achievements and recent advances on the description of the processes involved in erosion, deposition and transport of sediment in open-channels.

Section 13.2 is devoted to the characterization of the origins of fluvial sediments. Section 13.3 addresses the initiation of sediment transport. A theoretical model for the entrainment of a single particle on a granular bed is developed and expected values of the involved parameters are also estimated. The conditions for the derivation of formulas for incipient motion are then discussed. The principles of mechanics of bedload transport are discussed in Sect. 13.4. In Sect. 13.5 techniques for measuring sediment-related quantities are presented and discussed. The text is ended by a recapitulation of the main points.

13.2 Origins of Sediment

Reflecting conventional methods for measurements, sediment load of a stream is traditionally divided into two different modes of transport, bedload and suspended load. Bedload consists of particles moving in contact with the bed while suspended load refers to particles moving in the water column supported by buoyancy and by upwardly directed turbulent currents. Another division that provides insight into channel stability and morphology—and that accounts for the different origins of sediment—is the distinction between wash load and bed material load. Wash load is the relatively fine material (characteristic diameters smaller than those of fine sand) that moves directly throughout the reach in suspension without being deposited in the main channel; it is normally originated in the basin by hillslope erosion. Bed material load (coarse material), moving in suspension or as bedload, results from local processes of erosion from and deposition on channel bed and banks although substantial amounts of material may enter the stream by landslides or rock slides.

In mountain regions, sediment is delivered to channels from tributaries, bank erosion, surface erosion on slopes and mass movements. These are called external sources of sediment. In mountain streams, where channels are not buffered by a floodplain, mass movements and bank erosion are the primary sources of sediment into channels. Sediment from mass movement events enters stream channels from adjacent slopes or upstream tributaries. It may be deposit immediately and dam the channel for a more or less prolonged period, then to be remobilized and moved downstream gradually by fluvial processes (Sutherland et al. 2002). Sediment entering the channel as a mass movement event may travel along the channel as a debris flow for some distance and be deposited in low gradient reaches or behind obstacles to form sediment wedges (e.g., Roberts and Church 1986). Parts of the episodic deposited sediment in the channel are stored for a relatively long period of time to form persistent, complex channel morphology (e.g., Church 2006; Lisle et al. 1997). Bank erosion is a chronic sediment source especially into channel adjacent to developed floodplains or valley flats.

Channel response to external sediment supply largely depends on flood magnitude and sequence and the interaction among the major mobilization processes along the channel. Due to sediment supply history, flow events of the same magnitude may create different channel morphology and sediment mobility (e.g., Buffington and Montgomery 1999; Lisle et al. 2000). For example, Ryan (2001) reported that whereas streams with a relatively large sediment supply typically have fine surface sediments and high sediment transport rates, relatively low sediment transport rates and well armoured bed surface were observed for streams with relatively low sediment supply. These observations are consistent with experimental findings of Parker et al. (1982), Dietrich et al. (1989), Church et al. (1998) and Hassan and Church (2000).

Internal sources of sediment supply are located within the stream channel and banks. Bed material is supplied from sediment stored in pools, bars, behind obstructions such as logs and boulders, including rarely moved keystones and large wood (e.g., Whittaker 1987; Lisle and Hilton 1992). In steep channels, sediment supply from behind obstructions may occur by the displacement of the obstruction, suddenly releasing large volumes of sediment (Sidle 1988; Smith et al. 1993a, b). The localized release of large sediment stores results in the movement of sediment in waves or pulses. Swanson et al. (1982a, b) noted that the volume of temporarily stored material is up to an order of magnitude larger than the annual export of total sediment. Therefore, even if there is no sediment supply from external sources, changes in sediment storage can cause major changes in sediment transport rate (Hassan et al. 2008; Hassan and Zimmermann 2011). Furthermore, sediment storage within channel can also delay the spread of sediment waves and could produce a hysteresis effect on transport (Meade 1985; Hassan et al. 2014).

The magnitude of the hysteresis loop appears to depend on the amount of stored sediment and position along the channel (Bogen 1980; Hassan et al. 2014). However, the temporal and spatial variation in the amount of within-channel storage depends largely on the supply from external sources (e.g., Swanson et al. 1982a; Benda 1990; Hassan et al. 2008).

13.3 Initiation of Sediment Motion

13.3.1 Introduction

Initiation of sediment motion is a classic problem of sediment and fluid mechanics that has been studied at wide range of scales. We define analysis at grain scale as that concerning the statics of a single grain or a small number of grains, subjected to hydrodynamic forces, normally generated by turbulent flows, and to contacts with a limited number of neighbouring grains. At this scale, initiation of motion is a deterministic problem: a single grain is set in motion once the areal integral of normal and tangential forces per unit area acting on its surface produces a destabilizing hydrodynamic force larger than the stabilizing force resulting from its weight and contacts with other grains. This notion can be generalised for small groups of grains provided that momentum transfer per unit time from other moving grains in the mean flow direction is computed as a destabilizing force. A sample of experimental, theoretical and field works that addressed the problem of characterization of incipient motion at grain scale include Fernandez Luque and van Beek (1976), Yalin (1977, Chap. 4), Fenton and Abbott (1977), Wiberg and Smith (1987), Kirchner et al. (1990), Bridge and Bennet (1992), Buffington et al. (1992), Seminara et al. (2002) or Göğüş and Defne (2005).

By analysis at channel scale, we mean the investigation of a reach of the stream, sufficiently large to encompass a large number of sediment grains but sufficiently small not to experience important variations in key hydrodynamic variables. At this scale, it is a matter of simple observation that there is a threshold below which particle motion passes undetected in the bed reach. A first attempt to systematize this observational result was Kramer's (1935) classification of bed movement.

The first class is simply no movement and the second class—weak bedload—is described as the stage when some particles are "in motion, in isolated spots, and in countable numbers." At a colloquial level, threshold conditions may be identified with Kramer's (1935) weak bedload conditions.

The proper mathematical formulation of the threshold (including the debate on its very existence in the case of turbulent flows) in terms of hydrodynamic variables and sediment parameters is however more complex. The hydrodynamic conditions that correspond to this threshold for a given bed mixture have been the aim of several studies, and have been termed *critical flow* (Shields 1936), *incipient motion* (e.g., Buffington and Montgomery 1997) or *sediment threshold* (Komar and Miller 1973; Dey and Papanicolaou 2008). Other approaches include the definition a reference bedload transport rate (Shields 1936; Paintal 1971a; Parker and Klingeman 1982) below which finite sediment transport occurs but at very low rates. This approach has been used by Wilcock and McArdell (1993), Wilcock and Crowe (2003), among others.

Most formulae for total bedload transport rates include a parameter or a function using threshold conditions (e.g., Meyer-Peter and Müller 1948; Ashida and Michiue 1972; Fernandez-Luque and van Beek 1976) expressed in varying forms of Eq. (13.4). However, the wide variety of formulations, even for uniform sediment (Buffington and Montgomery 1997), indicates that flows at incipient motion conditions are not yet sufficiently well described. It can also be argued that, in turbulent flows, there is no sound foundation for an incipient motion condition as the probability of entrainment of any given exposed grain is never exactly zero, for infinite time, given the stochastic nature of hydrodynamic actions (e.g., Einstein 1950; Paintal 1971b).

For poorly sorted mixtures of sediment the problem acquires further complexity. Extensive experimental and field work by, among others, Wilcock and McArdell (1993, 1997), Hassan and Church (2001) or Church and Hassan (2002) have shown that there is no single critical flow for which all particle sizes of a given bulk mixture initiate motion and that it becomes necessary to consider the effects of bed surface structuring and armouring.

13.3.2 Dimensional Analysis

The relevant variables to describe static equilibrium threshold for a single particle of grain size d_i involve geometrical characteristics of the sediment grains, geometrical descriptors of grain positioning, properties of the minerals that compose the grains, fluid properties, variables that describe the flow in the vicinity of the grain and the acceleration of gravity. Symbolically one has

$$F\left(d_i, \{C_k\}, \{\phi_k\}, \psi, \rho^{(w)}, \mu^{(w)}, u_p, g(\rho^{(g)} - \rho^{(w)})\right) = 0 \qquad (13.1)$$

where d_i is the grain sieving diameter (defined as the width of the smallest square mesh through which the grain is able to pass and, if the particle is approximately elliptical, the length of the intermediate axis), $\{C_k\}$ is a set of shape parameters that characterize the non-sphericity of the grain, $\{\phi_k\}$ is a set of angles that describe the orientation of axes of the particle, ψ is the skin friction angle of the individual grain, $\rho^{(g)}$ and $\rho^{(w)}$ are the grain and fluid densities respectively, $\mu^{(w)}$ is the fluid viscosity, u_p is the longitudinal velocity of the fluid in front of a grain (a "local" velocity, not the depth-averaged flow velocity) and g is the acceleration of gravity. Variables $\left(d_i, \{C_k\}, \{\phi_k\}, \rho^{(w)}, \mu^{(w)}, u_p\right)$ determine both horizontal and vertical hydrodynamic forces; note however that the actual functional dependence is different. The submerged weight of the grain is determined by $\left(d_i, \{C_k\}, \left(\rho^{(s)} - \rho^{(w)}\right)g\right)$. Note that, in the analysis of grain threshold of movement cast as a limit static equilibrium problem, the grain density is not an independent variable: its role is to determine the buoyant weight, together with the acceleration of gravity and the volume of the grain. It would be an independent variable only if the grain was in motion, in which case there would be an inertial force depending on its mass. Since the inertial force is zero, $\rho^{(g)}$ and g cannot be separated in Eq. (13.1). Reactions at the supporting points are determined by the weight and hydrodynamic forces and by the friction angle characteristic of the grain skin roughness.

The non-dimensional form of Eq. (13.1) can be obtained applying the Vaschy-Buckingham's theorem choosing the grain sieving diameter d_i, the flow velocity u_p and the fluid density $\rho^{(w)}$ as dimensionally independent base:

$$\Phi\left(\{C_k\}, \{\phi_k\}, \psi, \frac{u_p d}{\nu^{(w)}}, \frac{u_p^2}{g(s-1)d_i}\right) = 0 \tag{13.2}$$

where $(s-1) = \left(\rho^{(g)} - \rho^{(w)}\right)/\rho^{(w)}$ is the submerged specific gravity of the grain and $\nu^{(w)} = \mu^{(w)}/\rho^{(w)}$ is the fluid kinematic viscosity. Equation (13.2) can be, for instance, employed to guide an empirical investigation of the local threshold flow velocity above which the grain is entrained, whose non-dimensional form is

$$\frac{u_p}{\sqrt{g(s-1)d_i}} = \Phi_u\left(\{C_k\}, \{\phi_k\}, \psi, \frac{u_p d_i}{\nu^{(w)}}\right) \tag{13.3}$$

where Φ_u is a general function of grain shape factors, grain positioning, skin friction and Reynolds number $u_p d_i/\nu^{(w)}$. It is clear that the relevant flow parameter to which a sensitivity analysis is due is the Reynolds number $u_p d/\nu^{(w)}$. For an isolated particle, there is a range of values of this Reynolds number (2×10^2 to 10^5, White 2011) for which the ratio of drag to lift forces becomes constant. If it is assumed that the same holds for grains exposed on a bed surface (e.g., Wiberg and Smith 1987; Göğüş and Defne 2005), the threshold velocity becomes a function of grain shape, submerged specific gravity of sediment, skin roughness and positioning.

The relation between the local threshold velocity u_p and the mean flow velocity cannot be expressed with grain-scale variables as it involves channel-scale issues related to the spatial variability of the non-dimensional parameters involved in Eq. (13.3).

At channel scale, and in a steady and uniform fluid flow, the non-dimensional relation between the volumetric sediment transport rate per unit width, q_s, and flow, fluid and sediment variables is

$$\Psi\left(\{C_k^{(j)}\},\left\{\frac{\gamma^{(j)}}{d^j}\right\},(s-1),\frac{\rho^{(w)}u_* d}{\mu^{(w)}},\frac{h}{d},\frac{q_s}{d\sqrt{gd(s-1)}},\frac{u_*^2}{gd(s-1)}\right)=0 \quad (13.4)$$

where d is a characteristic diameter of the bed granular mixture, for instance the median diameter of the substrate $d_{50}^{(sb)}$, $\left\{C_l^{(k)}\right\}$ represent a set of l shape parameters of grains of size-fraction k, $\{\gamma^{(j)}\}$ represents a characterization of the grain-size distribution through its first j-order moments, $\rho^{(g)}$ is the grain density, h is the flow depth, $u_* = \sqrt{\tau_0/\rho^{(w)}}$ is the friction velocity where, in a uniform flow, $\tau_0 = \rho^{(w)}hgJ$ is the bed shear stress and J is the channel slope (see Yalin 1977).

Equation (13.4) states that the non-dimensional sediment transport rate

$$\phi_s = \frac{q_s}{d\sqrt{gd(s-1)}} \quad (13.5)$$

is a function of grain shape, grain-size distribution, specific gravity, relative depth $Z = h/d$, the Reynolds number

$$X = \frac{\rho^{(w)}u_* d}{\mu^{(w)}}, \quad (13.6)$$

which expresses the influence of viscosity and the mobility number, best known as the Shields parameter,

$$Y = \frac{u_*^2}{gd(s-1)}, \quad (13.7)$$

which expresses the influence of the submerged grain weight of the grain.

At incipient motion conditions one has $\phi_s \approx 0$. In this case Z and the specific gravity of sediment grains are not relevant parameters[1] and Eq. (13.4) becomes

$$\frac{u_{*_c}^2}{gd(s-1)} = \Psi_c\left(\{C_k^{(j)}\},\left\{\frac{\gamma^{(j)}}{d^j}\right\},\frac{\rho^{(w)}u_{*_c}d}{\mu^{(w)}}\right)=0 \quad (13.8)$$

---

[1]Note that gravity appears in Eq. (13.4) through terms u_* and $g(\rho^{(g)}-\rho^{(w)})$; in the latter case it allows for separating clearly the effects of grain weight, directly depending of $g(\rho^{(g)}-\rho^{(w)})$, and inertial effects, depending only on $\rho^{(g)}$ (Yalin 1977).

Equation (13.8) expresses the functional relation between mean flow and granular mixture properties at the onset of generalised sediment motion, as opposed to Eq. (13.3) which expresses limit static equilibrium of a single grain in terms of a relation between its properties and flow in its near vicinity. For a given grain-size distribution and given type of particles, Eq. (13.8) becomes

$$Y_c = \Psi_c(X_c) \tag{13.9}$$

where $Y_c = \frac{u_{*c}^2}{gd(s-1)}$ and $X_c = \frac{\rho^{(w)} u_{*c} d}{\mu^{(w)}}$ are relative to the critical flow conditions expressed by u_{*c} for which incipient motion occurs. This was the starting point of the influential work of Shields (1936). Using narrowly graded sediment, he conducted flume experiments to investigate particle initiation of motion in streams. He determined that, for particles larger than 5 mm, corresponding to large values of X, Y is independent of X and about 0.06. For smaller grains, viscous effects would be relevant and two other ranges of values of X could be determined. The importance of Shields work was first recognised by Rouse (1939) as well as its main shortcomings, namely the fact that it did not include grain shape effects and is not applicable to poorly-sorted sediment mixtures.

13.3.3 Entrainment of Individual Grains: Theoretical Formulation

In this section we analyse the conditions for the onset of motion of a single grain contained in a sediment mixture of different grain sizes. Given the forces exerted by the fluid on a particle, i.e., drag and lift forces, its weight and the resistive forces due to the contacts with the surrounding grain, a force and a moment balances is established to determine the conditions for the onset of motion of the particle.

Consider a particle at rest in a hydraulically rough bed, as depicted in Fig. 13.1. The longitudinal bed gradient, tan(α), is small and there are no relevant bed forms. The sieving diameter, d_i, is equal to its horizontal projection (Fig. 13.1, plan view).

While at rest, the particle is supported by, at least, three grains. At the static limit (onset of motion) it is assumed that there are reacting forces in only two upfront supporting points. If the hydrodynamic actions increase, the particle will either (i) roll out of its position, (ii) slide over the upfront supporting grains or (iii) perform a jump. In any of these cases, the direction of the particle, as it leaves its static position, is normal to the pivoting axis (which may not be horizontal, i.e. $\gamma \neq 0$ in general, as seen in Fig. 13.1, front view). Evidence drawn from gravel bed experiments (Parker and Andrews 1985; Kovacs and Parker 1994; Nikora et al. 2002; Seminara et al. 2002; Parker et al. 2003) suggests that angle δ between the direction of the flow and the direction of displacement is, on average, small but not zero (Fig. 13.1, plan view). Hence, drag and lift should be considered in the direction normal to the pivoting angle.

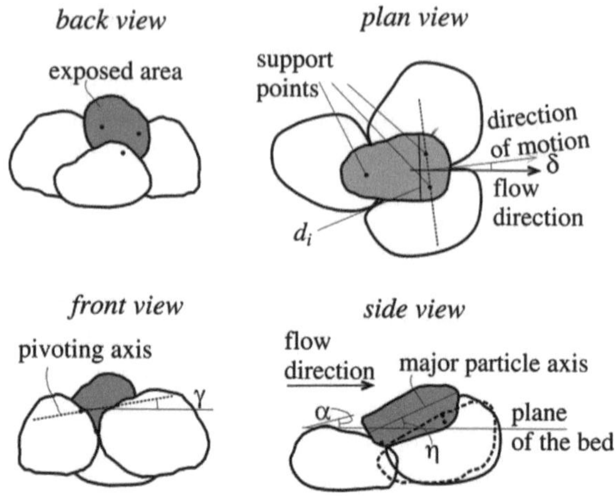

Fig. 13.1 Positioning of an individual sand or gravel particle on the surface of a granular bed composed of a poorly-sorted sediment mixture

A generally overlooked aspect (Carling et al. 1992 exempted) is that the major axis of the particle is tilted η relatively to the plane of the bed, itself tilted α relatively to the horizontal plane (Fig. 13.1, side view). If the particles are pronouncedly non-spherical, the angle η plays a major role in the stability of the particle since it determines its protrusion, defined as the maximum height of the particle above the mean bed elevation (Fenton and Abbott 1977).

In order to determine the limit static equilibrium conditions Yalin (1977), Wiberg and Smith (1987), Kirchner et al. (1990) or Valyrakis et al. (2011), among others, formulated the problem in terms of equilibrium of forces. Bridge and Bennet (1992), Fernandez Luque and van Beek (1976) or Aleixo and Maia (2006), among other, opted for equilibrium of moments while Ling (1995), Cheng and Chiew (1998) or Sun and Donahue (2000) formulated the problem for both (the latter two just for lift). Figure 13.2 shows the balances of forces (left) and moments (right) for a particle at the onset of motion.

If the particle performs a jump or if it slides out of its position, force equilibrium is broken before moment equilibrium. In this case the fundamental positioning parameters are the support angle φ and the skin friction angle ψ (Fig. 13.2, left).

Angle φ develops between the plane of bed and the plane that is normal to the particle surface at the contact points, herein *support plane*. This is a construct for which is necessary to assume that the pivoting axis is nearly horizontal ($\gamma \approx 0$) and that the tangential reactions **T** at each of the supporting points can be combined in a single force. Angle ψ is such that $|\mathbf{T}| = |\mathbf{N}| \tan(\psi)$ where **N** is the normal reaction at the supporting points.

If the particle rolls out of its position, equilibrium of moments is broken before equilibrium of forces; the most relevant positioning parameter is θ_1, the angle between the direction normal to the bed and the plane that encompasses the point of application of the hydrodynamic force and the pivoting axis (Fig. 13.2, right).

Angles φ and θ_1 are often mistakenly taken as the same (Li and Komar 1986; Kirchner et al. 1990; Fernandez Luque and van Beek 1976), which is true only if

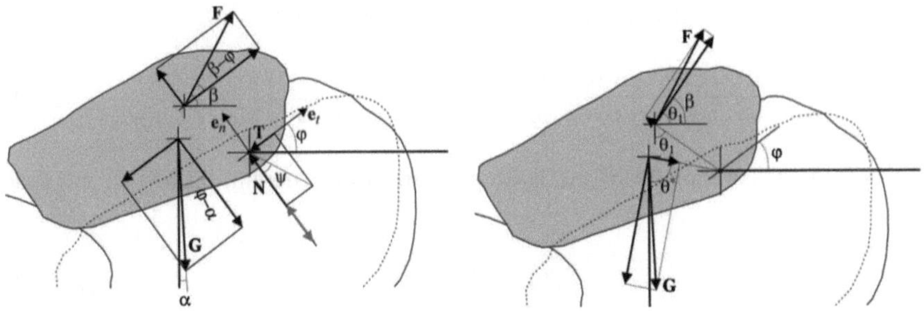

Fig. 13.2 Equilibrium of forces (*left*) and of moments (*right*) for the particle shown in Fig. 13.1

the particle is exactly spherical. Furthermore, Kirchner et al. (1990) erroneously refers to φ as the "friction angle" and, as a matter of fact, measures $\varphi + \psi$ while intending to measure φ. It is noted that ψ is the friction angle at the scale of the particle while $\varphi + \psi$ may be interpreted as a 'bed friction angle'.

Limit equilibrium conditions based on the balance of the forces shown in Fig. 13.2(left) can be formulated in the following manner. The first step is to find the projection of the forces in the direction of the displacement. Equilibrium of forces in the direction normal to the support plane reads as follows

$$N + F\sin(\beta - \varphi) - G\cos(\varphi - \alpha) = 0 \tag{13.10}$$

where $N \equiv |\mathbf{N}|$, $F \equiv |\mathbf{F}|$ is the absolute value of the hydrodynamic force (lift and drag), $G \equiv |\mathbf{G}|$ is the absolute value of the weight of the particle and β is the angle between the hydrodynamic force and horizontal plane such that $\tan(\beta) = F_L/F_D$ where F_D and F_L are the horizontal and vertical components of the hydrodynamic force, herein designated drag and lift forces, respectively.

Equilibrium of forces in the direction tangential to the support plane reads

$$-T + F\cos(\beta - \varphi) - G\sin(\varphi - \alpha) = 0 \tag{13.11}$$

where $T \equiv |\mathbf{T}|$. Considering that

$$T = N\tan(\psi) \tag{13.12}$$

combining Eqs. (13.10) and (13.12), introducing the result in Eq. (13.11) and solving for F one obtains

$$F = G\frac{\sin(\varphi - \alpha) + \cos(\varphi - \alpha)\tan(\psi)}{\cos(\beta - \varphi) + \sin(\beta - \varphi)\tan(\psi)} \tag{13.13}$$

Introducing trigonometric relations, Eq. (13.13) becomes

$$F = G\frac{\cos(\alpha)(\sin(\varphi) + \cos(\varphi)\tan(\psi)) - \sin(\alpha)(\cos(\varphi) - \sin(\varphi)\tan(\psi))}{\cos(\beta)(\cos(\varphi) + \tan(\beta)\sin(\varphi) + \tan(\beta)\cos(\varphi)\tan(\psi) - \sin(\varphi)\tan(\psi))} \quad (13.14)$$

The horizontal component of the hydrodynamic force, herein defined as the drag force, is

$$F_D = F\cos(\beta) \quad (13.15)$$

Introducing Eq. (13.14) in Eq. (13.15) and dividing by $\cos(\varphi)$ one obtains

$$F_D = G\frac{\cos(\alpha)(\tan(\varphi) + \tan(\psi)) - \sin(\alpha)(1 - \tan(\varphi)\tan(\psi))}{1 - \tan(\varphi)\tan(\psi) + \tan(\beta)(\tan(\varphi) + \tan(\psi))} \quad (13.16)$$

The usual parameterization of the drag and lift forces, the latter being the vertical component of the hydrodynamic force, are respectively [recall that the above manipulations concerned forces in the direction of the displacement, hence the factor $\cos(\delta)$]

$$F_D = \frac{1}{2}\rho^{(w)} C_D C_e C_0 \pi \frac{d_i^2}{4} u_p^2 \cos(\delta) \quad (13.17)$$

and

$$F_L = \frac{1}{2}\rho^{(w)} C_L C_e C_0 \pi \frac{d_i^2}{4} u_p^2 \quad (13.18)$$

where u_p is the near-bed horizontal velocity registered in front of the particle when forces are at limit equilibrium conditions, herein designated the competent velocity for individual particle entrainment, C_D and C_L are the drag and lift coefficients, C_e is an exposure coefficient, defined as the fraction of the visible projected area of the grain considering that a part of it may be hidden by an upstream grain, and C_0 is a shape parameter such that $C_0 = \max\{C_3 \cos\eta, C_1 \sin\eta\}$ (see Fig. 13.1) where $C_3 = d_m/d_i$ $C_1 = d_M/d_i$, being d_m and d_M respectively, the minor and major particle axis. It should be noticed that u_p is normally taken as stream-wise velocity; no major errors are committed if the channel slope is small. Note also that the definition of C_1 and C_3 presuppose that the particles are approximately elliptical and that the problem of the projected area is simplified. The correct projected area is a more complex problem whose full solution can be consulted in e.g., Binggeli (1980).

Since $\tan(\beta) = F_L/F_D$, one has

$$\tan(\beta) = \frac{C_L}{C_D\cos(\delta)} \quad (13.19)$$

The weight of the particle is

$$G = \rho^{(w)}(s-1)gC_V\pi\frac{d_i^3}{6} \tag{13.20}$$

where C_V, which can be back-calculated from (13.20), is a shape coefficient. Introducing Eqs. (13.17), (13.19) and (13.20) in Eq. (13.16) one obtains the specific form of Eq. (13.3) for force limit equilibrium:

$$\frac{u_p^2}{(s-1)gd_i} = \frac{4}{3}\frac{C_V}{C_D C_e C_O}\frac{\cos(\alpha)(\tan(\varphi)+\tan(\psi))-\sin(\alpha)(1-\tan(\varphi)\tan(\psi))}{\cos(\delta)(1-\tan(\varphi)\tan(\psi))+\frac{C_L}{C_D}(\tan(\varphi)+\tan(\psi))} \tag{13.21}$$

In the case of equilibrium of moments, limit equilibrium conditions expressing the geometrical conditions shown in Fig. 13.2, assuming that lift and drag are applied in the same point, read:

$$Fb_F\cos(\beta-\theta_1) - Gb_G\sin(\theta^*) = 0 \tag{13.22}$$

where $\theta^* = \theta - \alpha$ is the angle between the direction of the weight and the plane that encompasses the centre of mass of the particle and the pivoting axis, θ_1 is the angle between the direction normal to the bed and the plane that encompasses the point of application of the hydrodynamic force (Fig. 13.2, right) and the pivoting axis (b_F and b_G) are the lever arms of the hydrodynamic action and of the weight, respectively. It was pointed out by Chepil (1959), Fernandez Luque and van Beek (1976), Bridge and Bennet (1992), that the point of application of drag and lift may be different from the centre of mass of the particle. In fact, the point of application of drag may be different from that of lift, an aspect that will be overlooked in this study. It is implicit the forces are expressed in the direction of the displacement.

Considering that $\cos(\beta-\theta_1) = \cos(\beta)\sin(\theta_1) + \sin(\beta)\cos(\theta_1)$ Eq. (13.22) becomes

$$F = G\frac{b_G}{b_F}\frac{\sin(\theta^*)}{\cos(\beta)(\cos(\theta_1)+\tan(\beta)\sin(\theta_1))} \tag{13.23}$$

Introducing Eq. (13.15) in Eq. (13.23) one rewrites the latter as

$$F_D = G\frac{b_G}{b_F}\frac{\sin(\theta^*)/\cos(\theta_1)}{1+\tan(\beta)\tan(\theta_1)}$$

and, given Eqs. (13.17), (13.20) and (13.19), the specific form of Eq. (13.3) for limit equilibrium of moments becomes

$$\frac{u_p^2}{(s-1)gd_i} = \frac{C_V}{C_D C_e C_O}\frac{b_G}{b_F}\frac{\sin(\theta^*)/\cos(\theta_1)}{\cos(\delta)+\frac{C_L}{C_D}\tan(\theta_1)} \tag{13.24}$$

Equations (13.21) and (13.24) configure a model of entrainment of an individual particle expressed in terms of competent velocities. When the flow velocity in the near vicinity of the sediment grain, for instance in front of it and at the elevation of its centre of mass, is equal to u_p that sediment grain will be displaced. The dependence of the Reynolds number $u_p d_i/\nu^{(w)}$ is indirect, through the ratio C_L/C_D and through C_D. Assuming that most flows have $u_p d_i/\nu^{(w)}$ ranging between 2×10^2 to 10^5, C_L/C_D and C_D have a narrow range of variation and are not dominant terms. On the contrary, the normalised velocity u_p is much influenced by protrusion (quantified by C_0) and by exposure (C_e), being inversely proportional to both. Other major influences are angles φ, θ_1, and ψ.

In order to make this model operational, it is necessary to quantify its geometrical and dynamic parameters. The shape parameters, C_V, C_1 and C_3 were subjected to direct measurement and evaluation. A sample of over 200 particles of fine gravel was employed. The mean sieving diameter of the sample was 4.3 and the extreme values were 1.2 mm and 9.2 mm. The results are seen in Fig. 13.3.

The support angle, φ, was studied by Kirchner et al. (1990), in the wake of Miller and Byrne (1966), by tilting a frozen segment of a water worked bed where a particle was randomly placed. As explained before, it is envisaged that this method provides good estimates for the sum $\varphi + \psi$. From their work, and assuming that the probability density function of φ is symmetrical for each size fraction, the expected value of φ in degrees (°) is

$$\varphi = 55.2\left(\frac{d_i}{d_{50}}\right)^{-0.307} - \psi$$

The inclination of the particle, expressed by angle η in degrees, was estimated for a limited number of particles of the water worked beds described by Ferreira (2005, Chap. 2); it was obtained

$$\eta = \begin{cases} 3333(d_i - 0.001) & \text{if } d_i < 0.007 \text{ m} \\ 20 & \text{if } d_i > 0.007 \text{ m} \end{cases}$$

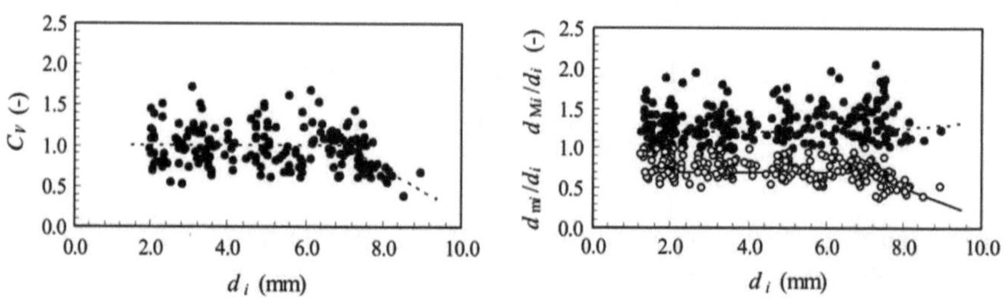

Fig. 13.3 Shape coefficients. Left panel: measured values of C_V (*black circle*) and mean values as a function of sieving diameter (*dashed line*). Right panel: measured values of $C_1(d_M/d_i)$ (*black circle*) and respective mean values (*dashed line*) and measured values of $C_3(d_m/d_i)$ (*open circle*) and expected mean values (*black line*)

Angle θ can be estimated as a function of φ. For an ellipsoid one has $\tan(\varphi') = (d_m/d_M)^2\tan(\theta')$ where $\theta = \theta' + \eta$ and $\varphi = \varphi' + \eta$, which, considering $K = (d_M/d_m) = (C_3/C_1)^2$, leads to

$$\tan(\theta) = \frac{\tan(\varphi)(K + \tan^2(\eta)) + \tan(\eta)(1 - K)}{1 + \tan(\eta)(\tan(\varphi)(1 - K) + \tan(\eta))} \tag{13.25}$$

The exposure coefficient C_e depends on the height of the crest of the particle relatively to the crest of the particle immediately upstream (Kirchner et al. 1990). It does not account for particle hiding as a result of placement inside the wake of a larger grain or cluster of grains. The uncertainty in this parameter is large. It is considered that exposure is uniformly distributed between 0 and 1 for grain-sizes smaller than 0.007 m and uniformly distributed between 0 and 0.5 for grain-sizes larger than 0.007 m.

Drag and lift coefficients were subjected to much attention in previous studies. It is assumed that the Reynolds number is sufficiently large to render constant both coefficients. Assuming that the range of $u_p d_i/\nu^{(w)}$ is compatible with constant drag and lift coefficients, it is assumed $C_D = 0.4$ (White 2011) and $C_L = 0.2$ Wiberg and Smith (1987). The ratio C_L/C_D is consistent with the direct measurements of Chepil (1961) who found that for exposed particles larger than 3 mm that lie above local mean bed elevation the ratio should be lower than 0.75. For smaller grain sizes, there is high uncertainty regarding the values of this ratio as empirical evidence is scarce.

The impact of angles δ and α is small. Without loss of generality, and considering the uncertainties already introduced, it is considered that $\delta = \alpha = 0$. From $\alpha = 0$ one obtains $\theta^* = \theta$. Given the lack of information in the moment arms of drag, lift and weight, it is assumed that $b_G/b_F = 1$ and that $\theta_1 = \theta$.

The particle friction angle, ψ, depends primarily on the micro-texture of the particle and, hence, varies with the type of constitutive minerals and the degree of abrasion. Preliminary experiments on gravel-sized particles indicate that this angle can vary from 5°, for mechanically polished feldspate particles, to 30° for crushed granite rock.

Introducing the expressions and values of the intervening parameters in Eqs. (13.21) and (13.24) the Froude-like parameter

$$Fr_p = \frac{u_p^2}{(s - 1)gd_i}$$

can be calculated for any given size fraction characterized by d_i. The results for $\psi = 15°$ are shown in Fig. 13.4. In can be concluded that:

(i) small particles (up to 2 mm), because of their high values of φ, are susceptible to be entrained by failure of moment equilibrium. They will roll out of its pocket and, given their size, are likely to be found in saltation;

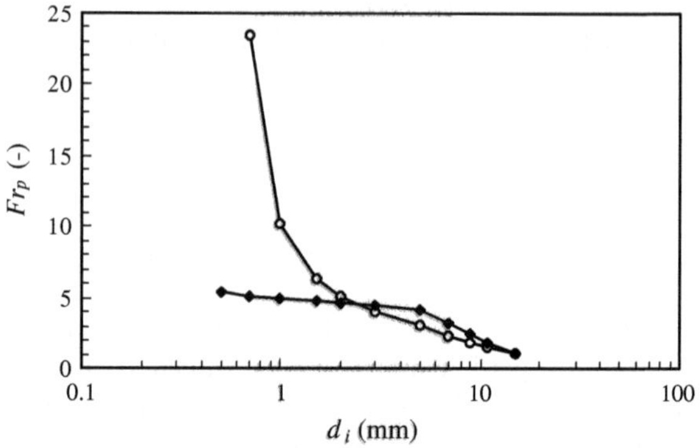

Fig. 13.4 Values of Fr_p computed from Eqs. (13.21) (—o—) and (13.24) (—◆—) for particles sized between 0.5 and 15 mm (coarse sand and fine gravel). Friction angle $\psi = 15°$

(ii) intermediate sized particles are susceptible to slide out of its locations by failure of force equilibrium. Larger particles may be set in motion by rolling due to failure of moment equilibrium. This is especially true for rounder particles;

(iii) the value of Fr_p for fine and intermediate sized particles is approximately constant. Hence, in these sizes, the square of the competent velocity of a grain, with 'average' shape and positioning characteristics, is approximately proportional to its diameter d_i;

(iv) for larger particles, the relation between u_p^2 and d_i is not a direct proportionality. In this model, the decreased resistance of larger grains is not so much due to greater exposure but fundamentally because it is assumed that larger particles have smaller supporting angles, φ. This effect does not owe to protrusion (sensu Fenton and Abbott 1977).

It is now necessary to understand the relation between the competent velocity for entrainment of a grain and the concept of incipient motion at channel scale. In a sufficiently large bed area, Fr_p is a random variable defined through the distributions of the parameters involved in Eqs. (13.21) and (13.24). The near-bed velocity field is also a random variable; these are the two main premises to cast incipient motion as a probabilistic problem.

13.3.4 Incipient Motion at Reach Scale. A Probabilistic Approach

For poorly sorted sediment mixtures, field work (e.g., Ramette and Heuzel 1962; Fahnestock 1963; Novak 1973; Church 1978; Bradley and Mears 1980; Parker et al. 1982; Andrews 1983; Carling 1983; Reid and Frostick 1984; Andrews and

Erman 1986; Church and Hassan 2002) and laboratory studies (e.g., Fenton and Abbott 1977; Parker and Klingeman 1982; White and Day 1982; Proffitt and Sutherland 1983; Wilcock and Southard 1988; Wilcock and McArdell 1993, 1997; Buffington and Montgomery 1997) show that Y_c [Eqs. (13.8) and (13.9)] ranges between 0.01 and 0.1 and even more. Such wide variations obtained for both natural and experimental gravels have been explained by the effect of sheltering and hiding of different grain sizes in poorly sorted mixed sediment, relative protrusion, pivot angle, channel morphology, bed surface structuring, and the interaction among grains. In a natural stream, in fact, the combined effect of all of these surface characteristics is likely to determine the threshold of movement of the bed sediment. Per se, each of these effects does not fully explain the wide range of values of Y_c obtained for natural sediments.

To account for this variability, the problem of incipient motion at reach scale can be cast in probabilistic terms. This has been attempted since, at least, Gessler (1965), who found that the probability for a grain to remain static on a bed surface was about 0.5 for $Y = Y_c$ and about 0.95 for $Y = 0.5Y_c$. More recently Wu and Chou (2003) derived a theoretical model of critical entrainment probability as a function of X_c, explicitly differencing rolling and lifting probabilities. However, they assume that particles are subjected to the mean flow velocity, given by the logarithmic law, which is not the case in hydraulically rough beds (Ferreira et al. 2010). The following text describes a probabilistic analysis of incipient motion at reach scale, based on experimental evidence and on the theoretical model for the entrainment of a single particle described in Sect. 1.3. Each exposed grain[2] in the bed, has a hydrodynamic threshold force. This force depends only on its geometrical arrangement and on the specific frictional interaction with its neighbours, as expressed in Eqs. (13.14) and (13.23). If, at a given time, the flow field generates normal and tangential stresses such that, integrated on surface of a grain, originate a force larger than its threshold force that grain will be entrained; otherwise it will remain static. Note that we are not taking into consideration the effect of the duration of flow event, see Valyrakis et al. (2011). Alternatively, one may express this principle in terms of flow velocities in the vicinity of a grain and competent velocities for the entrainment of a single grain given by Eqs. (13.21) or (13.24). Repeating this analysis over the totality of exposed grains, one obtains the number of grains that are set in motion at a given instant.

At this level, the analysis is still deterministic: each exposed grain experiences an action either smaller or larger than its threshold. Yet, given the stochastic nature of the turbulent flow, one may interpret such instantaneous analysis as one realization of a stochastic process. Repeating the analysis for a sufficiently large succession of instants one subjects the same bed location to the whole spectrum of velocities (and thus of hydrodynamic forces). Should each realization of the turbulent flow be accomplished with the same bed (same grains in the same positions) an empirical

[2]We define exposed grain as that which, to be able to move, would not have to disturb the position of its neighbours (we ignore, for now, the possibility of collective entrainment).

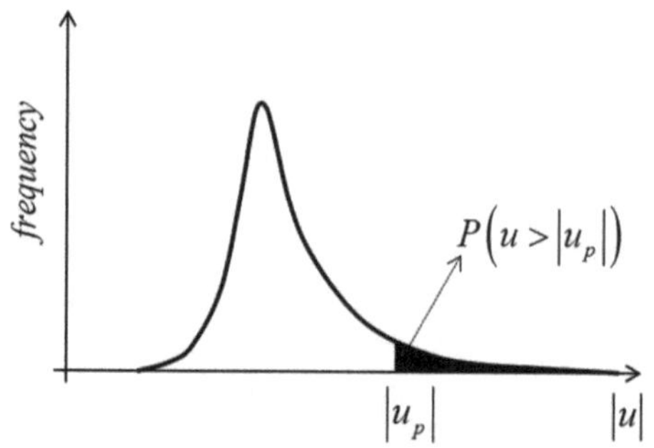

Fig. 13.5 Probability density of the flow velocity in the near-vicinity of a given grain, $f_u(u|\mathbf{x}_n)$. The probability of entrainment of the grain or, approximately, the probability of entrainment at a given bed location is the numerical value of the shaded area

probability of entrainment could be assigned to a given location as the probability of the instantaneous near-grain velocity exceeding its competent velocity,

$$P(u(x_n, y_n) > u_{p_n}) = \int_{u_{pn}}^{+\infty} f_u(u|\mathbf{x}_n)\mathrm{d}u \tag{13.26}$$

where the index n designates the grain, $\mathbf{x}_n = (x_n, y_n)$ are its longitudinal and lateral coordinates and $f_u(u|\mathbf{x}_n)$ is the empirical probability density function (PDF) of the instantaneous (turbulent) velocity in its near upstream vicinity. If $f_u(u|\mathbf{x}_n)$ is known, the probability of entrainment of a given grain is the shaded area depicted in Figure 13.5.[3]

The competent velocity varies from grain to grain—hence being written $u_p(\mathbf{x}_n)$—a direct consequence of the variation of parameters in Eqs. (13.21) or (13.24). At reach scale, one can view the competent velocity as a random variable and organize the values of $u_p(\mathbf{x}_n)$ into a probability distribution function $f_{u_p}(u_p)$, given that a bed composed of a poorly-sorted granular mixture exhibits a sufficiently large number of arrangements of grains.

Also, the PDF $f_u(u|\mathbf{x}_n)$ varies with the location $\mathbf{x}_n = (x_n, y_n)$, influenced by the bed micro-topography in the upstream vicinity of the grain. An example is shown in Fig. 13.6. Grains A and B have the same skin roughness, exposure coefficient, protrusion and pivoting angle. Assuming that the ratio C_L/C_D does not change significantly both grains should have approximately the same ratio $|\boldsymbol{F}|/|\boldsymbol{G}|$

[3]The ensemble analysis underlying Eq. (13.26) is not possible in practice since it is not feasible to subject the exact same bed to a succession of turbulent flows. The practical way to generate a succession of flow realizations is to expose the bed to a turbulent flow over a sufficiently long but finite time interval, so that all relevant turbulent scales are included. Hence, in practical terms, an empirical probability of entrainment can be obtained for a set of bed locations, coincident with the position of a set of grains that may be substituted over the course of time.

[Eqs. (13.14) or (13.23)]. Also, in absolute terms, the threshold competent velocity for the larger grain A is larger than that of (the smaller) grain B [Eqs. (13.21) or (13.24)]. However, micro-topography upstream of grain B contributes to reduce the flow velocity in its near vicinity. Thus, it may be the case that grain A, in spite of its greater size, has a larger probability of entrainment than grain B.[4]

A first consequence of the spatial variability of $u_p(\mathbf{x}_n)$ and of $f_u(u|\mathbf{x}_n)$ is that the probability of entrainment characteristic of a given bed location

$$P(u(\mathbf{x}_n) > u_p(\mathbf{x}_n)) = \int_{u_p(\mathbf{x}_n)}^{+\infty} f_u(u|\mathbf{x}_n)\mathrm{d}u$$

is not constant in the bed reach. Hence the analysis of the entrainment of the entire sequence of exposed grains in that bed reach is not a Bernoulli statistically random process. A second consequence is that it makes sense to speak about joint probability of local bed micro-topography features and entrainment of single grains. Hence, one can the interpret probability shown in Fig. 13.6 as the conditional probability of entrainment of a single grain given a specific bed morphology feature, for instance a wake of a cluster or an exposed region. Note that the quantification of the frequency of these micro-topography features should be susceptible to be parameterized employing grain diameter, grain-size mixture, mean flow parameters, fluid properties and gravity.

An empirical probability of entrainment, which will naturally account for size-selective transport, can be calculated at the scale of the bed reach, using the PDFs $f_u(u|\mathbf{x}_n)$ and $f_{u_p}(u_p)$ and the concept of joint probability of entrainment and bed micro-topography features.

One must first divide the bed reach into several classes of local morphological conditions. These classes are arbitrary but should be representative of phenomena susceptible to be characterized empirically. Following the example shown in

[4]The view of sediment transport as a purely size selective process has been rejected (Parker and Klingeman 1982; Komar 1987). Deterministic bedload discharge formulas that include a dependence of a power of $\theta - \theta_c$ require corrections to either the Shields or the critical Shields parameters when applied to the transport of size fractions. Hiding-exposure coefficients have been proposed to reduce/increase $\theta_k - \theta_{ck}$ for given size fractions d_k smaller/larger than a reference diameter (Egiazaroff 1965; Ashida and Michiue 1972). As seen in the example of Fig. 13.6, hiding and exposure describe different phenomena. The former is related to local modifications of the flow field which, for smaller size fractions may signify reduced mobility due to sheltering or decreased near bed turbulence (Nelson et al. 1995). The latter expresses the positioning of the grain relatively to its immediate neighbours. Increased exposure and decreased support angle are common of larger size fractions (Sect. 13.3 and, e.g., Komar and Li 1988). Due to this combined effect, in the limit, equal mobility of different size fractions can be nearly attained for gravel mixtures (Parker and Klingeman 1982; Parker et al. 1982), although the range of conditions for this to happen has been shown to be relatively narrow (Wathen et al. 1995; Parker and Toro-Escobar 2002). Hiding/exposure coefficients thus account, in a lumped way, for the differential grain response regarding entrainment, depending on the particular location/diameter of the particle relatively to their neighbours.

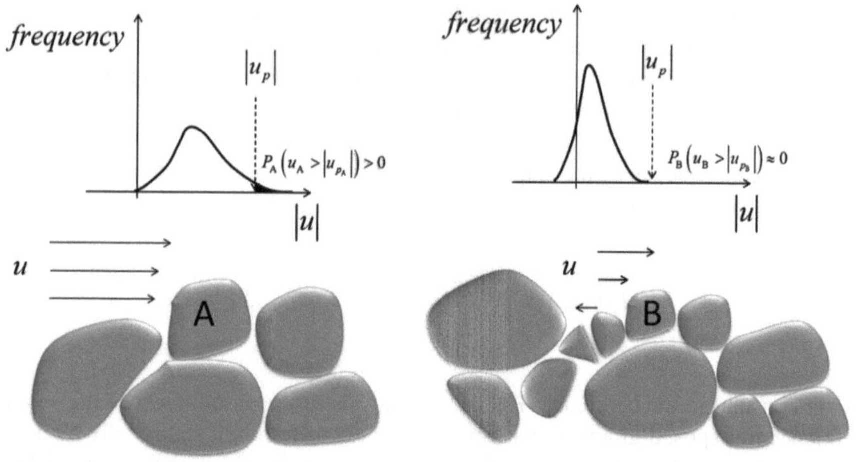

Fig. 13.6 Stream-wise vertical section of the bed depicting geometrically similar grains subjected to different flow turbulence fields. The influence of upstream micro-topography results in a greater probability of entrainment of the larger grain

Fig. 13.6, a class can group the grains of a given size fraction that have clusters in their upstream vicinity; one other class may group the grains of another size fraction that are more exposed. Other classes may be imposed but it is crucial that a frequency p_M can be assigned to each class M.

For each class, a conditional PDF of the competent velocity $f_{u_p}(u_p|M)$ is obtained, through Eqs. (13.21) and (13.24), from the PDFs of the intervening parameters. For the same class, the probability density function of the instantaneous turbulent velocities $f_u(u|M)$ can be obtained from judicious laboratory or field work, for instance collecting time series in the locations of typical bed features. From these probability densities, the empirical conditional probability of entrainment of given class M can be obtained as

$$P(E|M) = \int_{-\infty}^{+\infty} P(u > u_p|M)\, f_{u_p}(u_p|M) \mathrm{d}u_p \tag{13.27}$$

where $P(u > u_p|M) = \int_{u_p}^{+\infty} f_u(u|M)\,\mathrm{d}u$. Employing a frequentist interpretation of probability one may say that, in an actual bed reach subjected to a succession of N (turbulent) flows, Eq. (13.27) states the fraction $N\,P(E|M)$ is the number of flows in which the grains of class M are entrained.

A joint probability of class M and entrainment of grains of that class is given by the product $P(E|M)p_M$. Hence, the reach scale empirical probability of entrainment is the marginal probability

$$P(E) = \sum_M P(E|M)p_M \tag{13.28}$$

since the classes M are mutually exclusive (Feller 1968, p. 116).

A given flow is said to be in incipient motion conditions if the marginal probability expressed in Eq. (13.28) is equal to an arbitrarily specified small threshold value:

$$P(E) = P_\varepsilon \tag{13.29}$$

Incipient motion conditions can also be expressed in terms of bedload transport rates. Following Einstein (1950), the volumetric bedload discharge per unit width, q_s, can be expressed as

$$q_s = E\ell_s \tag{13.30}$$

where ℓ_s is the mean displacement length of entrained grains and E is the entrainment rate, in volume of granular material per unit bed area and per unit time, calculated from the probability of entrainment as

$$E = \frac{P(E)d(1-\lambda)}{2T} \tag{13.31}$$

where d is a characteristic diameter of grains on the bed surface, λ is the bed porosity and T is the integral length scale of the longitudinal velocity at the given elevation above the bed.

The displacement length has been the object of considerable research, including Einstein (1937) or Hu et al. (1992). Note that in the definition of E it is assumed that $2T$ is the period of the larger velocity fluctuations. It is implicit that the probability of entrainment, which increases with the size of the time interval employed in the stability analysis, becomes sufficiently close to its asymptotic value $P(E)$ for time intervals longer than $2T$. Multiples of this interval would render the same value because the (statistically) same hydrodynamics actions would be applied on the grain.

The entrainment rate at threshold conditions is thus

$$\varepsilon = \frac{P_\varepsilon d(1-\lambda)}{2T} \tag{13.32}$$

and the bedload discharge characteristic of incipient motion conditions becomes

$$q_{\varepsilon} = \varepsilon \ell_s \tag{13.33}$$

An example of calculation is seen next. The gravel bed data of Ferreira et al. (2012), obtained in 12 m long 40 cm wide channel under uniform flow conditions and sediment recirculation, are used to identify the flow that can be considered as incipient motion conditions. Tests T1, T2 and T8 of that database were picked for analysis in this text. The main parameters of the tests are reproduced in Table 13.1.

Variables involved in Table 13.1 are S_0, the channel slope, Q the total discharge, $Fr = U/\sqrt{gh}$, the Froude number where U is the depth-averaged mean flow velocity, $d_{50}^{(srf)}$ and $d_{90}^{(srf)}$, the median diameter of sediment in the surface layer and the diameter for which 90 % of the mass of the surface layer is finer, $Y_{50}^{(sb)}$, the Shields parameter [definition (13.7)] calculated with the median diameter of the substrate ($d_{50}^{(sb)}$), X, the boundary Reynolds number (calculated from definition (13.6) with $d \equiv k_s$ where k_s is a roughness scale defined as in Ferreira et al. (2012) and ϕ_s is given by definition (13.5).

For simplification, the classes needed to apply Eq. (13.27) were simply the grain sizes of the employed mixture of fine gravel. The grain-size distribution, with geometric standard deviation of 1.7, can be seen in Ferreira et al. (2012). The probability density function of instantaneous velocities was derived from Laser Doppler Anemometry data at the lowest measuring position, roughly at the elevation of the crests of the more exposed grains.

The probability density function of the competent velocities for each class $f_{u_p}(u_p|M)$ was obtained, through equations, from the empirical distribution of support angle φ determined by Kirchner et al. (1990) for grains of classes $d_i/d_{50}^{(sb)} = 0.45$ and $d_i/d_{50}^{(sb)} = 2.10$ (the empirical distributions of the remaining classes were determined by interpolation). Such distribution acknowledges that smaller grains tend to have higher values of angle φ. Values of the exposure coefficient C_e of each class were determined from the formula proposed by Kirchner et al. (1990). Values of the shape coefficients were determined from a large collection of grains (Fig. 13.3). Friction angle was $\psi = 10°$ for all classes.

Table 13.1 Main variables and parameters that characterize the laboratory tests

Test	S_0 (–)	Q (l/s)	Fr (–)	$d_{50}^{(srf)}$ (mm)	$d_{90}^{(srf)}$ (mm)	$Y_{50}^{(sb)}$ (–)	X (–)	ϕ_s (–)
T1	0.0014	14.3	0.43	3.2	6.0	0.027	221	3.3×10^{-7}
T2	0.0031	13.2	0.60	3.3	6.1	0.046	324	1.9×10^{-5}
T8	0.0046	18.5	0.76	3.7	6.1	0.067	376	3.3×10^{-3}

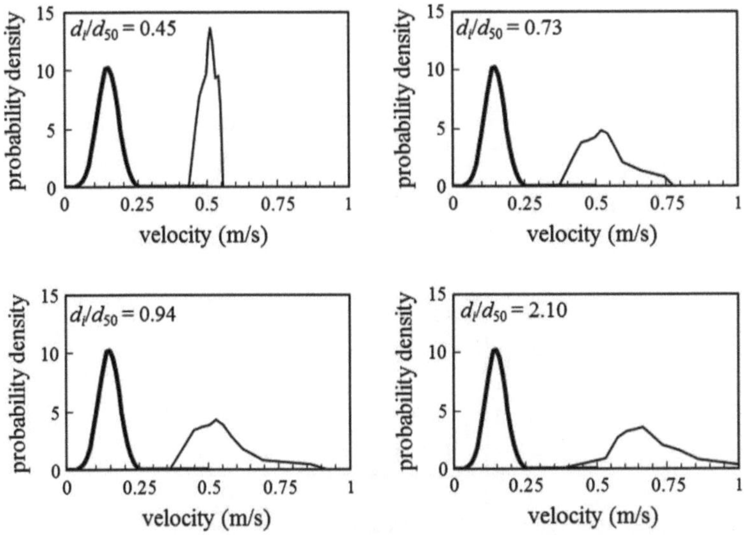

Fig. 13.7 Probability density functions of instantaneous velocities (*thick black line*) and competent velocities (*thin black line*) for 4 classes of grains in the bed, discriminated by size. Data of Test T1

All other variables were parameterized to φ, as is the case of θ, or included as averaged values, as was the case of η.

The PDFs $f_{u_p}(u_p|M)$ and $f_u(u|M)$ are shown in Fig. 13.7 for the classes $d_i/d_{50}^{(sb)} = \{0.45; 0.73; 0.94; 2.10\}$ of test T1 (Table 13.1). The same PDFs are shown for class $d_i/d_{50}^{(sb)} = \{2.10\}$ of tests T1, T2 and T3 in Fig. 13.8. The PDF of $f_u(u|M)$ is the thick black line in the figures. It is striking that the flow velocities are not normally distributed; a positive skewness is evident, especially in the test with the larger value of Shields parameter, as seen in Fig. 13.8c). The standard deviation of the velocity distribution is a measure of the turbulence intensity; there is a necessary positive correlation between its value and the value of the Shields parameter, which is apparent in Fig. 13.8.

Figure 13.7 illustrates that for Test T1 there is practically no overlapping between the distributions of flow velocities and competent velocities for all classes. The computed probability of entrainment was zero. This can be explained by the low value of the dimensionless critical shear stress (Table 13.1).

In the case of tests T2 and T8, Fig. 13.8 shows that there is some overlapping of probability density function, mostly in the classes of diameters equal and larger than the $d_{50}^{(sb)}$. Probabilities of entrainment were computed for each class with Eq. (13.27). Bedload discharges were also computed for each class according to the fractional counterpart of Eq. (13.30)

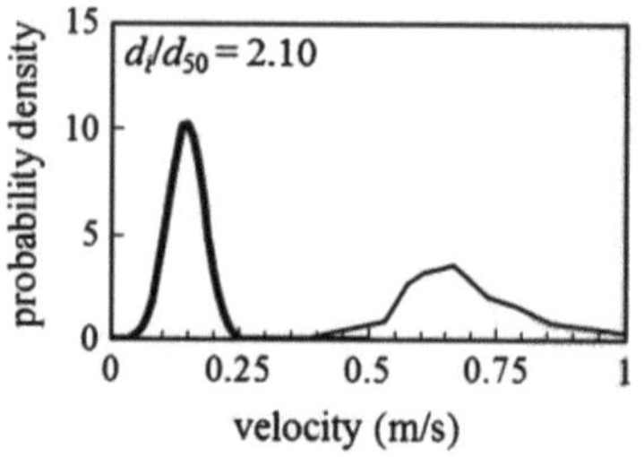

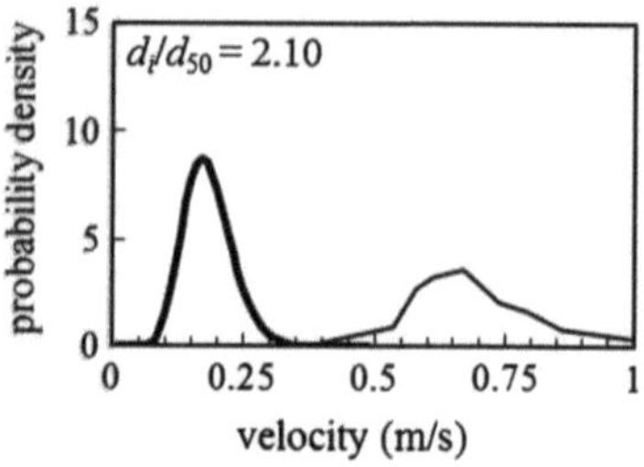

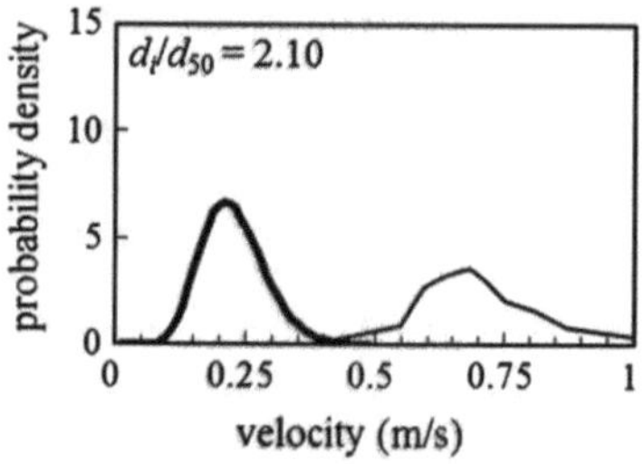

Fig. 13.8 Probability density functions of instantaneous velocities (*thick black line*) and competent velocities (*thin black line*) for class $d_i/d_{50}^{(sb)} = \{2.10\}$. Data of tests **a** T1, **b** T2 and **c** T8

$$q_{sk} = E_k \ell_{sk} \tag{13.34}$$

where q_{sk} is the bedload discharge in volume per unit width of size fraction k, E_k is the entrainment rate of that size fraction, calculated from the probability of entrainment as $E_k = P(E|k)p_k(1-\lambda)d/(2T)$ and ℓ_{sk} is the mean displacement length of grains of that size fraction, computed, in this case, with the formula for saltation in rough beds proposed by Hu et al. (1992).

Calculation for T1 rendered a zero probability of entrainment while the measured time-averaged discharge was the very small number of about 20 grains (with the median diameter) per hour. In Test T2, the calculated bedload rate was about 90 equivalent $d_{50}^{(sb)}$ grains per minute, more than four times larger the measured averaged bedload transport rate (about 20 grains per minute). In the case of Test T8, the computed bedload rate is about 100 equivalent grains per second, while the measured rate is 59 grains per second. Given the uncertainties involved in the parameters of the model the agreement is acceptable.

If incipient motion conditions were defined as the flow that generates a bedload transport rate of, say, 0.1 grains per second, this probabilistic approach would indicate that Test T1 is below threshold conditions, that Test T2 is slightly above threshold conditions and that Test T8 is under conditions of generalised transport.

The above probabilistic considerations allow for an objective definition of incipient motion conditions in the following sense: once the threshold value ε [Eq. (13.32)] is specified, whether a flow is above or below, threshold conditions become specified with no intervention from the observer. However, a decision must be made regarding the value of ε.

To avoid such decision, critical bed shear stresses have been obtained by extrapolation from $\phi_s(Y)$ curves. According to Vanoni (1966), Paintal (1971a) or Buffington (1999), among others, this was the procedure followed in the seminal work of Shields (1936). A regression law was fitted to measured pairs of bed shear stress and bedload transport rates; the critical bed shear stress was simply the bed shear stress for which the best fit law crossed the axis of zero bedload transport rate.[5] This definition presupposes a separation between high, and thus relevant (for practical engineering purposes), transport rates from indifferent ones.

13.3.5 Advanced Issues Concerning Mountain River Morphologies

The problem of incipient motion is more complex in coarse bedded streams with cascade, step-pool and riffle-pool morphologies. In this case the shear stress responsible for particle entrainment is a very small fraction of the total bed shear stress (e.g., Zimmermann et al. 2010). One of the outstanding characteristics of these streams affecting sediment entrainment and transport is the high boundary roughness from not only the large particles but also the frequent obstructions and rough banks that make up a large proportion of the boundary. In these streams, particle entrainment largely depends on the location within the local bed morphology, i.e., riffle, pool, step. Furthermore, large proportions of the bed remain stable during small and medium size events while during large events (larger than 20 years return period) most of the bed is mobile including boulder and keystones in steps (e.g., Lauffer and Sommer 1982; Jaeggi 1995; Zimmermann and Church 2001; Curran and Wilcock 2005; Zimmermann et al. 2010). For steep channels, the Shields' function may be modified to take into account the particle's downslope weight (e.g., Graf 1971; Mizuyama 1977; Rickenmann 1990, 2001; Lamb et al. 2011). For example, Solari and Parker (2000) reported that small particles are more stable than large particles at slopes >2 %. Due to the difficulty in obtaining reliable measure of the water depth/hydraulic geometry of steep streams with step-pool morphology and boulder-bed channels, field studies often report water discharge, and so the Shields function is converted to a discharge-based relation (e.g., Schoklitsch 1962; Bettess 1984; Ferguson 1994; Lenzi et al. 1999; Lenzi and D'Agostino 1999) or stream power (Gomez and Church 1989). However, stream power depends on water depth.

[5]Kennedy (1995) believed that Shields (1936) used Kramer's (1935) category of weak-countable transport conditions to define the incipient motion conditions, thus reducing subjectivity to the notion of "countable".

13.4 Mechanics of Bedload Transport

13.4.1 Definition of Bedload Transport Rate

Sediment transport concerns the motion of granular material. It is the general name for kinematic and dynamic phenomena occurring when granular material is picked up from a granular bed and set in motion by a continuous fluid phase or is triggered to move together with a fluid phase. The fundamental phenomenon occurring during the movement of coarse bed material by fluid is the dispersion of individual particles. The volumetric sediment transport rate, or sediment discharge, per unit width, can be defined as the flux

$$q_s = \frac{1}{b}\int_S \boldsymbol{u}^{(g)} \cdot \boldsymbol{n}\,\mathrm{d}S \tag{13.35}$$

where $\boldsymbol{u}^{(g)}$ is the velocity of the granular phase, $\boldsymbol{n}$ is the outward unit normal vector to the arbitrary surface S (Fig. 13.9) and b is a length calculated along surface S. As pointed out by Furbish et al. (2012a) this definition is not very practical as it requires the evaluation of the (spatially discontinuous) granular velocity field. It can be achieved by analysing high-speed video footage of relatively low bedload rates. An example is shown in Fig. 13.10: a sequence of images acquired at 300 frames per second of 5 mm glass beads passing through a rectangular cross-section section normal to the bed. From this sequence, the actual velocity field of the granular phase can be calculated, as well as the area of solid material.

In this case, video analysis allows for the determination of the velocity of each grain as it crosses a virtual surface. Since this velocity is considered uniform over the cross-sectional area of the grain, Eq. (13.35) is actually calculated as

$$q_s(t_0 + m\Delta t) = \frac{1}{b}\sum_{n=1}^{N(t_0+m\Delta t)} A_n^{(g)}\boldsymbol{u}_n^{(g)} \cdot \boldsymbol{n}, \tag{13.36}$$

where N is the number of grains that intercept S in the sampling instant $t_0 + m\Delta t$, m is the number of the video frame under analysis, $A_n^{(g)}$ is the area resulting from the intersection of surface S and grain n,

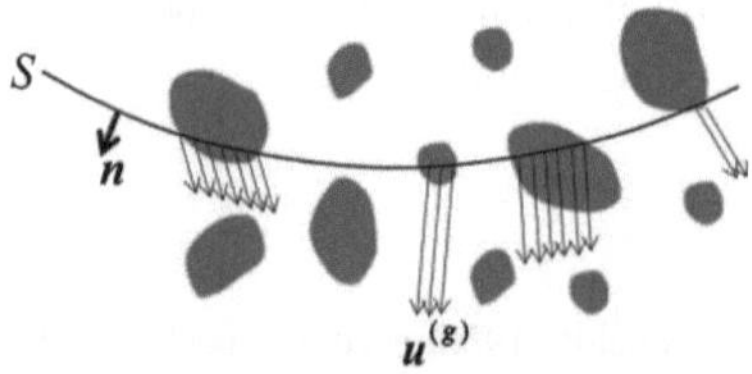

Fig. 13.9 General definition of volumetric transport rate: variables included in Eq. (13.35)

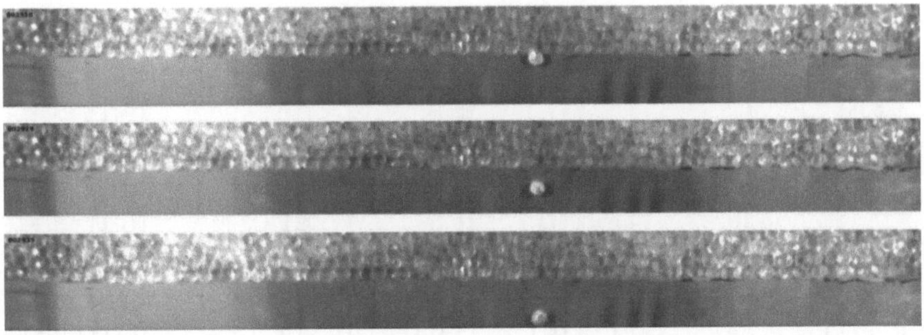

Fig. 13.10 Sequence of images distanced 33 ms of a particle moving as bedload. Particles are glass beads with 5 mm diameter. Channel is 0.405 m wide, recirculating water and granular material. Channel slope is 0.00385, Froude number is $Fr = 0.67$, Reynolds number [definition (13.6)] is $X = 288$, dimensionless transport rate is $\phi_s = 6.7 \times 10^4$

$$\boldsymbol{u}_n^{(g)} = \frac{\boldsymbol{x}(t_0 + m\Delta t) - \boldsymbol{x}(t_0 + (m-1)\Delta t)}{\Delta t}$$

is the velocity of grain n and $\boldsymbol{x}$ is the position of a fixed point of grain n.

A concentration of the granular phase can also be obtained from the video analysis. In this case, as the channel cross section is rectangular, it is $c(t_0 + m\Delta t) = (1/(bh)) \sum_{n=1}^{N(t_0+m\Delta t)} A_n^{(g)}$ where h is the flow depth.

A 5 s long time series of bedload discharge per unit width, taken from the same flow depicted in Fig. 13.10, is presented in Fig. 13.11. The stochastic nature of q_s at weak bedload rates is evident, itself a consequence of the stochastic character of the rates of particle entrainment and deposition. The former results from randomness in particle positioning, especially in poorly sorted grain-size mixtures (e.g., Church and Hassan 2002) and from the particular randomness associated to the magnitude and orientation of hydrodynamic actions, determined by local flow turbulence (see Sect. 13.3). Particle deposition has been less studied but has been shown to depend on the grain-size distribution of surface material and bed microtopography (Drake et al. 1988), themselves depending on turbulent flow variables (Nelson et al. 1995; Ferreira et al. 2012).

For relatively low transport rates, such as that depicted in Fig. 13.11, and well-sorted bed material, the distribution of intervals between particle arrivals (inter-arrival intervals) is approximately exponential (Einstein 1937), although other distributions may constitute better fits, including the Poisson's distribution (Turowski 2010). For higher transport rates, self-organization into bedload sheets and reach scale morphological features like alternate bars or pools and riffles result in a particle motion not amenable to a description by homogeneous random processes (Hassan et al. 1991).

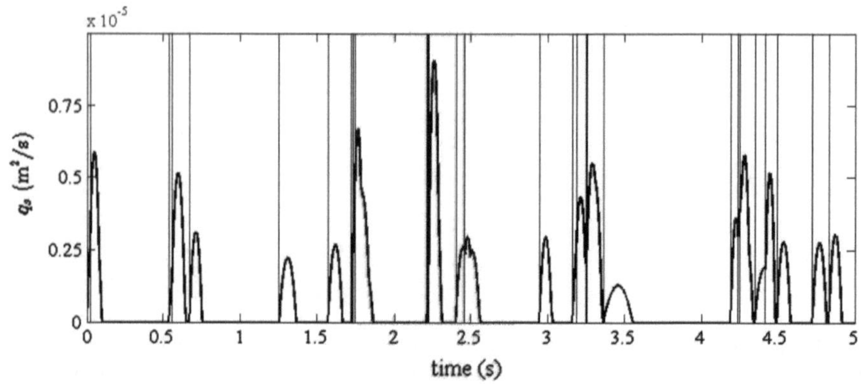

Fig. 13.11 Time series of q_s as directly calculated from definition (13.35). Sample of 5 s of the flow depicted in Fig. 13.10. Vertical thin lines mark the detection instant

Under equilibrium conditions for which flow is steady and nearly-uniform spatial distribution of moving particles (or active particles, Furbish et al. 2012a), ensemble-averaging over a large number of events of instantaneous bedload transport rates, one obtains

$$\bar{q}_s = U^{(g)} H^{(g)} \tag{13.37}$$

where $U^{(g)}$ is the ensemble-averaged grain velocity and $H^{(g)}$ is the volume of moving grains per unit bed area. If the channel is sufficiently wide to be approximated by a rectangular cross-section, equation above is equivalent to $\bar{q}_s = U^{(g)}Ch$ where the grain concentration is the ratio of the volume of granular material and of total volume. If ensemble-averaging is substituted by a time-average of quasi-continuous records of bedload transport rates, such as the series shown in Fig. 13.11, the averaging window has to be chosen with some care in order to obtain a smooth flux $\bar{q}_s$ (Ballio et al. 2014). Averaging windows of the order of magnitude of the mean inter-arrival time result in time-fluctuating averaged values. For instance, averaging the time series of Fig. 13.11 with a window of 1.0 s would produce a much more stable average flux than with a window of 0.25 s.

Still under near-equilibrium conditions, the mean bedload discharge can be expressed by Einstein (1950) formulation, already presented in Sect. 1.4:

$$\bar{q}_s = E\ell_s$$

where, it is recalled, ℓ_s is the mean displacement length, i.e. the mean length travelled by individual grains between the instants of entrainment and of distrainment (the latter defined as the instant for which the grains finds again a stable position in the bed surface) and E is the entrainment rate (volume of granular material per unit bed are per unit time), of which Eq. (13.31) is one particular definition. Equation (13.30) states a conservation equation of the volume of particles under the hypothesis entrainment and deposition rates are in equilibrium.

The volumetric rate of sediment transport in streams can also be expressed as:

$$\bar{q}_s = V_b D_s (1 - \lambda) \tag{13.38}$$

where V_b is the virtual rate of travel velocity ($V_b = \frac{\bar{L}}{\Delta t}$); $\bar{L}$ is the mean travel distance, D_s is the depth of active layer. Hence, correlation of the volumetric transport rate with mean flow characteristics is equivalent to seeking consistent correlations of travel distance, virtual rate of travel velocity, and depth of active layer. Equation (13.38) can be applied to the total or fractional bed material load. Although, Eq. (13.38) is simple, the quantification of these variables proves to be difficult. The above formulation represents a time and space averaged formulation of the discrete approach given by Eq. (13.36).

The difficulty in quantifying the virtual rate of travel velocity has to do with the interment nature of movement of individual particles in streams (see Fig. 13.11). The motion of grains is not continuous, but consists rather of a series of steps and rest periods. As it has been described in the previous section, see Sect. 4.3), the movement of individual stones appears to be a statistically random phenomenon (Einstein 1937). The patterns of particle dispersion in streams are due to irregularities in sediment transport and the local nature of the scour and fill. Factors of irregularities of sediment transport include flow turbulence, wide range of particle size and shape, bed surface arrangements, and channel morphology. These factors are superimposed to cause complicated dispersion with particle sorting. At present, it is difficult to deal with them simultaneously. The quantification of sediment transport using tracers differs with respect the procedure followed in Sect. 4.3 in the sense that the stochastic nature of the flow and thus of the particle entrainment is transferred to the virtual rate of travel velocity. This approach considers the physical consequence of flow on the particle entrainment, i.e. the movement of a particle. When the number of individual tracers increases, this movement, expressed as the travel distance of a particle and integrated according to Eq. (13.36), can also be seen as a part of a statistical distribution of the bedload transport rate for a given flow conditions. In other words, the stochastic nature of the particle location and its entrainment due to the time fluctuations of the turbulent flow is converted to a lumped statistic distribution of the sediment transport rate which is, indeed, the final outcome of the interactions between the water flow and bed surface particles and their distribution.

13.4.2 Step Length, Rest Period and Travel Distance

Einstein (1937) assumed that the step length and the rest period are independent, negative exponential variates. Subsequent work in sand bed river suggested that the step length may be best described by gamma probability function (e.g., Grigg 1970;

Nakagawa and Tsujimoto 1976). The step length and rest period are both stochastic parameters that scattered over wide range of values. Their distribution depends on flow conditions, bed material composition (texture and structure), channel morphology (bars, riffle-pool, step-pool), and bedforms. Bed surface structures and well developed channel morphology (e.g., bars) are likely to trap sediment and result in relatively long rest periods. On the other hand, an increase in the flow strength is likely to increase the step length and reduce the rest period duration. Field evidence shows that the distribution of step length and rest period follows exponential or gamma distributions (Schmidt and Ergenzinger 1992; Habersack 2001; McNamara and Borden 2004). In order to estimate sediment transport rates, Einstein (1950) assumed that the mean step length is about 100 grain diameters. Although, mean step length larger than $100d$ has been observed (Schmidt and Ergenzinger 1992; Habersack 2001), there is a need for more field and experimental research on the topic. As usually the movement period is much shorter than the rest period, the rest period and step length are especially important in order to express the bedload transport rate as a time series.

To solve Eq. (13.38) we need information about the travel distance of particles for a flood event or a time interval. The travel distance of particles, has been the focus of field, experimental and theoretical research. The distribution of travel distance of individual particles for a given number of steps is normally taken as gamma variate (although an exponential distribution may be used for simplicity, Hill et al. 2010). After an interval of time Δt, the travel distribution is a compound Poisson process (Einstein 1937). The field and experimental results of Sayre and Hubbell (1965) confirms Einstein's model for particle travel distance. In gravel bed rivers, studies showed that the gamma probability function and Einstein-Hubbell-Sayre model fit the data reasonably particularly one-peaked relatively short flow events during which the number of steps that particles are likely to preform is relatively low (Hassan et al. 1991; Hassan and Church 1992; Gintz et al. 1996; Wong et al. 2007). Poor performance of the models was observed for multi-peak long floods and complex channel morphologies that include well developed bars (Hassan and Church 1992; Gintz et al. 1996). Studying the influence of bars on the step length of particles, Pyrce and Ashmore (2003a, b, 2005) reported a departure from the gamma distribution which was due to the modulating effect of channel bars and riffle-pool on the travel distance. They demonstrated that the travel distance of particles corresponds to mean spacing of bars and riffle and pool morphologies. Working on steep channels, Lamarre and Roy (2008) demonstrated the strong effect of channel morphology on travel distance.

Recently, the stochastic approach of Einstein has been questioned such as the definition of characteristic time of entrainment and deposition rates, fluctuation in the fluid movement, and abnormal diffusion of particles (e.g., Ancey et al. 2008). For example, using sediment transport data from gravel bed streams, Nikora et al. (2002) suggested three scales (temporal and spatial) with different diffusion regimes to describe the longitudinal and transverse diffusion of bedload particles. The first is the local range (ballistic diffusion) which corresponds to ballistic particles between two successive collisions with static bed grains. The second is the intermediate range (normal or anomalous diffusion) which corresponds to particle trajectory between two successive periods of rest. Finally, the global range (subdiffusion) which is corresponds with particle trajectory that consists with many intermediate range particle trajectories. Most traditional research on the topic has assumed that the distributions of the step length and rest period have a well-defined mean and characteristic amount of variability suggesting that the advection-dispersion equation should describe the movement of sediment in streams (e.g., Bradley et al. 2010; Ganti et al. 2010; Hassan et al. 2013). Furthermore, over a very wide range of scales and when the pattern of particle displacement is controlled by multiplicity of mechanisms, Ganti et al. (2010) argued that particle advection/dispersion may no longer be described by classical advection/dispersion model which is based on constant downstream dispersion. In a variety of settings, fractional advection dispersion equations have been used to model environmental transport over a wide range of scales (e.g., Schumer et al. 2003; Bradley et al. 2010; Foufoula-Georgiou and Stark 2010). Ganti et al. (2010) proposed a theoretical model to study bedload dispersion on the streambed. They restricted their model to describe a set of situations that might lead to anomalous dispersion of tracers. For a mixture of sediment, Ganti et al. (2010) discussed how a heavy-tailed power law density function of travel distances can result from the marginal density of two thin-tailed densities: a conditional exponential density of step travel distance given particle size with a gamma distribution of grain sizes. To explore the applicability of heavy-tailed distributed step lengths, Bradley et al. (2010) reanalyzed the classic data collected by Sayre and Hubbell (1965) for sand bed rivers. Bradley et al. (2010) developed a model similar to the one proposed by Sayre and Hubbell assuming a heavy-tailed distribution of particle step lengths. To improve the performance of their model, they partitioned the tracers into detectable mobile phase and undetectable immobile phase. They concluded that fractional-order models match the observed plume shape and growth rates better than classical step length models. These results contrast Nikora et al. (2002) suggestion that anomalous super-diffusive exists as a preasymptotic local range within which the correlated particle motion dominates transport. Flume experiments conducted by Martin et al. (2012) yielded super-diffusive step length distributions that persisted after sediment was well mixed in the flume. On the other hand, Furbish et al. (2012b) suggested that the anomalous particle movement behavior may be the result of periodicities in the particle motion rather than super-diffusive step length. Recently, Roseberry et al. (2012) showed that the step length distribution is more exponential like than heavy tailed.

13.4.3 Mean Travel Distance and Virtual Travel Velocity

One way to estimate travel distance of each fraction is to characterize the relative mobility of grains is to quantify the scaled average distance moved by tracers in each size class during a single flood as a function of grain size (Church and Hassan 1992; Hassan and Church 1992). For example, the scaled travel distance was found to decrease with the increasing grain size for coarse grains (Church and Hassan 1992; Hassan and Church 1992; Ferguson and Wathen 1998; Ferguson et al. 2002; Haschenburger 2013). Wilcock (1997a) confirmed the relation between the scaled travel distance and particle size but asserted that such relation applies for the condition when material finer than the median size of the bed material is fully mobile while the coarser fractions are within the state of partial mobility. Recently, MacVicar and Roy (2011) showed that a power law function describes the decline in the travel distance with particle size for free surface particles in forced riffle-pool morphology of the Mores Creek.

The relation between the V_b and the flow properties is another practical relation for the estimation of the sediment transport rate using the dispersion parameters of the tracer particles. The concept was introduced by Einstein in 1937; Stelczer (1981) considered the virtual travel velocity of particle as a great contribution to the description of the process of bedload transport in streams. Sayre and Hubbell (1965) used the concept to estimate sediment transport rates in a sand bed river and flume experiments. However, Sayre and Hubbell (1965) show a wide scatter in the V_b even under the same flow conditions. This is likely due to the increase in the number of stationary particles because of the increase in the buried proportion of the tracer particles (Ferguson and Wathen 1998; Bradley et al. 2010; Haschenburger 2011b). Relating both the virtual travel velocity and travel distance to stream power calculated for the maximum peak flow or the total stream power above the threshold for movement yielded scatter data with weak correlation (Hassan et al. 1992; Haschenburger and Church 1998; Schneider et al. 2014). One of the main controls on the virtual velocity is the burial depth which is likely to slow down the downstream dispersion of tracer particles (e.g., Ferguson and Wathen 1998; Ferguson et al. 2002). Burial depth is likely to determine the timing and the duration of movement of particles within the active layer, deeply buried particles are likely to move later during the event and for shorter duration (Hassan and Church 1994; DeVries et al. 2001; Wong et al. 2007).

An implementation of Eq. (13.38) requires some knowledge of the depth of the active layer and particle. The depth of the active layer has been inferred from the probability distribution of depth of burial of tracers as well as from direct measurements of scour and fill over the event scale (Hassan 1990; Hassan and Church 1994; Wilcock 1997b; Haschenburger 1999, 2013; DeVries 2002). The distribution of burial depth of coarse bed material tends to follow an exponential distribution for a single-peaked flood (Hassan and Church 1994). Furthermore, Hassan and Church (1994) showed that a single event can bury the tracer particles up to five times the median size of the bed material.

13.4.4 Thickness of Active Layer

Information on the thickness of the mobile bedload layer with characteristics of particle displacement and dispersion enable the estimation of the volume of bedload transport during a flood event (Hassan and Ergenzinger 2003; Haschenburger and Church 1998; Houbrechts et al. 2012; Laronne et al. 1992; Sear (1996); Wong et al. 2007). The thickness of the layer influence sediment transport model mass balancing and prediction of degradation and aggradation in streams (DeVries 2002). Wilcock et al. (1996) related the thickness of the active layer to the local bed shear stress and the d_{90} of the bed material. They reported that particle exchange associated with bedload transport occurred to a thickness of $1.7d_{90}$ at a dimensionless shear stress of 0.035. Field and experimental research linked the thickness of the active layer to the flow strength in streams (Hassan et al. 1992; Wilcock and McArdell 1997; Haschenburger and Church 1998; Wong et al. 2007; Houbrechts et al. 2012; Schneider et al. 2014). Houbrechts et al. (2012), for example, showed that the thickness of the active layer positively correlated with the specific stream power estimated for the peak flow of the mobilizing event. However, limited field evidence suggests a value of $2d_{90}$ as an upper limit of the active layer (Wilcock et al. 1996; DeVries 2002; Haschenburger 2011a, 2012). Based on the probability distribution associated with bed elevation fluctuations and structure functions for variation in the entrainment and deposition rates with depth below the mean active, Parker et al. (2000) developed vertically continuous version of the Exner equation of sediment continuity for multiple grain sizes.

13.4.5 Vertical Mixing of Particles

Marked tracer particles offer a unique approach for studying of the dispersion of sediment within the active bed layer, bed surface composition and evolution. As part of the fluvial sediment dispersion process, part of the surface particles vertically exchange to the subsurface while a proportion of the subsurface material become exposed on the surface due to scour and fill. The vertical exchange within the active layer involves both active mobile gravels and passive stationary stones (Hassan 1990; Haschenburger 2011a, 2012). The vertical mixing of sediment and burial depth of tracer particles is likely to be accomplished in most cases by random scour and fill in gravel bed streams. This is not the case in sand bed stream where bedforms such dunes and antidunes dominate the channel morphology. Vertical sorting in the dune regime has been shown to be too complex to explain in terms of active layer (Ribberink 1987).

As mentioned before, the depth of the active layer depends on flow magnitude and duration, bed surface texture and structure (e.g., armouring) and channel morphology (Hassan and Church 1994; Haschenburger 1999; DeVries 2002; Wong et al. 2007). Particle entrainment, waiting time, and travel distance depend on the relative location of the particles within the active layer; deeply buried particles are likely to be entrained late in the flood, move short distances and have a relatively long waiting time.

At the event scale, the burial depth of individual particle varies with no simple relation to particle size (Hassan 1990; Hassan and Church 1994; Ferguson and

Wathen 1998; Ferguson et al. 2002; Gottesfeld et al. 2004; Hassan et al. 2008). Single-peaked and short flow events yielded exponential distribution of burial depth of the individual particles, whereas, multi-peak and long events resulted in uniform distributions (Hassan and Church 1994). Observing the movement of gravels over sand bed stream, Hassan et al. (1999) reported that the gamma distribution fit the burial distribution of individual particles. This could be due to the dunes that dominated the bed channel during floods (Carling et al. 2000). Using on flume experiments, Wong et al. (2007) reported an increase in the depth of exchange and bed mobility as the flow magnitude or duration increases; an outcome supported by field observations (e.g., Schick et al. 1987; Hassan 1990; Haschenburger 1999). Finally, using a long-term data set that span 17 years, Haschenburger (2011b, 2013), evaluated the depth of burial and vertical mixing over a long flood series. The vertical mixing of particles through the monitoring reaches was very rapid; within one to eight events, the mean burial depth reached two times the thickness of the bed surface layer (Haschenburger 2011a). A consistent pattern of particle exchange was achieved after the tracers particles become well mixed within the sediment exchange layer (Haschenburger 2011a, 2012).

Few vertical mixing models have been developed for both sand and gravel bed streams. Crickmore and Lean (1962) developed a two layer vertical mixing model assuming that the upper "half layer" on the zone of motion moves more frequently than particles in the lower half. They assumed that when motion occurs, there is a complete mixing between the layers. Full mixing and uniform vertical distribution of particles within the active layer are the main assumptions in Sayre and Hubbell (1965) model. In gravel bed rivers, Schick et al. (1987) developed a two layer model, surface and subsurface, for mixing of coarse bedload. The time for equilibrium exchange depends largely on the level of armouring and the magnitude of the mobilizing events (Schick et al. 1987). Using the extensive Carnation Creek and Allt Dubhaig data sets, Haschenburger (2012) successfully applied Schick et al. (1987) two layers model. The model that Hassan and Church (1994) developed consists of ten layers in which part of the particles move and change their vertical location while the rest remains in the same layer. The model is based on the assumption that both particle entrainment and travel are exponentially distributed. The model yielded an exponential decline in the vertical distribution for small events and well armoured surfaces. After many events, the model yielded a uniform distribution. Based on extensive data set collected from the Allt Dubhaig, Scotland, Ferguson and Hoey (2002) developed Markov model of exchanges between three layers of sediment. An equilibrium of exchange was achieved between three and five years for the study creek.

13.4.6 Characteristics of Sediment Transport at Reach-Scale

Detailed descriptions of the spatial and temporal variations in sediment transport are prerequisite information for understanding the sediment transport physics and for the development of predictive models. Short-term variations in the sediment transport rate have been documented in the field. At the single event scale, the sediment transport rate on the rising limb of the hydrograph is often higher than that on the falling limb (clockwise hysteresis). This variation is attributed to changes in sediment supplied and within channel storage between floods which is mobilized as soon as flows increase. A counter clockwise hysteresis in the sediment transport rate is typical of well-packed and armoured stream bed where little sediment moves until the armoured layer is disrupted. After the disruption of the armour layer, usually at peak flow, relatively large amounts of sediment are transported during the falling limb of the hydrograph (Milhous 1973; Reid et al. 1985). However, due to temporal changes in sediment storage and supply, both types of hysteresis could be observed in the same stream (Gomi and Sidle 2003). In addition to sediment supply and the partial destruction of armouring, it has been suggested that short term variations result from the differential pattern of scour and fill in relation to channel morphology, the fluctuations in flow velocity, the development and destruction of small scale bed forms such as pebble clusters, stone cells, and micro-scale eddies developing around large obstacles (Reid et al. 1985; Hassan and Reid 1990), and episodic sediment supply (Benda and Dunne 1997; Church et al. 1998).

At a longer time scale, variation in the sediment transport is associated with sediment supply from slopes and banks and sediment storage in bars and pools. For example, Dietrich et al. (1989) and Ryan (2001) linked the temporal variations in sediment transport to the sediment supply from outside the channel system. Inputs from mass movement increase the sediment supply to channels and hence dictate the sediment transport in streams. Large inputs of sediment from mass movement events may have a long-term influence on the sediment transport rates and texture and may result in complex channel morphology. Deposition of large inputs of sediment from mass movement events may create a wedge of sediment that will persist for decades to centuries and might result in the movement of sediment in pluses or waves along the channel (e.g., Benda 1990; Roberts and Church 1986; Miller and Benda 2000; Benda and Dunne 1997).

Based on flume and field measurements, three stages of sediment transport have been identified: (1) phase I consists of overpassing sand, (2) phase II occurs during the partial mobility of sediment, and (3) phase III occurs during the full mobility conditions (e.g., Jackson and Beschta 1982; Andrews 1983; Ashworth and Ferguson 1989; Wilcock and McArdell 1993, 1997; Zimmerman and Church 2001). In phase I fine sediment (mainly sand) move over static bed (no local scouring); phase II occurs during high flows when coarser material is locally entrained; and phase III occurs during the heights flows under conditions of full mobility of all sizes on the bed surface. In natural streams, phase II is the most dominant stage for the mobilization of local coarse sediment while phase III occurs during relatively extreme events. Therefore, phases II and II are likely to influence the channel morphology. However, the relative importance of the three phases largely depends on the degree of bed surface armouring/structuring and the sediment supply (amount and texture) into the channel. Working in St. Louis Creek, a boulder-cobble stream, Ryan and Troendle (1996) related the sediment transport phase to the flow conditions relative to the bankfull discharge. They reported that phase I was in effect at half bankfull discharge, and phase II was observed between 70 % and 100 % bankfull discharge. Since they did not observed phase III in study, we speculate that this phase is associated with a flow beyond bankfull discharge.

13.4.7 Bedload Capacity Formulas

A large quantity of bedload capacity formulas have been proposed since the first "modern" effort of DuBoys, in 1879 (review in Mavies et al. 1935). Its derivation entails pure theoretical arguments (e.g., Einstein 1950), best fit analysis between normalised sediment transport rates, measured in the field or in laboratory flumes, and non-dimensional parameters deduced by dimensional analysis (e.g., Ackers and White 1973) or a combination of both (van Rijn 1984). In any case, the parameters of the formulas must be calibrated with experimental data and validated with independent data.

Below is a non-exhaustive set of topics for discussion on existing bedload capacity predictors.

- In general, capacity formulas incorporate average hydraulic conditions and express mean transport rates or concentrations. However, sediment transport is highly intermittent in time and heterogeneous in space. Hence, mean values from collected samples may not represent the actual transport rate due to scale issues (Ballio et al. 2014); data collection should take place over a time interval larger than the characteristic time scale of the bedload fluctuations. In field studies, pit traps (Milhous 1973) provide adequate information while in the laboratory image analysis of video-footages (Roseberry et al. 2012) may represent an alternative.
- Bedload formulas can be classified as probabilistic (e.g., Einstein 1950) or deterministic (e.g., Bagnold 1956). Some explicitly include physics of bedload transport at grain-scale. For instance, classic probabilistic formulas of Einstein (1950) or Paintal (1971b) embed upscaling arguments to express reach scale motion based on criteria for single particle entrainment. Other formulas acknowledge that particles move preferentially when subjected to high-velocity turbulent bursts (Nelson et al. 1995; Cao 1997; Ferreira et al. 2009). Such flow structures must penetrate the laminar sub-layer (for low values of X, Gyr and Schmid 1997) or deep into the region between troughs and crests of roughness elements (Lacey and Roy 2007; Ferreira et al. 2010). Yet, some classic deterministic formulas include physics only indirectly in the correlation between mean flow non-dimensional flow parameters and mean non-dimensional transport rates (Meyer-Peter and Müller 1948; Brownlie 1981). Attempts to transform these predictors into stochastic formulas for poorly sorted sediment mixtures have been undertaken (Kleinhans and van Rijn 2002).

- Most capacity formulas are developed and calibrated from specific conditions of shear stress, discharge, bed conditions. The basic rule is to use the one whose conditions best approximate those of the problem at hand. Note that river, but most have been calibrated using the same data set (e.g., Oak Creek–Parker and Klingman 1982; Wilcock and Crowe 2003; Parker 2008) and many have been validated with laboratory data only (e.g., Wu and Yang 2004; McEwan et al. 2004; Ferreira et al. 2009).

Two examples of capacity bedload predictors are presented next, as examples of probabilistic (Sun and Donahue 2000) and deterministic (Wilcock and Crowe 2003) predictors.

Closely following Einstein (1950), Sun and Donahue (2000) employed probability concepts, mechanical formulations and experimental data to develop and validate a bedload transport capacity function for poorly sorted sediment mixtures.

The exchange of sediment particles between resting and active states is modelled with a continuous-time Markov process. The dimensionless particle average velocity and duration of single-step motion are functions of the Shields parameter for each size fraction. The probability of incipient motion for *k*th grain-size fraction, α_k, was derived considering random properties of forces and moments acting on a particle of that grain-size. They obtained:

$$\alpha_k = 1 - \frac{1}{\sqrt{2\pi}} \int_{-2.7\left(\sqrt{0.0822\psi_k}+1\right)}^{2.7\left(\sqrt{0.0822\psi_k}-1\right)} e^{-0.5x^2}\,\mathrm{d}x \tag{13.39}$$

Flow intensity is defined as the reciprocal of the grain Shields parameter, corrected by a shelter-exposure coefficient:

$$\psi_k = \frac{\rho^{(w)}(s-1)gd_k}{\tau_0'}\left(\frac{\widehat{d}}{d_k}\right)^{0.5}\sigma_g^{-0.25}$$

where $\tau_0' = \rho^{(w)} g R_h' J$ is the bed shear stress associated to grain roughness, R_h' is the hydraulic radius associated to grain roughness, σ_g is the geometric standard deviation (see, e.g., Garde and Ranga Raju 1985) and $\widehat{d}$ is the mean diameter. Including the estimates for particle average velocity, duration of single-step motion and probability of entrainment [Eq. (13.39)] in the probabilistic equation of mass flux Sun and Donahue (2000) obtained the dimensionless bedload capacity formula for the *k*th grain size:

$$\phi_{s_k} = 0.3 f_k \frac{\alpha_k}{\psi_k^{3/4}(1-\alpha_k)} \tag{13.40}$$

where f_k is the fraction of the grain size in the bed (although it is not clear if the authors meant the substrate or the surface). The key disadvantage of this model is that the transitional probabilities are only suitable for discrete-time Markov process. Validation of this formula took place with total transport data, not fractional bed-load rates.

Wilcock (1998) and Wilcock and Kenworthy (2002) introduced the role played by the fraction of sand in the bed surface enhancing bedload transport of a mixture of composed of two fractions of sand and gravel. These previous works were extended to poorly-sorted mixtures of a finite number N of grain classes by Wilcock and Crowe (2003). They proposed a surface-based bedload approach (Parker 1990; Parker 2008) in which the fraction of sand on the surface is explicitly included. The "all or nothing" drawbacks of considering a critical shear stress for a given grain size d_k are overcome by considering a reference dimensionless shear stress Y_{rk} associated to a very low dimensionless sediment transport ($W^* = 0.002$). The formulation proposed is expressed in terms of dimensionless fraction sediment transport rate for the kth grain size W_k^* :

$$W_k^* = \frac{(s-1)gq_{sk}}{F_k u_*^3} \tag{13.41}$$

where F_k is the surface fraction of the kth fraction and q_{sk} is the volumetric sediment transport rate per unit channel width for the ith grain size. Note that q_{sk} drops to zero when $F_k = 0$, i.e. when a particular grain size is not present on the bed surface. The total sediment transport q_s is obtained after summing up the partial contributions q_{sk} of all grain sizes in the mixture.

The dimensionless sediment transport rate W_k^* is obtained by:

$$W_k^* = \begin{cases} 0.002\vartheta^{7.5}, & \vartheta < 1.35 \\ 14\left(1 - \frac{0.894}{\vartheta^{0.5}}\right)^{4.5}, & \vartheta \geq 1.35 \end{cases} \tag{13.42}$$

where ϕ is defined as the hiding/exposure function such that

$$\vartheta = \frac{Y_{sg}}{Y_{ssrg}} \left(\frac{d_k}{d_g}\right)^{-b} \tag{13.43}$$

where d_g is the geometric mean size of the mixture, Y_{sg} is the dimensionless shear stress (Shields number) associated to d_g and Y_{ssrg} is the reference dimensionless shear stress associated to d_g which is obtained by

$$Y_{ssrg} = 0.021 + 0.015\, e^{-20F_s} \tag{13.44}$$

where F_s is the fraction of the sand in the surface. Lastly, the exponent b is computed by

$$b = \frac{0.69}{1 + e^{\left(1.5 - \frac{d_k}{d_g}\right)}}$$

Given a flow condition (characterized by the boundary shear stress τ_0 or a surrogate as u_*), and the grain size distribution of the mixture (represented by the surface fractions F_k, its geometric mean size d_g and the sand fraction F_s), the reference shear stress of the mixture is calculated by (13.44). After evaluating the exponent b (which varies for each grain size), the hiding/exposure function is calculated by Eq. (13.43). This result is plugged into (13.42) to obtain W_k^* and finally the fractional transport rate is obtained by solving (13.41) for q_{sk}.

The total sediment transport rate for both above predictors is

$$q_s = \sum_{k=1}^{N} q_{sk} \tag{13.45}$$

13.5 Measuring Sediment Related Quantities

Reliable measurements of sediment transport in streams, especially bed material load, are one of the hurdles in the development and the calibrations of transport models. The focus of the short review here is on bed load because of the relative importance of this mode of transport to channel stability and morphology and difficulty in obtaining reliable field measurements. Furthermore, bed load occurs under relatively high flow conditions and for a relatively short period of time and therefore difficult to measure. Wide range of methods has been used to measure bed load including sampling devices, pit traps, or magnetic detection devices, and observe weir ponds and tracers. This is particularly true in steep cobble-boulder channels in which the operation of bed load sampling devices is difficult.

Bed load sampling devices is the most common used method to estimate sediment transport rate and texture in streams. Over the last 100 year, a large number of sediment transport samplers have been developed. The Helley-Smith sampler is the most commonly used, however, it was designed for use in coarse sand and granule-gravel beds (Helley and Smith 1971) and is known to produce biased results in coarse materials (Sterling and Church 2002). Sediment transport samplers provide site specific or cross-sectional averages of transport rate and sediment texture. To represent spatial and temporal variations and in order to derive a reasonable estimates in sediment transport, a large number of samples is needed, which is labour-intensive and expensive task (e.g., Emmett 1980; Hubbell et al. 1985; Ryan and Emmett 2002; Gray et al. 2010). Although the Helley-Smith sampler has been widely used, especially in the US, there is little agreement as to what is considered to be a valid sampling scheme in any fluvial system (Ryan and Troendle 1996).

The best sampler for obtaining true sediment transport rates in gravel bed rivers is unknown.

Weir ponds and pit traps have been deployed successfully in small streams wide range of environments (e.g., Milhous 1973; Hayward 1980; Reid et al. 1980; Sawada et al. 1983, 1985; Kuhnle 1992; Reid and Laronne 1995). Traps are assumed to yield relatively good estimates of sediment transport rate and event bed load transport (Reid et al. 1980; Hassan and Church 2001). Weir ponds provide volumetric total bed material transport between surveys and particle size distribution of the integrated sample (e.g., Ryan and Troendle 1997; Gomi and Sidle 2003). Data of this nature provide a record of long-term sediment yield but the information is limited in temporal detail (Troendle et al. 1996).

Surrogate methods (e.g., active and passive sensors) for the estimation of sediment transport have been the focus of much research in the last few years. A comprehensive review of the surrogate methods can be found in Gray et al. (2010). The active sensors include acoustic Doppler current profilers (ADCPs), sonar, radar, and smart sensors. Passive sensors include geophones (pipes or plates) in direct contact with the streambed, hydrophones deployed in the water column, impact columns, and magnetic detection. The most developed ADCP for sand and geophones for gravel (Gray et al. 2010). However, most of these methods require both laboratory and field calibration.

Tracers can help solve some of the difficulties of obtaining reliable estimates of sediment transport using sampling devices. Hassan and Ergenzinger (2003) provide a comprehensive review of the available techniques. Passive and active tracers can be used to estimate volumetric sediment transport, three-dimensional dispersion of sediment, and flow competence, but their deployment and recovery is labour-intensive. Using particle kinematics, Wong et al. (2007) (see also Wilcock 1997a) detailed three methods to estimate sediment transport using tracers. In a comprehensive study Houbrechts et al. (2012) compared sediment transport estimates calculated using tracers data with data from Helley-Smith and dredged material show that the tracers' method yielded poor results. Therefore, practical relation for the estimation of sediment transport using individual grain movement characteristics is still an open question.

13.6 Summary and Recapitulation

This chapter summarizes some of the most important issues with regards the transport of sediment as bedload in fluvial systems. The theoretical functional relations describing both the initiation of motion and the sediment transport are introduced. These relations depend on the basic variables that characterize the mean flow properties and the sediment parameters. With regards the classical problem of the initiation of motion of particles, it has been treated both at a grain level (accounting for a force and moment balances acting on a grain) and from channel perspective. The first approach represents a particular version of the aforementioned functional relations. The second approach introduces the resulting expressions for the initiation of motion at a grain level and integrates them at a reach scale accounting for the stochastic nature of flow. This perspective results in a strategy to analyse the onset of motion at river reach scale that encompasses the deterministic response of a particle to be entrained or stay at rest depending on the forces and moments balances acting on it with the nature of the flow, which in turn is the cause of the particle movements. This procedure has been applied to a set of laboratory experiments. Finally, another way of quantifying the sediment transport rate in rivers is presented. This approach consists of tracking the travel distances of set of particles (tracers) on the river bed. By doing this sediment transport rate that, in the end, results from the interactions between the flow and the particles on the river bed surface, the nature of which has demonstrated to have an important stochastic behaviour, is converted to a lumped statistic distribution. Finally, some field and laboratory techniques for measuring sediment transport, accounting for its inherent fluctuations are introduced.

Acknowledgments First author acknowledges that this work was partially funded by FEDER, program COMPETE, and by national funds through Portuguese Foundation for Science and Technology (FCT) project RECI/ECM-HID/0371/2012.

References

Ackers P, White WR (1973) Sediment transport: a new approach and analysis. J Hydraul Div 99 (HY11):2041–2060

Aleixo R, Maia R (2006) The beginning of sediment transport—a different approach. In: Ferreira RML, Alves ECTL, Leal JGAB, Cardoso AH (eds) River flow 2006. Balkema, Lisse, pp 553–563

Ancey C, Davison AC, Böhm T, Jodeau M, Frey P (2008) Entrainment and motion of coarse particles in a shallow water stream down a steep slope. J Fluid Mech 595:83–114

Andrews ED (1983) Entrainment of gravel from naturally sorted riverbed materials. Geol Soc Am Bull 94:1225–1231

Andrews ED, Erman DC (1986) Persistence in the size distribution of surficial bed material during an extreme snowmelt flood. Water Resour Res 22:191–197

Ashida K, Michiue M (1972) Study on hydraulic resistance and bedload transport rate in alluvial streams. Trans Jpn Soc Civ Eng 206:59–69

Ashworth PJ, Ferguson RI (1989) Size-selective entrainment of bed-load in gravel bed streams. Water Resour Res 25:627–634

Bagnold RA (1954) Experiments on a gravity-free dispersion of large solid spheres in a Newtonian fluid under shear. Proc R Soc Lond A 225:49–63

Bagnold, R. A. (1956). The flow of cohesionless grains in fluids. Phil. Trans. Roy. Soc. London, A, 249: 964

Ballio F, Nikora V, Coleman SE (2014) On the definition of solid discharge in hydro-environment research and applications. J Hydraul Res 52(2):173–184

Benda L (1990) The influence of debris flows on channels and valley floors in the Oregon Coast Range, U.S.A. Earth Surf Proc Land 15:457–466

Benda L, Dunne T (1997) Stochastic forcing of sediment routing and storage in channel networks. Water Resour Res 33(12):2849–2863

Bettess R (1984) Initiation of sediment transport in gravel streams. Proc Instn Civ Eng 77(Part 2, Technical Note 407):79–88
Binggeli B (1980) On the intrinsic shape of elliptical galaxies. Astron Astrophys 82(3):289–294
Bogen J (1980) The hysteresis effect of sediment transport systems. Norsk Geogr Tidsskr 34:45–54
Bradley DN, Tucker GE, Benson DA (2010) Fractional dispersion in a sand bed river. J Geophys Res 115:F00A09. doi:10.1029/2009JF001268
Bradley WC, Mears AI (1980) Calculations of flows needed to transport coarse fraction of boulder creek alluvium at boulder, Colorado—summary. Geolog Soc Am Bull 91:135–138
Bridge JS, Bennet SJ (1992) A model for the entrainment and transport of sediment grains of mixed sizes, shapes and densities. Water Resour Res 28(2):337–363
Brownlie WR (1981) Prediction of flow depth and sediment discharge in open channels. W.M. Keck Laboratory of Hydraulics and Water Resources, Report no. KH-R-43A, California Institute of Technology, Pasadena, California
Buffington JM (1999) The legend of shields. J Hydraul Eng 125(4):376–387
Buffington JM, Montgomery DR (1997) A systematic analysis of eight decades of incipient motion studies with special reference to gravel bedded rivers. Water Resour Res 33(8):1993–2029
Buffington JM, Montgomery DR (1999) Effects of hydraulic roughness on surface textures of gravel-bed rivers. Water Resour Res 35. doi:10.1029/1999WR900138
Buffinton JM, Dietrich WE, Kirchner JW (1992) Friction angle measurements on a naturally formed gravel strembed. Implications for critical boundary shear stress. Water Resour Res 28 (2):411–425
Cao Z (1997) Turbulent bursting-based sediment entrainment functions. J Hydraul Eng 123 (3):233–236. doi:10.1061/(ASCE)0733-9429(1997)123:3(233)
Carling PA (1983) Threshold of coarse sediment transport in broad and narrow natural streams. Earth Surf Proc Land 8(1):1–18
Carling PA, Kelsey A, Glaister MS (1992) Effect of bed roughness, particle shape and orientation on initial motion criteria. In: Billy P, Day RD, Thorne CR, Taconi P (eds) Dinamics of gravel-bed river. Wiley, New York, pp 23–37
Carling PA, Golz E, Orr HG, Radecki-Pawlik A (2000) The morphodynamics of fluvial sand dunes in the River Rhine near Mainz, Germany. Sedimentol Morphol Sedimentol 47:227–252. doi:10.1046/j.1365-3091.2000.00290.x
Cheng N-S, Chiew Y-M (1998) Pickup probability for sediment entrainment. J Hydraul Eng 124 (2):232–235. doi:10.1061/(ASCE)0733-9429(1998)124:2(232)
Chepil WS (1959) Equilibrium of soil grains at the threshold of movement by wind. Soil Sci Soc Am J 23(6):422–428
Chepil WS (1961) The use of spheres to measure lift and drag on wind-eroded soil grains. Proc Soil Sci Soc 25:343–345
Church M (2006) Bed material transport and the morphology of alluvial river channels. Annu Rev Earth Planet Sci 34:325–354
Church M, Hassan MA (1992) Size and distance of travel of unconstrained clasts on a streambed. Water Resour Res 28(1):299–303
Church M, Hassan MA (2002) Mobility of bed material in Harris Creek. Water Resour Res 38:19-1–19-12. doi:10.1029/2001WR000753
Church M (1978) Palaeohydraulic reconstructions from a Holocene valley fill. In: Miall AD (ed) Fluvial sedimentology. Canadian Society of Petroleum Geologists, Mem. 5, Calgary, Canada, pp 743–772
Church M, Hassan MA, Wolcott JF (1998) Stabilizing self-organized structures in gravel-bed stream channels: Field and experimental observations. Water Resour Res 34(11):3169–3179
Crickmore MJ, Lean GH (1962) The measurement of sand transport by means of radioactive tracers. Proc R Soc London Ser A 266:402–421
Curran JC, Wilcock PR (2005) Characteristic dimensions of the step-pool bed configuration: An experimental study. Water Resour Res 41:W02030. doi:10.1029/2004WR003568

DeVries PE, Burges SJ, Daigneau J (2001) Measurements of the temporal progression of scour in a pool-riffle sequence in a gravel bed stream using an electronic scour monitor. Water Resour Res 37:2805–2816

DeVries P (2002) Bedload layer thickness and disturbance depth in gravel bed streams. J Hydraul Eng 128:983–991

Dey S, Papanicolaou AN (2008) Sediment threshold under stream flow: a state-of-the-art review. Water Eng KSCE J Civ Eng 12:45–60

Dietrich WE, Kirchner JW, Ikeda H, Iseya F (1989) Sediment supply and the development of the coarse surface layer in gravel-bedded rivers. Nature 340:215–217

Drake TG, Shreve RL, Dietrich WE, Whiting PJ, Leopold LB (1988) Bed load transport of fine gravel observed by motion-picture photography. J Fluid Mech 192:193–217

Egiazaroff IV (1965) Calculation of non-uniform sediment concentrations. J Hydraul Eng 91 (4):225–248

Einstein HA (1950) The bed-load function for sediment transportation in open channel flows. Soil Conserv. Serv. Tech. Bull. 1026, US Department of Agriculture, 73 pp

Einstein HA (1937) Bedload transport as a probability problem. Ph.D. thesis, Colorado State University (In Sedimentation, Shen, H.W. (ed.) 1972, App. C)

Emmett, W.W. (1980). A field calibration of the sediment-trapping characteristics of the Helley-Smith bedload sampler: U. S. geological survey. Professional paper 1139, 44 p

Ettema R, Mutel CF (2004) Hans Albert Einstein: innovation and compromise in formulating sediment transport by rivers. J Hydraul Eng 130(6):477–487

Fahnestock RK (1963) Morphology and hydrology of a glacial stream—White River, Mount Rainier, Washington. U.S. geological survey professional paper 422A, 70 pp

Feller W (1968) An introduction to probability theory and its applications, vol 1, 3rd edn. Willey, New York

Fenton JD, Abbott JE (1977) Initial movement of grains in a stream bed: the effects of relative protrusion. Proc R Soc A 352:523–527

Ferguson RI, Hoey TB (2002) Long-term slowdown of river tracer pebbles: generic models and implications for interpreting short-term tracer studies. Water Resour Res 38(8):17-1–17-11

Ferguson RI (1994) Critical discharge for entrainment of poorly sorted gravel. Earth Surf Proc Land 19:179–186

Ferguson RI, Wathen SJ (1998) Tracer-pebble movement along a concave river profile: virtual velocity in relation to grain size and shear stress. Water Resour Res 34:2031–2038

Ferguson RI, Bloomer DJ, Hoey TB, Werritty A (2002) Mobility of river tracer pebbles over different timescales. Water Resour Res 38:1045. doi:10.1029/2001WR000254

Fernandez Luque R, van Beek R (1976) Erosion and transport of bed-load sediment. J Hydraul Res 14(2):127–144

Ferreira RML (2005) River morphodynamics and sediment transport. Conceptual model and solutions. PhD thesis, IST, TULisbon

Ferreira RML, Franca MJ, Leal JGAB, Cardoso AH (2009) Mathematical modelling of shallow flows: closure models drawn from grain-scale mechanics of sediment transport and flow hydrodynamics. Can J Civ Eng 36(10):1605–1621. doi:10.1139/L09-033

Ferreira RML, Ferreira L, Ricardo AM, Franca MJ (2010) Impacts of sand transport on flow variables and dissolved oxygen in gravel-bed streams suitable for salmonid spawning. River Res Appl 26(4):414–438. doi:10.1002/rra.1307

Ferreira RML, Franca MJ, Leal JGAB, Cardoso AH (2012) Flow over rough mobile beds: Friction factor and vertical distribution of the longitudinal mean velocity. Water Resour Res 48(5): W05529, 1944–7973. doi:10.1029/2011WR011126

Foufoula-Georgiou E, Stark C (2010) Introduction to special section on stochastic transport and emergent scaling on earth's surface: rethinking geomorphic transport—stochastic theories, broad scales of motion and nonlocality. J Geophys Res 115. doi:10.1029/2010JF001661

Furbish DJ, Haff PK, Roseberry JC, Schmeeckle MW (2012a) A probabilistic description of the bed load sediment flux: 1. Theory. J of Geophys Res 117:F03031. doi:10.1029/2012JF002352

Furbish DJ, Ball AE, Schmeeckle MW (2012b) A probabilistic description of the bed load sediment flux: 4. Fickian diffusion at low transport rates. J Geophys Res Earth Surf 117

Ganti V, Meerschaert MM, Foufoula-Georgiou E, Viparelli E, Parker G (2010) Normal and anomalous diffusion of gravel tracer particles in rivers. J Geophys Res Earth Surf 115:F00A12. Doi:http://dx.doi.org/10.1029/2008JF001222

Garde RJ, Ranga Raju KG (1985) Mechanics of sediment transportation and alluvial stream problems, 2nd edn. Willey Eastern Limited, New Delhi

Gessler J (1965) The beginning of bedload movement of mixtures investigated as natural armouring in channels. Technical report no. 69, The Laboratory of Hydraulic Research and Soil Mechanics, Swiss Federal Institute of Technology, Zurich, (translation by W. M. Keck Laboratory of Hydraulics and Water Resources, California Inst. Technology)

Gintz D, Hassan MA, Schmidt K-H (1996) Frequency and magnitude of bedload transport in a mountain river. Earth Surf Proc Land 21:433–446

Göğüş M, Defne Z (2005) Effect of shape in incipient motion of large solitary particles. J Hydraul Eng 131(1):38–45

Gomez B, Church M (1989) An assessment of bed load sediment transport formulae for gravel bed rivers. Water Resour Res 25:1161–1186

Gomi T, Sidle RC (2003) Bed load transport in managed steep-gradient headwater streams of Southeastern Alaska. Water Resour Res 39, doi:10.1029/2003WR002440

Gottesfeld AS, Hassan MA, Tunnicliffe JF, Poirier RW (2004) Sediment dispersion in fish bearing streams: the influence of floods and spawning salmon. J Am Water Resour Assoc 40: 1071–1086

Graf WH (1971) Hydraulics of sediment transport. McGraw_Hill Book Co., New York

Gray JR, Laronne JB, Marr JDG (2010) Bedload-surrogate monitoring technologies. U.S. geological survey scientific investigations report 2010–5091

(Am Soc Civ Eng Proc)

Gyr A, Schmid A (1997) Turbulent flows over smooth erodible sand beds in flumes. J Hydraul Res 35(4):525–544

Habersack HM (2001) Radio-tracking gravel particles in a large braided river in New Zealand: a field test of the stochastic theory of bed load transport proposed by Einstein. Hydrol Process 15:377–391

Haschenburger JK (2012) On gravel exchange in natural channels. In: Church M, Biron PA, Roy A (eds) Gravel-bed river: processes, tools, environments. Wiley, New York, pp 56–67

Haschenburger JK, Church M (1998) Bed material transport estimated from the virtual velocity of sediment. Earth Surf Proc Land 23:791–808

Haschenburger JK (1999) A probability model of scour and fill depths in gravel-bed channels. Water Resour Res 35:2857–2869

Haschenburger JK (2011a) The rate of fluvial gravel dispersion. Geophys Res Lett 38(L24403). doi:10.1029/2011GL049928

Haschenburger JK (2011b) Vertical mixing of gravel over a long flood series. Earth Surf Proc Land 36:1044–1058

Haschenburger JK (2013) Tracing river gravels: insights into dispersion from a long-term field experiment. Geomorphology 200:121–131

Hassan MA, Church M (2000) Experiments on surface structure and partial sediment transport on a gravel bed. Water Resour Res 36:1885–1895

Hassan MA, Zimmermann AM (2011) Channel response and recovery to changes in sediment supply. In: Church M, Biron PA, Roy A (eds) Gravel-bed river: processes, tools, environments, pp 464–473 (Invited contribution)

Hassan MA, Church M, Schick AP (1991) Distance of movement of coarse particles in gravel bed streams. Water Resour Res 27:503–511

Hassan M, Smith B, Hogan D, Luzi D, Zimmermann A, Eaton B (2008) Sediment storage and transport in coarse bed streams: scale considerations. In: Habersack H, Piegay H, Rinaldi M (eds) Gravel-bed rivers VI: from process understanding to river restoration. Elsevier, Amsterdam, pp 473–496

Hassan MA (1990) Scour, fill, and burial depth of coarse material in gravel bed streams. Earth Surf Proc Land 15:341–356
Hassan MA, Church M (1992) The movement of individual grains on the streambed. In: Billi P, Hey RE, Thorne CR, Tacconi P (eds) Dynamics of gravel bed rivers. Wiley, New York, pp 159–175
Hassan MA, Church M (2001) Rating bedload transport in Harris Creek: seasonal and spatial variation over a cobble-gravel bed. Water Resour Res 37:813–825
Hassan MA, Church M (1994) Vertical mixing of coarse particles in gravel bed rivers: a kinematic model. Water Resour Res 30:1173–1185
Hassan MA, Ergenzinger P (2003) Tracers in fluvial geomorphology. In: Kondolf GM (ed) Tools in fluvial geomorphology. Wiley, Chichester
Hassan MA, Reid I (1990) The influence of microform bed roughness elements on flow and sediment transport in gravel bed rivers. Earth Surf Proc Land 15:739–750
Hassan MA, Church M, Ashworth PJ (1992) Virtual rate and mean distance of travel of individual clasts in gravel-bed channels. Earth Surf Proc Land 17:617–628
Hassan MA, Robinson SVJ, Voepel H, Lewis J, Lisle TE (2014) Modeling temporal trends in bedload transport in gravel-bed streams using hierarchical mixed-effects models. Geomorphology 219:260–269
Hassan MA, Schick AP, Shaw PA (1999) The transport of gravel in an ephemeral sandbed river. Earth Surf Proc Land 24:623–640
Hassan MA, Voepel H, Schumer R, Parker G, Fraccarollo L (2013) Displacement characteristics of coarse fluvial bed sediment. J Geophys Res Earth Surf 118:155–165
Hayward JA (1980) Hydrology and stream sediments in a mountain catchment. Special Publication 17, Tussock Grasslands and Mountain Lands Institutes, Lincoln College, Canterbury, New Zealnad, 236 pp
Helley EJ, Smith W (1971) The development and calibration of a pressure difference bed load sampler. U.S. geological survey, open file report, 18 pp
Hill KM, DellAngelo L, Meerschaert MM (2010) Heavy-tailed travel distance in gravel bed transport: An exploratory enquiry. J Geophys Res 115:F00A14. doi:10.1029/2009JF001276
Houbrechts G, Campenhout JV, Levecq Y, Hallot E, Peeters A, Petit F (2012) Comparison of methods for quantifying active layer dynamics and bedload discharge in armoured gravel-bed rivers. Earth Surf Proc Land 37:1501–1517
Hu C, Hui Y, Xia Z (1992) Experimental study on saltation of solid grains in flowing water. Int J Sediment Res 7(2):23–51
Hubbell DW, Stevens HH, Skinner JV, Beverage JP (1985) New approach to calibrating bed load samplers. J Hydraul Eng 111:677–694 (American Society of Civil Engineering)
Iverson R (1997) The physics of debris flows. Rev Geophys 35(3):245–296
Jackson WL, Beschta RL (1982) A model of two-phase bedload transport in an Oregon coast range stream. Earth Surf Proc Land 7:517–527
Jaeggi MNR (1995) Sediment transport in mountain rivers—a review. In: International conference on erosion control, 25–26 August, 1995, The Sabo Society of Japan, keynote address, 13 pp
Kennedy JF (1995) The albert shields story. J Hydraul Eng 121:766–772
Kirchner JW, Dietrich WE, Iseya F, Ikeda H (1990) The variability of critical shear stress, friction angle, and grain protrusion in water-worked sediments. Sedimentology 37:647–672
Kleinhans M, van Rijn L (2002) Stochastic prediction of sediment transport in sand-gravel bed rivers. J Hydraul Eng 128:412–425. doi:10.1061/(ASCE)0733-9429(2002)128:4(412) (Special Issue: Stochastic Hydraulics and Sediment Transport)
Komar PD (1987) Selective grain entrainment by a current from a bed of mixed sizes: a reanalysis. J Sediment Petrol 57(2):203–211
Komar PD, Miller MC (1973) The threshold of sediment movement under oscilatorry water waves. J Sediment Petrol 43(4):1101
Komar PD, Li MZ (1988) Application of grain pivoting and sliding analyses to selective entrainment of gravel and to flow competence evaluations. Sedimentology 35:681–695

Kovacs A, Parker G (1994) A new vectorial bedload formulation and its application to the time evolution of straight river channels. J Fluid Mech 267:153–183

Kramer H (1935) Sand mixtures and sand movement in fluvial models. Trans Am Soc Civ Eng 100:798–878

Kuhnle RA (1992) Bedload transport during rising and falling stages on two small streams. Earth Surf Proc Land 17:191–197

Lacey RWJ, Roy AG (2007) A comparative study of the turbulent flow field with and without a pebble cluster in a gravel bed river. Water Resour Res 43:W05502. doi:10.1029/2006WR005027

Lamarre H, Roy AG (2008) The role of morphology on the displacement of particles in a step–pool river system. Geomorphology 99:270–279

Lamb MP, Scheingross JS, Amidon WH, Swanson E, Limaye A (2011) A model for fire-induced sediment yield by dry ravel in steep landscapes. J Geophys Res: Earth Surf 116(F3)

Laronne JB, Reid I, Yitshak Y, Frostick LE (1992) Recording bedload discharge in a semiarid channel, Nahal Yatir, Israel. In: Bogen J, Walling DE, Day TJ (eds) Erosion and sediment transport monitoring programmes in river basins, vol 210. IAHS Publication, Wallingford, pp 79–86

Lauffer H, Sommer N (1982) Studies on sediment transport in mountain streams of the eastern Alps. In: Proceeding of 14th congress international commission on large dams, Rio de Janeiro, pp 431–453

Lenzi MA, D'Agostino V (1999) Bedload transport in the instrumented catchment of the Rio Cordon, Part II: analysis of bedload rate. CATENA 36:191–204

Lenzi MA, D'Agostino V, Billi P (1999) Bedload transport in the instrumented catchment of Rio Cordon, part I: analysis of bedload records, conditions and threshold of bedload entrainment. Catena 36:171–190

Li Z, Komar PD (1986) Laboratory measurements of pivoting angles for applications to selective e ntrainment of gravel in a current. Sedimentology 33(3):413–423. doi:10.1111/j.1365-3091.1986.tb00545.x

Ling C-H (1995) Criteria for incipient motion of spherical sediment particles. J Hydraul Eng 121 (6):472–478. doi:10.1061/(ASCE)0733-9429(1995)121:6(472)

Lisle TE, Nelson JM, Pitlick J, Madej MA, Barkett BL (2000) Variability of bed mobility in natural, gravel-bed channels and adjustments to sediment load at local and reach scales. Water Resour Res 36(12):3743–3755

Lisle TE, Pizzuto JE, Ikeda H, Iseya F, Kodama Y (1997) Evolution of a sediment wave in an experimental channel. Water Resour Res 33(8):1971–1981

Lisle TE, Hilton S (1992) The volume of fine sediment in pools: an index of sediment supply in gravel bed streams. Water Resour Bull 28:371–383

Ma H, Heyman J, Fu X, Mettra F, Ancey C, Parker G (2014) Bed load transport over a broad range of timescales: determination of three regimes of fluctuations. J Geophys Res Earth Surf 119 (12):2169–9011. doi:10.1002/2014JF003308

MacVicar BJ, Roy AG (2011) Sediment mobility in a forced riffle-pool. Geomorphology 125: 445–456

Martin R, Jerolmack D, Schumer R (2012) The physical basis for anomalous diffusion in bed load transport. J Geophys Res 117:F01018, 18 pp. doi:10.1029/2011JF002075

Mavies W, Ho C-H, Tu Y-C (1935) The transportation of detritus by flowing water–I. University of Iowa Studies in Engineering, Bulletin 5, 54 pp

McEwan I, Sørensen M, Heald J, Tait S, Cunningham G, Goring D, Willetts B (2004) Probabilistic modeling of bed-load composition. J Hydraul Eng 130(2):129–139. doi:10.1061/(ASCE)0733-9429(2004)130:2(129)

McNamara JP, Borden C (2004) Observations on the movement of coarse gravel using implanted motion-sensing radio transmitters. Hydrol Process 18(10):1871–1884

Meade RH (1985) Wavelike movement of bedload sediment, East Fork River, Wyoming. Environ Geol Water Sci 7:215–255

Meyer-Peter E, Müller R (1948) Formulas for bed-load transport. In: Proceedings of 2nd meeting, IAHR, Stockholm, Sweden, pp 39–64

Milhous, R.T. (1973). Sediment transport in a gravel-bottomed stream, PhD thesis, Oregon State University, Corvallis, Oregon, 232 pp

Miller DJ, Benda LE (2000) Effects of punctuated sediment supply on valley-floor landforms and sediment transport. Geolog Soc Am Bull 112:1814–1824

Miller RL, Byrne RJ (1966) The angle of repose for a single grain on a fixed rough bed. Sedimentology 6(4):303–314. doi:10.1111/j.1365-3091.1966.tb01897.x

Mizuyama T (1977) Bedload transport in steep channels. Ph.D. thesis, Kyoto University, Japan, 118 pp

Nakagawa H, Tsujimoto T (1976) Experimental Study on Evolution of Sand Waves from an Initially Flattened Bed. Disaster Prev Res Inst Ann 19:289–309

Nelson JM, Shreve RL, McLean SR, Drake TG (1995) Role of near-bed turbulence in bedload transport and bedform mechanics. Water Resour Res 31:2071–2086

Nikora V, Habersack H, Huber T, McEwan I (2002) On bed particle diffusion in gravel bed flows under weak bed load transport. Water Resour Res 38(6):1–9

Novak ID (1973) Predicting coarse sediment transport: the Hjulstrom curve revisited. In: Morisawa M (ed) Fluvial geomorphology. Binghampton New York State University, New York, pp 13–26

Paintal AS (1971a) Concept of critical shear stress in loose boundary open channels. J Hydraul Res 9(1):91–113

Paintal AS (1971b) A stochastic model of bed load transport. J Hydraul Res 9(4):527–554

Parker G (1990) Surface bedload transport relation for gravel rivers. J Hydraul Res 28(4):417–436

Parker G (2008) Transport of gravel and sediment mixtures. In: García MH (ed) Sedimentation engineering. Processes, measurements, modeling and practice, ASCE, Reston, pp 165–251

Parker G, Andrews E (1985) Sorting of bed load sediment by flow in meander bends. Water Resour Res 21(9):1361–1373

Parker G, Klingeman PC (1982) On why gravel-bed streams are paved. Water Resour Res 18:1049–1423

Parker G, Toro-Escobar CM (2002) Equal mobility of gravel in streams: the remains of the day. Water Resour Res 38(11):1264. doi:10.1029/2001WR000669

Parker G, Paola C, Leclair S (2000) Probabilistic form of Exner equation of sediment continuity for mixtures with no active layer. J Hydraul Eng 126:818–826

Parker G, Klingeman PC, McLean DG (1982) Bedload and size distribution in paved gravel-bed streams. J Hydraul Div 108(HY4):544–571

Parker G, Seminara G, Solari L (2003) Bed load at low shields stress on arbitrarily sloping beds: alternative entrainment formulation. Water Resour Res 39(7):1183

Proffitt GJ, Sutherland AJ (1983) Transport of non-uniform sediment. J Hydraul Res IAHR 21 (1):33–43

Pyrce RS, Ashmore PE (2003a) Particle path length distributions in meandering gravel-bed streams: results from physical models. Earth Surf Proc Land 28:951–966

Pyrce RS, Ashmore PE (2003b) The relation between particle path length distributions and channel morphology in gravel-bed streams. Geomorphology 56:167–187

Pyrce RS, Ashmore PE (2005) Bedload path length and point bar development in gravel-bed river models. Sedimentology 52:839–857

Ramette M, Heuzel M (1962) Le Rhône à Lyon. Etude de l'entraînement des galets à l'aide de traceurs radioactifs. Essai poor l'extension de la formule de Mayer-Peter au domaines du charriage partiel. Houille Blanche, pp 289–399

Reid I, Frostick LE (1984) Particle interaction and its effect on the thresholds of initial and final beldoad motion in coarse alluvial channels. Can Soc Pet Geo 19:61–68 (Koster EH, Steel RJ (eds) Sedimentary of gravels and conglomerates)

Reid I, Laronne JB (1995) Bedload sediment transport in an ephemeral stream and a comparison with seasonal and perennial counterparts. Water Resour Res 31:773–781

Reid I, Frostick LE, Lyman JL (1985) The incidence and nature of bedload transport during flood flows in coarse-grained alluvial channels. Earth Surf Proc Land 10:33–44

Reid I, Layman JT, Frostick LE (1980) The continuous measurement of bedload discharge. J Hydraul Res 18:243–249

Ribberink JS (1987) Mathematical modeling of one-dimensional morphological changes in rivers with non-uniform sediment. Ph.D. thesis, Delft University of Technology, 200 p

Rickenmann D (1990) Bedload capacity of slurry flows at steep slopes. In: Mitteilung der Versuchsanstalt fur Wasserbau, Hydrologie und Glaziologie, Rep. 103, Eidg. Tech. Hochsch, Zurich, Zurich, Switzerland, 249 pp

Rickenmann D (2001) Comparison of bedload transport in torrents and gravel bed streams. Water Resour Res 37:3295–3305

Roberts RJ, Church M (1986) The sediment budget in severely disturbed watersheds, Queen Charlotte Ranges, British Columbia. Can J Earth Sci 16:1092–1106

Roseberry JC, Schmeeckle MW, Furbish DJ (2012) A probabilistic description of the bed load sediment flux: 2. Particle activity and motions. J Geophys Res Earth Surf 117

Rouse H (1939) An analysis of sediment transportation in light of fluid turbulence. Report no. SCS-TP-25, Sediment Division, U.S. Department of Agriculture, Soil Conservation Service, Washington, DC

Ryan SE, Emmett WW (2002) The nature of flow and sediment movement in Little Granite Creek near Bondurant, Wyoming. U. S. Forest Service, General technical report RMRS-GTR -90, 48 p

Ryan SE (2001) The influence of sediment supply on rates of bed load transport: a case study of three streams on the San Juan National Forest. In: Proceeding of the seventh federal interagency sedimentation conference, March 25–29, 2001, Reno, Nevada, pp III-48–III-54

Ryan SE, Troendle CA (1996) Bed load transport patterns in coarse-grained channels under varying conditions of flows. In: Proceeding of the sixth federal interagency sedimentation conference on sedimentation technologies for management of natural resources in the 21st century, vol 2, March 10–14, 1996, Las Vegas, pp VI-22–VI-27

Ryan SE, Troendle CA (1997) Measuring bedload in coarse-grained mountain channels procedures, problems, and recommendations. In: Water resources education, training, and practice. Opportunities for the next century, American Water Resources Association, Keystone, Colorado, pp 949–958

Sawada T, Ashida K, Takahashi T (1983) Relationship between channel pattern and sediment transport in a steep gravel bed river. Z Geomorph NF Suppl Bd 16:55–66

Sawada T, Ashida K, Takahashi T (1985) Sediment transport in mountain basins. Paper presented at international symposium on erosion, debris flow and disaster prevention, Tsukuba, Japan

Sayre WW, Hubbell DW (1965) Transport and dispersion of labeled bed material. U.S. Geology survey professional paper 433-C, North Loup River, Nebraska, 48 pp

Schick AP, Hassan MA, Lekach J (1987) A vertical exchange model for coarse bedload movement: numerical considerations. Catena Suppl 10:73–83

Schmidt K-H, Ergenzinger P (1992) Bedload entrainment, travel lengths, step lengths, rest periods—studied with passive (iron, magnetic) and active (radio) tracer techniques. Earth Surf Process Land 17:147–165

Schneider JM, Turowski JM, Rickenmann D, Hegglin R, Arrigo S, Mao L, Kirchner JW (2014) Scaling relationships between bed load volumes, transport distances, and stream power in steep mountain channels. J Geophys Res Earth Surf 119:533–549. doi:10.1002/2013JF002874

Schoklitsch A (1962) Handbuch des Wasserbaues, 3rd edn. Springer, Vienna

Schumer R, Benson DA, Meerschaert MM, Baeumer B (2003) Fractal mobile/immobile solute transport. Water Resour Res 39:1296. doi:10.1029/2003WR002141

Sear DA (1996) Sediment transport processes in pool-riffle sequences. Earth Surf Proc Land 21:241–262

Seminara G, Parker G, Solari L (2002) Bed load at low shields stress on arbitrarily sloping beds: failure of the Bagnolg hypothesis. Water Resour Res 38(11):1249

Shields A (1936) Anwendung der Aehnlichkeitsmechanik und der Turbulenzforschung auf die Geschiebebewegung, Mitt. Preuss. Versuchsanst.Wasserbau Schiffbau; english translation by

W. P. Ott and J. C. van Uchelen, 36 pp, U.S. Dep. of Agric. Soil Conser. Serv. Coop. Lab., California, Institute of Technology, Pasadena
Sidle RC (1988) Bedload transport regime of a small forested stream. Water Resour Res 24: 207–218
Smith RD, Sidle RC, Porter PE (1993a) Effects on bedload transport of experimental removal of woody debris from a forest gravel-bed stream. Earth Surf Proc Land 18:455–468
Smith RD, Sidle RC, Porter PE, Noel JR (1993b) Effects of experimental removal of woody debris on the channel morphology of a forest, gravel bed stream. J Hydrol 152:153–178
Solari L, Parker G (2000) The curious case of mobility reversal in sediment mixtures. J Hydraul Eng 126:185–197
Stelczer K (1981) Bedload transport: theory and practice. Littleton, Colorado, Water Resources Publication, 295 pp
Sterling S, Church M (2002) Sediment trapping characteristics of a pit trap and a Helley-Smith sampler in a cobble-gravel bed river. Water Resour Res 38:19-1–19-11. doi:10.1029/2000WR000052
Sun Z, Donahue J (2000) Statistically derived bed load formula for any fraction of nonuniform sediment. J Hydraul Eng ASCE 126(2):105–111. doi:10.1061/(ASCE)0733-9429(2000)126:2(105)
Sutherland DG, Ball MH, Hilton SJ, Lisle TE (2002) Evolution of a landslide-induced sediment wave in the Navarro River, California. Geol Soc Am Bull 114:1036–1048
Swanson FJ, Janda RJ, Dunne T, Swanston DN (eds) (1982a). Sediment budgets and routing in forested drainage basins. United States Department of Agriculture, Forest Service, General technical report PNW-141, 165 pp
Swanson FJ, Janda RJ, Dunne T (1982b) Summary: sediment budgets and routing in forested drainage basins. In: Swanson FJ, Janda RJ, Dunne T, Swanston DN (eds) Sediment budgets and routing in forested drainage basins. United States Department of Agriculture, General Technical Report Number PNW-141, pp 157–165
Troendle CA, Nankervis JM, Ryan SE (1996) Sediment transport from small, steep-gradient watersheds in Colorado and Wyoming. In: Sixth federal interagency sedimentation conference on sedimentation technologies for management of natural resources in the 21st century, March 10–14, 1996, Las Vegas, NV, pp IX-39–IX-45
Turowski JM (2010) Probability distributions of bed load transport rates: a new derivation and comparison with field data. Water Resour Res 46:W08501. doi:10.1029/2009WR008488
Valyrakis M, Diplas P, Dancey CL (2011) Entrainment of coarse grains in turbulent flows: an extreme value theory approach. Water Resour Res 47(9):W09512. doi:10.1029/2010WR010236
Van Rijn LC (1984) Sediment transport, part I: bed load transport. J Hydraul Eng 110(10): 1431–1456
Vanoni VA (1966) Sediment transportation mechanics: Initiation of motion. J Hydraul Div ASCE 92(2):291–314
Vanoni VA (1984) Fifty years of sedimentation. J Hydraul Eng 110(8):1022–1057
Wathen SJ, Ferguson RI, Hoey TB, Werrity A (1995) Unequal mobility of sand and gravel in weakly bimodal river sediments. Water Resour Res 31(8):287–296. doi:10.1029/95WR01229
White FM (2011) Fluid mechanics, 7th edn. McGaw-Hill International Editions
White WR, Day TJ (1982) Transport of graded gravel bed material. In: Hey RD, Bathurst JC, Thorne CR (eds) Gravel-bed rivers, fluvial processes, engineering and management. Wiley, New York, pp 181–223
Whittaker JG (1987) Sediment transport in step-pool streams. In: Thorne CR, Bathurst JC, Hey RD (eds) Sediment transport in gravel-bed rivers. Wiley, Chichester, pp 545–579
Wiberg PL, Smith JD (1987) Calculations of the critical shear stress for motion of uniform and heterogeneous sediments. Water Resour Res 23:1471–1480
Wilcock PR, Crowe JC (2003) Surface-based tranport model for mixed-size sediment. J Hydraul Eng 129(2):120–128. doi:10.1061/(ASCE)0733-9429(2003)129:2(120)

Wilcock PR (1998) Two-fraction model of initial sediment motion in gravel-bed rivers. Science 280(410). doi:10.1126/science.280.5362.410
Wilcock PR, Kenworthy ST (2002) A two-fraction model for the transport of sand/gravel mixtures. Water Resour Res 38(10):1194. doi:10.1029/2001WR000684
Wilcock PR, Southard JB (1988) Experimental study of incipient motion in mixed-size sediment. Water Resour Res 24(7):1137–1151
Wilcock PR (1997a) The components of fractional transport rate. Water Resour Res 33(1):247–258. doi:10.1029/96WR02666
Wilcock PR, McArdell BW (1993) Surface-based fractional transport rates: mobilization thresholds and partial transport of a sand-gravel sediment. Water Resour Res 29(4):1297–1312. doi:10.1029/92WR02748
Wilcock PR, McArdell BW (1997) Partial transport of a sand/gravel sediment. Water Resour Res 33(1):235–245. doi:10.1029/96WR02672
Wilcock PR (1997b) Entrainment, displacement and transport of tracer gravels. Earth Surf Proc Land 22:1125–1138
Wilcock PR, Barta AF, Shea CC, Kondolf GM, Matthews WVG, Pitlick JC (1996) Observations of flow and sediment entrainment on a large gravel-bed river. Water Resour Res 32:2897–2909
Williams GR et al. (1937). Selected bibliography on erosion and silt movement. Water supply paper 797, US Department of Interior Geological Survey
Wong M, Parker G, DeVries P, Brown TM, Burges SJ (2007) Experiments on dispersion of tracer stones under lower-regime plane-bed equilibrium bed load transport. Water Resour Res 43: W03440. doi:10.1029/2006WR005172
Wu F-C, Yang K-H (2004) A stochastic partial transport model for mixed-size sediment: application to assessment of fractional mobility. Water Resour Res 40:W04501. doi:10.1029/2003WR002256
Wu F-C, Chou Y-J (2003) Rolling and lifting probabilities for sediment entrainment. J Hydraul Eng 129(2):110–119. doi:10.1061/(ASCE)0733-9429(2003)129:2(110)
Yalin MS (1977) Mechanics of sediment transport. 2nd edn. Pergamon Press, Oxford
Zimmerman A, Church M (2001) Channel morphology, gradient profiles and bed stresses during spring runoff in a step-pool channel. Geomorphology 40:311–328
Zimmermann AE, Church M, Hassan MA (2010) Step-pool stability: testing the jammed state hypothesis. J Geophys Res 115:F02008. doi:10.1029/2009JF001365

Chapter 14
Recent Advances from Research on Meandering and Directions for Future Work

Ana Maria Ferreira da Silva

Abstract This paper presents a review of the present understanding of the kinematics of meandering flow, and its relationship to bed deformation as well as downstream migration and lateral expansion of meander loops. Taking into account the conditions prevailing in natural, low-land alluvial meandering rivers, the paper focuses primarily on the behaviour of streams having "large" values of width-to-depth ratio. The present review is preceded by a brief description of meandering defining geometric characteristics, as these are invoked throughout this manuscript. The paper is also used as an opportunity to outline future directions for research. These involve matters related to the topics under consideration that remain obscure and which, in the writer's view, constitute subjects particularly worthwhile as future research topics for their scientific as well as practical significance. More specifically, these concern the nature and analytical formulation of meander wavelength; the value of width-to-depth ratio beyond which the effect of cross-circulation becomes of secondary importance where the meandering bed deformation is concerned; and, finally, the unification of present methods of determination of meandering planimetric evolution with the principle of self-formation of alluvial streams as expressed by regime theory.

Keywords Meandering rivers · Flow kinematics · Bed deformation · Bank deformation · Morphological evolution

14.1 Introduction

Following Leopold and Wolman (1957), streams are commonly classified as straight, meandering and braiding, with meandering being by far the plan shape most frequently acquired by natural streams. As such, meandering has been

A.M.F. da Silva (✉)
Department of Civil Engineering, Queen's University, Kingston, Ontario K7L 3N6 Canada
e-mail: amsilva@queensu.ca

P. Rowiński and A. Radecki-Pawlik (eds.), *Rivers—Physical, Fluvial and Environmental Processes*, GeoPlanet: Earth and Planetary Sciences,
DOI 10.1007/978-3-319-17719-9_14

attracting the attention of scientists and engineers for a long time, with the first systematic studies on the topic dating from the later part of the 19th century (see, e.g., Thomson 1879; de Leliavsky 1894; Jefferson 1902; see also Fargue 1908; Engels 1926). An excellent review of this early research was provided by Leliavsky (1959). Since the publication of the just mentioned works, and especially over the past 60 years, meandering became the object of intensive research. This has resulted in a voluminous body of literature covering different aspects of meandering, with contributions by specialists from a variety of fields such as physical geography and geomorphology, fluid mechanics and fluvial hydraulics, and including theoretical and numerical research, as well as laboratory and field investigations.

While much debate is still ongoing on several aspects of meandering, a great deal of progress has been achieved in the understanding of the nature of meandering flow, and how this determines bed deformation as well as meandering downstream migration and lateral expansion patterns. This paper focuses primarily on these aspects. Its objective is to provide a synopsis of the state-of-the-art, and at the same time outline matters related to the topics under consideration that remain obscure and which, in the writer's view, form particularly worthwhile subjects for future research, for strictly scientific reasons as well as reasons related to river engineering practice.

Taking into account the conditions prevailing in natural alluvial meandering streams, the paper centres on research specifically addressing the behaviour of streams having "large" values of width-to-depth ratio. The matters under consideration in this paper can be quite involved, as often different phenomena co-exist in meandering streams. To illustrate this point, consider the nature of the deformed bed in a meandering stream. In addition to the characteristic large scale pool-bar formations (or complexes) invariably occurring once per meander loop and strictly caused by the meandering of the stream, bed forms due to other reasons, such as ripples, dunes and bars may be superimposed on them. This paper, however, is restricted to those essential aspects or features of alluvial meandering streams that are always present, and thus can be viewed as forming the essence of the meandering phenomenon.

14.2 Defining Geometric Characteristics of Meandering Streams

In nature, different meandering streams exhibit different flow widths and depths, grain size distributions, flow regimes, etc., and in any given stream the conditions vary to a smaller or larger extent, from one meander loop to another. Yet, despite their differences, meandering streams share some essential, defining geometric properties. Since these properties will be invoked throughout this paper, it seems worthwhile, for the sake of clarity, to devote this section to their description. The following is also used by the writer as an opportunity to discuss existing formulations of meander wavelength and related views.

14.2.1 Meander Wavelength

(i) Following the works by Inglis (1947), Leopold and Wolman (1957) and Zeller (1967), among others, it has become clear that the meander wavelength Λ_M (see the definition sketch in Fig. 14.1) is directly proportional to the flow width B, i.e., that

$$\Lambda_M = n_M B \tag{14.1}$$

where n_M stands for the proportionality factor.
The realization of this fact has, over the years, prompted a number of authors to produce plots of meander wavelength versus B using data from various sources. To the best knowledge of the writer, the most recent of such plots are by Yalin and da Silva (2001) and Soar and Thorne (2001) (see Fig. 14.2a, b). Observe that Fig. 14.2a, which was produced by adapting and extending Fig. 13.12 in Garde and Raju (1977), includes data not only from alluvial streams, but also from meltwater channels on ice and meanders of the Gulf Stream. These data are from Leopold et al. (1964), who appear to have been the first to realize that "the meander pattern of meltwater channels on the surface of glaciers have nearly identical geometry to the meander bends in rivers" and that "the geometry in plan view of meanders in the Gulf Stream is also similar to that of rivers". It should be noted that, as pointed out by Leopold et al. (1964), p. 302, the "meandering channels on ice are formed without any sediment load or point-bar construction by sediment deposition", and that the meanders on the Gulf Stream too occur "… without debris load

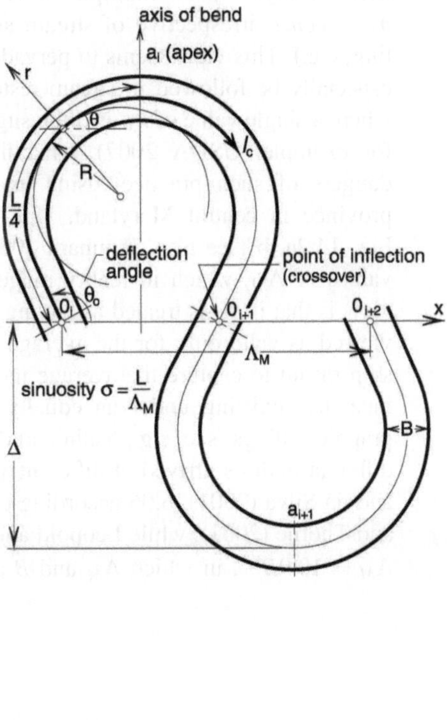

Fig. 14.1 Definition sketch

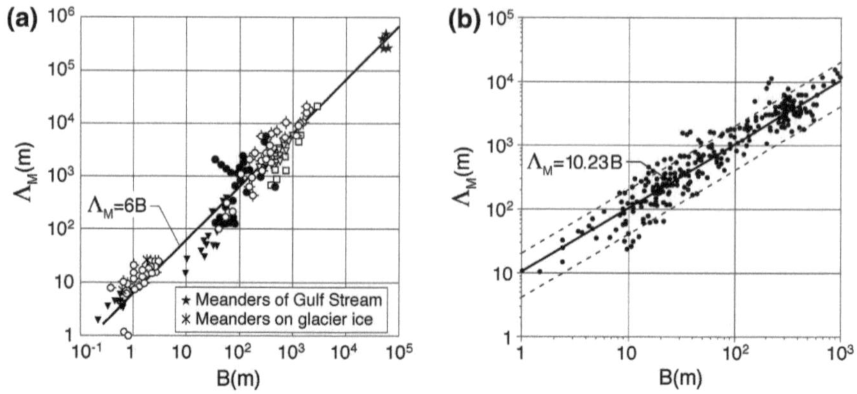

Fig. 14.2 Plots of meander wavelength Λ_M versus flow width B: **a** Adapted from Yalin and da Silva (2001); **b** Adapted from Soar and Thorne (2001) (*dashed lines mark* $\Lambda_M = 4B$ and $\Lambda_M = 20B$)

and, in this instance, without confining banks". Considering this, Yalin (1992), p. 161, defined meandering as a "*self-induced* plan deformation of a stream that is (ideally) periodic and anti-symmetrical with respect to an axis, x say, which may or may not be exactly straight"—where the term self-induced is used to imply that the deformation is induced by the stream itself, as opposed to being forced upon the stream by its environment. The term "meandering" is used throughout this paper in the sense implied by this definition.

(ii) Two different views regarding the nature of Eq. (14.1) emerge from the literature, either explicitly or implicitly. The first view is that Λ_M is theoretically (i.e., ideally) a *fixed* multiple of channel widths (or, in other words, that $n_M = const$, irrespective of stream size, flow conditions, physiographic setting, etc.). This view seems to pervade present river engineering practice, and especially be followed in stream restoration and re-naturalization guidelines, where a single value of n_M is often suggested for all practical applications (see, for example, USDA 2007). Rinaldi and Johnson (1997) pointed out the dangers of such practice, using as examples streams from the Piedmont province in central Maryland, U.S.A. The just mentioned work, just like Fig. 14.2a, b (see also Seminara 2006), highlights the large variation in the values of Λ_M, which in reality range from $\approx 4B$ to $\approx 20B$. Thus, the second view is that if n_M is treated as having a constant value, then Eq. (14.1) can be viewed as valid only for the average meander wavelength, i.e., as a relationship meant to capture the average meander wavelength of a large number of meanders existing under an equally large set of flow conditions and geographic settings (see, e.g., Yalin and da Silva 2001). In this sense, the fact that different authors arrived at different values of n_M [e.g., 6 according to Yalin and da Silva (2001), 6.06 according to Inglis (1947), 10.23 according to Soar and Thorne (2001); while Leopold and Wolman (1960) proposed the equation $\Lambda_M = 10.9B^{1.1}$, in which Λ_M and B are in feet] is merely a reflection of the

fact that the datasets they used to determine the average value of n_M are different, and not large enough to eliminate the effect of sample size when determining the 'true' average value of this proportionality factor.
Yalin and da Silva (2001) attributed the variation in values of Λ_M to the strong "random element" in natural meandering streams. However, while this certainly plays a role in the 'scatter' exhibited by the data around the 'best fit' lines in Fig. 14.2a, b, a possibility that cannot be excluded is that the scatter is also a reflection of n_M being a function of some variables—even if this function is not yet known. To further discuss this point, consider the case of alternate bars, also known as single-row or one-row bars, and which, as is well-known, are closely related to meandering. A recent plot of measured values of Λ_a/B versus B/D due to Boraey (2014), and including all the readily available data from the literature, is shown in Fig. 14.3 (for references to the data sources in these figures, see Boraey 2014 or, alternatively, Boraey and da Silva 2014). Here Λ_a is alternate bar length and D is representative grain size ($=D_{50}$). Figure 14.3a corresponds to the case of alternate bars produced under rough turbulent flows; Fig. 14.3b, to that of bars produced by flows in the transitional regime of turbulent flow (note that the reason for using B/D as abscissa in these plots is irrelevant to the present discussion). Figure 14.3a, b clearly illustrate the fact that the length of alternate bars too, varies within a wide range [$\approx 4B$–$18B$, on the basis of Fig. 14.3a, b, which is in agreement with previous statements by Ikeda (1984), Rhoads and Welford (1991), Welford (1993) and Bridge (2003)]. Yet, from the independent works by Ikeda (1984) and Boraey and da Silva (2014) (see also Boraey 2014), it follows that alternate bar length is best expressed as $\Lambda_a/B = n_a$, where n_a is not a constant, but rather a function of B/h and h/D (even if the exact form of this function is still under investigation). It thus seems only natural to expect the proportionality factor n_M in Eq. (14.1) to be a function of some variables. It would be particularly worthwhile to explore this possibility in the future, especially for its implications for river engineering practice.

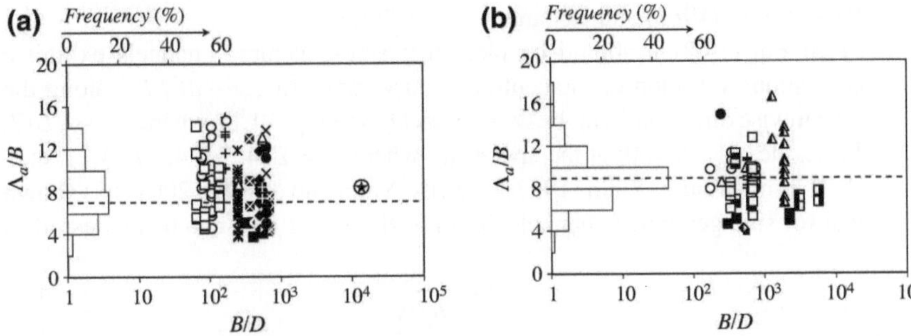

Fig. 14.3 Plots of normalized alternate bar length Λ_a/B versus B/D (after Boraey 2014): **a** bars produced under rough turbulent flows; **b** bars produced under transitional flows

14.2.2 Plan Shape of a Meandering Stream

(i) As demonstrated by Leopold and Langbein (1966), Langbein and Leopold (1966), the centrelines of natural, regular meandering streams closely follow the "sine-generated curve", given by

$$\theta = \theta_0 \cos\left(2\pi \frac{l_c}{L}\right) \tag{14.2}$$

(see Fig. 14.1 for the meaning of symbols in this equation).
As is well-known, Eq. (14.2) was introduced by the aforementioned authors as an approximation to the closed form integral resulting from the probabilistic derivation of meander path by von Schelling (1951), which rested on the postulate that for a given length between two points, A and B (two consecutive crossovers, say), a meander loop will acquire the shape corresponding to the minimum overall (average) curvature. Yalin (1992) succeeded in deriving Eq. (14.2) on a non-probabilistic and continuous basis, by treating the meander path as a isoperimetric variational problem based on the postulate that if a river is to turn in a meander loop having a given average curvature (squared), then this loop must be such that the average rate of change (squared) of its curvature is minimum. More recently, classical calculus of variations was also used by Movshovitz-Hadar and Shmukler (2006) to formulate the meander path. These authors, however, based their derivation on von Schelling's assumption of minimum overall curvature. [It should be noted that the term 'irregular meandering stream' appears to be generally used in the literature to describe any stream whose plan shape noticeably deviates from the sine-generated curve. While the deviations can be the result of external factors such as terrain irregularities or geological constraints, it appears that in the absence of such factors there are nonetheless some definable (deterministic) patterns in the shape of irregular meandering streams (see e.g. Kinoshita 1961; Hasegawa 1983), intrinsic to the phenomenon itself and requiring a physical explanation. This topic, however, is outside of the scope of the present paper, and the reader is referred to the works on the topic by, among others, Ikeda et al. (1981), Parker et al. (1982) and Seminara et al. (2001).]

(ii) From Eq. (14.2) it should be clear that sine-generated channels exhibit a continuous variation of the centreline curvature $1/R$ $(= -d\theta/dl_c)$ along the streamwise direction l_c: at the crossovers O_i (see Fig. 14.1), where $l_c = 0, L/2$, L, ..., etc., $|1/R| = 0$; at the apexes a_i, where $l_c = L/4$, $3L/4$, $5L/4$, ... etc., $|1/R|$ is maximum. Yalin (1992) (see also Yalin and da Silva 2001) has shown that for sine-generated channels the sinuosity $\sigma = L/\Lambda_M$ can be expressed as

$$\sigma = \frac{1}{J_0(\theta_0)}, \tag{14.3}$$

while the dimensionless curvature B/R along the centreline can be expressed as

$$\frac{B}{R} = [\theta_0 J_0(\theta_0)] \sin\left(2\pi \frac{l_c}{L}\right) \tag{14.4}$$

(or as a multiple of the right hand side of Eq. (14.4) if $\Lambda_M/B \neq 2\pi$). Here $J_0(\theta_0)$ is the Bessel function of first order and zero-th kind of θ_0. This equation yields for the apex sections

$$\frac{B}{R_a} = \theta_0 J_0(\theta_0). \tag{14.5}$$

Observe from Eqs. (14.3) and (14.5), whose graphs are shown in Fig. 14.4, that both σ and B/R_a are uniquely determined by θ_0. Note also that the maximum possible value of θ_0, corresponding to $J_0(\theta_0) = 0$ and $L, \sigma \to \infty$, is $\approx 138° = 2.41$ rad. However, in practice, this can never occur, for when θ_0 reaches the value $\approx 126° = 2.20$ rad ($\sigma \approx 8.5$), the meander loops come into contact with each other and the meandering pattern is destroyed. From Eq. (14.5) (see also Fig. 14.4) it follows that the largest values of (dimensionless) curvature at the apex occur for intermediate values of θ_0 ($\theta_0 \approx 70°$), and that B/R_a then gradually decreases with the increment of deviation of θ_0 from $\approx 70°$.

(iii) As follows from subsequent sections, the streamwise variation in stream curvature exhibited by freely meandering natural streams as reflected in the sine-generated curve, and the fact that B/R_a is of the nature just described, have far-reaching implications where flow patterns, modes and rates of bed and bank deformation, etc., are concerned. It is therefore highly advantageous to capture these properties when idealizing meandering streams for the purpose of investigating the phenomenon. Yet, from the 1960s to the 1990s, laboratory research on meandering was almost invariably carried out in single circular bends (of various central angles) or sequences of adjacent circular bends (for lists and reviews of such works see, e.g., Chang 1988; Yalin and da Silva 2001). More recent works carried out in circular bends include Blanckaert and de Vriend (2004), Zolezzi et al. (2005), Blanckaert

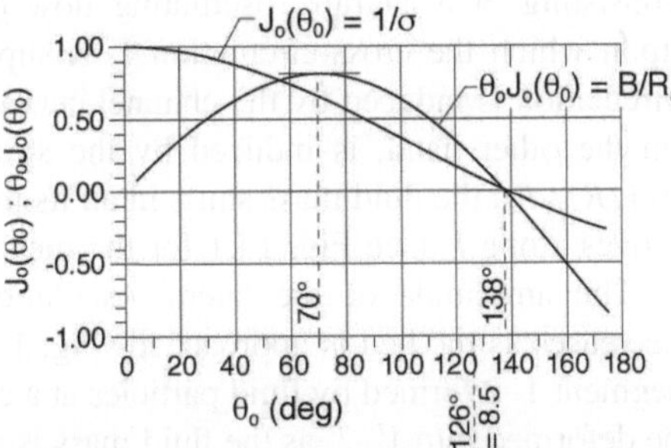

Fig. 14.4 Plots of $J_0(\theta_0)$ and $\theta_0 J_0(\theta_0)$ versus θ_0

(2010), Kashyap et al. (2012), among others. It is not until relatively recently that more realistic idealizations of meandering streams, and in particular the sine-generated curve, started to be frequently used in research. These include the works by Hooke (1974), Hasegawa (1983) (see also Hasegawa and Yamaoka 1983), Ikeda and Nishimura (1986), Whiting and Dietrich (1993a, b), da Silva (1995), Termini (1996), da Silva et al. (2006), da Silva and El-Tahawy (2008), Binns and da Silva (2009, 2011), Termini (2009), Termini and Piraino (2011), Xu and Bai (2013). [It appears that, so far, only Hasegawa (1983) and Abad and Garcia (2009a, b) used Kinoshita curves as idealizations of irregular meandering streams.]

The contribution of the research on circular bends to the overall understanding of processes in meandering rivers should not be minimized. However, as pointed out by Yalin (1992), it is also important to keep in mind that such research has significant limitations where explaining the behaviour of most natural, freely meandering rivers is concerned. In subsequent sections, the focus is on results of research carried out in sine-generated streams.

14.3 Meandering Flow Kinematics

In order not to encumber the explanations below, throughout this section it will be assumed that the meandering streambed is flat (i.e., the bed cross-sectional profiles are horizontal). If the streambed is movable, then the flat bed is assumed to represent the conditions at the beginning of an experiment (i.e., the "flat initial bed" at time t = 0).

14.3.1 General

As is well-known, a meandering flow manifests itself as a three-dimensional helicoidal fluid motion (Engelund 1974; Smith and McLean 1984; Chang 1988; Nelson and Smith 1989). Following Yalin (1992), this motion can be viewed as consisting of a laterally oscillating flow (the "convective base"; see Fig. 14.5a) upon which the cross-circulation Γ is superimposed (see Fig. 14.5b). The cross-circulation is induced by the channel curvature $1/R$. The lateral oscillation of flow, on the other hand, is induced by the streamwise variation of channel curvature $d(1/R)/dl_c$: the fluid mass shifts in all its thickness h periodically left and right as it moves along l_c (see Fig. 14.1 for the meaning of l_c).

The amplitude of the lateral oscillation of streamlines (due to $d(1/R)/dl_c$) decreases as the bed is approached (Fig. 14.5a). As a result, a vertical straight line segment 1–2 formed by fluid particles at a certain instant, one second later, say, will be deformed into 1′–2′ as the fluid mass is shifted sideways. The related radial flow

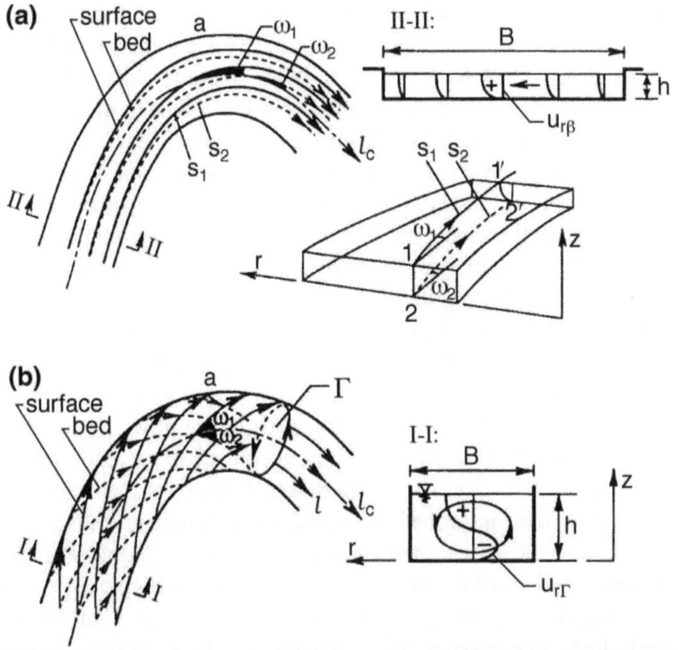

Fig. 14.5 Schematic representation of components of flow in a meandering channel (after Yalin 1992): **a** laterally oscillating flow induced by the streamwise variation in channel curvature; **b** cross-circulatory motion

velocities, $u_{r\beta 1}$ and $u_{r\beta 2}$, are in the same r-direction; and the deviation angles at the free surface, ω_1, and near the bed, ω_2,are of the same sign.

In contrast to this, cross-circulation produces a $u_{r\Gamma}$-diagram along z consisting of one negative part and one positive part having equal areas (Fig. 14.5b). The radial velocities $u_{r\Gamma 1}$ (at the free surface) and $u_{r\Gamma 2}$ (near the bed) are in the opposite r-directions, and therefore the deviation angles ω_1 and ω_2 of its streamlines s_1 and s_2 are also of opposite sign. The vertical average of a $u_{r\Gamma}$-diagram is zero (Nelson and Smith 1989; Yalin 1992). As pointed out by Yalin (1992), p. 191, "at any location in the flow the u_r-diagram is the sum of two components: the circulatory $u_{r\Gamma}$-diagram, and the translatory $u_{r\beta}$-diagram".

14.3.2 The Convective "Base"

The vertically-averaged streamlines of the laterally oscillating flow are between the free surface and near bed streamlines s_1 and s_2 (see Fig. 14.5a). Since the deviation between these is small, it follows that the plan behaviour of the lateral oscillations of flow is very adequately reflected by the vertically-averaged counterpart of the actual three-dimensional flow.

Let us then consider the vertically-averaged flow in a meandering channel over a flat bed. From the considerations above, it should be clear that the streamlines s of this flow must necessarily form, in flow plan, laterally adjacent convergence-divergence zones (in short, [CD]'s), as shown in Fig. 14.6. In this figure, s_* is the streamline separating two laterally adjacent (outer and inner) convergence-divergence flow zones, each conveying half the flow rate Q. In sine-generated streams, the pattern formed by the streamlines in plan view varies periodically along l_c. Hence the length of each convergence-divergence flow zone ([CD]) is $L/2$. At any point P, let $\overline{\omega}$ be the deviation angle between the vertically-averaged streamline passing through P and the longitudinal coordinate line l, as shown in Fig. 14.6. Clearly, at the sections coinciding either with the upstream-end or the downstream-end of a [CD], $\overline{\omega}$ is equal to zero. The sign of $\overline{\omega}$ remains either positive or negative within each [CD]; and it alternates periodically along l_c between positive and negative (positive in a $L/2$-long divergence-convergence flow zone as in Fig. 14.6, then negative for the subsequent $L/2$-long convergence-divergence zone, etc.).

It appears that the first systematic experimental study on the nature of the vertically-averaged flow over a flat bed in sine-generated channels was carried out by da Silva (1995), who used two channels having $\theta_0 = 30°$ and $110°$ and typifying "small" and "large" values of θ_0. In these experiments, $B = 0.40$ m, $h_{av} \approx 3$ cm, and $D = 2.2$ mm; $\Lambda_M/B = 2\pi$, $B/h_{av} \approx 13$, $(c_f)_{av} \approx 11$. It was assumed that B/h_{av} was sufficiently large to represent "wide" channels (where the term "wide" is used in the sense of Sect. 14.4.1.2(i)). Here h_{av} is channel-averaged flow depth, D is average size of the bed material and $(c_f)_{av}$ is (channel-averaged) dimensionless (Chézy) flow friction factor, determined from Eq. (14.1) in Yalin and da Silva (2001). It should be noted that the results of the velocity measurements in the 110° channel in these experiments are consistent with the free surface flow velocity measurements that had been previously carried out Whiting and Dietrich (1993b) in a 100° channel as well as the measurements of the vertically-averaged flow subsequently carried out by Termini (1996) in a 110° channel. These experiments made it clear that streams having small and large values of θ_0 possess distinctly different locations of their convergence-divergence zones. But this means that the location of

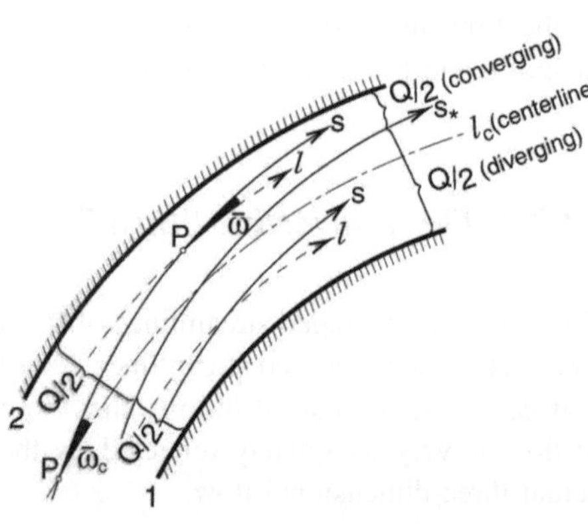

Fig. 14.6 Laterally adjacent convergence-divergence flow zones in a meandering flow

the [CD]'s in flow plan is not of a standard nature, but rather strongly dependent on (at least) θ_0. Considering this, da Silva et al. (2006) extended the previous measurements in da Silva (1995) by including also channels representing intermediate values of sinuosity ($\theta_0 = 50°$, 70° and 90°), with the goal of revealing experimentally how the location in the flow plan of the [CD]'s changes as θ_0 increases from "small" to "large". As examples, the velocity fields measured by da Silva (1995) in the channels having $\theta_0 = 30°$ and 110°, and that resulting from the measurements reported by da Silva et al. (2006) in their 70° channel are shown in Fig. 14.7.

From the aforementioned experiments, it follows that the plan configuration of the [CD]'s varies as shown in the schematic Fig. 14.8a–c, where the meandering channels are "straightened" for the sake of simplicity (consider only the top figure in each of Fig. 14.8a–c, where "CONV" and "DIV" indicate the zones of convergence and divergence of flow, respectively; and ξ_{c0} is the distance, normalized by L, from the crossover O_i to the upstream-end of the [CD] shown; the bottom figures, showing erosion-deposition zones will be discussed later, in Sect. 14.4.1.1):

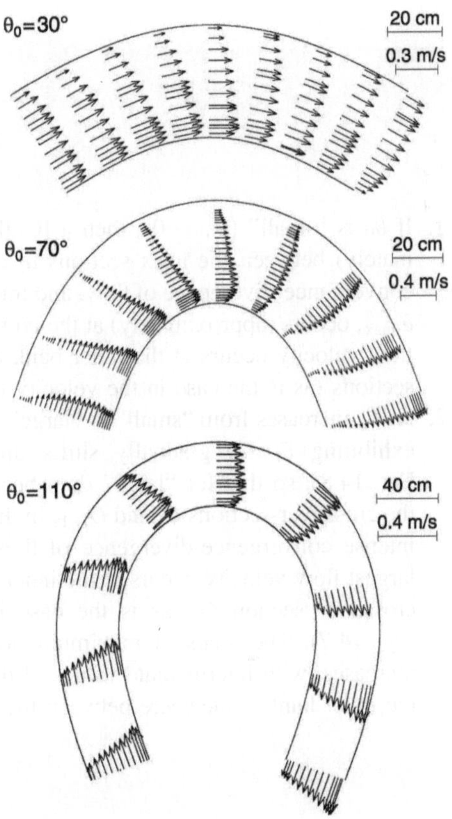

Fig. 14.7 Measured vertically-averaged flow velocity fields in sine-generated channels having $\theta_0 = 30°$ and 110° (adapted from da Silva 1995) and $\theta_0 = 70°$ (produced from the measurements reported by da Silva et al. 2006)

Fig. 14.8 Schematic representation of the location of flow convergence-divergence zones ([CD]'s) and erosion-deposition zones ([ED]'s) in sine-generated streams with varying values of θ_0. Meandering channels are "straightened" for the sake of simplicity

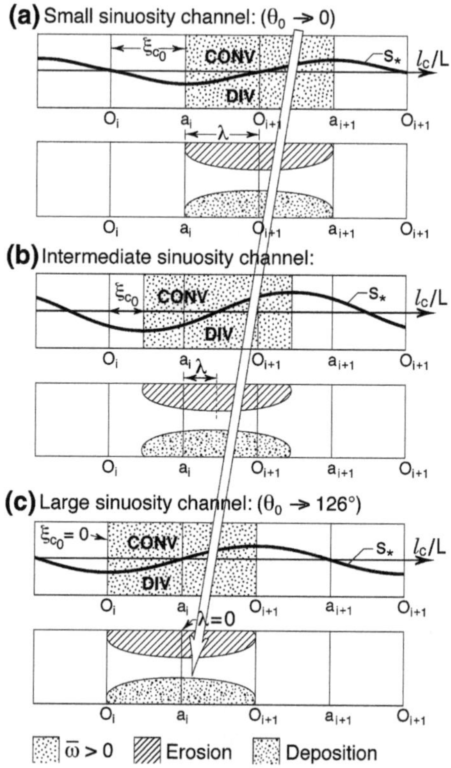

1. If θ_0 is "small" ($\theta_0 \to 0$), then a [CD] exhibiting $\overline{\varpi} > 0$ is situated (approximately) between the apex-sections a_i and a_{i+1}. In this case, the most intense convergence-divergence of flow, and thus the maximum value of $\overline{\varpi}$, here termed $\overline{\varpi}_{max}$, occurs (approximately) at the crossover section O_{i+1}; while the maximum flow velocity occurs at the inner bank (or in its proximity) very near the apex sections (as is the case in the velocity field for $\theta_0 = 30°$ in Fig. 14.7).
2. As θ_0 increases from "small" to "large", then the location of the analogous [CD] exhibiting $\overline{\varpi} > 0$ gradually shifts upstream (as implied by the arrow in Fig. 14.8), so that for "large" θ_0 it becomes situated (approximately) between the crossover-sections O_i and O_{i+1}. In the latter case, $\overline{\varpi} = 0$ at O_i, with the most intense convergence-divergence of flow occurring at the apex-section a_i; the largest flow velocity occurs at the inner bank (or in its proximity) very near the crossover section O_i (as is the case in the velocity field for $\theta_0 = 110°$ in Fig. 14.7). The locus of maximum flow velocity thus shifts upstream as θ_0 increases, with intermediate values of θ_0 exhibiting the largest flow velocity at the inner bank somewhere between the sections O_i and a_i.

(Note that because of bank effect, in reality the largest flow velocity does not occur exactly at the inner bank—hence the reason for the inclusion of '(or in its proximity)' in the statements above.)

The plan patterns of the streamlines of the vertically-averaged flow corresponding to small and large values of θ_0 are as shown in the schematic Fig. 14.17a, b, respectively. The conditions above can also be inferred from Fig. 14.9, showing a plot of the values of ξ_{c0} versus θ_0 resulting from the flow velocity measurements by da Silva (1995) and da Silva et al. (2006).

Figure 14.10 shows the graph of the values of $(\overline{\omega}_c)_{\max}$ versus θ_0 resulting also from the just mentioned experiments. Here $(\overline{\omega}_c)_{\max}$ stands for the maximum value of $\overline{\omega}_c$ found along the channel, $\overline{\omega}_c$ itself denoting the deviation angle $\overline{\omega}$ at points P falling on the channel centreline. For a given set of experimental conditions, as the deviation of θ_0 from $\approx 70°$ decreases and B/R_a $(=\theta_0 J_0(\theta_0))$ increases (see Sect. 14.2.2), the vertically-averaged flow must necessarily become "stronger" (in the sense that super-elevation increases, velocity gradients increase, the amplitude of oscillation of the streamline s_* (as defined earlier) around the channel centreline increases, etc. Therefore, $(\overline{\omega}_c)_{\max}$ must also increase. This explains why, in Fig. 14.10, $(\overline{\omega}_c)_{\max}$ reaches a maximum for $\theta_0 \approx 70° = 1.22$ rad.

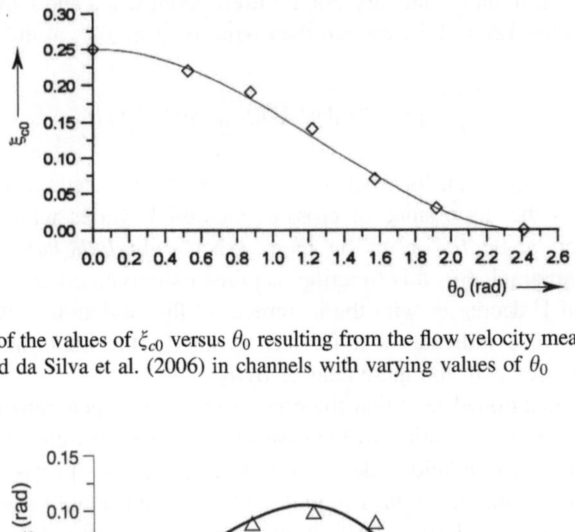

Fig. 14.9 Plot of the values of ξ_{c0} versus θ_0 resulting from the flow velocity measurements by da Silva (1995) and da Silva et al. (2006) in channels with varying values of θ_0

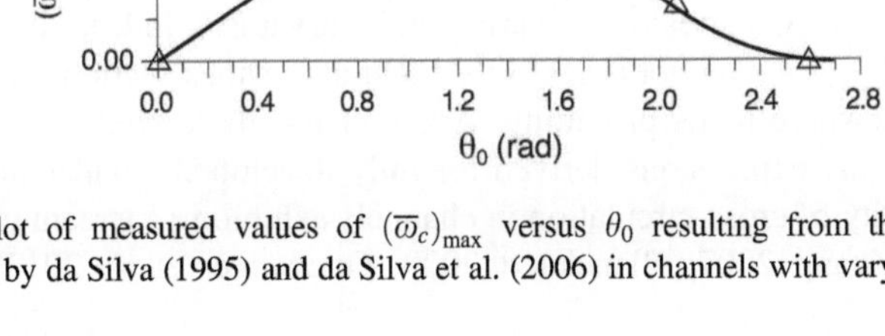

Fig. 14.10 Plot of measured values of $(\overline{\omega}_c)_{\max}$ versus θ_0 resulting from the flow velocity measurements by da Silva (1995) and da Silva et al. (2006) in channels with varying values of θ_0

No systematic experimental studies to investigate the dependency of the location in flow plan of the [CD]'s on variables other than θ_0, and in particular Λ_M/B, B/h_{av} and $(c_f)_{av}$, have been carried out to date. However, there are sufficient indications that the effect of Λ_M/B and $(c_f)_{av}$ in determining the location of the [CD]'s is relatively minor when compared to that of θ_0 (for more on the topic, see da Silva 1995; Zhang 2007; da Silva and El-Tahawy 2008). The effect of B/h_{av} is discussed later in Sect. 14.4.1.2.

14.3.3 The Cross-Circulatory Motion

Consider now the cross-circulation Γ. In earlier works on the topic (see, e.g., Chang 1988) it has been demonstrated that for the case of fully developed flow in circular bends, the related radial velocity $u_{r\Gamma}$ is best expressed as $u_{r\Gamma}/\overline{u} = const(h/B)(B/R)\cdot\varphi(z/h, c_f)$, where $\overline{u}$ is the longitudinal component of the local vertically-averaged flow velocity $\overline{U}$, h is flow depth ($\approx h_{av}$), and $\varphi(z/h, c_f)$ is a function that varies depending on distance from the channel bottom z (where z is measured vertically) and the friction factor c_f. Adopting this equation for sine-generated channels, as invariably done in the current literature (see, e.g., Jia and Wang 1999; Kassem and Chaudhry 2002; Olsen 2003; Chen and Duan 2006), and taking into account Eq. (14.5), we can then write for $(u_{r\Gamma}/\overline{u})_a$ at the apex-section:

$$\left(\frac{u_{r\Gamma}}{\overline{u}}\right)_a \approx const[\theta_0 J_0(\theta_0)]\frac{h}{B}\cdot\varphi(z/h, c_f). \qquad (14.6)$$

This relation indicates that for a given flow in a sine-generated channel, the value of $u_{r\Gamma}$. and thus the importance of cross-circulation Γ varies with the channel's initial deflection angle θ_0 or, to be more exact, with $\theta_0 J_0(\theta_0)$. As shown in Sect. 14.2.2, paragraph (ii), this function acquires its maximum at $\theta_0 \approx 70°$. Thus the relevance of Γ decreases with the increment of the deviation of θ_0 from $\approx 70°$. Equation (14.6) also indicates that for any given θ_0, the importance of Γ progressively decreases with the increment of B/h_{av}.

It should be mentioned here that the present practice of generalizing the earlier expressions for cross-circulation to the case of sine-generated channels is entirely due to the lack of formulations developed specifically for the case of the latter channels. From this practice, it thus should not be concluded that the expressions of cross-circulatory motion derived for fully developed circular flows are valid as they stand for the case of sine-generated channels (or, to that matter, other types of channels involving a streamwise variation in curvature). Indeed, in the latter the largest value of $u_{r\Gamma}/\overline{u}$ may not necessarily occur at the apex sections (Chang 1988), but rather somewhere in its proximity. Additionally, there seems to be a gradual realization that the expressions derived for fully developed circular flow overestimate the intensity of cross-circulation in channels exhibiting a streamwise variation in curvature (Zhang 2007; Blanckaert 2009; Ottevanger et al. 2012). For these reasons, Eq. (14.6) cannot be viewed as exact for the case of sine-generated channels. However, these matters are inconsequential where the conclusions of the discussion in the previous paragraph are concerned.

14.4 Meandering Morphodynamics: Deformation of Bed and Banks

14.4.1 Bed Deformation

According to Yalin and da Silva (2001), "perhaps the most pervasive and lasting misconception inherited from the literature stemming from the earlier laboratory research on meandering, is the exaggerated importance of cross-circulation (in determining the bed geometry and bank erosion-deposition patterns in meanders)". Indeed, still today any "radial manifestation" of a meandering stream or its boundaries (including both bed and bank deformation) is very often routinely attributed to the cross-circulation—no matter what the value of θ_0 and B/h_{av} might be. "The importance of cross-circulation in determining the geometry of river beds in meanders has been over-emphasized for many years, and it will take some time to bring the significance of such flow patterns into proper perspective" (Hooke 1980). This aspect is further elaborated below, by considering the bed erosion/deposition patterns in meandering streams and how these come into being.

14.4.1.1 Nature of Deformed Bed

In agreement with the current approach, it will be assumed in the following that the initial surface of the movable bed is flat: its slope along the stream centreline l_c is the channel slope S_c, its slope in the radial direction being zero. An experimental "run" commences at time $t = 0$. With the passage of time, the bed progressively deforms until the equilibrium (or developed) state is achieved at time $t = T_b$. The bed remains invariant for any $t \geq T_b$, T_b being the duration of development of the equilibrium bed topography.

As is well-known, the bed development consists of the growth, in the vertical z-direction, of laterally adjacent deposition bars and erosion pools (see as examples Fig. 14.17c, d, showing the developed beds measured by Losiyevskii [as reported by Makaveyvev (1975)] and Jackson (1975), respectively. In the case of sine-generated streams, each pool-bar complex forms a large-scale erosion-deposition zone having the length $L/2$ (which, for short, henceforth will be referred to as a [ED]). The location in flow plan of the [ED]'s remains nearly invariant as pools and bars grow in their magnitude (see Binns and da Silva 2011; Binns 2012).

Just like the location in flow plan of the [CD]'s, that of the [ED]'s is also not of a standard nature, but rather dependent on stream geometry and flow conditions. From their analysis of laboratory data in sine-generated streams, da Silva and El-Tahawy (2008) found that the location in flow plan of the $L/2$-long erosion-deposition zones is strongly dependent on θ_0. In order to characterize this location within a meander loop, da Silva and El-Tahawy (2008) introduced the quantity λ, defined as the distance (normalized by the meander length L) from the apex a_i to the mid-section of the [ED] where deposition occurs at the inner bank and erosion at the outer bank as shown in the schematic Fig. 14.8. Here the loop O_iO_{i+1} is used as reference. Note from Fig. 14.8 that the [ED] under consideration may be only partially located within the meander loop or fully located within it. The plot of the values of λ versus θ_0 produced by the just mentioned authors on the basis of data from the movable bed experiments carried out by Losiyevskii [as reported by Makaveyvev (1975)], Hasegawa (1983), Whiting and Dietrich (1993a, b), Termini (1996), as well as data from their own experiments, is shown in Fig. 14.11. In this figure, the solid line represents the general trend of the data; the data correspond to $\approx 10 \leq B/h_{av} \leq \approx 30$ and $(c_f)_{av} \approx 14.5$, with the (channel-averaged) dimensionless (Chézy) flow friction factor $(c_f)_{av}$ reflecting the influence of the granular skin roughness as well as of the grain size Reynolds number Re_*. A plot similar to Fig. 14.11 can be found in Whiting and Dietrich (1993b), the difference being that the latter authors considered the distance measured along the channel centreline to the cross-section of the deepest point of the pool, instead of λ. The results, however, are perfectly in agreement with the findings of da Silva and El-Tahawy (2008). Note that the graph in Fig. 14.11 implies that if θ_0 is "small", then the most pronounced erosion-deposition occurs around the crossovers O_{i+1} (as in Fig. 14.8a; see also Fig. 14.17c later on); if θ_0 is "large", it occurs around the apex-sections a_i (as in Fig. 14.8c); and if θ_0 is intermediate, then it occurs in an intermediate location with regard to those for small and large values of θ_0 (as in Fig. 14.8b; see also Fig. 14.17d). Observe also that the trend of the solid line in Fig. 14.11 is similar to that best fitting the plot of ξ_{c0} versus θ_0 in Fig. 14.9—which implies that the location of the [ED]'s is nearly identical to that of the [CD]'s of the initial flow. As previously pointed out by da Silva et al. (2006), this means that the information needed for a flow to subsequently generate its bed surface is already locked in the structure of the meandering flow at its initial stage (at $t = 0$, when the bed is still flat).

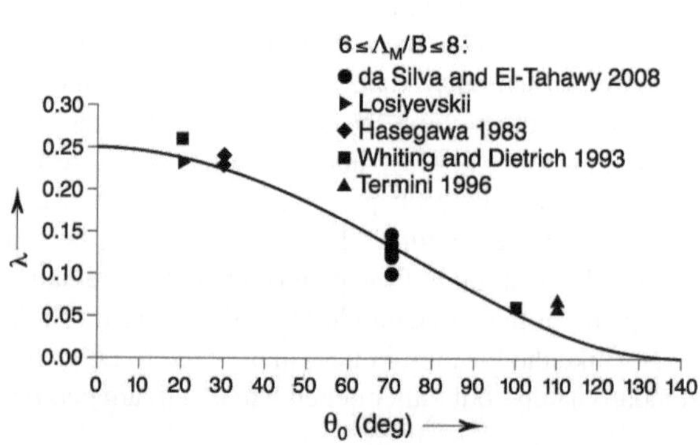

Fig. 14.11 Plot of values of λ versus θ_0 (after da Silva and El-Tahawy 2008)

14.4.1.2 Mechanics of Bed Deformation

(i) From the considerations in Sect. 14.3, it follows that there are two mechanisms responsible for bed deformation in meandering streams, namely convective acceleration/deceleration of flow and cross-circulation. The first of these can be explained on the basis of the sediment transport continuity equation, namely $(1-p)\partial z_b/\partial t = -\nabla \cdot \mathbf{q}_s$, where p is the porosity of bed material, z_b is the bed elevation measured with regard to an arbitrary datum, t is time and $\nabla \cdot \mathbf{q}_s$ is the divergence of the specific (per unit flow width) volumetric bed-load rate vector $\mathbf{q}_s$. As is well-known, the local sediment transport rate $\mathbf{q}_s$ is a strongly increasing function of the vertically-averaged flow velocity vector $\overline{\mathbf{U}}$ (at the same location). Therefore, the convective variation of $\overline{\mathbf{U}}$ in a flow zone inevitably causes the corresponding variation of $\mathbf{q}_s$ in that zone, i.e., it causes the scalar $\nabla \cdot \mathbf{q}_s$ to acquire a non-zero value. From the sediment transport continuity equation, it thus follows that the zones of the downward and upward bed displacements (i.e., the erosion and deposition zones) must coincide with the zones of convective acceleration and deceleration of flow, respectively. But if each [ED] is brought into being by a corresponding [CD], then the length of each [ED] must be the same as that of each [CD], viz $L/2$. Moreover, the location in flow plan of the [ED]'s must necessarily be approximately in coincidence with that of the [CD]'s, as implied by the schematic Fig. 14.8 (more on the topic in da Silva et al. 2006). Cross-circulation, on the other hand, will induce the growth of a pool-bar complex that is invariably located around the apex (see Fig. 14.12). That is, in contrast to the erosion-deposition patterns due to the convective behaviour of the flow, cross-circulation (by itself) is associated with a "standard" erosion-deposition pattern.
The convective behaviour of flow induced by the streamwise variation in curvature is always present, irrespective of the value of B/h_{av}; the intensity of cross-circulation, as follows from Sect. 14.3.3, depends on B/h_{av} (and also θ_0). But this means that for any θ_0, the extent to which Γ will play a role in the deformation of the bed must necessarily depend on the value of B/h_{av}. It is therefore only natural to expect cross-circulation to play an important role in determining the shape of the bed provided that B/h_{av} is sufficiently "small"; and to expect its role to eventually become negligible provided that B/h_{av} is sufficiently "large".

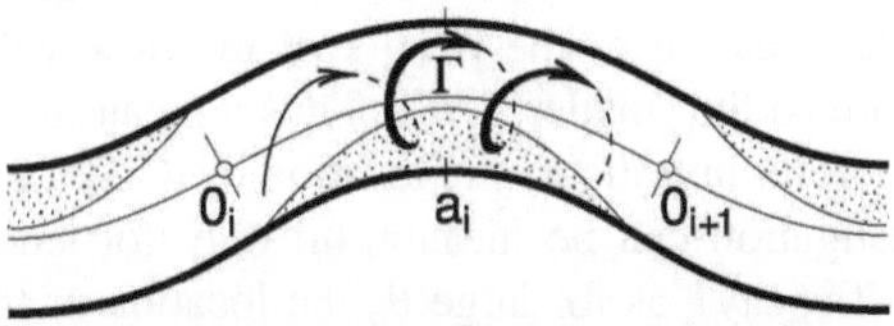

Fig. 14.12 Standard erosion-deposition pattern due to cross-circulation

(ii) On the basis of the content of the previous paragraph, it should be clear that the erosion-deposition patterns described in Sect. 14.4.1.1 can be very satisfactorily explained by attributing them primarily to the convective behaviour of flow arising from the streamwise variation in curvature, with cross-circulation playing only a rather secondary role. In fact, for large low-land meandering rivers, whose values of B/h_{av} can be as high as 100–120 (see, e.g., Fig. 14.3 in Ahmari and da Silva 2011) and where the effect of cross-circulation has since long been deemed only of secondary importance (Matthes 1941; Makaveyvev 1975; Hooke 1974), the convective behaviour of flow is the only means of explaining the resulting erosion-deposition patterns. Such explanation is also consistent with field observations, such as those by Matthes (1941) for the Mississippi River, that the bar in any meander loop is formed mainly by material eroded from the loop immediately upstream and transported downstream along the *same* side of the river; as well as similar observations by Hooke (1974), Whiting and Dietrich (1993a, b), da Silva and El-Tahawy (2008) and Binns and da Silva (2009) in their sine-generated laboratory channels.

(iii) The considerations so far in this paper suggest that the effect of cross-circulation becomes significant only for comparatively small values of B/h_{av}. Yet, no systematic research has been carried out to date to determine, for any given θ_0, what exactly is the "critical value" of B/h_{av} beyond which the effect of cross-circulation becomes of secondary importance where the bed topography is concerned. Such a research would be particularly worthwhile, as it would provide a useful guideline regarding the selection of the numerical hydro and morphological platform to be used in river engineering studies (3D versus 2D numerical models). Considering this, in the following the matter of the "critical value" of B/h_{av} is further explored.

For the present purposes, let us return to the considerations in Sect. 14.3.2. There, for the sake of simplicity, the convective behaviour of flow was entirely attributed to the streamwise variation in curvature. In reality, however, as demonstrated by Kalkwijk and de Vriend (1980), de Vriend (1981) (see also Struiksma et al. 1985; Blanckaert and de Vriend 2003), cross-circulation itself leads to the transport of momentum from the inner to the outer bank as flow moves around a bend. But this means that the convective behaviour of a flow where cross-circulation is negligible ("very large" B/h_{av}) and strictly produced by the streamwise variation in curvature should be somewhat different from that in the same channel but under conditions where cross-circulation is not negligible. We postulate here that, in comparison to the case where cross-circulation is negligible, as the effect of the latter increases the location of the [CD]'s of the flow will be shifted further upstream, and the more so the smaller B/h_{av}. Thus a means to establish the critical value of B/h_{av} would be to investigate the location of [CD]'s for decreasing values of B/h_{av}. Such investigation can be meaningful only for small and intermediate values of θ_0 ($\theta_0 \leq \approx 70°$, say), as for large θ_0 the location of the [CD]'s is around the apex no matter what the value of B/h_{av}.

In addition to affecting the convective flow patterns, and as follows from paragraph (i) above, the cross-circulatory motion can directly act on the grains forming the bed by moving them from the outer to the inner bank. That is, cross-circulation can determine the bed topography (or play a role in determining it) indirectly via its action on the convective flow patterns as well as directly through its effect on sediment transport patterns. Through this combined effect, one would expect the location of the [ED]'s too, to be shifted further upstream in a case where cross-circulation is not negligible when compared to one where it is negligible—and the more so, the smaller B/h_{av}. Hence, an alternative means to establish the critical values of B/h_{av} would be to investigate differences in the location of the [ED]'s for decreasing values of B/h_{av}. Note that this would not be meaningful for the case of large values of θ_0, as in this case the pool-bar complexes are invariably located around the apexes, no matter whether solely caused by the convective behaviour of flow due to the streamwise variation in curvature, solely by cross-circulation, or a combination of the two. However, in this case, there should be detectable differences in the shape of the pool-bar complexes. That is, for large values of θ_0, one would need to search for differences in the shape of the pool-bar complexes, instead of their location.

The above hypothesis was tested in da Silva et al. (2008), by plotting the measured values of λ resulting from two series of laboratory experiments in a 70° sine-generated channel, as shown in Fig. 14.13. Observe from this figure that λ remains approximately constant as long as $B/h_{av} > \approx 20$, but for $B/h_{av} < \approx 20$ it indeed gradually decreases with B/h_{av}. The experiments did not include tests with $B/h_{av} < 10$, but the data trend indicates that for such values of B/h_{av} the location in flow plan of the [ED]'s would become significantly different from that in streams having $B/h_{av} > \approx 20$.

The aforementioned suggests that for $\theta_0 = 70°$, the critical value of B/h_{av} beyond which the effect of cross-circulation becomes of secondary importance where the bed deformation is concerned is ≈ 20.

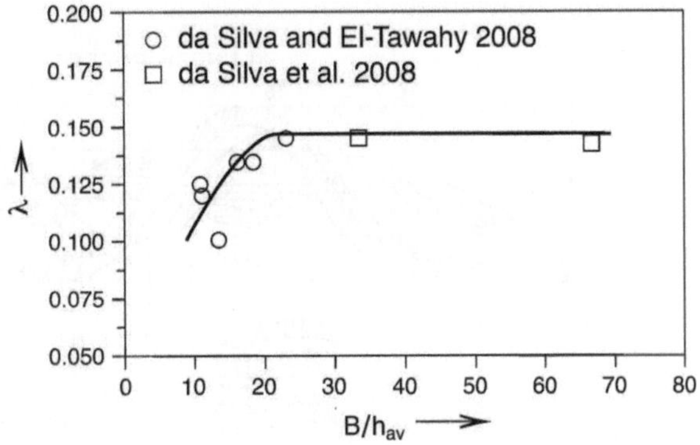

Fig. 14.13 Plot of values of λ measured in a 70° channel for decreasing values of B/h_{av} (after da Silva et al. 2008)

14.4.2 Bank Deformation

14.4.2.1 Downstream Migration and Lateral Expansion of Meander Loops

In the previous section, the streams were treated as having rigid (or virtually rigid) banks. However, as is well known, and as illustrated by the photo in Fig. 14.14, at the same time that the bed deforms, the banks deform too—and as a result, the meander loops evolve in plan with the passage of time. Such evolution occurs though (simultaneous) downstream migration and loop expansion as exemplified by Fig. 14.15, showing the results of one of the laboratory runs by Friedkin (1945). From field measurements in European and American rivers, but especially from

Fig. 14.14 Erosion of outer bank in the Hardebek-Brokelander Au River, in northern Germany (courtesy of Dr. M. Hassan Nasermoaddeli, BAW, Hamburg, Germany)

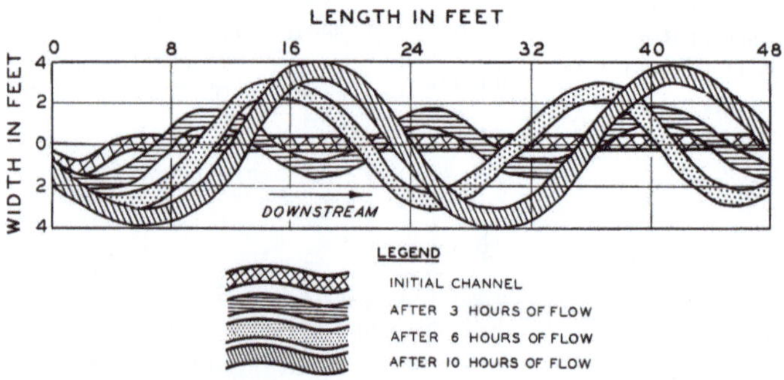

Fig. 14.15 Evolution of a laboratory meandering stream through downstream migration and lateral expansion (experiment by Friedkin 1945)

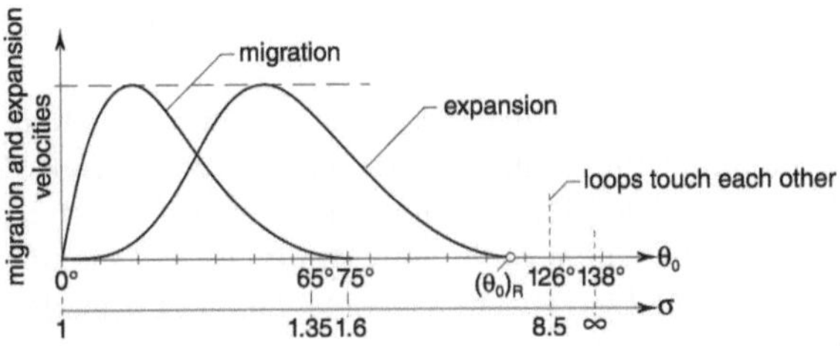

Fig. 14.16 Schematic graph showing the variation with θ_0 of (normalized) downstream migration and lateral expansion velocities of meandering streams (after Kondratiev et al. 1982)

field surveys carried out over long periods of time in Russian rivers including the Dnieper, Oka, Irtish, etc. (compiled and analysed by Kondratiev et al. 1982), it follows that the (normalized) migration velocity and the expansion speed of freely meandering rivers varies with θ_0 as shown in the schematic Fig. 14.16. "At the early stages (small θ_0), it is the downstream migration of the meander waves which is mainly observable, at the latter stages (large θ_0), it is their expansion which dominates" (Kondratiev et al. 1982, p. 108).

From the aforementioned, it follows that the patterns of migration and expansion described above are also most satisfactorily explained on the basis of the convective behaviour of the flow. The banks thus are eroded mostly in those locations where the bed adjacent to them is eroded; with a similar situation applying to deposition. This explains why we have mainly migration for "small" θ_0, mainly expansion for "large" θ_0 (see Fig. 14.17), and a combination of migration and expansion for "intermediate" values of θ_0.

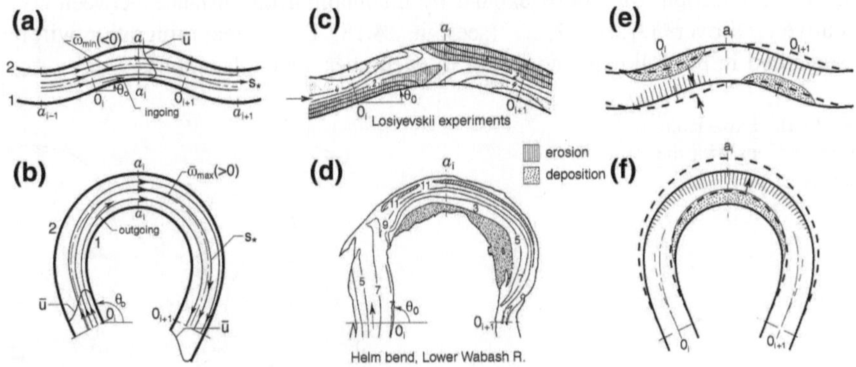

Fig. 14.17 Patterns of convective flow, bed erosion-deposition and planimetric downstream migration and lateral expansion in streams having "small" θ_0 (**a**, **c** and **e**) and "large" θ_0 (**b**, **d** and **f**). Figures are schematic representations, except (**c**) and (**d**): **c** measured by Losiyevskii (as reported by Makaveyvev 1975); **d** measured by Jackson (1975)

14.4.2.2 Computation of Meandering Planimetric Evolution

Recent research on meandering is marked by substantial efforts to extend previous models (Nelson and Smith 1989; Shimizu and Itakura 1989; Struiksma and Crosato 1989; Jia and Wang 1999, etc.) for the computation of bed deformation, invariably achieved by coupling the solution of the flow continuity and momentum equations with a sediment transport solver, also to the computation of bank deformation, with the goal of capturing the plan development of meandering streams (see, e.g., Nagata et al. 2000; Darby et al. 2002; Olsen 2003; Duan and Julien 2005; Chen and Duan 2006; Rüther and Olsen 2007; Zolezzi et al. 2012; Nasermoaddeli 2012; Motta et al. 2012). Only a brief outline of the rather involved process of bankline shifting was given in the previous section. No mention was made to bank geotechnical failure, near-bank flow structures, evaluation of bank shear stress and sediment transport, interaction between bed and bank, etc.—all of which are not yet completely understood and/or not satisfactorily described mathematically. The problem of developing physically-based, reliable numerical models for the simulation and prediction of the planimetric evolution of meandering rivers is compounded by the spatial and temporal scales involved. Planimetric adjustments of equilibrium in a meandering stream can range from a matter of hours/days in a laboratory stream, years in a small creek, to many decades (and sometimes centuries) in large rivers. An insightful discussion of some of the existing challenges has been presented by Mosselman (1995) in a paper which, despite substantial progress in the field over the past ten years, nonetheless remains nowadays as up-to-date as when it was first published (see also Crosato 2007). A comprehensive review of post-1995 research efforts has been presented by Nasermoaddeli (2012).

Yet, a different matter that has not yet been addressed emerges if we consider the time-growth of meander loops in the light of regime development, as done below. Taking into account that loops expand by maintaining the distance between consecutive crossovers O_1, O_2, O_3, ... (see Fig. 14.18), downstream migration will be disregarded in the following so as not to encumber the explanations.

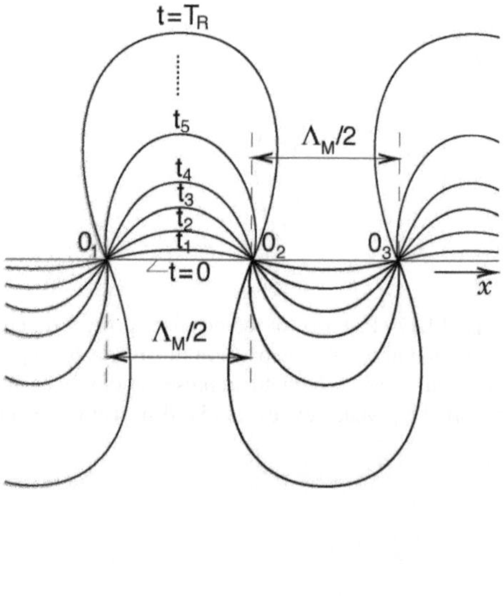

Fig. 14.18 Expansion of meander loops with the passage of time

14.4.2.3 Regime Development and Time-Growth of Meander Loops

In their classical treatments, meandering and regime channels used to be regarded as independent fluvial phenomena. The first suggestions that the phenomena mentioned may not really be independent are due to Bettess and White (1983), Chang (1988). An outline of the time-growth of meander loops in the light of the regime-trend following Yalin (1992), Yalin and da Silva (2001) (see also da Silva 2009) is given below.

(i) Before entering the topic of meandering, it seems in order to first clarify the nature of changes that a stream undergoes when striving to achieve a stable state. For this purpose, and in agreement with convention in classical regime theory, let us consider an experiment which starts at $t = 0$ in a straight initial channel excavated in an alluvial valley. The slope S_0 of the initial channel is the same as the valley slope S_v, i.e. $S_0 = S_v$. It is assumed that the granular material and fluid are specified, that the flow rate Q is given ($Q = Q_{bf} = const$, Q_{bf} being the bankfull flow rate), and that the conditions are such that sediment can be transported. It is also assumed that the initial channel (B_0, h_0, S_0) is such that the formation of the regime channel (B_R, h_R, S_R) is possible. The duration of formation of the regime channel is T_R. The laboratory research (see, e.g., Ackers 1964; Leopold and Wolman 1957) indicates that the variation of the flow width B, the flow depth h, and the slope S during T_R takes place as shown in the schematic Fig. 14.19. In the (very short) part $\hat{T}_0$ of T_R, B and h vary substantially, while S remains nearly constant ($S \approx S_0$); no regime development as such takes place. The part $\hat{T}_0$ of T_R is merely the duration needed to alter (the arbitrary) B_0 and h_0 into such $\hat{B}_0$ ($\approx B_R$) and $\hat{h}_0$, say, which are in equilibrium with the existing $S \approx S_0$ and which, together with S_0, are able to convey the given flow rate Q. The regime development in the proper sense takes place only after the *adjustment period* $\hat{T}_0$.

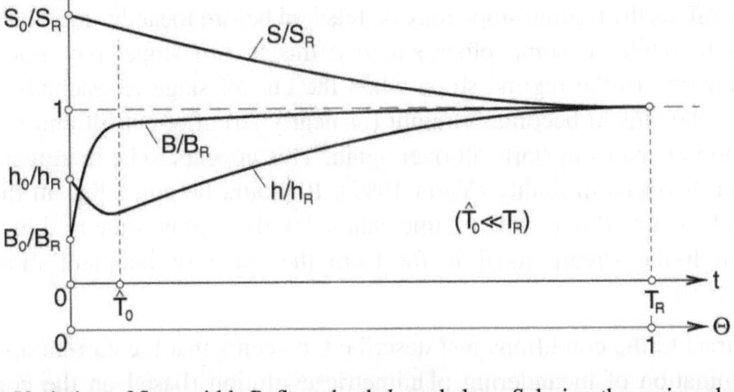

Fig. 14.19 Variation of flow width B, flow depth h, and slope S during the regime development. Subscripts 0 and R mark the initial and regime (or stable) states, respectively

(ii) According to the contemporary rational approaches to regime, the regime development is a process in which the stream appropriately alters its channel so that a certain energy-related quantity, A_* say, may be minimized. Although different authors proposed different quantities as A_* [e.g., according to Chang (1988), $A_* = \gamma QS$; according to Yang et al. (1981), $A_* = u_{av}S$; according to Jia (1990), Yalin (1992), $A_* = Fr$, where $Fr \sim S$; according to Yalin and da Silva (2001), $A_* = u_{av}$ (see also da Silva 2009, 2013)], almost invariably A_* is such that its minimization can only be achieved through the decrement of the slope. This is in agreement with the aforementioned experimental observations. Here γ is specific weight of fluid, u_{av} is average flow velocity, and Fr is Froude number.
Clearly, the decrement of the slope (from S_0 to S_R) can only be realized either by degradation-aggradation or by meandering (for the expansion of meander loops (see Fig. 14.18), i.e., the increment of their length, means the decrement of the channel slope)—or by a combination of both. The development stops, and thus the expansion of meander loops stops, at $t = T_R$ when $S = S_R$. In the case of large sand-bed rivers, the regime development is accomplished primarily by meandering. This is because the regime slope of large sand-bed rivers is usually rather small and, as pointed out by Chang (1988), "reduction of channel slope through incision would require tremendous degradation. For these reasons, the river channel usually adjusts by developing a flatter slope through meandering" (Chang 1988, p. 313). [Note that the matter of how exactly a straight stream initiates its meandering is to be considered and treated as a topic independent from the time-growth of meander loops (see, e.g., da Silva 2006). Accordingly, in the considerations below, the initial stream is to be viewed as a stream which already exhibits an incipient meandering at $t = \hat{T}_0$].
(iii) Consider now a stream whose initial slope (valley slope) is S_V, and which acquires at $t = T_R$ its regime slope S_R. Since the valley slope is arbitrary (i.e., it can take any value), the sinuosity of the stream at the regime state, namely $\sigma_R = S_V/S_R$, can also take any value (from small to large). But this means that, according to regime theory as outlined above, not all streams evolve to cut-off, as the regime slope may be reached before meandering loops begin to touch; while in some other streams, the stream slope may not yet have decreased to the regime slope when the cut-off stage is reached—in which case the stream becomes straight (or nearly so) after cut-off and the process of loop expansion starts all over again. This appears to be in agreement with what happens in reality (Yalin 1992). It should be noted that in the case of sand streams ("live-bed" regime channels), the regime state is achieved even though the stream itself is far from the stage of incipient motion (i.e., $q_s \gg 0$).

In contrast to the conditions just described, it seems that the current approach to the determination of meandering planimetric evolution (based on the coupling of the solution of the flow continuity and momentum equations with bed and bank sediment transport solvers) will always result in loop expansion until cut-off (unless, of course, the stream reaches the critical stage of initiation of motion before that). This, as previously pointed out by Yalin and da Silva (2001), suggests that the current approach is not complete as it stands. According to the just mentioned authors, it should be augmented so as to take into account also the conditions present because of the regime development. The matter is complex, but it is clear that there is a need to unify the two approaches, as the contradiction that results from their application cannot be ignored.

14.5 Conclusions

The present paper focuses on the kinematics of flow in sine-generated streams and its relationship to bed and bank erosion-deposition patterns.

It is shown that in the case of large values of width-to-depth ratio the meandering pool-bar complexes and the modes of bank deformation (downstream migration and lateral expansion) can be very satisfactorily explained by attributing them primarily to the convective behaviour of flow due to the streamwise variation in curvature. Two possible methods for the determination of the "critical value" of B/h_{av} beyond which cross-circulation becomes of secondary importance where the bed deformation is concerned, are discussed. Application of one of these methods to data from existing laboratory experiments suggests that for $\theta_0 = 70°$, the critical value of B/h_{av} is $\approx$20. This result, however, cannot be extrapolated for different values of θ_0, as the critical value of B/h_{av} should itself be a function of θ_0.

The time-growth of meander loops is analysed in view of regime theory. It is shown that this and the current approach to the determination of meandering planimetric evolution based on the flow kinematic and sediment transport equations may, in many instances, lead to different values of the maximum value of sinuosity that a stream may reach. Such a contradiction highlights the need to consolidate the two approaches. Even though no attempt has been made so far in this direction, this appears as a particularly worthwhile line of research, with the potential to greatly deepen our understanding of the fluvial system.

References

Abad JD, Garcia MH (2009a) Experiments in a high-amplitude Kinoshita meandering channel: 1. Implications of bend orientation on mean and turbulent flow structure. Water Resour Res 45: W02401. doi:10.1029/2008WR007016

Abad JD, Garcia MH (2009b) Experiments in a high-amplitude Kinoshita meandering channel: 2. Implications of bend orientation on bed morphodynamics. Water Resour Res 45:W02402. doi:10.1029/2008WR007017

Ackers P (1964) Experiments on small streams in alluvium. J Hydraul Div ASCE 90(HY4):1–37

Ahmari H, da Silva AMF (2011) Regions of bars, meandering and braiding in da Silva and Yalin's plan. J Hydr Res 49(6):718–727

Bettess R, White WR (1983) Meandering and braiding of alluvial channels. Proc Instn Civ Eng Part 2 75(3):525–538

Binns AD (2012) Nature and time-scale of bed morphological adjustments toward equilibrium in meandering streams: an experimental study. Ph.D. thesis, Queen's University, Kingston, Canada

Binns AD, da Silva AMF (2009) On the quantification of the bed development time of alluvial meandering streams. J Hydr Eng 135(5):350–360

Binns AD, da Silva AMF (2011) Rate of growth and other features of the temporal development of pool-bar complexes in meandering streams. J Hydr Eng 137(12):1565–1575

Blanckaert K (2009) Saturation of curvature-induced secondary flow, energy losses, and turbulence in sharp open-channel bends: laboratory experiments, analysis, and modeling. J Geophys Res 114:F03015. doi:10.1029/2008JF001137

Blanckaert K (2010) Topographic steering, flow recirculation, velocity redistribution, and bed topography in sharp meander bends. Water Resour Res 46:W09506. doi:10.1029/2009WR008303

Blanckaert K, de Vriend HJ (2003) Nonlinear modeling of mean flow redistribution in curved open channels. Water Resour Res 39(12):1375. doi:10.1029/2003WR002068

Blanckaert K, de Vriend HJ (2004) Secondary flow in sharp open-channel bends. J Fluid Mech 498:353–380

Boraey AA (2014) Alternate bars under steady-state flows: time of development and geometric characteristics. Ph.D. thesis, Queen's University, Kingston, Canada

Boraey AA, da Silva, AMF (2014) A new equation for alternate bar length. In: Schleiss AJ, de Cesare G, Franca MJ, Pfister M (eds) Proceedings of river flow 2014, 7th international conference on fluvial hydraulics, Lausanne, Switzerland, 3–5 Sept 2014. CRC Press, Taylor & Francis Group, London. ISBN 9781138026742, 1195-1202

Bridge JS (2003) Rivers and floodplains: forms, processes and the sedimentary record. Blackwell, Oxford

Chang HH (1988) Fluvial processes in river engineering. Wiley, New York

Chen D, Duan JD (2006) Simulating sine-generated meandering channel evolution with an analytical model. J Hydr Res 44(3):363–373

Crosato A (2007) Effects of smoothing and regridding in numerical meander migration models. Water Resour Res 43:W01401. doi:10.1029/2006WR005087

da Silva AMF (1995) Turbulent flow in sine-generated meandering channels. Ph.D. thesis, Queen's University, Kingston, Canada

da Silva AMF (2006) On why and how do rivers meander. J Hydr Res 44(5):579–590

da Silva AMF (2009) On the stable geometry of self-forming alluvial channels: theory and practical application. Can J Civil Eng 36(10):1667–1679

da Silva AMF (2013) River self-formation and regime channel geometry in the light of thermodynamic principles. In: Proceedings of 35th IAHR congress, Chengdu, China, 9–13 Sept, 12 pp

da Silva AMF, El-Tahawy T (2008) On the location in flow plan of erosion-deposition zones in sine-generated meandering streams. J Hydr Res 46(Extra Issue 1):49–60

da Silva AMF, El-Tahawy T, Tape WD (2006) Variation of flow pattern with sinuosity in sine-generated meandering channels. J Hydr Eng 132(10):1003–1014

da Silva AMF, Holzwarth S, Pasche E, El-Tahawy T (2008) Bed topography of alluvial meandering streams under varying width-to-depth-ratios. In: Proceedings of river flow 2008, 4th international conference on fluvial hydraulics, Cesme-Izmir, Turkey, 3–5 Sept 2008, pp 1363–1372

Darby SE, Alabyan AM, Van de Wiel MJ (2002) Numerical simulation of bank erosion and channel migration in meandering rivers. Water Resour Res 38(9):1163. doi:10.1029/2001WR000602 de Leliavsky N (1894) Currents in streams and the formation of stream beds. In: Sixth international congress on internal navigation, The Hague, The Netherlands

de Vriend HH (1981) Velocity redistribution in curved rectangular channels. J Fluid Mech 107:423–439. doi:10.1017/S0022112081001833

Duan JG, Julien PY (2005) Numerical simulation of the inception of channel meandering. Earth Surf Proc Land 30:1093–1110. doi:10.1002/esp.1264

Engels H (1926) Movement of sedimentary materials in river bends. In: Freeman JR (ed) Hydraulic laboratory practice. ASME, New York

Engelund F (1974) Flow and bed topography in channel bends. J Hydr Div 100(11):1631–1648

Fargue L (1908) La forme du lit des rivières à fond mobile. Gauthier-Villars, Paris

Friedkin JF (1945) A laboratory study of the meandering of alluvial rivers. U.S. Waterways Experiment Station, Vicksburg, Mississippi

Garde RJ, Raju KGR (1977) Mechanics of sediment transportation and alluvial stream problems. Wiley Eastern, New Delhi

Hasegawa K (1983) Hydraulic research on planimetric forms, bed topographies and flow in alluvial rivers. Ph.D. thesis (in Japanese), Hokkaido University, Sapporo, Japan

Hasegawa K, Yamaoka I (1983) A study on flows and bed topographies in meandering channels. In: Elliott CM (ed) Proceedings of conference on Rivers'83. River meandering. ASCE, New York

Hooke RL (1974) Distribution of sediment transport and shear stress in a meander bend, Rept. 30. Uppsala University Naturgeografiska Inst., 58

Hooke RL (1980) Shear-stress distribution in stable channel bends, discussion. J Hydraul Div 106 (HY7):1271–1272

Ikeda S (1984) Prediction of alternate bar wavelength and height. J Hydr Eng 100(4):371–386

Ikeda S, Nishimura T (1986) Flow and bed profile in meandering sand-silt rivers. J Hydr Eng 112 (7):562–579

Ikeda I, Parker G, Sawai K (1981) Bend theory of river meanders. Part 1. Linear development. J Fluid Mech 112:363–377

Inglis CC (1947) Meanders and their bearing on river training, Maritime and Waterways Engrg. Div., Inst. Civ. Eng., London

Jackson RJ (1975) Velocity-bed-form-texture patterns of meander bends in the lower Wabash river of Illinois and Indiana. Geol Soc Am Bull 86:1511–1522

Jefferson M (1902) The limiting width of meander belts. Natl Geogr Mag, 372–384

Jia Y (1990) Minimum Froude number and the equilibrium of alluvial sand rivers. Earth Surf Proc Land 15:199–209

Jia Y, Wang SSY (1999) Numerical model for channel flow and morphological change studies. J Hydr Eng 125(9):924–933

Kalkwijk JPTh, de Vriend HJ (1980) Computation of the flow in shallow river bends. J Hydr Res 18(4):327–342

Kashyap S, Constantinescu G, Rennie CD, Post G, Townsend R (2012) Influence of channel aspect ratio and curvature on flow, secondary circulation and bed shear stress in a bend. J Hydr Eng 138(12):1045–1059

Kassem AK, Chaudhry MH (2002) Numerical modeling of bed evolution in channel bends. J Hydr Eng 128(5):507–514

Kinoshita R (1961) Investigation of channel deformation in Ishikari river. Rep. Bureau of Resources, Department of Science and Technology, Japan

Kondratiev N, Popov I, Snishchenko B (1982) Foundations of hydromorphological theory of fluvial processes (in Russian). Gidrometeoizdat, Leningrad

Langbein WB, Leopold LB (1966) River meanders—theory of minimum variance. U.S. Geol. Survey Prof. Paper 422-H, 1–15

Leliavsky S (1959) An introduction to fluvial hydraulics. Constable and Company, Russell Square

Leopold LB, Langbein WB (1966) River meanders. Sci Am 214(6):60–70

Leopold LB, Wolman MG (1957) River channel patterns: braided, meandering and straight. U.S. Geol. Survey Professional Paper 282-B, 39–73

Leopold LB, Wolman MG (1960) River meanders. Geol Soc Am Bull 71(6):769–793

Leopold LB, Wolman MG, Miller JP (1964) Fluvial processes in geomorphology. W.H. Freeman, San Francisco

Makaveyvev NI (1975) River bed and erosion in its basin. Press of the Academy of Sciences of the USSR, Moscow

Matthes GH (1941) Basic aspects of stream meanders. Trans. Am. Geophys. Union, pp 632–636

Mosselman E (1995) A review of mathematical models of river planform changes. Earth Surf Proc Land 20:661–670

Motta D, Abad JD, Langendoen EJ, Garcia MH (2012) A simplified 2D model for meander migration with physically-based bank evolution. Geomorphology 163–164:10–25

Movshovitz-Hadar N, Shmukler A (2006) River meandering and a mathematical model of this phenomenon. Physica Plus, Online magazine of the Israel Physical Society (IPS), Issue No. 7. www.physicaplus.org.il

Nagata N, Hosoda T, Muramoto Y (2000) Numerical analysis of river channel processes with bank erosion. J Hydr Eng 126(4):243–252

Nasermoaddeli MH (2012) Bank erosion in alluvial rivers with non-cohesive soil in unsteady flow. Hamburger Wasserbauschriften 14, TuTech Verlag, Hamburg, Germany

Nelson JM, Smith JD (1989) Evolution and stability of erodible channel beds. In: Ikeda S, Parker G (eds) River meandering. Water Resour. Monograph, Am. Geophys. Union, 12, 321–378

Olsen NRB (2003) Three-dimensional CFD modeling of self-forming meandering channel. J Hydr Eng 129(5):366–372

Ottevanger W, Blanckaert K, Uijttewaal WSJ (2012) Processes governing the flow redistribution in sharp river bends. Geomorphology 163–164:45–55

Parker G, Sawai K, Ikeda S (1982) Bend theory of river meanders, part 2. Nonlinear deformation of finite-amplitude bends. J Fluid Mech 115:303–314

Rhoads BL, Welford MR (1991) Initiation of river meandering. Prog Phys Geo 15(2):127–256

Rinaldi M, Johnson P (1997) Characterization of stream meanders for stream restoration. J Hydr Eng 123(6):567–570

Rüther N, Olsen NRB (2007) Modelling free-forming meander evolution in a laboratory channel using three-dimensional computational fluid dynamics. Geomorphology 89:308–319. doi:10.1016/j.geomorph.2006.12.009

Seminara G (2006) Meanders. J Fluid Mech 554:271–297

Seminara G, Zolezzi G, Tubino M, Zardi D (2001) Downstream and upstream influence in river meandering. Part 2. Planimetric development. J Fluid Mech 438:213–230

Shimizu Y, Itakura T (1989) Calculation of bed variation in alluvial channels. J Hydr Eng 115 (3):367–384

Smith JD, McLean SR (1984) A model for flow in meandering streams. Water Resour Res 20 (9):1301–1315

Soar PJ, Thorne CR (2001) Channel restoration design for meandering rivers. Report ERDC/CHL CR-01-1, ERDC, US Army Corps of Engineers, Vicksburg, Mississippi

Struiksma N, Crosato A (1989) Analysis of a 2D bed topography model for rivers. In: Ikeda S, Parker G (eds) River meandering. Water Resour. Monograph, Am. Geophys. Union, 12, 153–180

Struiksma N, Olesen KW, Flokstra C, de Vriend HJ (1985) Bed deformation in curved alluvial channels. J Hydr Res 23(1):57–79

Termini D (1996) Evolution of a meandering channel with an initial flat bed: theoretical and experimental study of the channel bed and the initial kinematic characteristics of flow. Ph.D. thesis (In Italian), Department of Hydraulic Engineering and Environmental Applications, University of Palermo, Italy

Termini D (2009) Experimental observations of flow and bed processes in large-amplitude meandering flume. J Hydr Eng 135(7):575–587

Termini D, Piraino M (2011) Experimental analysis of cross-sectional flow motion in a large amplitude meandering bend. Earth Surf Proc Land 36(2):244–256

Thomson J (1879) On the flow of water round river bends. Proceedings of Institution of Mechanical Engineers, vol 6, Aug

USDA (2007) Stream restoration design, part 654, National Engineering Handbook, U.S. Department of Agriculture

Von Schelling H (1951) Most frequent particle path in a plane. Trans Am Geophys Union 32: 222–226

Welford MR (1993) Field evaluation of empirical equations in straight alluvial channels. Phys Geogr 14(6):581–598

Whiting PJ, Dietrich WE (1993a) Experimental studies of bed topography and flow patterns in large-amplitude meanders. 1. Observations. Water Resour Res 29(11):3605–3614

Whiting PJ, Dietrich WE (1993b) Experimental studies of bed topography and flow patterns in large-amplitude meanders. 2. Mechanisms. Water Resour Res 29(11):3615–3622

Xu D, Bai Y (2013) Experimental study on the bed topography evolution in alluvial meandering rivers with various sinuousnesses. J Hydro Env Res 7(2):92–102

Yalin MS (1992) River mechanics. Pergamon Press, Oxford

Yalin MS, da Silva AMF (2001) Fluvial processes, IAHR Monograph, IAHR, Delft, The Netherlands

Yang CT, Song CCS, Woldenberg MJ (1981) Hydraulic geometry and minimum rate of energy dissipation. Water Resour Res 17:1014–1018

Zeller J (1967) Meandering channels in Switzerland. In: International symposium on river morphology, Bern, IASH

Zhang Y (2007) On the computation of flow and bed deformation in alluvial meandering streams. Ph.D. thesis, Queen's University, Kingston, Canada

Zolezzi G, Guala M, Termini D, Seminara G (2005) Experimental observations of upstream overdeepening. J Fluid Mech 531:191–219

Zolezzi G, Luchi R, Tubino M (2012) Modeling morphodynamic processes in meandering rivers with spatial width variations. Rev Geophys, 50, RG4005, doi: 10.1029/2012RG000392

Chapter 15
Fluvial Processes in Braided Rivers

Nicola Surian

Abstract Braided rivers are characterized by an unstable network of multiple channels and very active channel processes. They can be found in different climate regions (e.g. from glacial areas to arid regions) and physiographic settings (e.g. from steep mountain areas to low coastal plains). This chapter aims to summarize the present knowledge about braided rivers and to point out some gaps that still require further research. The first part of the chapter focuses on channel processes while the second part deals with channel changes through time. As for the first part, the following processes are illustrated: bar formation and development; processes at bifurcations and confluences; lateral mobility; bedload transport; role of vegetation on river morphodynamics. The second part deals with the evolution of braided rivers, mainly in response to human alteration of fluvial systems, over some decades up to some centuries. The historical perspective is crucial to understand present morphology and processes, as well as to assess future channel evolution. Understanding about braided river morphology and processes has significantly increased over the last three decades or so, through physical modeling, field observations, and, to a lesser extent, numerical modeling. On the other hand, several open questions still remain and future research should address aspects such as metrics used to characterized braided rivers, measurements of flow and sediment transport, and prediction of future channel evolution.

Keywords Braided morphodynamics · Bars · Bedload transport · Lateral mobility · Channel pattern · Channel changes

---

N. Surian (✉)
Department of Geosciences, University of Padova, Via G. Gradenigo 6, 35131 Padova, Italy
e-mail: nicola.surian@unipd.it

P. Rowiński and A. Radecki-Pawlik (eds.), *Rivers—Physical, Fluvial and Environmental Processes*, GeoPlanet: Earth and Planetary Sciences,
DOI 10.1007/978-3-319-17719-9_15

15.1 Introduction

Braided rivers have a peculiar morphology and are characterized by very active channel processes. Braided morphology is characterized by an unstable network of multiple channels separated by ephemeral bars. Channel processes include frequent bar development and migration, intense bedload transport, high rates of bank retreat. Braiding processes occur when there are non-cohesive materials (e.g. gravel and sand), high sediment supply, and absence or limited lateral confinement.

Braided rivers can be found in different climate regions, e.g. from glacial areas to arid regions, and in very different physiographic settings, e.g. from steep mountain areas to low coastal plains. This means that braiding occurs in a very wide range of river scales, from small streams with drainage basins of few km^2 (e.g. in proglacial environments) to very large low gradient rivers draining basins over 10^5 km^2 (e.g. the Brahmaputra River). Consequently, sediment size of braided rivers ranges from cobbles to silt. Because high sediment supply is a key ingredient for occurrence of braiding, the spatial extent of braiding along a river and its persistence through time may change also over relatively short time periods (i.e. few decades or centuries) in relation to natural causes (e.g. occurrence of large floods, earthquakes, volcanic eruptions) and/or human modifications within the fluvial system (e.g. dams, land use changes, sediment mining) (e.g. Surian and Rinaldi 2003; Piégay et al. 2009).

Understanding of braided river morphology and processes has developed through physical modeling, field observations, and, to a lesser extent, numerical modeling (Ashmore 2013). Physical modeling has been very effective for investigation of braided morphology and morphodynamics over more than thirty years (e.g. Ashmore 1982; Young and Davies 1991; Hoey and Sutherland 1991). Field studies are not easy to carry out in a braided system but remote sensing and recent technologies (e.g. LiDAR, TLS) have improved our capability of observation in the field (Lane et al. 2003; Bertoldi et al. 2011b; Brasington et al. 2012). Insights about braiding derive also from numerical modeling (e.g. Murray and Paola 1994; Nicholas 2000; Thomas et al. 2007; Ziliani et al. 2013), but still several issues limit the use of such models to braided rivers.

The aim of this chapter is to summarize the present knowledge about braided rivers and, at the same time, to point out some open questions that still require further research. In the first section of the chapter the morphological characters of braided rivers are illustrated. The second section deals with the different processes that take place in a braided system. In the third section it is described how braided rivers may change through time, specifically in response to human alteration of fluvial systems. The last section deals with thresholds between channel patterns.

15.2 Braided River Morphology

Although the chapter focuses on processes, this section is intended to give a clear description of morphological features in braided rivers and clarify terminology commonly used to describe such features. Aerial photographs and satellite images have been commonly used to describe the main elements of braided river planform (e.g. Surian 1999; Piégay et al. 2009; Fotherby 2009; Bertoldi et al. 2011a; Belletti et al. 2014), while, more recently, new technologies have allowed construction of Digital Elevation Models and, therefore, more comprehensive description of braided morphology (e.g. Wheaton et al. 2013; Lallias-Tacon et al. 2014; Williams et al. 2014; Javernick et al. 2014) (Fig. 15.1).

The braidplain, also called "braid belt" or "braided channel belt" (e.g. Lane 2006; Ashmore 2013), is composed by different bar types (e.g. unit bars, complex braid bars), main and secondary channels, commonly called anabranches, and, in some cases, islands. Bars are the main depositional features and are exposed most of the time. Anabranches, representing the lowest topographic features, can carry water most of the time in more humid environments, but can be also dry for most of the year in arid climates or in small streams. Islands are commonly the highest

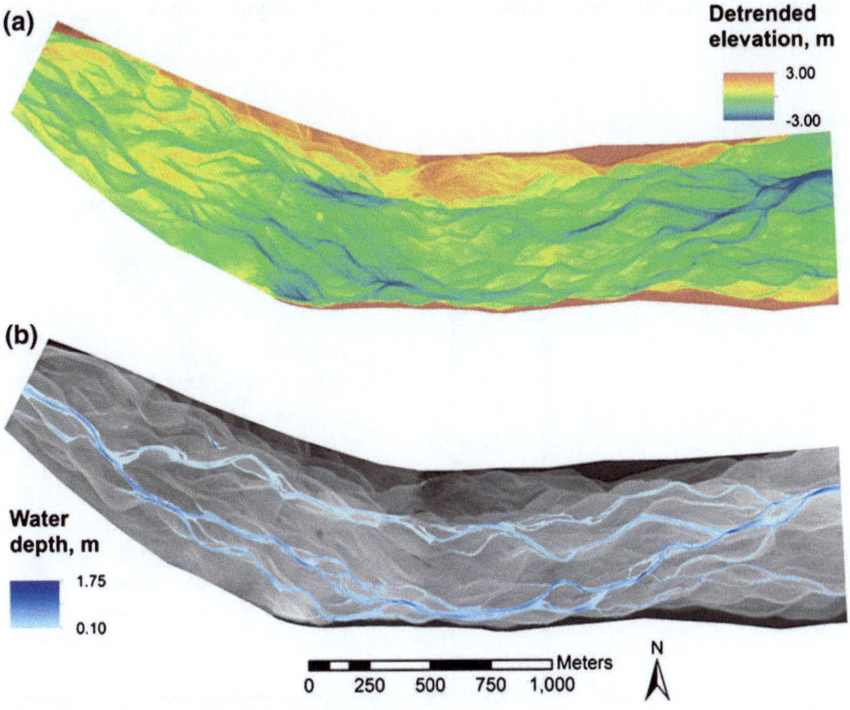

Fig. 15.1 Detrended digital elevation model (**a**) and bathymetric map (**b**) of the Ahuriri River, New Zealand (from Javernick et al. 2014)

topographic features displaying high vegetation cover, that is a significant portion occupied by shrubs and trees.

Braiding intensity, or degree of braiding, is a basic morphological property analogous to sinuosity in single-thread channels. Several indices have been used to measure braiding intensity, but the two most commonly used are the total sinuosity index (e.g. Hong and Davies 1979; Robertson-Rintoul and Richards 1993) and the channel count index (e.g. Howard et al. 1970; Ashmore 1991a). According to Egozi and Ashmore (2008), the channel count index gives the best combination of rapid measurements, precision, and range of sources from which the index can be reliably carried out. The estimate of this index implies counting of wet anabranches in a minimum number of cross sections which should be spaced no further apart than the average wetted width of the river (Egozi and Ashmore 2008). It is worth noting that braiding intensity is flow stage dependent; therefore, intensity should be measured at different flow stages or be referred to an index discharge.

Channel bifurcations and confluences are other basic morphological features of braided rivers. Hundey and Ashmore (2009) found that there is a linear relationship between channel width and the distance between anabranch confluences and downstream bifurcation. In a physical model, confluence-bifurcation lengths turned out to be 4–5 times the width of the main channel (Fig. 15.2). These findings, as well as other morphological observations, have suggested that braided rivers may be self-similar fractals (Sapozhnikov and Foufoula-Georgiou 1996).

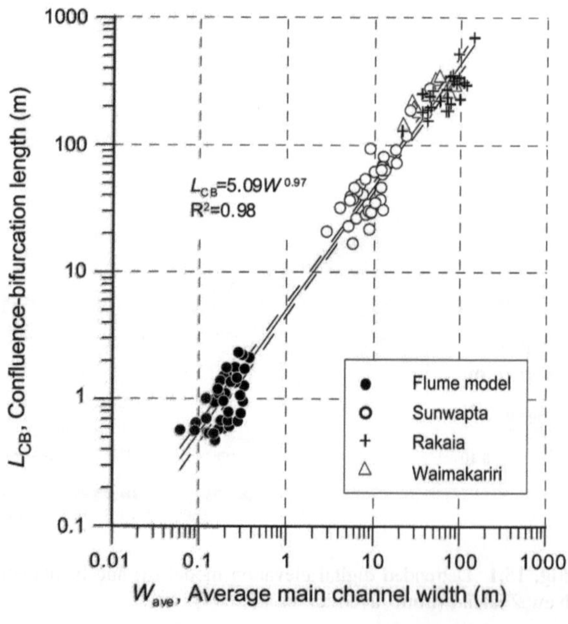

Fig. 15.2 Relation between the average width of the main channel and confluence-bifurcation length (from Hundey and Ashmore 2009)

15.3 Processes in Braided Streams

15.3.1 Bar Formation and Development

Bars are fundamental features in braided rivers because (i) their formation has a key role in development of braided pattern and (ii) presence and development of bars is strictly connected to other processes, such as channel bifurcations and confluences and lateral channel mobility. Bar formation and development of braiding has been observed in laboratory experiments (e.g. Ashmore 1982; Ikeda 1984). Such experiments have shown that there are two different manners for development of braiding, from alternate bars or from multiple row bars (Fig. 15.3). Braiding develops by cutoff of alternate bars (see upper sketch in Fig. 15.3) and bifurcation around mid-channel bars. Simple lobate bars that develop by such manners are commonly called "unit bars" (Smith 1974; Ashmore 1991b) which have distinct

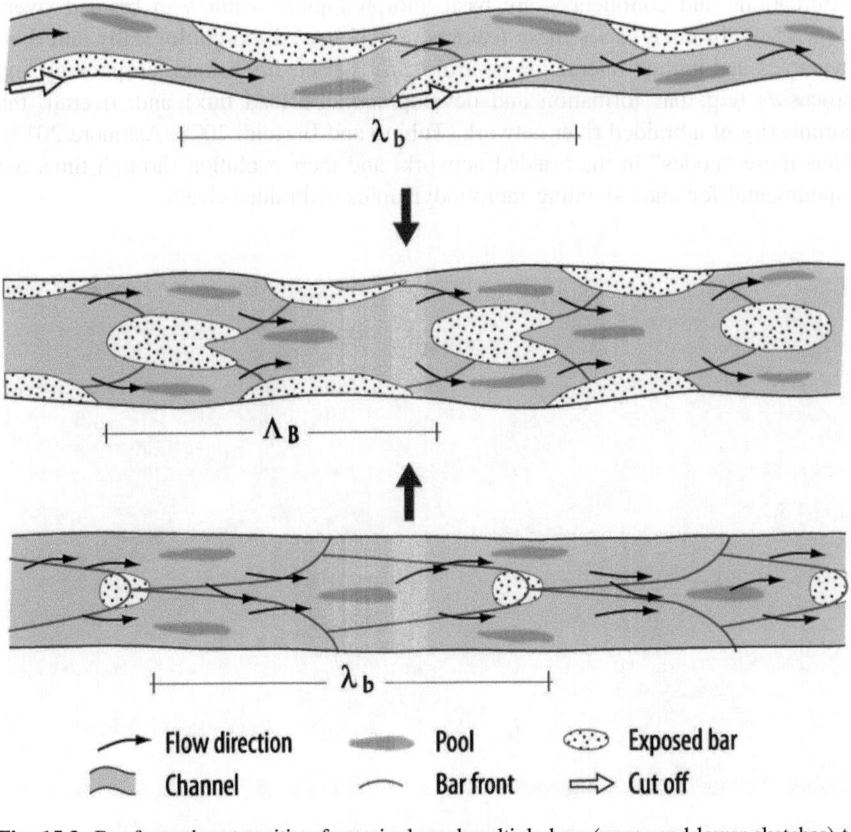

Fig. 15.3 Bar formation: transition from single and multiple bars (upper and lower sketches) to braided channel morphology (from Ashmore 2009)

downstream margins, with avalanche faces, especially in sand-bed rivers (Bridge 1993).

As braiding becomes more developed, large central or lateral bars are formed by gradual accretion of unit bars or bedload sheets. Sheets are formed when deposition is less rapid and/or more continuous (Bridge 2003). In sand-bed rivers, large bars are formed by accretion of low-amplitude dunes. These large bars are called "braid bars" (e.g. Ashmore 2013) or "compound bars" (e.g. Bridge 2003; Bridge and Lunt 2006). Braid bars are commonly built through a series of depositional and erosional events. These large bars migrate laterally and downstream but their migration rates are normally considerably much less than those in unit bars (Bridge 2003). Less mobile compound bars have been called "forced" bars, in contrast to "free" bars which are more mobile.

15.3.2 Bifurcations and Confluences

Bifurcations and confluences are basic morphological features in braided rivers (Fig. 15.4). In many cases these features are connected morphologically and their dynamics and mutual interaction control many aspects of channel morphology and processes (e.g. bar formation and development, bedload flux) and, overall, the complexity of a braided river network (Tubino and Bertoldi 2008; Ashmore 2013). Thus these "nodes" in the braided network, and their evolution through time, are fundamental for understanding morphodynamics of braided rivers.

Fig. 15.4 Bifurcations and confluences in the Rakaia River (New Zealand); braidplain width is about 900 m (from Google Earth)

Bifurcations control the partition of flow and sediment discharge in the anabranches of a braided network. How bifurcations are generated and how they evolve have been studied by laboratory (e.g. Ashmore 1991b; Federici and Paola 2003, Bertoldi and Tubino 2007) and field (e.g. Ferguson et al. 1992; Richardson and Thorne 2001; Zolezzi et al. 2006) observations and modeling approaches (e.g. Bolla Pittaluga et al. 2003; Jagers 2003). Bifurcations may be symmetric or asymmetric. In the first case both anabranches convey flow and transport bed material, while in asymmetric bifurcations only one anabranch transports bed material and one or both anabranches convey water. In the latter case upstream bar migration may cause a switch of activity between the anabranches. In gravel-bed rivers, where Shields stress is relatively low, bifurcations evolve toward asymmetrical configurations (e.g. Zolezzi et al. 2006), while in sand-bed rivers, where Shields stress is higher, they may evolve to a symmetrical configuration (e.g. Best et al. 2007). Zolezzi et al. (2006) have analyzed seven bifurcations in two gravel-bed rivers (the Ridanna Creek, in Italy, and the Sunwapta River, in Canada). They observed asymmetrical configurations which are reflected by (i) an unbalanced water distribution and different width of the downstream branches, (ii) the presence of a transverse step at the bifurcation, and (iii) the lateral shift of the main branch, the one carrying more water, towards the external bank, where most of erosion takes place. The fact that only one branch is morphologically active, at least at low-intermediate flow stages, is crucial for understanding braided morphodynamics. In fact the distinction between total braiding intensity and active braiding intensity (e.g. Egozi and Ashmore 2009) and the concept of active channel width (e.g. Ashmore et al. 2011) have been increasingly used over the last few years.

Confluences are other fundamental "nodes" within the braided network. These features represent a transfer zone between upstream lateral erosion sites and downstream sedimentation sites (i.e. bars). Some confluences display a simple symmetrical Y shape but in most cases their morphology is more complicated. Morphology of confluences varies in relation to several factors including total and relative discharges of the confluent anabranches, bedload delivery to the confluence, boundary shear stress, particle size distribution and confluence angle (Ashmore 1993). Confluence complexity may increase also in relation to the number of confluent anabranches.

Typical morphological features, flow structures and processes of confluence have been commonly described referring to the simple Y form (e.g. Mosley 1976; Ashmore and Parker 1983; Best 1986; Ashmore et al. 1992). This type of confluence is characterized by a deep scour at its center and mid-channel bar downstream. Scour size and depth is primarily controlled by confluence angle and discharge of the confluent anabranches (Ashmore 1993, 2013). Scour depth is larger in high-angle confluences with similar discharge in the confluent anabranches, while it is minimal if there is a large difference in discharge in the two anabranches, regardless of confluence angle (Mosley 1976; Ashmore and Parker 1983; Best 1986; Ashmore and Gardner 2008). The center of the confluence is also characterized by secondary flow dominated by a double helical circulation with a downward component in the shear zone and divergence at the bed (Ashmore et al.

1992). Strong secondary flows lead to size sorting in the thalweg and rapid lateral fining (Ashmore and Gardner 2008). Often one, or both, confluent anabranches build an avalanche face bar in the upstream end of the confluence. If the anabranches are of different size, the avalanche face may prograde into the confluence forcing lateral or downstream migration of the scour hole (Best 1988; Ashmore 1993).

15.3.3 Lateral Mobility

Lateral mobility, commonly characterized by high rates of bank retreat, is a fundamental process in braided river. The process is strictly linked with the processes described above (i.e. bar formation and development; bifurcations and confluences) and, it is worth to remember, to the presence of non-cohesive materials (i.e. gravel, sand, silt). Bank erosion occurs at specific sites along the river, in relation to the presence of the main anabranches, confluences, and sharp bends (Mosselman 2006). Commonly the magnitude of erosion is directly correlated with discharge, although in some cases, e.g. the right bank of the Brahmaputra River (Sarker et al. 2014), a weak relation between peak discharges and bank erosion was found. It is worth noting that bank erosion does not require high magnitude discharges and, therefore, it is effective also during frequent flood events (e.g. Mosselman 2006; Surian et al. 2009a).

Bank erosion rates can be very high in large braided rivers, a good example being the Brahmaputra River, in Bangladesh, which has a catchment area of about 573,000 km^2, mean annual flood of about 70,000 m^3 s^{-1}, and banks made of fine sand and silt (Thorne et al. 1993; Sarker and Thorne 2006). In this river, bank retreat is of the order of hundreds of meters per year (Mosselman 2006; Mount et al. 2013) and it can exceed locally 1 or even 2 km/year (Sarker et al. 2014) (Fig. 15.5). Lower, but still remarkable, erosion rates have been measured in gravel-bed rivers. In the Tagliamento River, with catchment area of 2580 km^2 and mean annual flood of about 1600 m^3 s^{-1}, bank erosion can be up to tens of meters per year. For instance, bank retreat was up to 90 m over a period of 16 months characterized by a flood with a recurrence interval of 12 year and some other floods of lower magnitude (Surian et al. 2009a) (Fig. 15.6).

The intense lateral mobility of braided rivers can represent a major issue for river basin managers. This is the case of the Brahmaputra River where bank erosion has remarkable societal and economical impact (i.e. loss of agricultural land, destruction of houses) (Mosselman 2006; Sarker et al. 2014). Piegay et al. (2006) pointed out that there is not a unique solution to managing braided river. Besides human safety and protection of economic interest, management should take into account existing or potential ecological values and the evolutionary trajectory of the river. As for this latter aspect, Ziliani and Surian (2012) found that lateral mobility is crucial for rivers that undergo severe channel adjustments (e.g. narrowing and

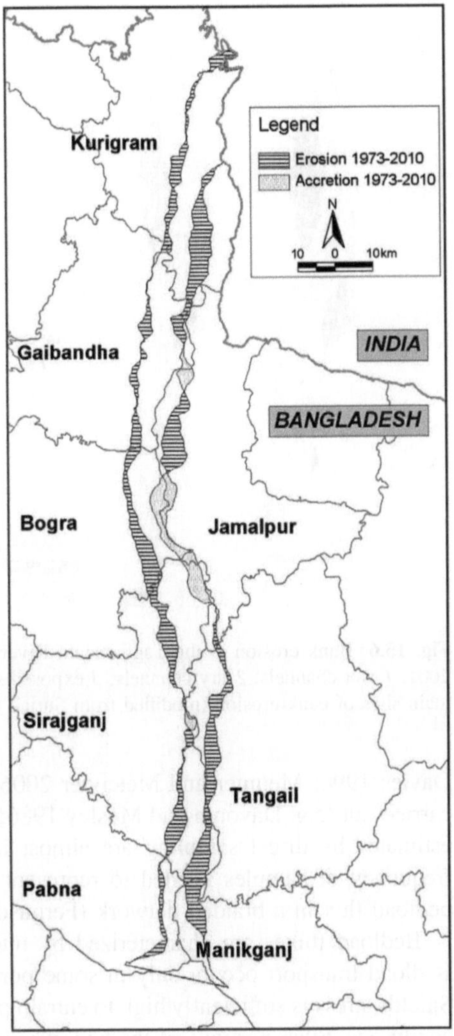

Fig. 15.5 Lateral mobility along the Brahmaputra River (Bangladesh) between 1973 and 2010 (from Sarker et al. 2014)

incision); therefore, the mobility should not be prevented unless for specific reasons (e.g. human safety).

15.3.4 Sediment Transport

Sediment transport, specifically bedload transport, is a key process to understand morphodynamics in braided rivers. Insights about bedload fluxes come mainly from flume experiments (e.g. Ashmore 1988; Hoey and Sutherland 1991; Warburton and

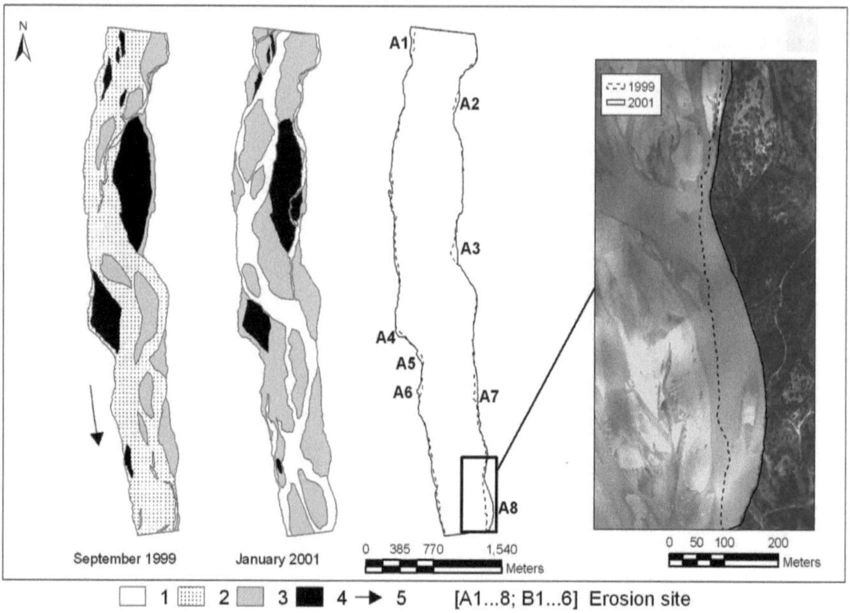

Fig. 15.6 Bank erosion in the Tagliamento River (Italy) between September 1999 and January 2001. *1* wet channels; *2* dry channels; *3* exposed sediments; *4* islands; *5* flow direction; *A1–A8*: main sites of bank erosion (modified from Surian et al. 2009a)

Davies 1994; Meunier and Méteivier 2006). Some field measurements have been carried out (e.g. Davoren and Mosley 1986; Ferguson et al. 1992), but reliable field estimates by direct sampling are almost impossible because of the number and frequency of samples needed to represent the spatial and temporal variations of bedload flux in a braided network (Ferguson et al. 1992; Bertoldi et al. 2009a).

Bedload fluxes are characterized by intense spatial and temporal fluctuations. Bedload transport occurs only in some portions of the whole cross-section where Shields stress is sufficiently high to entrain particle motion and to maintain transport conditions. Such portions of the bed, where sediment transport and morphological changes occur, define the "active width" which has been shown to be an useful element of braided river morphodynamics (Ashmore et al. 2011) (Fig. 15.7). Active width can be derived by parameters such as braiding intensity and wetted width and it can be used to predict bedload flux (Egozi and Ashmore 2009; Ashmore et al. 2011).

Besides spatial variations, remarkable temporal fluctuations of bedload flux at a given discharge have been measured. Most of measurements have been carried out in physical models, but some also in the field, e.g. in the Sunwapta River in Canada (Ashmore 1988; Hoey and Sutherland 1991; Ashmore 2013). Temporal fluctuations may deviate by a factor up to 10 relative to the mean bedload flux (Fig. 15.8). These

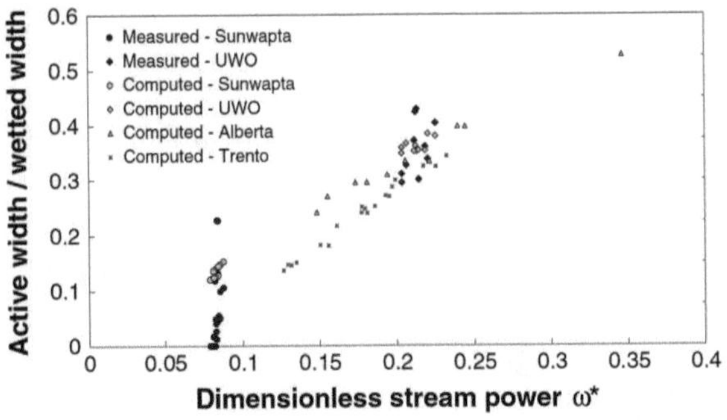

Fig. 15.7 Relationship between active width and dimensionless stream power (from Ashmore et al. 2011)

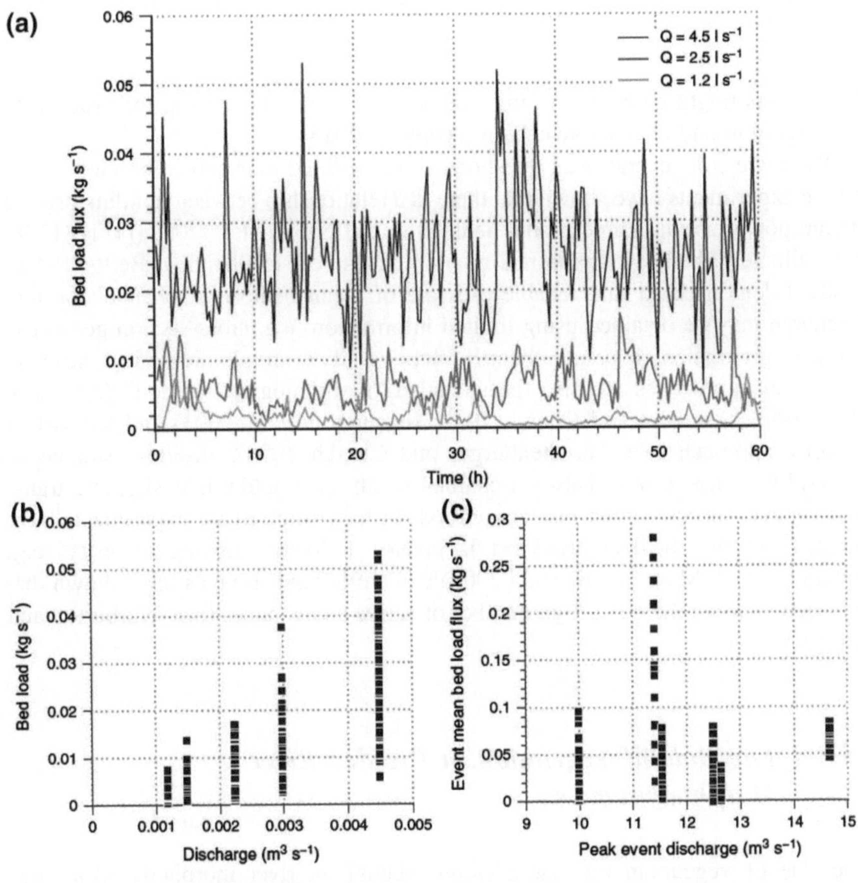

Fig. 15.8 Bedload transport fluctuations: **a** time series of fluctuations in bedload flux; **b** range of bedload variations, at constant discharge in physical models with a range of discharges; **c** morphological estimates of bedload flux in the Sunwapta River (Canada) over a series of daily meltwater hydrographs (from Ashmore 2013)

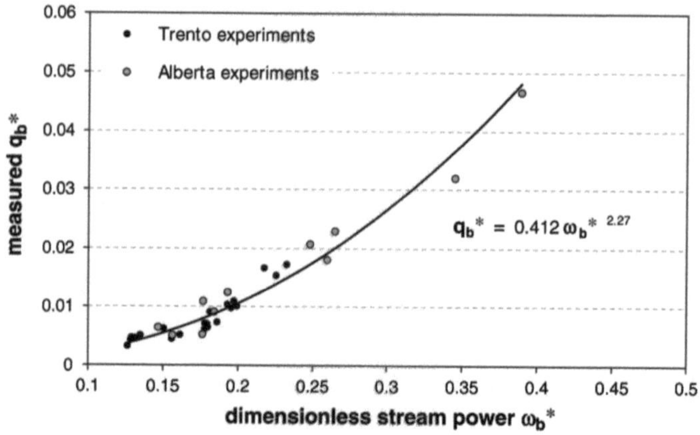

Fig. 15.9 Sediment flux as a function of dimensionless stream power; data from flume experiments (from Bertoldi et al. 2009a)

fluctuations might be related to the overall river morphodynamics, for instance to bar migration and channel switching (Ashmore 2013).

There are still several open questions about bedload transport in braided rivers. Flume experiments have shown that there is a relationship between nondimensional stream power and nondimensional bedload flux (Bertoldi et al. 2009a) (Fig. 15.9), but estimate of bedload flux in real rivers remains very challenging. Bertoldi et al. (2009a) demonstrated that reliable estimate of mean bedload at channel forming discharge may be obtained using limited information (e.g. cross-section geometry, average grain size, average channel slope) and a simple numerical scheme. Promising approaches are also the so-called "morphological method" (e.g. Lane et al. 1995; Ashmore and Church 1998; Brasington et al. 2003) and the virtual velocity approach (e.g. Haschenburger and Church 1998). Besides estimate of bedload flux, the relation between channel width, or braiding intensity, and transport capacity needs further research efforts. In fact, some flume experiments have pointed out that bedload transport is greater at lower braiding intensity (e.g. Ashmore 1988; Marti and Bezzola 2006) while other experiments have shown that it is greater when there is a high number of active anabranches (e.g. Warburton and Davies 1994).

15.3.5 The Role of Vegetation in Braided River Morphodynamics

The role of vegetation (i.e. large wood, plants) on river morphodynamics has received increasing attention over the last 10–15 years and it is now clearer than in the past that vegetation is a key component of fluvial systems (e.g. Gurnell 2014).

According to the extent of vegetation patches (i.e. islands) braided river can be classified as bar-braided with occasional islands (island area/gravel area <0.25), island-braided (island area/gravel area >0.25 and <0.5) or heavily island braided (island area/gravel area >0.5) (Gurnell et al. 2000) (Fig. 15.10).

Island formation and dynamics have been widely studied in the Tagliamento River, a high energy braided system in northeastern Italy (e.g. Gurnell et al. 2001; Gurnell and Petts 2006; Surian et al. 2015). Several key aspects have emerged from those studies on the Tagliamento. It was recognized that there is a threshold of unit stream power below which island can develop and persist (Gurnell and Petts 2006)

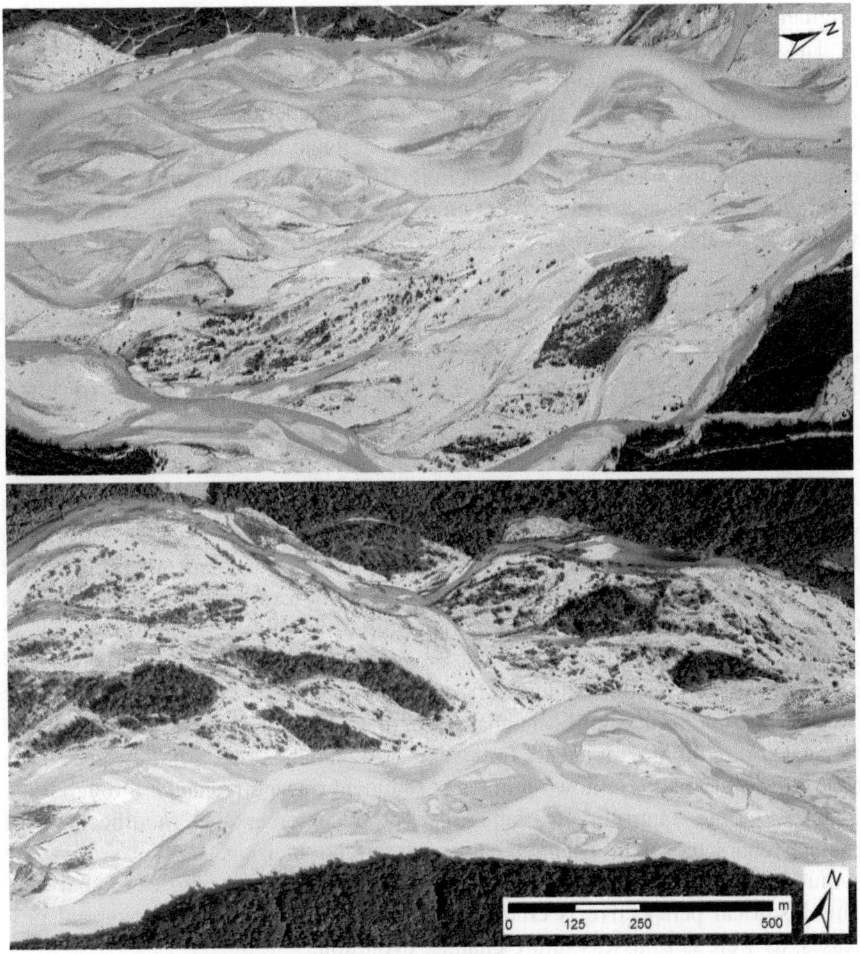

Fig. 15.10 Islands in the Tagliamento River (Italy): both reaches can be classified as bar-braided with occasional islands (island area/gravel area <0.25), although a significant difference in vegetation extent does exist between the two reaches. The reach with more islands (*lower sketch*) turned out to be less dynamic than the other one (modified from Surian et al. 2015)

and the fundamental role of large wood for island formation. Commonly pioneer islands start their formation around a deposited tree (i.e. living wood) that regenerates. This implies that seedling germination and growth is not, in most cases, the main process for vegetation establishment. Tree growing and trapping of fine sediments make islands relatively stable features which can persist for some years within the channel. Erosion and development of islands is strictly related to flow regime. Surian et al. (2015) have recently documented in the Tagliamento River that significant vegetation erosion is determined by relatively frequent floods (i.e. floods with a recurrence interval of 1–2.5 years); therefore, erosion is not controlled solely by large floods. This implies that vegetation turnover is rapid and few islands persist for several years. For instance, only 10 % of islands persists in the Tagliamento for more than 18–19 years (Surian et al. 2015).

The role of vegetation on braided morphodynamics has been analyzed in flume experiments (e.g. Gran and Paola 2001; Coulthard 2005; Tal and Paola 2010). Alteration of flow regime, specifically reduction of floods which are capable of vegetation erosion, produces vegetation expansion and may lead to channel pattern change (i.e. from braided to single-thread). Braided rivers affected by strong flow regulation, for instance below dams, may undergo dramatic changes because of vegetation establishment and its effects on the overall river morphodynamics (Hicks et al. 2008).

15.4 Braided Morphology and Processes in the Long Term Perspective

As illustrated in the previous sections, braided rivers are very dynamic systems which change their morphology frequently. Although some processes are more frequent than others (e.g. bedload transport in the main anabranches occurs more frequently than vegetation erosion), commonly significant morphological changes take place on one year scale, even without the occurrence of large floods. These changes, due to channel shifting, bar erosion or migration, bank retreat, island establishment or erosion, produce a frequent re-assemblage of morphological units but this may not imply significant changes in the main characters of the systems (i.e. channel width, braiding intensity or the extent of islands). This is not the case if the controlling variables, i.e. flow regime and sediment transport, change through time. Changes in the controlling variables can produce remarkable modifications of channel characters, e.g. channel narrowing or/and incision. In this section the evolution of braided rivers over some decades up to some centuries is analyzed. This historical perspective is crucial to understand present morphology and processes, as well as to assess future channel evolution.

The work by Gurnell et al. (2009) gives an overview of braided river evolution in Europe over the last centuries. Braided rivers are very common in the Alps and are found, to a lesser extent, in other European mountain regions, such as the

Apennines, the Pyrenees, the Carpathians, the Caucasus, the mountainous regions of Scandinavia and some upland areas of the United Kingdom. Braided rivers have undergone dramatic changes in many parts of Europe over the last few centuries. Many studies have documented that a major phase of braiding reduction took place during the 20th century (e.g. Winterbottom 2000; Muhar et al. 2008; Piégay et al. 2009; Surian et al. 2009b), but some studies suggest that this followed several centuries (approximately from the 14th to the 19th century) characterized by braiding expansion. An example of this morphological evolution (i.e. braiding expansion followed by braiding reduction) comes from the Polish Carpathians (Klimek 1987; Wyzga 1993, 2008), where the natural occurrence of multi-thread rivers was quite limited and largely concentrated in the Tatra Mountains during most of the Holocene. However, the increasing impact of human activities on slopes (deforestation, agriculture and pastoral activities on hillslopes) in the past increased sediment delivery to channels, inducing aggradation and braiding. The change of Carpathian rivers towards braiding gradually extended downstream, particularly during the Little Ice Age when it was probably enhanced by increased precipitation. During the 20th century, this trend was reversed under the influence of intense river channelization works, a marked decrease in human activities on hillslopes and extensive hillslope reforestation. Currently, the occurrence of braided pattern is once again largely confined to rivers draining the Tatra Mountains.

A marked phase of braiding expansion commencing in the Middle Ages, has also been recognized in the Alps and particularly in several French rivers (Bravard 1989; Piegay et al. 2006), and was associated with catchment-wide pressures induced by an increasing rural population. Widespread deforestation occurred as agriculture expanded onto more marginal, often steeply-sloping, land, and over-grazing was common on this cleared land as a result of relatively high stocking densities. These catchment-scale human activities increased runoff and sediment supply to river channels, leading to an increase in the magnitude and frequency of floods and the quantities of sediment that they transported, although Bravard (1989) has also attributed increases in sediment supply and braiding expansion to changes in climate, particularly during the Little Ice Age.

From the end of the 19th century and throughout the 20th century, human activities dramatically altered river dynamics so that the extent of braided and transitional river reaches were significantly reduced throughout Europe. For example, during the 20th century, the total length of braided reaches decreased by 70 and 95 %, respectively, in France and Austria (Habersack and Piégay 2008; Muhar et al. 2008). Besides, in those reaches where braided morphology has persisted braiding intensity have commonly decreased significantly (Fig. 15.11). These morphological changes were caused by a range of human activities that affected both drainage basins and river channels, such as land-use changes, torrent-control works, dam construction, flow regulation, channelization, and sediment mining. Complex sets of these activities altered flow regimes and sediment fluxes, as well as channel boundary conditions. In particular, a decrease in bedload supply to river channels has been identified as the driving factor of many channel adjustments (e.g. Kondolf et al. 2002; Surian et al. 2009b). In some rivers a cumulative effect of

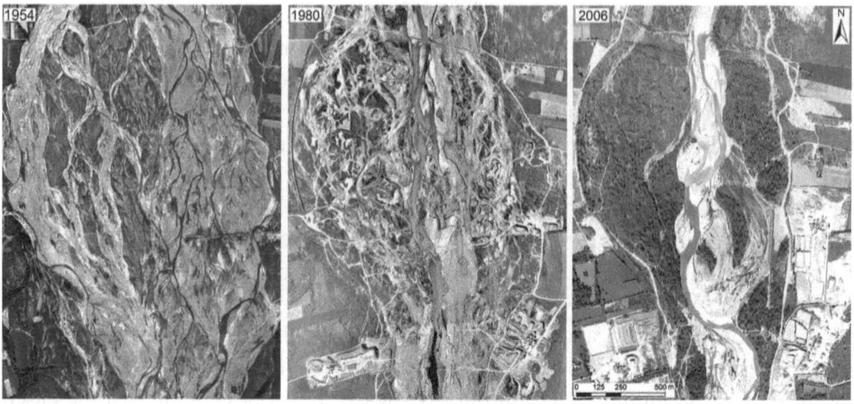

Fig. 15.11 Decrease of braiding intensity in the Trebbia River (Italy) between 1954 and 2006. The 1980 aerial photo shows channel morphology when mining activity was very intense (from Surian et al. 2011)

many different human activities has occurred, whereas in others one or two activities are thought to be responsible for channel changes. Furthermore, there are cases where basin-scale activities are identified as the major cause of channel change (e.g. land-use change, afforestation, torrent-control works; see Beguería et al. 2006; Kondolf et al. 2007) whereas in other cases reach scale interventions were more significant (e.g. sediment mining; see Peiry 1987; Surian et al. 2009b).

Historical analysis is very instructive to understand present morphology and predict future changes. Specifically, reconstruction of evolutionary trajectories needs to be coupled with the analysis of controlling factors (Fig. 15.12). This approach enables inferences to be made about future channel evolution (e.g. Surian et al. 2009b; Ziliani and Surian 2012). Prediction of future evolution has several practical implications because it may represent a key tool to guide management strategies. This said, we should be aware that prediction of channel morphology has inherent limitations since braided rivers are very complex systems and results of any model (e.g. conceptual, physical, analytical or numerical model) are affected by a degree of uncertainty.

15.5 Braided Morphology and Transition Between Channel Patterns

In the previous section it was illustrated how braided rivers may change notably their morphology over few centuries or even few decades. These changes may imply a decrease of braiding intensity but also more drastic changes, for instance from braided to single-thread channel pattern (Surian and Rinaldi 2003; Piégay et al. 2009; Gurnell et al. 2009). Is it possible to identify clear thresholds between

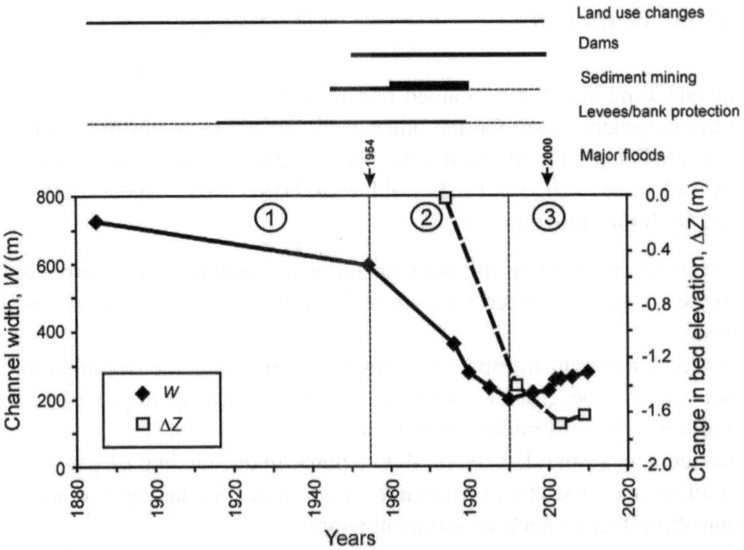

Fig. 15.12 Trends of width and bed-level adjustments in the Trebbia River (Italy) and their relationship with human impacts and main flood events. Human impacts: *horizontal bars* indicate temporal interval and relative intensity of the different impacts (from Bollati et al. 2014)

channel patterns? Which are the main factors that control channel pattern in alluvial streams? Questions of this type have been addressed by several authors over the last six decades or so (e.g. Leopold and Wolman 1957; Schumm and Khan 1972; Carson 1984; Chang 1985; Van den Berg 1995; Eaton et al. 2010; Mueller and Pitlick 2014). Since the classical work by Leopold and Wolman (1957), where distinction between braided, straight and meandering streams was approached taking into account discharge and channel slope, progresses have been made in two ways. First, other variables, besides discharge and channel slope, have been considered to explain thresholds between channel patterns, in particular bank strength (e.g. Carson 1984; Eaton et al. 2010) and sediment size (e.g. Van den Berg 1995). Second, there has been a shift from qualitative towards quantitative classifications. While qualitative or descriptive classifications are based on morphological variables (e.g. channel sinuosity, channel division), quantitative classifications are based on equations of flow and sediment transport, therefore attempting to relate channel pattern to fluvial processes (Eaton et al. 2010). Using a quantitative approach Eaton et al. (2010) recognized three distinct channel patterns (single-thread, anabranching, and braided channels) and showed that prediction of channel pattern is a three variable problem. They used channel slope, dimensionless discharge and relative bank strength to define thresholds between channel patterns.

15.6 Final Remarks

Our understanding about how braided rivers work has significantly increased over the last three decades or so. On the other hand, several open questions still remain and future challenges for research have been recently pointed out during a workshop held in France in June 2014 (Bertoldi et al. 2014). Some aspects that should be addressed by future research are:

(i) Metrics to characterize the wide spectrum of braided river morphology and processes: there is a need for both revising traditional metrics and using new ones;
(ii) Flow and sediment transport measurements: few data are available for "real" rivers, it will be crucial obtaining a larger dataset to improve our overall comprehension of braiding processes;
(iii) Management of braided rivers: this requires understanding of past and recent evolution and prediction of possible future trends, by taking into account key controlling factors such as sediment supply.

As for tools and methods, observations and measurements in braided rivers have been already facilitated by technologies such as airborne and terrestrial LiDAR and multibeam echo-sounding, but other emerging technologies, such as Structure from Motion could enhance our capability of measuring and understanding braiding processes. Finally, a great challenge is represented by numerical models. These models require improvements in reproducing later mobility, but there is also a need for a better integration between numerical modeling and the other approaches, specifically with insights coming from field observations and flume experiments.

Acknowledgments The author gratefully acknowledge funding from Fondazione CARIPARO (project "Linking geomorphological processes and vegetation dynamics in gravel-bed rivers") which supported his research on braided rivers over the period 2010–2013. I thank Robert Ettema for his comments and helpful suggestions.

References

Ashmore PE (1982) Laboratory modelling of gravel braided stream morphology. Earth Surf Process Land 7:201–225

Ashmore PE (1988) Bed load transport in braided gravel-bed stream models. Earth Surf Process Land 13:677–695

Ashmore P (1991a) Channel morphology and bed load pulses in braided, gravel-bed streams. Geografiska Annaler, Series A 73 A:37–52

Ashmore PE (1991b) How do gravel-bed rivers braid? Can J Earth Sci 28:326–341

Ashmore P (1993) Anabranch confluence kinetics and sedimentation processes in gravel-braided streams. In: Best JL, Bristow CS (eds) Braided rivers, vol 75. Geological Society of London, London, pp 129–146 (special publication)

Ashmore P (2009) Intensity and characteristics length of braided channel patterns. Can J Civ Eng 36:1656–1666

Ashmore P (2013) Morphology and dynamics of braided rivers. In: Shroder I (ed) Treatise on geomorphology, vol 9. Academic Press, Waltham, pp 289–312
Ashmore P, Church M (1998) Sediment transport and river morphology: a paradigm for study. In: Klingeman PC, Beschta RL, Komar PD, Bradley JB (eds) Gravel-bed rivers in the environment. Water Resources: Highlands Ranch, Colorado, pp 115–148
Ashmore P, Gardner JT (2008) Unconfined confluences in braided rivers. In: Rice S, Roy A, Rhoads BL (eds) River confluences, tributaries and the fluvial network. Wiley, Chichester, pp 119–147
Ashmore P, Parker G (1983) Confluence scour in coarse braided streams. Water Resour Res 19:392–402
Ashmore PE, Ferguson RI, Prestegaard KL, Ashworth PJ, Paola C (1992) Secondary flow in anabranch confluences of a braided, gravel-bed stream. Earth Surf Proc Land 17:299–311
Ashmore P, Bertoldi W, Gardner JT (2011) Active width of gravel-bed braided rivers. Earth Surf Proc Land 36:1510–1521
Beguería SJ, López-Moreno I, Gómez-Villar A, Rubio V, Lana-Renault N, García-Ruiz JM (2006) Fluvial adjustments to soil erosion and plant cover changes in the central Spanish Pyrenees. Geografiska Annaler: Series A Phys Geogr 88–3:177–186
Belletti B, Dufour S, Piégay H (2014) Regional assessment of the multi-decadal changes in braided riverscapes following large floods (example of 12 reaches in South East of France). Adv Geosci 37:57–71
Bertoldi W, Tubino M (2007) River bifurcations: experimental observations on equilibrium configurations. Water Resour Res 43
Bertoldi W, Ashmore P, Tubino M (2009a) A method for estimating the mean bed load flux in braided rivers. Geomorphology 103:330–340
Bertoldi W, Zanoni L, Tubino M (2009b) Planform dynamics of braided streams. Earth Surf Proc Land 34:547–557
Bertoldi W, Drake NA, Gurnell AM (2011a) Interactions between river flows and colonizing vegetation on a braided river: exploring spatial and temporal dynamics in riparian vegetation cover using satellite data. Earth Surf Process Land 36:1474–1486
Bertoldi W, Gurnell AM, Drake NA (2011b) The topographic signature of vegetation development along a braided river: results of a combined analysis of airborne lidar, color air photographs, and ground measurements. Water Resour Res 47
Bertoldi W, Recking A, Surian N (2014) Braiding rivers: state of the art and future challenges. EOS 95:381
Best JL (1986) The morphology of river channel confluences. Prog Phys Geogr 10:157–174
Best JL (1988) Sediment transport and bed morphology at river channel confluences. Sedimentology 35:481–498
Best JL, Ashworth PJ, Sarker MH, Roden JE (2007) The Brahmaputra-Jamuna River, Bangladesh. In: Gupta A (ed) Large rivers: geomorphology and management. Wiley, Chichester, pp 395–434
Bolla Pittaluga M, Repetto R, Tubino M (2003) Channel bifurcation in braided rivers: equilibrium configurations and stability. Water Resour Res 39
Bollati IM, Pellegrini L, Rinaldi M, Duci G, Pelfini M (2014) Reach-scale morphological adjustments and stages of channel evolution: the case of the Trebbia River (northern Italy). Geomorphology 221:176–186
Brasington J, Langham J, Rumsby B (2003) Methodological sensitivity of morphometric estimates of coarse fluvial sediment transport. Geomorphology 53:299–316
Brasington J, Vericat D, Rychkov I (2012) Modeling river bed morphology, roughness, and surface sedimentology using high resolution terrestrial laser scanning. Water Resour Res 48
Bravard JP (1989) La metamorphose des rivieres des Alpes francaises a la fin du Moyen-Age et a l'epoque moderne. Bulletin de la Societe Geographie de Liege 25:145–157
Bridge JS (1993) The interaction between channel geometry, water flow, sediment transport and deposition in braided rivers. In: Best JL, Bristow CS (eds) Braided rivers. Geological Society of London, London, pp 75:13–71 (special publication)

Bridge JS (2003) Rivers and floodplains: forms, processes and sedimentary record. Blackwell, Oxford

Bridge JS, Lunt IA (2006) Depositional models of braided rivers. In: Sambrook Smith GH, Best JL, Bristow CS, Petts GE (eds) Braided rivers: process, deposits, ecology and management. International Association of Sedimentologists, vol 36. Blackwell, Oxford, pp 11–50 (special publication)

Carson MA (1984) The meandering-braided river threshold: a reappraisal. J Hydrol 73:315–334

Chang HH (1985) River morphology and thresholds. J Hydraul Eng 111:503–519

Coulthard TJ (2005) Effects of vegetation on braided stream pattern and dynamics. Water Resour Res 41:1–9

Davoren A, Mosley MP (1986) Observations of bedload movement, bar development and sediment supply in the braided Ohau River. Earth Surf Proc Land 11:643–652

Eaton BC, Millar RG, Davidson S (2010) Channel patterns: braided, anabranching, and single-thread. Geomorphology 120:353–364

Egozi R, Ashmore P (2008) Defining and measuring braiding intensity. Earth Surf Process Land 33:2121–2138

Egozi R, Ashmore P (2009) Experimental analysis of braided channel pattern response to increased discharge. J Geophys Res F Earth Surf 114

Federici B, Paola C (2003) Dynamics of channel bifurcations in noncohesive sediments. Water Resour Res 39

Ferguson RI, Ashmore PE, Ashworth PJ, Paola C, Prestegaard KL (1992) Measurements in a braided river chute and lobe: 1. Flow pattern, sediment transport, and channel change. Water Resour Res 28:1877–1886

Fotherby LM (2009) Valley confinement as a factor of braided river pattern for the Platte River. Geomorphology 103:562–576

Gran K, Paola C (2001) Riparian vegetation controls on braided stream dynamics. Water Resour Res 37:3275–3283

Gurnell AM (2014) Plants as river system engineers. Earth Surf Proc Land. doi:10.1002/esp.3397

Gurnell A, Petts G (2006) Trees as riparian engineers: the Tagliamento River, Italy. Earth Surf Process Land 31:1558–1574

Gurnell AM, Petts GE, Harris N, Ward JV, Tockner K, Edwards PJ, Kollmann J (2000) Large wood retention in river channels: the case of the Fiume Tagliamento, Italy. Earth Surf Process Land 25:255–275

Gurnell AM, Petts GE, Hannah DM, Smith BPG, Edwards PJ, Kollmann J, Ward JV, Tockner K (2001) Riparian vegetation and island formation along the gravel-bed Fiume Tagliamento, Italy. Earth Surf Proc Land 26:31–62

Gurnell A, Surian N, Zanoni L (2009) Multi-thread river channels: a perspective on changing European alpine river systems. Aquatic Sci 71:253–265

Habersack H, Piégay H (2008) River restoration in the Alps and their surroundings: past experience and future challenges. In: Habersack H, Piégay H, Rinaldi M (eds) Gravel-bed rivers VI: from process understanding to river restoration, developments in earth surface processes. Elsevier, Amsterdam, pp 703–738

Haschenburger JK, Church M (1998) Bed material transport estimated from the virtual velocity of sediment. Earth Surf Process Land 23:791–808

Hicks DM, Duncan MJ, Lane ST, Tal M, Westway R (2008) Contemporary morphological change in braided gravel-bed rivers: new developments from field and laboratory studies, with particular reference to the influence of riparian vegetation. In: Habersack H, Piégay H, Rinaldi M (eds) Gravel-bed rivers VI: from process understanding to river restoration. Developments in earth surface processes. Elsevier, Amsterdam, pp 557–584

Hoey TB, Sutherland AJ (1991) Channel morphology and bedload pulses in braided rivers: a laboratory study. Earth Surf Proc Land 16:447–462

Hong LB, Davies TRH (1979) A study of stream braiding. Bull Geol Soc Am 90:1839–1859

Howard AD, Keetch ME, Vincent CL (1970) Topological and geometrical properties of braided streams. Water Resour Res 6:1674–1688

Hundey EJ, Ashmore PE (2009) Length scale of braided river morphology. Water Resour Res 45
Ikeda S (1984) Prediction of alternate bars wavelength and height. J Hydraul Eng 110:371–386
Jagers HRA (2003) Modeling planform changes of braided rivers. PhD thesis, University of Twente
Javernick L, Brasington J, Caruso B (2014) Modeling the topography of shallow braided rivers using Structure-from-Motion photogrammetry. Geomorphology 213:166–182
Klimek K (1987) Man's impact on fluvial processes in the Polish Western Carpathians. Geogr Ann 69A:221–226
Kondolf GM, Piégay H, Landon N (2002) Channel response to increased and decreased bedload supply from land use change: contrasts between two catchments. Geomorphology 45:35–51
Kondolf GM, Piégay H, Landon N (2007) Changes in the riparian zone of the lower Eygues River, France, since 1830. Landscape Ecol 22:367–384
Lallias-Tacon S, Liébault F, Piégay H (2014) Step by step error assessment in braided river sediment budget using airborne LiDAR data. Geomorphology 214:307–323
Lane SN (2006) Approaching the system-scale understanding of braided river behaviour. In: Sambrook Smith GH, Best JL, Bristow CS, Petts GE (eds) Braided rivers: process, deposits, ecology and management, vol 36. International Association of Sedimentologists, Blackwell, Oxford, pp 107–135 (special publication)
Lane SN, Richards KS, Chandler JH (1995) Morphological estimation of the time-integrated bed load transport rate. Water Resour Res 31:761–772
Lane SN, Westaway RM, Hicks DM (2003) Estimation of erosion and deposition volumes in a large, gravel-bed, braided river using synoptic remote sensing. Earth Surf Process Land 28:249–271
Leopold LB, Wolman MG (1957) River channel pattern; braided, meandering and straight. Professional paper 282-B. US Geological Survey, Washington
Marti C, Bezzola GR (2006) Bed load transport in braided gravel-bed rivers. In: Sambrook Smith GH, Best JL, Bristow CS, Petts GE (eds) Braided rivers: process, deposits, ecology and management, vol 36. International Association of Sedimentologists Blackwell, Oxford, pp 199–215 (special publication)
Meunier P, Méteivier F (2006) Sediment transport in a micro-scale braided stream: from grain size to reach scale. In: Sambrook Smith GH, Best JL, Bristow CS, Petts GE (eds) Braided rivers: process, deposits, ecology and management, vol 36. International Association of Sedimentologists Blackwell, Oxford, pp 217–231 (special publication)
Mosley MP (1976) An experimental study of channel confluences. J Geol 84:535–562
Mosselman E (2006) Bank protection and river training along the braided Brahmaputra-Jamuna River, Bangladesh. In: Sambrook Smith GH, Best JL, Bristow CS, Petts GE (eds) Braided rivers: process, deposits, ecology and management, International Association of Sedimentologists Special Publication 36, Blackwell, Oxford, 277–287
Mount NJ, Tate NJ, Sarker MH, Thorne CR (2013) Evolutionary, multi-scale analysis of river bank line retreat using continuous wavelet transforms: Jamuna River, Bangladesh. Geomorphology 183:82–95
Mueller ER, Pitlick J (2014) Sediment supply and channel morphology in mountain river systems: 2. Single thread to braided transitions. J Geophys Res Earth Surf 119:1516–1541
Muhar S, Jungwirth M, Unfer G, Wiesner C, Poppe M, Schmutz S, Hohensinner S, Habersack H (2008) Restoring riverine landscapes at the Drau River: successes and deficits in the context of ecological integrity. In: Habersack H, Piégay H, Rinaldi M (eds) Gravel-bed rivers VI: from process understanding to river restoration. Developments in earth surface processes. Elsevier, Amsterdam, pp 779–803
Murray AB, Paola C (1994) A cellular model of braided rivers. Nature 371:54–57
Nicholas AP (2000) Modelling bedload yield braided gravel bed rivers. Geomorphology 36:89–106
Peiry JL (1987) Channel degradation in the middle Arve river, France. Regulated Rivers Res Manage 1:183–188

Piegay H, Grant G, Nakamura F, Trustrum N. (2006) Braided river management: from assessment of river behaviour to improved sustainable development. In: Sambrook Smith GH, Best JL, Bristow CS, Petts GE (eds) Braided rivers: process, deposits, ecology and management, vol 36. International Association of Sedimentologists, Blackwell, Oxford, pp 257–275 (special publication)

Piégay H, Alber A, Slater L, Bourdin L (2009) Census and typology of braided rivers in the French Alps. Aquat Sci 71:371–388

Richardson WR, Thorne CR (2001) Multiple thread flow and channel bifurcation in a braided river: Brahmaputra-Jamuna River, Bangladesh. Geomorphology 38:185–196

Robertson-Rintoul MSE, Richards KS (1993) Braided-channel pattern and palaeohydrology using an index of total sinuosity. In: Best JL, Bristow CS (eds) Braided rivers. Geological Society of London, London, pp 75:113–118 (special publication)

Sapozhnikov V, Foufoula-Georgiou E (1996) Self-affinity in braided rivers. Water Resour Res 32:1429–1439

Sarker MH, Thorne CR (2006) Morphological response of the Brahmaputra-Padma-Lower Meghna river system to the Assam earthquake of 1950. In: Sambrook Smith GH, Best JL, Bristow CS, Petts GE (eds) Braided rivers: process, deposits, ecology and management, vol 36. International Association of Sedimentologists, Blackwell, Oxford, pp 289–310 (special publication)

Sarker MH, Thorne CR, Aktar MN, Ferdous MR (2014) Morpho-dynamics of the Brahmaputra-Jamuna River, Bangladesh. Geomorphology 215:45–59

Schumm SA, Khan HR (1972) Experimental study of channel patterns. Geol Soc Am Bull 83:1755–1770

Smith ND (1974) Sedimentology and bar formation in the upper Kicking Horse River, a braided meltwater stream. J Geol 82:205–223

Surian N (1999) Channel changes due to river regulation: the case of the Piave River, Italy. Earth Surf Process Land 24:1135–1151

Surian N, Rinaldi M (2003) Morphological response to river engineering and management in alluvial channels in Italy. Geomorphology 50:307–326

Surian N, Mao L, Giacomin M, Ziliani L (2009a) Morphological effects of different channel-forming discharges in a gravel-bed river. Earth Surf Process Land 34:1093–1107

Surian N, Ziliani L, Comiti F, Lenzi MA, Mao L (2009b) Channel adjustments and alteration of sediment fluxes in gravel-bed rivers of north-eastern Italy: potentials and limitations for channel recovery. River Res Appl 25:551–567

Surian N, Rinaldi M, Pellegrini L (2011) Channel adjustments and implications for river management and restoration. Geografia Fisica e Dinamica Quaternaria 34:145–152

Surian N, Barban M, Ziliani L, Monegato G, Bertoldi W, Comiti F (2015) Vegetation turnover in a braided river: frequency and effectiveness of floods of different magnitude. Earth Surf Process Land 40:542–558 doi:10.1002/esp.3660

Tal M, Paola C (2010) Effects of vegetation on channel morphodynamics: results and insights from laboratory experiments. Earth Surf Process Land 35:1014–1028

Thomas R, Nicholas AP, Quine TA (2007) Cellular modelling as a tool for interpreting historic braided river evolution. Geomorphology 90:302–317

Thorne CR, Russell APG, Alam MK (1993) Planform pattern and channel evolution of the Brahmaputra River, Bangladesh. In: Best JL, Bristow CS (eds) Braided rivers, vol 75. Geological Society of London, London, pp 257–276 (special publication)

Tubino M, Bertoldi W (2008) Bifurcations in gravel-bed streams. In: Habersack H, Piégay H, Rinaldi M (eds) Gravel-bed rivers VI: from process understanding to river restoration. Developments in earth surface processes. Elsevier, Amsterdam, pp 133–159

Van den Berg JH (1995) Prediction of alluvial channel pattern of perennial rivers. Geomorphology 12:259–279

Warburton J, Davies T (1994) Variability of bedload transport and channel morphology in a braided river hydraulic model. Earth Surf Process Land 19:403–421

Wheaton JM, Brasington J, Darby SE, Kasprak A, Sear D, Vericat D (2013) Morphodynamic signatures of braiding mechanisms as expressed through change in sediment storage in a gravel-bed river. J Geophys Res Earth Surf 118:759–779

Williams RD, Brasington J, Vericat D, Hicks DM (2014) Hyperscale terrain modelling of braided rivers: fusing mobile terrestrial laser scanning and optical bathymetric mapping. Earth Surf Process Land 39:167–183

Winterbottom SJ (2000) Medium and short-term channel planform changes on the Rivers Tay and Tummel, Scotland. Geomorphology 34:195–208

Wyzga B (1993) River response to channel regulation: case study of the Raba River, Carpathians, Poland. Earth Surf Process Land 18:541–556

Wyzga B (2008) A review on channel incision in the Polish Carpathian rivers during the 20th century. In: Habersack H, Piégay H, Rinaldi M (eds) Gravel-bed Rivers VI: from process understanding to river restoration. Developments in earth surface processes. Elsevier, Amsterdam, pp 525–553

Young WJ, Davies TRH (1991) Bedload transport processes in a braided gravel-bed river model. Earth Surf Process Land 16:499–511

Ziliani L, Surian N (2012) Evolutionary trajectory of channel morphology and controlling factors in a large gravel-bed river. Geomorphology 173–174:104–117. doi:10.1016/j.geomorph.2012.06.001

Ziliani L, Surian N, Coulthard TJ, Tarantola S (2013) Reduced-complexity modeling of braided rivers: assessing model performance by sensitivity analysis, calibration, and validation. J Geophys Res Earth Surf 118:2243–2262. doi:10.1002/jgrf.20154

Zolezzi G, Bertoldi W, Tubino M (2006) Morphological analysis and prediction of channel bifurcations. In: Sambrook Smith GH, Best JL, Bristow CS, Petts GE (eds) Braided rivers: process, deposits, ecology and management, vol 36. International Association of Sedimentologists, Blackwell, Oxford, pp 233–256 (special publication)

[illegible]

Chapter 16
Lagrangian Modelling of Saltating Sediment Transport: A Review

Robert J. Bialik

Abstract One hundred years of research on the saltation in rivers, both experimental and numerical, has allowed for a fairly good improvement of our knowledge of the physics of the saltation process. Lagrangian modelling has played a huge role in this field and has made it possible to apply the knowledge obtained in the analysis of processes associated with the movement of sediment particles. The present paper briefly reviews the current state-of-the-art of the Lagrangian modelling of saltating grains in open channels and highlights recent findings in three areas in which employment of the Lagrangian models of saltation improve our understanding of sediment transport in rivers, namely: initial motion of saltating grains, diffusion of particles and calculation of the bedload transport rate. The particular challenges in all of these research areas are discussed and future ways forward are presented.

Keywords Entrainment · Numerical simulations · Particle diffusion · Saltation · Sediment transport · Turbulence structures

16.1 Introduction

Saltation as defined by Niño (1995) and further emphasized by Niño and Garcia (1998a, b) is "*the unsuspended transport of particles over a granular bed by a fluid flow, in the form of consecutive hops that nonetheless keep the particles within the near-bed region, which is governed mainly by the action of hydrodynamic forces that carry the particles through the flow, the downward pull of gravity, and the particles' collision with the bed, which transfers their streamwise momentum into upward momentum, thus sustaining the saltation motion*". Generally, saltation,

R.J. Bialik (✉)
Institute of Geophysics, Polish Academy of Sciences, Ks. Janusza 64, 01-452 Warsaw, Poland
e-mail: rbialik@igf.edu.pl

P. Rowiński and A. Radecki-Pawlik (eds.), *Rivers—Physical, Fluvial and Environmental Processes*, GeoPlanet: Earth and Planetary Sciences,
DOI 10.1007/978-3-319-17719-9_16

besides sliding and rolling, is one of the forms of sediment transport in bedload and appears if the shear stress exceeds some critical value, allowing particles to jump downstream.

The term 'saltation' originates from the Latin verb 'saltatus', the past participle of 'saltare', which means to leap or to dance, frequentative of 'salire' meaning to jump. As a curiosity, to the best of the author's knowledge, this term was used for the first time in 1646 by Sir Thomas Browne in his magnum opus "*Pseudodoxia Epidemica*" when he attempted to characterize the grasshopper; we can read: "*The word Grashopper*[1] *itself is improper, and the term of Grashopper not appliable* (see footnote 1) *unto the Cicada; for therein the organs of motion are not contrived for* ***saltation****, nor are the hinder legs of such extension, as is observable in salient animals, and such as move by leaping*". The first known use of the term 'saltation' associated with the transport of sediment was made by McGee (1908), who claimed that "(…) *Transportation may be regarded as the general movement of earth matter seaward by streams*; *it comprises carriage of material* (*a*) *in solution*, (*b*) *in suspension, and* (*c*) *in what may be denoted* ***saltation*** (...)". However, Gilbert (1914) was the first to use the word 'saltation' in a very similar sense to that in which it is used today, stating that "*If the bed is uneven, the particle usually does not retain continuous contact but makes leaps, and the process is then called* ***saltation****, an expressive name introduce by McGee*".

Until the early 1980s, research on the saltation motion of particles had been mostly based on experiments (Einstein 1942; Bagnold 1956, 1973; Gordon et al. 1972; Francis 1973; Abbot and Francis 1977; Murphy and Hooshiari 1982) which mainly focused on analysis of the velocities of the saltating particles. It should be noted, however, that they established a basis for the construction of later models. However, only the dynamic development of and open access to computers in the 1980s and 1990s allowed for the first significant step in the understanding of saltation through the possibility of carrying out simulations based on theoretical and numerical models (Wiberg and Smith 1985, 1989; Sekine and Kikkawa 1992; Lee and Hsu 1994; Niño and Garcia 1994, 1998b; Rowiński 1995). Note that only models on saltation in flowing water are considered here. However, it is important to mention three works on saltation in the air condition on the Earth (McEwan and Willetts 1991) and the Martian (White et al. 1976 and White 1979) surfaces as these papers present models which were one of the first used to analyse the different processes, including entrainment and diffusion of particles in open channels.

The advantage of these new, Lagrangian models of sediment transport, compared to the existing ones, lies in the possibility of their derivation from the general equations of hydrodynamics. Moreover, each of the models' terms has a physical meaning that was previously often neglected. All of these models are based on the equation of a single spherical particle in a flowing fluid. Two studies on this phenomenon are particularly important: Tchen (1947), who summarized the works of Basset (1888), Boussinesq (1903) and Ossen (1927), describing the BBO

[1]The original spellings are used here.

equation (after the first letters of the names of the authors), well-known from fluid mechanics, which was formulated for the sphere in a stationary liquid; the second work was done by Maxey and Riley (1983) wherein, based on an analysis similar to that reported by Corrsin and Lumley (1956), the authors showed the detailed derivation of the equations describing the behaviour of such particles in flowing fluid. Since then, a number of modifications to this equation have been proposed, the most commonly quoted being that given by Mei et al. (1991), which in the form presented in Bialik (2011a) is shown here:

$$\overbrace{m_p \frac{\mathbf{du}_p}{dt}}^{\text{inertia}} = \overbrace{m_f \mathbf{g}}^{\text{bouyancy force}} - \overbrace{m_p \mathbf{g}}^{\text{gravity force}} + \overbrace{\frac{1}{2}\rho C_D A_D u_r \mathbf{u}_r}^{\text{drag force}} + \overbrace{m_f C_m \frac{\partial \mathbf{u}_r}{\partial t}}^{\text{added mass force}}$$
$$+ \underbrace{\frac{1}{2}\rho C_L A_D \left(|\mathbf{u}_r|_T^2 - |\mathbf{u}_r|_B^2 \right) \mathbf{e}_z}_{\text{lift force}} + \underbrace{\frac{3}{2} d_p \sqrt[2]{\rho \mu \pi} \int_0^t \frac{\mathbf{du}_r}{d\tau} \frac{d\tau}{\sqrt{t-\tau}}}_{\text{Basset force}}, \tag{16.1}$$

where u_p is the particle velocity vector, $\mathbf{u}_r = \mathbf{u}_f - \mathbf{u}_p$ denotes the relative water and particle velocity vector, d_p is the particle diameter, C_D is the drag coefficient, C_L stands for the lift coefficient, and C_m denotes the virtual mass coefficient, which for spherically shaped particles is equal to 0.5, A_D is the cross-section area, m_p and m_f are particle and water masses, respectively, ρ is water density, t is the time of saltation, $\mathbf{g}$ stands for acceleration of gravity, μ is dynamical viscosity and τ is characteristic time. Subscript B denotes the particle bottom and subscript T is the particle top and $\mathbf{e}_z$ is a unit vector in the direction perpendicular to the relative velocity. It should be noted that the Maxey and Riley (1983) as well as the BBO equations are valid only for very small particle Reynolds number $Re_p \ll 1$, whereas Eq. (16.1) used by Bialik (2011a, b) is valid for finite Re_p. In addition, Eq. (16.1) should be supplemented with the trajectory equation, equation for the particle rotation and with the sub-model for the flow velocity, which for simplicity in such models has usually been described by the logarithmic law. However, ideally, the velocity field should be resolved and coupled with the particle motion, as done for example by Ji et al. (2014). Figure 16.1 shows sample trajectories of particle motion obtained with use of the Bialik (2011a, b) model and by solving Eq. (16.1).

Initially, the main purpose of constructing the Lagrangian models was the accurate calculation of the amount of transported sediment, mostly in bedload. However, it soon turned out that it was only possible to calculate the several trajectories of particles, which was insufficient for calculation of the bedload rate, so the goal was transformed into improving the models, such that they might be used further when the development of computers allowed for their employment. The focus was on better description of the shape of a single trajectory through more accurate assessment of the impact of individual processes such as: collisions between particles (Sørensen and McEwan 1996; Schmeeckle et al. 2001; Lee et al. 2002; Yeganeh-Bakhtiary et al. 2009; Bialik 2010, 2011a; Moreno and Bombardelli

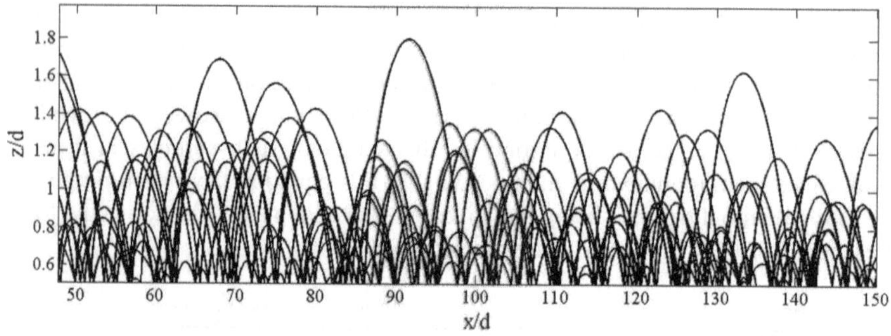

Fig. 16.1 2-D sample trajectories of particle motion

2012), particles' collisions with the channel bottom (Sekine and Kikkawa 1992; Niño and Garcia 1994; Lukerchenko et al. 2006; Bialik 2011b, Kharlamova and Vlasak 2015), taking into account the three-dimensional nature of the particles' motion (Lee et al. 2006; Lukerchenko et al. 2009a; Bialik et al. 2012) or a better approximation of the different forces acting on the flowing particles, such as: Basset force (Bombardelli et al. 2008; Lukerchenko et al. 2012), lift force (Lee and Hsu 1996; Zou et al. 2007; Lukerchenko et al. 2008) and drifting force (Lee 1987; Lee and Wiesler 1987; Czernuszenko 1998) due to turbulent diffusion, which, with some exceptions (Rowiński and Lee 1993; Rowiński 1995), has usually been neglected in the Lagrangian models of sediment transport. It should be noted that all of these models mostly skip the idea of their creation, namely their use for the evaluation of the phenomena associated with the movement of particles near the bed region.

The present review paper aims to achieve two goals: (1) to present the current state-of-the-art of the Lagrangian modelling of saltating grains in open channels; and (2) to highlight recent findings in three areas, in which the employment of such models improves our understanding of sediment transport in rivers, namely: initial motion of saltating grains, diffusion of particles and calculation of the bedload transport rate.

16.2 Initiation of Saltating Grains Movement

In order to find the particle trajectory using the Lagrangian models of saltation, the initial conditions, which determine the initial position of the particle as well as its velocity, are required. Abbott and Francis (1977) were among the first who suggested, based on the experimental data, that the particles' velocity in both the streamwise u_p and vertical w_p directions should be of approximately $2u_*$, where u_* is the shear velocity. In addition, if it is assumed that initially the particle lies on the channel bottom and that its movement is three-dimensional, then such conditions are as follows: $x_p(t_0) = 0$, $y_p(t_0) = 0$, $z_p(t_0) = 0.5\text{d}$, $u_p(t_0) = 2u_*$, $v_p(t_0) = 0$ and $w_p(t_0) = 2u_*$. However, the initial velocities of the particle are sometimes treated as parameters for the model calibration (i.e. Lee et al. 2000) and further several of the particles' first jumps should be discarded for the statistical analysis. This is due to the influence of the initial conditions on the height and length of saltation. Niño and Garcia (1998b) suggested that in order to draw any conclusions on this issue, the first 5–20 jumps should be omitted, which means that the information about the initial movement of the particle is lost and, therefore, the processes that play an important role in the particles' entrainment would be neglected in the Lagrangian modelling of saltation. Bialik (2011b) confirmed the findings of Niño and Garcia (1998b) by comparative analysis between the commonly used and the following initial conditions: $x_p(t_0) = 0$, $z_p(t_0) = 0.5d$, $u_p(t_0) = 0$, and $w_p(t_0) = 0$, which were employed in the 2D Lagrangian model of saltation. The model of Bialik (2011b) takes into account the fluctuation of the flow velocity which causes non-moving particles to begin to jump, which allows for analysis of the local range of particles' diffusion (this issue will be discussed in Sect. 16.3) and of the processes responsible for the particles' entrainment.

According to McEwan and Willets (1991) who investigated aeolian saltation, the moving grains may be considered as made up from two components: aerodynamically entrained grains and impact generated grains. Further, McEwan et al. (1999) developed this approach, adapting it to the fluvial conditions, suggesting that the number of uniform grains N_e entrained per unit area per unit time is equal to

$$N_e = \alpha_* \left[(\tau_*)_f^b - (\tau_*)_{cr} \right], \tag{16.2}$$

where α_* denotes the coefficient which determines the entrainment rate, $(\tau_*)_f^b$ is the dimensionless fluid shear stress at the channel bottom, and $(\tau_*)_{\mathrm{cr}}$ is the dimensionless critical fluid shear stress, defined by the Shield's criterion. They used the 'Aberdeen' Lagrangian model of saltation for the calculation of the bedload transport rate (this issue will be discussed in detail in Sect. 16.4) for different values of entrainment coefficient α_*. Based on the results of simulations, they showed that the fluid shear stress at the channel bottom and the transport rate are dependent on the sediment availability, parameters which depend on the entrainment coefficient. It should be mentioned that the McEwan and Willets (1991) entrainment model is proportional to the first power of the difference between the shear stress and the critical fluid shear stress, whereas other models specify a highly non-linear power law with larger exponents, at least 1.5, as it was proposed for example by Garcia and Parker (1991).

Wiberg and Smith (1987) carried out an analysis of the forces acting on the sediment particle, which were taken into account in their model of saltation (Wiberg and Smith 1985), in order to find the value of the critical shear stress of noncohesive sediment ζ_*. They proposed an expression which omits the critical shear velocity $(u_*)_{cr}$ and depends only on the grain size D, sediment ρ_s and fluid ρ densities and viscosity v, which allow for the calculation of the critical shear stress for any given grain in a particular fluid environment:

$$\zeta_* = \frac{d_p^3}{\nu^2} \frac{(\rho_s - \rho)}{\rho} g, \tag{16.3}$$

It should be noted that both McEwan and Willets (1991) and Wiberg and Smith (1987) mostly focused on the saltation occurring in air conditions and, due to the low density of this fluid, argued that the main role in the sediment particles entrainment is played by forces associated with their collisions with the bed. It seems, however, that in the case of the saltation in rivers, more important in sediment transport and the incipient motion of grains are near-bed turbulence structures, and in particular those associated with the fluctuation in streamwise velocity (e.g., Drake et al. 1988; Nelson et al. 1995).

Bialik (2013) used the previously proposed 3D Lagrangian model of saltation (Bialik et al. 2012) to analyse the influence of the near-bed turbulence structures on the motion of the saltating particles. One of the most important ingredients in this model was the generator for the flow velocity field originally proposed by Nikora et al. (2001), which is based on the Monte Carlo simulation of time series of velocity fluctuation. For details, readers are referred to the original works. Three values of particle mobility parameters $K = 1.1$; 1.5; 2.5 and three values of relative sizes of saltating particles $d/D = 0.5$; 1; 2 were used in the simulations, where d denotes a flying particle and D is the bottom particle. Figure 16.2 summarizes all of the obtained results which are presented in a histogram of time fractions of four quadrants. This analysis has been usually used (i.e. Lu and Willmarth 1973; Nezu and Nakagawa 1993; Czernuszenko and Rowiński 2008) to determine important instantaneous Reynolds stress and four regions (quadrants) may be detected in dependence on the signs of streamwise (u′) and vertical (w′) velocity fluctuations and then by checking fractional Reynolds stresses belonging to those particular regions. Thus, outward interactions (Q_1) are sought when ($u' > 0$, $w' > 0$), ejections (Q_2) when ($u' < 0$, $w' > 0$), inward interactions (Q_3) when ($u' < 0$, $w' < 0$) and sweeps (Q_4) when ($u' > 0$, $w' < 0$). The simulation results supported earlier

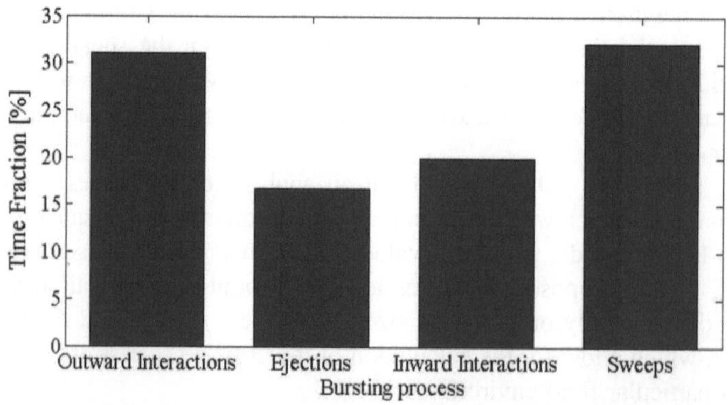

Fig. 16.2 Cumulative results of all simulation showing time fractions of four quadrants

experimental observations, showing that the initiation of sediment movement is strongly related with fluctuation of the positive streamwise velocity and, despite the more common existence of ejections in fluid flow, it was presented that the outward interactions and sweeps play a crucial role in the entrainment of saltating particles. Moreover, the results suggested that ejections have the smallest part in this process. More details on the quadrant analysis technique and the results can be found in Bialik (2013).

Entrainment and the movement of particles in the form of saltation is actually mostly a result of a combination of two forces: drag and lift. Their impact in a particular phase of the particles' movement determines the shape of the particles' trajectory. Particularly striking is the dominant role of the sweeps events in the initial motion of the grains, which was confirmed by experimental (Cameron 2006) and numerical (Bialik 2013) studies. The negative values of the vertical velocity decrease the lift force and should thus reduce the entrainment, but the results of simulations and experiments indicate a completely different role of sweeps events. Therefore, further attention is required to clarify the contribution of the individual forces being exerted on the particle and the role of the coherent structures in the particles' entrainment.

16.3 Diffusion of Saltating Particles

The diffusive nature of the bedload particles has been observed for almost 50 years (e.g., Sayre and Hubbell 1965; Drake et al. 1988). However, only the combination of experimental studies (e.g., Habersack 2001; Nikora et al. 2002; Campagnol et al. 2013) and Lagrangian models of saltation (e.g., Nikora et al. 2001; Lukerchenko et al. 2009a, b; Bialik 2011b; Bialik et al. 2012, 2015; Moreno and Bombardelli 2012) have allowed for the improvement of our knowledge on this issue and the building of the theory of particles' diffusion in bedload. Figure 16.3 shows

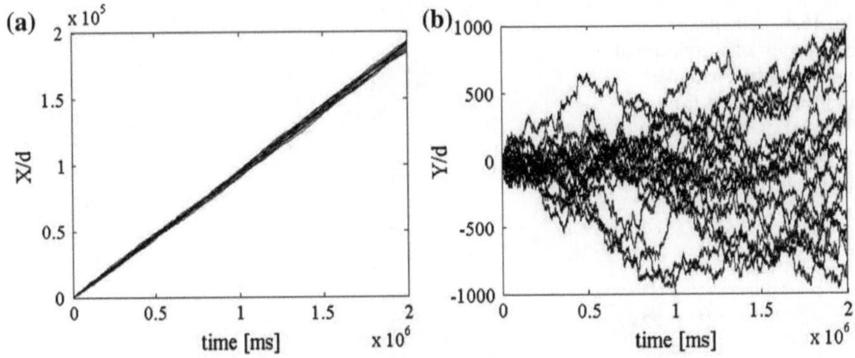

Fig. 16.3 Particle trajectories: **a** $X(t)/d$, and **b** $Y(t)/d$

examples of temporal changes of the particles' positions based on the simulations carried out using the Bialik et al. (2012) model. It should be noted that the diffusion phenomenon is easily visible in this figure.

Nikora et al. (2001, 2002) presented a conceptual model for the diffusion of saltating/rolling particles which consists of at least three ranges of scales: local, intermediate and global. They suggested that the local diffusion is ballistic and is related to particles' single jump; diffusion in the intermediate range, which consists of many jumps, could be normal or anomalous (super- or sub-diffusion); and the global range of particles' diffusion, which consists of many intermediate ranges, is sub-diffusive. Figure 16.4 schematically summarizes the model of Nikora et al. (2001, 2002).

On the other hand, Lukerchenko et al. (2009a, b) proposed the concept of a bundle or fascicle of particle trajectories starting from the same point. Generally, the main difference in comparison to the model proposed by Nikora et al. (2001, 2002) was that they introduced the disperse angle, which was defined as the angle between the bundle boundary and its axis, and found that this angle is linearly dependent on the lateral channel slope, whereas the model of Nikora et al. (2001, 2002) is based on the variances of particle positions. Lukerchenko et al. (2009a, b) showed that the value of the angle increases when the particle diameter increases and the shear velocity decreases.

According to Nikora et al. (2002), who suggested that the boundaries between diffusion ranges should probably depend on the Shields parameter and the relative sizes between particles, Bialik et al. (2012) investigated the effects of the particle mobility parameter, bed topography, and turbulence on the particles' diffusion. They found that the spread of saltating particles in the lateral direction depends on turbulence, whereas in the transversal direction it depends more on the relative sizes of the particles. Because they found a different value of the boundary between the intermediate and global ranges (based on the simulations results they suggested that it should be more than $tu_*/d \sim 800$) from previous studies (for example, Nikora et al. (2002) suggested that it should be equal to $tu_*/d \sim 15$), they speculated that it

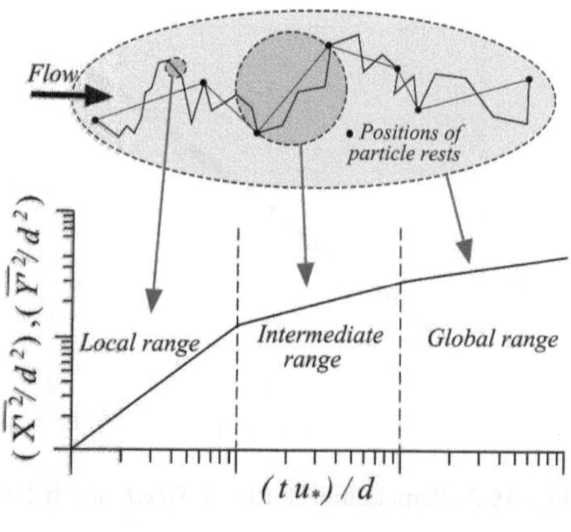

Fig. 16.4 Schematic presentation of a conceptual model of a bed particles' diffusion (adapted from Nikora et al. 2001, 2002)

is thus necessary to look for other important factors affecting the movement of particles. Moreno and Bombardelli (2012) showed that sudden changes in the particles' velocity and acceleration which appear due to particle–particle collisions introduce additional transverse diffusion, while in the streamwise direction diffusive behaviour due to this process does not exist. On the other hand, in the paper by Bialik et al. (2015) it was shown that two different time scales for defining the local–intermediate and intermediate–global boundaries should be considered. The first one is controlled by turbulence, as suggested by Niño (1995), Nikora et al. (2002) or Bialik et al. (2012), and the second one depends on the duration of the resting period of the particles, which is the most important factor controlling the spread of particles in the global range of diffusion. Moreover, a new method of data normalization based on the mean resting time was proposed in this paper, allowing for the identification of the diffusive behaviour of particles at different temporal scales. Figure 16.5 shows the change in time of the variance of particle longitudinal position obtained with use of the Bialik et al. (2012) model. At least three diffusion ranges are visible and the results confirm the Nikora et al.'s (2001, 2002) theoretical approach.

Recently, there has been a noticeable tendency for the diffusion of sediment particles to be analysed with the use of the Random Walk Models (i.e., Bradley et al. 2010; Tucker and Bradley 2010; Roseberry et al. 2012; Zhang et al. 2012). Generally, in such models the particle step lengths and resting periods are described by the probability distributions, but the physics of the individual particle jump is missing. This approach allows for the analysis of particles diffusion in the very large temporal scales (longer than the resting time) which, in the case of Lagrangian models based on Newton's second law, is still not achievable due to computer

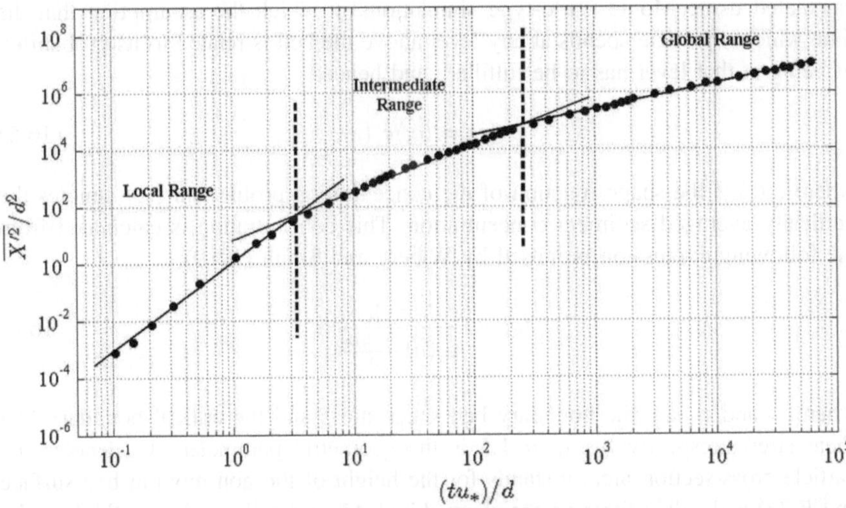

Fig. 16.5 Time evolution of the second order moments of particle longitudinal position

limitations. However, the use of both methods shows some contradictory results, for example, concerning the determination of the boundaries between the diffusion regimes or identification of the type of regime, in particular in the global scales. Further study is therefore required in order to clarify the differences between the results obtained when using these two approaches.

16.4 Bedload Transport Rate

The flux of the bedload sediment q_s is given by the product of the downstream particle velocity and sediment concentration in the bedload layer and hence

$$q_s = \int_{\eta}^{H_s} c_s(z) u_p(z) dz, \tag{16.4}$$

where $c_s(z)$ is the vertical distribution of the particles' concentration, $u_p(z)$ stands for the vertical distribution of the particles' velocity in the streamwise direction, z denotes the distance from the channel bottom, η is the level of the non-moving bed surface and H_s is the maximum height of the saltation. Wiberg and Smith (1989) suggested that the saltation models allow us to obtain most of the parameters required to calculate the bedload transport rate and, in order to prove this statement, they used the saltation model described in Wiberg and Smith (1985). In addition, they proposed a method for the description of the distributions of the particles' concentration, which cannot be directly obtained from the saltation model. Generally, as noted in Bialik and Czernuszenko (2013), these distributions may be calculated using Monte Carlo-type simulations in which the assumption that the time which a particle spends at any level above the bed is related to its probability of being at that level has to be fulfilled, and hence:

$$c_s(z) = \langle c_s \rangle c_*(z), \tag{16.5}$$

where $c_*(z)$ is the shape function of the concentration profile and $\langle c_s \rangle$ denotes the vertically averaged sediment concentration. This concentration is calculated from the following expression proposed by Wiberg and Smith (1989):

$$\langle c_s \rangle = \frac{\tau_* - \tau_{cr}}{c_*(\eta) \frac{F_D(\eta) \alpha_D}{A_D}}, \tag{16.6}$$

where τ_* and τ_{cr} are the boundary bed shear stress and the critical boundary bed shear stress, respectively, α_D = 1.5 is the geometric parameter, A_D denotes the particle cross-section area, η stands for the height of the non-moving bed surface and $F_D(\eta)$ is the drag force on the channel bed. More details on this method can be found in Wiberg and Smith (1989) or Bialik (2011a).

Wiberg and Smith (1989) showed that the results of the bedload transport rate obtained with the proposed method fit very well to the experimental data and, compared with the traditional equations (i.e., Meyer-Peter and Muller (1948) or Fernandez-Luque and van Beek (1976)) do not require the measurement of any flow or sediment coefficients. In addition, the results may be obtained directly from the simulations, which was the great advantage of this method. The authors also claimed that their proposal may be a good starting point for a more complex solution that could take into account more components of the sediment transport in the bedload. For example, Bialik and Czernuszenko (2013), based on the Wiberg and Smith (1989) approach, verified positively the hypothesis "*that the distributions of particles' velocity and sediment concentration, instead of averaged characteristics of these values, should be taken into account in the calculation of sediment transport rate*". Moreover, they investigated the influence of the effect of inter-particle collisions on the bedload transport and confirmed the earlier suggestions of Lee et al. (2002), who speculated that the bedload transport rate obtained based on the saltation modelling may be slightly overestimated due to neglect of the bed forms and especially the inter-particle collisions between saltating particles. The inclusion of the latter effect, as shown by Bialik and Czernuszenko (2013) in the saltation model, reduces the overestimation of the sediment transport rate, especially for gravel bed particles.

On the other hand, Niño and Garcia (1994, 1998b) proposed the formula for calculation of the bedload transport rate, based on the numerically simulated saltation characteristics:

$$\phi = 43.0(\tau_* - \tau_{cr})(\sqrt{\tau_*} - \sqrt{\tau_{cr}}), \tag{16.7}$$

where ϕ stands for the non-dimensional bedload transport, τ_* denotes the non-dimensional bed shear stress and τ_{cr} is the non-dimensional critical boundary shear stress. Equation (16.7) is quite similar in structure to that of Engelund and Fredsøe (1976); however, the obtained results were three times higher than those obtained using the Engelund and Fredsøe (1976) formulae. They explained this difference by the estimation of the Lagrangian dynamics friction coefficient calculated from the saltation model, which differs from the dynamic friction coefficient used in the derivation of the bedload model. They suggested that the proposed formula may be valid only for very high particle concentrations and small particles diameters.

The application of Lagrangian models of saltation for the calculation of the bedload sediment transport rate, although not as popular as expected when creating these models (as mentioned in the introduction), has certainly provided important information about this process. However, as emphasized by most authors (i.e., Wiberg and Smith 1989; Niño and Garcia 1994, 1998b; Lee et al. 2002), these models tend to overestimate the obtained results. They pointed out that this problem may have different underlying reasons, namely the omission of the collisions between the saltating particles (Lee et al. 2002) or the need to determine the magnitude of the shear stresses exerted on the bed by the particles (Niño and Garcia 1994). Lajeunesse et al. (2010) and Roseberry et al. (2012) pointed out that the velocities of most of the particles moving in the bedload are considerably smaller than if they only moved in the form of saltation, which seems to be the most important problem in the application of these models when calculating the bedload transport rate. In addition, there is still no clear answer as to what is the percentage of particle motion in bedload transport that moves in the form of saltation. There is no doubt that the Lagrangian models of saltation will meet the expectations placed on them only when the above concerns have been clarified.

16.5 Summary

Lagrangian modelling of saltation from the very beginning of its application has had both its supporters and opponents. Although many are critical of such an approach, there is no doubt that it has brought great results allowing for the improvement of knowledge of the bedload transport. In this paper the achievements following from the application of Lagrangian modelling of saltation have been highlighted regarding three areas: initial motion of saltating grains, diffusion of particles and calculation of bedload transport rate. The particular challenges in all of these research areas and future ways forward were presented in this review paper. It should be noted here that the topic is not limited only to these issues, however. One of the most important breakthroughs would be a transition from models describing the motion of a single spherical particle to models of particles representing more natural shapes. The research questions presented here, and many others, are still awaiting answers. Therefore, well-established researchers, students and practitioners will find in the Lagrangian approach a useful tool that should allow for further significant achievements in sediment transport mechanics.

Acknowledgments This work was supported within statutory activities No. 3841/E-41/S/2014 of the Ministry of Science and Higher Education of Poland. I am very grateful to the Editors of this book: Professors Paweł Rowiński and Artur Radecki-Pawlik for their invitation to write this chapter. I appreciate the thoughtful discussions with many colleagues with whom I have been privileged to interact on this subject. In particular, I wish to thank to my family, and Professors: Francesco Ballio, Włodzimierz Czernuszenko, Willi Hager, Jarosław Napiórkowski, Vladimir Nikora and Paweł Rowiński.

References

Abbott JE, Francis JRD (1977) Saltation and suspension trajectories of solid grains in a water stream. Phil Trans R Soc Lond A 284(1321):225–254

Bagnold RA (1956) The flow of cohesionless grains in fluid. Phil Trans R Soc Lond A 249: 235–297

Bagnold RA (1973) The nature of saltation and of bedload transport in water. Proc R Soc Lond A 332(1591):473–504

Basset AB (1888) A treatise on hydrodynamics: with numerous examples. Deighton, Bell and Co., Cambridge; George Bell and Sons, London, p 264

Bialik RJ (2010) Modeling of the saltating particles motion in the river flow and the bedload sediment transport. Ph.D. thesis, Institute of Geophysics, Polish Academy of Sciences, Warsaw, (in Polish)

Bialik RJ (2011a) Particle–particle collision in Lagrangian modeling of saltating grains. J Hydraul Res 49(1):23–31

Bialik RJ (2011b) Numerical study of saltation of non-uniform grains. J Hydraul Res 49(5):697–701

Bialik RJ, Nikora VI, Rowiński PM (2012) 3D Lagrangian modelling of saltating particles diffusion in turbulent water flow. Acta Geophys 60(6):1639–1660

Bialik RJ (2013) Numerical study of near-bed turbulence structures influence on the initiation of saltating grains movement. J Hydrol Hydromech 61(3):202–207

Bialik RJ, Czernuszenko W (2013) On the numerical analysis of bedload transport of saltating grains. Int J Sediment Res 28(3):413–420

Bialik RJ, Nikora VI, Karpiński M, Rowiński PM (2015) Diffusion of bedload particles in open-channels flows: distribution of travel times and second-order statistics of particle trajectories. Environ Fluid Mech (on-line first)

Bombardelli FA, González AE, Niño YI (2008) Computation of the particle basset force with a fractional-derivative approach. J Hydraul Eng ASCE 134(10):1513–1520

Boussinesq J (1903) Théorie analytique de la chaleur. Gauthier-Villars, Imprimeur-Libraire, Paris

Bradley DN, Tucker GE, Benson DA (2010) Fractional dispersion in a sand bed river. J Geophys Res 115:F00A09

Browne T (1646) Pseudodoxia Epidemica. Of the picture of a Grashopper 5(3):274–275

Cameron SM (2006) Near-boundary flow structure and particle entrainment. Ph.D. thesis, University of Auckland, New Zealand

Campagnol J, Radice A, Nokes R, Bulankina V, Lescova A, Ballio F (2013) Lagrangian analysis of bedload sediment motion: database contribution. J Hydraul Res 51(5):589–596

Corrsin S, Lumley J (1956) On the equation of motion for a particle in turbulent fluid. Appl Sci Res 6(2–3):114–116

Czernuszenko W (1998) The drift velocity concept for sediment-laden flows. J Hydraul Eng 124 (10):1026–1033

Czernuszenko W, Rowiński PM (2008) Shear stress statistics in a compound channel flow. Arch Hydro-Engin Environ Mech 55(1–2):3–27

Drake TG, Shreve RL, Dietrich WE, Whiting PJ, Leopold LB (1988) Bedload transport of fine gravel observed by motion-picture photography. J Fluid Mech 192:193–217

Engelund F, Fredsøe J (1976) A sediment transport model for straight alluvial channels. Nord Hydrol 7:293–306

Einstein HA (1942) Formula for the transportation of bedload. Trans ASCE 107:561–577

Fernandez-Luque R, van Beek R (1976) Erosion and transport of bedload sediment. J Hydraul Res 14(2):127–144

Francis JRD (1973) Experiments on the motion of solitary grains along the bed of a water stream. Proc R Soc Lond A 332:443–471

Garcia M, Parker G (1991) Entrainment of bed sediment into suspension. J Hydraul Eng 117 (4):414–435

Gilbert GK (1914) The transportation of debris by running water. US Geol Surv Prof Pap 89:263

Gordon R, Carmichael JB, Isackson FJ (1972) Saltation of plastic balls in a one-dimensional flume. Water Resour Res 8(2):444–459

Habersack HM (2001) Radio-tracking gravel particles in a large braided river in New Zealand: a field test of the stochastic theory of bedload transport proposed by Einstein. Hydrol Process 15 (3):377–391

Ji C, Munjiza A, Avital E, Xu D, Williams J (2014) Saltation of particles in turbulent channel flow. Phys Rev E 89:052202

Kharlamova IS, Vlasak P (2015) Dependence of saltation characteristics on bed organisation in numerical simulation. Geosci J 9(1):177–184

Lajeunesse E, Malverti L, Charru F (2010) Bedload transport in turbulent flow at the grain scale: experiments and modeling. J Geophys Res 115:F04001
Lee SL (1987) A unified theory on particle transport in a turbulent dilute two-phase suspension flow II. Int J Multiphase Flow 13(1):137–144
Lee SL, Wiesler MA (1987) Theory on transverse migration of particles in a turbulent two-phase suspension flow due to turbulent diffusion I. Int J Multiphase Flow 13(1):99–111
Lee HY, Hsu IS (1994) Investigation of saltating particles motions. J Hydraul Eng 120(7):831–845
Lee HY, Hsu IS (1996) Particle spinning motion during saltating particles. J Hydraul Eng 122 (10):587–589
Lee HY, Chen YH, You JY, Lin YT (2000) Investigation of continuous bedload saltating process. J Hydraul Eng 126(9):691–700
Lee HY, You JY, Lin YT (2002) Continuous saltating process of multiple sediment particles. J Hydraul Eng 128(4):443–450
Lee HY, Lin YT, Chen YH, You JY, Wang HW (2006) On three-dimensional continuous saltating process of sediment particles near the channel bed. J Hydraul Res 44(3):374–389
Lu SS, Willmarth WW (1973) Measurements of the structure of the Reynolds stress in a turbulent boundary layer. J Fluid Mech 60:481–571
Lukerchenko N, Chara Z, Vlasak P (2006) 2D numerical model of particle-bed collision in fluid-particle flows over bed. J Hydraul Res 44(1):70–78
Lukerchenko N, Kvurt Y, Kharlamov A, Chara Z, Vlasak P (2008) Experimental evaluation of the drag force and drag torque acting on a rotating spherical particle moving in fluid. J Hydrol Hydromech 56(2):88–94
Lukerchenko N, Piatsevich S, Chara Z, Vlasak P (2009a) 3D numerical model of the spherical particle saltation in a channel with a rough fixed bed. J Hydrol Hydromech 57(2):100–112
Lukerchenko N, Piatsevich S, Chara Z, Vlasak P (2009b) Numerical model of spherical particle saltation in a channel with a transversely tilted rough bed. J Hydrol Hydromech 57(3):182–190
Lukerchenko N, Dolansky J, Vlasak P (2012) Basset force in numerical models of saltation. J Hydrol Hydromech 60(4):277–287
Maxey MR, Riley JJ (1983) Equation of motion of a small rigid sphere in a nonuniform flow. Phys Fluids 26(4):883–889
McEwan IK, Willetts BB (1991) Numerical model of the saltation cloud. Acta Mech Suppl 1:53–66
McEwan IK, Jefcoate BJ, Willetts BB (1999) The grain-fluid interaction as a self-stabilizing mechanism in fluvial bedload transport. Sedimentology 46:407–416
McGee WJ (1908) Outlines of hydrology. Geol Soc America Bull 19:193–220
Mei R, Adrian RJ, Hanratty T (1991) Particle dispersion in isotropic turbulence under Stokes drag and Basset force with gravitational settling. J Fluid Mech 225:481–495
Meyer-Peter E, Muller R (1948) Formulas for bedload transport. Proc 2nd Int IAHR Congress
Moreno PM, Bombardelli FA (2012) 3D numerical simulation of particle–particle collisions in saltation mode near stream beds. Acta Geophys 60(6):1661–1688
Murphy PJ, Hooshiari H (1982) Saltation in water dynamics. J Hydraul Div 108(11):1251–1267
Nelson JM, Shreve RL, McLean SR, Drake TG (1995) Role of near-bed turbulence structure in bedload transport and form mechanics. Water Resour Res 31(8):2071–2086
Nezu I, Nakagawa H (1993) Turbulence in open-channel flows. Balkema, Netherlads
Nikora V, Heald J, Goring D, McEwan I (2001) Diffusion of saltating particles in unidirectional water flow over a rough granular bed. J Phys A: Math Gen 34:743–749
Nikora V, Habersack H, Huber T, McEwan I (2002) On bed particle diffusion in gravel bed flows under weak bedload transport. Water Resour Res 38(6):17-1–17-9
Niño Y (1995) Particle motion in the near wall region of a turbulent open channel flow: implications for bedload transport by saltation and sediment entrainment into suspension. Ph. D. thesis, University of Illinois at Urban-Champaign, Urbana, Illinois
Niño Y, Garcia M (1994) Gravel saltation 2. Modelling. Water Resour Res 30(6):1915–1924
Niño Y, Garcia M (1998a) Experiments on saltation of sand in water. J Hydraul Eng 124 (10):1014–1025

Niño Y, Garcia M (1998b) Using Lagrangian particle saltation observations for bedload sediment transport modelling. Hydrol Process 12(8):1197–1218

Ossen CW (1927) Neure methoden und ergebnisse in der hydrodynamik. Akademische Verlagsgesellschaft, Leipzig, p 337

Rowiński PM, Lee SL (1993) Equilibrium profile of suspended sediment concentration. Acta Geophys Pol 41(2):163–176

Rowiński PM (1995) Transport of solid particles in turbulent water flow. Ph.D. thesis, Institute of Geophysics, Polish Academy of Sciences, Warsaw, (in Polish)

Roseberry JC, Schmeeckle MW, Furbish DJ (2012) A probabilistic description of the bedload sediment flux: 2. Particle activity and motions. J Geophys Res 117:F03032

Sayre W, Hubbell D (1965) Transport and dispersion of labeled bed material, North Loup River, Nebraska. US Geol Surv Prof Pap 433(C):1–48

Schmeeckle MW, Nelson JM, Pitlick J, Bennett JP (2001) Interparticle collision of natural sediment grains in water. Water Resour Res 37(9):2377–2391

Sekine M, Kikkawa H (1992) Mechanics of saltating grains. II. J Hydraul Eng 118(4):536–558

Sørensen M, McEwan IK (1996) On the effect of mid-air collisions on aeolian saltation. Sedimentology 43(1):65–76

Tchen CM (1947) Mean value and correlation problems connected with the motion of small particles suspended in a turbulent fluid. Ph.D. thesis, The Hague, Marinus Nijhoff

Tucker GE, Bradley DN (2010) Trouble with diffusion: reassessing hillslope erosion laws with a particle-based model. J Geophys Res 115:F00A10

White BR (1979) Soil transport by winds on Mars. J Geophys Res 84(B9):4643–4651

White BR, Greeley R, Iversen JD, Pollack JB (1976) Estimated grain saltation in a Martian atmosphere. J Geophys Res 81(32):5643–5650

Wiberg PL, Smith JD (1985) A theoretical model for saltating grains in water. J Geophys Res 90 (C4):7341–7354

Wiberg PL, Smith JD (1987) Calculation of the critical shear stress for motion of uniform and heterogeneous sediments. Water Resour Res 23(8):1471–1480

Wiberg PL, Smith JD (1989) Model for calculating bedload transport of sediment. J Hydraul Eng 115(1):101–123

Yeganeh-Bakhtiary A, Shabani B, Gotoh H, Wang SSY (2009) A three-dimensional distinct element model for bedload transport. J Hydraul Res 47(2):203–212

Zhang Y, Meerschaert MM, Packman AI (2012) Linking fluvial bed sediment transport across scales. Geophys Res Lett 39:L20404

Zou XY, Cheng H, Zhang CL, Zhao YZ (2007) Effects of the Magnus and Saffman forces on the saltation trajectories of sand grain. Geomorphology 90(1–2):11–22

Chapter 17
Runoff Generation and Soil Erosion After Forest Fires from the Slopes to the Rivers at a Basin Scale

Antonio Velasco and Xavier Úbeda

Abstract In many parts of the world, forest fires are very common. Fires burn more or less severely the vegetation and even the most organic parts of the forest soil. Fire can modify the physical properties of soil; due to the combustion of vegetation and organic matter, some chemical substances appear and ashes are deposited in the soil surface in various quantities. During the combustion, there is a production of gases, which go to the atmosphere but some of them go deep into the soil and condensate around the soil particles. Due to these important changes resulting from fires, the rainfall arrives at the soil surface in a different way, the infiltration capacity changes due to hydrophobicity, the runoff generation can increase and also the movement of fine and coarse soil particles takes place. All these changes are important for the water and sediment production at the slopes and the consequences can remain during months or years at a basin scale.

Keywords Soil properties · Hydrophobicity · Infiltration capacity · Ashes

17.1 Introduction

Many environments of the world are not stranger to the problem of forest fires (Carretero et al. 2003; Ferreira-Leite et al. 2013). We do not believe the fire to be an enemy to fight, because we share the argument that many of our landscapes, for example the Mediterranean landscapes (Ferreira-Leite et al. 2013) have been shaped by fire. It is a mistake to want to reach the total fire suppression, taking examples like in the Unites States, where this policy of "fire suppression" proved to be a big mistake (Úbeda and Outeiro 2008) and caused ones of the biggest forest fires ever seen. So, when we talk about the "fire problem", we use different terms to define the

A. Velasco · X. Úbeda (✉)
GRAM (Grup de Recerca Ambiental Mediterrània). Department of Physical Geography, University of Barcelona, Montalegre 6, 08001 Barcelona, Spain
e-mail: xubeda@ub.edu

P. Rowiński and A. Radecki-Pawlik (eds.), *Rivers—Physical, Fluvial and Environmental Processes*, GeoPlanet: Earth and Planetary Sciences,
DOI 10.1007/978-3-319-17719-9_17

fires which burned hundreds of hectares, sometimes emphasizing the importance of the burned area with the term "Large Forest Fires", and sometimes focusing on the fire intensity with the term "High Intensity Fires". In this paper we will use the term Large Forest Fire; the features of such fires are, among others: very high fire intensities, impossibility of directly fighting by the teams of extinction, and large burned areas.

Forest history of the last 50 years included: (1) rural depopulation, which has led to an accumulation of forest fuel, many times dry and dead; (2) an increase in forest cover and continuity, vertically and horizontally; (3) a proliferation of urban-rural interface; and (4) the risk of periods of drought, with very hot summers and very low humidity. Consequently, the risk of having a Large Forest Fire will surely be repeated in the world.

We define the basin as a closed system of input and output of water. Entries can be either in the form of liquid precipitation (rain) and solid (snow) and outputs are usually given through the main channel (river, stream, creek, etc.). The basin is the basic unit used for hydrographic studies, allowing measurements and balance of inputs and outputs of water. The study of a basin includes soil, water fluxes and potential aquifers. The surface of the basin is expressed in km^2. Water catchment system for the various elements present in the basin and the movement of water and its status changes is called the hydrologic cycle. Usually a percentage of the precipitation leaves the main channel of the river, as part of the water is retained by the soil, vegetation, and groundwater or can evaporate into the atmosphere, either directly or indirectly.

An important factor that determines the amount of water that leaves the basin is the different land use (farming, forestry, urban, etc.). Depending of the land use, the soil will have a different capacity to retain and infiltrate water. Also soil characteristics (composition, porosity, structure), and even the type of vegetation cover as well as the type of rain and the topography play an important role. Only a percentage of the water that precipitates generates runoff that arrives later in the channels reaching the outlet of the basin to another river, lake, reservoir or the sea. The amount of water generated as runoff varies throughout the year, depending on factors such as rainfall or vegetation existing in the basin.

Some effects of fire on hydrology are as follows: reduced infiltration rates and increased runoff (De Bano et al. 1967), increased total flow and peak flow (Brown 1972; Lavabre et al. 1993) and accelerated soil erosion and consequently increased sediment load of the river (Brown 1972; Batalla and Sala 1998). Fire effects on hydrological processes depend on many factors: features (fire intensity and severity), the type of soil, the rainfall (amount and intensity), topography, the basin area affected, the vegetation characteristics or the time elapsed after the fire (Lopez 1999).

Thus, the hydrological processes that may be affected by fire are basically the intercepting rainfall infiltration and the surface runoff (Batalla 2002). When a basin is affected by a forest fire, the whole system is altered (Sala and Rubio 1994), as the fire quickly destroys the vegetation, leaving a bare soil (if the fire intensity is high enough). The disappearance of the vegetation cover is an important lack of soil protection against rainfall drops. In case of forest fires, the intensity of fire plays a critical role in the degree of affectation of the soil. Another important element is the litter (leaves and branches of fallen trees which are decomposing over time), which has several functions in hydrological terms:

1. Retains a portion of the moisture.
2. Controls water infiltration into deeper layers of soil
3. Protects soil aggregates from the direct impact of raindrops.

If the fire is of a low or medium intensity, part of this litter or vegetation cover can remain and make the soil to be not exposed to rain. On the other hand, if the fire intensity is high, vegetation, litter and even part of the organic matter can disappear for combustion. This loss can destroy the soil structure and permeability, increasing erosion (Badia and Úbeda 2013).

When forest fires take place, they generate a layer of ash and organic residues. This layer can sometimes protect the soil from erosion during a variable period of time (Cerdà and Doerr 2008; Zavala et al. 2009). When this layer disappears, the soil is bare and rainwater will impact directly on it, fragmenting aggregates, generating splash, and the particles of the soil will be able filling the pores and sealing the soil surface (Inbar et al. 1998). This sealing of the surfaces can increase the surface runoff, and, with it, the possible increase of soil erosion (Badia and Úbeda 2013). This post-fire response is very common in the Mediterranean environments since fires are frequent in summer (dry season) and the first autumn storms are very intense. However, in the Atlantic climate the rainfall is more even throughout the year (no dry season), and vegetation recovers faster.

If a fire burns at a very high temperature, it can affect certain properties of the mineral soil horizons (Badia et al. 2011). The accumulation of ash from burning can alter the hydrological behavior of the basin. The accumulation of ash can form hydrophobic surfaces, reducing water infiltration capacity of the soil and increase surface runoff. The high rates of runoff and erosion will be present until it recovers or vegetation recovery makes a protection of the soil in front of heavy rains and reduces the runoff generation (Stoof 2011).

17.2 Rainfall Interception

The severe reduction of vegetation cover alters the processes of interception and evapotranspiration and seriously affects the hydrological cycle (Fig. 17.1). Interception is the process by which land cover (vegetation, organic matter, etc.) interrupts the fall of precipitation on the soil and thus reduces the impact of raindrops on the soil surface (Tiedemann et al. 1979). This loosening of the soil, caused by the disappearance of the vegetation cover, together with surface runoff is a major component of the erosion processes.

The effectiveness of cover against soil erosion is proportional to the amount of existing vegetation cover (Osborn 1954). Interception loss in a rain forest that has

Fig. 17.1 Burnt slope without canopy protection (Photography by Xavier Úbeda)

not undergone any forest fire during an episode of heavy rainfall is very low (2–10 %). However, in a forest where the vegetation has been removed by fire, the effectiveness of interception of rain drops, increasing the potential for runoff and subsequently the potential soil evaporation due to increased insolation and ventilation (Batalla 2002).

Soto et al. (1994) reported a 70 % reduction in the interception after prescribed burning in low intensity in a study area in Galicia. It is generally believed that nearly all of interception is completely eliminated with high fire intensity, but that against low or medium intensity fires interception is much less affected (Batalla 2002).

17.3 Soil Hydrophobicity

The hydrophobicity or water repellence of soil is an important factor that affects the processes of runoff and infiltration and is a key component in hydrological models (Doerr et al. 2003; Doerr and Shakesby 2009).

One of the effects of fire on soils is usually caused by increased hydrophobicity or water repellence. This effect is studied since the late 1960s (DeBano et al. 1970). The repellence to water is a property of the soil that reduces the infiltration capacity and has important hydrological and geomorphological consequences. This effect usually occurs after forest fires but is also associated with various types of soils and climates in different parts of the world (Jordan et al. 2010). The accumulation of ash as a result of forest fires, as well as the volatilization of organic compounds during combustion and, once the fire around the condensation of soil aggregates, can help to increase this to hydrophobicity provided the temperature of the forest fire is between 200 and 250 °C (Osborn et al. 1964) or weaken and even destroy it if the temperature in the soil is up to 270–300 °C (DeBano et al. 1976; Nayaka 1982). These temperature ranges are not static, but vary depending on the type of soil. There are many factors involved in the development of soil hydrophobicity. Two basic factors are met during combustion: temperature and residence time (Jordan et al. 2010). Many authors have found that the water repellence in soils can be induced after suffering high temperatures (DeBano and Krammer 1966; DeBano 2000b). DeBano et al. (1970) and Savage (1974) found that fire can induce water repellence in soils that previously did not show it, as well as reduce it or increase it on soils that were hydrophobic. These changes are related to the combustion temperature, residence time of heat, the amount and type of fuel, soil moisture prior to the fire, soil type, etc. (Doerr et al. 2000). However, many studies show water repellence in soils naturally depending on the moisture content or the soil temperature (Witter et al. 1991; Ritsema et al. 1993; Dekker and Ritsema 1994; Moral et al. 2002). An example of an unburnt soil that is hydrophobic is shown in Fig. 17.2.

The repellence to water causes a reduction in the adsorption of water by the soil. Thus, the layer of soil that acquires this property has to offer strong resistance to wetting, causing decreased infiltration of water accumulated on the surface during a time period that can range from a few seconds to days or weeks (King 1981; DeBano 1981; Dekker and Ritsema 1994; Doerr and Thomas 2000). This effect has been observed in different soil types and different climates and vegetation types worldwide (Wallis and Horne 1992; DeBano 2000a; Doerr et al. 2000; Jaramillo et al. 2000). However, DeBano et al. (1970) and Wallis et al. (1993) found that this

Fig. 17.2 An extremely hydrophobic sandy soil sample. Slovakia (Photography by Jorge Mataix-Solera)

effect is usually more pronounced in acidic soils with sandy texture because they have a lower specific surface area.

The decreased rates of water infiltration in the soil as a result of hydrological and geomorphological hydrophobicity have immediate consequences, to which must be added other effects on the soil and survival of plants (House 1991; York 1993). By decreasing the rate of infiltration into the soil surface, the water repellence helps reduce runoff generation time and intensify surface flow, causing other serious consequences such as an increased risk of erosion or accelerated washing of soil (Imeson et al. 1992; Shakesby et al. 1993; Ritsema et al. 1993, 1997; Doerr and Shakesby 2009). However, water repellence is not always negative because it can increase the structural stability (Blanco-Canqui and Lal 2009) or carbon uptake (Urbanek et al. 2007).

Examples of repellence to water have been collected since 1917 (Schantz and Piemeisel 1917). However, this property has not been studied more intensively until the second half of the twentieth century, with searches of Jamison (1947), Van't Woudt (1959) and Domingo (1950). In the 1960s and 1970s the number of scientific publications dealing this phenomenon grew, emphasizing its strong relationship with forest fires (DeBano 1981). During the later years, the study of hydrophobicity has been continued with increasing frequency because it was shown to be a field of study with great significance for the research into the relationship between forest fires and soil. Thanks to research carried out until now, we know that hydrophobicity is a process much more widespread than previously thought and is a very normal phenomenon in a variety of soils (Doerr et al. 2000). In fact, it has been studied on every continent (except Antarctica), in areas with seasonal tropical climates or subarctic, in soils with different uses, in sandy and clay soils or soils with a very different amount of moisture (Wallis and Horne 1992; Doerr et al. 2000; Dekker and Ritsema 1996). Several studies have shown that, over time, the hydrophobicity of the soil decreases due to the rupture of hydrophobic components formed during heating of the soil. The duration of this phenomenon, however, has not been determined accurately. It can be present for days or years. Dyrness (1976) found that hydrophobicity of a soil after a fire in Oregon high intensity persisted for six years. Instead, Huffman et al. (2001) observed that hydrophobicity in a ponderosa pine forest persisted for 22 months. Velasco (2013) found that in a soil burnt 18 years ago there was not any kind of hydrophobicity, even in parts of the burnt area that was burnt at high intensity fire.

17.4 Soil Infiltration Capacity

Infiltration can be defined as the amount of water that can be absorbed by the soil in a given time. If supplies more water than soil can infiltrate, the result is runoff generation. Discontinuities of the surface hydrophobicity and discontinuities of hydrophobic/hydrophilic patches on the soil will determine whether flow occurs localized or more widespread in the area. Several studies have shown that usually there is no runoff generation in soils with hydrophilic surface in forest soils (Tsukamoto 1961; Dunne and Black 1970) due to these, soils usually have enough capacity to support infiltration in front of high intensity rainfalls.

Several factors affect infiltration (soil cover, vegetation type, the physico-chemical characteristics, etc.) and these can be altered severely by fire, causing a decrease in infiltration and therefore an increase in runoff. Arend (1941) observed a reduction in the infiltration rate of 2 cm per hour in plots burned annually for six years, compared to non-burned areas in the highlands of the Ozarks (Missouri). Krammer and DeBano (1965) observed that the fire created hydrophobic soil layers in areas of "chaparral vegetation" in Southern California.

When the rate of infiltration or infiltration capacity of the soil has been surpassed by the amount of rainfall, irrigation or by the rate of snowmelt, there is runoff generation. Several variables affect the rate of runoff: the intensity and duration of rainfall, slope and if the soil is frozen or not (Wisler and Brater 1959). DeByle and Packer (1972), in a study in eastern Montana, found that surface runoff caused by snowmelt was eight times higher in burned areas compared with control areas. Campbell et al. (1977) described a 66 % reduction in the rate of infiltration into soil occupied by a pine forest in Arizona, resulting in a 800 % increase in the flow of the stream of watershed burned in the first wet season after the fire. Although the problems caused by hydrophobicity are more pronounced in areas of "chaparral", the problem is not confined to the arid zones. Dyrness (1976) showed increases in hydrophobicity in the first 23 cm of soil in the first five years after the fire in a pine forest in Oregon which generated a decrease in the infiltration capacity. Other studies indicate that the physicochemical properties and infiltration capacity of the soil are little affected by low to moderate intensity fires (Hudson et al. 1983).

The process of water logging and infiltration in hydrophilic and hydrophobic soils was studied in detail by Wallach and Jortzick (2008). They observed that while the water quickly penetrated in hydrophilic soils, hydrophobic soils tended to expand over the surface. They found that the time of water penetration into hydrophobic soils could be controlled by the contact time between soil particles and water, as well as the pressure of the column of water above the soil surface. DeBano (1971), in a laboratory experiment after destroying the repellence to water in a soil by heat treatment, showed that the infiltration capacity of the soil increased up to 25 times. Wallis et al. (1990) also observed in the laboratory that the infiltration capacity of the hydrophilic soil was up to 6 times greater than the infiltration capacity in a hydrophobic soil.

17.5 Runoff Generation

Increases in runoff in burned areas have been observed particularly in Mediterranean areas, ranging from 11 % (Anderson 1976) to 300 % (Nasseri 1988) in large basins, or from 800 % in small basins (Campbell et al. 1977) up to 50,000 % at a scale plot (Inbar et al. 1998). Scott and Van Wyk (1990) observed increases of 200 % and up to 300 % in discharge peaks in the year following a wildfire in a basin of South Africa. López and Batalla (2001) described an increase of 30 % in runoff and 120 % in flood peaks for the same rainfall 6 months after a wildfire. Eight months after the fire the discharge values started to decrease, although they were still higher than the values before the wildfire.

Lavabre et al. (1993) studied the hydrological effects of a fire that burned 85 % of the oaks in a Mediterranean experimental basin in different time scales. The annual runoff increased by 25 % for the same annual precipitation, due to reduced evapotranspiration for the burning of the vegetation cover. The frequency of floods also grew. In these cases, the return time to the values of the pre-fire runoff conditions depends largely on the rate of recovery of vegetation cover and vegetation characteristics interception (Batalla 2002).

Jordán et al. (2008) analyzed the hydrological response of hydrophobic and non-hydrophobic soils in the Mediterranean area and found that the time to have runoff generation was lower and the rate of runoff increased with increasing degree of water repellence. They also found that the relationship between hydrophobicity with other soil properties and discontinuities of each contribute to a heterogeneous spatial distribution of the response of soil erosion. Jordan et al. (2009) analyzed the relationship between water repellence and soil loss by rainfall simulation in volcanic areas of Mexico, and observed that the hydrological response is related to the degree of hydrophobicity of the studied soils. Although there were no significant correlations between the persistence of water repellence and the rate of runoff, it was observed that the surface flow was more intense in soil water repellent soils than in hydrophilic ones.

Despite the fact that fire can modify the soil properties and provoke an accelerated runoff, it is known that dry soils show a high degree of repellence (Witter et al. 1991; Ritsema and Dekker 1994) and that implies that the runoff coefficient tends to increase when soil moisture is low. It is also necessary to take into account that water repellence depends largely on type and quantity of rains in a hydrological year (Crockford et al. 1991; Miyata et al. 2007), which may affect the runoff coefficient. Although several authors have shown a strong relationship between the impact of repellence to water and runoff generation (Osborn et al. 1964; Cerdà et al. 1998; Benito et al. 2003; Leighton-Boyce et al. 2007), the precise role of water repellence is difficult to determine, since it is very difficult to isolate it from the rest of the physical properties and moisture content of the soil (Miyata et al. 2007).

17.6 Soil Erosion

Fire affects soil systems in terms of the physical, chemical and biological properties and with this, the response to the hydrological and erosive rainfall is also altered. Thus, fires directly influence erosion processes by eliminating vegetation cover and produce changes in the structure, nutrient and soil water balance. Forest fires alter erosion processes, affect soil degradation and change land geomorphology because of accelerated erosion rates, transport and deposition of materials outside the soil system (Cerdà and Bodí 2007).

Erosion is a process of denudation of the soil by extraction, transport and deposition of material by various agents. It is a natural process where erosion rates allow the recovery and development of different lands. This phenomenon has led to the formation of the most fertile soils in the world, located in floodplains, river deltas and terraces built by the deposition of materials from the mountains. It is therefore a sustainable phenomenon dependent on natural conditions of each ecosystem (Cerdà and Bodí 2007).

Where human activity in the area is old, for example in the Mediterranean basin, erosion rates are high because the vegetation has been reduced and degraded soil. This is due to the exploitation of the soil by man (use of fire as a tool to lighten or remove the vegetation, fertilize the soil, etc.) (Cerdà and Bodí 2007).

Although the rates of soil erosion in water repellent soils are usually lower under dense vegetation cover (Cerdà and Doerr 2007), they can be very high in areas where natural vegetation has declined or disappeared as a result of human pressure or wildfires (Scott 1993; Shakesby et al. 1993; Úbeda and Sala 1998; Cerdà and Doerr 2007; Jordán et al. 2009). Irregularities in seasonal climates, like the Mediterranean, enhance the risk of soil erosion, especially at the beginning of the rainy season, which is, when the soil moisture is very low and the degree of hydrophobicity is very high (Osborn et al. 1964; Cerdà et al. 1998; Doerr et al. 2003; Jordán et al. 2008).

In sandy soils of the coast of Holland, Witter et al. (1991) demonstrated that the loss of sediment from runoff water increased almost 20 times in soil water repellent soils in front hydrophilic soils, although the relationship between the two parameters could not be isolated from other factors such as soil moisture and organic matter content.

Through experiments of rainfall simulations, Jordán et al. (2009) determined a good correlation between the rate of runoff and soil loss. They found an increase of 40 % in average soil water repellent compared to hydrophilic. The role of water repellence on soil erosion is very important in the case of soils affected by fires of high intensity, where infiltration can be reduced drastically (Campbell et al. 1977; Rulli and Rosso 2007). In the case of low-intensity fires, the impact is smaller, recovery time and the properties prior to the fire have more to do than the recovery of the vegetation cover (Hudson et al. 1983). In the case of Mediterranean soils, the vegetation recovery and regeneration of the layer of leaf litter and soil characteristics prior to the fire can be quite rapid (Fuentes et al. 1994; Cerdà et al. 1998), demonstrating the strong adaptation of these ecosystems to natural fires (Naveh and Lieberman 1984).

In Table 17.1 there some data about erosion rates in slopes after forest fire at plot scale.

Table 17.1 shows only few experiments to calculate erosion rates, but they are illustrative of the different results at plot scale. It is also clear how the Unites States, Australia and Mediterranean countries have developed the main part of the research

Table 17.1 Erosion rates at plot scale on the burnt slopes in some forest after fires

Erosion rate (T/ha/year)	Country	Author	Year
0.81	USA	Meginnis	1935
165	USA	Hendricks and Johnson	1944
7.61	USA	Copley et al.	1944
0.51	USA	Ursic	1970
3.85	Scotland	Kinako and Gimingham	1980
49.03	Portugal	Lourenço and Bento-Gonçalves	1990
21.75	Spain	Marqués	1991
41.46	Portugal	Terry	1994
11.10	Spain	Andreu et al.	1994
24.05	Spain	Úbeda and Sala	1998
16.34	France	Martin	1996
4.00	USA	Benavides and MacDonald	2001
17.6	Portugal	Nunes et al.	2013

of this topic, due to the recurrence of this phenomenon and the importance of soil losses due to forest fires.

17.7 Repercussions on the Hydrological Cycle

Rainfall, when infiltrates into the soil, fills the pores of surface horizons. The infiltration capacity depends on the speed water can move through the profile. Infiltration decreases rapidly in direct relation to soil depth, and is much higher in forest soils than in agriculture ones. Permeability is defined by large pores through which water can move by gravity. One important reason favouring the permeability of the forests soils is the continuous addition of organic matter to the forest soils as a result of falling leaves, branches, etc., which contributes to the fact that its structure is more granular; this addition varies with the type of forest throughout the year (Mataix-Solera and Guerrero 2007). The unaltered forest soils have the maximum infiltration capacity.

In addition to the infiltration and surface runoff caused by the removal of vegetation cover after forest fires, certain effects on the physic-chemical properties of the soil contribute to this alteration of the hydrological cycle in the affected area. The formation of hydrophobic substances and their impact on the reduction of soil permeability and increased runoff, due to the direct loss of vegetation cover and other indirect effects, are the most important (Mataix-Solera and Guerrero 2007). After forest fires, erosion studies on quantification revealed significant impact on the geomorphology of the soils surfaces (Moody and Martin 2001). The vegetation determines the runoff generation as it controls the distribution of infiltration rates (Cerdà 1995). The natural recovery of vegetation is usually slower in the sunny slopes than of shadow ones, and this factor is also important in controlling erosion rates in the months following the fire (Cerdà et al. 1995).

After a forest fire, with increasing hydrophobicity of the soil, changes in infiltration and runoff generation, and if vegetation cover is not recovered before the first heavy rains arrive, the soil can be eroded in a much accelerated way.

This rain hitting against the bare soil contributes significantly to the destruction of aggregates. Finer fractions clogged the pores, reducing the infiltration rate (Ela et al. 1992), increasing surface runoff (Swanson 1981), loss of nutrients and also increasing the erosion surface (Úbeda et al. 1990; Andreu et al. 1994). The stability of aggregates that are on the surface is very important. Unstable aggregates form a crust that inhibits the movement of water and air in soil (Mataix-Solera and Guerrero 2007).

Slopes with steep gradients can accelerate erosion process, sweeping away much of the soil and exposing the underlying rock. Some authors have investigated how important is the slope gradient and orientation and they have found relationships between these slope variables and erosion rates in areas affected by wildfires (Llovet and Ponce 1996; Andreu et al. 2001).

17.8 Final Remarks

It is interesting how, once the researchers have known about the effects of fires on soils and how the variables (hydrophobicity, infiltration capacity, slopes, time, etc.) participate in the increase of soil erosion, the main part of these researchers in the last decades of the XX century focused their investigations on how to mitigate the erosion rates.

Nowadays there are works which try to minimize the effects of the fire on soils, once the fire took place, incorporating mulch on the soil to benefice the infiltration capacity and avoid splash erosion, until the barriers are installed in determined spots of the slopes where the generation of accelerated soil erosion is more likely.

Today, researchers and forest managers have understood that "prevention" is the only way to avoid great effects on soil erosion after forest fires. It is even cheaper, economically, to concentrate resources in prevention than control the erosion after forest fires. The forest management is the key for the prevention. Create forest structures that in case of forest fires do not develop High Intensity Fires, which are responsible for generating high erosion rates, and forest mosaic so as to avoid the forest masses connection which could develop fires to burn large extents of landscape.

Acknowledgments The authors want to thank the Spanish Project from the Ministerio de Economía y Competitividad: *Evaluacion de la calidad del suelo, control de la erosion y recuperacion de la cubierta vegetal en distintos escenarios y manejos postincendio* (CGL2013-47862-C2-2-R).

References

Anderson HW (1976) Fire effects on water supply, floods and sedimentation. In: Tall timbers, fire ecology conference, n° 15, pp 249–260

Andreu V, Imeson AC, Rubio JL (2001) Temporal changes in soil aggregates and water erosion after a wildfire in a Mediterranean pine forest. Catena 44:69–84

Andreu V, Rubio JL, Forteza J, Cerní R (1994) Long term effects of forest fires on soil erosion and nutrient losses. In: Sala M, Rubio JL (eds) Soil erosion as a consequence of forest fires. Geoforma Ediciones, Logroño, pp 79–90

Arend JL (1941) Infiltration rates of forest soils in the Missouri Ozarks as affects by weed burning and litter removal. J Forestry 39:726–728

Badía D, Úbeda X (2013) Consecuencias hidrológicas y erosivas de un incendio forestal. Ficha técnica. FGR2013/05

Badía D, Martí C, Charte R (2011) Soil erosion and conservation measures in semiarid ecosystems affected by wildfires. In: Godone D, Stanchi S (eds) Soil erosion studies. InTech Open Access Publisher, Rijeka, pp 87–110

Batalla RJ (2002) Hydrological implications of forest fires: an overview. In: Pardini G, Pintó G (eds) Fire, landscape and biodiversity: an appraisal of the effects and effectiveness. Lectures form the 3rd International Summer School on the Environment. Institut de Medi Ambient, Universitat de Girona, pp 99–115

Batalla RJ, Sala M (1998) Changes on sediment and dissolved load after wildland fire in a Mediterranean river basin. XXIII General Assembly of the European Geophysical Society, Niza, France, April 1998

Benavides J, MacDonald LH (2001) Post-fire runoff and erosion from simulated rainfall on small plots, Colorado front range. Hydrol Process 15(15):2931–2952

Benito E, Santiago JL, de Blas E, Varela ME (2003) Deforestation of water-repellent soils in Galicia (NW Spain): effects on surface runoff and erosion under simulated rainfall. Earth Surf Proc Land 28:145–155

Blanco-Canqui H, Lal R (2009) Extent of soil water repellency under long term and till soils. Geoderma 149:171–180

Brown JAH (1972) Hydrologic effects of a bushfire in a catchment in south-eastern New South Wales. J Hydrol 15:77–96

Campbell RE, Baker MB, Folliott PF (1977) Wildfire effects on a ponderosa pine ecosystem: an Arizona case study. USDA Forest Service Papers, RM-191, Rocky Mountain Forest and Range Experimental Station, United States Department of Agriculture, Forest Service, Fort Collins, CO

Carretero JM, Guzmán E, Martínez M, Úbeda X (2003) Els incendis forestals a l'àrea del Garraf-Castelldefels. II Premi: Castelldefels Sostenible, 148 pp

Cerdà A (1995) Influencia de la exposición sobre la producción de sedimentos y escorrentías en ambientes semiáridos. Cuadernos de Investigación Geográfica, 20–21. Logroño

Cerdà A, Bodí MB (2007) Erosión hídrica en suelos afectados por incendios forestales. In: Mataix-Solera J (ed) Incendios Forestales, Suelos y Erosión Hídrica. Caja Mediterráneo, CEMACAM Font Roja-Alcoy, Alicante, pp 71–117

Cerdà A, Doerr SH (2007) Soil wettability, runoff and erodibility of major dry-Mediterranean land use types on calcareous soils. Hydrol Process 21:2325–2336

Cerdà A, Doerr SH (2008) The effect of ash and needle cover on surface runoff and erosion in the immediate post-fire period. Catena 74:256–263

Cerdà A, Imeson AC, Calvo A (1995) Fire and aspect induced differences on the erodibility and hydrology of soils at La Costera, Valencia, Southeast Spain. Catena 24:289–304

Cerdà A, Schnabel S, Ceballos A, Gomez-Amelia D (1998) Soil hydrological response under simulated rainfall in the Dehesa land System Extremadura, SW Spain. Under drought conditions. Earth Surf Proc Land 23:195–209

Copley TL, Forrest AL, McCall AG, Bell FG (1944) Investigations in erosion control and reclamation of eroded land at the Central Piedmont Conservation Experiment Station. Statesville, N.C.U.S. Depart Agric Technol Bull 873:66

Crockford S, Topalidis S, Richardson DP (1991) Water repellency in a dry sclerophyll forest-measurement and processes. Hydrol Process 5:405–420

DeBano LF (1971) The effect of hydrophobic substances on water movement in soil during infiltration. Soil Sci Soc Am Proc 35:340–343

DeBano LF (1981) Water repellent soils: a state-of-the-art. General Technical Report, PSW-46. United States Department of Agriculture, Forest Service, Berkeley, CA

DeBano LF (2000a) Water repellency in soils: a historical overview. J Hydrol 231232:4–32

DeBano LF (2000b) The role of fire and soil heating on water repellency in wildland environments: Aa review. J Hydrol 231–232:195–206

DeBano LF, Krammer JS (1966) Water repellent soils and their relation to wildfire temperatures. Bull Int Assoc Sci Hydrol 2:14–19

DeBano LF, Mann LD, Hamilton DA (1970) Translocation of hydrophobic substances into soil by burning orgànic litter. Soil Sci Soc Am Proc 34:130–133

DeBano LF, Osborne JF, Krames JF, Letey J (1967) Soil wettability and wetting agents. Our corrent knowledge of the problema. USDA, For. Serv., Res. Pap, PSW-43, Berkeley, USA

DeBano LF, Savage SM, Hamilton DA (1976) The transfer of heat and hydrophobic substances during burning. Soil Sci Soc Am J 40:779–782

DeByle NV, Packer PE (1972) Plant nutrient and soil losses in overland flow from burned forest clearcuts. In: Watersheds in transition, Proceedings of the American Water Resources Association and Colorado State University Symposium, pp 296–307

Dekker LW, Ritsema CJ (1994) How water moves in a water-repellent sandy soil. I. Potential and actual water repellency. Water Resour Res 20:2507–2517

Dekker LW, Ritsema CJ (1996) Variation in water content and wetting patterns in Dutch water repellent peaty clay and clayey peat soils. Catena 28:89–105

Doerr SH, Shakesby RA (2009) Soil water repellency. Principles, causes and relevance in fire-affected environments. In: Cerdà A, Mataix-Solera J (eds) Efectos de los incendios forestales sobre los suelos en España. El estado de la cuestión visto por los científicos espanyoles. Universitat de València. Valencia

Doerr SH, Thomas AD (2000) The role of soil moisture in controlling water repellency: new evidence from forest soils in Portugal. J Hydrol 231–232:134–147

Doerr SH, Shakesby RA, Walsh RPD (2000) Soil water repellency: its causes, characteristics and hydrogeomorphological significance. Earth Sci Rev 51:33–65

Doerr SH, Ferreira AJD, Walsh RPD, Shakesby RA, Leighton-Boyce G, Coelho COA (2003) Soil water repellency as a potential parameter in rainfall-runoff modelling: experimental evidence at point to catchment scales from Portugal. Hydrol Process 17:363–377

Domingo WR (1950) Some notes on irreversibly dried difficulty wettable soils. Landbouwkund tijdschr 62:250–260

Dunne T, Black R (1970) An experimental investigation of runoff production in permeable soils. Water Resour Res 6:478–490

Dyrness CT (1976) Effect of wildfire on soil wettability in the High Cascades of Oregon. USDA For. Serv. Res. Pap. PNW-202, Pacific Northwest Forest and Range Exp. Stn. Portland, Oreg., 18 p

Ela SD, Gupta SC, Rawis JW (1992) Macropore and surface seal interactions affecting water infiltration into soil. Soil Sci Soc Am J 56:714–721

Ferreira-Leite F, Bento-Gonçalves A, Lourenço L, Úbeda X, Vieira A (2013) Grandes Incêndios Florestais em Portugal Continental como resultado das perturbações nos regimes de Fogo no Mundo Mediterrâneo. Silva Lusitana, nº Especial: 1–9, 129–144

Fuentes ER, Segura AM, Holmgren M (1994) Are the response of matorral shrubs diferent from those in an ecosystem with a reputed fire history? In: Moreno JM, Oechel WC (eds) The role of fire in Mediterranean-type ecosystems. Springer, New York

Hendricks BA, Johnson JM (1944) Effects of fire on steep mountain slopes in Central Arizona. J Forest 24:568–571

House MG (1991) Selected committee enquiry into land conservation. Legislative Assembly, Perth, Western Australia

Hudson J, Kellman M, Sanmugadas K, Alvarado C (1983) Prescribed burning *Pinus occarpa* in Honduras. For Ecol Manage 5:269–281

Huffmann EL, MacDonald LH, Stednick JD (2001) Strength and persistence of fire-induced hydrophobicity under Ponderosa and lodgepole pine, Colorado Front Range. Hydrol Process 15:2877–2892

Imeson AC, Verstraten JM, Van Mullingen EJ, Sevink J (1992) The effects of fire and water repellency on infiltration and runoff under Mediterranean type forests. Catena 19:345–361

Inbar M, Tamir M, Wittenberg L (1998) Runoff and erosion processes after a forest fire in Mount Carmel, a Mediterranean area. Geomorphology 24:17–33

Jamison VC (1947) Resistance of wetting in the surface of sandy soils under citrus trees in central Florida and its effect upon penetration and the efficiency of irrigation. Soil Sci Soc Am Proc 11:103–109

Jaramillo DF, Dekker LW, Ritsema CJ, Hendrickx JMH (2000) Occurrence of soil water repellency in arid and humid climates. J Hydrol 231–232:105–111

Jordán A, Zavala LM, Bellinfante N (2008) Heterogenity in soil hydrological response from diferent land cover types in southern Spain. Catena 74:137–143

Jordán A, Zavala LM, Nava AL, Alanís N (2009) Ocurence and hydrological effects of water repellency in diferent soil and land use types in Mexican volcanic highlands. Catena 79:60–71

Jordán A, Zavala LM, González FA, Bárcenas-Moreno G, Mataix-Solera J (2010) Repelencia al agua en suelos afectados por incendios: métodos sencillos de determinación e interpretación. In: Cerdà A, Jordán A (eds) Actualización en métodos y técnicas para el estudio de los suelos afectados por incendios forestales. Càtedra de Divulgació de la Ciència. Universitat de València, pp 145–183

Kinako PD, Gimingham CH (1980) Heather burning and soil erosion on upland heaths in Scotland. J Environ Manage 10:277–284

King PM (1981) Comparison of methods for measuring severity of water repellence of sandy soils and assessment of some factors that affect its measurement. Aust J Soil Res 19:275–285

Krammer JS, De Bano LF (1965) Soil wettability: a neglected factor in watershed management. Water Resour Res 1:283–286

Lavabre J, Sempere D, Cernesson F (1993) Changes in the hydrological response of a small Mediterranean basin a year after a wildfire. J Hydrol 142:273–299

Leighton-Boyce G, Doerr SH, Shakesby RA, Walsh RPD (2007) Quantifying the impact of soil water repellency on overland flow generation and erosion: a new approach using rainfall simulation and wetting agent on in situ soil. Hydrol Process 21:2337–2345

Llovet J, Ponce JM (1996) Evolución temporal en la producción de sedimentos de zonas afectadas por incendios forestales: efecto de la orientación. First European Conference on Erosion Control. Sitges-Barcelona

López R (1999) Consecuencias hidrológicas de los incendios forestales. Temas de Ingeniería Hidrológica para Forestales. Cap. VII, ETSEA, Universitat de Lleida

López R, Batalla RJ (2001) Análisis del comportamiento hidrológico de la Cuenca Mediterrània de Arbúcies antes y después de un incendio forestal, III Congreso Forestal Español, Granada

Lourenço L, Bento-Gonçalves A (1990) The study and measurement of surface flow and soil erosion on slopes affected by forest fires in the Serra da Lousã. In: Proceedings of international conference on forest fire research, Coimbra, pp C.05-1–C.05-13

Marqués MA (1991) Soil erosion research: experimental plots on agricultural and burnt environments near Barcelona. In: Sala M, Rubio JL, García-Ruiz, JM (eds) Soil erosion studies in Spain, Geofroma Ediciones, pp 153–164

Martin CL (1996) L'érosion hydrique à l'échele de la parcelle et d'un petit bassin versant après incendie de fôret dans le Massif des Maures. Étude et Gestion des sols 3(3):179–192

Mataix-Solera J, Guerrero C (2007) Efectos de los incendios forestales sobre las propiedades edáficas. In: Mataix-Solera J (ed) Incendios forestales, suelos y erosión hídrica). Caja Mediterráneo CEMACAM Font Roja-Alcoi, 196 pp

Meginnis HG (1935) Effect of cover on surface runoff and erosion in the loess in uplands of Mississippi. US Department of Agriculture, Circ. 347, 15 p
Miyata S, Kosugi K, Gomi T, Onda Y, Mizuyama T (2007) Surface runoff as affected by soil water repellency in a Japanese cypress forest. Hydrol Process 21:2365–2376
Moody JA, Martin DA (2001) Initial hydrologic and geomorphic response following a wildfire in the Colorado front range. Earth Surf Proc Land 26:1049–1070
Moral FJ, Giráldez JV, Laguna AM (2002) La hidrofobia en los suelos arenosos del parque natural de Doñana: caracterización y distribución. Ingeniería del Agua 9:37–50
Nasseri I (1988) Frequency of floods from a burned chaparral watershed, in Proceeding of the syposium on fire and watershed management. General Technical Report PSW-18, USDA (Forest Service). Berkeley, California., RM-191. Rock Mountain Forest ad Range Experimental Station, Fort Collins, Colorado
Naveh Z, Lieberman A (1984) Landscape ecology. Theory and applications. Springer, New York
Nayaka N (1982) Water repellency of soils. Jpn Agric Res Q 6:24–28
Nunes JP, Marisa Santos J, Bernard-Jannin L, Keizer JJ (2013) Comparing erosion rates in burnt forests and agricultural fields for a mountain catchment in NW Iberia. EGU General Assembly 2013, Vienna, Austria, id. EGU2013-11514, 7–12 April 2013
Osborn B (1954) Effectiveness of cover in reducing soil splash by raindrop impact. J Soil Water Conserv 9:70–76
Osborn JR, Pelishek RE, Krammes JS, Letey J (1964) Soil wettability as a factor in erodibility. Soil Sci Soc Am Proc 28:294–295
Ritsema CJ, Dekker LW (1994) How water moves in a water repellent sandy soil: 2. Dynamics of fingered flow. Water Resour Res 30:2519–2531
Ritsema CJ, Dekker LW, Hendricks JMH, Hamminga W (1993) Preferential flow mechanism in a water repellent sany soil. Water Resour Res 29:2183–2193
Ritsema CJ, Dekker LW, Heijs AWJ (1997) Three-dimensional fingered flow patterns in a water repellent sandy field soil. Soil Sci 162:79–90
Rulli MC, Rosso R (2007) Modeling catchment erosion after wildfires in the San Gabriel Mountains of Southern California. Geophys Res Lett 32:L19401
Sala M, Rubio JL (1994) Soil erosion as consequence of forest fires. Geoforma Ediciones, Logroño
Savage SM (1974) Mechanism of fire-induced water repellency in soils. Soil Sci Soc Am Proc 38:652–657
Schantz EC, Piemeisel FJ (1917) Fungus fairy rings in Eastern Colorado and their effect on vegetation. J Agric Res 11:191–245
Scott DF (1993) The hydrological effects of fire n South African mountains catchments. J Hydrol 150:409–432
Scott DF, Van Wyk DB (1990) The effects of fire on soil water repellency, catchment sediment yields and streamflow. In: Van Wilgen BW et al. (eds) Fire in South African Mountain Fynbos. Springer, Berlin, pp 216–239
Shakesby RA, Coelho COA, Ferreira AD, Terry JP, Walsh RPD (1993) Wildfire impacts on soil erosion and hydrology in wet Mediterranean forest, Portugal. Int J Wildland Fire 3:95–110
Soto B, Lasanta R, Benito E, Pérez R, Díaz-Ferros F (1994) Runoff and erosion from burnt soils in Northwest Spain. In: Sala M, Rubio JL (eds) Soil erosion as a consequence of forest fires. Geoforma Ediciones, Logroño, pp 91–98
Stoof C (2011) Fire effects on soil and hydrology. Thesis, Wageningen University, 182 p
Swanson FJ (1981) Fires and geomorphic processes. In: Fires regimes and ecosystems conference, Honolulu. Gen. Tech. Rep. WO-26 USDA, Washington, DC, pp 401–420
Terry JP (1994) Soil loss from erosion plots of differing post-fire forest cover, Portugal. In: Sala M, Rubio JL (eds) Soil erosion as consequence of forest fires. Geoforma Ediciones, Logroño, pp 133–148
Tiedemann AR, Conrad CE, Dieterich JH, Hornbeck JW, Megahan WF, Viereck LA, Wade DD (1979) Effects of fire on water. A state-of-knowledge review. General Technical Report WO-I0, USDA, Forest Service, Washington DC
Tsukamoto Y (1961) An experiment on sub-surface flow. J Japn Forest Soc 43:62–67

Úbeda X, Sala M, Imeson AC (1990) Variaciones en la variabilidad y consistencia de un suelo forestal después de haber sido sometido a un incendio. I Reunión Nacional de Geomorfología. Sociedad Nacional de Geomorfología, pp 677–685
Úbeda X, Sala M (1998) Variation in runoff and erosion in three areas with different fire intensities. Geoökodynamik 19:179–188
Úbeda X, Outeiro L (2008) Physical and Chemicals effects of fire on soil. In: Cerdà A, Robichaud P (eds) Fire effects on soils and restoration strategies. Science Publishers, Enfield, pp 105–132
Urbanek E, Hallet P, Feeney D, Horn R (2007) Water repellency and distribution of hydrophilic and hydrophobic compounds in soil aggregates from different tillage systems. Geoderma 140:147–155
Ursic SJ (1970) Hydrologic effects of prescribed burning and deadening upland hardwoods in northern Mississippi. USDA For. Serv. Res. Pap. SO-54, 15 p
Van't Woudt BD (1959) Particle coatings affecting the wettability of soils. J Geophys Res 64:263–267
Velasco A (2013) Efectes a llarg terme d'un incendi forestal en algunes propietats del sòl. 1994–2012. Degree in Geography. University of Barcelona. 61 pp
Wallach R, Jortzick C (2008) Unstable fingir-like flow in water-repellent soils during wetting and redistribution—the case of a point water source. J Hydrol 351:26–41
Wallis MG, Horne DJ (1992) Soil water repellency. Adv Soil Sci 20:91–146
Wallis MG, Horne DJ, McAuliffe KW (1990) A study of water repellency and its amelioration in a yellow Brown sand: 1. Severity of water repellency and the effects of wetting and abrasion. N Z J Agric Res 33:139–144
Wallis MG, Horne DJ, Palmer AS (1993) Water repellency in a New Zealand development sequence of yellow-brown sands. Aust J Soil Res 31:641–654
Wisler CO, Brater EF (1959) Hydrology. Wiley, New York. 408 p
Witter JV, Jungerius PD, Ten Harkel MJ (1991) Modelling water erosion and the impact of water repellency. Catena 18:115–124
York CA (1993) A questionnaire survey of dry patch on golf courses in the United Kingdom. J Sports Turf Res 69:20–26
Zavala LM, Jordán A, Gil J, Bellinfante N, Pain C (2009) Intact ash and charred litter reduces susceptibility to rain splash erosion post-wildfire. Earth Surf Proc Land 34:1522–1532

Part III
Environmental Processes

Chapter 18
Ecological Thresholds and Resilience in Streams

Robert H. Hilderbrand and Ryan M. Utz

Abstract Ecological thresholds and resilience are powerful heuristics for understanding how lotic ecosystems change. Ecosystems may exist in self-organized states based on their taxonomic composition or the range of ecosystem functions, which are influenced by environmental drivers such as thermal or hydrologic regimes, channel morphology, and availability of nutrients. Changes in these underlying drivers may exceed an ecosystem's ability to maintain its characteristic attributes and shift the system into alternative states of organization, which are often regarded as degraded or undesired. The boundaries where transitions occur are known as ecological thresholds and often show a rapid ecosystem response across a relatively small change in the environmental driver. Resilient ecosystems have the capacity to retain attributes in the face of disturbances. However, at some disturbance magnitude an ecosystem may become altered, and the magnitude necessary for a regime shift decreases as resilience declines. While ecological resilience remains largely metaphorical in lotic ecosystems, we describe some approaches for identification and assessment.

Keywords Ecological resilience · Ecological thresholds · Headwater streams · Land use · Disturbance

---

R.H. Hilderbrand (✉)
Appalachian Laboratory, University of Maryland Center for Environmental Science, 301 Braddock Road, Frostburg, MD 21532, USA
e-mail: rhilderbrand@al.umces.edu

R.M. Utz
Falk School of Sustainability, Chatham University, 1 Woodland Road, Pittsburgh, PA 15232, USA

P. Rowiński and A. Radecki-Pawlik (eds.), *Rivers—Physical, Fluvial and Environmental Processes*, GeoPlanet: Earth and Planetary Sciences,
DOI 10.1007/978-3-319-17719-9_18

18.1 Introduction

As discussed in previous chapters, complex, nonlinear processes may operate at multiple spatial and temporal scales to simultaneously control physical and chemical properties of running waters. Examining the biological attributes of riverine systems increases the complexity of the picture significantly because lotic organisms and the processes they affect are simultaneously influenced by the geomorphic, thermal, hydrologic, and chemical properties of the systems in which they reside. Even the shape of lotic ecosystems presents a challenge to organisms: the linear-dendritic structure and continuous downstream flow of river networks renders dispersal, a critical process for organisms, difficult. Streams also integrate many processes that occur in the terrestrial portion of their watersheds (Hynes 1975; Allan and Johnson 1997), and legacies of land use or other modifications can remain within streams for decades even when watersheds return to a system resembling a pre-disturbance state (Harding et al. 1998). In addition to being literally immersed in the products of upstream and upslope drivers, lotic organisms must cope with disturbance regimes perhaps without parallel in terrestrial and marine systems, from catastrophic floods that can completely reorder their physical habitat to droughts that may cause their environment to effectively disappear. Thus a high degree of environmental variability is the norm for most stream ecosystems.

Lotic organisms have evolved numerous strategies to persist and thrive in such highly dynamic ecosystems. As a prime example, floods may appear to be catastrophic events, but organisms possess diverse adaptations to cope, and many species rely on flooding to maintain habitats necessary to fulfill life history requirements (Lytle and Poff 2004). Some aquatic insects can recognize cloud cover to detect eminent high flow events and move to refuge habitat (Bogan and Lytle 2011), while others have evolved flattened or streamlined body shapes to minimize drag (Alba-Tercedor 2008). In the opposite extreme, drought and desiccation may seem intuitively disastrous, but many organisms have adapted to cope with this stressor as well. For instance, the embryonic life stage of some invertebrates may enter a state of diagenesis to allow persistence of populations through decades-long dry periods (Jenkins and Boulton 2007), while others use atmospheric or ionic signals of impending drought to cue migration towards perennial stream reaches (Lytle et al. 2008). Adaptive traits for surviving dynamic riverine environments are by no means limited to multicellular organisms. Seemingly sessile benthic diatoms disperse within lotic ecosystems by riding currents until a favorable patch is encountered (Stevenson and Peterson 1991).

Even well adapted organisms possess limited tolerance to extreme events, chronically elevated physicochemical stress, or simply changing environmental conditions that at some point no longer allow for persistence. Floods and droughts can be severe enough to eliminate all individuals in populations of fishes or invertebrates (Adams and Warren 2005; Roghair et al. 2002; Bogan and Lytle 2011). Chronically elevated temperatures or chemical stressors may also induce local extirpations. Among the most pervasive threats to lotic organisms are landscape-scale stressors such as agricultural and urban development, both of which influence multiple physicochemical stressors in receiving waters and frequently cause widespread biodiversity loss (Allan 2004). Ecological changes are not limited to state variables: natural or anthropogenic change can fundamentally alter biogeochemical processes such as nutrient retention and whole-system metabolism (Gibbins 2015, this issue). Periphyton growth may increase nonlinearly if concentrations of key chemical constituents of lotic ecosystems, such as inorganic nitrogen or phosphorous, exceed critical values (Dodds et al. 2002). When such restructuring events or responses exhibit large shifts across a narrow range of an environmental driver, they are known as ecological thresholds (Fig. 18.1). While many ecological responses are nonlinear, not all are necessarily thresholds. Rather it is the rapidity of change in an ecological attribute over a small change in the driver variable that constitutes a threshold.

Ecological thresholds have parallels to abiotic components of rivers in that they can appear at multiple spatial or temporal scales (Church 2002). Two examples of geomorphic thresholds in flowing water systems include exceeding the shear stress that induces sediment transport, which represents a threshold at a fine temporal resolution, and the transition from a braided to a single-channel system when sediment supply diminishes—an example of a long-term regime shift. Ecological thresholds exhibit similar heterogeneity. A localized flood may temporarily extirpate a population of organisms until migration from nearby reaches recolonizes the disturbed habitat. At the opposite end of the spectrum, assemblages of fishes and macroinvertebrates may entirely turn over when stream networks fed by glaciers transition from turbid to clear water subsequent to glacial retreat.

Not all ecosystem-environment relationships within riverine systems behave in threshold fashion. In fact, most interactions probably do not result in rapid nonlinear changes. While the documented number of threshold relationships is

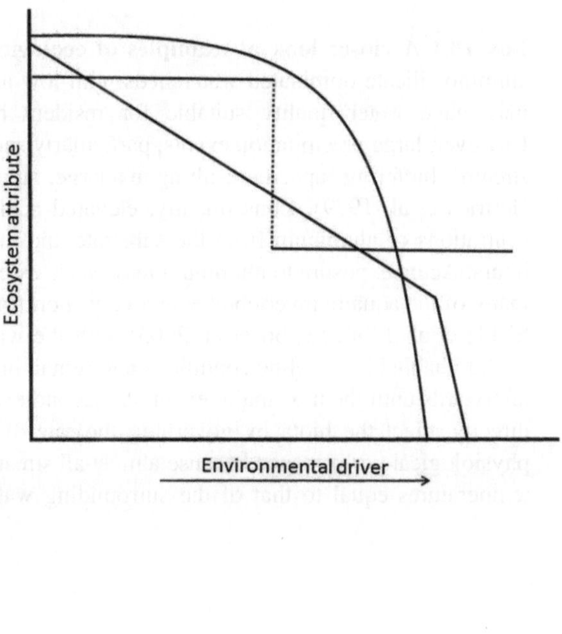

Fig. 18.1 Examples of ecological thresholds as an ecosystem attribute responds across the range of an environmental driver. The direction of the response can also be upwards

increasing, their continued relative rarity in riverine systems calls their importance into question (Groffman et al. 2006). However, where thresholds exist, they are rarely known until crossed, and many such relationships may yet be discovered.

18.2 Thresholds and the Challenge of Scale

Where threshold responses exist, few exhibit consistency among species or locations because the response scale to a gradient of change is often highly idiosyncratic. Detrimental conditions for one species may be tolerable or even beneficial to another. For example, aquatic insects requiring clean, stable substrates may be extirpated by fine (≤2 mm) sediment deposition, while populations of burrowing taxa may grow under such conditions. Even for drivers that trigger a threshold response, the inflection point will likely be site-, region-, or taxon-specific. Thus, a single, all-encompassing threshold value is rarely applicable to multiple biological attributes excepting catastrophic events such as a toxic chemical spill or volcanic eruption. Such inconsistencies among thresholds largely result from the highly variable conditions under which stream biota have evolved and the substantial heterogeneity in physical and chemical templates among basins.

As with most environment-ecological interactions, biotic responses to environmental change can unfold at different spatial and temporal scales (Box 18.1). The effects can be immediate, as happens when physiological thresholds are exceeded and organisms die, whereas other threshold responses may transpire over one or more generations when changing conditions trigger sublethal effects such as depressed reproductive output. In the latter case, rapid nonlinear reproductive declines may occur, but one or more generations are required for the result to be expressed at the population level.

Box 18.1 A closer look at examples of ecological thresholds Streams in alumino-silicate dominated watersheds with low acid neutralizing capacities may have water quality suitable for resident biota almost year round. However, large precipitation events, particularly snowmelt, may overwhelm a stream's buffering capacity resulting in a large, rapid drop in surface water pH (Jeffries et al. 1979). Consequently, elevated acidity may leach lethal concentrations of aluminum from the substrate and into streams for only a few hours. Acute exposure to aluminum may cause extensive mortality to fish and many of the aquatic invertebrates in a very short time (Baker et al. 1996, Van Sickle et al. 1996; Lepori et al. 2003) with the water quality later returning back to habitable, baseline conditions and remaining suitable for long periods afterwards until the next major event. As second example, water temperatures directly affect the biota by governing the rate of enzymatic processes and physiological performance because almost all stream-dwelling biota maintain temperatures equal to that of the surrounding water. Temperatures directly influence the rates of processes as diverse as nutrient uptake (Rasmussen et al. 2011) and enzymatic activity (Ward and Stanford 1982). Aquatic biota have upper thermal limits that behave as thresholds above which a species rarely survives (Quinn et al. 1994; Beitinger et al. 2000). Many coldwater species, such as salmonids, have difficulty surviving temperatures exceeding 25 °C, while many warm water stream fishes have thermal thresholds around 30 °C (Allan 1995). Geomorphic changes to streams such as channel widening will allow greater solar-conductive water heating and consequently may change thermal regimes to exceed thresholds. Thermal regime shifts can produce acute responses from rapid heating or chronic, longer-term responses if elevated temperatures are not lethal but rather increase physiological stress. As a final example, the effects of fine sediment dynamics in streams may influence biological processes over multiple time scales. As seen in earlier chapters, sediment size heterogeneity is both present and necessary for channel equilibrium and maintenance. A lack of fine particles can result in channel armoring, particularly downstream of dams where the channel may be sediment-starved. However, excessive fine sediment in the channel has been regarded as one of the most pervasive biological stressors in stream ecosystems (Waters 1995). Fine sediments exceeding around 20 % areal coverage of the streambed produces threshold responses in some benthic macroinvertebrate communities (Burdon et al. 2013). For salmonid embryos incubating in the hyporheic zone, fine sediments may be particularly lethal: volumetric percentages of fine particles exceeding 10 % typically cause high mortality (reviewed in Jensen et al. (2009)), which may result in a multi-year time lag in population declines. Fine sediment deposition may also reduce hydraulic conductivity between the stream and its hyporheos, which may account for substantial portions of the total whole stream system metabolism (Fellows et al. 2001). We expect declines in stream metabolism to become rapid and nonlinear as some yet to be defined porosity threshold is exceeded, and the steepness of the decline proportional to hyporheic depth.

The question of whether thresholds are detected can be contingent on the response scale of the observations. For example, threshold responses of tolerance to landscape-scale changes associated with urbanization in the form of pavement and rooftops (collectively referred to as impervious surface area, hereafter ISA) are evident for many benthic insects (Utz et al. 2009; King and Baker 2010). In many cases, populations of sensitive organisms may disappear at very low levels of ISA within a watershed. The most sensitive taxa are lost at about 3 % watershed ISA while nearly half of all resident taxa may be extirpated by 15–20 % ISA, although site-specific values may vary among geoclimatic settings (Utz et al. 2009) and the degree of ISA connectivity to stream channels (Collier and Clements 2011). However, aggregating individual taxa responses by considering total taxonomic richness rather than individual taxa produces a linear decline across the same ISA gradient (Fig. 7 in Utz et al. 2009). Although taxa richness is a common community-level metric, other metrics exhibit threshold responses to ISA (Hilderbrand et al. 2009; King et al. 2011). Thus, the choice of both taxonomic resolution and how it is measured can substantially influence whether or not a threshold is detected.

18.3 Thresholds and the Challenge of Covariates

Ecological thresholds have tremendous allure both as heuristic devices to understand how streams behave and as potentially powerful tools for environmental management. Unfortunately, thresholds are often assumed to be static such that maintaining an environmental driver below a critical value can preserve stream ecosystem structure and function. In theory, thresholds can be used to optimize the degree to which a system may be altered and still maintain ecological function. In practice, however, numerous interacting factors often distort or shift threshold boundaries. Failure to recognize such limitations through adherence to static values can result in considerable damage to ecosystems. For example, early research on the effects of urbanization on streams noted negative changes in ecosystem structure and function as watershed ISA levels exceeded around 10–15 % (Klein 1979; Schueler 1995). The 10 % value grew in notoriety and became recommended for land use planning activities (US EPA 2004). Multiple subsequent studies, however, have detected significant changes in species richness and community structure at watershed ISA levels well below 10 %.

Because each stream represents a unique combination of physical, chemical, climatic, and historical legacies, analyses of ecological thresholds have many co-varying attributes that interact to affect biota. Thus even where threshold responses are evident, factors that may not be directly related to the stressor of interest can distort threshold responses. For example, stream biota face particular challenges with osmoregulation because of a need to retain a higher salt balance than the surrounding freshwater. Most aquatic organisms have evolved to retain salts in water with low ionic concentrations, but elevated concentrations overwhelm their physiological tolerance and may become lethal. High concentrations of anionic salts (Cl^-, SO_4^{2-}) in particular can result in large-scale biodiversity loss over a relatively small gradient such as downstream of coal mining activities (Garcia-Criado et al. 1999; Pond et al. 2008). However, toxicity at a given anionic salt concentration decreases substantially as water hardness increases (Mount et al. 1997; Soucek and Kennedy 2005), making the use of a single threshold for all streams dubious. Similar blurring of threshold boundaries occur when atypical geologic settings result in a cool, stable water temperature regime, which allows typically sensitive species such as brook trout (*Salvelinus fontinalis*) to persist in streams with watershed ISA above levels that cause brook trout extirpation elsewhere (Stranko et al. 2008). Thus, planning for system change using rigid, static approaches may inadvertently shift the system into an undesired state if the threshold is not completely understood.

Because of the multidimensional and contingent nature of many thresholds, the exact mechanism affecting a stream ecosystem is often unknown or not measured. Thus cause and effect relationships with univariate drivers are rare in many ecosystems. When a particular environmental driver causes many disparate physicochemical changes, surrogates representing the combined effects are often used. An example involves the use of watershed urbanization or agricultural land uses as a surrogate for the numerous physicochemical stressors associated with landscape-scale anthropogenic change. The combined effects of urbanization is akin to death by a thousand cuts; urbanization is associated with increases in fine sediment, bed armoring, elevated specific conductance, higher and more variable water temperatures, increased flood frequency and magnitude, greater concentrations of toxic metals, and numerous other stressors (Paul and Meyer 2001). In an urban stream, biological structure or processes may be responding to any or all of the above-mentioned stressors, but rarely do sufficient data exist to identify the specific mechanism for ecological changes while controlling for all of the other potential drivers. A given biological attribute of interest could be responding directly to elevated salt concentrations or indirectly to an altered hydrologic regime and consequential changes in channel geomorphology.

Although we have cautioned about the uncertainties and potential misapplication of using ecological thresholds for management, they have value beyond heuristics. The fact that not every stream responds exactly the same way or to the same value of a driver known to elicit a response does not negate the threshold. Rather, the concept provides a range within which change is likely to occur (Schueler et al. 2009), and effective management will recognize the need to mitigate or plan for uncertainties within such ranges. Ecological thresholds also provide the opportunity to develop futures scenarios and forecast ecological changes (e.g., Van Sickle et al. 2004; Hilderbrand et al. 2010). While futures scenarios can be conducted for non-threshold responses as well, the results are often instructive when large ecosystem changes are forecasted to result from relatively small changes to input drivers. Such exercises may also be used to better understand the concept of ecological resilience, which can be lost when thresholds are crossed.

18.4 Ecological Resilience and Alternate States

In a resilient stream ecosystem, key ecological attributes such as species composition or the range of ecosystem functions will remain consistent over long time scales or return to pre-disturbance states following perturbations. Resilience can have many meanings such as the related concept of engineering resilience (Pimm 1984), which refers to the amount of time required for an ecosystem attribute to return to conditions prior to a disturbance. While ecosystem recovery time is an important attribute of ecological resilience, it is incomplete. Ecological resilience as we use it in this chapter is the degree to which an ecosystem can absorb or withstand environmental stress or disturbance and still maintain self-organization (i.e., characteristic structure and function; (Holling 1973; Gunderson and Holling 2001). Our treatment hereafter will be with respect to lotic ecosystems and may not follow all of the classical definitions, as streams and rivers possess different attributes and behavior relative to the systems where the foundational concepts of ecological resilience were developed (shallow lakes, terrestrial, and socioeconomic ecosystems). Thus, we may raise as many questions as answers, and we encourage the reader to identify similarities and differences and to challenge our perceptions because ecological resilience can be highly abstract. Several excellent, more comprehensive treatments place ecological resilience in a broader context (Holling 1973; Carpenter et al. 2001; Gunderson and Holling 2001).

Exceeding the limits of resilience and crossing an ecological threshold may shift a system from one ecosystem state into an alternate state (Fig. 18.2). State shifts often entail transitioning from a desired to an undesired state. Figure 18.2 depicts an urbanizing stream with the ball-in-cup model illustrating resilience and alternative

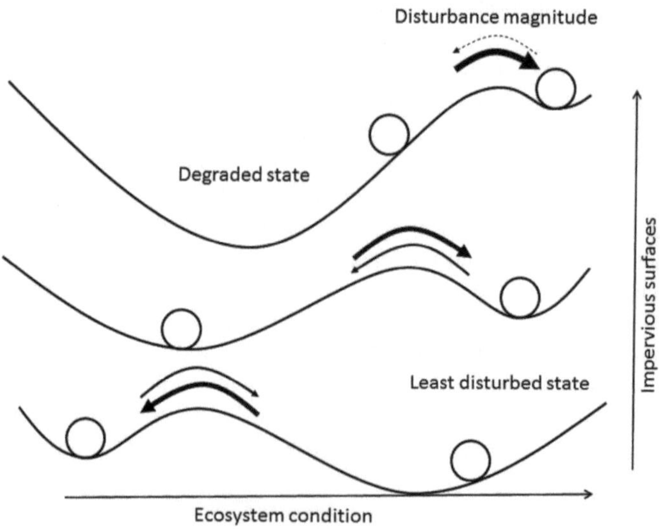

Fig. 18.2 Lotic ecosystems may exist as multiple alternate states, or domains, (*troughs*) separated by thresholds (*ridges*) dividing the states. An ecosystem at a specific point (*open circles*) may have variation in its attributes, but still belong within a state having substantially different attributes, such as taxonomic composition or nutrient processing rates, from other states. Ecological resilience is represented by the width of the domain, while engineering resilience is the depth or steepness of the sides. As the underlying drivers change, as can happen due to increases in impervious surfaces in the watershed, some domains will begin to lose resilience, while others may increase. As the resilience decreases, progressively smaller disturbances (*arrows*) can shift the stream from one alternate state into another. These regime shifts may be semi-permanent on human time scales if the drivers have been substantially altered or there are no dispersal paths for organisms to recolonize after extirpations

stable states as developed by Holling et al. (1995) and used later by Beisner et al. (2003); Scheffer and Carpenter (2003). This heuristic device can represent both biological community and ecosystem frameworks. With respect to biological communities, ecosystem drivers are relatively constant, and the community composition changes mostly from internal, biologically controlled drivers such as competition or predation. In the ecosystem perspective, the system responds to changes in the underlying environmental drivers (Scheffer et al. 2001). Both community and ecosystem frameworks are possible within lotic ecosystems because the relative importance of external (abiotic) factors versus internal (biotic interactions) factors forms a continuum along which communities may be structured (Zalewski and Naiman 1985). Streams dominated by changing environmental conditions are more externally controlled by abiotic variables, while streams in more stable environments tend to be structured more internally by biotic interactions.

Changes in ecosystem state, also known as regime shifts, occur when the resilience of an ecosystem has been exceeded. Resilience, state changes, and thresholds are closely intertwined because thresholds represent boundaries between ecosystem states. While not couched explicitly as alternative states, the lotic literature is rich with examples because the concept is foundational to biomonitoring and assessment. As the physical, chemical, or thermal conditions of a stream change beyond levels observed due to natural variation, detectable changes in taxonomic composition ensue (e.g., Brenden et al. 2008) or ecological functions no longer exhibit expected behavior (Dodds 2006; Earl et al. 2006). Thus with a changing environment, the same section of a stream or river can fundamentally change ecosystem attributes.

Many examples of transitions between alternative states involve rapid, catastrophic shifts (e.g., Carpenter et al. 1999; Scheffer et al. 2001), but loss of resilience may also prove gradual as the drivers change slowly or taxa are sequentially lost. In such cases, the alternate states may behave similarly to how taxonomists define closely related species—differences may be clear at the extremes, but the exact point where the change occurs is uncertain. As detailed below, we believe that more gradual changes are the more common mode of state change in streams because lotic ecosystems are highly influenced by many environmental variables and disturbance regimes (Resh et al. 1988).

Multiple pathways can cause ecosystem state shifts by altering the underlying drivers (Beisner et al. 2003). Changes to any number of drivers such as sediment supply, altered thermal regime, or channel confinement and their interactions can fundamentally change the environmental conditions and result in loss of ecosystem structure or function. Excellent examples involve rivers where a dam regulates discharge, and the resulting altered discharge regime alters hydro-geomorphic attributes such as sediment delivery, resulting in ecological regime shifts (Poff et al. 1997). The dynamic, disturbance-prone nature of lotic systems that exerts strong external controls on community structure in streams may also promote the formation of alternative states (Didham and Watts 2005). Seasonal and annual variation in temperature, discharge, and sediment dynamics probably discourage the development of tightly coupled biotic interactions in many, but not all, riverine systems. Rather, biota must respond to a constantly changing and challenging physical fluid environment, and species sort along their preferred environmental conditions. As long as the collective suite of abiotic drivers remains within certain bounds, an ecosystem will likely remain loosely organized based on the existing physical habitat template (sensu Southwood 1977; Poff and Ward 1990) and distinguishable from states associated with degraded conditions. However, easily identified, discrete states seem less common than assortments of species driven by a few common taxa.

Disturbances that directly affect the actual organisms or processes may also produce shifts to alternate states. Examples include population losses resulting from a chemical spill, harvest, disease, extreme drought, or other event that causes excessive mortality without the ability to recover or recolonize (e.g., Franssen et al. 2006; Bogan and Lytle 2011). State shifts may also arise from internal controls due to biotic interactions, particularly when a top predator or keystone species is affected (e.g. Power et al. 1985; Bechara et al. 1992; Carpenter et al. 2011). In the case of Power et al. (1985), the exclusion of largemouth bass (*Micropterus salmoides*), a piscivorous predator, shifted a stream ecosystem state from one with substantial periphyton production to one with clean substrates. The regime shift occurred because largemouth bass preyed on small herbivorous grazing fish (central stoneroller, *Capostoma anomallum*), and the exclusion of largemouth bass released these herbivores to feed in areas they would normally avoid in the presence of their predators. Reintroducing largemouth bass reverted the system back to one with heavy periphyton production. In another case study, the catadromous, top predator American Eel (*Anguilla rostrada*) seems to exert significant influence on the lower trophic levels (Stranko et al. 2014). Large dams preclude eel migrations and result in changes in the densities of the benthic fish and macroinvertebrates.

18.5 Assessing Resilience

Assessing ecological resilience has proven challenging, and the science is still in an early state. Advances in resilience theory have greatly outpaced applications (Thrush et al. 2009). The disparity may exist because measuring ecological resilience requires us to identify a specific ecosystem component and the stressor against which resilience is assessed as well as the necessary time scale of natural recovery to the stressor (Carpenter et al. 2001). Not all ecosystem components respond similarly to a given driver nor are we always able to extract the effect of a single driver from contingent interactions. Thus most advances in assessing resilience have been theoretical (e.g., Petchey and Gaston 2009) or based on simulation models (e.g., Carpenter and Brock 2006). Rarely has resilience been assessed on empirical data (but see Carpenter et al. 2011). Emerging evidence suggests that systems approaching a threshold and regime shift will exhibit higher variance in response (Carpenter and Brock 2006) or a critical slowing down of recovery time after disturbance (Scheffer et al. 2009; Drake and Griffen 2010; Lindegren et al. 2012; Veraart et al. 2012). However, non-catastrophic state shifts may also exhibit similar behaviors (Kéfi et al. 2013). Nonetheless, such information can be quite useful for predicting when a system is undergoing change regardless of its speed. Ideally, several years or decades of data will exist on multiple ecosystem attributes and drivers, but such data are extremely rare. Thus, resilience is often assessed through retrospective analysis and inferred from large datasets not explicitly collected specifically for resilience assessments and often using space-for-time substitutions (Box 18.2). Multiple ecological thresholds may need to be crossed in order to measure resilience. With sufficient time and data however, advances in identifying ecological thresholds, alternate states, and resilience are likely to emerge.

Box 18.2 Measuring ecological resilience Ecological resilience is an abstract concept. We therefore provide a more detailed example of an attempt at its quantification. Our research group (Hilderbrand and Raesly 2009) has studied ecological thresholds, alternate states, and resilience by retrospectively analyzing a large, long-term dataset on headwater streams collected by the Maryland Biological Stream Survey (Klauda et al. 1998). We focus specifically on the benthic macroinvertebrate community because it is an important measure of stream ecosystem health known to be sensitive to changes in channel structure, water quality, and general degradation (Resh and Jackson 1993). Streams in the least disturbed watersheds clustered into two identifiable and slightly overlapping alternate states (Fig. 18.3). The ellipsoids define sites in similarity space that are not statistically different from the least disturbed reference sites in terms of benthic macroinvertebrate community composition. Sites outside of either ellipsoid are statistically different from either reference state and may reside in a separate alternate state (not shown due to complexity). The boundary of each ellipsoid represents the edge of the domain occupied by each alternate state and may be construed as a threshold between alternate states. Environmental stressors at sites residing within a reference state can now be quantified and compared against stressor levels for sites outside of the ellipsoid. The community is resilient to a specific stressor when stressor levels do not differ among alternate states. In contrast differences in stressor levels among alternate states indicate sensitivity to the stressor, and the point where the community changes can be calculated. Using this approach, many potential stressors can be assessed against an ecosystem attribute. Using the approach described above is imperfect because statistical criteria are applied to demarcate a boundary, but the approach incorporates many of the important attributes of alternate states, thresholds, and resilience. The use of multiple least-disturbed sites to define the ellipsoid implicitly allows for the wide variation in community structure observed in nature and quantifies the presence of one or more alternate states. The approach also provides boundaries for use in

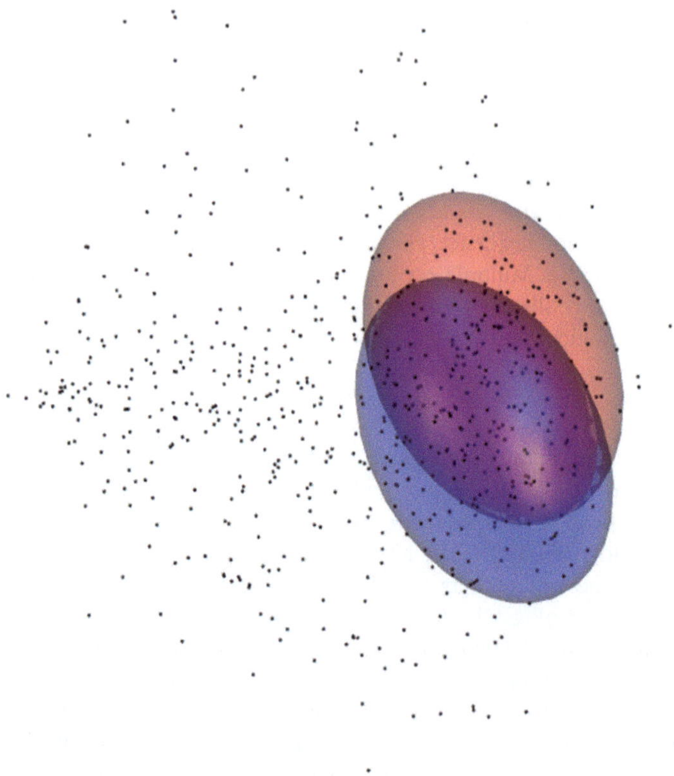

Fig. 18.3 Alternate states derived from nonmetric nondimensional scaling of benthic macroinvertebrate communities occurring in least disturbed watersheds. Ellipsoids define the boundaries of alternate ecological states identified in the least disturbed watersheds. Sites falling outside of an ellipsoid have benthic macroinvertebrate communities that are statistically different from the alternate state representing least disturbed conditions. In this example, there are two general community states for least disturbed systems. The edges of the ellipsoids represent boundaries of the alternate states. Axes have been removed for better visual clarity

determining if a state has been exited. Measuring the range of a given stressor across all sites falling within a reference ellipsoid is akin to measuring the width of the domain within which the alternate state resides with respect to that stressor and operationalizes the theory into something measureable. Thus we believe that many insights can be derived from retrospective analysis of the monitoring data that many agencies are charged with.

The choice of ecological response used to quantify resilience may exert a large influence due to the high degree of variability among rates of ecological processes. System aspects governed by microbial activity (e.g., nutrient uptake, stream metabolism) may respond very rapidly because microbes reproduce rapidly and consequently an event or driver that alters the microbial community may respond or recover over short time scales. In contrast, longer-lived organisms, such as fish and freshwater mussels, may take much longer to respond or recover from a disturbance. A long recovery time may be mistaken for a loss of resilience when it simply reflects the time required for reproduction or dispersal. Until sufficient time has passed however, we may be unable to separate a regime shift and loss of resilience from a slow recovery trajectory.

18.6 Maintenance, Recovery, and Restoration of Resilience

Resilient ecosystems have the ability to recover from disturbances of differing magnitudes. Small disturbances may temporarily alter relative abundances of organisms or alter processing rates. As the magnitude of a disturbance increases to the point where taxa are locally extirpated, connections within the hydrologic network may be required for recolonization and a return to a recognizable, pre-disturbance state. Thus, maintaining resilient ecosystems in the face of larger perturbations requires the integration of increasingly larger spatial and temporal scales.

Across scales, resilience tends to decline with a loss of natural variation in environmental drivers and is often the result of human interventions in trying to manage or control the environment, such as with channel and floodplain alterations to limit flooding (Holling and Meffe 1996). As resilience decreases, smaller disturbances may cause regime shifts (Fig. 18.2). In contrast, maintaining variation in the external drivers tends to promote a larger range of ecosystem functions and greater biodiversity, thus increasing systemic capacity to adapt to, and recover from, changes without fundamentally altering the underlying structure. Such adaptive capacity is thought to be critical in maintaining resilient systems (Gunderson 2000). Unfortunately, designing resilient stream ecosystems for alternate states found in non-degraded watersheds can be very difficult. Invariably, 'managing' a system for resilience by focusing on an incomplete suite of drivers will neglect important and unknown drivers, likely reducing overall resilience. Thus, the best way to maintain resilience is to promote its retention by allowing physical and chemical attributes of the system to experience as much of the range of natural conditions as is possible.

Sometimes however, resilience is not a desired quality and may inhibit what we consider the proper functioning of lotic ecosystems. Some watersheds are so highly altered that only the most tolerant lotic biota are capable of surviving. These ecosystems tend to contain a small subset of the regional biodiversity and have highly altered ecosystem functions. However, they can be highly resilient systems because the magnitude of change in the drivers required to shift to a more biodiverse or ecologically functional state is nearly insurmountable (Bernhardt and

Palmer 2011). These systems often exhibit hysteresis—the lack of reversion even after some suitable conditions are restored to a pre-disturbance state.

The visual and heuristic power of thresholds as tipping points and the resilience that prevents regime shifts naturally leads to a desire to integrate the concepts in restoration projects (Irvine 2012). If riverine ecosystems can shift into a different state due to changes in the underlying drivers, then the process should be reversible with directed management actions. However, because most ecological relationships are highly multidimensional the perceived ability to restore a system back to its predisturbance state is usually naïve (Hilderbrand et al. 2005). The hysteresis exhibited by many degraded streams makes reversion back to the original, or a less degraded, ecosystem state difficult or improbable because of fundamentally altered drivers or the inability of the biota to recolonize and persist even after repeated reintroductions. Despite thousands of restoration projects and billions of dollars spent on stream restoration (Hassett et al. 2005), we have little evidence of successfully restoring stream ecosystems back to an undisturbed state (Bernhardt and Palmer 2011; Stranko et al. 2012); post-restoration monitoring has largely been insufficient to determine if ecosystem attributes have moved into a more desirable condition (Palmer 2008). Nonetheless, many activities that highly degrade lotic ecosystems continue to assume that all aspects of a stream can be restored or even re-created in newly constructed channels (Palmer et al. 2010).

18.7 Conclusion

Thresholds and resilience have the potential to serve as powerful heuristics for understanding stream ecosystems and how they respond as the world around them changes. The utility of identifying ecological thresholds and quantifying ecological resilience for managing and maintaining the quality of our streams in an increasingly human modified world is tremendous. However, the allure can also result in problems because we cannot anticipate every interaction. We also cannot know how the hydrologic cycle and the resulting thermal and geomorphic adjustments in each river system will change as the climate and precipitation patterns change. Thus, these concepts should be primarily viewed more as guiding principles rather than as static decision points for management. The potential to over-manage watersheds (Holling and Meffe 1996) could easily end in a Sisyphus Complex (Hilderbrand et al. 2005) where a seeming endless suite of actions are initiated in response to topical symptoms rather than identifying and addressing the underlying changes. A difficult, but more long-term view will allow for change in such a way as to not dramatically alter the underlying processes and constraints under which each stream and its watershed evolved. Thus, we too must be resilient in our ongoing efforts to understand and ensure the ecological resilience and adaptive capacity of streams.

References

Adams SB, Warren ML Jr (2005) Recolonization by warmwater fishes and crayfishes after severe drought in upper coastal plain hill streams. Trans Am Fish Soc 134:1173–1192

Alba-Tercedor J (2008) Aquatic macroinvertebrates. In: Ziglio G, Siligardi M, Flaim G. Biological monitoring of rivers. John Wiley, New York, pp 71–87

Allan JD, Johnson LB (1997) Catchment-scale analysis of aquatic ecosystems. Freshwater Biol 37:107–111

Allan JD (1995) Stream ecology: structure and function of running waters. Chapman and Hall, New York

Allan JD (2004) Landscapes and riverscapes: the influence of land use on stream ecosystems. Annu Rev Ecol Syst 35:257–284

Baker JP, Van Sickle J, Gagen CJ, DeWalle DR Jr, Sharpe WF, Carline RF, Baldigo BP, Murdoch PS, Bath DW, Kretser WA, Simonin HA, Wigington PJ (1996) Episodic acidification of small streams in the sortheastern United States: effect on fish populations. Ecol Applic 6:422–437

Bechara JA, Moreau G, Planas D (1992) Top-down effects of brook trout (*Salvelinus fontinalis*) in a boreal forest stream. Can J Fish Aquat Sci 49:2093–2103

Besiner BE, Haydon DT, Cuddington K (2003) Alternative stable states in ecology. Front Ecol Envir 1:376–382

Beitinger TL, Bennett WA, McCauley RW (2000) Temperature tolerances of North American freshwater fishes exposed to dynamic changes in temperature. Env Biol Fish 58:237–275

Bernhardt ES, Palmer MA (2011) River restoration: the fuzzy logic of repairing reaches to reverse catchment scale degradation. Ecol App 21:1926–1931

Bogan MT, Lytle DA (2011) Severe drought drives novel community trajectories in desert streams. Freshwat Biol 56:2070–2081

Brenden TO, Wang L, Su A (2008) Quantitative identification of disturbance thresholds in support of aquatic resource management. Env Man 42:821–832

Burdon FJ, McIntosh AR, Harding JS (2013) Habitat loss drives threshold response of benthic macroinvertebrate communities to deposited sediment in agricultural streams. Ecol Applic 23:1036–1047

Carpenter SR, Ludwig D, Brock WA (1999) Management of eutrophication for lakes subject to potentially irreversible change. Ecol Applic 9:751–771

Carpenter S, Walker B, Anderies JM, Abel N (2001) From metaphor to measurement: resilience of what to what? Ecosystems 4:765–781

Carpenter SR, Cole JJ, Pace ML, Batt R, Brock WA, Cline T, Coloso J, Hodgson JR, Kitchell JF, Seekell DA, Smith L, Weidel B (2011) Early warnings of regime shifts: a whole-ecosystem experiment. Science 332:1079–1082

Carpenter SR, Brock WA (2006) Rising variance: a leading indicator of ecological transition. Ecol Lett 9:311–318

Church M (2002) Geomorphic thresholds in riverine landscapes. Freshwat Biol 47:541–557

Collier KJ, Clements BL (2011) Influences of catchment and corridor impervious surfaces on urban stream macroinvertebrate communities at multiple spatial scales. Hydrobiologia 664:35–50

Didham RK, Watts CH (2005) Are systems with strong underlying abiotic regimes more likely to exhibit alternative stable states? Oikos 110:409–416

Dodds WK, Smith VH, Lohman K (2002) Nitrogen and phosphorus relationships to benthic algal biomass in temperate streams. Can J Fish Aquat Sci 59:865–874

Dodds WK (2006) Eutrophication and trophic state in rivers and streams. Limnol Oceanogr 51:671–680

Drake JM, Griffen BD (2010) Early warning signals of extinction in deteriorating environments. Nature 467:456–459

Earl SR, Valett HM, Webster JR (2006) Nitrogen saturation in stream ecosystems. Ecology 87:3140–3151

Fellows CS, Valett HM, Dahm CN (2001) Whole-stream metabolism in two montane streams: contribution of the hyporheic zone. Limnol Oceanogr 46:523–531

Franssen NR, Gido KB, Guy CS, Tripe JA, Shrank SJ, Strakosh TR, Bertrand KN, Franssen CM, Pitts KL, Paukert CP (2006) Effects of floods on fish assemblages in an intermittent prairie stream. Freshwater Biol 51:2072–2086

Garcia-Criado F, Tome A, Vega FJ, Antolin C (1999) Performance of some diversity and biotic indices in rivers affected by coal mining in northwestern Spain. Hydrobiol 394:209–217

Gibbins C (2015) Coupling biological and physical processes: the ecological significance of river channel hydraulics and fluvial dynamics, this issue

Groffman PM, Baron JS, Blett T, Gold AJ, Goodman I, Gunderson LH, Levinson BM, Palmer MA, Paerl HW, Peterson GD, Poff NL, Rejeski DW, Reynolds JF, Turner MG, Weathers KC, Wiens J (2006) Ecological thresholds: the key to successful environmental management or an important concept with no practical application? Ecosystems 9:1–13

Gunderson LH (2000) Ecological resilience- in theory and application. Annu Rev Ecol Syst 31:425–439

Gunderson L, Holling CS (eds) (2001) Panarchy: understanding transformations in human and natural systems. Island Press, Washington, DC

Harding JS, Benfield EF, Bolstad PV, Helfman GS, Jones EBD (1998) Stream biodiversity: the ghost of land use past. P Natl Acad Sci USA 95:14843–14847

Hassett B, Palmer MA, Bernhardt ES, Smith S, Carr J, Hart D (2005) Status and trends of river and stream restoration in the Chesapeake bay watershed. Front Ecol Env 3:259–267

Hilderbrand RH, Watts AC, Randle AM (2005) The myths of restoration ecology. Ecol Soc 10 (1):19. (online) url: http://www.ecologyandsociety.org/vol10/iss1/art19/

Hilderbrand RH, Raesly RL (2009) Quantifying ecological thresholds and resilience in stream ecosystems. Final report to U.S. Environmental Protection Agency, National Center for Environmental Research, Science to Achieve Results (STAR) Program (RD-83244401-2)

Hilderbrand RH, Utz RM, Stranko SA, Raesly RM (2010) Applying thresholds to forecast potential biodiversity loss from human development. J N Am Benthol Soc 29:1009–1016

Holling CS (1973) Resilience and stability of ecological systems. Annu Rev Ecol Syst 4:1–23

Holling CS, Meffe GK (1996) Command and control and the pathology of natural resource management. Conserv Biol 10:328–337

Holling CS, Schindler DW, Walker BW, Roughgarden J (1995) Biodiversity in the functioning of ecosystems: an ecological synthesis. In: Perrings C, Mäler KG, Folke C, Holling CS, Jansson BO (eds) Biodiversity loss: economic and ecological issues. Cambridge University Press, Cambridge, pp 44–83

Hynes HBN (1975) The stream and its valley. Verh Int Theoret Ang Limnol 19:1–15

Irvine K (2012) The tragedy of the threshold: revising perceptions for aquatic conservation. Aquatic Conserv: Mar Freshw Ecosyst 22:705–711

Jeffries DS, Cox CM, Dillon PJ (1979) Depression of pH in lakes and streams in central Ontario during snowmelt. J Fish Res Board Can 36:640–646

Jenkins KM, Boulton AJ (2007) Detecting impacts and setting restoration targets in arid-zone rivers: aquatic micro-invertebrate responses to reduced floodplain inundation. J App Ecol 44:823–832

Jensen DW, Steel EA, Fullerton AH, Pess GR (2009) Impact of fine sediment on egg-to-fry survival of Pacific salmon: a meta-analysis of published studies. Rev Fish Sci 17:348–359

Kéfi S, Dakos V, Scheffer M, van Nes EH, Rietkerk M (2013) Early warning signals also precede non-catastrophic transitions. Oikos 122:641–648

King RS, Baker ME (2010) Considerations for analyzing ecological community thresholds in response to anthropogenic environmental gradients. J N Am Benthol Soc 29:998–1008

King RS, Baker ME, Kazyak PF, Weller DE (2011) How novel is too novel? Stream community thresholds at exceptionally low levels of catchment urbanization. Ecol Applic 21:1659–1678

Klauda R, Kazyak P, Stranko S, Roth N, Chaillou J (1998) Maryland biological stream survey: a state agency program to assess the impact of anthropogenic stresses on stream habitat quality and biota. Environ Monit Assess 51:299–316

Klein RD (1979) Urbanization and stream quality impairment. Wat Resour Bull 15:948–963

Lepori F, Barbieri A, Ormerod SJ (2003) Effects of episodic acidification on macroinvertebrate assemblages in Swiss alpine streams. Freshw Biol 48:1873–1885

Lindegren M, Dakos V, Groger JP, Gardmark A, Kornilovs G, Otto SA, Mollmann C (2012) Early detection of ecosystem regime shifts: a multiple method evaluation for managment application. PLoS ONE 7:e38410

Lytle DA, Poff NL (2004) Adaptation to natural flow regimes. Trends Ecol Evol 19:94–100

Lytle DA, Olden JD, McMullen LE (2008) Drought-escape behaviors of aquatic insects may be adaptations to highly variable flow regimes characteristic of desert rivers. Southwest Nat 53:399–402

Mount DR, Gulley DD, Hockett R, Garrison TD, Evans JM (1997) Statistical models to predict the toxicity of major ions to *Ceriodaphnia dubia*, *Daphnia magna*, and *Pimephales promelas* (fathead minnows). Environ Toxicol Chem 16:2009–2019

Palmer MA (2008) Reforming watershed restoration: science in need of application and applications in need of science. Estuaries Coasts 32:1–17

Palmer MA, Bernhardt ES, Schlesinger WH, Eshleman KN, Foufoula-Georgiou E, Hendryx MS, Lemly AD, Likens GE, Loucks OL, Power ME, White PS, Wilcock PR (2010) Mountaintop mining consequences. Science 327:148–149

Paul MJ, Meyer JL (2001) Streams in the urban landscape. Annu Rev Ecol Syst 32:333–365

Petchey OL, Gaston KJ (2009) Effects on ecosystem resilience of biodiversity, extinctions, and the structure of the regional species pool. Theor Ecol 2:177–187

Pimm SL (1984) The complexity and stability of ecosystems. Nature 307:321–326

Poff NL, Ward JV (1990) Physical habitat template of lotic systems: recovery in the context of historical pattern of spatiotemporal heterogeneity. Env Man 14:629–645

Poff NL, Allan JD, Bain MB, Karr JR, Prestegaard KL, Richter BD, Sparks RE, Stromberg JC (1997) The natural flow regime. Bioscience 47:769–784

Pond GJ, Passmore ME, Borsuk FA, Reynolds L, Rose CJ (2008) Downstream effects of mountaintop coal mining: comparing biological conditions using family- and genus-level macroinvertebrate bioassessment tools. J N Am Benthol Soc 27:717–737

Power ME, Matthews WJ, Stewart AJ (1985) Grazing minnows, piscivorous bass, and stream algae: dynamics of a strong interaction. Ecology 66:811–814

Quinn JM, Steele GL, Hickey CW, Vickers ML (1994) Upper thermal tolerances of twelve New Zealand stream invertebrates. N Zeal J Mar Freshwat Res 28:391–397

Rasmussen JJ, Baatrup-Pedersen A, Riis T, Friberg N (2011) Stream ecosystem properties and processes along a temperature gradient. Aquat Ecol 45:231–242

Resh VH, Brown AV, Covich AP, Gurtz ME, Li HW, Minshall GW, Reice SR, Sheldon AL, Wallace JB, Wissmar RC (1988) The role of disturbance in stream ecology. J N Am Benthol Soc 7:433–455

Resh VH, Jackson JK (1993) Rapid assessment approaches to biomonitoring using benthic macroinvertebrates. In: Rosenberg DM, Resh VH (eds) Freshwater Biomonitoring and benthic macroinvertebrates. Chapman & Hall, New York, pp 195–233

Roghair CN, Dolloff CA, Underwood MK (2002) Response of a brook trout population and instream habitat to a catastrophic flood and debris flow. Trans Am Fish Soc 131:718–730

Scheffer MS, Carpenter SR, Foley JA, Folke C, Walker B (2001) Catastrophic shifts in ecosystems. Nature 413:591–596

Scheffer M, Bascompte J, Brock WA, Brovkin V, Carpenter SR, Dakos V, Held H, van Nes EH, Rietkerk M, Sugihara G (2009) Early-warning signals for critical transitions Nature 461:53–59

Scheffer MS, Carpenter SR (2003) Catastrophic regime shifts in ecosystems: linking theory to observation. TREE 18:648–656

Schueler TR (1995) The peculiarities of perviousness. Watershed Prot Techn 2:233–238

Schueler TR, Fraley-McNeal L, Cappiella K (2009) Is impervious cover still important? Review of recent research. J Hydrol Eng 14:309–315
Soucek DJ, Kennedy AJ (2005) Effects of hardness, chloride, and acclimation on the acute toxicity of sulfate to freshwater invertebrates. Environ Toxicol Chem 24:1204–1210
Stevenson RJ, Peterson CG (1991) Emigration and immigration can be important determinants of benthic diatom assemblages in streams. Freshwat Biol 26:279–294
Stranko SA, Hilderbrand RH, Morgan RP II, Staley MW, Becker AJ, Roseberry-Lincoln A, Perry ES, Jacobson PT (2008) Brook trout declines with land cover and temperature changes in Maryland. N Am J Fish Manage 28:1223–1232
Stranko SA, Hilderbrand RH, Palmer MA (2012) Comparing the fish and benthic macroinvertebrate diversity of restored urban streams to reference streams. Restor Ecol 20:747–755
Stranko SA, Ashton MJ, Hilderbrand RH, Weglein SL, Kazyak DC, Kilian JV (2014) Fish and benthic macroinvertebrate densities in small streams with and without American Eel. Trans Am Fish Soc 143:700–708
Southwood TRE (1977) Habitat, the templet for ecological strategies? J Anim Ecol 46:337–365
Thrush SF, Hewitt JE, Dayton PK, Coco G, Lohrer AM, Norkko A, Norkko J, Chiantore M (2009) Forecasting the limits of resilience: integrating empirical research with theory. Proc R Soc B 276:3209–3217
U.S. Environmental Protection Agency (2004) Protecting water resources with smart growth (#EPA231-R-04-002). Washington, DC
Utz RM, Hilderbrand RH, Boward DM (2009) Identifying regional differences in threshold responses of aquatic invertebrates to land cover gradients. Ecol Indic 9:556–567
Van Sickle J, Baker JP, Simonin HA, Baldigo BP, Kretser WA, Sharpe WF (1996) Episodic acidification of small streams in northeastern United States: fish mortality in field bioassays. Ecol APplic 6:408–421
Van Sickle J, Baker J, Herlihy A, Bayley P, Gregory S, Haggerty P, Ashkenas L, Li J (2004) Projecting the biological condition of streams under alternative scenarios of human land use. Ecol Applic 14:368–380
Veraart AJ, Faassen EJ, Dakos V, van Nes EH, Lurling M, Scheffer M (2012) Recovery rates reflect distance to a tipping point in a living system. Nature 481:357–359
Ward JV, Stanford JA (1982) Thermal responses in the evolutionary ecology of aquatic insects. Ann Rev Entomol 27:97–117
Waters TF (1995) Sediment in streams: sources, biological effects, and control. American Fisheries Society Monograph, USA
Zalewski M, Naiman RJ (1985) The regulation of riverine fish communities by a continuum of abiotic-biotic factors. In: Alabaster JS (ed) Habitat modification and freshwater fisheries. Butterworths, Boston, pp 3–9

Chapter 19
Coupling Biological and Physical Processes: The Ecological Significance of River Channel Hydraulics and Fluvial Dynamics

Chris Gibbins

Abstract This chapter focuses on how flow and associated geomorphic processes influence river and stream invertebrates. It stresses the importance of flows and fluvial dynamics across a range of scales, discussing how they influence individual organisms, populations and whole communities. It considers not just benthic larvae, but how flow and sedimentological conditions influence other life stages. Consideration is also given to how organisms influence habitat—so called 'habitat engineering'. The chapter argues that a detailed understanding of species' ecologies is needed if we are to understand precisely why they are affected by flow conditions and to allow us to manage rivers sustainably.

Keywords Invertebrates · Hydraulics · Sediments · Larvae · Adult oviposition

19.1 Introduction and Scope

The forces of flowing water create the tune to which, in order to persist, stream-dwelling organisms must be able to dance. Understanding how they persist has been a central focus for freshwater ecology for decades, with research providing insights into an array of morphological, physiological, behavioural and life history traits that allow organisms to survive and reproduce in what are often high-energy and extremely variable environments. This chapter provides an insight into the ways in which flowing water influences benthic organisms. Flow has a direct effect on them, but it also influences them in a number of indirect ways, perhaps most notably as a result of the way in which it affects the sedimentological and geomorphological characteristics of river channels. The chapter deals also with some of these indirect influences. Flow-biota interactions are not uni-directional, so the chapter also discusses how organisms influence river bed conditions.

C. Gibbins (✉)
Northern Rivers Institute, School of Geosciences, The University of Aberdeen, Aberdeen AB24 3UF, UK
e-mail: c.gibbins@abdn.ac.uk

P. Rowiński and A. Radecki-Pawlik (eds.), *Rivers—Physical, Fluvial and Environmental Processes*, GeoPlanet: Earth and Planetary Sciences,
DOI 10.1007/978-3-319-17719-9_19

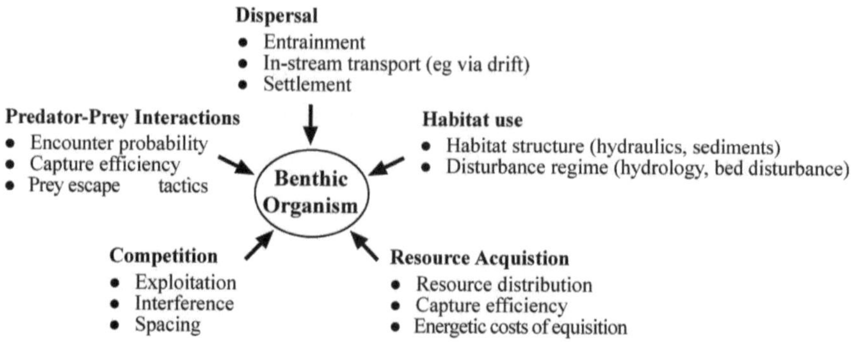

Fig. 19.1 The ecological processes influencing benthic invertebrates and which are affected by flow. These processes influence the performance, distribution and abundance of invertebrates. Modified from Hart and Finelli (1999)

There is a large and rapidly expanding literature on flow-biota interactions and a number of extensive and insightful reviews already exist (e.g. Hart and Finelli 1999; Gurnell 2013). Rather than providing another general review, the current chapter focuses on the need to connect knowledge of how flow influences individual organisms to the implications of this for whole populations. The chapter discusses flow influences where individuals of a given species occur, how it affects their dispersal and what it means for their life histories. It stresses the importance of considering multiple life stages and the need to include scales that are relevant both to individuals and whole populations in scientific studies. The chapter draws mainly on literature related to benthic macro-invertebrates (Fig. 19.1). These small and apparently fragile organisms are an important but challenging group to study. They are important because they play a range of functional roles in river ecosystems. For example, some are primary consumers which in turn are preyed upon by higher trophic groups such as fish and birds, and so they play an important role in energy transfer. Invertebrates are mostly rather small, so collecting data at the scales relevant to individuals poses many challenges, especially in field studies. Population level processes may be played out at a different (larger) scale, so it is important that studies which attempt to understand the controls on things such as birth or immigration rates characterise habitat at an appropriate scale.

19.2 Multi-scale Habitat Influences on Invertebrates

19.2.1 Physical and Ecological Links Across Scales

An individual *Baetis* mayfly nymph (i.e. the larval life-stage) will no longer be able to hold on to the surface of a pebble if instantaneous shear stress exceeds some critical threshold. Once entrained, how far it drifts downstream and where it settles are also a function of hydraulic characteristics, but characteristics that are better represented over slightly larger spatio-temporal scales than those which caused its entrainment. It may experience multiple entrainment and settlement episodes over the time it spends in the larval stage, so flow conditions over scales of tens or hundreds of meters and over weeks or months are relevant. However, other factors operating over much greater spatio-temporal scales will have dictated the location, within the stream network, of the pebble it was on. *Baetis* occur in patches of bed where substrate is of a suitable size and where their food and oxygen requirements can be met. The presence of suitable substrate at the microhabitat or patch scale is dictated by variables that influence spatial variation in sediment supply and transport. Depending on the volume and size distribution of sediment being supplied to a reach from upstream and the channel's competence to convey this material downstream, different reach types (e.g. pool-riffle, braided) develop. Supply and transport capacity correlate strongly with reach gradient, so reaches with certain gradients are more likely to provide more suitable sediment for *Baetis* than others; consequently we can expect to find relationships between *Baetis* abundance and reach-scale attributes such as gradient, even though individual animals are not influenced directly by the gradient of the patch of bed they are occupying.

Channel evolution is driven by certain high flow events, so not all of the flow regime is important in creating the geomorphic and sedimentological conditions required by *Baetis*. Therefore, while it is conceivable that an individual nymph may never experience the annual high flow events (e.g. a nymph in the summer cohort may emerge as an adult and complete its life-cycle before the winter floods) it nevertheless has been influenced by them because if the way they influence the distribution of suitable and unsuitable benthic habitat. Montgomery and Buffington (1997) provide a framework within which the hierarchical controls on physical, habitat-forming processes can be understood. An application of their framework to understand the hierarchy of scales influencing benthic invertebrate community structure is provided by Buendia et al. (2013a).

19.2.2 Adaptations to Flow

Three topics are prominent in the literature on benthic invertebrates: (i) the morphological adaptations of organisms to flow, (ii) the importance of substrate for community structure, and (iii) drift. In the 1980s and 1990s disturbance emerged as a prominent theme, and is a theme which connects these three topics. Disturbance is addressed by Hilderbrand and Utz in this volume and is the subject of numerous reviews (e.g. recently by Stanley et al. 2010) so is not discussed explicitly here. However, as much of the material in this chapter deals with how organisms cope with flow extremes and related fluvial dynamics, the material presented herein relates implicitly to disturbance. Aquatic organisms experience a very difference world to terrestrial ones, and those inhabiting flowing (lotic) water experience a rather different world to those inhabiting standing (lentic) water. The mechanics of moving or even standing still in lotic environments have had a profound influence on all aquatic organisms, but especially small ones such as invertebrates. Lancaster and Downes (2013) provide a very useful recent synthesis of this subject.

Water is approximately 50 times more viscous than air, so movement in water is more difficult. Viscosity is temperature dependant, so the viscous forces experienced by stream-dwelling organisms change with stream temperature. Water molecules at the surface are strongly attracted to each other, but not to the air molecules above. This creates what is termed surface tension, which is apparent as a 'film' at the water surface. Surface tension allows some small organisms (notably Hempitera) to skate or walk on the water surface; such organisms are typically more abundant in lentic water, but they also occur in lotic environments, often in marginal (channel edge) locations. Surface walking requires organisms to be within a certain size (weight) range; too heavy and the surface film will not be able to support them, and too light and they will not have the requisite force to manipulate the surface film (such manipulation is needed for movement; Hu et al. 2003). Surface walking also requires the evolution of morphological traits which produce water-repellent leg surfaces. Such traits include waxy coatings and dense coverings of hairs (Lancaster and Downes 2013). Many invertebrates need to pass through the water surface to complete their life cycle. For instance, adult mayflies (Ephemeroptera) of some species need to pass down through the water surface in order to deposit their eggs, while the larvae later pass in the opposite direction in order to metamorphose into adults. Others, such as adult beetles (Coleoptera), move across the water surface more frequently.

The mass of water creates a pressure on surfaces or objects below. Organisms inhabiting deep oceans and lakes experience considerable pressure because of the mass of water above. This pressure is lower in the shallower conditions typically associated with streams and rivers and so, on its own, is less relevant in lotic systems. However, other forces and hydraulic conditions created by the interaction of water mass and its movement downstream exert a key influence on individual invertebrates. The now classic work of Hynes (1970) and Statzner and Holm (1989) provide important reviews of this topic.

The Reynolds number (Re) is a useful descriptor which helps explain how a fluid behaves on meeting an object. Re is a function of (i) kinematic viscosity (V), the ratio of viscosity to the fluid's density, (ii) the velocity of the water (U), and (iii) the length of the object (*l*). Re helps describe changes in flow from laminar to turbulent and, most importantly, because *I* is the length of the object of interest, it helps explain how the flow that an organism experiences is a function of its size. Re is also relevant because it influences drag. Pressure drag (drag related to differences in pressure at the front and rear of an object) tends to be most important at high Re. Different body forms are effective in different Re; for instance, an organism with a rounded, relatively wide front-end and a long, tapering rear-end will be well adapted to life at high Re due to reduced pressure drag. Conversely, this body form confers no advantage at low Re.

Benthic invertebrates have evolved an array of different structures to help them deal with flow forces, including suckers, friction pads, a variety of hooks and claws

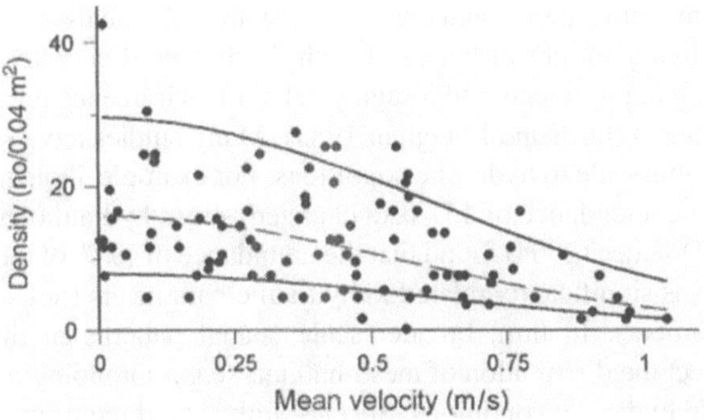

Fig. 19.2 Velocity use by Heptageniidae mayflies. The *fitted lines* represent limiting response models fitted using quantile regression. *Source* adapted from Lancaster and Beylea (2006)

and seteose structures (e.g. Ditsche-Kuru et al. 2010); other organisms (e.g. blackfly and caddisfly larvae) use silk or sticky secretions to avoid being swept downstream. Although the subject of much early attention (e.g. Steinmann 1907), this topic has not garnered much in the way of recent interest. Consequently, most existing studies are now rather old and there remains much scope for application of new high resolution technologies to characterise flow around different lotic invertebrate body forms, as has been done with some marine organisms (e.g. Stamhius 2006). Lancaster and Downes (2013) provide a thoughtful discussion of some of the problems of interpreting the functional role of particular body forms as adaptations to flow. They suggest that some unsubstantiated, misleading or incorrect assertions have been perpetuated in the literature through repeat citation. For instance, the body flatness seen in Heptageniidae mayflies is regularly interpreted as an adaptation to living in high velocity environments; however, flatness may actually increase lift, and it seems that some Heptageniidae occur primarily in lower velocity microhabitats (Lancaster and Beylea 2006; Fig. 19.2) as they may be less able than other organisms to cope with high velocities (Gibbins et al. 2010).

Life would be relatively simple if all invertebrates had to do was avoid being swept downstream. However, they also have to feed and this may require them to move around and/or occupy exposed areas. Many species graze algae and so have to spend part of their time on the top of exposed bed surfaces, placing them at risk of displacement. Some species feed by filtering suspended organic material from the water column, so organisms such as blackflies (Simuliidae) occupy exposed, high velocity locations in order to filter-feed. Flow also effects filter-feeders in other ways, as ingestion rates are often limited by low seston flux rates at low velocities, while high drag may impair the performance of feeding apparatus at high velocities (Morin et al. 1988). Other organisms are very mobile and risk dislodgement while foraging. Elliott (2003) assessed the distances travelled by individual invertebrates and found that predators moved most; distances travelled per 24 h period were mostly rather short in both upstream and downstream directions, with patterns for all species best described by an inverse power function. Flow has been found to influence the movement of caddisfly larvae, with individuals crawling less frequently and moving shorter distances at high discharges (Lancaster et al. 2006).

These feeding requirements and foraging behaviours influence microhabitat-scale distribution patterns (Orth and Maughan 1983). Many studies have sought to relate distributions at this scale to hydraulic conditions. For example, Doledec et al. (2007) illustrate how the abundance of 151 taxa changed across hydraulic gradients, while Merigoux and Doledec (2004) found that the abundance of 77 % of the taxa included in their study was significantly related to hydraulic parameters (notably shear stress and Froude number). In turn, broader scale spatial patterns in distribution and abundance reflect the distribution of meso-habitats (e.g. morphological units) which provide suitable hydraulic conditions. Species with shared requirements, or at least overlap in some of their habitat requirements, therefore tend to co-occur, leading to distinct communities associated with particular habitats; the different invertebrate assemblages of riffles and pools is an obvious and well known example of this (see Books et al. 2005 and references therein).

19.3 Linking Individual to Population Level Processes

19.3.1 The Role of Drift

Drift is the downstream movement of invertebrates suspended within the water column. Its prominence in the literature is due to its ubiquity and its importance. Drift is central to the life history of many invertebrates, as it is a primary means of longer distance movement and dispersal by larvae—their movement into and out of stream reaches. Increased drift is also a direct ecological response to physical disturbances such as floods: drifters are at best displaced, and at worst injured or killed. However, drift is also a mechanism for the re-colonisation of denuded areas once the disturbance has passed, and so is an important means of population recovery. Through its effects on mortality, immigration and emigration, drift therefore plays an important role in population dynamics.

Early work elucidated the temporal rhythms to drift, drift distances and time spent in the drift, as well as the relationships between the downstream drift of larvae and the movements of adults (e.g. Elliott 2003 and the review by Brittain and Eikland 1988). Drift occurs for different reasons at different times, leading to the distinction between voluntary (=behavioural) and involuntary drift. Flow forces are a primary driver of involuntary drift, while the search for food, avoidance of competitors and escape from predators are causes of behavioural drift. However, irrespective of the initial cue, flow conditions influence time spent in the drift and the downstream distance travelled by drifters. Drift increases during periods of elevated discharge, as greater numbers of invertebrates are sheared from the bed. The term 'catastrophic drift' is applied to the marked increase in involuntary drift observed during floods. Statzner (1984) hypothesised that during floods the major increase in drift occurs once the bed begins to move, and more recent laboratory

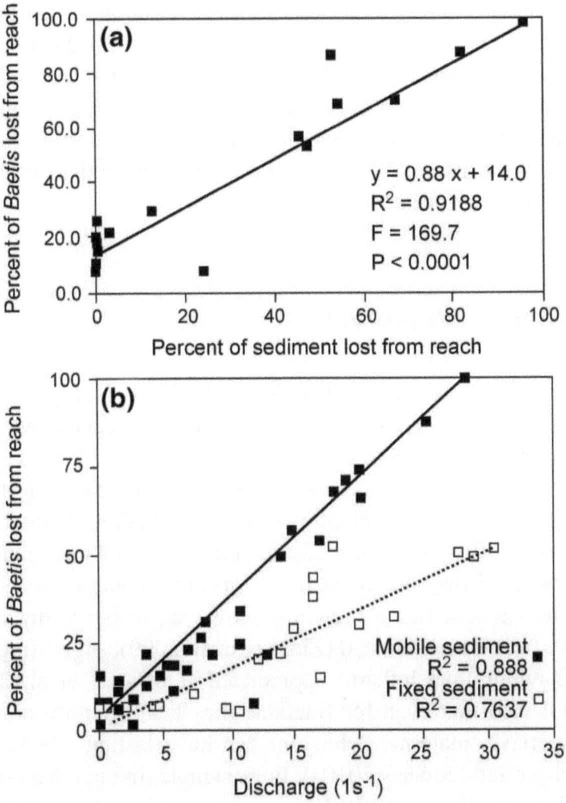

Fig. 19.3 Relations between the number of *Baetis* lost from the bed of laboratory flume and sediment stability. The *upper plot* (**a**) indicates that, in general, a higher rate of loss of animals occurs as the bed become increasingly unstable. The *lower plot* (**b**) indicates the magnitude of the effect of bed instability, over and above that of the direct shear of animals from the bed surface. The same bed grain-size distributions and range of discharges were used, but in one set of experiments the sediment was fixed to the flume bed while in the other it was not fixed, and hence was entrained as a function of discharge and associated hydraulic conditions. *Source* adapted from Gibbins et al. (2005)

studies (Gibbins et al. 2005) have confirmed both that the rate of loss of animals from the bed parallels sediment loss (Fig. 19.3a), and that fewer animals are lost from stable beds than mobile ones (Fig. 19.3b). Evidence from field experimentation suggests that a major upward inflection in the drift response to increasing discharge occurs at the same point that the critical entrainment threshold of the sediment is exceeded (Gibbins et al. 2007a). What is interesting is that the rates of sediment transport associated with catastrophic drift are not necessarily great, suggesting that the hydrological events responsible for ecological disturbance are lower in magnitude than those responsible for geomorphic disturbance (Gibbins et al. 2007b).

Lower levels of involuntary drift occur during conditions when the bed remains stable. Based on flume experiments, Blanckaert et al. (2013) found that, unlike sediment particles, dislodgement of invertebrates is related not only to the peak values of turbulent flow forcing but also to the temporal fluctuations in this flow. They also found that the dominant temporal fluctuations are not due to local turbulent structures of the size of the invertebrate, but to those that scale with the flow depth and are inherited from upstream.

19.3.2 The Role of Oviposition

In lotic systems the distribution of individual larvae is typically patchy and non-random, and historically has been assumed to be dictated by their dispersal, either through drift or crawling. The initial distribution of larvae (neonates) is related to the behaviour of their mother and, for those species that deposit eggs in clusters in specific locations, is a function of the distribution of suitable oviposition sites (Peckarsky et al. 2000). Historically the effects of this maternal behaviour on larval distribution have been assumed to be transient, being erased by subsequent larval movements via drift and/or crawling. However, dispersal distances for some groups may have been overestimated (Zimmer et al. 2009), suggesting a greater role of maternal behaviour than hitherto appreciated (Lancaster et al. 2008). Several studies have indicated that even for Baetidae mayflies, a family normally considered highly dispersive, maternal behaviour can have lasting effects on larval distribution (Lancaster and Downes 2010a). For example, the numbers of early instars of *Baetis rhodani*, and the caddisfly *Rhyacophila dorsalis* (which shares the same oviposition strategy), have been shown to be positively associated with the abundance of suitable oviposition sites; riffles with the most oviposition sites were found to have more early instar *B. rhodani and R. dorsalis* (Lancaster and Downes 2014). Such findings are now challenging long-held assumptions about how larval hydraulic and sedimentary habitat requirements must dictate spatial patterns in their abundance; at least for some species, it appears that the requirements of ovipositing females play a more prominent role.

Female Baetidae oviposit on large cobbles and boulders projecting above the water surface. However, not all projecting cobbles and boulders appear equally suitable. Peckarsky et al. (2000) found that, over the three years of their study, females oviposited on less than 10 % of available cobbles/boulders, and in some years single large boulders held more than half of the total egg masses observed. The best predictors of the presence of egg masses appear to vary between streams, but in general large clasts located in places where there is water splash associated with high velocities are used most frequently (Encalada and Peckarsky 2006); the area exposed above the water surface, and hence the area available for females to land on, appears an important criteria for females selecting oviposition sites (Encalada and Peckarsky 2006). Thus, the fluvial processes which dictate the abundance and distribution of suitably large substrate, as well as the hydrological

Plate 19.1 A male *Calopteryx virgo* (Beautiful Demoiselle) holds station on a protruding boulder. Such boulders are important for Demoiselles as they are prominent positions used by males to signal their ownership of a particular territory, and as a base for feeding sorties. Boulders are also important for other groups; for example, female Baetidae mayflies use them for egg deposition. *Photo* Chris Gibbins, Catalunya, Spain

regimes which dictate whether (as a function of water depth) potentially suitable boulders are available to females at the times of the year that they are on the wing, may exert a critical influence on the success of egg laying. Work on this topic is also beginning to show that variation in neonate recruitment, linked to adult oviposition behaviour, can alter larval population sizes and potentially mask the effects of processes acting subsequently on larvae. It therefore helps illustrate that flows and fluvial processes are important for more than just larvae, and that caution is needed when interpreting patterns of larval abundance solely within the context of larval hydraulic habitat requirements (Plate 19.1).

19.3.3 The Influence of Habitat Complexity and Simplification

Flow and substrate interact and create habitat that varies in character and complexity across a range of spatial scales. Cummins and Lauf (1969) provided an early review of the effects of substrate on benthic organisms. Much of this early, and indeed some later literature, attempted to understand the extent to which substrate characteristics (such as clast shape, surface texture and pore space volume) influence the diversity of invertebrate assemblages (review by Hynes 1970). However, different empirical studies led to contradictory conclusions (e.g. see Wise and Molles 1979).

Habitat heterogeneity and complexity may promote species diversity, so increasingly sophisticated means of quantifying these elements have been employed in the search for general patterns. Complexity refers to the abundance of distinct, individual physical elements (e.g. crevices); habitats become more complex as more elements (e.g. more crevices) are added. Heterogeneity refers to the number of different elements present; habitats become more heterogeneous the more elements there are (vertical elements such as plant stems present along with crevices). McCoy and Bell (1991) proposed a three-dimensional conceptualisation of habitat structure, with heterogeneity and complexity each being an axis. Their third axis relates to the spatial scale of interest and helps represent the idea that both heterogeneity and complexity are scale dependent.

At small spatial scales, the presence of mineral substrates of different sizes, shapes and surface texture creates bed architectural complexity and heterogeneity. By manipulating complexity and heterogeneity, Downes et al. (1998) were able to show how habitat structure, quantified by adding elements to increase both complexity and heterogeneity, influences invertebrate diversity. However, more recent work has found that habitat heterogeneity may not influence diversity and abundance, and that the effects of complexity can be weak (Barnes et al. 2013). The latter authors suggest that habitat influences invertebrates through factors other than physical structure, and so advocate a re-appraisal of the processes involved. Clearly, our ability to detect habitat effects depends on the methods used to characterise habitat structure. Thus, more than 40 years after Cummins' early review there remains considerable uncertainty about how habitat structure influences diversity, with generalizable rules not yet agreed (Barnes et al. 2013).

Although the effects of habitat complexity have proven difficult to assess, it is clear that simplification of habitat can lead to reduced diversity. For example, large volumes of fine sediment, resulting from either natural or anthropogenic processes, can homogenise river bed habitat and impact invertebrates (reviewed by Wood and Armitage 1997). Fine sediment smothers riverbed micro-topography and clogs interstitial spaces, and can alter bed stability as a result of changes in particle cohesion and entrainment thresholds. In gravel bed rivers, relatively few taxa occur in locations dominated by fine sediment, and as a result assemblages in such locations are dominated by a small number of species characterised by particular ecological traits (Buendia et al. 2013b). At larger scales, whole reaches and sub-catchments may show community patterns that reflect differences in how much of the sediment matrix is composed of fine material (Buendia et al. 2013b). Temporal patterns related to the combined effects of flow and fine sediment may also be evident; for example, in a river receiving large volumes of fine sediment as a result of natural geological features and runoff processes (Plate 19.2), invertebrate communities tracked seasonal rhythms of fine sediment accumulation, rhythms which themselves were dictated by seasonal changes in flood magnitude and

Plate 19.2 The River Isabena, Catalunya, Spain. Due to highly erodible Marls in its catchment, this river carries a very high fine sediment load. In places where this fine material settles out, it can smother the bed and homogenise physical habitat conditions. Relatively few organisms can cope with such conditions, so invertebrate species richness can be low. *Photo* Chris Gibbins

frequency (Buendia et al. 2013c; Fig. 19.4). Both alpha and beta diversity changed over time as a result of the gradual increase in fine sediment accumulated on the bed and which progressively homogenised habitat.

19.4 Species Impacts on Habitat—Habitat Engineering

Habitat and ecosystem engineering by river organisms is well known, although most evidence comes from rather obvious examples. The North American beaver, *Castor Canadensis,* has long been considered a model ecosystem engineer, with numerous studies documenting the physical, chemical and ecological impacts of beavers across a range of spatial scales (Anderson et al. 2007; Hammerson 1994). Recently, Fuller and Peckarsky (2011) documented how beaver ponds influence stream temperatures and, in turn, the size of mature female *Baetis bicaudatus.* Female size is significant in Baetidae population dynamics, as large females are more fecund than small ones. An indication of the magnitude of habitat engineering by fish can be gained from work on mass spawning salmonids. Hassan et al. (2008), for example, found that salmonids were responsible for around half of the annual bedload yield and influenced river bed morphology over the whole year (i.e. long after spawning ceased). Other families of fish have been shown to be significant

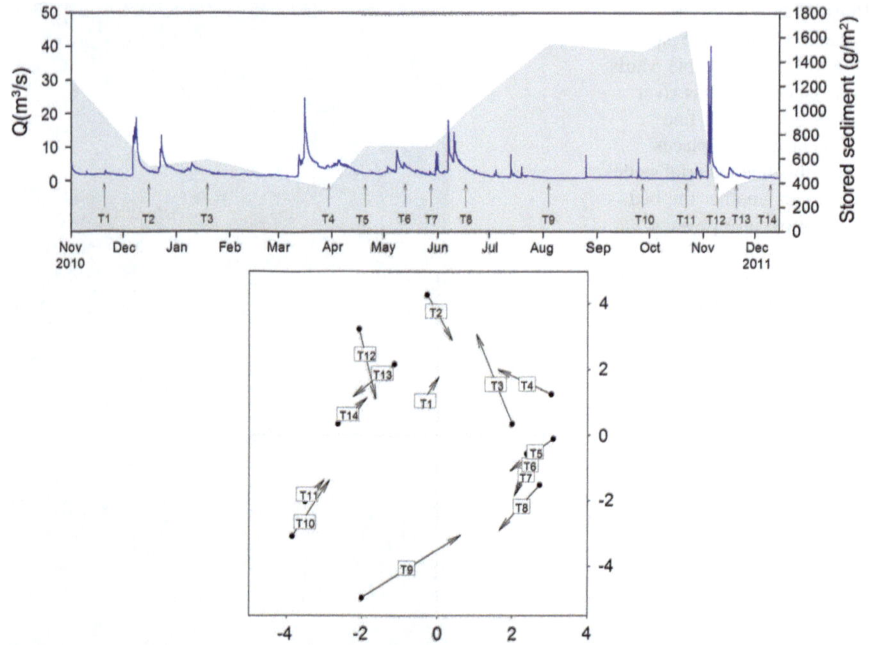

Fig. 19.4 Flow and fine sediment dynamics in the River Isabena (Catalunya, Spain) and invertebrate community responses. The *upper panel* shows discharge and the amount of fine sediment stored on the bed of the study reach. Numbers prefixed T relate to dates on which invertebrate samples were collected. The *lower panel* shows the centroids (•) of invertebrate sample replicates (n = 25–30 on each occasion), ordinated using Co-inertia Analysis (Co-IA). The Co-IA indicated a strong correspondence between invertebrate patterns and the sedimentary and hydrological conditions over the study period. *Arrows* indicate changes in community structure between sampling occasions (T1–14). Adapted from Buendia et al. (2013c)

bioturbators of river-bed habitat, due to their feeding action (Statzner et al. 2003; Statzner and Sagnes 2008).

Rather less obvious but potentially important (especially when considered cumulatively) are the engineering effects of numerous invertebrates. Different invertebrates alter habitat in different ways, with some altering hydraulic conditions and others bed sediments. Crayfish have been shown to influence both coarse and fine sediment in rivers. Their feeding and movement, especially the aggressive interactions between males, can disturb gravel substrates. Their burrowing behaviour affects bed architecture, while as they walk they alter the bed by re-orientating material and changing friction angles between grains (Johnson et al. 2011). Such changes have been estimated to reduce critical shear stress by up to 75 %, with reductions leading to increased base-flow transport of sand grains by up to 2 kg m^{-2} d^{-1} and of gravels by up to 4 kg m^{-2} d^{-1} (Statzner et al. 2003; Statzner and Peltret 2006). Most studies have been conducted in laboratory settings, but Harvey et al. (2014) demonstrated that crayfish activity can result in pulses of suspended sediment in natural river channels at night-time, the period when they are most active.

Many smaller invertebrates, including numerous insects, worms and molluscs, engineer bed sediments in a variety of ways. Statzner (2012) provided a review of their effects, classified according to whether they have bioturbation or consolidation impacts on bed sediments. Feeding, borrowing, walking and building tubes or cases all have bioturbation effects; in their active search for prey, for example, large predatory stoneflies remove sand grains from within the bed matrix (Zanetell and Peckarsky 1996). Such effects are not trivial: Statzner et al. (1996) estimated that *Dinocras cephalotes* may erode 200–400 kg m^{-2} y^{-1} of sand. Conversely, various taxa help consolidate sediments. Some caddisflies, moths and dipterans construct cases or tubes from sand grains bound together with silk. In addition, some taxa use silk as a safety thread or pad when moving or resting, glueing the silk to coarser bed material (Malmqvist et al. 2004). These silks act as bridges of varying strength between particles, and may last for long periods (up to one month; see Statzner 2012). Caddisfly larvae of the family Hydropsychidae use silk for several purposes: as a safely thread, to fix gravel together for their larval retreat, to create nets (which capture food) between coarse material, and to bind together gravel for pupal cases. Hydropsychids can occur at high densities, and estimates of the length of silk produced by them exceed hundreds of km y^{-1} m^{-2} (see Statzner 2012). The effect of these silks on bed stability may be considerable, because of how they increase critical shear stress (Johnson et al. 2009).

Different species are responsible for consolidation and bioturbation of bed material and, as they may occur in different locations, can influence spatial patterns of bed stability/instability within river reaches. Statzner (2012) illustrated schematically the magnitude of the effects of a number of co-occurring bioturbators and consolidators in a hypothetical reach (Fig. 19.5). While his model did not include all taxa and necessitated a degree of upscaling and extrapolation, it nonetheless gives a thought-provoking insight into the potential cumulative influence of organisms in lotic environments. It also provides a framework for understanding how the disappearance of native engineers or the introduction of alien ones (e.g. zebra mussels *Dreissina polymorpha* or signal crayflish *Pacifastacus leniusculus*) may impact fluvial processes (Johnson et al. 2011; Harvey 2014).

19.5 Conclusions: Opportunities and Challenges for Future Research

It should be evident from the material discussed above that understanding species ecologies, behaviours and life histories is central to understanding the ecological importance of flow. Work which fails to do this runs the risk of being merely descriptive (Lancaster and Downes 2010b) and so may be of limited value in helping to guide sustainable river management or aid species conservation.

The trait-based approach provides potentially powerful insights into the ecological mechanisms underpinning observed distribution and abundance patterns

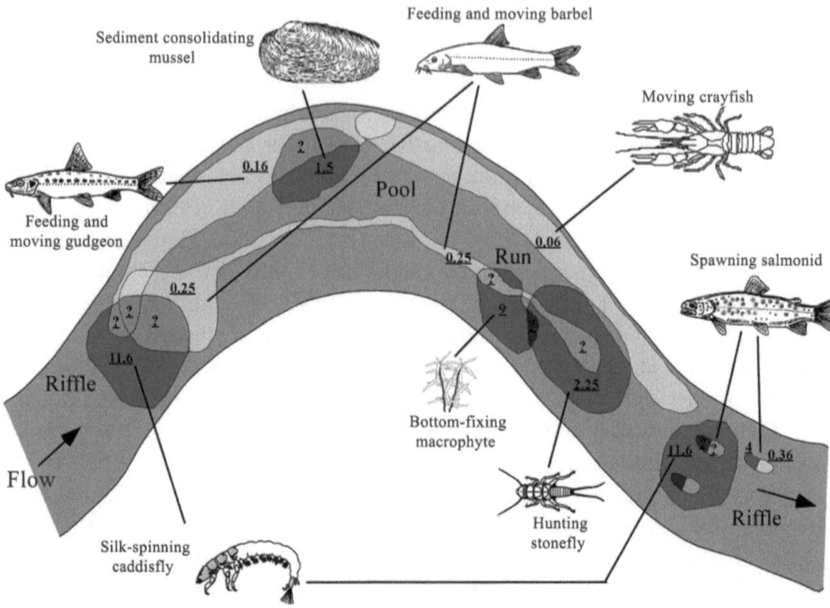

Fig. 19.5 Potential effects of engineers on habitat conditions in a stream reach. The diagram illustrates how by bioturbating or consolidating sediments, organisms alter the discharge needed for sediment entrainment (critical discharge, Q_c). Intermediate *grey shading* indicates areas unaffected by the organisms shown. *Lighter* and *darker shading* indicated respectively the decrease or increase in Q_c as a result of organisms; e.g. through its consolidations effects, the mussel increases Q_c by a factor of 1.5 in areas where it occurs. *Question marks* indicate where uncertainty exists over the net effects of species in locations where they co-occur. Adapted from Statzner (2012)

(e.g. a recent example by Buendia et al. 2013b). For this reason, traits are increasingly being incorporated into monitoring and assessment tools. Trait-based approaches are not new, but they are assuming greater significance because of the need to properly assess human impacts on aquatic environments. Opportunities also now exist for much more realistic characterisation of hydraulic and sedimentary habitat, providing a much better platform for understanding abundance-environment relationships. For instance, high resolution techniques such as particle image and acoustic Doppler velocimetry (PIV and ADV respectively) can provide hydraulic data at the scales relevant to individual organisms, not just in laboratory settings but in the field (Smith et al. 2012; Cameron et al. 2013). Similarly, developments in survey and data interrogation techniques allow seamless characterisation of river bed architecture from the patch to the reach scales (e.g. Terrestrial Laser scanning; Brasington 2010; aerial photogrammetry; Vericat et al. 2009). Quinlan et al. (2014) provide a recent review of how such technologies can provide insights into the interactions between the endangered freshwater pearl mussel *Margaritifera margaritifera* and its habitat, and thus provide a robust basis for habitat management.

While there are opportunities to move forward at increased pace, there are also significant challenges faced by aquatic scientists. Thomas et al. (2014) provide a summary of the issues and difficulties associated with understanding flow-biota interactions. They conclude by stressing that the results of studies carried out over short durations or at conditions of low to moderate hydraulic forcing cannot easily be extrapolated to longer timescales or larger (catastrophic) events and, therefore, that there is an urgent need for larger and more flexible research facilities. In particular such facilities need to be 'hybrids', capable of experimentation with both abiotic and biotic variables; when married with carefully planned and executed experiments, such facilities should further our understanding of the coupling between physical and ecological processes in streams and rivers.

Acknowledgments I am indebted to the many collaborators I have been fortunate enough to work with. I would particularly like to thank Ramon Batalla, Cristina Buendia, Iain Malcolm and Damia Vericat.

References

Anderson CB, Rosemond AD (2007) Ecosystem engineering by invasive exotic beavers reduces in-stream diversity and enhances ecosystem function in Cape Horn, Chile. Oecologia 154:141–153

Barnes JB, Vaughan IP, Ormerod SJ (2013) Reappraising the effects of habitat structure on river macroinvertebrates. Freshw Biol 58:2154–2167

Blanckaert K, Garcia XF, Ricardo AM, Chen Q, Pusch MT (2013) The role of turbulence in the hydraulic environment of benthic macroinvertebrates. Ecohydrology 6:700–717

Brasington J (2010) From grain to floodplain: hyperscale models of braided rivers. J Hydraul Res 48(53 Suppl.):4

Brittain JE, Eikeland TJ (1988) Invertebrate drift—a review. Hydrobiologia 166:77–93

Brooks AJ, Haeusler T, Reinfelds I, Williams S (2005) Hydraulic microhabitats and the distribution of macroinvertebrates in riffles. Freshw Biol 50:331–344

Buendia C, Gibbins CN, Vericat D, Batalla RJ (2013a) Reach and catchment scale influences on invertebrate assemblages in a river with naturally high fine sediment loads. Limnologia 42:362–370

Buendia C, Gibbins CN, Vericat D, Batalla RJ, Douglas A (2013b) Detecting the structural and functional impacts of fine sediment on stream invertebrates. Ecol Ind 25:184–196

Buendia C, Gibbins CN, Vericat D, Batalla RJ (2013c) Effects of flow and fine sediment dynamics on spatial and temporal turnover of stream invertebrate assemblages. Ecohydraulics. doi:10.1002/eco.1443

Cameron SM, Nikora VI, Albayrak I, Miler O, Stewart S, Siniscalchi F (2013) Interactions between aquatic plants and turbulent flow: a field study using stereoscopic PIV. J Fluid Mech 732:345–372

Cummins K, Lauf J (1969) The influence of substrate particle size on the microdistribution of stream macrobenthos. Hydrobiologia 34:145–181

Ditsche-Kuru P, Koop JHE, Gorb SN (2010) New insights into a life in current: do the gill lamellae of *Epeorus assimilis* and *Iron alpicola* larvae function as friction or sucker pads? J Exp Biol 213:1950–1959

Dodelec S, Lamouroux N, Fushs U, Merigoux S (2007) Modelling the hydraulic preferences of benthic macroinvertebrates in small European streams. Freshw Biol 52:145–164

Downes B, Lake PS, Schreiber ESG, Glaister A (1998) Habitat structure and regulation of local species diversity in a stony, upland stream. Ecol Monogr 68:237–257

Elliott JM (2003) A comparative study of the dispersal of 10 species of stream invertebrates. Freshw Biol 48:1652–1668

Encalada AC, Peckarsky BL (2006) Selective oviposition of the mayfly *Baetis bicaudatus*. Oecologia 148:526–537

Fuller MR, Peckarsky BL (2011) Ecosystem engineering by beavers affects mayfly life histories. Freshw Biol 56:969–979

Gibbins CN, Scott E, Soulsby C, McEwan I (2005) The relationship between sediment mobilisation and the entry of *Baetis* mayflies into the drift in a laboratory flume. Hydrobiologia 533:115–122

Gibbins CN, Batalla RJ, Vericat D (2007a) Shaking and moving: sediment movement triggers mass drift of invertebrates. Can J Fish Aquat Sci 64:1–5

Gibbins CN, Vericat D, Batalla RJ (2007b) When is stream invertebrate drift catastrophic? Freshw Biol 52:2369–2384

Gibbins CN, Batalla RJ, Vericat D (2010) Invertebrate drift and benthic exhaustion during disturbance: response of mayflies to increasing shear stress and river-bed instability. River Res Appl 26:499–510

Gurnell A (2013) Plants as river system engineers. Earth Surf Process Landf 39:4–25

Hammerson GA (1994) Beaver (*Castor canadensis*): ecosystem alterations, management, and monitoring. Nat Areas J 14:44–57

Hart DD, Finelli CM (1999) Physical-biological coupling in streams: the pervasive effects of flow on benthic organisms. Annu Rev Ecol Syst 30:363–395

Harvey J, Alexander James Henshaw AJ, Moorhouse TP, Clifford N, Holah H, Grey J, Macdonald DW (2014). Invasive crayfish as drivers of fine sediment dynamics in rivers: field and laboratory evidence. Earth Surf Process Landf 39:259–271

Hassan MA, Gottesfeld AS, Montgomery DR, Tunnicliffe JF, Clarke GKC, Wynn G, Jones-Cox H, Poirier R, MacIsaac E (2008) Salmon-driven bed load transport and bed morphology in mountain streams. Geophys Res Lett 35:LO4405

Hu DL, Chan B, Bush JWM (2003) The hydrodynamics of water strider locomotion. Nature 424:663–666

Hynes HBN (1970) The ecology of running waters. University of Toronto Press, Canada

Johnson ME, Reid I, Rice SJ, Wood P (2009) Stabilisation of fine gravels by net-spinning caddisfly larvae. Earth Surf Process Landf 34:413–423

Johnson ME, Rice SJ, Reid I (2011) Increase in coarse sediment transport associated with disturbance of gravel river beds by signal crayfish (*Pacifastacus leniusculus*). Earth Surf Process Landf 36:1680–1692

Lancaster J, Belyea LR (2006) Defining the limits to local density: alternative views of abundance-environment relationships. Freshw Biol 51:783–796

Lancaster J, Downes B (2010a) Lasting effects of maternal behaviour on the distribution of a quintessential disperser in an advection-dominated system. J Anim Ecol 163:373–384

Lancaster J, Downes B (2010b) Linking the hydraulic world of individual organisms to ecological processes putting the ecology into ecohydraulics. River Res Appl 26:385–403

Lancaster J, Downes B (2013) Aquatic entomology. Oxford University Press, Oxford

Lancaster J, Downes BJ (2014) Maternal behaviours may explain riffle-scale variations in some stream insect populations. Freshw Biol 59:502–513. doi:10.1111/fwb.12281

Lancaster J, Buffin-Bélanger T, Reid I, Rice S (2006) Flow- and substratum-mediated movement by a stream insect. Freshw Biol 51:1053–1069. doi:10.1111/j.1365-2427.2006.01554.x

Lancaster J, Downes B, Arnold A (2008) Environmental constraints on oviposition may limit density in a stream insect at multiple scales. Oecologia 163:373–384

Malmqvist B, Adler PH, Kuusela K, Merritt RW, Wotton RS (2004) Blackflies in the boral biome, key organisms in both terrestrial and aquatic environments: a review. Ecoscience 11:187–200

McCoy ED, Bell SS (1991) Habitat structure: the evolution and classification of a complex topic. In: Bell SS, McCoy ED, Mushinsky HR (eds) Habitat structure: the physical arrangement of objects in space. Chapman Hall, London, pp 3–27

Merigoux S, Dodelec S (2004) Hydraulic requirements of stream communities: a case study of invertebrates. Freshw Biol 49:600–613

Montgomery DR, Buffington JM (1997) Channel reach morphology in mountain drainage basins. Geol Soc Am Bull 109:596–611

Morin A, Back C, Chalifour A, Boisvert J, Peters RH (1988) Empirical models predicting ingestion rates of blackfly larvae. Can J Fish Aquat Sci 45:1711–1719

Orth DJ, Maughan OE (1983) Microhabitat preferences of benthic fauna in a woodland stream. Hydrobiologia 106:157–168

Peckarsky BL, Taylor BW, Caudill CC (2000) Hydrologic and behavioural constraints on oviposition of stream insects: implications for adult dispersal. Oecologia 125:186–200

Quinlan E, Gibbins CN, Malcolm IA, Batalla RJ, Vericat D, Hastie L (2014) A review of the physical habitat requirements and research priorities needed to underpin conservation of the endangered freshwater pearl mussel *Margaritifera margaritifera*. Aquat Conserv Mar Freshw Ecosyst. doi:10.10002/acq.2484

Smith M, Vericat D, Gibbins CN (2012) Through-water terrestrial laser scanning of gravel beds at the plot scale. Earth Surf Process Landf 37:411–421

Stamhius EJ (2006) Basics and principles of particle image velocimetry (PIV) for mapping biogenic and biologically relevant flows. Aquat Ecol 40:463–479

Stanley E, Powers SM, Noah NR (2010) The evolving legacy of disturbance in stream ecology: concepts, contributions and coming challenges. J North Am Benthol Soc 29:67–83

Statzner B (2012) Geomorphological implications of engineering bed sediments by lotic animals. Geomorphology 157–158:49–65

Statzner B, Fuchs U, Higler LWG (1996) Sand erosion by mobile predaceous stream insects: implications for ecology and hydrology. Water Resour Res 32:2279–2287

Statzner B, Holm TF (1989) Morphological adaptation of shape to flow: microcurrents around lotic macroinvertebrates with known reynolds numbers at quasi-natural flow conditions. Oecologia 78:145–157

Statzner B, Peltret O (2006) Assessing potential abiotic and biotic complications of crayfish-induced gravel transport in experimental streams. Geomorphology 74:245–256

Statzner B, Sagnes P (2008) Crayfish and fish as bioturbators of streambed sediments: assessing joint effects of species with different mechanistic abilities. Geomorphology 93:267–287

Statzner B, Dejoux C, Elouard J-M (1984) Field experiments on the relationship between drift and benthic densities of aquatic insects in a tropical stream (Ivory Coast) 1 introduction: review of drift literature, methods and experimental conditions. Rev Trop Hydrobiol 17:319–334

Statzner B, Peltret O, Tomanova S (2003a) Crayfish as geomorphic agents and ecosystem engineers: effect of a biomass gradient on baseflow and flood-induced transport of gravel and sand in experimental streams. Freshw Biol 48:147–163

Statzner B, Sagnes P, Champagne J-Y, Viboud S (2003b) Contribution of benthic fish to the patch dynamics of gravel and sand transport in streams. Water Resour Res 39:1309–1325. doi:10.1029/2003WR002270

Steinman P (1907) Die Tierwelt der Gebirgsbäche. Eine faunistische-biologische Studie. Ann Biol Lacustre 2:30–150

Thomas RE, Johnson MF, Frostick LE, Parsons DB, Boumac TJ, Dijkstrad JT, Eiffe O, Gobert S, Henry PE, Kemp P, Stuart J, Mclelland SJ, Frederic Y, Moulin FY, Myrhaugg D, Neytsg A, Pauli M, Ellis W, Puijalonj S, Rice SJ, Stanicak P, Tagliapietral D, Talm M, Tørumn A, Vousdoukasio MI (2014) Physical modelling of water, fauna and flora: knowledge gaps, avenues for future research and infrastructural needs. J Hydraul Res. doi:10.1080/00221686.2013.876453

Vericat D, Brasington J, Cowie M, Wheaton J (2009) Accuracy assessment of aerial photographs acquired using lighter-than-air blimps: low-cost tools for mapping river corridors. River Res Appl 15:985–1000

Wise DH, Molles MC Jr (1979) Colonisation of artificial substrates by stream insects: influence of substrate size and diversity. Hydrobiologia 65:69–74

Wood PJ, Armitage PD (1997) Biological effects of fine sediment in the lotic environment. Environ Manage 21:203–217

Zanatell BA, Peckarsky BL (1996) Stoneflies as ecological engineers—hungry predators reduce fine sediments in stream beds. Freshw Biol 36:569–577

Zimmer RK, Fingerut JT, Zimmer CA (2009) Dispersal pathways, seed rains and the dynamics of larval behaviour. Ecology 90:1933–1947

Chapter 20
Why Do We Need Bankfull and Dominant Discharges?

Artur Radecki-Pawlik

Abstract Within scientific hydrological, geomorphological and also engineering literature the two discharges that very often demand attention of hydrologists as well as ecologists and river engineers and water catchment managers are, respectively, the dominant and bankfull discharges. What are they? Do we really need them and why? The present paper explains the main ways of the dominant and bankfull discharges determinations and calculations. As far as the dominant discharge is concerned, the most important methods to find its value are: the Rzanicyn, the Debski, the Lambor, the Marlette and Walker, the Wolman and Miller, and the Makkavieiev methods. When considering bankfull discharge, the most important concepts and definitions as well as methods of its determination are: the Williams, the Wolman, the Schumm and Brown, the Riley, the Woodyer as well as the Radecki-Pawlik and Skalski methods. All those concepts and methods are described. Additionally, examples of bankfull calculations are given from one of the Carpathians rivers to show the difference obtained in bankfull values using different methods. Both morphometric and biological methods are used to determine bankfull. Special attention is given to biological approaches of bankfull methods since the Water Framework Directive gives priority to biological measures and findings along river and stream reaches.

Keywords Dominant discharge · Bankfull · Mountain stream · River · Water framework directive

20.1 Introduction

Within modern scientific literature dealing mostly with hydrology, catchment management as well as fluvial geomorphology, two key discharges are very often reported: the dominant and the bankfull discharges. Also many river engineers, especially those who cooperate with forest engineers, ecologists and geomorphol-

A. Radecki-Pawlik (✉)
Department of Hydraulic Engineering and Geotechnics, Environmental Engineering and Geodesy Faculty, Agricultural University of Cracow, Cracow, Poland
e-mail: rmradeck@cyf-kr.edu.pl

P. Rowiński and A. Radecki-Pawlik (eds.), *Rivers—Physical, Fluvial and Environmental Processes*, GeoPlanet: Earth and Planetary Sciences,
DOI 10.1007/978-3-319-17719-9_20

ogists are paying an attention due to the variety of variables of the dominant discharge and bankfull, since bankfull values are helpful when designing river training practices. The values of the two discharges are even more important when talking about river rehabilitation processes as well as environmentally-sensitive training of the river. As far as dominant discharge is concerned, one wants quite often to describe the river regime with one single discharge value. Such a discharge might be responsible for most changes in the river channel morphology in terms of its geometry and, at the same time, during such a discharge the largest amount of sediment might be transported. In general, the bed-forming processes are due to a variety of variables, but discharges are one of the most important. Often the peak or the catastrophic discharges transport the largest amount of sediment. But such spectacular events occur quite rarely within river channels. From the channel forming process point of view, it is important to consider discharges which are large enough to initiate the sediment movement (when the water depth is higher than the critical depth which allows the river bed material to move) and have a sufficient frequency at the same time too. It is because those discharges are really responsible for river channel shaping processes. So, the concept of dominant discharge (called, for the matter of simplicity, Qdd) was introduced, which could be defined as a discharge which determines the size of the river channel cross-section or a size of river channel pattern at the particular location, which also depends on the character and quantity of the sediment transported as well as the composition of river banks and bed material (Thomas and Goudie 2000). In the literature, one can find the dominant discharge (or sometimes the bed-forming flow) defined in many different ways. At the same time, in the literature one can find the term bankfull discharge. Bankfull (called, for the matter of simplicity, shortly Qb) very often has been assumed to be a singular discharge or critical channel forming flow which is important in determining the size and shape of the river channel and it has therefore be regarded as equivalent to Qdd (Thomas and Goudie 2000). Also it might be described as the maximum amount of water a specific channel can carry before water overflows the stream bank and causes flooding. However, as was pointed out by Radecki-Pawlik (2002), one should rather look for the value of Qb within the range of discharges, then select one value, which does not necessarily relate to channel shaping process within a particular river reach and even one cross-section. As mentioned by Williams (1978), bankfull is something more than just one-reference discharge, but rather it marks the conditions of incipient flooding. In this latter respect, bankfull is also of great interest to river engineers and catchment managers and planners (Nixon 1959). If so, in the present view of the Fifth European Water Directive, bankfull could be a useful tool which might be used by river engineers, but it is also becoming an important factor for ecologists and biologists since they are interested in river changes from the point of view of local ecosystems which are located within the corridor of a river. Finally, it has to be said that bankfull information is not including the flow which is very often running through the gravel bed of some alluvial rivers and it might be as large as up to 60 % of the total flow (Carling et al. 2006; Lach and Wyżga 2002). But it is another story which might be included in bankfull investigations later and it is not within the scope of the present chapter, although must be mentioned since many scientists and practitioners are rising up that issue, especially in countries with large scarcity of water.

When in the early 1990s I started my work at the Univeristy of Agriculture in Krakow after many years of being employed before as the hydraulic structures and river training designer in "Hydroprojekt" Designing Office in Krakow, the first task I met was to be acquainted with bankfull and dominant discharges. I must say that at that time only a few people had any information on that topic in Poland. So I started to dig deeply in the problem. I found very soon that the concepts of bankfull and dominant discharges are useful not only for scientists but first of all for practitioners, designers and engineers from river/stream management. It is also quite important for me, since very often, while working in the designing office before, I was just given one single discharge to determine the designing flood for establishing, for example, dimensions of engineering structures. It was the value of the discharge of a given probability. Just having only this number I was supposed to decide on dimensions of river channel trained or dimensions of check dams, bank revetments or river drop hydraulic structures. For me this information was not sufficient. I had a number but I did not "feel" the river. I had the number but I did not know anything about river channel processes occurring in the particular cross-section or along the river distance I was working with. Now it was a great discovery time: Bankfull (!); Dominant discharge (!). Two additional values showing me exactly what is going on in the river or stream cross-section in terms of fluvial processes. That is probably the answer why I needed, and still need, bankfull and dominant discharge value and why others may require to find those values and to know how to use and interpret them. But the problem, especially in Poland, is different. Many people do not want to know bankfull and dominant. One could ask simply: why? The calculation of bankfull or dominant is not so difficult and the advantage of having their values seems to be justified. So what? The answer why some do not need those values is pretty simple: since if you have those values, you need to firstly include river processes in your project and you have to know something about them. If you forget about dominant and bankfull you are just playing with discharges from hydrological calculations (based on probability) which, when used separately, very often overestimate the hydraulic structures dimensions. People simply do not want to waste time on that, because, although the calculation process of bankfull and dominant is maybe not very sophisticated, it needs time and quite a bit of knowledge about river forming processes. Many professional engineers need to see the river much deeper than through one number, they need to feel the river—beautifully about it writes Luna Leopold in "The view of the river" (2005)—the man I never met in my life but as far as rivers are concerned I owe him a lot. But at the end of the day, the consequences of such a way of thinking are rather sad and we can see them unfortunately along many rivers in my country. Anyway—I trust that a new way of thinking for river engineers is coming (for the last 23 years I am publishing in Polish journals about bankfull and dominant making the two concepts popular). Thus, the following paper presents some definitions of the dominant discharge (Qdd) and bankfull discharge (Qb) as well as some chosen methods of their determination showing how to deal with channel fluvial geomorphological processes knowing bankfull and dominant discharges.

20.2 Methodology

20.2.1 Dominant Discharge

It is believed that Schaffernak (1922) introduced the term "dominant discharge" firstly. He defined it as a discharge which, during a long period of time, transports the maximum amount of sediment. There are several other definitions of Qdd that appeared later in scientific papers: discharge which has the significant influence on forming the channel geometry (Debski 1967); the 1.58 year flow on the annual series used as a statistical definition of bank-full discharge (Durry 1973); the range of flows which, over a period of time, transports the most bed-load or bed material load (Marlette and Walker 1968); the effective discharge—the increment of discharge that transports the largest fraction of annual sediment load over a period of years (Andrews 1988); the most significant discharge influencing alluvial channel capacity through morphological adjustment which transported most bed sediment for its given frequency of occurrence (Wolman and Miller 1960); the constant flood over a period of time which has the same influence for a channel geometry such as the variable short-time discharge (the same scouring and accumulation) (Rzanicyn 1960); the discharge which transports most bed sediment in a stream that is close to steady-state conditions (Carling 1988). So, there are many different ways of defining and thus determining Qdd. The most important are listed below. Schaffernak (1922) and Raczyński (1977) suggested that having a frequency of discharges curve (f) as well as a sediment transport rate (T) we can find the products of these two curves. The maximum of the products is the dominant discharge:

$$Qdd = (fT)_{\max}$$

Lambor (1971) shows a similar formula but the transport rate is replaced with the rating curve and Qdd with the stage for a dominant discharge Hdd (Fig. 20.1):

$$Hdd = (fQ)_{\max}$$

Makkavieiev (1966) gives his own formula for calculating Qdd values for the rivers of Western Europe. He suggested that bankfull is a discharge which is responsible for the basic shape of a river:

$$Qdd = Q_{\min}(1 + K_f \frac{I_w}{I_n})$$

where: Q_m—mean low flood over a period of time (m^3/s), Q—mean maximum flood over a period of time (m^3/s), I_w—mean slope during flood over a period of time, I_n—mean slope during low flood over a period of time; K_f is a coefficient: $K_f = (\frac{K_n}{65})^3$ for Latochin factor L less than 1, $K_f = (\frac{K_n}{75})^3$ for Latochin factor L greater than 1; K_n is a coefficient defined as

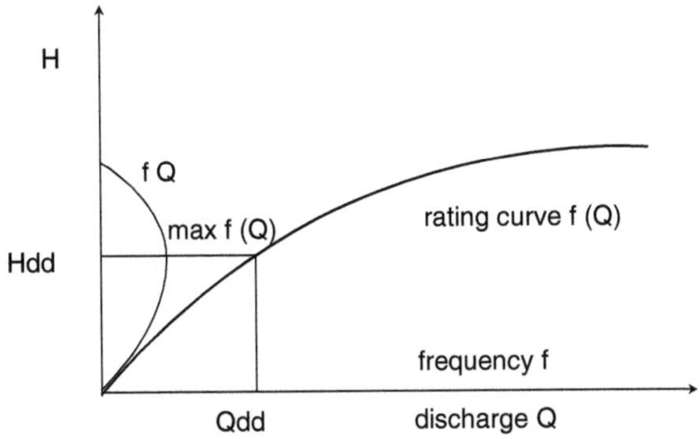

Fig. 20.1 Lambor's concept of determination of the dominant discharge Qdd

$$K_n = \frac{(Q - Q_m)\ 100}{Q + Q_m}$$

and the Latochin factor is $L = \frac{d}{i}$, where $d = d_{50}$ (the 50-percent diameter of bedload from a grain-size curve) and i the mean slope. Rzanicyn (1960) treats the flood-time as the most important for the channel forming process. His graphic method uses the flood hydrograph, the intensity-channel-forming curve and the cumulative intensity-channel-forming curve. Having the equation to calculate the intensity coefficient as well as the graphic example, we are able to find out Qdd according to Rzanicyn:

$$S = \frac{hi}{d}\left(\frac{v_1^3}{v_2^3} - 1\right)$$

where: h—mean depth of a cross-section (m), S—mean slope between two cross-sections, $d = d_{50}$ is a grain size value, v_1—mean velocity upstream (m/s), v_2—mean velocity downstream (m/s) (Fig. 20.2).

Schoklitsch (1952), see also Mordziński (1972) is looking for dominant discharge value using his own function for sediment transport (this is the empirical relation merely between values given there but not the typical equation):

$$T = f\left(\frac{7000}{d^{0.5}}\, i^{1.5}(Q - Q_0)\right)$$

where: $d = d_{50}$, i—mean slope, Q—discharge (m^3/s), Q_0—discharge for initial the motion of sediment (m^3/s).

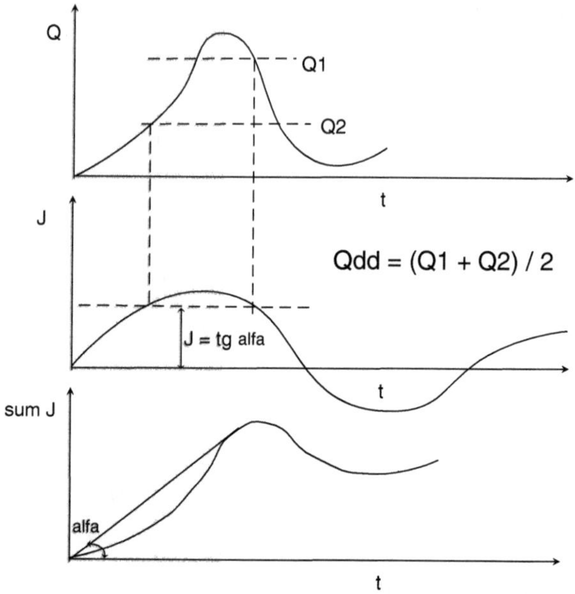

Fig. 20.2 Rzanicyn's concept of determination of the dominant discharge Qdd

The author states that the dominant discharge stage is the maximum of two products:

$$Hdd = f(Q - Q_0)_{\max}$$

where f is the frequency of discharges in days.

Wierzbicki (1965) sees the dominant discharge connected with the mean annual flow:

$$Qdd = (1.03 - 1.07)\ Q_{mean}$$

Finally, the Wolman and Miller (1960) concept of finding out the dominant discharge emphasized that the most significant discharge influencing alluvial channel capacity through morphological adjustment was the one which transported most of bed sediment for its given frequency of occurrence (Fig. 20.3).

In Polish literature, the concept of dominant discharge is described by Mordziński (1972, 1986) who is giving also case studies of its determination. Also Radecki-Pawlik (2002) put together many definitions of Qdd, trying to analyse some of them.

For clear interpretation and understanding as well as for better practical use, the above information on dominant discharge is gathered in Table 20.1.

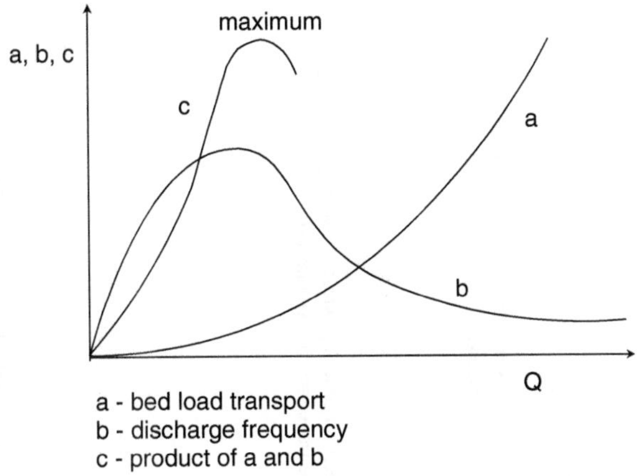

Fig. 20.3 Wolman and Miller's concept of determination of the dominant discharge Qdd

20.2.2 Bankfull

To identify the bankfull line in the field before the assessment of its value one needs to know what means 'floodplain'. With river morphology in the field, usually as the concave bank recedes due to erosion and the point bar builds outward from the concave bank, the river/stream channel width remains the same. The progressive growth of point bars forms a flat surface called the floodplain. This floodplain is built primarily by point bar extension—and in some instances deposition by overbank adds to the floodplain level (Leopold 2005). Thus, knowing what is floodplain—it has to be said that the level of it is the elevation of the top of river channel banks. When the channel is flowing full, the water surface is at floodplain level and the flow rate is the bankfull discharge. There are many papers in which bankfull is especially mentioned (Lenzi et al. 2006; Wu et al. 2008; Navratil et al. 2006; Adams and Spotila 2005; Wohl et al. 2010; Carpenter and Taylor 2007; Pike and Scatena 2010; Navratil and Albert 2010; Radecki-Pawlik and Skalski 2008a, b; Skalski et al. 2012; Jiménez and Wohl 2011; Shamir et al. 2013; Soar and Thorne 2011; Agouridis et al. 2011; Pitlick et al. 2013; Frandofer and Lehotský 2013), but in the present paper the concentration is focused on the most classical ways of viewing it.

Thus, there are two main ways of how authors define bankfull. The first group of definitions describes bankfull in terms of the geometry of a cross-section; the second group of definitions describes it as a bankfull discharge in terms of volume of water. Within the first group one can find methods which are associated with pure morphometry, and here we would find the Wolman method and the Riley method (the so-called morphometric methods), and one can find methods associated with biological findings in the cross section, and it would be the Woodyer method and Radecki-Pawlik and Skalski method (the so-called biological methods).

Table 20.1 Dominant discharge (Q_{dd}) determination methods

Author	Equation and/or determination method	Comments
1	2	3
Schoklitsch (1952)	$G = \frac{7000}{d^{0,5}} \cdot J^{1,5}(Q - Q_0)$ [kg/sek] where: d—grain diameter D_{50} (mm), J—average inclination of water table [0,5], Q—flow rate at particular water stage (m^3/s) Q_0—limit flow at which dragging begins, calculated from the formula given by: Peter: $Q_0 = 0.0592 \frac{d^{3/2}}{J^{3/2}}$ $[m^3/sek/m]$, Favre: $Q_o = 0.0683 \frac{d^{3/2}}{J^{3/2}}$ $[m^3/sek/m]$	Q_{dd} is calculated as a maximal value of $(Q - Q_0)$ n, where n daily frequency
Lambor (1971)	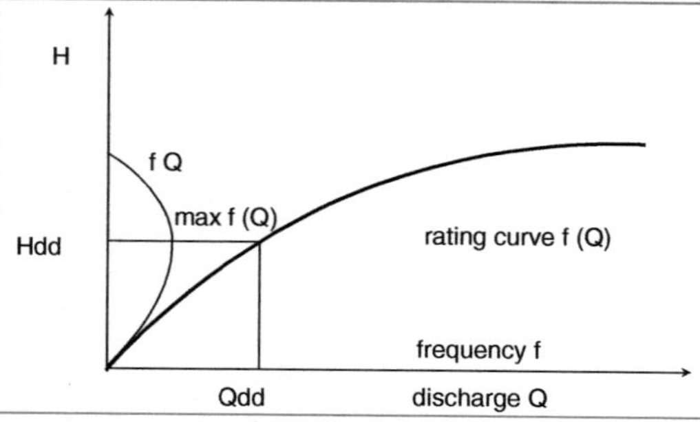	Q_{dd} is determined as a product of discharge value, Q, and discharge frequency (maximum QT)
Makka-viejev (1955)	$Q_f = Q_m\left(1 + K_f \frac{I_w}{I_n}\right)^2$ $[m^3/sek]$ where: Q_f—discharge corresponding with Q_{dd} [m^3/s], Q_m—mean minimum multiannual discharge (m^3/s), I_w—average multiannual inclination along given river stretch at high water stages (–), I_n—average multiannual fall along given river stretch at low water stages, K_f—coefficient calculated from Lochtin's formula: Łochtin's: $K_f = \left(\frac{K_N}{65}\right)^3$, $K_f = \left(\frac{K_N}{75}\right)^3$ (–) where: K_N—coefficient of non-uniformity calculated from: $K_N = \frac{(Q - Q_m)100}{Q + Q_m}$ (-)	Equation valid for the East European Plain

(continued)

Table 20.1 (continued)

Author	Equation and/or determination method	Comments
van Berghholm (1969)	$K = \frac{nGp}{t + \Delta t}$ where: $K = Q_{dd}$ (m^3/s), *t*—depth (cm), Δt—depth change in a particular time interval *n*—discharge duration (days), *G*—dragging intensity (cm^3/s), *p*—coefficient calculated from: $p = 1 + \left(\frac{0.25}{G(bd)^{3/2}}\right)$ where: *b*—width of water-table (m), *d*—grain diameter corresponding with grain-size distribution d_{50} (mm)	Known as the "Dutch method"
Rzanicyn (1960)	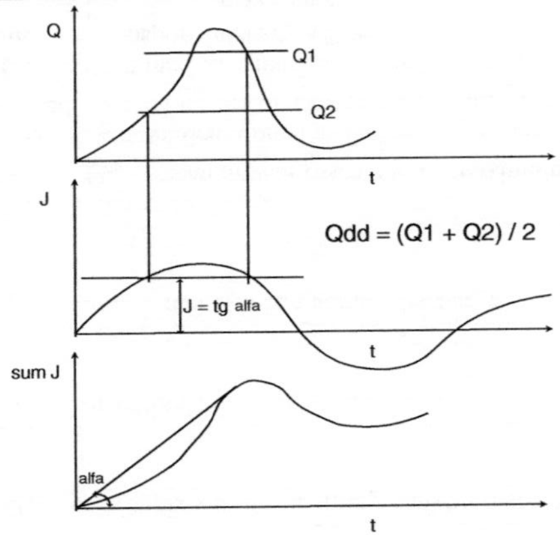 $J = \frac{hi}{g}\left(\frac{V_1^3}{V_2^3} - 1\right)$ where: *J*—intensity index of bed-forming process (–), *h*—average depth of river-bed (m) *i*—adjusted longitude fall along a given stretch (%), *d*—grain diameter corresponding with grain-size distribution d_{50} (m), V_1—average velocity in upper part of water stretch (m/s), V_2—average velocity in lower part of water stretch (m/s)	Q_{dd}—taken from the diagram given below, representing discharge, *Q*, versus its duration, *t*
Bray (1972)	$Q_{dd} = Q_{50\ \%}$ (2-years water)	Discharge analysis of 71 gravel-bed

(continued)

Table 20.1 (continued)

Author	Equation and/or determination method	Comments
Pickap and Warner (1976)	Q_{dd}—corresponds to value of discharge occurring every 1.15–1.40 year, on average; equivalent to $Q_{75\ \%}$	Discharge analysis of streams belonging to the Cumberland river basin
Mordziński (1986)	Q_{dd}—corresponds to 0.92–1.14 of mean annual discharge	Discharge of 3 rivers in control in Poland (Nida, Prosna and Pilica) have been studied
Laval-Makkaviejev (1949)	Q_{dd}—corresponds to mean spring bankful discharge	Simplified method
Głuszkow-Antropovskij (1969)	Q_{dd}—corresponds to bankfull discharge	Simplified method
Romaszin-Antropovskij (1969)	Q_{dd}—corresponds to mean annual discharge	Simplified method
Wierzbicki (1965)	$Q_{dd} = (1.03 - 1.07)\ Q_{śr}$ where: $Q_{śr}$—mean annual discharge of maximum waters	Simplified method
Marlette and Walker (1968)	Q_{dd}—discharge determined as follows: at a stage higher than this discharge, transport of bed material occurs and more than a half of the transported material is displaced	Simplified method
Wolman and Miller (1960)	Q_{dd}—determined as a maximal value of the product of discharge value and bed material transport intensity value 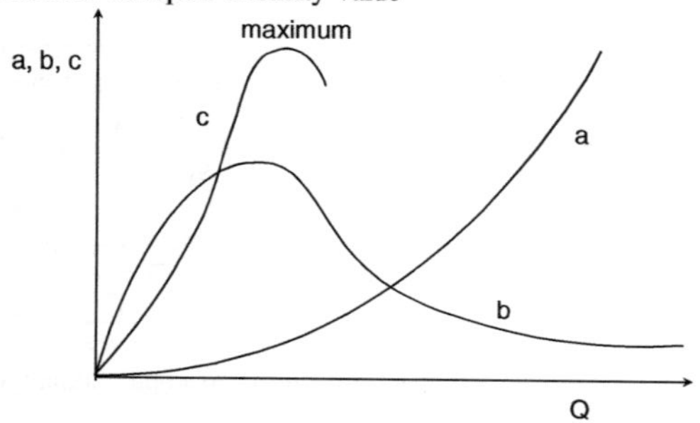 a—bed load transport b—discharge frequency c—product of *a* and *b*	Method tested during the analysis of river and stream discharges in the Polish Carpathian rivers, ex.: the Raba

20.2.2.1 Wolman Method

A classical example of the first group of methods describing bankfull in terms of the geometry of a cross-section is the Wolman method. The Wolman method consists of plotting at a-station hydraulic geometry relations between discharge Q and such flow features as cross-sectional area A, water surface width W, and mean depth D (=A/W) at the cross-section of interest. Bankfull width and/or area are determined from field measurements, and bankfull discharge is read from the appropriate hydraulic geometry graph. Alternatively, a channel cross-sectional survey can provide values of A, W, and D with stage. Stage is plotted against any one of A, W, or D (or W/D), and a break in the trend of plotted points usually identifies the bankfull level (Wolman 1955; Carling 1988). Wolman suggests that the bankfull stage is a stage at which the ratio of channel width to channel depth is at a minimum. From a set of widths for given depths (Fig. 20.4) at each cross-section, the minimum value of a ratio is:

$$R = \frac{W(i)}{D(i)}$$

20.2.2.2 Riley Method

Riley method also classifies according to morphometric methods and consists of plotting at-a-station hydraulic geometry relations between discharge Q and such flow features as cross-sectional area A, water surface width W, and mean depth D (=A/W) at the cross-section of interest. In 1972 Riley proposed the concept of a bench index BI:

$$BI = \frac{W(i) - W(i+1)}{D(i) - D(i+1)}$$

where W and D are defined above and i = 1, 2, 3…(n − 1)-th measurements.

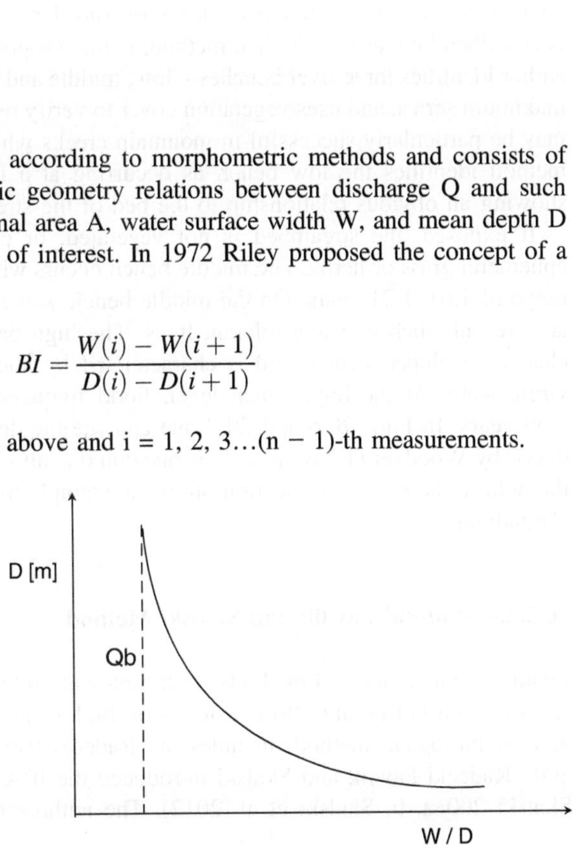

Fig. 20.4 The sketch for Wolman's method of bankfull determination

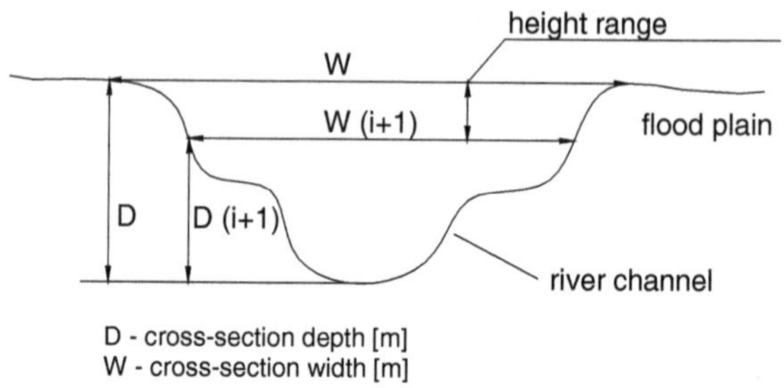

Fig. 20.5 Sketch for Riley's method of bankfull determination

The bench index BI (Fig. 20.5) plotted against depth D for decreasing values of depth within a channel shows a marked peak value near the actual bankfull stage (Riley 1972).

20.2.2.3 Woodyer Method

The next example of the first group of methods describing bankfull in terms of the geometry of a cross-section also takes into consideration biological findings, and so is described here as a biological method, is that proposed by Woodyer (1968). The author identifies three river benches—low, middle and high—in terms of the annual maximum series, and uses vegetation cover to verify results (Fig. 20.6). His method may be particularly successful in mountain creeks where plants are abundant. The method identifies the low bench as occurring at a low stage only, and usually showing an obvious relationship to the bed of the stream.

If exposed, the streambed is not vegetated, or carries only a thin cover of ephemeral grass or herbs. The middle bench occurs where flood frequency is in the range of 1.01–1.21 years. On the middle bench, some larger species of vegetation are present, such as water-tolerant trees. The high bench is the widest and most clearly developed bench, and is characterised by abundant tree cover in a near-virgin state. At the high bench level, flood frequency varies between 1.24 and 2.69 years. In Figs. 20.6 and 20.7 one can see the details of three benches introduced by Woodyer (1968) as well as one could analyse the detailed plants found in the field in the river cross section on given example from one of the streams in the Carpathians.

20.2.2.4 Radecki-Pawlik and Skalski Method

Finally, the first group of methods describing bankfull in terms of the geometry of a cross-section but taking into consideration biological findings, so again described here as biological method, includes the Radecki-Pawlik and Skalski method. In 2008 Radecki-Pawlik and Skalski introduced the IBA index (Radecki-Pawlik and Skalski 2008a, b; Skalski et al. 2012). The authors postulate that is possible to

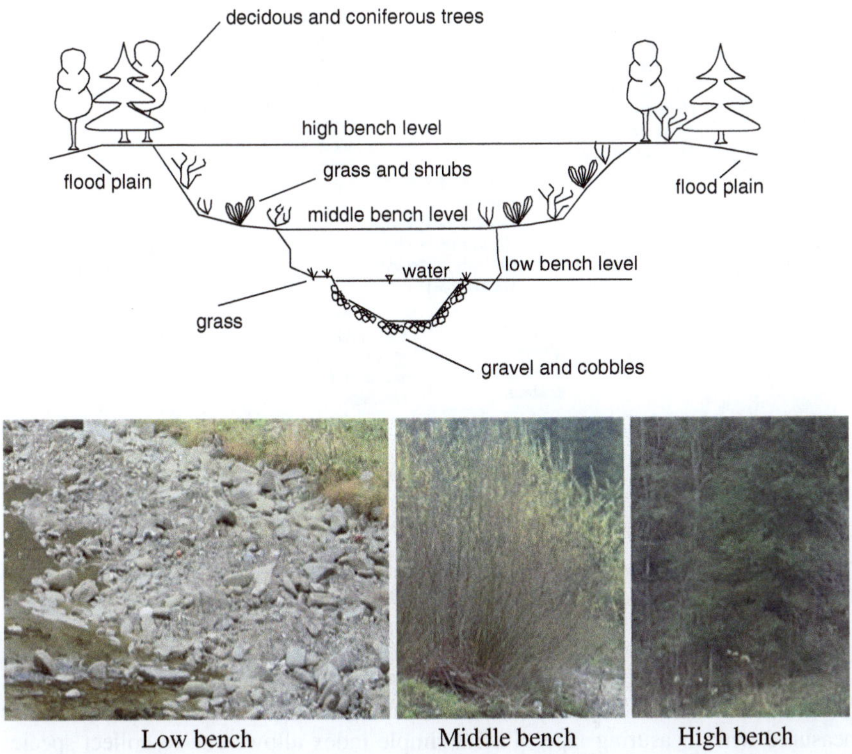

Fig. 20.6 Sketch for Woodyer's method of bankfull determination (*Photo* A. Radecki-Pawlik)

predict the value of the bankfull discharge knowing the proportions of dimensions of dominant species in a given bench of the river cross-section. To calculate the bankfull level, the authors introduced the so-called Invertebrate Bankfull Method (IBA).

They hypothesized that the distribution of particular size group in the riverine invertebrate community will change with frequency of floods. Since it is known the size of most central European Riverine ground beetles varies between 1 and 35 mm (Skalski et al. 2012), thus in more disturbed habitats the proportion of the smallest toward medium sized beetles will be good predictor and can be applied as an indicator of habitat changes (flood frequency). In this manner it is proposed that the formula of bankfull assessment can be tested with the particular example of the Ochotnica River research cross-section in the Polish Carpathians:

$$\textbf{\textit{IBA index}} = \textbf{\textit{abundance of}}\ 1\text{–}6\,\text{mm}\ \textbf{\textit{specimens}}/\textbf{\textit{abundance of}}\ 18\text{–}24\,\text{mm}\ \textbf{\textit{specimens}}$$

In that way the authors proposed (similar to Woodyer method) to introduce three benches: high, with IBA < 1; middle, with IBA ≥ 1; and low, where IBA $\gg 1$.

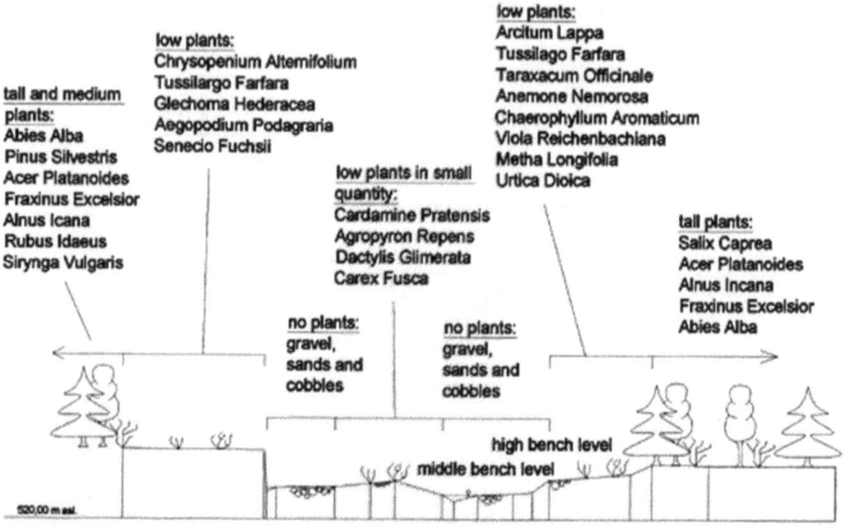

Fig. 20.7 The example sketch for Woodyer's method of bankfull determination from the Skawica river cross-section (modified from Radecki-Pawlik 2002)

The animals (riverine ground beetles) are trapped in Berber traps (Fig. 20.8; Photo 20.1a, b)—(Radecki-Pawlik and Skalski 2008a, b), and their bodies are measured with measuring tape. Such a simple index allows one to collect species within any cross-section, measure them and classify into size groups. Next, having the value of the index and the level/altitude of a particular class/community occurrence and also at the same time having the level of the calculated t-year flood for the particular cross section, a bankfull level could be easy to asses from the rating curve. This method seems to be in very good accordance with the Water Framework Directive of EU philosophy where one can see that conditions of streams and rivers depend mostly on fauna living there as well as hydromorphology. In this context, the use of this method is very important, giving river engineers and catchment managers the real view of living river cross-section. Also when planning any hydraulic structures this method might be used as a reference method indicating the real changes of fauna abundance and exact habitat in the river reach or cross-section which would influence the dimensions of hydraulics structures, especially in terms of their height (altitude).

The second group of methods defines bankfull as a bankfull discharge in terms of the volume of water. The best known are here the Gauckler-Manning method (also known as Schumm and Pickup and Warner method) and Williams's method.

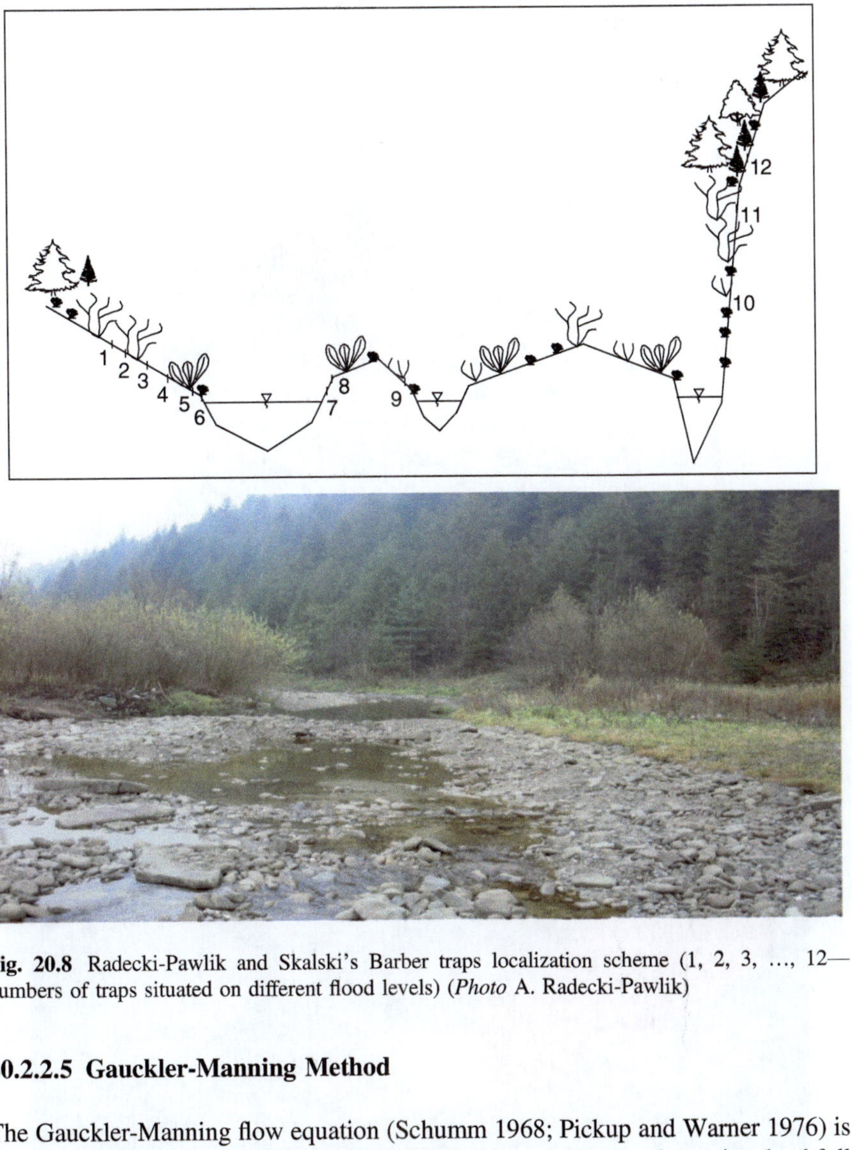

Fig. 20.8 Radecki-Pawlik and Skalski's Barber traps localization scheme (1, 2, 3, ..., 12—numbers of traps situated on different flood levels) (*Photo* A. Radecki-Pawlik)

20.2.2.5 Gauckler-Manning Method

The Gauckler-Manning flow equation (Schumm 1968; Pickup and Warner 1976) is one of several equations to be used directly at a station to determine bankfull discharge. The equation as presented here is:

$$Q_b = (1.0/n)A_b D_b^{2/3} S^{1/2}$$

The required variables are therefore the resistance coefficient n (Te Chow 1959), the bankfull flow area A_b, the bank-full depth D_b, and the water surface slope S. Investigators obtain the resistance coefficient either by estimating bankfull conditions, or by actually measuring a lower discharge, computing the coefficient, and

Photo 20.1 **a** Installation of Barber traps for IBA Radecki-Pawlik and Skalski bankfull method (2008a, b)—the river bank, bench 1 (*Photo* A. Radecki-Pawlik). **b** Installation of Barber trap for IBA Radecki-Pawlik and Skalski bankfull method (2008a, b)—the trap (*Photo* A. Radecki-Pawlik)

assuming that 'n' for the same coefficient applies at bankfull stage. Bankfull geometric properties are determined from a further measurement of cross-sections. When one cross-section is used, the resulting Q_b applies to that section only. Q_b for a reach can be obtained by using several cross-sections with the slope-area method. In the field, the slope of the energy gradient is often taken as the bed or water surface slope. This slope should be measured in the field, but is sometimes obtained from a topographic map (Williams 1978).

20.2.2.6 Williams Method

The second well known method which defines bankfull as discharge in terms of volume of water is the Williams method. Williams (1978) proposed an equation for bankfull, after investigating fifty-one different rivers. Williams' final regression relationship was (warning—this is not a typical mathematical equation, but empirical relationship merely between A and S to give Q using observed data):

$$Q_b = 4.0\, A_b^{1.21} S^{0.28}$$

where Q_b is in cubic meters per second, A_b is in square meters, and S is dimensionless.

After all what was said above about floodplain and bankfull one can notice that in order to determine a bankfull, some bankfull indicators are very often needed. And again—to identify the bankfull line in the field, before calculating or/and assessing its value with the above-mentioned methods we need to find principal indicators in the field. These are: (1) Point bars: the point bar is the sloping surface that extends into the river channel from the convex bank of a curve. The top of the point bar is at the level of the floodplain. (2) Bank slope: In stream channels with natural riparian areas and a low, flat floodplain, the bankfull edge is located at the edge of this plain. Often the floodplain will slope down very gradually and then more abruptly. This abrupt slope-break is usually a good indicator. (3) Topographic break: There is usually a topographic break at the bankfull. The stream bank may change from a sloping bar to a vertical bank. It may change from vertical bank to horizontal plane on top of floodplain. The change of topography may be as subtle as a change in slope of the bank. (4) Vegetation: The bankfull edge is often indicated by a demarcation line between lower areas that are either bare or have aquatic and annual vegetation, and higher areas with perennial vegetation such as ferns, shrubs, and trees. Therefore, one has to look at vegetation. (5) Animals—river ground beetles: The bankfull edge is often indicated by a demarcation line between lower areas with small ground beetles and higher areas with bigger specimens, so the IBA index is useful. (6) Point bars and bank undercuts: Often on the inside of meander bends, the stream will build up a bar of sediment from the eddy current created by the bend; the top of such a bar is the minimum height of bankfull. Similarly, on the outside of such bends, the stream will often undercut the bank and expose root mats. If one could reach up beneath this mat, he can estimate the upper extent of the undercut. This would also be the minimum height of bankfull. (7) Lines on boulders/bedrock: If one is in a steep with no clear floodplain, he has to look for the highest mineral-stain line or the lowest line of lichen or moss on stable rock. All those indicators might help to identify the bankfull line in the field. (8) Material size: Bankfull is often registered by a change in the size distribution of materials at the surface, from fine gravel to cobbles or from sand to gravel. It can change from fine to coarse or from coarse to fine, but the change is common. (9) Debris deposits: One can notice a subtle change in the debris deposited between rocks, such as the amount of leaves, needles, seeds or other organic debris. Such indicators might be the confirmation of a bankfull.

Finally in this section it is worthy to mention two more things about the bankfull to help in its identification. The first is that all investigations from the field show that bankfull discharge has a recurrence interval in the range of 1.0–2.5 years (Leopold 2005). In that context, the value of 1.5-year appears to be a quite reasonable average for bankfull. The second thing is that it is worthy of note that river channel width at bankfull is closely related to bankfull discharge over a wide range of channels and regional climates (also in Leopold 2005). Therefore, it might be worthy to find the relation of channel width to discharge and analyse it with other data from different rivers in the plotted diagrams. At the same time, in the field the measurement of width is not subject to large errors even if the elevation of the bankfull condition is not correct. The width is not much different whether the elevation chosen is bankfull or somewhat below bankfull. Depth, in contrast, is very sensitive to an exact choice of bankfull elevation.

20.3 Case Study

Instead of a typical summary, I would like to present, avoiding the already given thoughts, the importance of using different methods to obtain a bankfull. In Table 20.2, there are presented the results of bankfull assessment for one of the Carpathian rivers, the Ochotnica River in the Polish Carpathians, to give a view on the methodology and outputs. The bankfull discharge values listed in Table 20.2 were obtained using different morphometric and biological methods.

When briefly discussing the obtained values of bankfull discharge for the considered research cross-section, as it was postulated by Radecki-Pawlik (2002), one should expect it to lie just between values of bankfull discharges which were found using many methods—in the present case by abiotic and biotic methods. At the same time, there is a need to have just one value of bankfull to which other values of discharges can be referred to the particular river cross-section. Here, in the particular case of the Ochotnica River research cross section (Table 20.2), there is a need to exclude from consideration some results of bankfull obtained from some of the employed methods. Firstly, rejected are the smallest values of the Woodyer method since those values are too small to be responsible for any river morphology changes in the context of t-year values as well as because the shear stresses which can be

Table 20.2 Bankfull values for the Ochotnica River obtained using different morphometric and biological methods (modified from Radecki-Pawlik and Skalski 2008a, b)

Bankfull discharge values in a cross-section within the investigated research reach, Q_b ($m^3\ s^{-1}$) according to different methods of calculation					
Wolman	Riley	The Gauckler-Manning (Schumm, Brown and Warner)	Williams	Woodyer	IBA method (Invertebrate Bankfull Assessment method)
Minimum W/D index 4.10	First index value 8.88 local index value 4.10	(Manning's n = 0.02-0.03 range) n = 0.02 minimum 23.37 n = 0.025 average 18.67 n = 0.03 maximum 15.58	Williams' equation 19.94	High bench 40.47 Middle bench 11.10 Low bench 3.20	Bench 3 $I > 1$ 38.40 Bench 2 $I \leq 1$ 11.10 − 38.40 central value 24.75 Bench 1 $I \ll 1$ 4.10 − 11.10 central value 7.60

predicted under such discharges are not big enough to initiate an entrainment. Also excluded are the local maximum Riley and Wolman numbers as they seem to be good indicators for bankfull within only one-braid channel but when dealing with a multi-braided stream cross-section (which we have in the case of the Ochotnica) the values of discharge obtained here are too small. Thus to find the range of discharges where the expected bankfull value should lie, the first Woodyer terrace value, the average value of Schumm, Brown and Warner number (Manning-Gaucler method), the first index of Riley value, the Williams number and the middle bench value of the Woodyer biotic method (the second bench value of Woodyer) are considered. Finally, bench 2 was considered with resepct to the IBA method with its central value of other discharges where one could expect to find bankfull. So, calculated are the central values of discharge within all ranges of bankfull discharges mentioned above, which were referred to different abiotic and biotic methods and next their arithmetic average is calculated, which is $Q = 18.4\ m^3\ s^{-1}$. That value is termed the bankfull value to which one could refer as to one number $Q_b = 18.4\ m^3\ s^{-1}$. This number is bigger than $Q_{50} = 11.3\ m^3\ s^{-1}$ (the two-year flood) for that cross section but close to $Q_{25} = 21.2\ m^3\ s^{-1}$, which follows the general statements of bankfull definition given by some authors, for example Pickup and Warner (1976). Also, the obtained $Q_b = 18.4\ m^3\ s^{-1}$ is very close to $Q = 18.6\ m^3\ s^{-1}$ which is the Shumm, Brown and Warner bankfull value for n = 0.025, which also seems to be the best described roughness coefficient of the considered Ochotnica River cross-section (from field observations). Finally the obtained $Q_b = 18.4\ m^3\ s^{-1}$ confirms the present hydrological situation in that part of the Polish Carpathians where big floods were experienced in 1997, 2000 and 2003 and when were experienced such values of discharges in the Ochotnica River. They were marked in the field as nearly or just above the bankfull but still conforming to the channel shape. The discharge $Q_{20} = 40.0\ m^3\ s^{-1}$ was noted at the Tylmanowa gauge station for the area of the catchment $A = 108\ km^2$ as a flood noted on the Ochotnica River which is an equivalent of $Q_{20} = 23.9\ m^3\ s^{-1}$ for the research cross section described in the present paper (calculated using hydrological analogous Fall method (Radecki-Pawlik 1995). To recapitulate: the value of bankfull $Q_b = 18.4\ m^3\ s^{-1}$ calculated within the research cross section employing biotic and abiotic methods seems to be the best for recommendation to any planners, managers and developers but first of all to river engineers who would like to undertake any training works within the considered cross-section of the Ochotnica. This value turns out to be a consensus number when we employed abiotic and biotic methods to calculate bankfull. It looks reliable from the point of view of hydrology and geomorphology but also it covers the biological requirements. It also is very important because it gives understanding of river channel forming processes here and is in line with main thought of River Framework Directive of EU where including fauna factor to hydromorphological parameters of a particular river/stream cross-section is of prior importance. In my opinion it is most of all the prior importance for our world rivers.

Acknowledgments I would like to give my greatest thanks to Prof. Paul Carling for his valuable comments to that paper encouraging me to publish it, as well as for editing the text of this publication. Also, having the chance to write it here, I would like to say thank you to Prof. Carling for many years of cooperation, for his great knowledge thanks to which I could better understand rivers and for his personality which always was towards help and brilliant pieces of advice.

References

Adams RK, Spotila JA (2005) The form and function of headwater streams based on field and modeling investigations in the Southern Appalachian Mountains. Earth Surf Proc Land 30 (12):1521–1546

Agouridis C, Brockman R, Workman S, Ormsbee L (2011) Bankfull hydraulic geometry relationships for the inner and outer bluegrass regions of Kentucky. Water 2011(3):923–948

Andrews E (1988) Effective and bankfull discharges of streams in the Yampa River Basin, Colorado and Wyoming. J Hydrol 46

Antropovskij WI (1969) Gidromorfologiczieskije zavisimisti i ich dalniejszeje rozviatje. Trudy GGI, vyp. 169,Gidrometeorozat

Berdenis van Berlekom AH (1969) The role of rivers to mankind. Conference: selected problems from thetheory from simulation of hydrodynamic phenomena. Jablonna, June, 1969.

Bray DI (1972) Generalised regime-type analysis of Alberta Rivers. PhD thesis, University of Alberta, Edmonton,Canada.

Carling P (1988) The concept of dominant discharge applied to two gravel-bed streams in relation to channel stability thresholds. Earth Surf Proc Land 13:355–367

Carling P, Whitcombe L, Benson I, Hankin B, Radecki-Pawlik A (2006) A new method to determine interstitial flow patterns in flume studies of sub-aqueous gravel bedforms such as fish nests. River Res Appl 22(6):691–701

Carpenter TM, Taylor SV (2007) Surveying flash flood response in mountain streams. EOS Trans Am Geophys Union 88(66):69–80

Dębski K (1967) Regulacja rzek—cz.1 (River training, part 1) PWN, Warsaw
Dury G (1973) Magnitude frequency analysis and channel morphometry. In: Morisawa M (ed) Fluvial geomorphology. State University of Binghamton, pp 99–121
Frandofer F, Lehotský M (2013) Stupňovitá štruktúra dna koryta vodného toku s výrazným pozdĺžnym sklonom (na príklade horného toku rieky topľa). Geografický časopis (Geogr J) 65 (2):141–159
Jiménez MA, Wohl E (2011) Solute transport modeling using morphological parameters of step-pool reaches. Water Resour Res 49:1345–13459
Lach J, Wyżga B (2002) Channel incision and flow increase of the upper Wisłoka River, southern Poland, subsequent to the reafforestation of its catchment. Earth Surf Proc Land 27(4):445–462
Lambor J (1971) Hydrologia inżynierska (Engineering hydrology). PWN, Warsaw
Lenzi MA, Mao L, Comiti F (2006) Effective discharge for sediment transport in a mountain river: computational approaches and geomorphic effectiveness. J Hydrol 326:257–276
Leopold LB (2005) A view of the river. Harvard University Press, Cambridge, p 298
Makkavieiev M (1966) Ruslovoj rezim riek i trassirowanie prorezej. Ak. Nauk. ZSRR, Moscow
Marlette R, Walker R (1968) Dominant discharges at the Plate-Missouri Confluence. J Waterways Harb Divis 94(1):23–32
Mordzinski S (1972) Przepływ kształtujący koryto i metody jego wyznaczania. Wiad Inst Melior i Użytków Zielonych X(4)
Mordziński S (1986) Wyznacznie przepływu kształtującego koryto na podstawie transportu rumowiska wleczonego. Gosp Wodna 9
Navratil O, Albert MB (2010) Non-linearity of reach hydraulic geometry relations. J Hydrol 388:280–290
Navratil O, Albert MB, Herouin E (2006) Determination of bankfull discharge magnitude and frequency: comparison of methods on 16 gravel-bed river reaches. Earth Surf Proc Land 31:1345–1363
Nixon MA (1959) Study on the bank-full discharges of rivers in England and Wales. Proc Inst Civil Eng 12:157–174
Pickup G, Warner RF (1976) Effects of hydrologic regime on magnitude and frequency of dominant discharge. J Hydrol 29:51–75
Pike AS, Scatena FN (2010) Riparian indicators of flow frequency in a tropical montane stream network. J Hydrol 382:72–87
Pitlick J, Marr J, Pizzuto J (2013) Width adjustment in experimental gravel-bed channels in response to overbank flows. J Geophys Res Earth Surf 118:553–570
Raczyński K. 1977. Projektowanie regulacji rzek i potoków górskich z uwzględnieniem ruchu rumowiska. Gosp Wodna 9
Radecki-Pawlik A (1995) WODA-v.2.0, a simple hydrological computer model to calculate the t-year flood. In: Wiezik B (ed) Hydrological processes in the catchment. University of Technology, Cracow, pp 131–141
Radecki-Pawlik A (2002) Bankfull discharge in mountain streams: theory and practice. Earth Surf Proc Land 27:115–123
Radecki-Pawlik A, Skalski T (2008) A new concept to determinate bankfull discharge using invertebrate communities—an example from the Ochotnica stream, Polish Carpathians. Electron J Pol Agric Univ (EJPAU) 11:1–13, part 1
Radecki-Pawlik A, Skalski T (2008b) Bankfull discharge determination using the new invertebrate bankfull assessment method. J Water Land Dev 12:145–154
Riley SJ (1972) A comparison of morphometric measures of bankfull. J Hydrol 17:23–31
Rzanicyn A (1960) Morfologiczskije i gidrologiczskije zakonomiernosti stroczenia riecznoj sieti. Ak. Nauk ZSRR, Lenningrad
Schaffernak F (1922) Neue grundlagen für die berechnung der geschiebefuhrung in fluszlaufen. Leipzig and Wien, Franc Deuticke
Schoklitsch A (1952) Der Grundbau: Handbuch fur Studium und Praxis, 2nd edn. Springer, Wien
Schumm SA (1968) River adjustment to altered hydrologic regimen. U.S. geological survey professional paper, vol 598, 65p

Shamir E, Ben-Moshe L, Ronen A (2013) Geomorphology-based index for detecting minimal flood stages in arid alluvial streams. Hydrol Earth Syst Sci 17:1021–1034

Skalski T, Kędzior R, Radecki-Pawlik A (2012) Riverine ground beetles as indicators of inundation frequency of mountain stream: a case study of the Ochotnica Stream, Southern Poland. Baltic J Coleopterol 12(2):117–126

Soar PJ, Thorne CR (2011) Design discharge for river restoration. Stream restoration in dynamic fluvial systems: scientific approaches, analyses, and tools geophysical monograph series 194, Copyright 2011 by the American Geophysical Union

Te Chow V (1959) Open-channel hydraulics. McGraw-Hill, New York, pp 108–114

Thomas DSG, Goudie A (2000) A dictionary of physical geography, 3rd edn. Blackwell Publishers Ltd, Malden

Wierzbicki J (1965) Elementy trasy regulacyjnej w świetle badań układu poziomego koryt rzecznych. Doctor's thesis, Warsaw University of Technology

Williams PG (1978) Bankfull discharge of rivers. Water Resour Res 14(6):1141–1154

Wohl E, Cenderelli DA, Dwire KA (2010) Large in stream wood studies: a call for common metrics. Earth Surf Proc Land 35:618–625

Wolman MG (1955) The natural channel of Brandywine Creek, Pensylvania. U.S. geological survey professional paper, vol 282, pp 86–109

Wolman M, Miller J (1960) Magnitude and frequency of forces in geomorphic process. J Geol 68, 54–74

Woodyer KD (1968) Bankfull frequency in rivers. J Hydrol 6:114–142

Wu B, Wang G, Xia J, Fu X, Zhang Y (2008) Response of bankfull discharge to discharge and sediment load in the Lower Yellow River. Geomorphology 100(3–4):366–376

Chapter 21
Hydrodynamics of Vegetated Channels

Jochen Aberle and Juha Järvelä

Abstract Hydrodynamics of vegetated channels and streams is a rapidly developing research area, and this chapter summarizes the current knowledge considering both aquatic and riparian zones. The benefit of an advanced parameterization of plant morphology and biomechanical properties is highlighted. For this purpose, the response of flexible and foliated plants and plant communities to the flow is illustrated, and advanced models for the determination of drag forces of flexible plants are described. Hydrodynamic processes governing flow patterns in vegetated flows are presented for submerged and emergent conditions considering spatial scales ranging from the leaf to the vegetated reach scale.

Keywords Rivers and floodplains · Aquatic vegetation · Riparian vegetation · Hydrodynamics · Flow field · Flow resistance · Modelling · Parameterization

21.1 Introduction

Vegetation represents a ubiquitous feature in river systems affecting many physical, chemical, and biological processes. In addition to the flow boundary conditions governed by the geomorphic setting of streams and channels, vegetation interacts with the flow thereby affecting the turbulent flow field, hydraulic resistance, and bed shear stress (e.g., Järvelä 2004; Nikora 2010a; Folkard 2011; Neary et al. 2012; Nepf 2012a; Aberle and Järvelä 2013) and consequently backwater profiles, sediment transport, and channel morphology (e.g., Yen 2002; Osterkamp et al. 2012; Yager and Schmeeckle 2013; Gurnell 2014). Moreover, flow-vegetation interaction

J. Aberle (✉)
Norwegian University of Science and Technology, Trondheim, Norway
e-mail: jochen.aberle@ntnu.no

J. Järvelä
Aalto University School of Engineering, Espoo, Finland
e-mail: juha.jarvela@aalto.fi

P. Rowiński and A. Radecki-Pawlik (eds.), *Rivers—Physical, Fluvial and Environmental Processes*, GeoPlanet: Earth and Planetary Sciences,
DOI 10.1007/978-3-319-17719-9_21

affects solute and particulate budgets (e.g., Shucksmith et al. 2010) as well as the shapes, sizes and spatial distributions of vegetated patches and mosaics and hence the habitat structure (e.g., Gurnell and Petts 2006; Gurnell et al. 2006). As a consequence, the hydrodynamics of vegetated channels has been in the focus of many scientific disciplines ranging from aquatic ecology to hydraulic engineering.

Vegetation growing in riverine areas is versatile in its properties, as illustrated in Fig. (21.1). Vascular plants growing in aquatic ecosystems are called macrophytes and can be classified in plants growing completely under the water surface (submersed aquatic vegetation; SAV), plants rooted in shallow water but protruding through the water surface (emergent aquatic vegetation; EAV), and floating vegetation which may or may not be anchored in the stream bed (e.g., Janauer et al. 2013). Macrophytes grow also in the interface zone between aquatic and terrestrial ecosystems, i.e. in the riparian zone, but the more common vegetation types growing between low- and high-water marks are woody trees, bushes and shrubs (Richardson et al. 2007).

The above classification reflects partly the classical hydraulic engineering submergence based vegetation classification in submerged, emergent, and floating vegetation, which takes into account that the flow field changes significantly when the flow depth exceeds the vegetation height or is affected by plants floating on the water surface (e.g., Folkard 2011; Plew 2011). However, a purely water

Fig. 21.1 Rough classification of vegetation types based on commonly observed conditions in the nature. Note that the visual impression is strongly dependent on the water level and season

depth-related classification neglects important plant characteristics such as plant shape and plant biomechanical properties.

Plants and plant communities subjected to a flow direct water to move around stems, branches, leaves, patches, etc. and are thus exposed to flow-induced forces in form of drag and lift. The drag force F_D (Eq. 21.1 defined in Fig. 21.2) results from viscous drag along the wetted plant surface and pressure drag from the pressure differences around the plants and can be parameterized by the fluid density ρ, the drag coefficient C_D, the characteristic plant area A_c, and the reference velocity U, often defined as the bulk velocity. For isolated rigid elements such as cylinders, C_D can be obtained easily using charts from handbooks relating C_D to the stem-Reynolds number $Re_S = Ud/\nu$ where d denotes the cylinder diameter and ν the kinematic viscosity of the water (e.g., Hoerner 1965; Schlichting and Gersten 2006). The characteristic area A_c can be defined by the projected area A_p, i.e. the area of the cylinder exposed to the flow corresponding for emergent conditions to $A_c = Hd$, where H denotes the water depth. However, such a parameterization is much more difficult for complex shaped flexible vegetation which reacts to the flow by hydrodynamic reconfiguration if the drag force is larger than plant-resistance forces (Fig. 21.2). In this case, both the projected area and C_D decrease with increasing flow velocity due to the reconfiguration of the plant, i.e. streamlining, reflecting the effort of the plant to minimize pressure drag (e.g., Vogel 1994; Sand-Jensen 2003; Nikora 2010a).

Most of the research on hydrodynamics of vegetated channels has been carried out with idealized plants (e.g. Aberle and Järvelä 2013) and focused on homogeneously distributed canopies. Such simplifications are seldom justified as vegetation characteristics vary amongst species and habitats (see Fig. 21.1). Therefore, an important step in the proper description of the hydrodynamics of vegetated channels and streams is the suitable characterization of morphological and biomechanical

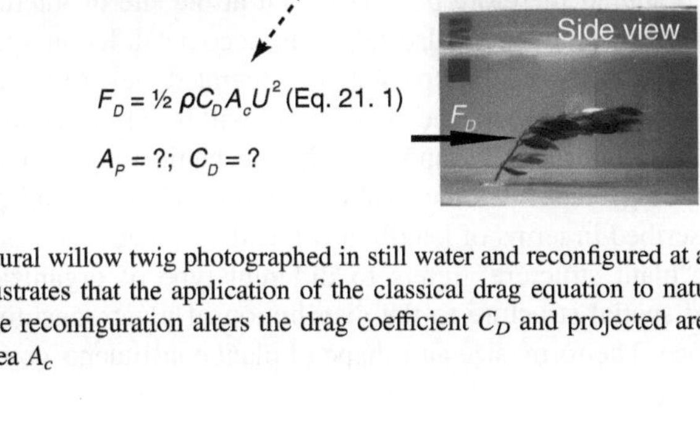

Fig. 21.2 A natural willow twig photographed in still water and reconfigured at a bulk velocity of $U = 0.5$ m/s illustrates that the application of the classical drag equation to natural vegetation is complicated. The reconfiguration alters the drag coefficient C_D and projected area A_p used as the characteristic area A_c

vegetation properties so that flow-induced and plant reaction forces can be parameterized at the plant scale, i.e. for individual plants, as well as for plant communities at the canopy, patch or patch mosaic scale (e.g., Nikora 2010a; Luhar and Nepf 2013). Furthermore, the quantification of the net impact of vegetation on flow properties and transport processes is aggravated by the complex multi-scale flow pattern arising through flow-vegetation interaction from the leaf to patch mosaic scale (e.g., Nikora 2010a; Nepf 2012a, b). An adequate framework for the investigation and the modelling of such multi-scale flow patterns is provided by the double-averaging methodology (DAM) which is based on the averaging of continuity, momentum, and other hydrodynamic equations in both time and space (e.g., Nikora et al. 2007a, b; Nikora and Rowinski 2008 and references therein). The DAM framework allows for up-scaling of physical interactions and mass transfer processes, as well as for considerations of mobile-boundary conditions (Nikora et al. 2013b) such as flexible vegetation. An in-depth consideration of this methodology is beyond the scope of this chapter, and the reader is referred to the aforementioned references for details.

The objective of the present chapter is to summarize the current knowledge on the hydrodynamics of vegetated channels and streams considering both aquatic and riparian zones. Recent review papers (e.g. Green 2005a; Nikora 2010a; Folkard 2011; Nepf 2012a, b; Aberle and Järvelä 2013; Curran and Hession 2013; Vargas-Luna et al. 2015) are complemented by reporting latest advances in this rapidly developing research area. Focus will be set on the plant characteristics and parameterization (Sect. 21.2) and their significance for the hydrodynamics of emergent and submerged flow situations on the canopy and patch scale with an overview on issues related to reach scale considerations (Sect. 21.3).

21.2 Plant Characteristics

Plants growing in aquatic and riparian habitats show significant differences with regard to their shape and biomechanical properties. Thus, a suitable parameterization of vegetation for hydrodynamic analyses must reflect these differences. Even though species-specific parameters are favorable, such parameters depend also on succession, seasonality and local habitat conditions (Puijalon et al. 2008; Thomas et al. 2014), and should therefore be determined at the site of interest. Moreover, vegetation parameterization must also take into account different spatial scales as the characteristics at larger scales represent an integrated view of the characteristics at smaller scales (Nikora 2010a). The latter aspect will be discussed in the following presentation of plant morphology and plant biomechanical parameters.

Plant morphology describes the visible structure of plants and plant communities and is described in terms of length, areal, and volumetric scales (Table 21.1). The notion of "plant structure" refers to different type of organization of plant constituents and spatial structure to the distribution of plant constituents in three-dimensional space. The form, size and shape of plant constituents correspond to the

Table 21.1 Summary of different parameters used for the description of plant morphology

Parameter	Symbol	Parameter	Symbol
Stem diameter (m)	d	Number of plants per uba[1] (m^{-2})	m
Plant height and width (m)	h, W	Volume of plants per uba (m)	m_v
Deflected canopy height (m)	h_w	Frontal area per uba (−)	λ_f
Spacing between plants[3] (m)	a_x, a_y	Wetted plant area per uba (−)	λ_w
Frontal (projected) area (m^2)	A_p	Frontal area per ucv[2] (m^{-1})	a
Leaf area (one sided) (m^2)	A_L	Leaf area index (−)	LAI
Stem area (m^2)	A_S	Solid volume fraction (−)	ϕ
Cross sectional area (m^2)	A_{cs}	Canopy porosity (−)	p
Plant, leaf, stem volume	V, V_L, V_S		

[1]unit bed area; [2]unit canopy volume; [3]here defined as spacing in (a_x) and transverse to flow direction (a_y)

geometrical structure, and the topological structure describes the decomposition of a plant into elementary constituents and determines which constituents are connected with each other (Sinoquet and Rivet 1997; Godin and Caraglio 1998). The most common parameters to describe an individual plant element in hydraulic engineering applications have been the stem diameter d and plant height h. Such a description reduces the plant basically to a cylindrical shape, and subsequently areal and volumetric measures such as frontal projected area A_p and plant volume V can be easily derived. However, the spatial structure of most natural plants differs significantly from idealized simple shapes as they are composed of different constituents such as stems, branches and leaves (e.g. Fig. 21.3). The growth structure is species specific and depends on succession and local habitat conditions and its a priori parameterization is therefore not straightforward. The vertical structure of vegetation has largely been neglected in hydraulic engineering applications by assuming a homogeneous structure over height although the actual width W of a plant may be much larger than for example the stem diameter. The latter has often been used to parameterize complex shaped vegetation elements (Aberle and Järvelä 2013 and references therein) but a more complete description of areal and

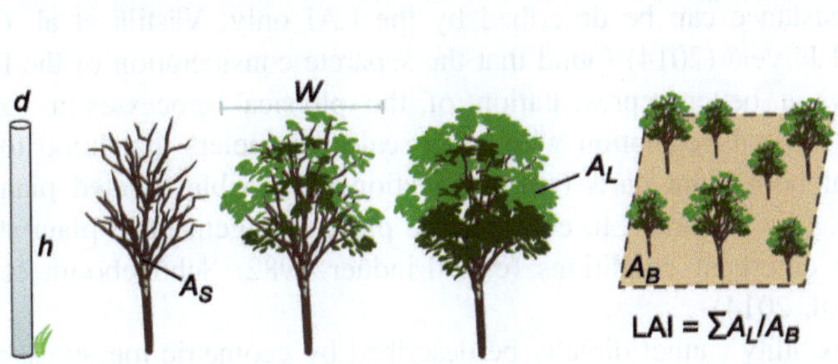

Fig. 21.3 Parameterization of riparian vegetation. For foliated vegetation, the canopy density can be described with the leaf area index LAI which is defined as the one-sided leaf area per unit bed area. For emergent conditions, this definition may be refined dependent on water level as the wetted one-sided leaf area per unit bed area

volumetric measures may include the separate determination of stem and one-sided leaf areas (A_S and A_L in Fig. 21.3) and volumes, respectively (Table 21.1), which may also be determined over height. In this context it must be emphasized that the vertical plant structure may have important implications for the vertical flow field through the generation of higher velocities in zones with lower plant mass, e.g., below the foliated zone in the case of riparian plants (Jalonen et al. 2012).

In hydraulic models, plant canopies are typically characterized by considering plant parameters per unit bed area or unit canopy volume. An important property is the canopy density for which different definitions such as the solid volume fraction ϕ, canopy porosity p, the frontal area per unit bed area (λ_f) or per unit canopy volume (a), respectively, are used in the literature (Table 21.1). For flow through non-bending vegetation such as erect reeds or trees where only the stems are exposed to the flow, the vegetation may be parameterized sufficiently by the stem diameter and vegetation density (e.g., m, λ_f, a in Table 21.1, see also Petryk and Bosmajian 1975; Nepf 1999). However, for foliated plants such a parameterization is inadequate. For such vegetation, the canopy density can be expressed in terms of the leaf area index (LAI) which is defined as the one-sided leaf area per unit bed area (LAI = $\Sigma A_L / A_B = m < A_L >$ with A_B = bed area and the angled brackets indicate spatial averaging; Fig. 21.3), which is thus slightly differently defined as λ_f or the wetted area per unit bed area (λ_w) as the plant stems and branches are not considered.

LAI has been used in a broad range of models in earth sciences, and various definitions of LAI have been introduced for different purposes. Hemi-surface LAI is closely related to one-sided LAI, with the difference that the same one-sided leaf area measure is projected onto a horizontal datum independent of ground slope. A third common definition is horizontally projected LAI that is typical in remote sensing applications. It is important that a definition of LAI is precisely addressed to make reported results comparable. LAI can be derived using various destructive and nondestructive methods, such as ripping and measuring foliage area (Jalonen et al. 2013; Jalonen and Järvelä 2014), or remote sensing methods such as terrestrial and airborne laser scanning and image analysis (e.g. Antonarakis et al. 2010; Jalonen et al. 2014, 2015) thus also allowing for a depth-related definition of LAI. As an alternative to assuming that the combined effect of vegetation density and foliage on flow resistance can be described by the LAI only, Västilä et al. (2013) and Västilä and Järvelä (2014) found that the separate consideration of the foliage and stem allows a better representation of the physical processes as opposed to describing foliated vegetation with plant-scale parameters that lump together the influence of both plant parts (see sub-section on flexible foliated plants below). In addition, it is important to consider the plant arrangement in plant stands, particularly in emergent conditions (e.g., Lindner 1982; Schoneboom et al. 2011; Ricardo et al. 2014).

Plant flexibility cannot directly be described by geometric measures, although it indirectly affects some of the parameters listed in Table 21.1. For example, a bending plant has a lower height than in still water (Fig. 21.2) and the flexibility is reflected by defining the deflected plant or canopy height (h_w). Similarly, the reconfiguration of plants affects A_p and canopy porosity p which hence become a function of flow regime, flow velocity and biomechanical properties, as discussed before.

Plant biomechanical properties define the ability of plants to reconfigure in the flow and hence govern plant-reaction forces. The most important plant biomechanical properties are the plant density ρ_p, Young's (bending) modulus E and the second moment of cross-sectional area I. Young's (bending) modulus E is a measure of stiffness characterizing stem flexibility under bending forces orthogonal to the plant stem. The product EI defines the flexural rigidity of an object describing its resistance while undergoing deformation. It should be noted that many natural plants cannot be assumed to have a simple cross-sectional shape, and thus it may be difficult to derive reliable and meaningful I values. The density ρ_p defines the gravity force of a plant ($F_G = \rho_p g V$, where g = acceleration due to gravity) and hence the submerged weight of the plant ($F_S = (\rho_p - \rho)gV$). It is not uncommon for aquatic plants that $\rho_p < \rho$ so that $F_S < 0$ represents an upward directed buoyancy force and therefore a plant-restoring force (e.g., Nepf 2012a). Plants with a very low flexural rigidity (e.g., buoyant plants) passively follow the flow and experience mainly viscous drag along the plant surface, whereas plants with higher flexural rigidity expose a resisting force to the flow (and react through bending), generating pressure drag and downstream vortices (Nikora 2010a).

Plant biomechanical properties are characterized by a high variability with regard to environment, species, and scales (Albayrak et al. 2014; Miler et al. 2014; Paul et al. 2014), and their implementation in studies of freshwater ecosystems and hydrodynamic considerations is not yet common (e.g., Nikora 2010a; Miler et al. 2011). Nonetheless, there exist studies that have focused on plant biomechanical properties by relating flexural rigidity to deflected plant height for grass and riparian vegetation (e.g., Kouwen 1992; Kouwen and Fathi-Moghadam 2000 and references therein) or biomechanical properties to macrophyte habitat (Miler et al. 2011, 2014). Other studies focused on the development of scaling criteria for vegetation (e.g., Ghisalberti and Nepf 2002) or developing analytical and numerical models for deformation of simple shaped flexible elements (e.g., Chen et al. 2011; Kubrak et al. 2012; Stone et al. 2013 and references therein).

Plant reaction to flow is governed by the interplay between flow forces and the plant-reaction forces. The governing flow forces are the drag and lift force which can be expressed in a dimensionless form via the drag coefficient C_D and the lift coefficient C_L which are a function of the object (plant) Reynolds-number Re_p (or Re_s for cylinders; Hoerner 1965; Schlichting and Gersten 2006). For a rigid plant and a flow situation where the frontal area is constant (i.e., completely submerged or constant water depth for emergent conditions) and the drag coefficient C_D does not vary with Re_p, F_D increases proportionally to the squared flow velocity U^2 according to Eq. (21.1).

A flexible plant, on the other hand, reconfigures with increasing U when the hydrodynamic force exceeds the plant-reaction force due to stiffness and buoyancy until the restoring force is equal to the drag force (e.g. Luhar and Nepf 2011). The reconfiguration and the associated decrease in the projected area is illustrated in Fig. 21.4 using photographs of a full scale tree towed at different velocities in

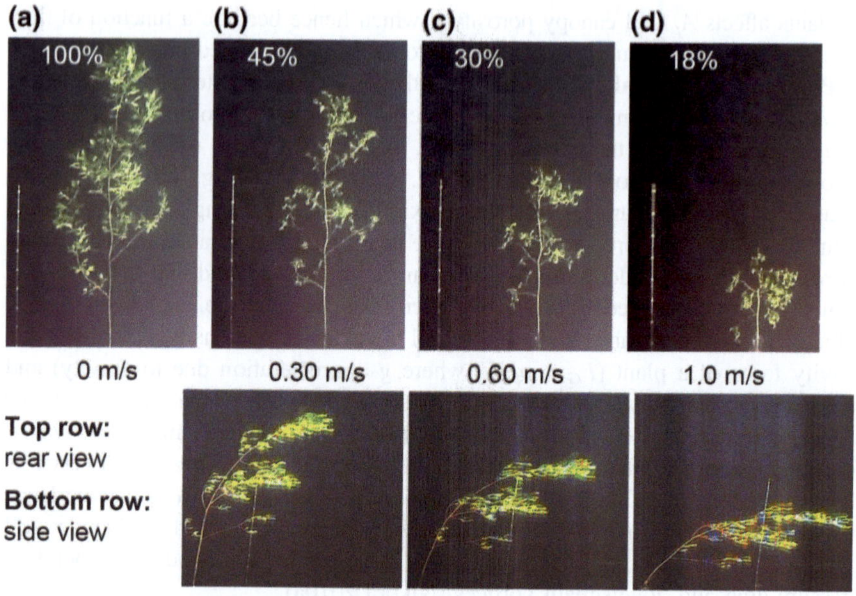

Fig. 21.4 Frontal projected area and side view of a submerged, 1.8 m tall Goat Willow exposed to different flow velocities. The percentages indicate the share of the projected area compared to the no-flow case (specimen SC7 from Jalonen and Järvelä 2014)

towing tank experiments carried out at Aalto University in Finland. In the example shown in Fig. 21.4 the frontal projected area reduced as much as 82 % compared to still water conditions. The reconfiguration results in parameterization difficulties for both the drag coefficient C_D and the characteristic area A_c as both parameters depend on the shape and porosity of the elements.

The use of different definitions for the characteristic flow velocity and area often aggravates a direct comparison between studies and may even lead to contradicting results (Statzner et al. 2006). An example is the use of different definitions of the characteristic area A_c in Eq. (21.1) to calculate the drag coefficient C_D from drag force measurements. Lumping both parameters into a combined C_DA_c parameter may be a possibility as only the combined effect of C_D and A_c is important (e.g. de Langre 2008). However, such an approach is not a complete solution as the lumped C_DA_c parameter still depends on the same variables as the individual parameters and cannot be determined a priori. A solution to this problem may be to consider non-changeable areas as reference area (e.g., Aberle and Järvelä 2013), such as the leaf and stem areas (Västilä and Järvelä 2014).

As both, C_D and the projected area A_p, decrease with increasing velocity until the plant reaches its maximum deformation, the drag growth with velocity is for flexible vegetation elements lower than for the usual rigid body proportionality, which has been shown in many studies on water and air flows (e.g., Vogel 1994; Järvelä 2002; de Langre 2008; Gosselin et al. 2010; Jalonen and Järvelä 2014 and references therein). The deviation of the F_D–U relationship from the quadratic law for flexible plants has been expressed in the literature in a more general way according to $F_D \propto U^{2+b}$, where b denotes the Vogel exponent (de Langre 2008). For rigid plants $b = 0$ whereas for natural plants and leaves b has been found to vary between −0.2 and −1.2 (e.g., Albayrak et al. 2011; de Langre et al. 2012; Aberle and Järvelä 2013; Jalonen and Järvelä 2014).

Taking the aforementioned forces into account, important force ratios can be defined to investigate plant deformation, such as the Cauchy number (describing the ratio of inertial to elastic forces), the buoyancy parameter B (ratio of buoyancy forces to elastic forces), the solid mass ratio ρ/ρ_p, and the reconfiguration number R comparing the drag of a flexible body to that of an equivalent rigid body at the same Reynolds number. Additionally, dimensionless geometrical parameters describing the plant structure may be defined, such as LAI or the slenderness ratio S_l which is the ratio of the length of a vertical column and its least radius of gyration (Chakrabarti 2002; Alben et al. 2002; de Langre 2008; Gosselin and de Langre 2011; Luhar and Nepf 2011). Additional dimensionless parameters may be defined to account for flexural rigidity at different plant scales and to enable the subdivision between tensile and bending plants (Nikora 2010a). Further investigation of these force ratios may provide guidance on how flexible vegetation elements may be scaled in laboratory investigations which currently remains an unresolved issue.

Until today, many studies focused on the investigation of plant reconfiguration in a time-averaged domain so that plant reconfiguration has often been determined by single snapshots and not been monitored over time. However, the dynamic interaction between plants and turbulent flow may result in a dynamic reconfiguration of the plants which becomes visible as flapping-like motions. Such motions have been found to be correlated with drag fluctuations and upstream turbulence (Siniscalchi and Nikora 2013). The drag fluctuations are closely related to ambient turbulence (Siniscalchi and Nikora 2012) and hence to large-scale eddies interacting with the plant. However, shear layer turbulence at the plant surface may also contribute to plant motion (e.g., Siniscalchi et al. 2012; Cameron et al. 2013; Albayrak et al. 2014; see also Sect. 21.3).

Flexible foliated plants and their reaction to the flow, as shown in Figs. 21.2 and 21.4, are discussed in this section to highlight the importance of foliage, as several studies have shown that foliage contributes significantly to the total drag exerted by plants (Vogel 1994; Järvelä 2002; James et al. 2008; Wilson et al. 2008; Dittrich et al. 2012; Jalonen and Järvelä 2014). This is of particular importance for riparian plants where leaves may account for a large portion of the total area or of the total mass (e.g. de Langre 2008). Moreover, foliated plants represent pervious bodies and flow passes through them, a feature which is known as bleed flow (e.g., Grant and Nickling 1998). However, due to the interplay between leaves and wooden plant parts, it is not possible to obtain the total drag of a leafy plant by simply multiplying the drag exerted by a single leaf with the number of leaves (Vogel 1994).

Leaves are highly flexible structures and the resulting drag at the leaf scale is affected by their shape, surface roughness, and biomechanical properties (Vogel 1994; Albayrak et al. 2011) indicating a species-specific hydrodynamic resistance even at the leaf scale (Vogel 2009). Studies with artificial and natural specimens at different scales (twigs and full-scale trees) showed that the contribution of foliage drag F_F to the total drag of a specimen F_D depends on flow velocity and involves significant natural variability (Aberle and Järvelä 2013 and references therein). Despite this variability, recent studies revealed a decreasing trend of the ratio F_F/F_D with increasing flow velocity which reaches an almost constant value for higher velocities. This indicates that leaves contribute more significantly to total drag at low velocities and that the contribution gradually decreases with increasing velocity which can be associated with the different biomechanical properties of the plants wooden parts and the leaves (Dittrich et al. 2012; Whittaker et al. 2013; Jalonen and Järvelä 2014).

Recent developments on the parameterization of the exponential force-velocity relationship for flexible woody vegetation have resulted in various models as shown in Table 21.2. Järvelä (2004) presented a model for estimating flow resistance using LAI = A_L/A_B as the key vegetative parameter. More recently, Jalonen et al. (2013) found that the bulk resistance could be estimated with better accuracy by replacing A_L/A_B with A_{tot}/A_B. In more detailed investigations, Västilä and Järvelä (2014) concluded that the leaf (A_L) and stem areas (A_S) are the key parameters to estimate the resistance, and proposed a model describing the foliage and stem with physically based parameters: drag coefficients, reconfiguration parameters, and leaf as well as frontal projected stem area per ground area. Whittaker et al. (2013) based their model on the Cauchy number and incorporated the tree volume and flexural rigidity EI in their analyses.

In a full-scale towing tank study, Jalonen and Järvelä (2014) collected a comprehensive dataset on drag forces and tree properties for comparing the suitability and reliability of different plant parameterizations for physically-based modeling applications. The results showed that at low velocities the stem drag of foliated trees reduced in comparison to the defoliated condition due to more efficient reconfiguration of the stem caused by the leaf mass. This implied that the actual foliage drag

Table 21.2 Parameterization of vegetated drag force in recent models based on the classical drag equation (Fig. 21.2)

Drag model	Notes	Reference
$F_D = \frac{1}{2}\rho \frac{C_{D\chi}}{U_\chi^\chi} A_L U^{2+\chi}$	A_L/A_B = LAI	Järvelä (2004)*
$F_D = \frac{1}{2}\rho K \left(U\sqrt{\frac{\rho VH}{EI}}\right)^\chi U^2$	K = empirical coefficient	Whittaker et al. (2013)
$F_D = \frac{1}{2}\rho \left(\frac{C_{D\chi,F}}{U_{\chi,F}^{\chi F}} A_L U^{2+\chi_F} + \frac{C_{D\chi,S}}{U_{\chi,S}^{\chi S}} A_S U^{2+\chi_S}\right)$	$F_D = F_F + F_S$	Västilä and Järvelä (2014)*
$F_D = \frac{1}{2}\rho \frac{C_{D\chi}}{U_\chi^\chi} A_{tot} U^{2+\chi}$	$A_{tot} = A_L + A_S$	Jalonen and Järvelä (2014)

*The equations were reformulated in terms of drag force by Jalonen and Järvelä (2014), as originally they were formulated to determine the vegetated friction factor f''. All these models use an exponential relationship with velocity as $F \propto U^{2+\chi}$, where the reconfiguration parameter χ equals the dimensionless Vogel exponent b. See Table 21.1 for parameter definitions

at low velocities can be somewhat larger than estimated by $F_F = F_D - F_S$. Variation in the ratio of leaf area to the stem area was found for trees of different sizes, and the share of leaf area appeared to increase for the smallest specimens. Evaluations with the new independent data by Jalonen and Järvelä (2014) revealed that the tested models (Table 21.2) were capable of producing reasonable results. The Järvelä (2004) model proved to be applicable for fully foliated trees and published parameter values for twigs were suitable for predicting the bulk drag at a broad range of velocities for trees of different scales. Models by Whittaker et al. (2013) and Västilä and Järvelä (2014) are suited for both foliated and defoliated conditions, but the piece-wise $F_S - U$ relationship can cause difficulties in estimating the stem drag and total $F_D = F_F + F_S$. Overall, the model predictions were dictated by the parameter values used rather than the model structure or plant scale (Jalonen and Järvelä 2014).

21.3 Hydrodynamic Processes

Following the above consideration at the plant scale, this section focuses on hydrodynamic processes at canopy scale with a subdivision in submerged ($H/h > 1$) and emergent ($H/h \leq 1$) flow situations, followed by considerations of floating, patch, and reach scale vegetation.

Canopy scale—submerged conditions For submerged conditions, the flow depth exceeds the (deflected) canopy height, i.e. $1 < H/h$. This general classification can be further refined by distinguishing between deeply submerged ($H/h > 10$) and shallow ($1 \leq H/h < 5$) conditions (Nepf and Vivoni 2000). Deeply submerged conditions resemble flows over terrestrial canopies where a large scale free surface layer dominates the flow field, although the vertical transport at the canopy interface is still dominated by canopy-scale vortices forming at the interface with the canopy (Nepf et al. 2012a, b and references therein). Deeply submerged canopies are rare in natural channels due to the constraint of sunlight, and thus most submerged aquatic canopies are characterized by $H/h < 5$ (e.g., Nepf 2012b). Nonetheless, deeply submerged flow conditions may occur in grass-lined channels or for canopies composed of buoyant tensile plants which are easily bent over during a flood so that they cover the bed surface, i.e. they will act as surface roughness.

The multi-scale flow patterns in a vegetated channel for shallow conditions are illustrated in Fig. 21.5 according to Nikora (2010a). The canopy interface represents a porous surface and therefore the flow field above the canopy can show boundary layer characteristics where the turbulent scales are governed by the depth scale (pattern #1 in Fig. 21.5). Although such a boundary layer is "perturbed" by the canopy, and hence different from a canonical boundary layer, the velocity distribution above the canopy has often been modelled using a logarithmic velocity distribution (Stephan and Gutknecht 2002 and references therein).

In close proximity of dense canopies ($C_Dah > 0.1$), the flow field is affected by a shear layer forming through the different velocities within and above the canopy

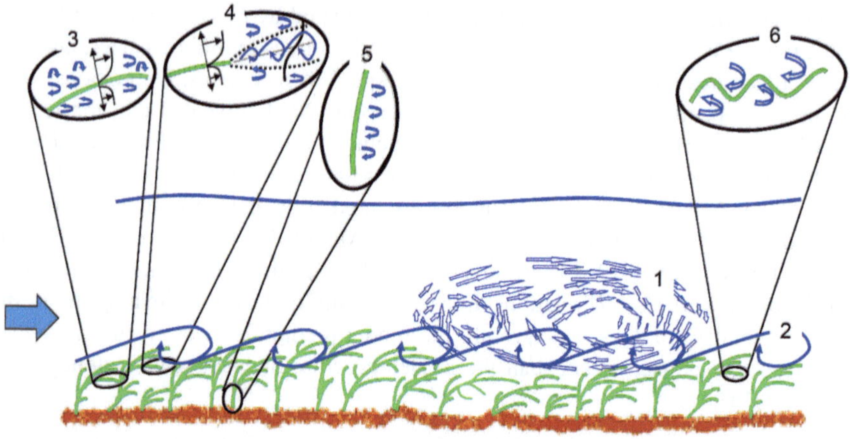

Fig. 21.5 Side view of multi-scale flow patterns in vegetated channels. Figure modified from Nikora (2010a)

(pattern #2, Fig. 21.5). The shear (or mixing) layer velocity profile layer is characterized by an inflection point near the canopy top (Fig. 21.6b, c) and its shape may be described by a hyperbolic tangent function (e.g. Chen et al. 2013 and references therein). The existence of an inflection point is the prerequisite for the generation of Kelvin-Helmholtz instabilities at the canopy interface that evolve into eddies dominating the vertical transport at the canopy interface (canopy-scale turbulence; pattern #2 in Fig. 21.5; e.g., Nepf 2012b and references therein). For $0.1 < C_Dah \approx< 0.2$, these mixing layer eddies penetrate the bed (Fig. 21.6b) governing the turbulence pattern throughout the canopy (Nepf and Ghisalberti 2008), but their penetration depth steadily reduces for larger canopy densities ($C_Dah > 0.2$; Fig. 21.6c). As a consequence, eddies no longer penetrate the bed and the level of ventilation of the canopy is reduced (Nikora 2010a). Moreover, for relative submergences $H/h < 2$, the size and strength of the mixing layer eddies is reduced due to the interaction with the water surface (Nepf and Vivoni 2000).

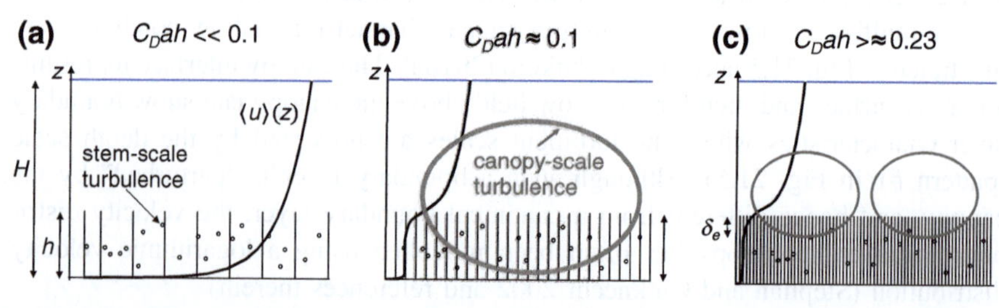

Fig. 21.6 Velocity distributions in a sparse canopy **(a)** and velocity distribution and eddy-penetration depth δ_e in dense canopies **(b and c)** (modified from Nepf 2012a)

The study of Okamoto and Nezu (2009) indicated a larger penetration of momentum into the canopy layer for a rigid than for a flexible canopy implying increased momentum absorption for flexible canopies. Canopies composed of flexible elements show often a waving motion, known as "monami" (Ackerman and Okubo 1993; Nepf 2012b and references therein). It has been generally accepted that this motion is caused by canopy scale vortices if the instantaneous drag associated with their passage is sufficient to depress individual plants (e.g. Okamoto and Nezu 2009). Thus, this motion may produce turbulence (pattern #6, Fig. 21.5). As the motion depends also on biomechanical properties, it remains to be investigated to what extent these properties affect mixing layer eddies and the "monami" (Nikora 2010a). We note that patterns #3 and #4 hypothesized by Nikora (2010a) in Fig. 21.5 are also related to flexible elements as they describe the generation of leaf scale boundary layer turbulence (pattern #3) resulting from flow around individual leaves as well as from small scale mixing-layers forming through different roughness of a leaf at the opposite leaf sides (leaf scale mixing layer turbulence, pattern #4).

For sparse vegetation densities ($C_D ah \ll 0.1$), the momentum absorption in the canopy is not large enough to generate an inflection point in the velocity distribution near the canopy and the velocity distribution follows a turbulent boundary layer profile (e.g. Nepf 2012b; Fig. 21.6a). However, turbulence within the canopy is still produced by both bed shear and canopy drag (generation of stem scale wake turbulence, pattern #5 in Fig. 21.5) and, as a consequence, it is not possible to predict turbulence levels in analogy to classical open-channel flows, i.e. where turbulence is generated only through bed shear (Nepf 2012a).

The flow through the canopy is for both sparse and dense densities affected by the form drag of the plant stems and branches causing flow separation and hence the generation of wake turbulence (stem scale turbulence; pattern #5 in Fig. 21.5). In regions of the canopy where both free-stream and near bed surface effects are negligible, the velocity distribution is a function of bed slope and vegetation density (Nikora et al. 2004; Lightbody and Nepf 2006). In case the vegetation is vertically homogeneously distributed and hence $C_D a$ = constant over depth, the velocity distribution in the canopy becomes uniform (Nikora et al. 2004). On the other hand, a vertically non-uniform density results in more complex shaped profiles due to different resistance imposed at different layers (e.g. Lightbody and Nepf 2006). Close to the bed, the flow is affected by the bed surface roughness resulting in the formation of a surface boundary layer.

There exist several approaches for the description of the velocity profile for submerged vegetation which can be broadly classified in "single-profile" and "segmented-profile" concepts (Nikora et al. 2013a and references therein) as well as analytical and numerical concepts. Single-profile concepts are based on the isolated consideration the aforementioned flow pattern #1 (perturbed boundary layer concept) or pattern #2 (mixing layer analogy), whereas the segmented-profile concept is based on the vertical segmentation of the velocity profile taking into account the aforementioned different physical principles. Analytical concepts have been derived using an eddy-viscosity model for the definition of the turbulent stress, and numerical approaches use first and second order turbulence closures or large eddy simulation (LES) (Nepf 2012a and references therein; Kubrak et al. 2008; Stoesser et al. 2009; King et al. 2012). Nikora et al. (2013a, b) introduced an approach in the form of a linear superposition of the velocity distribution in the canopy, mixing layer, boundary layer, and taking the Coles (1956) law of the wake into account. Although most of these approaches have been developed for canopies composed of rigid cylinders, some of them consider the flexibility of the elements (e.g., Kubrak et al. 2008; Dijkstra and Uittenbogaard 2010; Luhar and Nepf 2013).

Canopy scale—emergent conditions In emergent conditions, the flow in the canopy is driven by gravity and is dominated by the decay and spread characteristics of wakes forming at upstream elements (e.g., Li and Shen 1973; Lindner 1982; Wilkerson 2007), i.e. flow pattern #5 in Fig. 21.5 is dominating the hydrodynamic conditions. The length scale governing turbulence production is generally defined by the stem diameter d, or in very dense canopies where the spacing between stems is smaller than d, by the pore size (e.g., Nepf 2012b). The dominating spatial scales are therefore canopy density and stem diameter (Tanino and Nepf 2008; Ricardo et al. 2014). In general, the mean turbulence intensity increases with the solid volume fraction whereas the velocity reduces due to the increased form resistance. However, for very dense canopies, turbulence levels may be dampened and may be even lower than turbulence intensities in comparable non-vegetated flow conditions (e.g., Nepf 2012b). For low vegetation densities the flow behaves similarly to the flow around isolated elements, although the turbulence patterns are different from those of unobstructed open channel flows due to the effect of the wakes (Stoesser et al. 2010). The velocity distribution in emergent conditions can be described in the same way as for the vegetation layer in submerged conditions (see the preceding section) taking into account that free-stream effects are not present.

In order to express canopy resistance, it is necessary to consider the spatially averaged drag force. It has often been postulated that the drag coefficient of the elements in the canopy can be directly estimated using data derived for isolated objects. However, several studies showed that for dense canopies lower drag coefficients can be expected for a stem within a multistem array indicating the need to consider spatially averaged drag coefficients (e.g., Li and Shen 1973; Lindner 1982; Nepf 1999; Poggi et al. 2004). Herein it must be pointed out that contradictory results have been reported with regard to the density effect on drag coefficients (Aberle and Järvelä 2013 and references therein) which may be partially explained by differences in the experimental boundary conditions (array density and stem-Reynolds number), the definition of hydraulic variables (e.g., Statzner et al. 2006), and that drag forces were not measured directly in all these studies. For woody foliated plants, the spatially averaged drag force may be estimated by upscaling the drag force equations listed in Table 21.2, i.e. by considering plant characteristics per unit bed area and assuming that the Vogel exponent is scale-independent (e.g., Järvelä 2004; Jalonen et al. 2013; Västilä et al. 2013; Västilä and Järvelä 2014; Jalonen and Järvelä 2014).

Floating vegetation The hydrodynamic processes associated with floating vegetation, i.e. in situations where the flow is confined between the bottom and the floating vegetation elements, have been researched less than the other types of vegetation discussed above. For such conditions, the flow field is characterized by the bottom boundary layer, a shear layer at the interface to the floating vegetation, and an internal canopy layer. The results reported by Plew (2011) in a study with floating cylinders suggest that the bottom boundary layer height is inversely related to the canopy density $C_D a h_c$ and that the boundary layer is coupled with the shear layer. The latter has its maximum velocity at the top of the boundary layer and its minimum velocity in the canopy, and its size is hence limited by the distance between the canopy and bottom boundary layer. Due to the flow deceleration in the canopy, the velocity distribution is characterized by an inflection point near the bottom of the floating canopy and the flow structure in the canopy layer is governed by stem-scale turbulence (Plew 2011).

Patch-scale Under natural conditions, plants often form spatially heterogeneous communities—patches which together with non-colonized spaces, or spaces colonized by different types of vegetation, form irregular mosaics. Although the patchiness of aquatic vegetation is presently an important topic of ecological research (e.g., Nikora 2010a; Vandenbruwaene et al. 2011; Zong and Nepf 2011), its effect on flow structure and transport processes has been less examined. The occurrence of patches in channels may transform relatively two-dimensional open-channel flow into complex three-dimensional flows (Sukhodolov and Sukhodolova 2010; Siniscalchi et al. 2012). In fact, the flow patterns presented in Fig. 21.5 must be extended with regard to patches by the consideration of large-scale turbulence associated with flow separation and wakes at the patch scale (pattern #7, Fig. 21.7),

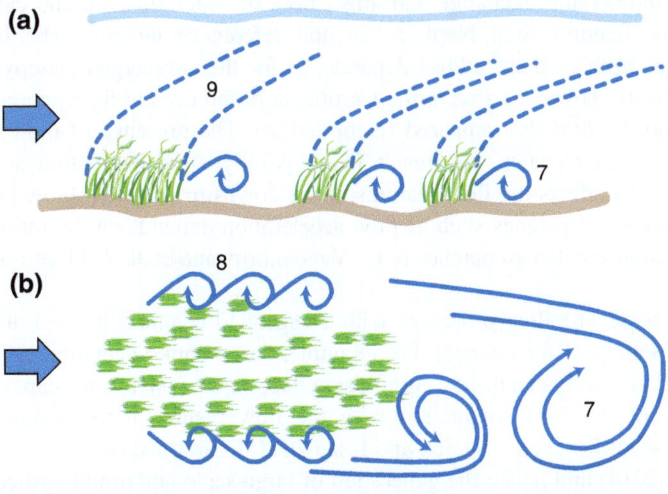

Fig. 21.7 Flow patterns at patch scale: **(a)** side view considering patch mosaic structure and **(b)** plan view at patch scale (modified from Nikora 2010b)

boundary layers and mixing layers developing at the patch side (pattern #8, Fig. 21.7b), as well as interacting vertical (pattern #9, Fig. 21.7a) and horizontal boundary layers at the patch mosaic scale (e.g., Nikora 2010b; Zong and Nepf 2010; Sukhodolov and Sukhodolova 2012).

Studies with submerged patches spanning the channel width showed that the upstream part of the patch diverts the flow upwards over the patch resulting in decelerating flow velocities in the canopy and flow acceleration above the patch. This velocity difference contributes to the formation of a shear layer (Ghisalberti and Nepf 2004; Sukhodolova and Sukhodolov 2012) enhancing vertical turbulent transport of momentum (Okamoto and Nezu 2013; Zeng and Li 2014). Moreover, such a flow feature suggests that plants at patch edges experience significantly larger drag than plants in the middle of the patch as they are exposed to larger flow velocities (Nikora 2010a; Siniscalchi et al. 2012).

The patch density and geometry are dominating factors for the turbulent flow field in and around the patches (e.g., Green 2005b; Sukhodolov and Sukhodolova 2012). Increasing the vegetation density results in faster development of velocity and turbulence inside the patch due to the larger resistance compared to sparse densities (Soulioutus and Prinos 2011). Moreover, these in-canopy flow features develop faster than the flow characteristics above the canopy (Zeng and Li 2014). Vegetated patches represent porous patches and this porosity affects the wake flow conditions. For example, the resulting wake from a porous patch is much longer compared to the wake generated by a similar solid obstruction as the bleed flow delays the onset of the von Karman vortex street (Nepf 2012a and references therein).

In the case of emergent patches, the flow is deflected sideways from the patch and a shear layer develops at the interface between the patch and the free flow (pattern #8 in Fig. 21.7b). The resulting horizontal mixing layer eddies dominate mass and momentum exchange and affect both the open channel and canopy turbulent flow features (e.g. Nepf 2012a and references therein). The horizontal penetration depth of these eddies depends, as for the submerged canopy case, on canopy density. However, due to the significant differences in flow geometry, both cases cannot be directly compared (Nepf 2012a). The presence of more than one patch, i.e. patch mosaics, can result in a hydrodynamic interaction so that the upstream patch affects the flow features of the downstream patch. Flow interaction between vegetated patches such as flow acceleration depends on the ratio of patch size and distance between patches (e.g., Vandenbruwane et al. 2011 and references therein).

Reach scale The flow processes within vegetated channels depend also on the geometric setting of the channel. For example, dense bank vegetation in a channel of compact geometry can have a significant effect on the conveyance capacity as the interaction of the bank vegetation with the flow results in the formation of a transverse shear flow (e.g., Sanjou and Nezu 2011; Czarnomski et al. 2012; de Lima and Izumi 2014) and hence the generation of large-scale horizontal vortices. These macro-turbulent structures decelerate the flow in the main channel significantly compared to the case with non-vegetated banks, and may lead to a significant reduction of the conveyance capacity while the conveyance in the vegetated zone is only partly affected (Mertens 2006). The enhanced momentum exchange between the main channel flow and the vegetated zone can be considered by introducing an apparent interface roughness which depends on the vegetation density (e.g., Mertens 2006; Pasche and Rouvé 1985).

In non-vegetated compound channels, flow resistance is already affected by the channel geometry (Shiono and Knight 1991; Knight 2013 and references therein) but the presence of vegetation has further consequences. In wide compound channels, bank vegetation affects the flow in the main channel only in spatially restricted area near the banks, and the flow resistance may be estimated by subdividing the overall flow field into the undisturbed main channel flow, the interface, and the vegetated floodplain region, respectively, taking into account vegetated and non-vegetated zones (Pasche and Rouvé 1985).

A common feature in many compound channels is the occurrence of trees at the floodplain edge, which can be interpreted as "one-line form roughness". Such vegetation stripes have a significant effect on the velocity distribution and result in decreased boundary shear stress and an increase in flow resistance (Dittrich et al. 2002; Sun and Shiono 2009), i.e. in a decrease of the conveyance capacity. Similarly, it is important to consider the main flow direction on floodplains in order to determine if vegetation stripes formed of shrubs and trees will be oriented in or transverse to the main flow direction on the floodplain thus affecting local energy losses (Klaasen and van Urk 1985; Dittrich and Aberle 2010).

Grass-lined channels Grassy type of vegetation is widely used as a protective liner in agricultural waterways, floodways, and emergency spillways. Thus a significant amount of practical and theoretical research is available on such linings (Järvelä 2005). For designing vegetated waterways, Palmer (1945) introduced the *n-UR* method relating Manning's n with the product of average velocity U and hydraulic radius R for various channel slopes and plant stands. The US Soil Conservation Service presented the method in a revised form (SCS 1954) making it popular in practise. Kouwen and Unny (1973) criticised the application of the method, since their experiments on flexible plastic roughness indicated that the resistance over such roughness is primarily a function of the relative roughness, defined as the ratio of the deflected plant height to the flow depth. They introduced a stiffness parameter mEI, flexural rigidity per unit area, which reflects the overall resistance to deformation of a plant stand as a result of a flow passing over it. Temple (1999) concluded that although alternate approaches for predicting vegetal flow resistance have been proposed, the *n-UR* method has remained the primary tool for practical application to grass-lined channel conditions. Conventional empirical methods are still widely used in practise (e.g., Wilson and Horritt 2002; Rhee et al. 2008), but it would be desirable to eventually replace them by less empirical relations.

21.4 Summary

This chapter summarized the current knowledge on the hydrodynamics of vegetated channels and streams taking into account both aquatic and riparian zones, i.e. in-stream vegetation as well as vegetation growing on banks and floodplains. Vegetation is an important feature in channels affecting the flow field across various spatial scales ranging from the leaf to the reach scale by imposing additional form resistance. The effect of vegetation on the flow field has often been studied in a simplified way by reducing complex shaped vegetation to simple shapes such as cylinders and neglecting plant biomechanical properties. However, these properties represent a critical control for the hydrodynamics of vegetated channels as they define the flow-vegetation interaction: if the flow force is larger than the plant resistance forces, the plant parts or whole plant will react to this force by streamlining to reduce the experienced drag which in turn affects the flow field and flow resistance.

The significance of plant characteristics was explored in Sect. 21.2 by critically reviewing conventional parameterization of vegetation. There is a clear need for advanced parameterization of vegetation characteristics considering plant morphology and biomechanical properties so that the interaction of natural plants and plant communities with the flow can be modelled more realistically. The significance of such parameters was highlighted by addressing how foliated plants react to the flow and how the drag force can be characterized while taking plant deformation into account. Hydrodynamic processes affecting the flow in vegetated channels were reviewed in Sect. 21.3 at different spatial scales using the classical hydraulic engineering classification of vegetation into emergent and submerged flow conditions. The dominating flow features that have been identified in studies with submerged and emergent canopies composed of rigid and flexible elements, floating canopies, and vegetation patches were described including brief considerations for interplay of the geometric setting of the channel. Overall, much effort has been devoted to solving the challenging flow-vegetation interaction problems, but there are still many research questions awaiting clarification, as discussed in the above review. Future research efforts should therefore focus on truly interdisciplinary research methodologies aiming at a broader incorporation of environmental hydraulics into studies of ecosystems and vice versa.

References

Aberle J, Järvelä J (2013) Flow resistance of emergent rigid and flexible floodplain vegetation. J Hydraul Res 51(1):33–45

Ackerman JD, Okubo A (1993) Reduced mixing in a marine macrophyte canopy. Funct Ecol 7:305–309

Albayrak I, Nikora V, Miler O, O'Hare M (2011) Flow–plant interactions at a leaf scale: effects of leaf shape, serration, roughness and flexural rigidity. Aquat Sci 74(2):267–286

Albayrak I, Nikora V, Miler O, O'Hare M (2014) Flow–plant interactions at leaf, stem and shoot scales: drag, turbulence, and biomechanics. Aquat Sci 76(2):269–294

Alben S, Shelley M, Zhang J (2002) Drag reduction through self-similar bending of a flexible body. Nature 420(6915):479–481

Antonarakis AS, Richards KS, Brasington J, Muller E (2010) Determining leaf area index and leafy tree roughness using terrestrial laser scanning. Water Resour Res 46(W06510). doi:10.1029/2009WR008318

Cameron SM, Nikora VI, Albayrak I, Miler O, Stewart M, Siniscalchi F (2013) Interactions between aquatic plants and turbulent flow: a field study using stereoscopic PIV. J Fluid Mech 732:345–372

Chakrabarti S (2002) The theory and practice of hydrodynamics and vibration. World Scientific, Singapore

Chen L, Stone MC, Acharya K, Steinhaus KA (2011) Mechanical analysis for emergent vegetation in flowing fluids. J Hydraul Res 49(6):766–774

Chen Z, Jiang C, Nepf H (2013) Flow adjustment at the leading edge of a submerged aquatic canopy. Water Resour Res 49(9):5537–5551

Coles D (1956) The law of the wake in the turbulent boundary layer. J Fluid Mech 1(2):191–226

Curran JC, Hession WC (2013) Vegetative impacts on hydraulics and sediment processes across the fluvial system. J Hydrol 505:364–376

Czarnomski N, Tullos D, Thomas R, Simon A (2012) Effects of vegetation canopy density and bank angle on near-bank patterns of turbulence and Reynolds stresses. J Hydraul Eng 138 (11):974–978

de Langre E (2008) Effects of wind on plants. Annu Rev Fluid Mech 40:141–168

de Langre E, Gutierrez A, Cossé J (2012) On the scaling of drag reduction by reconfiguration in plants. Comptes Rendus Mécanique 340(1–2):35–40

de Lima A, Izumi N (2014) Linear stability analysis of open-channel shear flow generated by vegetation. J Hydraul Eng 140(3):231–240

Dijkstra JT, Uittenbogaard RE (2010) Modeling the interaction between flow and highly flexible aquatic vegetation. Water Resour Res 46(W12547), 125. doi:10.11029/12010WR009246

Dittrich A, Aberle J (2010) Die Vegetation an Fliessgewässern aus dem Blickwinkel eines Hydraulikers. Ingenieurbiologie/Genie Biologique 3:37–42

Dittrich A, Koll K, Becker K. (2002) Resistance of local flood berm vegetation in compound channels. In: Proceedings of the 5th International Conference Hydro-Science and Engineering

Dittrich A, Aberle J, Schoneboom T (2012) Drag forces and flow resistance of flexible riparian vegetation. In: Rodi W, Uhlmann M (eds) Environmental fluid mechanics: memorial colloquium on environmental fluid mechanics in honour of Professor Gerhard H Jirka. CRC Press, London, pp 195–215

Folkard AM (2011) Vegetated flows in their environmental context: a review. Eng Comput Mech 164(EM1):3–24

Ghisalberti M, Nepf HM (2002) Mixing layers and coherent structures in vegetated aquatic flows. J Geophys Res 107(C2):3.1–3.11

Ghisalberti M, Nepf HM (2004) The limited growth of vegetated shear layers. Water Resour Res 40, doi:10.1029/2003WR002776

Godin C, Caraglio Y (1998) A multiscale model of plant topological structures. J Theor Biol 191 (1):1–46

Gosselin F, de Langre E (2011) Drag reduction by reconfiguration of a poroelastic system. J Fluid Struct 27(7):1111–1123

Gosselin F, de Langre E, Machado-Almeida B (2010) Drag reduction of flexible plates by reconfiguration. J Fluid Mech 650:319–341

Green JC (2005a) Modelling flow resistance in vegetated streams: review and development of new theory. Hydrolog Process 19:1245–1259

Green JC (2005b) Comparison of blockage factors in modelling the resistance of channels containing submerged macrophytes. River Res Appl 21:671–686

Grant PF, Nickling WG (1998) Direct field measurement of wind drag on vegetation for application to windbreak design and modelling. Land Degrad Dev 9:57–66

Gurnell A (2014) Plants as river system engineers. Earth Surf Proc Land 39(1):4–25

Gurnell A, Petts G (2006) Trees as riparian engineers: the Tagliamento River, Italy. Earth Surf Proc Land 31(12):1558–1574
Gurnell AM, van Oosterhout MP, de Vlieger B, Goodson JM (2006) Reach-scale interactions between aquatic plants and physical habitat: River Frome, Dorset. River Res Appl 22:667–680
Hoerner S (1965) Fluid-dynamic drag. Brick Town
Jalonen J, Järvelä J (2014) Estimation of drag forces caused by natural woody vegetation of different scales. J Hydrodyn 26(4):608–623. doi:10.1016/S1001-6058(14)60068-8
Jalonen J, Järvelä J, Aberle J (2012) Vegetated flows: drag force and velocity profiles for foliated plant stands. In: Murillo R (ed) Proceedings of river flow 2012, international conference on fluvial hydraulics, September 2012, San José 2012. Taylor & Francis, London
Jalonen J, Järvelä J, Aberle J (2013) Leaf area index as vegetation density measure for hydraulic analyses. J Hydraul Eng 139(5):461–469
Jalonen J, Järvelä J, Koivusalo H, Hyyppä H (2014) Deriving floodplain topography and vegetation characteristics for hydraulic engineering applications by means of terrestrial laser scanning. J Hydraul Eng 140(11):04014056. doi:10.1061/(ASCE)HY.1943-7900.0000928
Jalonen J, Järvelä J, Virtanen J-P, Vaaja M, Kurkela M, Hyyppä H (2015) Determining characteristic vegetation areas by terrestrial laser scanning for floodplain flow modeling. Water 7(2):420–437. doi:10.3390/w7020420
James CS, Goldbeck UK, Patini A, Jordanova AA (2008) Influence of foliage on flow resistance of emergent vegetation. J Hydraul Res 46(4):536–542
Janauer GA, Schmidt-Mumm U, Reckendorfer W (2013) Ecohydraulics and aquatic macrophytes: assessing the relationship in river floodplains. In: Maddock I, Harby A, Kemp P, Wood P (eds) Ecohydraulics—an integrated approach. Wiley, Chichester, pp 245–260
Järvelä J (2002) Flow resistance of flexible and stiff vegetation: a flume study with natural plants. J Hydrol 269:44–54
Järvelä J (2004) Determination of flow resistance caused by non-submerged woody vegetation. Int J River Basin Manage 2(1):61–70
Järvelä J (2005) Effect of submerged flexible vegetation on flow structure and resistance. J Hydrol 307(1–4):233–241
King AT, Tinoco RO, Cowen EA (2012) A k–ε turbulence model based on the scales of vertical shear and stem wakes valid for emergent and submerged vegetated flows. J Fluid Mech 701:1–39
Klaasen GJ, van Urk A (1985) Resistance to flow of floodplains with grasses and hedges. In: proceedings of 21st IAHR Congress, Melbourne, Australia
Knight DW (2013) River hydraulics—a view from midstream. J Hydraul Res 51(1):2–18
Kouwen N, Unny TE (1973) Flexible roughness in open channels. J Hydr Div 99(5):713–728
Kouwen N (1992) Modern approach to design of grassed channels. J Irrig Drain Eng 118(5):733–743
Kouwen N, Fathi-Moghadam M (2000) Friction factors for coniferous trees along rivers. J Hydraul Eng 126(10):732–740
Kubrak E, Kubrak J, Rowinski PM (2008) Vertical velocity distributions through and above submerged, flexible vegetation. Hydrological Sci J 53(4):905–920
Kubrak E, Kubrak J, Rowiński P (2012) Influence of a method of evaluation of the curvature of flexible vegetation elements on vertical distributions of flow velocities. Acta Geophys 60 (4):1098–1119
Li RM, Shen W (1973) Effect of tall vegetations on flow and sediment. J Hydraul Div 99 (HY5):793–814
Lightbody AF, Nepf HM (2006) Prediction of near-field shear dispersion in an emergent canopy with heterogeneous morphology. Environ Fluid Mech 6:477–488
Lindner K (1982) Der Strömungswiderstand von Pflanzenbeständen. Mitt. Leichtweiß-Institut für Wasserbau No. 75, Braunschweig, Technische Universität Braunschweig, Germany, in German
Luhar M, Nepf HM (2011) Flow-induced reconfiguration of buoyant and flexible aquatic vegetation. Limnol Oceanogr 56(6):2003–2017
Luhar M, Nepf HM (2013) From the blade scale to the reach scale: a characterization of aquatic vegetative drag. Adv Water Resour 51:305–316

Mertens W (2006) Hydraulisch-sedimentologische Berechnungen naturnah gestalteter Fliessgewässer. Deutsche Vereinigung für Wasserwirtschaft, Abwasser und Abfall e.V., Hennef, in German

Miler O, Albayrak I, Nikora V, O'Hare M (2011) Biomechanical properties of aquatic plants and their effects on plant–flow interactions in streams and rivers. DOI, Aquat Sci. doi:10.1007/s00027-00011-00188-00025

Miler O, Albayrak I, Nikora V, O'Hare M (2014) Biomechanical properties and morphological characteristics of lake and river plants: implications for adaptations to flow conditions. Aquat Sci. doi:10.1007/s00027-014-0347-6

Neary V, Constantinescu S, Bennett S, Diplas P (2012) Effects of vegetation on turbulence, sediment transport, and stream morphology. J Hydraul Eng 138(9):765–776

Nepf HM (1999) Drag, turbulence, and diffusion in flow through emergent vegetation. Water Resour Res 35(2):479–489

Nepf HM (2012a) Hydrodynamics of vegetated channels. J Hydraul Res 50(3):262–279

Nepf HM (2012b) Flow and transport in regions with aquatic vegetation. Annu Rev Fluid Mech 44:123–142

Nepf HM, Vivoni ER (2000) Flow structure in depth-limited, vegetated flow. J Geophys Res 105 (C12):28457–28557

Nepf H, Ghisalberti M (2008). Flow and transport in channels with submerged vegetation. Acta Geophys 56(3). doi:10.2478/s11600-11008-10017-y

Nikora N, Nikora V, O'Donoghue T (2013a) Velocity profiles in vegetated open-channel flows: Combined effects of multiple mechanisms. J Hydraul Eng 139(10):1021–1032

Nikora V (2010a) Hydrodynamics of aquatic ecosystems: an interface between ecology, biomechanics and environmental fluid mechanics. River Res Appl 26(4):367–384

Nikora V (2010b) Hydrodynamics of aquatic ecosystems: current state, challenges, and prospects. In: Proceedings of 17th Australasian fluid mechanics conference, Auckland New Zealand

Nikora VI, Rowinski PM (2008) Rough-bed flows in geophysical, environmental, and engineering systems: double-averaging approach and its applications. Acta Geophys 56(3):529–533

Nikora V, Koll K, McEwan I, McLean S, Dittrich A (2004) Velocity distribution in the roughness layer of rough-bed flows. J Hydraul Eng 130(10):1036–1042

Nikora V, McEwan I, McLean S, Coleman S, Pokrajac D, Walters R (2007a) Double-averaging concept for rough-bed open-channel and overland flows: theoretical background. J Hydraul Eng 133(8):873–883

Nikora V, McLean S, Coleman S, Pokrajac D, McEwan I, Campbell L, Aberle J, Clunie D, Ka Koll (2007b) Double-Averaging concept for rough-bed open-channel and overland flows: applications. J Hydraul Eng 133(8):884–895

Nikora V, Ballio F, Coleman S, Pokrajac D (2013b) Spatially averaged flows over mobile rough beds: Definitions, averaging theorems, and conservation Equations. J Hydraul Eng 139(8):803–811

Okamoto T, Nezu I (2009) Turbulence structure and "Monami" phenomena in flexible vegetated open-channel flows. J Hydraul Res 47(6):798–810

Okamoto TA, Nezu I (2013) Spatial evolution of coherent motions in finite-length vegetation patch flow. Env Fluid Mech 13:417–434

Osterkamp WR, Hupp CR, Stoffel M (2012) The interactions between vegetation and erosion: new directions for research at the interface of ecology and geomorphology. Earth Surf Proc Land 37:23–36

Palmer VJ (1945) A method for designing vegetated waterways. Agric Eng 26(12):16–520

Pasche E, Rouvé G (1985) Overbank flow with vegetatively roughened flood plains. J Hydraul Eng 111(9):1262–1278

Paul M, Henry PYT, Thomas RE (2014) Geometrical and mechanical properties of four species of northern European brown macroalgae. Coast Eng 84:73–80

Petryk S, Bosmajian G (1975) Analysis of flow through vegetation. J Hydraul Div 101(7):871–884

Plew D (2011) Depth-averaged drag coefficient for modeling flow through suspended canopies. J Hydraul Eng 137(2):234–247

Poggi D, Porporato A, Ridolfi L, Albertson JD, Katul GG (2004) The effect of vegetation density on canopy sub-layer turbulence. Bound-Layer Meteor 111:565–587

Puijalon S, Léna J-P, Riviére N, Champagne JY, Rostan JC, Bornette G (2008) Phenotypic plasticity in response to mechanical stress: hydrodynamic performance and fitness of 4 aquatic plant species. New Phytol 177:907–917

Rhee DS, Woo H, Kwon BA, Ahn HK (2008) Hydraulic resistance of some selected vegetation in open channel flows. River Res Appl 24:673–687

Ricardo AM, Koll K, Franca MJ, Schleiss AJ, Ferreira RML (2014) The terms of turbulent kinetic energy budget within random arrays of emergent cylinders. Water Resour Res 50(5):4131–4148

Richardson DM, Holmes PM, Esler KJ, Galatowitsch SM, Stromberg JC, Kirkman SP, Pyšek P, Hobbs RJ (2007) Riparian vegetation: degradation, alien plant invasions, and restoration prospects. Divers Distrib 13(1):126–139

Sand-Jensen K (2003) Drag and reconfiguration of freshwater macrophytes. Freshwat Biol 48 (2):271–283

Sanjou M, Nezu I (2011) Turbulence structure and concentration exchange property in compound open-channel flows with emergent trees on the floodplain edge. Int J River Basin Manag 9(3–4):181–193

Schlichting H, Gersten K (2006) Grenzschicht-Theorie. Springer, Berlin

Schoneboom T, Aberle J, Dittrich A (2011) Spatial variability, mean drag forces, and drag coefficients in an array of rigid cylinders. Exp Methods Hydrau Res Geoplanet: Earth Planet Sci 1:255–265. doi:10.1007/978-3-642-17475-9_18

SCS (1954) Handbook of channel design for soil and water conservation. Soil conservation service SCS-TP-61. U.S. Department of Agriculture, Washington, D.C

Shiono K, Knight DW (1991) Turbulent open-channel flows with variable depth across the channel. J Fluid Mech 222:617–646

Shucksmith JD, Bocall JB, Guymer I (2010) Effects of emergent and submerged natural vegetation on longitudinal mixing in open channel flow. Water Resour Res 46, W04504. doi:04510.01029/02008WR007657

Siniscalchi F, Nikora VI (2012) Flow-plant interactions in open-channel flows: a comparative analysis of five freshwater plant species. Water Resour Res 48(W05503). doi:10.1029/2011WR011557

Siniscalchi F, Nikora V (2013) Dynamic reconfiguration of aquatic plants and its interrelations with upstream turbulence and drag forces. J Hydraul Res 51(1):46–55

Siniscalchi F, Nikora VI, Aberle J (2012) Plant patch hydrodynamics in streams: mean flow, turbulence, and drag forces. Water Resour Res 48(W01513). doi:10.1029/2011WR011050

Sinoquet H, Rivet P (1997) Measurement and visualization of the architecture of an adult tree based on a three-dimensional digitising device. Trees 11(5):265–270

Souliotis D, Prinos P (2011) Effect of a vegetation patch on turbulent channel flow. J Hydraul Res 49(2):157–167

Statzner B, Lamoroux N, Nikora V, Sagnes P (2006) The debate about drag and reconfiguration of freshwater macrophytes: comparing results obtained by three recently discussed approaches. Freshwat Biol 51(11):2173–2183

Stephan U, Gutknecht D (2002) Hydraulic resistance of submerged flexible vegetation. J Hydrol 269:27–43

Stoesser T, Palau Salvador G, Rodi W, Diplas P (2009) Large eddy simulation of turbulent flow through submerged vegetation. Transport Porous Med 78:347–365

Stoesser T, Kim SJ, Diplas P (2010) Turbulent flow through idealized emergent vegetation. J Hydraul Eng 136(12):1003–1017

Stone MC, Chen L, McKay SK, Goreham J, Acharya K, Fischenich C, Stone AB (2013) Bending of submerged woody riparian vegetation as a function of hydraulic flow conditions. River Res Appl 29:195–205

Sukhodolov A, Sukhodolova T (2010) Case study: effect of submerged aquatic plants on turbulence structure in a lowland river. J Hydraul Eng 136(7):434–446

Sukhodolov AN, Sukhodolova TA (2012) Vegetated mixing layer around a finite-size patch of submerged plants: part 2. turbulence statistics and structures. Water Resour Res 48(12): W12506. doi:10.1029/2011WR011805

Sukhodolova TA, Sukhodolov AN (2012) Vegetated mixing layer around a finite-size patch of submerged plants: 1. theory and field experiments. Water Resour Res 48(10):W10533. doi:10.1029/2011WR011804

Sun X, Shiono K (2009) Flow resistance of one-line emergent vegetation along the floodplain edge of a compound open channel. Adv Water Resour 32:430–438

Tanino Y, Nepf H (2008) Lateral dispersion in random cylinder arrays at high Reynolds number. J Fluid Mech 600:339–371

Temple DM (1999) Flow resistance of grass-lined channel banks. Appl Eng Agric 15(2):129–133

Thomas RE, Johnson MF, Frostick LE, Parsons DR, Bouma TJ, Dijkstra JT, Eiff O, Gobert S, Henry PY, Kemp P, McLelland SJ, Moulin FY, Myrhaug D, Neyts A, Paul M, Penning WE, Puijalon S, Rice SP, Stanica A, Tagliapietra D, Tal M, Tørum A, Vousdoukas MI (2014) Physical modelling of water, fauna and flora: knowledge gaps, avenues for future research and infrastructural needs. J Hydraul Res. doi:10.1080/00221686.2013.876453

Vandenbruwaene W, Temmerman S, Bouma TJ, Klaassen PC, De Vries MB, Callaghan DP, van Steeg P, Dekker F, van Duren LA, Martini E, Balke T, Biermans G, Schoelynck J, Meire P (2011) Flow interaction with dynamic vegetation patches: implications for biogeomorphic evolution of a tidal landscape. J Geophys Res 116, F01008, doi:01010.01029/02010JF001788

Vargas-Luna A, Crosato A, Uijttewaal WSJ (2015) Effects of vegetation on flow and sediment transport: comparative analyses and validation of predicting models. Earth Surf Process Land 40:157–176. doi:10.1002/esp.3633

Västilä K, Järvelä J (2014) Modeling the flow resistance of woody vegetation using physically based properties of the foliage and stem. Water Resour Res 50(1):229–245

Västilä K, Järvelä J, Aberle J (2013) Characteristic reference areas for estimating flow resistance of natural foliated vegetation. J Hydrol 492:49–60

Vogel S (1994) Life in moving fluids: the physical biology of flow, 2nd edn. Princeton University Press, Princeton

Vogel S (2009) Leaves in the lowest and highest winds: temperature, force and shape. New Phytol 183(1):13–26

Whittaker P, Wilson C, Aberle J, Rauch HP, Xavier P (2013) A drag force model to incorporate the reconfiguration of full-scale riparian trees under hydrodynamic loading. J Hydraul Res 51 (5):569–580

Wilkerson GV (2007) Flow through trapezoidal and rectangular channels with rigid cylinders. J Hydraul Eng 133(5):521–533

Wilson CAME, Horrit MS (2002) Measuring the flow resistance of submerged grass. Hydrolog Process 16:2589–2598

Wilson CAME, Hoyt J, Schnauder I (2008) Impact of foliage on the drag force of vegetation in aquatic flows. J Hydraul Eng 134(7):885–891

Yager EM, Schmeeckle MW (2013) The influence of vegetation on turbulence and bed load transport. J Geophys Res Earth Surf 118(3):1585–1601

Yen BC (2002) Open channel flow resistance. J Hydraul Eng 128(1):20–39

Zeng C, Li CW (2014) Measurements and modeling of open-channel flows with finite semi-rigid vegetation patches. Env Fluid Mech 14(1):113–134

Zong L, Nepf H (2010) Flow and deposition in and around a finite patch of vegetation. Geomorphology 116:363–372

Zong L, Nepf H (2011) Spatial distribution of deposition within a patch of vegetation. Water Resour Res 47(W03516). doi:03510.01029/02010WR009516

Chapter 22
Hydraulic Influences on Dispersion and Reaeration in Rivers

J. Russell Manson and Steve G. Wallis

Abstract An important application of environmental hydraulics is the prediction of the fate and transport of dissolved oxygen within fluvial systems. For rivers this requires knowledge of the principle hydrological processes such as advection and dispersion and the physico-chemical process of re-aeration. Currently, in the absence of appropriate field measurements quantifying the mixing or aeration processes in a river, we rely on semi-empirical predictive equations that attempt to relate these processes to global flow and channel parameters. Although there is some theoretical justification for the form of these equations, they are not particularly successful even for channels of simple shape. As more complex channel shapes (e.g. two-stage flood relief channels) are tackled the equations become increasingly inappropriate. To help address this concern, the chapter proposes a theoretical approach for evaluating both the longitudinal dispersion coefficient and the re-aeration coefficient in channels of arbitrary shape that is based on integral formulations and which uses theoretical predictions of the transverse flow structure that are based on Shiono and Knight's (J Fluid Mech 222:617–646, 1991) momentum balance equation. The results for a simple channel (trapezoidal) are consistent with current knowledge, but they reveal unexpected patterns for a complex channel (two-stage, trapezoidal with active floodplains) that contains zones of distinctly different velocity and depth. The results also explore the role of the transverse turbulent transfer of momentum. For the simple channel, the dispersion coefficient was very small (being in the range 0–1 m^2/s for flow rates between 0 and 35 m^3/s and channel widths of approximately 15 m), and *increased* approximately linearly with flow rate. The influence of the transverse turbulent momentum exchange was relatively significant. For the complex channel, the dispersion coefficient was very large (being in the range 27,000–500 m^2/s for flow rates between 35 and 175 m^3/s and widths of approximately 55 m), and *decreased* with flow rate according to a power law with an exponent of about −4.7.

J.R. Manson (✉)
Stockton University, New Jersey, USA
e-mail: Russell.Manson@stockton.edu

S.G. Wallis
Heriot-Watt University, Edinburgh, UK

P. Rowiński and A. Radecki-Pawlik (eds.), *Rivers—Physical, Fluvial and Environmental Processes*, GeoPlanet: Earth and Planetary Sciences, DOI 10.1007/978-3-319-17719-9_22

The influence of the transverse turbulent momentum exchange was less than for the simple channel case. The predictions for both flow conditions are consistent with observed trends reported in Rutherford (River mixing. Wiley, Chichester, 1994). The very large dispersion coefficients found in the complex channel case could not be predicted using the existing semi-empirical equations proposed by Liu (J Environ Eng Div, Am Soc Civil Eng 103(EE1):59–69, 1977) and Deng et al. (J Hydraul Eng, Am Soc Civil Eng 127(11):919–927, 2001); neither could the rapid decrease with increasing flow rate. This is not surprising because the equations cannot represent the extremely strong transverse velocity shear that exists in these flows that contain zones of quite different velocity and depth. For the re-aeration coefficient in the simple channel we identified a power law decrease (exponent of about −0.5) with flow rate from about 40 to 10 per day up to the bank full condition. Once flows went over-bank the re-aeration coefficient jumped considerably (to about 100 per day) due to the small depths on the floodplains. It then reduced as a power law as flow rate increased (exponent of about −0.9). The influence of the transverse turbulent momentum exchange was not very significant for either channel case. Results from a semi-empirical equation proposed by Bennett and Rathbun (Reaeration in open-channel flow. United Sates Geological Survey, Washington, 74 pp, 1972) mirrored the computational results, but under-predicted the coefficient by about 50 % for both the simple and complex channel cases. Clearly, existing semi-empirical equations for the dispersion coefficient and the re-aeration coefficient should not be used for predicting non-conservative chemical transport for the over-bank case of a complex channel. A sensitivity analysis for the case of a steady oxygen demanding waste water discharge showed that the maximum dissolved oxygen sag and its location were insensitive to dispersion but were significantly sensitive to re-aeration for both channel cases. Hence, for this waste water scenario future work should focus on improving the prediction of re-aeration coefficients in both types of channel.

Keywords River longitudinal dispersion · Reaeration · Simple complex channel · Dissolved oxygen

22.1 Introduction

An important application of environmental hydraulics is the prediction of the fate and transport of natural and anthropogenic substances within fluvial systems (Rutherford 1994; Scott et al. 2003b; Ryan et al. 2004; Chin 2013) and therefore much effort has been expended developing mathematical models to aid this. The majority of these approaches have been based on constructing governing equations that describe mass transport within a moving fluid, followed by analytical or numerical solutions under specified boundary and/or initial conditions (Runkel and Chapra 1993; Manson and Wallis 2000; Cox et al. 2002). During the derivation of these equations, coefficients (theoretical or empirical) are introduced that characterise various physical processes, but application of these models is often hindered because it is difficult to predict, a priori, appropriate values of the required coefficients. When field observations are available recourse to optimisation techniques can yield the values of the coefficients that give a "best fit" to observations (Chapra and Wilcock 2000; Scott et al. 2003a; Rodrigues et al. 2013) but the coefficients are, of course, only valid for the hydraulic conditions prevailing at the time of measurement.

For the particular case of solute transport in rivers, a one-dimensional approach is usually adopted if interest focuses on the far field conditions following, for example, an accidental or controlled release of a pollutant (Fischer 1967) or if the time scales of the analysis are large enough such that a coarse grained spatial analysis is warranted (Demars et al. 2011). Experience suggests that there are three main physical mechanisms that need to be included in such a model, assuming the solute behaves conservatively (Fischer 1967; Rutherford 1994; Runkel 1998). These are: advection (describing longitudinal transport in the direction of flow), dispersion (describing bi-directional longitudinal spreading relative to the advection) and transient storage (describing temporary trapping of solute). In this approach a river is viewed as containing a main channel, in which advection and dispersion are the dominant mechanisms, and a surrounding zone encompassing parts of the river-bed and banks in which transient storage dominates (Runkel and Chapra 1993; Chapra and Wilcock 2000; Runkel 2002; Scott et al. 2003a; Ryan et al. 2004). Other mechanisms need to be included when dealing with non-conservative solutes. In the case of dissolved oxygen (the focus here) these include the oxidation rate of carbonaceous (and possible nitrogenous) compounds, the rate of reaeration from the atmosphere, the rate of photosynthesis in aquatic flora and the rate of respiration in aquatic fauna.

In contrast to much of the previous work in this field which has been concerned with simple channels, this chapter concerns complex channels, i.e. channels having two (or more) distinct flow zones. Many natural channels conform to this description as do many man-made channels. Figure 22.1 illustrates an idealised channel of this sort in which the flow is confined to the lower trapezoidal part of the cross-section during low flows (referred to as in-bank) but which occupies the complete cross-section during high flows (referred to as over-bank). Channels of

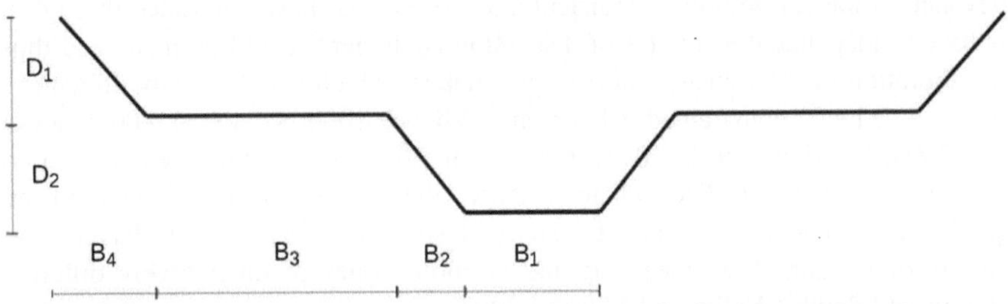

Fig. 22.1 Idealised channel cross-section

this shape are frequently constructed as flood protection measures (where they are often referred to as two-stage channels) and clearly reflect natural geomorphological conditions in which rivers spill from a main channel on to flood plains during high flow events. Although much work has been undertaken on the hydraulic and flow conveyance features of idealised complex channels (Shiono and Knight 1991; Cokljat and Younis 1995; Lambert and Sellin 1996; Ervine et al. 2000; Conway et al. 2012), such as that illustrated in Fig. 22.1, little work has been devoted to their mass transport or metabolic characteristics. Accepted knowledge of solute transport mechanisms in rivers (Fischer 1967; Fischer et al. 1979; Rutherford 1994) would suggest, however, that during periods of high flow the transverse velocity gradients between the main channel and the flood plains in this type of channel would play a pivotal role in the generation of dispersion. However, only a little previous work (Manson and Wallis 2004; Wallis and Manson 2008) has quantified the dispersive strength of this particular form of (multi-zone) transverse shear in such channels over a wide range of flows, including the transition from the simpler transverse velocity distribution found when the flow is confined to the main channel under low flow conditions. Similarly, as far as we are aware, no previous work has considered the consequences for reaeration rates of the sort of transverse variations in hydraulic conditions such as velocity and depth found in this type of channel.

The aim of this chapter, therefore, is to study these phenomena. A theoretical approach is adopted that allows longitudinal dispersion coefficients and global reaeration coefficients to be estimated from predictions of the transverse flow structure. Results are compared from two models of this flow structure that enable the importance of the transverse turbulent mixing of momentum to be identified. Furthermore, results are compared between over-bank conditions, when the channel behaves as a complex channel, and in-bank conditions, when the channel behaves as a simple channel. In both cases the predicted dispersion coefficients and reaeration coefficients are compared against existing documented trends provided by semi-empirical equations. Finally, the significance of the results is explored by undertaking a sensitivity analysis for a hypothetical water quality scenario in which the consequences of uncertainties in predicted dispersion coefficients and reaeration coefficients on dissolved oxygen sags downstream of a continuous release of oxygen demanding waste water are quantified.

Since the transverse shear is expected to be the dominant longitudinal mixing mechanism, transient storage is neglected from the analysis. The work does have relevance to the calibration of transient storage models, however, since through it improved independent estimates of dispersion coefficients could be made and thus the calibration of the transient storage parameters (which is notoriously difficult to do) could be better constrained (Chapra and Wilcock 2000; Scott et al. 2003a; Ryan et al. 2004; Rodrigues et al. 2013). Finally, although the illustrative example used here consists of an idealised channel shape, the concepts that are explored are equally relevant to the fate and transport of solutes in any natural channel that consists of a main flow zone and one or more zones of distinctively different velocity and depth (Wallis and Manson 2008).

22.2 Fate and Transport Theory of Riverine Solutes

One-dimensional solute fate and transport in steady, uniform flows in rivers may be described by the following mass conservation equation which accounts for: advection by the cross-sectional average longitudinal flow velocity; longitudinal dispersion (accounting for in-stream mixing due to the collective action of velocity shear, turbulent diffusion and secondary flows) (Fischer 1967; Fischer et al. 1979; Rutherford 1994); and possible losses or gains due to physical, biological or biochemical processes:

$$\frac{\partial C(x,t)}{\partial t} + U\frac{\partial C(x,t)}{\partial x} = D\frac{\partial^2 C(x,t)}{\partial x^2} \mp F(C(x,t)) \qquad (22.1)$$

In Eq. (22.1), C(x, t) is the cross-sectional average concentration of solute, U is the cross-sectional average longitudinal flow velocity, D is the longitudinal dispersion coefficient, x is the longitudinal spatial co-ordinate, t is time and F is a generic function describing solute losses or gains. Focussing attention on (a) dissolved oxygen as the solute of interest, and (b) the case of the recovery of depleted dissolved oxygen concentrations via reaeration from the atmosphere, Eq. (22.1) becomes:

$$\frac{\partial C(x,t)}{\partial t} + U\frac{\partial C(x,t)}{\partial x} = D\frac{\partial^2 C(x,t)}{\partial x^2} + K_a[C_s - C(x,t)] \qquad (22.2)$$

where K_a is the reaeration coefficient and C_s is the saturation concentration of dissolved oxygen in the river water. As long as $C(x, t) < C_s$, oxygen diffuses into the river from the atmosphere. Ignoring the possibility of super-saturated conditions (Schofield et al. 2012) occurring, C(x, t) never exceeds C_s. Since these equations describe a one-dimensional view, the coefficients U, D and K_a need to encapsulate any relevant multi-dimensional effects if the solutions to these equations are to be reliable. How this is achieved is described in the following three sub-sections, where the dominant theme is the distribution of the local longitudinal flow velocity within the river cross-section.

22.2.1 Advection

Assuming that the (steady) local longitudinal flow velocity is given by u(y,z) and that the local water depth is given by h(y), where y and z are the transverse and vertical spatial co-ordinates, the total volumetric flow (Q) and the cross-sectional area of flow (A) are defined as:

$$Q = \int u(y,z)dA \tag{22.3}$$

$$A = \int h(y)dz \tag{22.4}$$

where W is the top width of the flow cross-section (i.e. the width at the water surface). The rate of longitudinal advection, usually known as the cross-sectional average flow velocity, is simply defined as:

$$U = \frac{1}{A}\int u(y,z)dA = \frac{Q}{A} \tag{22.5}$$

22.2.2 Dispersion

The theoretical background to the nature of the dispersion term in Eqs. (22.1) and (22.2) is well established (Taylor 1954; Fischer 1967) for steady, uniform shear flows. The essential nature of the term is that longitudinal dispersion is modelled using a diffusion-like expression in which the rate of longitudinal spreading of a solute cloud is governed by the magnitude of the dispersion coefficient. Modelling dispersion in this way has the advantage of simplicity, but the dispersion coefficient is rather difficult to estimate a priori because dispersion arises from the interaction of two physical processes, namely differential longitudinal advection and cross-sectional mixing. The former encapsulates the transverse and vertical gradients of the longitudinal flow velocity [described by u(y, z)] that drive the longitudinal spreading while the latter continually re-distributes solute within the plane of the cross-section and tempers the longitudinal spreading. Both turbulent diffusion and secondary currents contribute to the cross-sectional mixing.

Despite the inherent difficulty of estimating the dispersion coefficient, the way in which the dispersion coefficient depends on differential longitudinal advection and cross-sectional mixing, under the Taylor/Fischer theory, is described by the following equation:

$$D = -\frac{1}{A}\int u'(y,z)f(y,z)\,dA \tag{22.6}$$

where $u'(y,z)$ is the local deviation of the longitudinal flow velocity from the cross-sectional average longitudinal flow velocity (i.e. $u'(y,z) = u(y,z) - U$) and f is a function that describes the cross-sectional distribution of the transverse and vertical mixing coefficients.

Since the work of Fischer (1967) in open channels, which was based on the earlier work of Taylor (1954) in pipes and of Elder (1959) in very wide open channels, it has generally been accepted that transverse differential longitudinal advection and transverse mixing dominate their vertical counterparts in the generation of longitudinal dispersion in natural watercourses, simply because rivers and streams are usually much wider than they are deep. The evaluation of D from Eq. (22.6) is then simplified because only the transverse distributions of depth-averaged longitudinal flow velocity and depth-averaged transverse mixing coefficient are required. Nevertheless, the practical application of this flow structure method is far from simple. Most published work suggests it requires the evaluation of the following triple integral (Jain 1976; Fischer et al. 1979; Rutherford 1994; Deng et al. 2001):

$$D = -\frac{1}{A}\int h(y)\overline{u}'(y)\int \frac{1}{\varepsilon_t(y)h(y)}\int h(y)\overline{u}'(y)dydydy \tag{22.7}$$

where $\varepsilon_t(y)$ is the local depth average transverse mixing coefficient, $\overline{u}'(y)$ is the local deviation of the depth average longitudinal flow velocity $\overline{u}(y)$ from the cross-sectional average longitudinal flow velocity (U) and the other symbols are as previously defined. Note, however, that the following alternative (but little-known) formulation exists (Smith 1976):

$$D = \frac{1}{A}\int \frac{1}{\varepsilon_t(y)h(y)}\left\{\int h(y)\overline{u}'(y)dy\right\}^2 dy \tag{22.8}$$

This is simpler to evaluate than Eq. (22.7) and may have some advantages if analytical expressions for h(y) and $\overline{u}'(y)$ are available because only two integrations are required. More usually, however, only discrete values of h(y) and $\overline{u}'(y)$ are available so that numerical integration is used: in which case either form is easily applied.

22.2.3 Reaeration

In contrast to dispersion, for which there is a single coherent well-established body of theory, there are several theoretical approaches for evaluating reaeration coefficients in rivers, (Dobbins 1964; Wilson and Macleod 1974; Melching and Flores 1999; Moog and Jirka 1998; Gualtieri and Gualtieri 2004; Raymond et al. 2012). Unfortunately, it is often difficult to decide on which of these theories is the most appropriate one to use for any particular location: also, the required data may not be available. Hence recourse is usually made to one of a great many empirical or semi-empirical equations that enable the coefficient to be estimated from commonly available hydraulic variables (Bennett and Rathbun 1972; Wilson and Macleod 1974; Rathbun 1977; Genereux and Hemond 1992; Moog and Jirka 1998; Melching and Flores 1999; Jain and Jha 2005; Aristegi et al. 2009; Haider et al. 2013). Interestingly, the danger of using an inappropriate predictor (e.g. because it is based on a theory valid for, or on observations taken under, radically different flow conditions to those being studied) is often overlooked when choosing one to use. Probably the most frequent combination of parameters used in these predictors is water depth, longitudinal flow velocity and longitudinal channel slope, so that many of them are of the following form:

$$k_a(y) = Eh(y)^q\tilde{u}(y)^p S_0^r \tag{22.9}$$

where $k_a(y)$ is the local reaeration coefficient, E is a numerical constant, S_0 is the longitudinal channel slope, h(y) is the water depth (as before), $\tilde{u}(y)$ is a representative local longitudinal flow velocity and p, q and r are numerical constants. Note that this equation will give k_a at a standard temperature and so this must be corrected for the prevailing temperature (Demars and Manson 2013). For a simple cross-section such as would pertain in Fig. 22.1 if the water depth were less than D_1, using the cross-sectional average depth and cross-sectional average longitudinal flow velocity in place of h(y) and $\tilde{u}(y)$, respectively, in one of these equations would probably be a justifiable approach, but the same is not true for the more complex flows which would pertain if the water depth were greater than D_1. Here, there are significantly different local water depths and local longitudinal flow velocities at different parts of the cross-section, so that the reaeration coefficient is likely to vary across the channel width. A simple, but effective, way to account for these transverse heterogeneities is to evaluate the reaeration coefficient locally and to then compute the width average value of the distributed coefficient using the following equation:

$$K_a = \frac{1}{W}\int k_a(y)dy \tag{22.10}$$

In the following section we discuss briefly some options for providing the velocity field required to implement the methods introduced above, and we describe the particular approach we used.

22.3 Theoretical Model for the Velocity Field

Several hydrodynamic models have been proposed for predicting the transverse distribution of the longitudinal flow velocity in channels. For example, three-dimensional numerical models have been developed (Krishnappan and Lau 1986; Naot et al. 1993; Lin and Shiono 1995; Cokljat and Younis 1995; Sofialids and Prionos 1999; Rameshwaran et al. 2013), however simpler two-dimensional models suffice for our purposes here in which we require the transverse distribution of depth-averaged longitudinal flow velocity. Keller and Rodi (1988) developed such a model that incorporated a two-equation model of turbulence while several workers have adopted simpler zero-equation approaches for turbulence (Shiono and Knight 1991; Lambert and Sellin 1996; Ervine et al. 2000). Even more simply, the transverse longitudinal flow velocity distribution may be obtained by the local application of any bed friction law. However, this approach ignores the effect of the transverse mixing of momentum by turbulence, so may be too simplistic.

The particular approach adopted here is based on the SKM model (Shiono and Knight 1991; Knight 2013), which provides a balance between containing sufficient detail whilst allowing solutions to be relatively easily obtained. In addition, a minor modification to it also enables the performance of the most simplistic approach introduced above to be easily assessed.

Following Shiono and Knight (1991), therefore, the two-dimensional depth-averaged shallow water equations may be simplified for the case of fully developed steady flow in a river of general cross-section. In the simplest case, the steady depth-averaged longitudinal momentum conservation equation can be written as:

$$\rho g S_o h(y) = \tau_b(y)\cos\theta(y) - \alpha\frac{d}{dy}\left(h\tau_{yx}(y)\right) \tag{22.11}$$

in which ρ is the fluid density, g is the gravitational acceleration, S_o is the channel bed slope, h(y) is the local depth (as before), $\tau_b(y)$ is the local longitudinal bed shear stress, θ(y) is the local transverse slope of the bed and $\tau_{yx}(y)$ is the local depth-averaged shear stress in the x-direction on the plane perpendicular to the y-direction. A binary constant, α, has been included here for convenience (the minor modification referred to earlier). If α is set to 0 then a transversely varying velocity profile results, based only on the local representation of longitudinal bed shear stress (the most simplistic method introduced above). However, when α is set to 1, the role of the transverse momentum exchange is included. Equation (22.11), which applies at all transverse locations across a river cross-section, describes a local balance between the component of fluid weight down the channel slope and the resisting longitudinal shear stresses on horizontal and vertical planes (caused by bed roughness and transverse turbulent momentum exchange, respectively). In Eq. (22.11) it has been assumed that the effect of transverse secondary flows can be ignored. Also note that for a transversely horizontal domain, in which h(y) is constant, θ(y) = 0.

Using a quadratic friction law the local longitudinal bed shear stress can be rewritten in terms of Manning's resistance coefficient, n(y), giving:

$$\tau_b(y) = \rho g\frac{\bar{u}(y)^2 n(y)^2}{h(y)^{1/3}} \tag{22.12}$$

where $\bar{u}(y)$ is the local depth-averaged longitudinal flow velocity (as before) and the hydraulic radius has been replaced with the local depth. This local application of a global resistance relationship mirrors the assumptions made by others who have undertaken similar work (Jain 1976; Deng et al. 2001, 2002).

Using the Boussinesq eddy-viscosity concept, the longitudinal shear stress caused by transverse turbulent momentum exchange can be represented in terms of the transverse gradient of local depth-averaged longitudinal flow velocity, as follows:

$$\tau_{yx}(y) = \rho\varepsilon_{yx}(y)\frac{d\bar{u}(y)}{dy} \tag{22.13}$$

where $\varepsilon_{yx}(y)$ is the eddy viscosity, which in the simplest case can be written as:

$$\varepsilon_{yx}(y) = \lambda h(y)u(y) \tag{22.14}$$

in which λ is a global dimensionless eddy viscosity and $u_*(y)$ is the local shear velocity. Combining Eqs. (22.11)–(22.14) gives:

$$S_o h(y) = \frac{\bar{u}(y)^2 n(y)^2 \cos\theta}{h(y)^{1/3}} - \frac{\alpha}{g}\frac{d}{dy}\left(\lambda h(y)^2 u(y)\frac{d\bar{u}(y)}{dy}\right) \tag{22.15}$$

The local shear velocity was evaluated by applying Eq. (22.12), assuming that locally $\tau_b(y) = \rho u_*(y)^2$. Thus:

$$u(y) = \frac{g^{1/2}\bar{u}(y)n(y)}{h(y)^{1/6}} \tag{22.16}$$

Equation (22.15) is non-linear in the depth average velocity because of the quadratic term and because the shear velocity has a linear dependency on it. However, the equation is readily solved for the depth average velocity using a numerical method, as described below.

22.4 Computational Details

22.4.1 Advection

The velocity field was computed by solving Eq. (22.15) within a one-dimensional computational domain extending between the channel banks, and located perpendicular to the longitudinal flow. The domain contained N computational points (nodes) uniformly distributed at a spacing of Δy between y = 0 and y = (N − 1)Δy. Equation (22.15) was discretised using central finite differences for the spatial derivatives, producing the following non-linear equation for the j-th computational node:

$$S_o h_j = \frac{\bar{u}_j^2 n_j^2 \cos\theta_j}{h_j^{1/3}} - \frac{\alpha}{g}\frac{1}{\Delta y}\left(\lambda(h^2u)_{j+1/2}\frac{\bar{u}_{j+1} - \bar{u}_j}{\Delta y} - \lambda(h^2u)_{j-1/2}\frac{\bar{u}_j - \bar{u}_{j-1}}{\Delta y}\right) \tag{22.17}$$

Since j extends from 1 to N there are potentially N equations in the system of equations describing the entire channel width. However, Eq. (22.17) need only be applied at the N-2 interior nodes because two boundary conditions are available (at j = 1 and j = N). At these boundary nodes $\overline{u}_j$ was set to zero since in the cases examined here the velocity (and depth) reduce simultaneously to zero at the boundaries. Values of the local shear velocity (mid-way between nodes) were calculated using Eq. (22.16). These calculations used (simple) mean values of velocity, depth and roughness derived from the surrounding nodal values.

The system of non-linear equations that was built up from Eq. (22.17) was linearized using Newton's method (Gerald and Wheatley 1994) and assembled into a matrix equation. This was solved by a direct method using a double-sweep approach for a tri-diagonal matrix (Wallis et al. 1989). Required values of local depth and local transverse bed slope were provided from the channel geometry of the case study (see later). Once the velocity field had been computed, Q was obtained from Eq. (22.3) using numerical integration.

22.4.2 Dispersion

The dispersion coefficient was evaluated using Eq. (22.7) into which the computed transverse distribution of depth-averaged longitudinal flow velocity was substituted. Assuming a neutrally buoyant solute and in the absence of secondary flows (see earlier), it was assumed that momentum and solute mix at the same rate (Rutherford 1994), although this is not supported by all work (Lin and Shiono 1995). Hence the depth-averaged transverse mixing coefficient was set equal to the eddy viscosity. Note that since Eq. (22.14) was used to specify the eddy viscosity, the transverse mixing of solute was treated in a truly local manner. Equation (22.7) was evaluated using numerical integration on the same computational grid that was used to evaluate the velocity field.

22.4.3 Reaeration Coefficient

The reaeration coefficient was evaluated locally using the following equation (Bennett and Rathbun 1972):

$$k_a(y) = 5.58h(y)^{-1.689}\overline{u}(y)^{0.607} \tag{22.18}$$

in which $k_a(y)$ has units of per day and is the value at a temperature of 20 °C. Clearly, it is a particular form of the general expression introduced earlier [Eq. (22.9)]. The rationale for choosing this equation from the many tens available was as follows. Haider et al. (2013) is one of the latest studies to evaluate the performance of predictive equations for reaeration rates. They tested 29 equations when simulating the spatial distribution of dissolved oxygen (DO) concentration in the Ravi River, Pakistan, using the classic DO sag model approach (Streeter and Phelps 1925) incorporating both carbonaceous and nitrogenous oxidation phases. Comparisons between observed and simulated DO concentrations were used to derive performance statistics for the 29 equations used. They concluded that equations containing only longitudinal flow velocity and water depth performed better than more complex equations. Of the 13 equations of this type, 5 were from flumes and 1 was from a mountainous river, and we considered these to be unsuitable for our application. The remaining 7 had velocity exponents in the range 0.5 to 1.0 and depth exponents in the range 1.4 to 1.89. Seeking to use a representative equation, we sought one that had exponents from the centre of these ranges, and Bennett and Rathbun's equation met this criterion quite well. Additionally in some preliminary tests an equally plausible equation (Bansal 1973) yielded similar values, so the choice of equation did not appear to be a particularly significant issue.

Local values of the reaeration coefficient were evaluated mid-way between the nodes of the computational grid used for predicting the velocity field. Required values of local water depth and local depth average longitudinal flow velocity were provided from the channel geometry of the case study and the computed velocity field, respectively. Finally, K_a was evaluated numerically from Eq. (22.10).

22.5 Application to Case Study

The channel shape shown in Fig. 22.1 was used to illustrate the phenomenon under study. The shape is based on a hypothetical benchmark channel suggested by Ackers (1992, 1993). The channel parameters shown in the figure took the following values: B_1 = 15 m; B_2 = 1.5 m; B_3 = 20 m; B_4 = 1.5 m; D_1 = 1.5 m; D_2 = 1.5 m. Manning's n was taken to be 0.05 for the main channel and 0.07 for the floodplains. The bed slope, S_o, was set to 0.003 and the global dimensionless transverse mixing coefficients (for momentum and solute) were assigned the value 0.16—a typical value for straight symmetric channels (Rutherford 1994).

Calculations were carried out for thirty-two main channel water depths from 0.09375 to 3 m inclusive, thus covering flow scenarios in both the in-bank (simple) and over-bank (complex) cases. For each water depth the flow rate, the dispersion coefficient and the (width average) reaeration coefficient were computed. For each flow depth at least four runs were undertaken with Δy being successively refined until the results became independent of Δy. Only when the results had converged were they deemed to be acceptable. These so-called grid independence tests are a necessary part of any numerical modelling study since it must be recognised that: (1) numerical errors are present in all numerical solutions and they must be minimized and (2) scale dependent physical processes are present in most systems and grid independence testing helps ensure that all important scales are captured. In this case capturing the shear layers was important.

In addition, estimates of the dispersion coefficient and the reaeration coefficient were made using cross-sectional average values of the basic hydraulic parameters. These so-called naïve values of the coefficients were typical of those that would be obtained if the transverse velocity distribution are unavailable: such is the case, for example, when using commercial one-dimensional river modelling software packages. For the reaeration coefficient we used Eq. (22.18) with cross-sectional average values in place of local values. Several options were available for the dispersion coefficient (Wallis and Manson 2004; Chin 2013). Similarly to the case of the reaeration coefficient, predictive equations for the dispersion coefficient typically use flow velocity and water depth, but flow width and shear velocity are also important. We chose the following two equations (Liu 1977; Deng et al. 2001) primarily because they performed well when predicting observed values in a previous study (Wallis and Manson 2004).

$$\text{Liu} \quad D = \frac{0.18 W^2 U^{0.5} U^{0.5}}{H} \tag{22.19}$$

$$\text{Deng et al.} \quad D = \frac{0.15 W^{1.67} U^2}{8 \varepsilon H^{0.67} U} \tag{22.20}$$

where

$$\varepsilon = 0.145 + \frac{1}{3520} \frac{W^{1.38} U}{H^{1.38} U} \tag{22.21}$$

Here W and U are as previously defined, H and U_* are cross-sectional average values of water depth and shear velocity and ε is a dimensionless transverse mixing coefficient.

22.6 Results

22.6.1 Advection

Transverse profiles of longitudinal flow velocity for a typical in-bank depth (simple channel shape) and a typical over-bank depth (complex channel shape) are shown in Fig. 22.2. For the former, the velocity is the same over the majority of the flow width with very narrow shear zones close to the banks. In contrast, for the latter the velocity in the main channel is about three times larger than the velocity on the flood plains and there are significant shear zones at the boundaries between the main channel and the flood plains. The results for the two values of α differ only in the shear zones. The way in which the cross-sectional average velocity varies with flow rate is shown in Fig. 22.3, where the velocity axis has a logarithmic scale. The transition between in-bank and over-bank flow occurs at a flow rate of about 35 m^3/s. For the in-bank

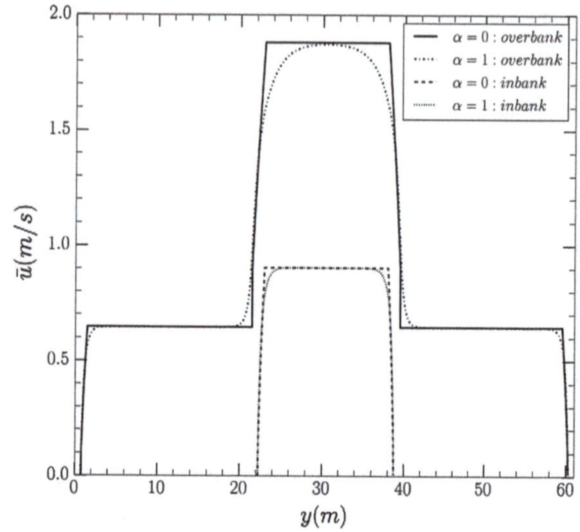

Fig. 22.2 Transverse velocity profile: in-bank, Q = 10.6 m^3/s, water level 0.75 m; over-bank, Q = 89.9 m^3/s, water level 2.25 m

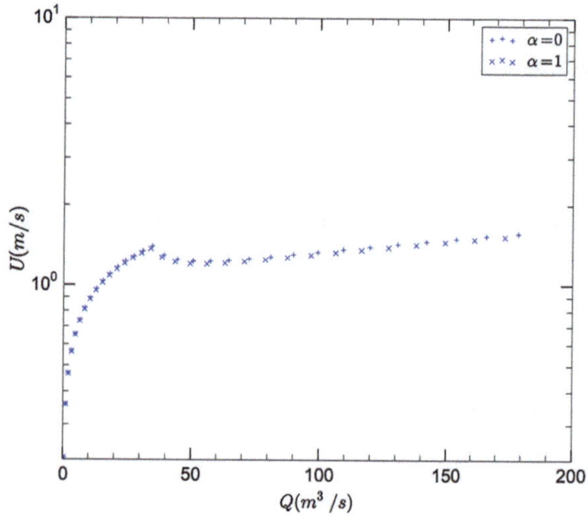

Fig. 22.3 Variation of cross-sectional average flow velocity with flow rate

case the velocity increases smoothly and rapidly with flow rate. For the over-bank case the velocity initially decreases (reflecting the presence of a large expanse of slowly moving water on the flood plains) and then increases smoothly but slowly. These trends are the same for both values of α. The effect of α, however, is noticeable when results for the same water level are compared. In particular, the inclusion of the transverse mixing of momentum (α = 1), produces a lower cross-sectional average velocity and lower flow rate compared to the α = 0 case. The smaller velocities in the α = 0 case can also be seen in the shear zones of the transverse profiles shown in Fig. 22.2.

22.6.2 Dispersion

The results for both values of α are shown in the form of dispersion coefficient plotted against flow rate in Fig. 22.4 for both the in-bank and over-bank cases (a logarithmic scale being used for dispersion). For the in-bank case the dispersion coefficient increases smoothly with increasing flow rate and is very small, being in the range 0–1 m^2/s for flows between 0 and 35 m^3/s. For both values of α the trend is well approximated by a linear relationship. Note this is not obvious from Fig. 22.4 due to the logarithmic scale for the y axis. When α = 1, the dispersion coefficient is larger than its value for α = 0 at a similar flow rate. This difference reduces from a factor of about 2.5 to a factor of about 1.6 as the flow rate increases. Both the size of the dispersion coefficients and the influence of α are supported by the predicted transverse velocity profiles that are shown in Fig. 22.2. Since dispersion is promoted by gradients in transverse velocity, the very narrow shear zones at the channel sides when α = 0 are only capable of inducing minimal dispersion. When α = 1 the shear zones are wider, causing an increased dispersion coefficient. However, the velocity is approximately constant over about three-quarters of the width of the channel, so that the dispersion coefficient remains very small. On the other hand the sensitivity of the dispersion coefficient to the shape of the velocity profile is clearly shown by the relatively minor differences between the profiles causing a significant enhancement of the coefficient (a factor of 2.4, for the case shown in Fig. 22.2). Note that these predicted dispersion coefficients are generally smaller than observed values in rivers of a similar size (Rutherford 1994). This is probably caused by the predicted velocity profiles being smoother than those encountered in the field.

For the over-bank case the results are significantly different. Not only is the magnitude of the dispersion coefficient markedly different, but so is the variation

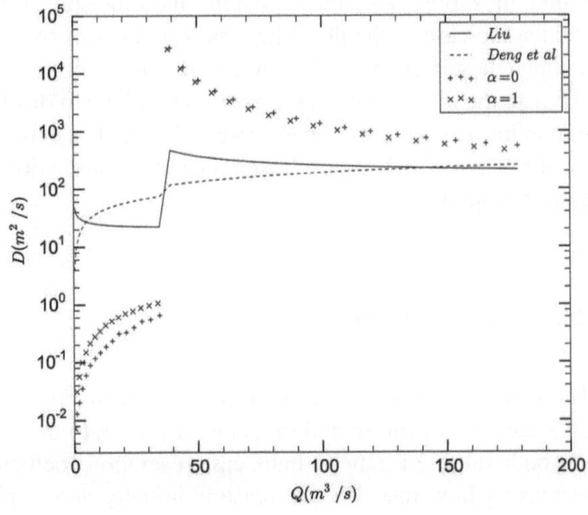

Fig. 22.4 Variation of dispersion coefficient with flow rate

with flow rate. In addition, the influence of α is reversed and the sensitivity of the results to the value of α is much reduced, see Fig. 22.3. As the flow moves from in-bank to over-bank there is a sharp and large increase in dispersion. The dispersion coefficients are now in the range 27,000–500 m^2/s for flow rates between 35 and 175 m^3/s, respectively—several orders of magnitude larger than for the in-bank case. The coefficients decrease smoothly and rapidly initially and then more slowly at higher flow rates. The overall trend is described by $D \propto Q^{-a}$ where a takes a value of approximately 4.7. Larger dispersion coefficients are now found when $\alpha = 0$ than when $\alpha = 1$, in contrast to the in-bank case, but the differences are now quite small, being about 10 % over the range of flow rate shown.

Clearly, the dispersion is dominated by the role of the strong shear that develops as the flow begins to occupy the small depth and velocity regions on the flood plains. This shear is easily recognised in the transverse velocity profiles shown in Fig. 22.2, which shows that the velocity in the main channel is about two times larger than the velocity in the shallow zones on either side for both values of α. The influence of the velocity shear decreases as the flow depth in the slow flowing region increases (increasing flow rate), but it retains its dominant role throughout the flow rate range considered. At all flow rates in the over-bank case the differences in the velocity profiles between $\alpha = 0$ and 1 are insignificant in comparison to the velocity gradients between the fast and slow flowing regions of the channel. Note that these predicted dispersion coefficients are generally much larger than observed values in rivers of a similar size (Rutherford 1994). This is probably because few if any of these measured values have been observed during flow conditions that are consistent with those considered here, i.e. channels containing zones of markedly different velocity and depth, and no secondary currents.

Figure 22.4 also shows dispersion coefficients obtained using the semi-empirical equations of Liu (1977) and Deng et al. (2001). For the in-bank case both semi-empirical equations significantly over-estimate the dispersion coefficient. Interestingly, they display opposite trends with increasing flow rate, which was also found in a previous study (Wallis and Manson 2004). Both equations show a sudden increase as the flow changes from in-bank to over-bank, with Liu's showing a much larger change. For over-bank flows both equations significantly under-estimate the dispersion coefficient and continue to display opposite trends. It is very important to realise that these results imply that existing semi-empirical equations for the longitudinal dispersion coefficient do not work for overbank flows in rivers with floodplains.

22.6.3 Reaeration

The results for both values of α are shown in Fig. 22.5 for both the in-bank and over-bank cases (presented in a similar format to the dispersion coefficient results). For both values of α the in-bank case reaeration coefficient decreases smoothly with increasing flow rate. The decrease is initially very rapid but it becomes slower as

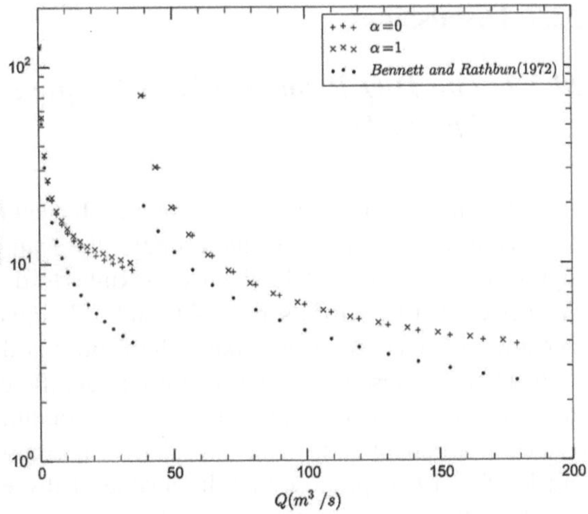

Fig. 22.5 Variation of reaeration coefficient with flow rate

bank full conditions are approached. At very low flow rates, the coefficient is very large due primarily to the presence of small values of water depths in Eq. (22.18). Otherwise, it takes values in the range 40–10 per day for flows between 0 and 35 m^3/s. The results for $\alpha = 1$ are a little larger than those for $\alpha = 0$ (about 3 %), but for both values of α the overall trend is described by $K_a \propto Q^{-b}$ where b takes a value of approximately 0.53.

The results are similar in form for the over-bank case, but there is a discontinuity as the flow condition changes from in-bank to over-bank. The very large values found when the flood plains carry very shallow water initially reduce rapidly and then more slowly as the flow rate increases. The influence of α is similar to the in-bank case but the sensitivity of the results to the value of α is reduced, see Fig. 22.5. The reaeration coefficients are now in the range 100–6 per day for flow rates between 35 and 175 m^3/s, respectively. The overall trend is described by $K_a \propto Q^{-b}$ where b takes a value of approximately 0.9. The reaeration appears to be dominated by the role of small depths and grows rapidly as the depth approaches zero; this phenomenon occurs both for the in-bank case (when the main channel depth is low) and for the over-bank case (when the floodplain depth is low).

Figure 22.5 also shows reaeration coefficients obtained using the semi-empirical equation of Bennett and Rathbun (1972) and global (cross-sectional averaged) values for velocity and depth. For both the in-bank and over-bank cases the semi-empirical equation under-estimates the reaeration coefficient and, as with the other results, there is a discontinuity as the flow changes from in-bank to over-bank conditions. Not unexpectedly, Bennett and Rathbun's equation is highly influenced by small depths. However, it differs from the semi-empirical equations for dispersion in the important regard that the correct trend in K_a versus Q is captured for both in-bank and over-bank cases.

22.7 Discussion

22.7.1 The D-Q Relationship in Simple and Complex Channels

The different behaviour of the dispersion coefficient between in-bank and over-bank flows is an important result. Interestingly, the nature of these theoretical results is supported by experimental evidence in Rutherford (1994). Firstly, for the simpler channel case, Rutherford's Figs. 4.6 and 4.7 suggest that dispersion coefficients increase with increasing flow rate. The former is directly comparable to the case studied here because it considers conditions at different flow rates in the same reach, for four rivers. In contrast, the latter shows conditions from a very wide range of reaches from about sixty rivers worldwide. It is also worth noting that McQuivey and Keefer (1974) propose that, by analogy between the diffusion of flood waves and the diffusion of solute, for simple channels D α Q. The results in Fig. 22.4 support this hypothesis.

Secondly, for the more complex channel case Rutherford's Fig. 5.14 shows dispersion coefficients decreasing rapidly with increasing flow rates in four reaches of a single river. The results presented in Fig. 22.4 support Rutherford's data and his explanation is resonant of the conditions in the case studied here, namely that bankside areas of shallow flow create strong transverse velocity shear at low flows, but as the flow increases the velocity in the shallows increases causing a reduction in the transverse shear and a concomitant reduction in the dispersion coefficient. Deng et al. (2001) also comment that in over-bank flows the dispersion coefficient may reduce with increasing flow rate. Interestingly, they do not comment on what may happen as the flow changes from in-bank to over-bank, but they do suggest that dispersion coefficients in compound (similar to complex herein) and more natural cross-sectional geometry channels are likely to be large.

It is evident that the magnitude of the dispersion coefficient and its relationship to flow rate is critically dependent on the flow regime. In a study of existing semi-empirical relationships for dispersion coefficients Wallis and Manson (2004) argue that none of them are able to account for the effects of the strong transverse shear due to velocity gradients occurring in complex channels. Hence, they are all likely to under-predict dispersion coefficients in such cases, probably by several orders of magnitude. Clearly, workers using one-dimensional models to predict solute transport in such conditions will find it extremely difficult to calibrate their models using existing predictors. At the very least, they must recognise the fundamental difference between simple and complex channels, namely that for the former dispersion coefficients tend to increase with increasing flow rate, but for the latter they tend to reduce. Also, there is an extremely rapid increase (over several orders of magnitude, indeed practically a discontinuity) in the case study used here as the flow begins to occupy the shallow flow depth regions adjacent to the main channel in the complex case. In more general terms, large dispersion coefficients would be expected in a complex channel, whenever the flow occupies more than one velocity zone.

22.7.2 The K_a-Q Relationship in Simple and Complex Channels

The most striking thing about the K_a relationship with discharge is the effect of small depths whether in the main channel at very low flow rates or when the flow has just gone over-bank and thus small depths pertain across the floodplains. Bennett and Rathbun's semi-empirical equation captures the correct trend in K_a even if using global (cross-sectional averaged) values for the hydraulic variables. However it does not seem to capture the correct magnitudes; the semi-empirical equation gives values that are about 40–60 % lower than the transversely integrated ones. This occurs because the lateral discretisation allows significantly increased local values of reaeration to be included in the integrated value. It is probably necessary to investigate further ways of computing k_a at low water depths.

22.7.3 The Effect of α

Interestingly, the inclusion of the transverse turbulent momentum exchange in Eq. (22.11) produces a more realistic transverse velocity profile, indicating that existing predictive equations for dispersion coefficients in simple channels that rely on the local application of a resistance equation to generate a velocity profile (Jain 1976; Deng et al. 2001, 2002) may not be as reliable as previously thought. Note that in the context of the present approach setting $\alpha = 0$ does not result in an exact equivalency with these previous approaches since the solutions of Eq. (22.3) obtained here use truly local values for both $u_*(y)$ and $\varepsilon_t(y)$, in contrast to the use of global values. None-the-less, the results presented here indicate that previous theoretical approaches (Jain 1976; Deng et al. 2001, 2002) may underestimate both D and K_a by artificially constraining the width of shear layers.

22.7.4 Some Caveats

The ideas discussed above raise some interesting ideas. Four issues in particular, however, should be borne in mind when considering the results, and their bearing on the predictions are undoubtedly in need of further research. Firstly, how appropriate is it to ignore secondary flows for either or both the simple and two-stage channel case? Secondly, is it appropriate to assume that the transverse mixing of momentum and solutes are the same? Thirdly, is it appropriate to assume that the non-dimensional transverse mixing coefficient is (a) the same for both simple and two-stage channel shapes and is (b) constant across either cross-sectional shape? Fourthly, what is the most appropriate way of representing transverse mixing if both mixing caused by turbulent diffusion and transverse dispersion need to be included? Interestingly, all these issues concern transverse mixing, and it should be noted that, when addressed, if they resulted in an increase in the strength of transverse mixing the predicted dispersion coefficients would decrease. Because the effect of α is not as pronounced in the K_a predictions these issues may be less important for them.

22.7.5 Sensitivity Analysis

Sensitivity analysis were undertaken in order to explore the consequences of uncertainties in the predicted dispersion coefficients and reaeration coefficients. A typical water quality scenario of the impact on dissolved oxygen downstream of a continuous release of waste water was considered using the well-known Streeter-Phelps analytical model including the effect of longitudinal dispersion (Thomann and Mueller 1987). Results were obtained for two base cases in the channel considered before, one each for in-bank and for over-bank conditions, followed by results for the same cases but with individual perturbations in the range −25 to +25 % in dispersion and reaeration coefficients. For both base cases the following parameter values were used: waste water loading rate of 3 kg/s; water temperature of 10 °C; de-oxygenation rate of 0.5 per day (at 10 °C); dissolved oxygen saturation concentration of 11.3 mg/l (at 10 °C). For the in-bank case, other base data (taken from the results presented earlier) were: flow rate of 10.5 m^3/s, flow velocity of 0.88 m/s, dispersion coefficient of 0.33 m^2/s, reaeration coefficient of 13.4 per day (equivalent to 17.0 per day at 20 °C). For the over-bank case, other base data were: flow rate of 87.5 m^3/s, flow velocity of 1.27 m/s, dispersion coefficient of 1493 m^2/s, reaeration coefficient of 5.63 per day (equivalent to 7.18 per day at 20 °C). Attention focused on the maximum dissolved oxygen deficit and its location.

For both flow cases, the results were very insensitive to dispersion coefficient (changes in both model outputs being <0.3 % in magnitude). Usually, dispersion is ignored in the Streeter-Phelps model and the insensitivity in the in-bank case where the dispersion coefficients were very small was expected. It was not, a priori, clear what would be found for the over-bank case where the dispersion coefficients were very large. However, these results show that under these extremely dispersive conditions the Streeter-Phelps model remains insensitive to dispersion.

Results showing the sensitivity of the model outputs to reaeration are shown in Fig. 22.6. Both model outputs show a slightly larger sensitivity to reaeration for the in-bank case (Fig. 22.6a) compared to the over-bank case (Fig. 22.6b). In general, however, for both flow cases the magnitudes of the changes in both model outputs are similar to the magnitude of the perturbation in reaeration, although the relationship is an inverse one (i.e. a positive perturbation in reaeration produces a negative change in the model outputs).

Fig. 22.6 **a** Sensitivity plot: in-bank case, Q = 10.5 m³/s, U = 0.88 m/s, D = 0.33 m²/s, K_a = 17.0/day. **b** Sensitivity plot: over-bank, Q = 87.5 m³/s, U = 1.27 m/s, D = 1493 m²/s, K_a = 7.2/day

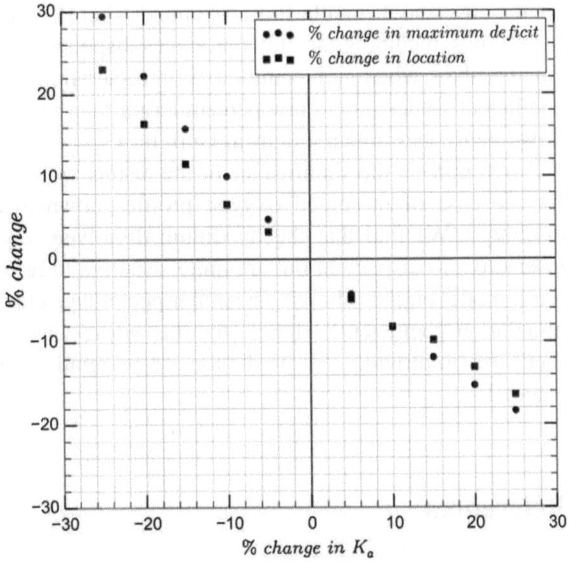

22.8 Conclusions

In this chapter we demonstrate that existing semi-empirical equations for computing longitudinal dispersion coefficients and re-aeration coefficients based on global (cross-sectional average) quantities, such as longitudinal flow velocity, water depth, top-width and shear velocity, are inappropriate for rivers with complex cross-sections, i.e. channels that contain zones of distinctly different velocity and depth. We achieve this by using an alternative approach based on a modified version of Shiono and Knight's (1991) model for the transverse distribution of longitudinal flow velocity and integral expressions for the dispersion and re-aeration coefficients.

The results for dispersion in a simple channel are consistent with current knowledge (dispersion increases with flow rate), but they reveal unexpected patterns for a complex channel (dispersion reduces with flow rate, having jumped by several orders of magnitude at the transition between channel types). The results for re-aeration in both channel types are consistent with each other and with current knowledge (re-aeration decreases with increasing flow rate), but there is a jump of about one order of magnitude at the transition. An important issue with many empirical equations for the re-aeration coefficient is that, with flow depth being in the denominator, the coefficient approaches infinity as the flow rate approaches zero. We suggest that an improved formulation for very low flow rates is required to overcome this difficulty.

A study of the importance of including the transverse mixing of momentum in the SKM model shows that more realistic transverse velocity profiles are obtained when it is included. For in-bank flows, including the effect produces dispersion coefficients about double those obtained when it is omitted. In contrast, for over-bank flows, including the effect produces dispersion coefficients that are about 10 % smaller than when it is omitted. For re-aeration, including the effect produces coefficient values only a few percent larger than when it is omitted, with over-bank flows being less affected than in-bank flows.

We report on the significance of the results by undertaking a sensitivity analysis for the case of dissolved oxygen profiles downstream of a steady waste water discharge. We find that the maximum dissolved oxygen sag and its location are insensitive to dispersion but are quite sensitive to re-aeration in both channel cases (both model outputs showing a similar percentage change to that used to perturb the re-aeration coefficient). Hence, for this waste water scenario future work should focus on improving the prediction of re-aeration coefficients in both types of channel.

References

Ackers P (1992) Hydraulic design of two-stage channels. Proc Inst Civil Eng, Water Marit Energy 96:247–257

Ackers P (1993) Flow formulae for straight two-stage channels. J Hydraul Res 31(4):509–531

Aristegi L, Izagirre O, Elosegi A (2009) Comparison of several methods to calculate reaeration in streams, and their effects on estimation of metabolism. Hydrobiologia 635:113–124

Bansal MK (1973) Atmospheric reaeration in natural streams. Water Res 7:769–782

Bennett JP, Rathbun RE (1972) Reaeration in open-channel flow. Professional paper 737, United Sates Geological Survey, Washington, 74 pp

Chapra SC, Wilcock RJ (2000) Transient storage and gas transfer in lowland streams. J Environ Eng, Am Soc Civil Eng 126(8):708–712

Chin DA (2013) Water-quality engineering in natural systems. Wiley, Hoboken

Cokljat D, Younis BA (1995) 2nd-order closure study of open-channel flows. J Hydraul Eng 121:2, 94–107

Conway P, O'Sullivan JJ, Lambert MF (2012) Stage–discharge prediction in straight compound channels using 3D numerical models. Proc Inst Civil Eng, Water Manage 166(1):3–15

Cox TJ, Rutherford JC, O'Sullivan MJ (2002) Towards modelling nutrient transport and uptake in lowland streams with transient storage. In: Proceedings of CSCE/ASCE international conference on environmental engineering, Niagara Falls, Canada

Demars BOL, Manson JR, Olafsson JS, Gislason GM, Gudmundsdottir R, Woodward G, Reiss J, Pichler D, Rasmussen JJ, Friberg N (2011) Temperature and the metabolic balance of streams. Freshw Biol 56:1106–1121

Demars BOL, Manson JR (2013) Temperature dependence of stream aeration coefficients and the effect of water turbulence: a critical review. Water Res 47:1–15

Deng Z-Q, Singh VP, Bengtsson L (2001) Longitudinal dispersion coefficient in straight rivers. J Hydraul Eng, Am Soc Civil Eng 127(11):919–927

Deng Z-Q, Bengtsson L, Singh VP, Adrian DD (2002) Longitudinal dispersion coefficient in single-channel streams. J Hydraul Eng, Am Soc Civil Eng 128(10):901–916

Dobbins WE (1964) BOD and oxygen relationships in streams. J Sanitary Eng Div, Am Soc Civil Eng 90(SA3):53–77

Elder JW (1959) The dispersion of marked fluid in turbulent shear flow. J Fluid Mech 5(4):544–560

Ervine DA, Babaeyan-Koopaei K, Sellin RHJ (2000) Two-dimensional solution for straight and meandering overbank flows. J Hydraul Eng, Am Soc Civil Eng 126(9):653–669

Fischer HB (1967) The mechanics of dispersion in natural streams. J Hydraul Div, Am Soc Civil Eng 93(HY6):187–215

Fischer HB, List EG, Koh RCY, Imberger J, Brooks NH (1979) Mixing in Inland and Coastal Waters. Academic Press, New York

Genereux DP, Hemond HF (1992) Determination of gas exchange rate constants for a small stream on Walker Branch Watershed, Tennessee. Water Resour Res 28(9):2365–2374

Gerald CF, Wheatley PO (1994) Applied numerical analysis. Addison-Wesley, California

Gualtieri C, Gualtieri P (2004) Turbulence-based models for gas transfer analysis with channel shape factor influence. Environ Fluid Mech 4:249–271

Haider H, Ali W, Haydar S (2013) Evaluation of various relationships of reaeration rate coefficient for modelling dissolved oxygen in a river with extreme flow variations in Pakistan. Hydrol Process 27:3949–3963

Jain SC (1976) Longitudinal dispersion coefficients for streams. J Environ Eng Div, Am Soc Civil Eng 102(EE2):465–474

Jain SK, Jha R (2005) Comparing the stream re-aeration coefficient estimated from ANN and empirical models. Hydrol Sci 50(6):1037–1052

Keller RJ, Rodi W (1988) Predictions of flow characteristics in main channel/flood plain flows. J Hydraul Res 26(4):425–441

Knight DW (2013) Hydraulic problems in flooding: from data to theory and from theory to practice. In: Rowinski P (ed) Experimental and computational solutions of hydraulic problems. Springer, Berlin

Krishnappan BG, Lau YL (1986) Turbulence modelling of flood plain flows. J Hydraul Eng, Am Soc Civil Eng 112(4):251–267

Lambert MF, Sellin RHJ (1996) Discharge prediction in straight compound channels using the mixing length concept. J Hydraul Res 34(3):381–394

Lin B, Shiono K (1995) Numerical modelling of solute transport in compound channel flows. J Hydraul Res 33(6):773–778

Liu H (1977) Predicting dispersion coefficient of streams. J Environ Eng Div, Am Soc Civil Eng 103(EE1):59–69

Manson JR, Wallis SG (2000) A conservative, semi-Lagrangian fate and transport model for fluvial systems: Part 1—theoretical development. Water Res 34(15):3769–3777

Manson JR, Wallis SG (2004) Fluvial mixing during floods. Geophys Res Lett 31:14, 4 pp

McQuivey RS, Keefer TN (1974) Simple method for predicting dispersion in streams. J Environ Eng Div, Am Soc Civil Eng 100(EE4):997–1011

Melching CS, Flores HE (1999) Reaeration equations derived from U.S. Geological Survey database. J Environ Eng, Am Soc Civil Eng 125:5, 407–414

Moog DB, Jirka GH (1998) Analysis of reaeration equations using mean multiplicative error. J Environ Eng, Am Soc Civil Eng 124(2):104–110

Naot D, Nezu I, Nakagawa H (1993) Hydrodynamic behaviour of compound rectangular open channel flow. J Hydraul Eng, Am Soc Civil Eng 119(3):390–408

Rameshwaran P, Naden P, Wilson CAME, Malki R, Shukla DR, Shiono K (2013) Inter-comparison and validation of computational fluid dynamics codes in two-stage meandering channel flows. Appl Math Model 37(20–21):8652–8672

Rathbun RE (1977) Reaeration coefficients of streams – State of the art. J Hydraulics Div, Am Soc Civil Eng 103(HY4):409–424

Raymond PA, Zappa CJ, Butman D, Bott TL, Potter J, Mulholland P, Laursen AE, McDowell WH, Newbold D (2012) Scaling the gas transfer velocity and hydraulic geometry in streams and small rivers. Limnol Oceanogr Fluids Environ 2:41–53

Rodrigues PPGW, González YM, de Sousa EP, Neto FDM (2013) Evaluation of dispersion parameters for River São Pedro, Brazil, by the simulated annealing method. Inverse Prob Sci Eng 1:34–51

Runkel RL, Chapra SC (1993) An efficient numerical solution of the transient storage equations for solute transport in small streams. Water Resour Res 29:211–215

Runkel RL (1998) One-dimensional transport with inflow and storage (OTIS)—a solute transport model for streams and rivers. Water-resources investigations report 98-4018, United Sates Geological Survey, 73 pp

Runkel RL (2002) A new metric for determining the importance of transient storage. J North Am Benthological Soc 21(4):529–543

Rutherford JC (1994) River mixing. Wiley, Chichester

Ryan RJ, Packman AI, Welty C (2004) Estimation of solute transport and storage parameters in a stream with anthropogenically produced unsteady flow and industrial bromide input. Water Resour Res 40:1, 6 pp

Schofield O, Roarty H, Saba G, Yi Xu, Kohut J, Glenn S, Manderson J, Oliver M (2012) Phytoplankton dynamics and bottom water oxygen during a large bloom in the summer of 2011. Oceans 1(6):14–19

Scott DT, Gooseff MN, Bencala KE, Runkel RL (2003a) Automated calibration of a stream solute transport model: implications for interpretation of biogeochemical parameters. J North Am Benthological Soc 22(4):492–510

Scott DT, Runkel RL, McKnight DM, Voelker BM, Kimball BA and Carraway ER (2003b) Transport and cycling of iron and hydrogen peroxide in a freshwater stream: Influence of organic acids. Water Resour Res 39:11, 1308, 14 pp

Shiono K, Knight DW (1991) Turbulent open-channel flows with variable depth across the channel. J Fluid Mech 222:617–646

Smith R (1976) Longitudinal dispersion of a buoyant contaminant in a shallow channel. J Fluid Mech 78:677–688

Sofialids D, Prionos P (1999) Turbulent flows in open channels with smooth and rough flood plains. J Hydraul Res 37(5):615–640

Streeter HW, Phelps EB (1925) A study of the pollution and natural purification of the Ohio river. III. Factors concerned in the phenomena of oxidation and reaeration. Public Health Bulletin no. 146, Reprinted by U.S. Department of Health, Education and Welfare, Public Health Service

Taylor GI (1954) The dispersion of matter in turbulent flow through a pipe. Proc R Soc Lond A223:446–468

Thomann RV, Mueller JA (1987) Principles of surface water quality modelling and control. Harper and Row, New York

Wallis SG, Manson JR (2004) Methods for predicting dispersion coefficients in rivers. Proc Inst Civil Eng, Water Manage 157(WM3):131–141

Wallis SG, Manson JR (2008) The influence of floodplain characteristics on longitudinal dispersion in a natural channel. Proceedings of hydropredict 2008, Prague, September, pp 135–138

Wallis SG, Crowther JM, Curran JC, Milne DP, Findlay JS (1989) Consideration of a one-dimensional transport model of the upper Clyde estuary. In: Advances in water modelling and measurement. BHRA Publications, pp 23–41

Wilson GT, Macleod N (1974) A critical appraisal of empirical equations and models for the prediction of the coefficient of reaeration of deoxygenated water. Water Res 8(6):341–366

Chapter 23
Exchange of Pollutants Between Rivers and the Surrounding Environment: Physical Processes, Modelling Approaches and Experimental Methods

M. Zaramella, A. Bottacin-Busolin, M. Tregnaghi and A. Marion

Abstract The fate of solute and pollutants is controlled by a broad number of different transport and storage mechanisms, ranging from simple processes (i.e. molecular diffusion, advection etc.) to more complex phenomena (i.e. evapotranspiration, groundwater flows, etc.). Different mathematical models, accounting for different exchange processes, have been developed and applied to specific experimental studies to assess transport and storage parameters. Experimental research focused on transport and retention processes induced by the transient storage in the dead zones, by the river bed topography and vegetation, by evapotranspiration. The analysis of these physical processes is generally conducted observing the behavior of solutes in field environments or in scaled laboratory models, using artificial or environmental tracers to track the fate of transported substances and assess transport and retention parameters. To improve the knowledge of pollutant exchange mechanism between a river and the surrounding environment, new experimental techniques focusing on long timescale retention and investigating the link between river biology and hydrodynamics are required. The development of new protocols for tracer tests design and the use of new specific tracers will open future research perspectives.

Keywords Hyporheic · Stream-aquifer · Sw-gw interactions · Surface water · Groundwater · River · Tracer tests

M. Zaramella (✉) · M. Tregnaghi · A. Marion
Department of Industrial Engineering, University of Padua, Via Marzolo 9, 35131 Padua, Italy
e-mail: mattia.zaramella@unipd.it

A. Bottacin-Busolin
School of Mechanical, Aerospace and Civil Engineering, The University of Manchester, M60 1QD Manchester, England

P. Rowiński and A. Radecki-Pawlik (eds.), *Rivers—Physical, Fluvial and Environmental Processes*, GeoPlanet: Earth and Planetary Sciences,
DOI 10.1007/978-3-319-17719-9_23

23.1 Introduction

The fate of pollutants in rivers is controlled by surface hydrodynamics and by exchange processes with the surrounding environment. A solute transported in a natural watercourse can be temporarily trapped in vegetated zones or in the sediment, follow deep flow paths in the porous medium and return to the surface water after some time. These solute trapping phenomena are generally referred to as "retention" processes or "transient storage", and have a major impact on the spatial and temporal distributions of nutrients and contaminants in fluvial systems. The characterization of these processes is important in many impact and vulnerability assessment studies, and for this reason they have been subject of extensive research during the last decades.

The hydrodynamic exchange with storage zones can be conceptually classified into short and long term retention. Short term retention is commonly due to surface dead zones, such as side pockets of recirculating water or vegetated zones, whereas long term retention is due to temporary storage of solutes in the upper part of the sediment bed, which is commonly referred to as the *hyporheic zone*. In addition to that, interactions between the stream water and the aquifer can occur over longer timescales. They are driven by gradients between the groundwater table and the river free surface, and depend on the hydraulic conductivity of the aquifer and the riverine morphology. Solute losses from river systems also occur as a consequence of evapotranspiration, which accounts for evaporation and plant transpiration from the water body and the soil.

The mass and momentum exchange between the water flowing in the river channel and the surrounding environment plays an important role in controlling the pollution of water bodies and connected ecological systems. The main transport mechanisms that determine the fate of nutrients and contaminants in rivers are briefly reviewed in Sect. 23.2. Section 23.3 reviews the main one-dimensional models proposed in the literature to represent the downstream propagation of solutes in streams and rivers; finally, Sect. 23.3.4 introduces the topic of field tracer tests and their use to calibrate transport models and characterize river contamination processes.

23.2 Physical Transport Processes

The conceptual description of transport processes depends on the scale of observation, which can range from the molecular scale to the watershed scale. Depending on the problem, it can be convenient to consider physical quantities averaged over certain spatial and temporal scales, or to operate a conceptual distinction between different parts of a continuum. This leads us to the identification of different physical processes that are relevant to different spatial and temporal scales, or physical domains. The main processes governing the fate of solutes in fluvial systems are described in the following sections.

23.2.1 Advection and Molecular Diffusion

Advection is a mechanical process by which a conserved physical quantity is transported in a fluid in motion. This mechanism is fully described by the amount of solute per unit volume (concentration $c(x,t)\left[\mathrm{M/L^{-3}}\right]$) and the velocity vector field $\boldsymbol{u} = (u, v, w)\left[\mathrm{LT^{-1}}\right]$ of the fluid defined at each point as a function of time.

If the substance behaves like a solute, that is, it has the same density as the medium or does not feel significant effect of its buoyant weight; its molecules are displaced along the direction of the local velocity vector, following the same path of the fluid particle as if they were part of the medium. This assumption allows advective transport to be modeled in a relatively simple way by writing the mass balance for an elementary control volume:

$$\frac{\partial c(x,t)}{\partial t} = -\boldsymbol{u} \cdot \nabla c(x,t). \tag{23.1}$$

Advection is always associated with molecular diffusion. This is the process by which matter is transported as a result of random molecular motions moving the net flux of matter from higher to lower concentrations. A quantitative description of the molecular diffusion process was given by Fick (1855) who expressed the corresponding net flux of the transported substance as equal to the concentration gradient multiplied by a physical property called molecular diffusivity or diffusion coefficient, indicated by $D_m\left[\mathrm{LT^{-2}}\right]$, leading to the second Fickian law:

$$\frac{\partial c(x,t)}{\partial t} = -D_m \nabla^2 \mathrm{c}(x,t). \tag{23.2}$$

Whenever the fluid is in motion, advection and diffusion processes act simultaneously. The total flux is thus given by the sum of the advective and diffusive fluxes and is expressed by the combination of (23.1) and (23.2):

$$\frac{\partial c(x,t)}{\partial t} = -\nabla \cdot \left[\mathrm{c(x,t)}\mathbf{u} - D_m \nabla \mathrm{c}(x,t)\right]. \tag{23.3}$$

23.2.2 Turbulent Diffusion

Molecular diffusion produced by Brownian motion is no longer the dominant diffusion mechanism when the flow velocity becomes fast enough to overcome viscous forces and the flow becomes turbulent. Under these conditions, diffusion is controlled by continuous displacement of fluid elements in all directions induced by turbulence. While molecular diffusion is isotropic, turbulent diffusion is typically different in each direction, as eddies are continuously stretched and deformed by the flow. Turbulent flows are usually modeled splitting the physical quantities into time-averaged mean values and fluctuations around the mean. After manipulation of the advection-diffusion Eq. (23.3), the time-averaged mass transport equation becomes:

$$\frac{\partial \bar{c}}{\partial t}+\bar{u}\frac{\partial \bar{c}}{\partial x}+\bar{v}\frac{\partial \bar{c}}{\partial y}+\bar{w}\frac{\partial \bar{c}}{\partial z}=\frac{\partial}{\partial x}\left(D_{xx}^{T}\frac{\partial \bar{c}}{\partial x}\right)+\frac{\partial}{\partial y}\left(D_{yy}^{T}\frac{\partial \bar{c}}{\partial y}\right)+\frac{\partial}{\partial z}\left(D_{zz}^{T}\frac{\partial \bar{c}}{\partial z}\right), \quad (23.4)$$

where D_{xx}^T, D_{yy}^T, and D_{zz}^T, are eddy diffusion coefficients in the three spatial directions, x, y and z, respectively, and the notation $(\bar{\square})$ denotes temporal average. The expression for the vertical dispersion coefficient $D_{zz}^T = 0.067u^*d$ in rivers can be derived from the logarithmic velocity profile where d is the water depth and u^* is the shear velocity. An approximate expression of the coefficient valid for uniform straight channels was empirically derived by Fischer et al. (1979) based on laboratory and field experiments: $D_{yy}^T = 0.15u^*d$, suggesting $D_{yy}^T = 0.6u^*d$ for irregular channels where geometrical variations enhance transverse mixing. For longitudinal mixing it can often be assumed that $D_{yy}^T = D_{xx}^T$. These relationships can be used to estimate the distance mix from an injection (for a lateral injection the distance is 4 times greater):

$$L_{mix} = 0.1\frac{Ub^2}{D_{xx}^T}, \quad (23.5)$$

where b is the channel width.

23.2.3 Longitudinal Dispersion

The effect of velocity gradients becomes apparent when spatial averaging of the physical quantities is carried out along with temporal averaging. For streams and rivers, it is often convenient to simplify the description of flow and mass transport by averaging flow quantities and solute concentration over the vertical direction (shallow water approach) or over a cross-section (one-dimensional approach). Depth averaging is justified when dealing with large rivers, estuaries and lagoons, based on the evidence that vertical mixing is usually much faster than lateral and longitudinal mixing, due to the limited extension of the domain in the vertical direction. The one-dimensional approach is justified when the extension of the domain in the transverse direction is small compared to the longitudinal dimension. This is the reason why the one-dimensional approach is commonly adopted for dispersion processes in open channel flows. In the case of cross-sectional averaging of the physical quantities, the mass balance equation is reduced to the one-dimensional form:

$$\frac{\partial C(x,t)}{\partial t} + U\frac{\partial C(x,t)}{\partial x} = \frac{1}{A(x,t)}\frac{\partial}{\partial x}\left(A(x,t)D_L(x,t)\frac{\partial C(x,t)}{\partial x}\right), \tag{23.6}$$

where C and U are the cross-sectional average concentration and flow velocity respectively, A is the flow cross-sectional area (L^2) and D_L is the longitudinal dispersion coefficient ($L^2\ T^{-1}$). The velocity gradients over the cross-section and turbulent transverse mixing play a competitive role in determining the spreading of solutes along the stream wise direction: while turbulent mixing decreases dispersion, velocity gradients enhance it. An approximated relationship for the longitudinal dispersion for streams with large width-to-depth ratios was suggested by Fischer (1975):

$$D_L = 0.011\frac{U^2 b^2}{du^*}. \tag{23.7}$$

23.2.4 Exchanges with the Surrounding Environment

Rivers are typically characterized by irregular cross-sections, and often feature lateral vegetated pockets of stagnant or recirculating water that act like dead zones (Fig. 23.1). Due to transverse mixing, the substances transported in the main channel, where the velocity are relatively higher and transverse mixing is faster, can enter such dead zones and be temporarily trapped before returning to the main channel. Similarly, the boundary of a river channel is a porous interface through which mass and momentum exchanges occur. Filtration through the porous boundary of a river bed leads the dissolved substances within the porous medium where sorption onto the sediments and other biogeochemical reactions may significantly affect their fate. The near stream region of the porous boundary that is directly affected by the concentration of solutes in the stream is called the hyporheic zone (Fig. 23.1), and is recognized to be an important transition environment for the evolution of a riverine ecosystem.

Surface dead zones and hyporheic zones can be envisioned as separate physical domains of immobile water that interact with a main flow domain identified as the main channel. The delay process resulting from mass and momentum exchanges with the immobile domains is called "transient storage", or retention. These physical exchange processes produce deviations from the asymptotic dispersion regimes represented by the advection-dispersion equation (Eq. 23.6), and can be taken into account by including an additional term in Eq. 23.6:

$$\frac{\partial C}{\partial t} + U\frac{\partial C}{\partial x} = \frac{1}{A}\frac{\partial}{\partial x}\left(AD_L\frac{\partial C}{\partial x}\right) + \frac{q}{A}(C_G - C) - \frac{P}{As}\Phi_s, \tag{23.8}$$

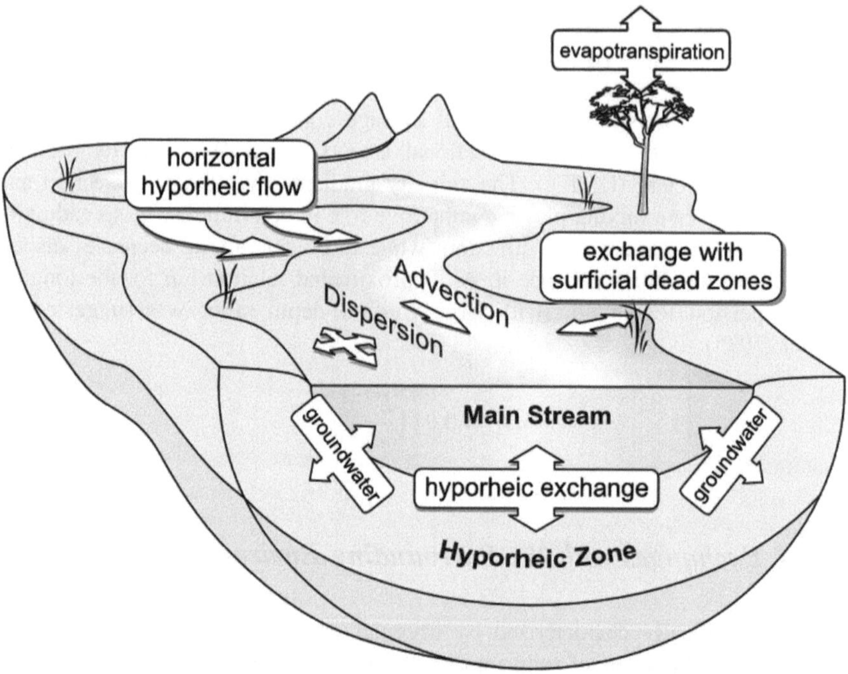

Fig. 23.1 Illustration of the transport processes acting in a river. The downstream transport of solutes is governed by advection and hydrodynamic dispersion in the main stream, and by mass exchanges with different retention zones. These include vertical exchanges with the underlying sediments, where adsorption process may take place; lateral exchanges with surficial dead zones, typically vegetated pockets; horizontal hyporheic flows induced by planimetric variation of the stream direction; groundwater flows; and evapotranspiration

where q is the groundwater flux per unit length ($L^2\ T^{-1}$), P is the wetted perimeter (L) and Φ_S ($M\ L^2\ T^{-1}$) denotes the exchange flux at the stream-storage zone interface. Various formulations for the term Φ_S will be discussed in Sect. 23.3.

23.2.4.1 The Role of Bed Topography

Streambed surface forms and obstructions (bedforms, meanders, obstacles, etc.) play a direct role in driving hyporheic exchange: the water pressure variations induced by the interaction between flow and topography induce flows into and out the bed sediments (Hendricks and White 1991; Harvey and Bencala 1993; Brunke and Gonser 1997; Cardenas et al. 2004; Storey et al. 2003; Savant et al. 1987; Thibodeaux and Boyle 1987; Franken et al. 2001). Bedforms with small wavelengths and amplitude produce effects on relatively small spatial and temporal scales. However, the resulting subsurface flows play a significant role in determining larger hyporheic flow patterns (Harvey and Bencala 1993).

The role of stream bedforms and roughness on hyporheic exchange processes has been studied in controlled laboratory studies (Marion et al. 2002; Packman et al. 2004; Elliott and Brooks 1997a, b; Salehin et al. 2003) and models have been proposed for homogeneous (Elliott and Brooks 1997a, b; Bottacin-Busolin and Marion 2010) and heterogeneous sediment beds (Marion et al. 2008a).

Wörman et al. (2002) modeled the hyporheic exchange in the Säva Brook, Sweden, coupling a longitudinal in-stream solute transport with a physically based representation of storage in the hyporheic zone.

23.2.4.2 The Effect of Vegetation

It is well known that vegetation increases flow resistance and can have considerable impact on turbulence and flow patterns (Nepf 1999). By decreasing the local flow velocity and producing complex wake structures, vegetation can generate trapping phenomena that significantly delay the downstream migration of a contaminant. Moreover, it is demonstrated that vegetation alters streambed permeability, with the introduction of organic matter to hyporheic sediments and intrusion of plant roots having a pronounced effect on solute transfer to the hyporheic zone (Packman and Salehin 2003). The presence of in-stream vegetation thereby reduces the exchange rates, increasing the release timescale of solute retention (Hendricks and White 1991; Harvey and Bencala 1993; Brunke and Gonser 1997; Cardenas et al. 2004). A number of studies have attempted to quantify the effect of submerged and emergent vegetation on the dispersion of solutes (Nepf 1999; Nepf et al. 2007; Shucksmith et al. 2011) and the storage effects associated with microbial biofilms (Battin et al. 2003; Bottacin-Busolin et al. 2009).

23.2.4.3 Evapotranspiration as a Driver of Hyporheic Exchange

Riparian vegetation plays an important role in driving hyporheic fluxes from the stream flow to the surrounding riparian environment (Jones et al. 2008; Duke et al. 2007; Loheide 2008; Harvey et al. 2005). Evapotranspiration (ET) is a process through which plants roots "pump" water from the riparian sediments exploiting capillarity forces. This water is then lost as vapor through stomata in leaves during diurnal hours. The "pumping" effect of ET drives fluxes of stream water to the sediments, enhancing the solute exchange processes from surface water to the surrounding hyporheic areas.

ET had a pronounced effect on solute residence times in the hyporheic zone (Larsen et al. 2014), diminishing them as a result of enhanced bidirectional flux across the interface between the hyporheic zone and surface water. This flux prevents hyporheic interstitial water from being caught by deep groundwaters, limiting the extent of the exchange layer and recalling upward water fluxes from the aquifer.

23.3 Models

Exchange processes can be studied individually by mathematical or numerical techniques. This can usually be done for idealized configurations by decoupling surface from subsurface flow. In field tracer tests, reach-integrated exchange processes are studied from the surface water perspective, by evaluating their effect as a whole on the solute concentration in the surface water. This approach is typically used in combination with one-dimensional transport models, often based on modified versions of the advection-dispersion equation. The use of transport models for determining tracer concentrations in the surface water, in combination with tracer experiments, represents the "surface water point of view". This approach is limited by the timescale and the magnitude of surface processes compared to surface-subsurface exchange mechanisms and aquifer interactions, and also by the modelling complexity of surface transport phenomena. These known limitations pushed researchers to investigate the reliability of different model formulations to quantify exchange processes from field tracer experiments (Harvey et al. 1996; Davis et al. 2000; Gooseff et al. 2013).

A review of the main 1-D transport models proposed in the literature is given in the following sections (a comprehensive review on hyporheic transport processes can be found in Boano et al. (2014)).

23.3.1 Transient Storage Model (TSM)

One of the most widely used transport models in field applications is the Transient Storage Model (TSM). The TSM was presented by Bencala and Walters (1983), although analogous formulations can be found in an earlier works (Hays et al. 1966; Nordin and Troutman 1980). This model has been widely applied to field experiments conducted both in small streams and large rivers (Bencala 1984; Castro and Hornberger 1991; Vallet et al. 1996; Mulholland et al. 1997; Harvey and Fuller 1998; Runkel et al. 1998; Choi et al. 1999; Fernald et al. 2001; Cheong and Seo 2003).

In the TSM, the net mass transfer from the main flow channel to the retention domains is assumed to be proportional to the difference of concentration between the surface water and a storage zone of constant cross-sectional area. The exchange flux at the stream-storage zone interface in (23.8) is assumed to be $\Phi_S = A/P\alpha(C_S - C_W)$. The mathematical formulation of the TSM for non-reactive solutes is then given by the following equations (Nordin and Troutman 1980; Bencala and Walters 1983; Czernuszenko and Rowinski 1997; Lees et al. 2000; De Smedt and Wierenga 2005; De Smedt 2006):

$$\frac{\partial C_W}{\partial t} + U\frac{\partial C_W}{\partial x} = D_L\frac{\partial^2 C_W}{\partial x^2} - \alpha(C_S - C_W); \tag{23.9}$$

$$\frac{\partial C_S}{\partial t} = -\alpha\frac{\mathrm{A}}{\mathrm{A_S}}(\mathrm{C_S} - \mathrm{C_W}), \tag{23.10}$$

where U is the mean flow velocity, α is a transfer coefficient, $\mathrm{A}/\mathrm{A_S}$ is the ratio of stream to storage cross-sectional areas, $\mathrm{C_W}$ is the in-stream solute concentration, and $\mathrm{C_S}$ is the concentration in the storage zone. A numerical solution of (10 was presented by Runkel and Chapra (1993), which formed the basis of their One-dimensional Transport with Inflow and Storage (OTIS), later extended by Runkel (1998) with a parameter estimation technique (OTIS-P).

23.3.2 The Aggregated Dead Zones Model (ADZ)

The ADZ model approach assumes that the river channel is divided into two interconnected regions (Davis and Atkinson 2000): a central core of flowing water exchanging with dead zones that are represented as immobile water domains around the bulk flow region. While the bulk flow region is characterized by pure advection, the dead zones account for both longitudinal dispersion and transient storage processes. The formulation of the model is derived from the TSM formulation, with $\mathrm{D_L} = 0$:

$$\frac{\partial C_W}{\partial t} + U\frac{\partial C_W}{\partial x} = -\alpha(C_S - C_W); \tag{23.11}$$

$$\frac{\partial C_S}{\partial t} = -\alpha\frac{A}{A_S}(C_S - C_W). \tag{23.12}$$

The model was found to be in agreement with experimental data where the effects of dead zones were predominant over dispersion (Davis and Atkinson 2000).

23.3.3 Diffusive Model

A model for solute transport of conservative species where hyporheic exchange is represented by a diffusive term was suggested by Jackman et al. (1984). This model assumes solute penetration into the bed to be a vertical diffusion process described by Fick's law. In the general Eq. (23.8), the flux at the stream-subsurface interface is assumed to be proportional to the vertical gradient of concentration in the bed:

$$\Phi_S(t,x) = -D_S \frac{\partial C_S(t,x,y)}{\partial y}, \tag{13}$$

where C_S is the concentration and D_S is the diffusion coefficient in the porous medium.

23.3.4 The Multi-rate Mass Transfer Approach (MRMT)

In the MRMT formulation, the flow domain is divided into mobile and immobile regions. In the mobile regions solute transport is modeled by the classical ADE, while immobile regions produce a delay in the migration of the tracer. The mobile domain is associated with the main flow channel, where the velocity is relatively higher and transverse mixing is faster, while the storage zones represent the immobile regions, where there is no significant propagation of the solute in the downstream direction. In the one-dimensional case, the migration of a solute in the mobile domain (i.e. the main flow channel) is described by the following mass balance equation (Haggerty et al. 2000):

$$\frac{\partial C_W}{\partial t} + \Gamma_S(x,t) = \frac{\partial}{\partial x}\left(D_L \frac{\partial C_W}{\partial x} - UC_W\right), \tag{23.14}$$

where $\Gamma_S(x,t)$ (M L^{-3} T^{-1})is a source-sink term for the mass exchange with the immobile sites. Comparison of Eqs. (23.8) and (23.14) yields $\Gamma_S(x,t) = -P/A$ $\Phi_S(x,t)$. The source-sink term can be expressed by variations of the concentrations in the immobile domains (van Genuchten and Wierenga 1976), but it is often more convenient to express it as a convolution integral, following Carrera et al. (1998) and assuming Γ_S to be independent of x. This is given by:

$$\Gamma_S(t) = C_W(x,t) * \frac{\partial f_M(t)}{\partial t} + C_W(x,t) f_M(0) - C_W(x,0) f_M(t), \tag{23.15}$$

where $f_M(t)$ is a memory function (T^{-1}) (Haggerty et al. 2000; Carrera et al. 1998):

$$f_M(t) = \int_0^\infty \alpha p_\alpha(\alpha) e^{-\alpha t} d\alpha. \tag{23.16}$$

The term $p_\alpha(\alpha)$ [T] is the probability density function of first order coefficients. Haggerty et al. (2000) proposed the following relation for $p_\alpha(\alpha)$, known as the "truncated power law density":

$$p_\alpha(\alpha) = \frac{\beta_{tot}}{\alpha_{max}^{\kappa-2} - \alpha_{min}^{\kappa-2}} \alpha^{\kappa-3}, \tag{23.17}$$

where α_{max} (T^{-1}) is the maximum rate coefficient, α_{min} (T^{-1}) is the minimum rate coefficient, κ is the exponent, and β_{tot} is the capacity coefficient (Haggerty and Gorelick 1995), the zeroth moment of the density function of rate coefficients $p_\alpha(\alpha)$.

23.3.5 Continuous Time Random Walk (CTRW)

In the conceptual framework of the continuous time random walk (Montroll and Weiss 1965; Scher and Lax 1973), the motion of solute molecules (or "particles") is envisioned as a sequence of displacements (or "jumps") of variable length and duration, considered as random variables. In this framework, the concentration of a solute at a given instant and position is derived statistically as the probability for a particle to occupy a specified position at a given time. The classical advection-dispersion equation (ADE) can be derived as a special case of a continuous time random walk in which every displacement has the same length and occurs in random directions at regular time intervals (Fischer et al. 1979). In the CTRW theory, the length and duration of particle jumps are random variables with joint probability density function (PDF) $\Psi(x,t)$, and marginal distributions and $\psi_T(t)$, respectively. This conceptualization of particle motion leads to the following generalized master equation (GME):

$$\frac{\partial C_W(x,t)}{\partial t} = -\int_0^t f_M(t-\tau)\left[U\frac{\partial C_W(x,t)}{\partial X} - D_L\frac{\partial^2 C_W(x,t)}{\partial x^2}\right]d\tau, \tag{23.18}$$

where:

$$U = \frac{1}{t}\int_{-\infty}^{+\infty} x\psi_L(x)dx, \tag{23.19}$$

$$D_L = \frac{1}{2t}\int_{-\infty}^{+\infty} x^2\psi_L(x)dx, \tag{23.20}$$

are the time-invariant velocity and longitudinal dispersion coefficients, respectively, over the averaging timescale $\bar{t}$. The function $f_M(t-\tau)$ is a memory function defined in Laplace domain as:

$$\tilde{f}_M(s) = s\bar{t}\frac{\tilde{\psi}_L(s)}{1 - \tilde{\psi}_L(s)}, \tag{23.21}$$

where $(\tilde{\square})$ denotes Laplace transform. The CTRW approach has been widely applied to study anomalous dispersion in fractured and heterogeneous media (Berkowitz and Scher 1995; Scher et al. 2002), and has been applied to the transport of solutes in streams (Boano et al. 2007).

23.3.6 Fractional Advection-Dispersion Equation (FADE)

A generalization of the classical advection-dispersion equation can be given using mathematical tools of fractional calculus. Using fractional order derivatives, Fick's law can be generalized to the form (Chaves 1998; Metzler and Klafter 2000; Schumer et al. 2001):

$$\Phi = -D_\varepsilon\left(\frac{1+\varsigma}{2}\partial_{+x}^{\varepsilon-1} + \frac{1-\varsigma}{2}\partial_{-x}^{\varepsilon-1}\right)C_W + UC_W, \tag{23.22}$$

where ε is the order of the fractional derivative, D_ε is the dispersion coefficient and $-1 \leq \varsigma \leq 1$ is a skewness parameter. For $\varepsilon = 1$ and $\varsigma = 0$ Eq. 23.22 reduces to the Fick law (23.2). Depending on the parameters ε, ς and D_ε, the residence time distributions predicted by the fractional advection-dispersion equation can be sensibly skewed and heavy tailed resembling those typically observed in natural streams. Applications of the fractional ADE has been reported by Deng et al. (2004, 2006) showing good agreement with experimental data.

23.3.7 A Generalized Residence Time Approach to Solute Transport in Rivers: The STIR Model

A conceptual model using distinct residence time distributions to represent surface and subsurface transport and a trapping probability distribution to represent stream-dead zone exchanges was proposed by Marion et al. (2008b). This residence time formulation formed the basis of what the authors called the STIR (Solute Transport In Rivers) model. In this conceptual framework, the river is represented as a system composed by distinct physical domains interacting with each other through mass exchanges. The river is conceptually divided into a main flow channel and different retention domains, such as surface dead zones and the hyporheic layer. By considering the total time spent by a solute particle in a river reach as the sum of the time spent in the main channel and the time spent in the storage zones, and using the summation rule of random variables, the residence time PDF of a solute particle in a stream segment of length x can be expressed as:

$$r(t;x) = \int_0^t r_w(t-\tau;x)\sum_{n=0}^{\infty} p(n|t-\tau)[\varphi(\tau)]^{*n} d\tau, \tag{23.23}$$

where $r_w(t;x)$ represents the residence time PDF in the main channel in absence of retention phenomena, $p(n|t)$ is the conditional probability of the number of trapping events, n, given the time t spent in the main channel, and $\varphi(t)$ is the residence time PDF in a storage domain for a single trapping event. In Eq. 23.23 the notation $[\square]^{*n}$ represents convolution power.

Equation 23.23 represents a generalization of the MRMT approach described in Sect. 23.3.4 in which the RTD describing the tracer migration in the mobile domain, as well as the probability distribution describing the solute transfer to the immobile domains, can take an arbitrary form. Assuming that the transport in the main channel is represented by the classical advection dispersion equation, and that the exchange with the immobile domains is a Poisson process, the STIR model can be shown to be equivalent to the following advection-dispersion-mass-transfer equation:

$$\begin{aligned}\frac{\partial C_W(x,t)}{\partial t} + \frac{Q}{A}\frac{\partial C_W(x,t)}{\partial x} &= D_W\frac{\partial^2 C_W(x,t)}{\partial x^2} \\ &- \alpha\left(C_W(x,t) - \int_0^t C_W(x,t-\tau)\varphi(\tau)d\tau\right)\end{aligned} \tag{23.24}$$

This is equivalent to (23.8) with:

$$\Phi_S = \frac{P}{A}\alpha\left(C_W(x,t) - \int_0^t C_W(x,t-\tau)\varphi(\tau)d\tau\right). \tag{23.25}$$

It is often reasonable to consider two main types of storage domains (Bottacin-Busolin et al. 2011), corresponding to surface and subsurface retention domains. Hence, assuming $\alpha = \alpha_1 + \alpha_2$:

$$\varphi(t) = \frac{1}{\alpha_1 + \alpha_2}[\alpha_1\varphi_1(t) + \alpha_2\varphi_2(t)]. \tag{23.26}$$

This formulation allows for faster and slower exchange processes to be represented separately. In particular, fast exchanges with surface dead zones can be represented by an exponential RTD, whereas other retention phenomena, such as hyporheic flows induced by bed-forms, bars and meanders can be represented by a physically-based model in terms of a transfer rate and a residence time distribution. Because the storage within surface dead zones and the storage within the sediments are generally characterized by very different timescales, this decomposition of transient storage is expected to yield values that are more representative of the physics of the processes.

23.4 Tracer Tests

Streams and rivers are complex environmental systems characterized by irregular geometry and heterogeneous properties. Field tracer tests provide a way to quantify reach-averaged transport quantities by measuring the solute concentrations at a section in response to a tracer injection upstream.

23.4.1 Tracers

A tracer is usually a soluble substance that follows the movement of water. Some of these substances are conservative, as they do not decay or react with any other compounds, and are not subject to sorption and deposition processes. Sometimes, non-conservative tracers can be used to study the physical or chemical properties of the surface water, the sediment bed or the storage zones. These tracers allow to investigate the behavior of solid particles in the water (i.e. suspended sediments, colloids, spores, etc.) or to understand the retardation or degradation potential of a substrate containing surface or sub-surface water. Tracers that are purposely introduced in the environment are called "artificial" (Ward et al. 1998) in opposition to environmental tracers, which are those substances or water properties that naturally (or anthropogenically) exist in the environment. Environmental tracers can be particular chemical compounds or water properties like conductivity or temperature, but also microbes or algae. The analysis of these tracers can be particularly useful when groundwater and surface water, with different tracer content, mix at the sediment-water interface. An example of environmental tracer is the use of radon-222 (Ellins et al. 1990; Cecil and Green 2000). Radon (^{222}Rn) is an environmental tracer released from radon-bearing rock. Radon concentrations are higher in groundwater than in surface water and can be used to identify groundwater inputs to a stream (Yoneda et al. 1991).

Anthropogenic environmental tracers result from their accidental release into the environment due to human activities. Frequently such tracers are widely distributed, have weak signals and result from prolonged accidental release. An example of anthropogenic environmental tracer is CFCs, which were released between the 1950s and 1980s and can be used to trace 50 year old water at a low detection limit (Busenberg and Plummer 1992; Plummer and Busenberg 2000).

Tracers must be chosen in order to obtain the best experimental control during an investigation. The addition of an artificial tracer to stream water may be used to provide improved understanding of hyporheic retention and natural attenuation processes. Radioisotopes like tritium (conservative, ^{3}H as tritiated water) and chromium (reactive, ^{51}Cr as Cr(III)) were used in field studies in the Säva Brook, Sweden (Johansson et al. 2001; Jonsson and Wörman 2001; Jonsson et al. 2003). By using different tracers at the same time, the investigators were able to detect the retardation of the non-conservative (reactive) tracer relative to the conservative (inert) tracer.

The sorption of strontium (Sr) and potassium (K) tracers in bed sediments was investigated by Bencala et al. (1983). Other researchers used caesium (Cs) to investigate the sorption processes in aquatic sediments (Nyffeler et al. 1984; Comans et al. 1991; Comans and Hockley 1992; Smith and Comans 1996). In these studies Caesium (Cs) was an environmental tracer originated from the 1986 Chernobyl accident in the Ukraine.

Fuller and Harvey (2000) used the reactive environmental tracer bromide (Br-) to track the fate of metals in the hyporheic zone of a stream contaminated by mining in Arizona, pointing out an active uptake and retardation of these metals by hyporheic flow-paths along the study reach.

Particular attention was devoted to the analysis of transport and retention of nitrate in the hyporheic zone (Triska et al. 1989a, b, 1990, 1993; Duff and Triska 1990). Chloride and bromide (inert tracers) were injected in the study reach and recoverd from the hyporheic zone by wells positioned at the side of the stream. The use of an inert solute provides information on the retardation and biological properties of the hyporheic zone. By injecting nitrate ($NaNO_3$) into the stream water and directly in the hyporheic sediments, it was demonstrated that biotic uptake within the hyporheic zone sediments acted as the major retention and storage mechanism. Groundwater and surface water mixing in the hyporheic zone was found to govern the denitrifying capacity of sediments, due to the different dissolved oxygen concentration (DOC) provided by the two sources of water.

Storey et al. (2003) used the ^{15}N isotope to describe denitrification processes in laboratory column experiments with sediments cores. Crenshaw et al. (2005) used the ^{15}N isotope in field studies to assess the denitrification process within the hyporheic zone.

Fluorescent dyes are an important group of tracers for hydrologic field studies in rivers and wetlands. Among a variety of commercial dyes available in a variety of colors, uranine, lissamine FF and rhodamine WT are the most used fluorescent water tracers (Wilson et al. 1986). Since fluorescent dyes are cost effective, highly soluble, low-toxic and easily detectable with specific fluorometers, they have been widely used for stream tracing applications, especially rhodamine WT (Atkinson and Davis 2000; Fernald et al. 2001; Bencala et al. 1983; Bottacin-Busolin et al. 2011).

Among the fluorescent dyes, particular attention should be devoted to "smart" tracers like resazurin. Resazurin is a weakly fluorescent, non-toxic, redox-sensitive phenoxazine dye. An important property of resazurin is the reduction to the strongly fluorescent resorufin under mildly reducing conditions, most commonly in the presence of aerobic respiration (Karakashev et al. 2003). These tracers have been recently used to analyze microbial activity in the storage zones (Haggerty et al. 2009).

Temperature is another environmental tracer largely used in groundwater studies. Groundwater has a relatively constant temperature compared to surface or stream waters (Constantz 1998). Conant (2004) investigated stream-groundwater exchange profiling water temperature through a depth profile of hyporheic sediments. Confirmation of the delineation of zones of groundwater discharge or recharge within the hyporheic zone derived from temperature measurements can be provided by the simultaneous injection of conservative solute tracers in the study reach. Becker et al. (2004) used heat as a tracer and measured the longitudinal variation of the stream discharge to quantify the stream-groundwater fluxes. The work of Anderson (2005) offers a comprehensive background review of the relevant literature. Recent studies have also attempted to provide a more accurate description of the transport in the hyporheic zones, by coupling the information from tracer tests with direct sampling of temperatures in the sediment bed (e.g. Lautz et al. 2010; Neilson et al. 2010a, b; Wörman et al. 2012). Neilson et al. (2010a, b) presented the formulation and calibration of a two-zone temperature and solute model which separates between surface and subsurface transient storage. The authors showed that the use of a multiobjective calibration method accounting for temperature and solute data at different locations at the same time produced a much better match between simulation results and observations compared to a single-objective procedure considering only solute concentrations or temperatures at single locations. A comprehensive review of heat as a tracer to quantify hyporheic flows has recently been given by Rau et al. (2014).

23.4.2 Designing Stream Tracer Tests to Quantify Retention Phenomena

Tracer tests are conducted by injecting a tracer in a study reach and measuring the temporal evolution of the tracer concentration downstream. Injection is usually performed with a constant (plateau) concentration of fixed duration to ensure proper dilution along the reach. The concentration distributions generated by the injection, known as breakthrough curves (BTCs), are measured over time at one or more sections downstream from the injection point, at a distance greater than the length scale of transverse mixing (23.5). BTCs contain information relevant to surface processes, namely advection and longitudinal dispersion, but also signatures of retention phenomena that produce deviations from the asymptotic dispersion regimes described by Taylor dispersion theory. Tracer BTCs carry signatures of complex stream-storage zone mixing and exchange dynamics, and the challenge of modeling and designing field tracer tests lies in the identification and the parameterization of these signatures. The major issue with their characterization is that the solute transport and retention phenomena are characterized by a continuum spectrum of timescales, and only processes characterized by significantly different timescales can be potentially separated. Furthermore, there are instrumental and experimental limits in the detection of tracer concentrations, which can dramatically hinder the analysis of long-term retention phenomena unless highly detectable tracers are used. Other factors that limit the characterization of long-term retention are the duration of the experiment and the presence of background noise in the measurements. In particular, the surface water point of view is generally blind to those processes characterized by small exchange fluxes and by temporal scales that are much longer than the duration of the experiment.

It must also be stressed that longitudinal dispersion leaves clear signatures only within a short distance from the upstream boundary section and that the fast exchange with the dead zones overwhelm the effect of surface mixing at longer distances (Davis and Atkinson 2000). In other words, longitudinal dispersion and fast exchange processes act as a bulk process over longer distances from the injection point, producing no deviations from the asymptotic dispersion behavior described by the advection-dispersion equation. This means that there are actual limits in the separability of transient storage phenomena, and that the limit depends on the scales of transverse mixing and the distance from the injection point. A nondimensional number that can be used to characterize the detectability of retention processes is the Damköhler number (Wagner and Harvey 1997). This can be expressed as:

$$DaI = \frac{(\alpha + 1/T)L}{u} \tag{23.27}$$

where α is the transfer rate, T is the average residence time in the storage zones, L is the length of the study reach, and u is the average flow velocity. The optimal DaI range is between 2 orders of magnitude of 1 (0.1–10), corresponding to peak storage sensitivity (Briggs et al. 2009). A DaI within this range indicates that there has been sufficient time for the exchange process to produce a visible effect on the BTCs, and that the exchange timescales are not too short compared to advective travel time. Under these conditions, the uncertainty in the estimated parameters of transient storage models is minimized.

A work on how signatures of different mechanism can be observed on BTC was presented by Bottacin-Busolin et al. (2011). BTCs where obtained with a controlled plateau injection of rhodamine WT in two different small Italian rivers, the Desturo and the Brenton canals. BTCs were analyzed using the STIR model with two exponential RTDs storage compartments (23.26). Comparison between the optimized model and BTCs is plotted in Fig. 23.2, using linear and logarithmic scale for the vertical axis (left and right column, respectively). A multiple domain solute transport model properly detects the different signatures of transient storage, and show that retention patterns at different timescales can be detected.

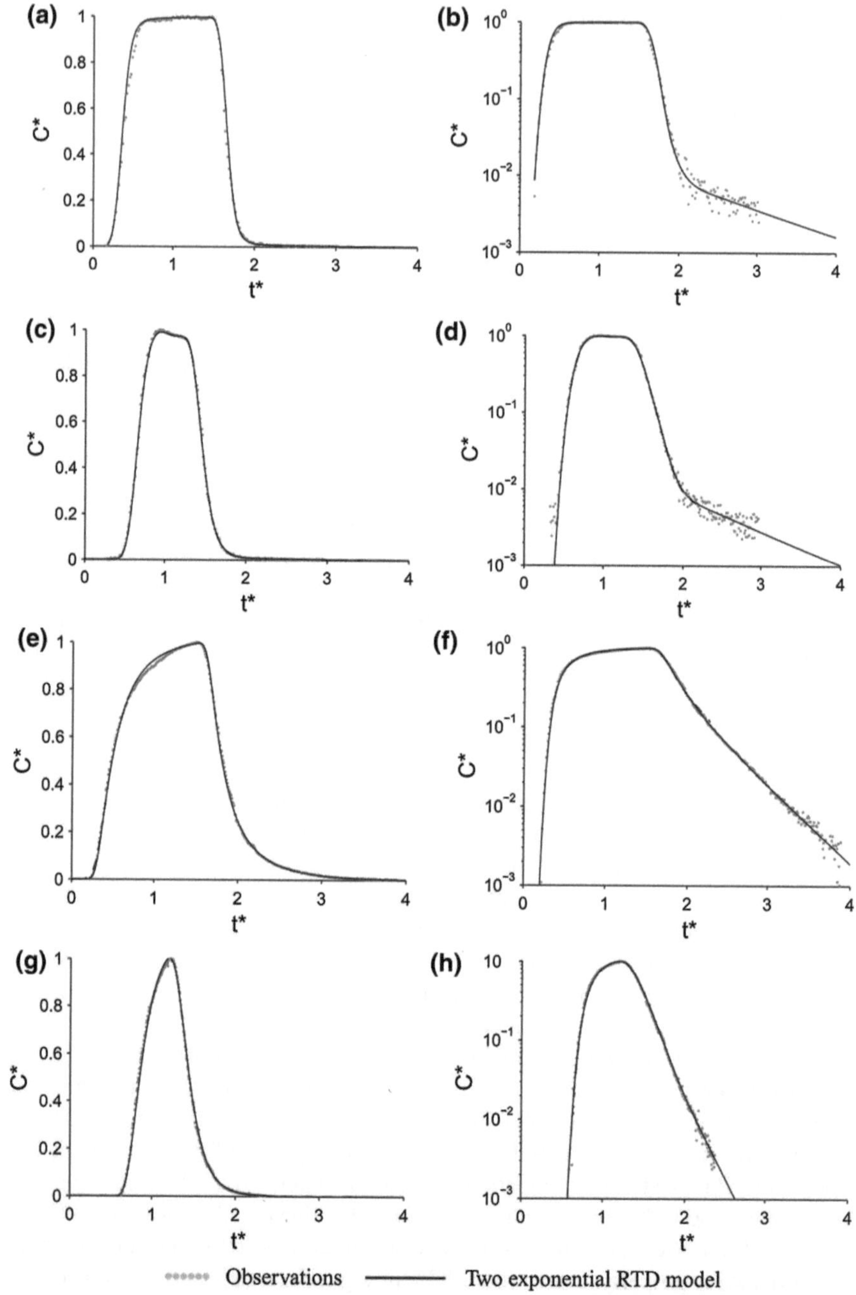

Fig. 23.2 Observed and simulated breakthrough curves for the Brenton and the Desturo rivers using a double exponential residence time distribution storage domain (figures modified from Bottacin-Busolin et al. 2011). Tails of concentration contains signature of transient storage that can be described by a multi-compartment model. **a** Brenton, Reach 1, **b** Brenton, Reach 1, **c** Brenton, Reach 2, **d** Brenton, Reach 2, **e** Desturo, Reach 1, **f** Desturo, Reach 1, **g** Desturo, Reach 3, **h** Desturo, Reach 3

23.4.3 *Limitations and Challenges in Inverse Modelling of Transport Processes*

Stream tracer tests can provide an estimate of reach-integrated transport quantities that summarize the effect of surface and subsurface transport fluxes. The major challenge in the design of tracer tests and the analysis of tracer test data is the separation between processes, which can be important in many impact assessment studies. While hyporheic retention produces signature on solute BTCs, the quantification of interfacial fluxes and residence times based on such signatures is a complex inverse modelling problem that is inherently ill posed. While the initial studies talked this problem by simple calibration of one-dimensional transport models based on the BTCs of passive tracers, the recent research has focused on the development of improved techniques for better conditioning the problem. This has included the use of smart tracers and combined solute-temperature measurements, or the direct measurement of solute fluxes and residence time in surface dead zones. At the same time, researchers have attempted to improve the capability of one-dimensional transport models to represent surface and subsurface transport processes. Furthermore, physically-based models of hyporheic exchange and surface transient storage have been proposed. However, despite significant advancements in the comprehension of transport phenomena in natural watercourses, our ability to measure and quantify surface and subsurface retention processes remains limited. Recent studies have shown that the parameters of the transient storage models can depend on the scale of the study reach (Bottacin-Busolin et al. 2011; Gooseff et al. 2013), and that moments of the breakthrough curves do not scale according to the predictions of the classical advection-dispersion theory, which is the basis of the most commonly used transport models (González-Pinzón and Haggerty 2013). The inconsistency between observations and model predictions points out an existing gap of knowledge in the hydrodynamics of mixing in the complex configurations found in natural watercourses. New conceptual understanding of river mixing processes is therefore needed in combination with new experimental techniques integrating tracer concentration data with velocity and bathymetry data. Such data can be used to directly model surface and subsurface transport dynamics, therefore reducing the number of parameters that require calibration or limiting their range of variability in inverse modeling procedures.

Acknowledgements This work was supported by the Research Executive Agency, through the 7th Framework Programme of the European Union, Support for Training and Career Development of Researchers (Marie Curie—FP7-PEOPLE-2012-ITN), which funded the Initial Training Network (ITN) HYTECH 'Hydrodynamic Transport in Ecologically Critical Heterogeneous Interfaces', N.316546.

References

Anderson MP (2005) Heat as a groundwater tracer—review paper. Ground Water 43(6):951–968

Atkinson TC, Davis PM (2000) Longitudinal dispersion in natural channels: l. Experimental results from the River Severn, U.K. Hydrol Earth Syst Sci 4:345–353. doi:10.5194/hess-4-345-2000

Battin TJ, Kaplan LA, Denis Newbold J, Hansen CME (2003) Contributions of microbial biofilms to ecosystem processes in stream mesocosms. Nature 426(6965):439–442. doi:10.1038/nature02152

Becker MW, Georgian T, Ambrose H, Siniscalchi J, Frederick K (2004) Estimating flow and flux of groundwater discharge using temperature and velocity. J Hydrol 296:221–233

Bencala KE (1984) Interactions of solutes and streambed sediment. II. A dynamic analysis of coupled hydrologic and chemical processes that determine solute transport. Water Resour Res 20(12):1804–1814

Bencala KE, Walters RA (1983) Simulation of solute transport in a mountain pool-and-riffle stream: a transient storage model. Water Resour Res 19(3):718–724

Bencala KE, Rathbun RE, Jackman AP, Kennedy VC, Zellweger GW, Avanzino RJ (1983) Rhodamine WT dye losses in a mountain stream environment. Water Resour Bull 19(6):943

Berkowitz B, Scher H (1995) On characterization of anomalous dispersion in porous and fractured media. Water Resour Res 31(6):1461–1466

Boano F, Packman AI, Cortis A, Revelli R, Ridolfi L (2007) A continuous time random walk approach to the stream transport of solutes. Water Resour Res 43(W10):425

Boano F, Harvey JW, Marion A, Packman Al, Revelli R, Ridolfi L, Worman A (2014) Hyporheic flow and transport processes: mechanisms, models, and biogeochemical implications, Review of Geophysics (in press)

Bottacin-Busolin A, Marion A (2010) Combined role of advective pumping and mechanical dispersion on time scales of bed form–induced hyporheic exchange. Water Resour Res 46(8): W08518. doi:10.1029/2009WR008892

Bottacin-Busolin A, Singer G, Zaramella M, Battin TJ, Marion A (2009) Effects of streambed morphology and biofilm growth on the transient storage of solutes. Environ Sci Technol 43 (19):7337–7342. doi:10.1021/es900852w

Bottacin-Busolin A, Marion A, Musner T, Tregnaghi M, Zaramella M (2011) Evidence of distinct contaminant transport patterns in rivers using tracer tests and a multiple domain retention model. Adv Water Resour 34(6):737–746. doi:10.1016/j.advwatres.2011.03.005

Briggs MA, Gooseff MN, Arp CD, Baker MA (2009) A method for estimating surface transient storage parameters for streams with concurrent hyporheic storage. Water Resour Res 45(4): W00D27. doi:10.1029/2008WR006959

Brunke M, Gonser T (1997) The ecological significance of exchange processes between rivers and groundwater. Freshw Biol 37:1–33

Busenberg E, Plummer LN (1992) Use of chlorofluorocarbons (CCl_3F and CCl_2F_2) as hydrologic tracers and age-dating tools: the alluvium and terrace system of central Oklahoma. Water Resour Res 28:2257–2284

Cardenas MB, Wilson JL, Zlotnik VA (2004) Impact of heterogeneity, bed forms, and stream curvature on sub-channel hyporheic exchange. Water Resour Res 40

Carrera J, Sanchez-Vila X, Benet I, Medina A, Galarza G, Guimer J (1998) On matrix diffusion: formulations, solution methods and qualitative effects. Hydrogeol J 6(1):178–190

Castro NM, Hornberger GM (1991) Surface-subsurface interactions in an alluviated mountain stream channel. Water Resour Res 27(7):1613–1621

Cecil DL, Green JR (2000) Radon-222. In: Cook P, Herczeg AL (eds) Environmental tracers in subsurface hydrology. Kluwer Academic Publishers, Dordrecht

Chaves AS (1998) A fractional diffusion equation to describe Lévy flights. Phys Lett A 239(1–2):13–16

Cheong TS, Seo IW (2003) Parameter estimation of the transient storage model by a routing method for river mixing processes. Water Resour Res Am Geophys Union 39(4):HWC 1-1-1-11

Choi J, Hulseapple SM, Conklin MH, Harvey JW (1999) Modeling CO_2 degassing and pH in a stream-aquifer system. J Hydrol 209(1–4):297–310

Comans RNJ, Hockley DE (1992) Kinetics of caesium sorption in illite. Geochim Cosmochim Acta 56:1157–1164

Comans RNJ, Haller M, De Preter P (1991) Sorption of caesium on illite: nonequilibrium behaviour and reversibility. Geochim Cosmochim Acta 55:433–440

Conant BJ (2004) Delineating and quantifying ground water discharge zones using streambed temperature. Ground Water 42(2):243–257

Constantz J (1998) Interaction between stream temperature, streamflow and groundwater exchanges in alpine streams. Water Resour Res 34(7):1609–1615

Crenshaw CL, Dahm CN, Sheibley RW, Grimm NB, Pershall AD (2005) Nitrogen dynamics in hyporheic zone sediments using $^{15}N–NO_3$ tracers. ASLO Aquatic Sci Meet. Abstract

Czernuszenko W, Rowinski PM (1997) Properties of the dead zone model of longitudinal dispersion in rivers. J Hydraul Res 35(4):491–504

Davis PM, Atkinson TC (2000) Longitudinal dispersion in natural channels: 3. An aggregated dead zone model applied to the River Severn, U.K. Hydrol Earth Syst Sci 4:373–381. doi:10.5194/hess-4-373-2000

Davis PM, Atkinson TC, Wigley TML (2000) Longitudinal dispersion in natural channels: 2. The roles of shear flow dispersion and dead zones in the River Severn, U.K. Hydrol Earth Syst Sci 4:355–371. doi:10.5194/hess-4-355-2000

De Smedt F (2006) Analytical solutions for transport of decaying solutes in rivers with transient storage. J Hydrol 330(3–4):672–680

De Smedt F, Wierenga PJ (2005) Analytical solution for solute transport resulting from instantaneous injection in streams with transient storage. J Hydrol 315(1–4):25–39

Deng Z-Q, Singh VP, Bengtsson L (2004) Numerical solution of fractional advection-dispersion equation. J Hydraul Eng 130(5):422–431

Deng Z-Q, Bengtsson L, Singh VP (2006) Parameter estimation for fractional dispersion model for rivers. Environ Fluid Mech 6(5):451–475

Duff JH, Triska FJ (1990) Denitrification in sediments from the hyporheic zone adjacent to a small forested stream. Can J Fish Aquat Sci 47:1140–1147

Duke JR, White JD, Allen PM, Muttiah RS (2007) Riparian influence on hyporheic-zone formation downstream of a small dam in the Blackland Prairie region of Texas. Hydrol Process 21(2):141–150

Ellins KK, Roman-Mas A, Lee R (1990) Using 222Rn to examine groundwater/surface discharge interaction in the Rio Grande De Manati, Puerto Rico. J Hydrol 115:319–341

Elliott AH, Brooks NH (1997a) Transfer of non-sorbing solutes to a streambed with bed forms: theory. Water Resour Res 33:123–136

Elliott AH, Brooks NH (1997b) Transfer of non-sorbing solutes to a streambed with bed forms: laboratory experiments. Water Resour Res 33:137–151

Fernald AG, Wigington PJ, Landers DH (2001) Transient storage and hyporheic flow along the Willamette River, Oregon: Field measurements and model estimates. Water Resour Res 37 (6):1681–1694

Fick A (1855) Über diffusion. Annalen der Physik 170:59

Fischer HB (1975) Discussion of "Simple method for predicting dispersion in streams". J Env Eng Div ASCE 101(3):435–455

Fischer HB, List JE, Koh CR, Imberger J, Brooks NH (1979) Mixing in inland and coastal waters. Academic, London

Franken RJM, Storey RG, Williams DD (2001) Biological, chemical and physical characteristics of downwelling and upwelling zones in the hyporheic zone of a northtemperate stream. Hydrobiologia 444:183–195

Fuller CC, Harvey JW (2000) Reactive uptake of trace metals in the hyporheic zone of a mining contaminated stream, Pinal Creek, Arizona. Environ Sci Technol 34:1150–1155

González-Pinzón R, Haggerty R (2013) An efficient method to estimate processing rates in streams. Water Resour Res 49(9):6096–6099. doi:10.1002/wrcr.20446

Gooseff MN, Briggs MA, Bencala KE, McGlynn BL, Scott DT (2013) Do transient storage parameters directly scale in longer, combined stream reaches? Reach length dependence of transient storage interpretations. J Hydrol 483:16–25. doi:10.1016/j.jhydrol.2012.12.046

Haggerty R, Gorelick SM (1995) Multiple-Rate mass transfer for modeling diffusion and surface reactions in media with Pore-Scale heterogeneity. Water Resour Res 31(10):2383–2400

Haggerty R, McKenna SA, Meigs LC (2000) On the late-time behavior of tracer test breakthrough curves. Water Resour Res 36(12):3467–3479

Haggerty R, Marti E, Argerich A, von Schiller D, Grimm NB (2009) Resazurin as a "smart" tracer for quantifying metabolically active transient storage in stream ecosystems. J Geophys Res 114:G03014. doi:10.1029/2008JG000942

Harvey JW, Bencala KE (1993) The effect of streambed topography on surface–subsurface water exchange in mountain catchments. Water Resour Res 29:88–99

Harvey JW, Fuller CC (1998) Effect of enhanced manganese oxidation in the hyporheic zone on basin-scale geochemical mass balance. Water Resour Res 34(4):623–636

Harvey JW, Wagner BJ, Bencala KE (1996) Evaluating the reliability of the stream tracer approach to characterize stream-subsurface water exchange. Water Resour Res 32(8):2441–2451

Harvey JW, Newlin JT, Saiers JE (2005) Solute transport and storage mechanisms in wetlands of the Everglades, South Florida. Water Resour Res 41:W05009. doi:10.01029/02004WR003507

Hays JR, Krenkel PA, Schnelle KBJ (1966) Mass transport mechanism in open channel flow. Tech Rep 8, Vanderbilt University, Nashville, Tennessee

Hendricks SP, White DS (1991) Physiochemical patterns within a hyporheic zone of a northern Michigan River, with comments on surface water patterns.Canadian. J Fish Aquat Sci 48:1645–1654

Jackman A, Walters R, Kennedy V (1984) Transport and concentration controls for chloride, strontium, potassium and lead in Uvas Creek, a small cobble-bed stream in Santa Clara County, California, USA: 2. Mathematical modeling. J Hydrol 75(1–4):111–141

Johansson H, Jonsson K, Forsman KJ, Wörman A (2001) Retention of conservative and sorptive solutes in streams—simultaneous tracer experiment. Sci Total Environ 266:229–238

Jones KL, Poole GC, Woessner WW, Vitale MV, Boer BR, O'Daniel SJ, Thomas SA, Geffen BA (2008) Geomorphology, hydrology, and aquatic vegetation drive seasonal hyporheic flow patterns across a gravel-dominated floodplain. Hydrol Process 22:2105–2113

Jonsson K, Wörman A (2001) Effect of sorption kinetics on the transport of solutes in streams. Sci Total Environ 266:239–247

Jonsson K, Johansson H, Wörman A (2003) Hyporheic exchange of reactive and conservative solutes in streams—tracer methodology and model interpretation. J Hydrol 278:152–171

Karakashev D, Galabova D, Simeonov I (2003) A simple and rapid test for differentiation of aerobic from anaerobic bacteria. World J Microbiol Biotechnol 19:233–238. doi:10.1023/A:1023674315047

Larsen LG, Harvey JW, Maglio MM (2014) Dynamic hyporheic exchange at intermediate timescales: testing the relative importance of evapotranspiration and flood pulses. Water Resour Res 50:318–335. doi:10.1002/2013WR014195

Lautz LK, Kranes NT, Siegel DI (2010) Heat tracing of heterogeneous hyporheic exchange adjacent to in-stream geomorphic features. Hydrol Process 24(21):3074–3086. doi:10.1002/hyp.7723

Lees MJ, Camacho LA, Chapra SC (2000) On the relationship of transient storage and aggregated dead zone models of longitudinal solute transport in streams. Water Resour Res 36(1):213–224

Loheide SP II (2008) A method for estimating subdaily evapotranspiration of shallow groundwater using diurnal water table fluctuations. Ecohydrology 1(1):59–66

Marion A, Bellinello M, Guymer I, Packman A (2002) Effect of bed form geometry on the penetration of nonreactive solutes into a streambed. Water Resour Res 38(10):27–1.

Marion A, Packman AI, Zaramella M, Bottacin-Busolin A (2008a) Hyporheic flows in stratified beds. Water Resour Res 44(9):W09433. doi:10.1029/2007WR006079

Marion A, Zaramella M, Bottacin-Busolin A (2008b) Solute transport in rivers with multiple storage zones: the STIR model. Water Resour Res 44(10):W10406. doi:10.1029/2008WR007037

Mulholland PJ, Marzolf ER, Webster JR, Hart DR, Hendricks SP (1997) Evidence that hyporheic zones increase hetrotrophic metabolism and phosphorus uptake in forest streams, Limnol. Oceanogr 42:443–451.

Neilson BT, Chapra SC, Stevens DK, Bandaragoda C (2010a) Two-zone transient storage modeling using temperature and solute data with multiobjective calibration: 1. Temperature. Water Resour Res 46(12):W12520. doi:10.1029/2009WR008756

Neilson BT, Stevens DK, Chapra SC, Bandaragoda C (2010b) Two-zone transient storage modeling using temperature and solute data with multiobjective calibration: 2. Temperature and solute. Water Resour Res 46(12):W12521. doi:10.1029/2009WR008759

Nepf HM (1999) Drag, turbulence, and diffusion in flow through emergent vegetation. Water Resour Res 35(2):479–489. doi:10.1029/1998WR900069

Nepf H, Ghisalberti M, White B, Murphy E (2007) Retention time and dispersion associated with submerged aquatic canopies. Water Resour Res 43(4):W04422. doi:10.1029/2006WR005362

Nordin CF, Troutman BM (1980) Longitudinal dispersion in rivers: the persistence of skewness in observed data. Water Resour Res 16:123–128

Nyffeler UP, Li YH, Santschi PH (1984) A kinetic approach to describe traceelement distribution between particles and solution in natural aquatic systems. Geochim Cosmochim Acta 48:1513–1522

Packman AI, Salehin M (2003) Relative roles of stream flow and sedimentary conditions in controlling hyporheic exchange. Hydrobiologia 494:291–297

Packman AI, Salehin M, Zaramella M (2004) Hyporheic exchange with gravel beds: basic hydrodynamic interactions and bedform-induced advective flows. J Hydraul Eng 130(7): 647–656

Plummer NL, Busenberg E (2000) Choloroflurocarbon. In: Cook P, Herczeg AL (eds) Environmental tracers in subsurface hydrology. Kluwer Academic Publishers, Dordrecht

Rau GC, Andersen MS, McCallum AM, Roshan H, Acworth RI (2014) Heat as a tracer to quantify water flow in near-surface sediments. Earth Sci Rev 129:40–58. doi:10.1016/j.earscirev.2013.10.015

Runkel RL (1998) One-dimensional transport with Inflow and Storage (OTIS): A solute transport model of streams and rivers, US Geol. Surv. Water Resour. Invest. Rep., pp 98–4018

Runkel RL, Chapra SC (1993) An efficient numerical solution of the transient storage equations for solute transport in small streams. Water Resour Res 29(1):211–215

Salehin M, Packman AI, Wörman A (2003) Comparison of transient storage in vegetated and unvegetated reaches of a small agricultural stream in Sweden: seasonal variation and anthropogenic manipulation. Adv Water Res 26:951–964

Savant SA, Reible DD, Thibodeaux LJ (1987) Convective transport within stable river sediments. Water Resour Res 23(9):763–1768.

Scher H, Margolin G, Berkowitz B (2002) Towards a unified framework for anomalous transport in heterogeneous media. Chem Phys 284(1–2):349–359

Schumer R, Benson DA, Meerschaert MM, Wheatcraft SW (2001) Eulerian derivation of the fractional advection-dispersion equation. J Contaminant Hydrol 48(1):69–88

Shucksmith JD, Boxall JB, Guymer I (2011) Determining longitudinal dispersion coefficients for submerged vegetated flow. Water Resour Res 47(10): n/a–n/a. doi:10.1029/2011WR010547

Smith JT, Comans RNJ (1996) Modelling the diffuse transport and remobilisation of 137 Cs in sediments: the effects of sorption kinetics and reversibility. Geochim Cosmochim Acta 60: 995–1004

Storey RG, Howard KWF, Williams DD (2003) Factors controlling riffle-scale hyporheic exchange flows and their seasonal changes in a gaining stream: a threedimensional groundwater flow model. Water Res Res 39(2):Art. No. 1034

Thibodeaux LJ, Boyle JD (1987) Bedform-generated convective transport in bottom sediment. Nature 325:341–343

Triska FJ, Kennedy VC, Avanzino RJ, Wellweger GW, Bencala KE (1989a) Retention and transport of nutrients in a third-order stream: channel processes. Ecology 70:1877–1892

Triska FJ, Kennedy VC, Avanzino RJ, Wellweger GW, Bencala KE (1989b) Retention and transport of nutrients in a third-order stream in Northwestern California: hyporheic processes. Ecology 70:1893–1905

Triska FJ, Duff JH, Avanzino RJ (1990) Influence of exchange flow between the channel and hyporheic zone on nitrate production in a small mountain stream. Can J Fish Aquat Sci 47:2099–2111

Triska FJ, Duff JH, Avanzino RJ (1993) Patterns of hydrological exchange and nutrient transformation in the hyporheic zone of a gravel bottom stream: examining terrestrial–aquatic linkages. Freshw Biol 29:259–274

Vallet HM, Morice JA, Dahm CN, Campana ME (1996) Parent lithology, surface-groundwater exchange, and nitrate retention in headwater streams. Limnol Oceanogr 41(2):333–345

van Genuchten MT, Wierenga PJ (1976) Mass transfer studies in Sorbing porous media I. Analytical solutions. Soil Sci Soc Am J 40(4):473–480

Wagner BJ, Harvey JW (1997) Experimental design for estimating parameters of rate-limited mass transfer: analysis of stream tracer studies. Water Resour Res 33(7):1731–1741. doi:10.1029/97WR01067

Ward RS, Williams AT, Barker JA, Brewerton LJ, Gale IN (1998) Groundwater tracer tests: a review and guidelines for their use in British aquifers. Environment Agency R&D Technical Report W160

Wilson JF, Cobb ED, Kilpatrick FA (1986) Fluorometric procedures for dye tracing. techniques of water-resources investigations of the United States Geological Survey, Chaper A12

Wörman A, Packman AI, Johansson H, Jonsson K (2002) Effect of flow-induced exchange in hyporheic zones on longitudinal transport of solutes in streams and rivers. Water Resour Res 38:1001

Wörman A, Riml J, Schmadel N, Neilson BT, Bottacin-Busolin A, Heavilin JE (2012) Spectral scaling of heat fluxes in streambed sediments. Geophys Res Lett 39(23):L23402. doi:10.1029/2012GL053922

Yoneda M, Inoue Y, Takine N (1991) Location of groundwater seepage points into a river by measurement of radon-222 concentration in water using activated charcoal passive collectors. J Hydrol 124(3–4):307–316

Chapter 24
Thermal Pollution in Rivers—Modelling of the Spread of Thermal Plumes

Monika B. Kalinowska and Paweł M. Rowiński

Abstract Modeling framework for stream temperature, especially after introducing substantial amount of heat pollution, is presented in this chapter. An overview of the mathematics and solution techniques suited for heat transfer quantification is given and the models presented range from 3D aimed at short distances towards 1D approach allowing for modeling of heat transfer over long distances. A special attention has been paid to the depth averaged two-dimensional models which are particularly useful when the fate of heat pollution such as the heat discharged from a steam power station is considered. The processes of exchange between the river water and river surrounding are also discussed. Examples of computational solutions are provided and discussed as well.

Keywords Heat transfer · Thermal pollution · Advection-dispersion equations · Thermal plume · Near field zone · Mid field zone · Far field zone

24.1 Introduction

A precondition for the proper understanding of the biology of an aquatic ecosystem is the cognition of thermal regime. For the aquatic life, the crucial factor is the presence of dissolved oxygen and the obvious observation is that as the temperature of water increases, its dissolved oxygen content decreases. This is a somewhat simplified picture; the interrelationship between water temperature and dissolved oxygen (DO) is a subject of many studies (Steele 1989) but recent data shows that under certain conditions the relationship between temperature and DO might have hysteretic pattern (Rajwa et al. 2014). Recently Demars et al. (2011) have studied the impact of temperature upon whole stream ecosystem respiration, gross primary productivity and the ecosystem production in Icelandic geothermal streams along a

M.B. Kalinowska (✉) · P.M. Rowiński
Institute of Geophysics, Polish Academy of Sciences, Ks. Janusza 64, Warsaw, Poland
e-mail: Monika.Kalinowska@igf.edu.pl

P. Rowiński and A. Radecki-Pawlik (eds.), *Rivers—Physical, Fluvial and Environmental Processes*, GeoPlanet: Earth and Planetary Sciences,
DOI 10.1007/978-3-319-17719-9_24

5–25 °C temperature gradient and demonstrated that temperature is an important (but not alone) driver of stream and rivers metabolism. Whatever the mechanism is, the increase of water temperature might have dramatic consequences. Elevated water temperature affects both the rate of reaeration (Demars and Manson 2013) and the rate of decomposition of organic wastes and hence affects dissolved oxygen. We do realize that metabolism requires oxygen; therefore, some species may be eliminated entirely if the water temperature rises. This single fact alone causes that the studies upon heat budget in a river are of crucial importance and cannot be omitted when the health of an aquatic ecosystem is judged.

Water has a unique feature, i.e. the capability to absorb thermal energy while experiencing only small changes in temperature, but this feature causes that most aquatic organisms have developed enzyme systems that operate in a narrow temperature range. So relatively small changes in natural ambient temperature might create substantial environmental problems. For example, in extreme cases an increase of water temperature can kill heat intolerant fish, but also plants, thereby disrupting the web of life dependent on the aquatic food chain. A tidbit in this respect (relevant for tropical climate) might be the fact that in case of crocodiles, alligators, several reptiles, such as some lizards and certain turtles, the temperature of egg incubation is the major factor determining sex (Murray 2002). Females are produced at one or both extremes of the range of viable incubation temperatures and the intermediate temperatures produce males so the increase or decrease of water temperature may disturb the sex ratio for those species, which plays a crucial role in their population dynamics and survival.

River temperature results from atmospheric conditions, its topography, water discharge and the streambed. The thermal regime of rivers is a subject of numerous studies (e.g. Caissie 2006 and the references given there); in this paper we will deal only with the situations that cause an abrupt change in thermal conditions in a river. One may think of various situations that might affect the thermal regime of a river. An obvious situation is the long-term climatic change that may create permanent changes in river water temperatures (e.g. Ficklin et al. 2013; Null et al. 2012), or on the other extreme an immediate heating of water during a fire in a forested watershed (Ice 2008)—in such a case elevation of stream temperature is caused by both heat of combustion and the removal of shade. Of crucial importance are various anthropogenic perturbations, such as water releases from dams or reductions in river flow due to, e.g., irrigation or finally the thermal effluents. In this paper we will concentrate on the latter which in most cases is an inevitable by-product of the cooling operations in power plants, petroleum refining, pulp and paper, iron and steel and chemical manufacturing industries. Fortunately such thermal loads may be treated as degradable due to the fact that heat is dissipated in receiving waters. Further in the paper such thermal loads that change ambient water temperature will be often called **thermal pollution**. When talking about thermal pollution one immediately thinks of elevated water temperatures caused by a variety of the above-mentioned factors but we should keep in mind the reverse, for example, the discharges of cold bottom water from deep-water reservoirs behind large dams, which might change the downstream biological communities

(Sherman 2000). In this study we will focus on the analyses of heat transfer, i.e. the processes that change heat in a defined water volume.

Although heat transfer in a turbulent flow can be of interest per se, most studies in relation to rivers are of applied nature, and their results may be used in decision making that may carry large ecological risks.

Scientific knowledge of heat transfer in turbulent flows has progressed to such an extent that we are now able to unambiguously determine the fate of a thermal plume in the stream. However, as it is going to be shown further, quantification of those processes is fraught with rather a lot of uncertainties.

The issues raised in the paper are not, of course, exhaustive but reflect the flavor of discussions provoked by mathematical modeling of the processes of transfer of heat in rivers and their further consequences. Although a substantial amount of working models for such analyses are available in literature, still much work is required to increase their credibility. Literature is rich in various studies of environmental consequences of the increase of stream temperature, however there are not too many in-depth studies that would show any substantial progress in methods allowing for much better quantification of heat transfer in rivers. It is fair to say that interdisciplinary studies that will make predictions of heating and cooling of stream waters more accurate are highly demanded.

24.2 Transport of Heat in Rivers

Conservation of heat in open channels yields the transport equation for heat, or rather, for the change of temperature which in general case should be described by three-dimensional (3D) transport equation (see, e.g., Szymkiewicz 2010; Whitaker 1983):

$$\frac{\partial T}{\partial t} = \nabla[(\mathbf{D_M} + \mathbf{D_T}) \cdot \nabla T] - \nabla[\mathbf{v} \cdot T] + Q; \tag{24.1}$$

where: t—time [s], $T(\mathbf{x}, t)$—time-averaged water temperature [°C], $\mathbf{x} = (x, y, z)$—position vector [m], x—longitudinal coordinate [m], y—transversal (lateral) coordinate [m], z—vertical coordinate [m], $\mathbf{v}(\mathbf{x}) = (v_x, v_y, v_z)$—time-averaged velocity vector [m/s], $\mathbf{D_M}(\mathbf{x})$—molecular heat diffusion tensor [m^2/s], $\mathbf{D_T}(\mathbf{x})$—turbulent heat diffusion tensor [m^2/s], Q—source function describing additional heating or cooling processes. Note that the values of components of molecular heat diffusion tensor are much smaller than those of the turbulent heat diffusion tensor, and therefore they are usually omitted. In principle, the process of heat transfer is the result of operation of two basic mechanisms: advection, which is the mechanism connected with the movement of water, and diffusion, which is an intrinsive mechanism of transporting heat in the direction of decreasing temperature.

To solve the transport equation (Eq. 24.1), a huge amount of information either from other models of from detailed measurements is required. On top of that one needs to know the exact 3D velocity field and the water depths. All that data is

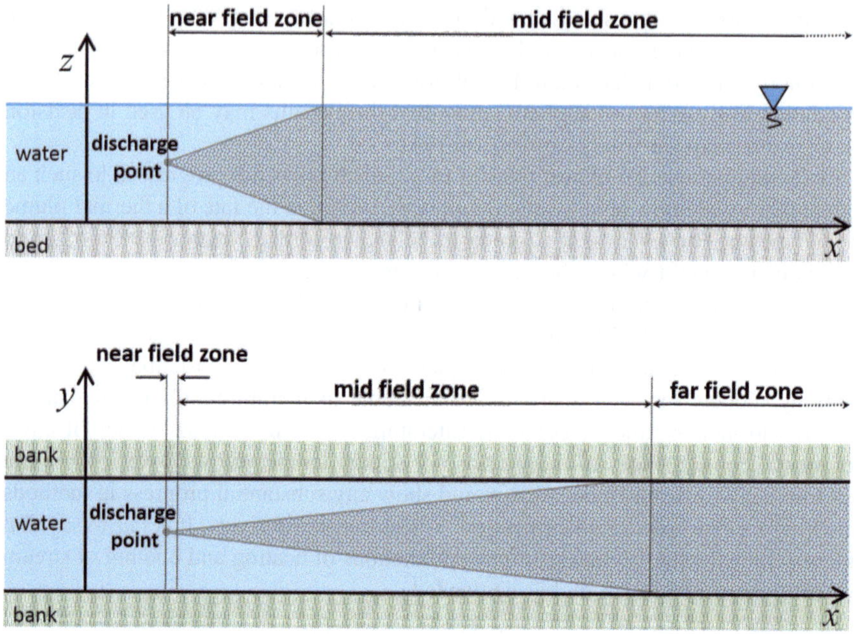

Fig. 24.1 Schematically presented characteristic mixing zones in rivers. *Top* subfigure—side view, *bottom* subfigure—plane view

usually difficult to obtain. The computational costs of the solution of Eq. 24.1 are also very high, so different simplifications should be considered in practice. The most obvious simplifications pertain to the reduction of the problem to two (2D) or even one dimension (1D).

In rivers we can distinguish three characteristic mixing zones (see Fig. 24.1) associated with the temporal and spatial scales of the heat transfer process. Each of them requires different descriptions and allows different simplifications. They are briefly described below.

- **Near field zone**—starting at the discharge point and continuing to the point of complete vertical mixing (Fig. 24.2).
- **Mid field zone**—stretching down the river until complete lateral mixing occurs (Figs. 24.3 and 24.4).
- **Far field zone**—starting after the complete mixing along the depth and width of the channel (Fig. 24.5).

Mixing in each zone should be described, respectively, in three-, two-, and one-dimension.

The first, near field zone—close to the discharge point—is usually very short. This is because most of rivers are shallow compared to their width and vertical mixing runs relatively fast. It breaks down thermal stratification and unlike in lentic

Fig. 24.2 Tracer test performed in a natural Narew river in north-east Poland in June 2005 (Rowiński et al. 2007, 2008); dye release (*top figure*), section directly behind the discharge (*bottom figure*)

waters, the temperature in rivers is nearly uniform from surface to bottom (Allan 1995; Rutherford et al. 1992). Complete vertical mixing takes maximally a few tens of water depth (see Jirka and Weitbrecht 2005). Additionally, information about the 3D temperature distribution in practical applications is usually not necessary. Therefore, the near field zone is often beyond practical considerations and treating the process of thermal pollution spreading as two-dimensional seems natural.

Fig. 24.3 Aerial photograph (ca. 1960) of an industrial discharge located near the middle of the regulated River Rhine upstream of Lake Constance (Bodensee); *photograph* courtesy of D. Vischer, Zurich; *source* Jirka and Weitbrecht (2005)

The equation describing the process in the 2D case is obtained by the averaging of the 3D equation (Eq. 24.1) along the depth. Detailed description of averaging process may be found in several papers (e.g., Kalinowska and Rowiński 2008; Rowiński 2002; Rutherford 1994; Szymkiewicz 2010). Then the 2D heat transport equation takes the following form (Kalinowska and Rowiński 2012; Rodi et al. 1981; Seo et al. 2010; Szymkiewicz 2010):

$$h\frac{\partial T}{\partial t} = \nabla(h\mathbf{D}\cdot\nabla T) - \nabla(h\mathbf{v}\cdot T) + Q; \tag{24.2}$$

Fig. 24.4 Tracer study of horizontal mixing in Missouri river in USA, *source* Holley (2001) (*left figure*); Rhodamine dye study of hydrology of stream-lake interactions (Spring Creek lake), *source* http://www.aslo.org/, author: Prof. Wayne Wurtsbaugh (*right figure*)

Fig. 24.5 Tracer test performed in a natural Narew river in north-east Poland in June 2005 (Rowiński et al. 2007, 2008); after complete mixing along the river width

where: t—time [s], $T(\mathbf{x}, t)$—depth-averaged water temperature [°C], $\mathbf{x} = (x, y)$—position vector [m], $h(\mathbf{x})$—local river depth [m], $\mathbf{v}(\mathbf{x}) = (v_x, v_y)$—depth-averaged velocity vector [m/s], $\mathbf{D}(\mathbf{x})$—heat dispersion tensor [m^2/s], Q—source function describing the heat exchange between the river and its environment and additional heating or cooling processes. The dispersion tensor includes four dispersion coefficients:

$$\mathbf{D}(\mathbf{x}) = \begin{bmatrix} D_{xx} & D_{xy} \\ D_{yx} & D_{yy} \end{bmatrix} \tag{24.3}$$

that appear in Eq. 24.2 as the result of depth-averaging. Components of turbulent heat diffusion tensor are usually included in the heat dispersion tensor components. Further simplifications of Eq. 24.2 usually relate to dispersion coefficients, but often used simplifications (involving the omission of the off-diagonal elements of dispersion tensor) are not admissible (see Rowiński and Kalinowska 2006). The heat dispersion tensor components should be properly determined, e.g., on the basis of the so-called longitudinal (D_L) and transversal (D_T) dispersion coefficients. Note that to simplify the notation in Eq. 24.2 we used the same symbols for temperature (T) and velocity vector (**v**) as in Eq. 24.1, but here they denote the depth-averaged values.

As the next step, we can average Eq. 24.2 along the channel width that gives us the 1D heat transport equation. In practical applications, such simplified equation is often considered. It can be very useful, but frequently its application is insufficient. The mid field zone (in which we should use the 2D heat transport equation) may extend over very long distance (see Fig. 24.3). For large rivers in an extreme case that distance may reach hundreds of kilometers (Endrizzi et al. 2002; Jirka and Weitbrecht 2005). Only after complete lateral mixing (in the far field zone) the 1D equation may reflect the real behavior of the thermal plume. For the steady flow, such equation can be written as follows (Szymkiewicz 2010):

$$\frac{\partial T}{\partial t} = \frac{1}{A}\frac{\partial}{\partial x}\left(AD\frac{\partial T}{\partial x}\right) - v_x\frac{\partial T}{\partial x} + Q; \tag{24.4}$$

where: t—time [s], x—downstream distance [m], $T(x, t)$—cross-sectionally averaged water temperature [°C], $v_x(x)$—cross-sectionally averaged river velocity [m/s], $D(x)$—longitudinal dispersion coefficient [m^2/s], Q—source function describing the heat exchange between the river and its surrounding and additional heating or cooling processes. Note that again to simplify the notation we use the same symbols: (T) denotes temperature and (v_x) denotes velocity, but herein they are cross-sectionally averaged values. For steady and uniform flow (A = const, v_x = const) and constant dispersion coefficient (D = const), Eq. (24.4) may be rewritten as:

$$\frac{\partial T}{\partial t} = D\frac{\partial^2 T}{\partial x^2} - v_x\frac{\partial T}{\partial x} + Q; \tag{24.5}$$

where: t—time [s], x—downstream distance [m], $T(x, t)$—cross-sectional averaged water temperature [°C], v_x—mean river velocity [m/s], D—longitudinal dispersion coefficient [m^2/s]. In the far field region the dispersion process is very often neglected since the most important role plays the exchange with the surrounding environment and then Eq. (24.5) is further simplified to:

$$\frac{\partial T}{\partial t} = -v_x \frac{\partial T}{\partial x} + Q. \tag{24.6}$$

24.2.1 Heat Exchange Between a River and Its Environment

The heat exchange between the river water and river surrounding should be included in the above equations (Eqs. 24.1–24.6) in the source function Q. The form of the source term very much depends on the question posed. Moreover, the decision which of the processes should be considered in the given equation must be made taking into account the process significance, the temporal and space scale of the process, and the availability of data necessary to calculate the heat exchange with suitable accuracy. Note that even if some of the necessary data can be obtained from local meteorological stations, usually conditions on the river banks differ significantly from those at the location of these meteorological stations. The river may be in a valley or may be sheltered from the wind and sun by a large number of trees, the conditions may also significantly vary along the river channel. Often, inaccurate calculations of the contribution of additional processes may introduce much larger errors in the final results then their omission. As a result, the decision whether and which processes should be taken into account is difficult. In practical applications, most of processes, especially if we are interested in the situation of water temperature changes following the introduction of thermal pollution to the river, can be neglected at short temporal and spatial scales. The situation is different if the time scale of the phenomenon is long and diurnal or seasonal changes are important. Then additional processes may play a meaningful role. In particular, the following processes may be significant: heat exchange between the water and air (Q_A), heat exchange between the water and river bottom and banks (Q_B), and heat exchange between the water and sediment (Q_S). They are schematically presented in Fig. 24.6. The heat may also be transferred advectively from upstream, tributaries, rainfall and groundwater flows (Evans et al. 1998) and influenced by heat production by biological and chemical processes (Joss and Resele 1987). Finally the source function Q will be a sum of all functions describing heating or cooling processes taken into account, e.g.:

$$Q = Q_A + Q_B + Q_S + \cdots \tag{24.7}$$

- **Heat exchange with the atmosphere**
  A huge amount of heat exchange between the water and its environment occurs through the water/air interface (Ashton 1986). The process is controlled by many mechanisms that depend on a number of meteorological factors and it has

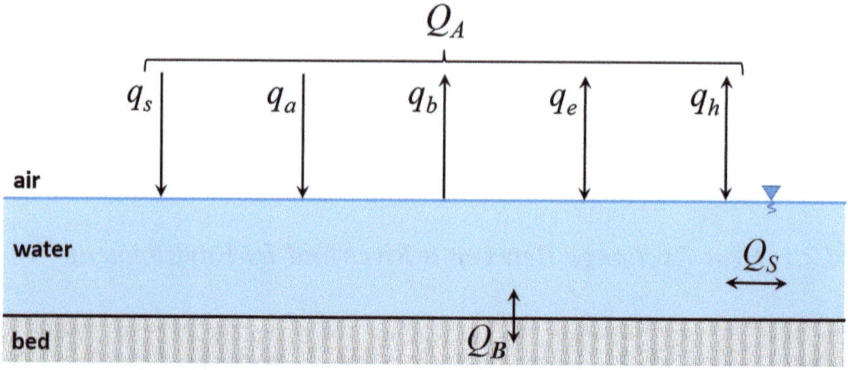

Fig. 24.6 Heat exchange between the river and its environment processes

been studied for a long time. The importance of the process may be different in different cases. We can omit it when the temporal scale of the analyzed case is small (of order of hours or less). It can be also omitted when we are interested in the difference between the temperature of ambient water and discharged heated water and this difference is not high (then the exchange of heat with the atmosphere proceeds in more or less the same way for the ambient and discharged water, and does not influence the difference between their temperatures). That means that the process may be neglected in the near field and mid field zones (unless the mixing across the channel is very poor and takes a very long time), but it should be considered in the far field zone. After complete vertical and horizontal mixing, additional cooling or heating may be strongly dependent on heat loss to the atmosphere. The exchange with the atmosphere must be included in all works related to the thermal regimes of rivers, in which the diurnal and seasonal changes are important. The heat flux through the water surface may by positive (heat is absorbed by the water from the atmosphere) or negative (heat is emitted from the water to the atmosphere). It results from the energy balance at the water/air interface. Taking into account the most important processes it may be written as (Edinger et al. 1974; Rutherford et al. 1993):

$$Q_A = q_s + q_a - q_b \pm q_e \pm q_h; \tag{24.8}$$

where: q_s—short-wave solar radiation, q_a—long-wave atmospheric radiation, q_b—long-wave back radiation (emitted by the water surface), q_e—evaporation and condensation, q_h—conduction. All components may be calculated based on water temperate, meteorological and hydrological data such as: wind speed, air temperature, cloud cover, bathymetric pressure, radiation, solar radiation, sun's altitude, humidity, or emissivity of water. There are different (more or less complicated) formulae available in the literature that can be used to compute particular components of Eq. 24.8 (see, e.g., Jurak 1978; Jurak and Wiśniewski 1989; Rutherford et al. 1993). For example, long wave back radiation is proportional to the fourth power of the surface water temperature (Rutherford et al. 1993). Finally, Eq. 24.8 becomes a complicated nonlinear function of water temperature, but in practice the formula is linearized to make the solution of heat transport equation simpler. For details, see (Jobson 1973; Rutherford et al. 1993).

- **Heat exchange with the river bed and banks**
  Heat exchange with the river bed and banks is negligible for most practical applications, except for shallow rivers (Benedini and Tsakiris 2013; Joss and Resele 1987). For such rivers it may by significant over long time scales, but its calculation is difficult and requires complex measurements. Its value anyway is small compared to the value of heat exchange between the water surface and atmosphere. Jurak and Wiśniewski (1989) post a table with the values of heat input or output from the river bed and banks for each month of year, depending on the north latitude and the averaged water depth, which may be used in practical cases. In the Northern Hemisphere we can notice the heat input from the bed to water in autumn-winter season and heat output from the bed to water in spring-summer season. According to studies performed by Evans et al. (1998) the river bed represented a significant energy sink during the summer months, being colder than the overlying water. We may also notice some energy sink during mid-day hours over a diurnal period and an energy source at night time. Nevertheless, in practical applications for deep rivers and near and mid field regions the heat exchange with the river bed and banks may be totally neglected. Note that in some cases, for example when the source of heated water is located near or on the river bank, the heat exchange with the river banks may be important.
- **Heat exchange with sediment**
  Similarly to the heat exchange with the bed and banks, the heat flux through the water-sediment interface is considered very small compared to the air-water interface (Edinger et al. 1974; Gu et al. 1998; Hockey et al. 1982) and is usually negligible in practical applications. Again it may be important especially in shallow water and in case of long time scales—over the time scales on the order of weeks and longer (Smith 2002).

24.3 Modelling of the Spread of Thermal Plumes

Modeling of spread of thermal plumes in rivers is usually necessary in practical applications concerning discharges of heated water into a river. Usually we are interested in possible threads to the environment from newly constructed objects (e.g. power plants). Special reports on environmental impact assessments are required by law in many countries. In such situations very often we deal with limited measurement data and at the same time many possible scenarios of pollution spreading must be taken into account. The Eqs. 24.1–24.6—describing the heat transport—have no analytical solutions for real boundary and initial conditions (for some very simple cases, analytical solutions are possible). In such situations, numerical models are indispensable. Nevertheless, choosing of the appropriate model is difficult and usually is a tradeoff between the results accuracy and costs that must be incurred. But it is not just about the computational costs and time, but also and maybe most of all about costs of obtaining the required number of model input data. More accurate models are usually more complicated and more time- and "data"-consuming. Simplified models solve the problem fast and with less data, nevertheless at the same time restrict the number of situations in which they may be applied. Models which only calculate the increase of water temperature above the natural river temperature are usually less demanding (Joss and Resele 1987). Not without significance is also whether the model is open source or commercial. Sometimes—for the more advanced users—access to the source code and the possibility of its modification may be also useful (this is especially important in scientific applications).

The first studies on the modeling of thermal pollution in rivers took place in the 1960s (Edinger and Geyer 1965); afterwards, many studies have been related to this problem (e.g. Chapra 2008; Dortch and Martin 1989; Fischer 1979; McCutcheon 1990; Thomann and Mueller 1987). Today, when restrictions on the computing power and data capacity are becoming smaller, the number of works and the available models is overwhelming. This makes the choice of the appropriate model even harder. Additionally, over the three different mixing zones described above, the heat transfer cannot be effectively simulated by a single model due to the wide range of space and time scales. To achieve the proper accuracy and resolution of results we should consider the use of different models in different mixing zones, taking into account the time and space scale of the process in the analyzed case. Examples of different models that can be used in each of mixing zones are presented below.

24.3.1 Near Field Zone—3D Modeling

As mentioned above, the near field zone is usually very short. Additionally, to meet receiving heated water temperature requirements, faster mixing may be achieved by discharging the heated water in several points instead of discharging it by a single nozzle. Different types of diffusers (see, e.g., Roberts 2011) may be also applied to accelerate mixing along the depth (often also along the width) of the channel. This further shortens the length of the near field zone. It is usually from ten to hundreds meters long and takes a few minutes and therefore in practical applications 3D models are usually not applied and we will not focus here on them. If it is necessary, for example, the following models can be considered:

- Visual Plumes (Frick 2004)—a simple, so-called plume model;
- CORMIX (Jirka et al. 1996)—near field expert system;

- Model presented by Tang et al. 2008—the Reynolds-averaged Navier-Stokes computational fluid dynamics model, developed for simulating initial mixing in the near field of thermal discharges at real-life geometrical configurations;
- THREETOX (Maderich et al. 2008)—3D numerical model developed for prediction of cooling water transport and mixing in the inland and coastal waters.

24.3.2 *Mid Field Zone—2D Modelling*

While the mixing along the depth is relatively fast, the mixing along the width can take a very long time, as pointed earlier. In the far field zone, the increase in temperature is usually small enough to fulfill all necessary conditions, thus the mid-field zone from environmental regulations point of view is the most important one. In practical applications, the two dimensional models have been shown to give reasonable results. Therefore, we will focus here on 2D modeling.

Although the number of two dimensional models is constantly increasing, not many of them can be used for streams or rivers. Existing models for reservoir and lakes, like, e.g., CE-QUAL-W2 (Cole and Buchak 1995), BETTER (Bender et al. 1990), are averaged laterally and may be suitable for a thermally stratified flow. Available depth-averaged models not always may be applied for natural rivers with complex geometries, with which we usually have to deal with in practice. They may take into account additional terms responsible for heat exchange with the atmosphere, with the bottom, etc., but usually they neglect the off-diagonal components of the dispersion tensor in the transport equation (Eq. 24.2), which may cause a huge error in the final results—see the example presented in Fig. 24.7. The problem has been discussed by the authors in the previous publications: (Kalinowska and Rowiński 2012; Rowiński and Kalinowska 2006). Such simplifications of dispersion tensor may be admissible for simple geometries, while the flow direction is more or less parallel to one of the axes of coordinates system, or while curvilinear coordinate system (see, e.g., Czernuszenko 1987, 1990) is applied. Therefore, to present an example of 2D modelling of thermal pollution spreading in the mid-field zone we have used a River Mixing Model (RivMix) that solves the full 2D heat transport equation (Eq. 24.2). This is the finite difference model which has been developed by the authors at the Institute of Geophysics, Polish Academy of Sciences, to simulate the transport of passive pollutants in rivers (Kalinowska and Rowiński 2008), and then adopted to simulate the thermal pollution spreading (Kalinowska and Rowinski 2012, 2014; Kalinowska et al. 2012). The use of adequate numerical scheme is very important, since the numerical errors may become huge for same cases—see the example in Fig. 24.8. Detailed description of presented example may be found in (Kalinowska and Rowiński 2008). The truncation errors caused by the numerical schemes that can be chosen to solve the model equations have been calculated and can be found in (Kalinowska and Rowiński 2007).

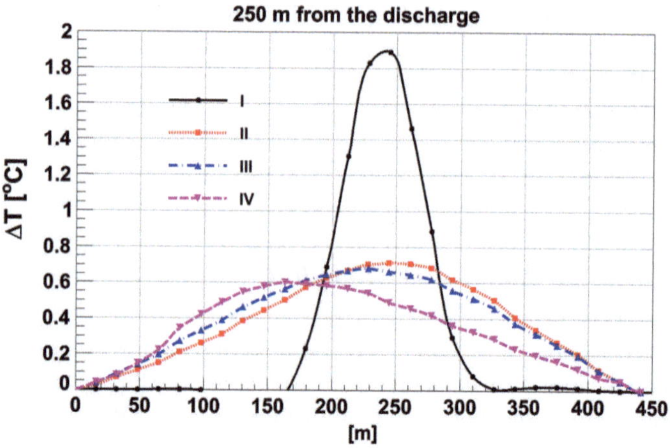

Fig. 24.7 Distribution of temperature increase (ΔT) in example case of the point-like continuous discharge along the cross-section located about 250 m from the discharge point: I—with the proper method of the dispersion tensor computation; II—with simplified method in which the off-diagonal elements of dispersion tensor D_{xy} and D_{yx} are omitted; III—with simplified method in which dispersion coefficients D_L and D_T are treated as a vector; IV—with simplified method in which the diagonal elements of dispersion tensor D_{xx} and D_{yy} are simply replaced by D_L and D_T, the off-diagonal elements are treated as 0. *Source* Kalinowska and Rowiński (2012)

To use the RivMix model, the following input data are necessary: the river geometry and the water depth, the two dimensional velocity field, the dispersion tensor, the initial and boundary conditions, the information about the sources of heated water, the simulation parameters (like, e.g., time step, grid spacing, simulation time).

The results of predictions of water temperature change for a real case study has been chosen to demonstrate the 2D modelling of thermal pollution spreading with application of RivMix model. The study has been done for an anticipated heated water jet that is planned to be continuously released in the lowland, urban part of the Vistula River in Poland after an existing thermal-electric power plant is reconstructed. The given reach of the Vistula River has semi-natural braided channel geometry and is protected by Nature 2000 protocol. The cooling water discharges and particularly their localization must be then designed to minimize the anticipated heat effects on local fish communities and environment. Different variants of warm water release have been analyzed. The results of the computed temperature distributions, for the selected case for the summer period, have been presented in Fig. 24.9. Computations have been made assuming constant intensity of 9 m^3/s of the discharged heated water with temperature 8 °C higher than the temperature of ambient river water (equal to 22.7 °C, which is the average temperature of natural water in summer in the analyzed river reach). Since we have been interested in the extreme conditions, the computations have been made for the mean low-flows of the river $Q = 200$ m^3/s. Additionally, the calculations have been performed for the lowest possible value of the flow, $Q = 100$ m^3/s. The values have

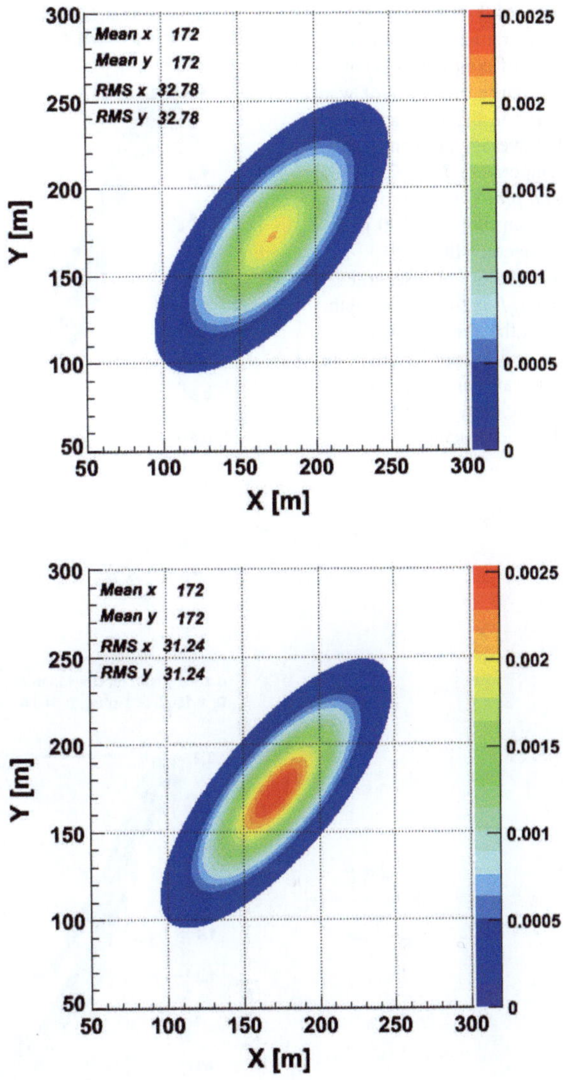

Fig. 24.8 Example of simulation results with the use of two different numerical schemes (*top panel*—Upwind scheme, *bottom panel*—Crank Nicolson scheme). Simulation parameters: $v_x = v_y = 0.106$ m/s, $D_{xx} = D_{yy} = 0.425$ m^2/s, $D_{xy} = D_{yx} = 0.325$ m^2/s, $x = y = 1$ m and $t = 0.5$ s. Mass ($M = 10$ arbitral units) has been injected instantaneously at a point $x_0 = 50$ m, $y_0 = 50$ m and time $t_0 = 0$ s. *Source* Kalinowska and Rowiński (2008)

been determined on the basis on the longstanding observations (century of observations), which is crucial for such studies. We can notice than in case of the lowest flow the increase in water temperature is larger close to the discharge point. Additional crucial aspect in such studies is the best choice of a discharge point and the way of heat release. An example of such studies is presented in Fig. 24.10. The increase in water temperature may be different depending on the way and location of discharge. Detailed description of the study, study area and presented variants may be found in (Kalinowska et al. 2012). Additional important aspects of 2D heat transport modeling and possible sources of uncertainty have been pointed out by the authors in (Kalinowska and Rowiński 2012).

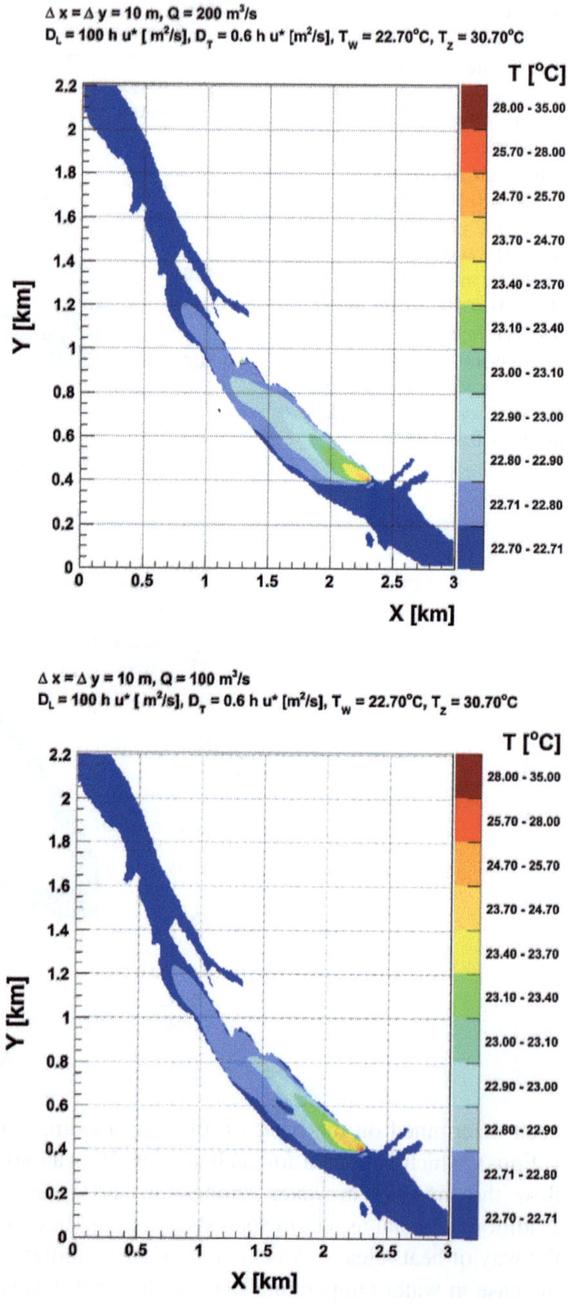

Fig. 24.9 Predicted 2D temperature distribution in case of continuous point discharge of 9 m^3/s of warm water in the summer time for the average ambient water temperature $T_W = 22.7$ °C. *Top plot* $Q = 200$ m^3/s, *bottom plot* $Q = 100$ m^3/s. T_Z denotes the water temperature at discharge point, h is the water depth, u^* is the shear velocity, D_L and D_T are longitudinal and transverse dispersion coefficients, Δx and Δy denote the grid spacing

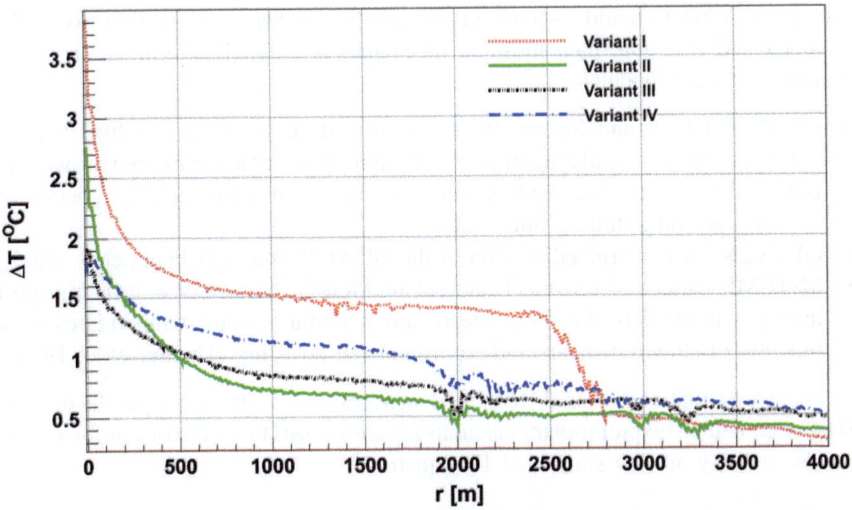

Fig. 24.10 Values of maximum increase of temperature within the distance r from the discharge point for different variants of way of discharge of heated water from the gas-stem power plant and discharge location for the real case study on the Vistula river in Poland. **Variant I**—point-like continuous discharge of 14 m^3/s of the heated water located in the middle of the channel; **Variant II**—point-like continuous discharge of 14 m^3/s of heated water on the left bank of the channel; **Variant III**—14 m long straight exit pipe containing nozzles for uniform distribution of heated water, i.e. for continuous discharge of 1 m^3/s of heated water per each 1 m of the exit pipe; **Variant IV**—28 m long straight exit pipe containing nozzles for uniform distribution of heated water, i.e. for continuous discharge of 0.5 m^3/s of heated water per each 1 m of the pipe. *Source* Kalinowska et al. (2012)

A few other 2D models used for the thermal pollution spreading are available, to mention, for example, RMA11—two/three Dimensional Finite Element Model for Water Quality Simulation (King 2003). The model may be used to simulate temperature with heat exchange with the atmosphere.

24.3.3 Far Field Zone—1D Modelling

One dimensional models may by very useful in practical applications, but very often they are used without proper justification for large rivers with complicated geometries, when 2D models have to be used to get reasonable results. The 1D approach may be used in the far field zone after complete vertical and horizontal mixing. Nevertheless, sometimes such approach is unavoidable even in mid field zone while there is no enough data to use the 2D models, but then results are often biased by very large errors.

There is a large number of 1D models available on the market. They are more or less complicated, but also they may be more or less accurate, and choosing of appropriate model should depend on the situation under consideration. Required accuracy and available data must be taken into account. Let us mention a few examples of used models:

- QUAL2E (River and Stream Water Quality Model)—developed by US EPA that is capable of simulating diurnal variations in water temperature (Brown and Barnwell 1987); the heat budget and temperature are simulated as a function of meteorology on a diurnal time scale;
- QUAL2K—a modernized version of the QUAL2E model (Chapra et al. 2005);
- SNTEMP (Stream Network Temperature Model)—steady-flow heat transport model that predicts the daily mean and maximum water temperatures as a function of stream distance and environmental heat flux (Theurer et al. 1984);

There are situations when there is no data necessary to run the proper 2D or even 1D model, and some preliminary calculations are a must. In such cases or for simple canals one may use the simplified 1D equation:

$$\rho c_p v_x h \frac{\mathrm{d}T}{\mathrm{d}x} = -k(T - T_\infty), \tag{24.9}$$

where: ρ—water density [kg/m^3], c_p—specific heat of water [J/(kg K)], v_x—mean river velocity [m/s], h—mean river depth [m], x—distance from the discharge point [m], T—water temperature [°C], T_∞—water temperature far away from discharge point [°C], k—heat exchange coefficient [W/m^2 K]. Such equation may give reasonable results after complete vertical and horizontal mixing, when additional heating or cooling proceeds mainly due to exchange through the water-air interface. Taking into account the initial condition:

$$T(x = 0) = T_0, \tag{24.10}$$

where: T_0—water temperature at discharge point; we obtain:

$$T(x) = T_\infty + (T_0 - T_\infty)\exp\left(\frac{-k}{\rho c_p UH}x\right). \tag{24.11}$$

T_0 may be calculated using the heat balance equation:

$$T_0 = \frac{Q_W T_W + Q_D T_D}{Q}, \tag{24.12}$$

where: T_W—the initial (natural) water temperature in the channel [°C], T_D—the temperature of discharged heated water [°C], Q_W—the water flow in the channel [m^3/s], Q_D—the heated water flow [m^3/s], $Q = Q_W + Q_D$—the resultant flow discharge [m^3/s]. We assume that the temperature of water far away from the discharge point is equal to the initial (natural) water temperature in the channel:

$$T_{\infty} = T_w. \quad (24.13)$$

Figure 24.11 presents water temperature as a function of distance below the discharge point of heated water for a channel in which h = 2.5 m, b = 30 m, $Q_W = 2\ m^3/s$. In such channel, the initial mixing should occur relatively fast. Then the additional cooling will proceed by heat exchange with the environment. Neglecting the heat exchange with the bottom and sediment, we take into account the heat exchange with the atmosphere by choosing the appropriate value of k coefficient. The heat exchange coefficient k will be here the most difficult factor to guess. We may assume it equal to (Rup 2006):

$$k = 100\,\mathrm{W/m^2\,K} \quad (24.14)$$

but in extreme conditions, in the absence of wind, it may fall to (Adams et al. 1981):

$$k = 5\,\mathrm{W/m^2\,K} \quad (24.15)$$

Therefore, in the example presented below results for different values of heat exchange coefficients have been calculated. They are plotted for ambient (natural) water temperature in the channel equal to: T_W = 22 °C. The heated water with temperature 7 °C higher then ambient water temperature (T_D = 29 °C) is discharged with constant intensity $Q_D = 0.5\ m^3/s$. We may observe the difference in the results

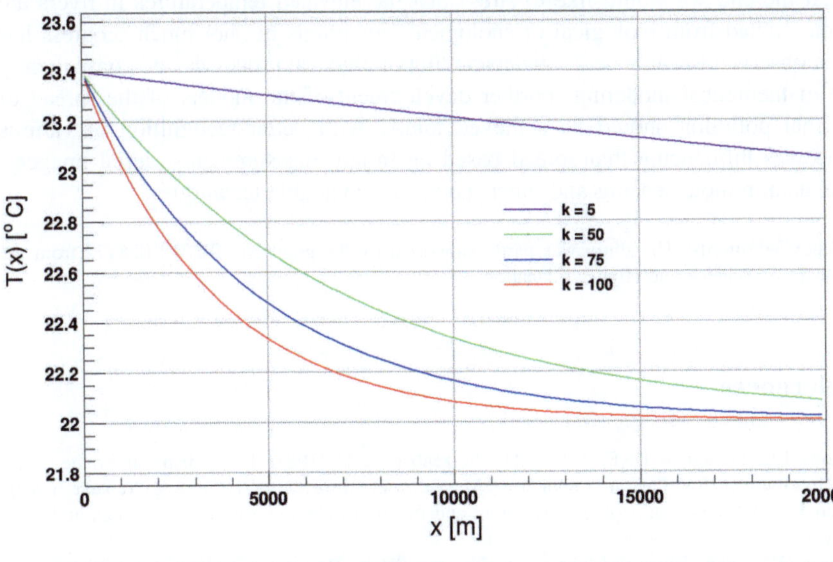

Fig. 24.11 Water temperature in the exemplary channel as a function of distance below the discharge point of heated water for different values of heat exchange coefficient k

depending on the k coefficient value. For the bigger k coefficient, the water temperature below the discharge point tends to the ambient water temperature much quicker than in case of small heat exchange coefficient.

24.4 Conclusions

The general theoretical framework presented in this chapter appears to be able to describe consistently a continuum of various settings for the heat transfer in a river ranging from short space scale (three dimensional approach) to long distances reflected by 1D approach. Special attention has been paid to the spread of thermal pollution and 2D approach has been favored as accurate enough and allowing to answer real life problems. In most of the presented examples, the authors used the RivMix model which was driven by the need for better computer-based tools for parts of Environmental Impact Assessment related to the fate of heat pollution in a river. At present a lot of data needed for proper description of heat transport are relatively scarce, mainly due to cost considerations and therefore numerous simplifications and assumptions have been discussed herein. We feel that fundamental research is still needed to ascertain the basic principles for the transport of heat in the turbulent surface water. Firstly, still much effort is needed in recognition of river turbulence and its relation to dispersion processes and also the exchange processes with the surrounding environment and their parameterization require further detailed studies.

At the end we would like to stress that the elevated temperatures in rivers are often studied from biological or ecological viewpoints but not much progress has been observed towards better understanding of physical processes and consequently the mathematical modeling. Further development of the models of the spread of thermal pollution should be achieved jointly with better recognition of various processes influencing that spread based on in situ measurements, aerial imagery, GIS data, remote sensing and other, nowadays available techniques.

Acknowledgments The study was partly supported by the grant No. IP2012 028772 from the Ministry of Science and Higher Education.

References

Adams EE, Harleman DRF, Jirka GH, Stolzenbach KD (1981) Heat, disposal in the water environment. RM Parsons Laboratory, Massachusetts Institute of Technology (course notes)

Allan JD (1995) Stream ecology—structure and function of running waters. Chapman and Hall, UK

Ashton GD (1986) River and lake ice engineering. Water Resources Publications, Littleton

Benedini M, Tsakiris G (2013) Water quality modelling for rivers and streams. Springer, Berlin

Bender MD, Hauser GE, Shiao MC, Proctor WD (1990) BETTER: A two-dimensional reservoir water quality model: technical reference and user's guide. Tennessee Valley Authority Engineering Laboratory, Norris, Tennessee

Brown LC, Barnwell TO (1987) The enhanced stream water quality models QUAL2E and QUAL2E-UNCAS: EPA/600/3-87-007. U.S. Environmental Protection Agency, Athens, p 189

Caissie D (2006) The thermal regime of rivers: a review. Freshw Biol 51(8):1365–2427

Chapra SC (2008) Surface water-quality modeling. Waveland Press, Salem

Chapra SC, Pelletier GJ, Tao H (2005) QUAL2K: a modeling framework for simulating river and stream water quality, version 2.02: documentation and users manual. Civil and Environmental Engineering Department, Tufts University, Medford

Cole TM, Buchak EM (1995) CE-QUAL-W2: a two-dimensional, laterally averaged, hydrodynamic and water quality model, version 2.0: user manual no. WES/IR/EL-95-1

Czernuszenko W (1987) Dispersion of pollutants in rivers. Hydrol Sci J 32(1):33–42

Czernuszenko W (1990) Dispersion of pollutants in flowing surface water. In: Encyclopedia of fluid mechanics, surface and groundwater flow phenomena, vol 10. Gulf Publishing Company, Houston, pp 119–168

Demars BOL, Manson JR, Olafsson JS, Gisalson GM, Friberg N (2011) Stream hydraulics and temperature determine the metabolism of geothermal Icelandic streams. Knowl Manage Aquat Ecosyst 402(05):1–17

Demars BOL, Manson JR (2013) Temperature dependence of stream aeration coefficients and the effect of water turbulence: a critical review. Water Res 47:1–15

Dortch MS, Martin JL (1989) Water quality modeling of regulated streams. In: Gore JA, Petts GE (eds) Alternatives in regulated river management. CRC Press, Florida, pp 63–90

Edinger JE, Geyer JC (1965) Heat exchange in the environment. Edison Electric Institute, New York

Edinger JE, Brady DK, Geyer JC (1974) Heat exchange and transport in the environment. Electric Power Research Institute, Palo Alto

Endrizzi S, Tubino M, Zolezzi G (2002) Lateral mixing in meandering channels: a theoretical approach. In: Bousmar D, Zech Y (eds) River flow 2002: proceedings of the international conference on fluvial hydraulics, Louvain-la-Neuve, Belgium. Swets & Zeitlinger, Lisse, 4–6 Sept 2002

Evans EC, McGregor GR, Petts GE (1998) River energy budgets with special reference to river bed processes. Hydrol Process 12:575–595

Ficklin DL, Stewart IT, Maurer EP (2013) Effects of climate change on stream temperature, dissolved oxygen, and sediment concentration in the Sierra Nevada in California. Water Resour Res 49(5):2765–2782

Fischer HB (1979) Mixing in inland and coastal waters. Academic press, New York

Frick WE (2004) Visual plumes mixing zone modeling software. Environ Model Softw 19 (7):645–654

Gu R, Montgomery S, Austin TA (1998) Quantifying the effects of stream discharge on summer river temperature. Hydrol Sci J 43(6):885–904

Hockey JB, Owens IF, Tapper NJ (1982) Empirical and theoretical models to isolate the effect of discharge on summer water temperatures in the Hurunui River. J Hydrol NZ 21:1–12

Holley ER (2001) Field tests for evaluating hydraulic transport processes in rivers. In: Rowinski PM, Napiórkowski JJ (eds) Water quality issues in the Upper Narew valley (Chap. 3). Publications of the Institute of Geophysics, Polish Academy of Sciences, Warsaw, pp 39–51

Ice GG (2008) Stream temperature and dissolved oxygen. In: Hydrological and biological responses to forest practices. Springer, Berlin, pp 37–54

Jirka GH, Weitbrecht V (2005) Mixing models for water quality management in rivers: continuous and instantaneous pollutant release. In: Czernuszenko W, Rowiński PM (eds) Water quality hazards and dispersion of pollutants (Chap. 1). Springer, USA, pp 1–34

Jirka GH, Doneker RL, Hinton SW (1996) User's manual for CORMIX: a hydrodynamic mixing zone model and decision support system for pollutant discharges into surface waters: prepared for USEPA Office of science and technology, cooperative agreement no. CX824847-010

Jobson HE (1973) The dissipation of excess heat from water systems. Am Soc Civ Eng 99(1):89–103
Joss J, Resele G (1987) Mathematical modelling of the heat exchange between a river and the atmosphere. Bound-Layer Meteorol 41(1–4):27–40
Jurak D (1978) Heat and mass exchange coefficient for open water surface. In: Modelling the water quality of the hydrological cycle: proceedings of the Baden symposium. International Association of Hydrological Sciences, Washington, DC
Jurak D, Wiśniewski A (1989) Simple model of thermal pollution in rivers. Working paper WP-89-066, International Institute for Applied Systems Analysis
Kalinowska M, Rowiński PM (2007) Truncation errors of selected finite difference methods for 2D advection-diffusion equation with mixed derivatives. Acta Geophys 55:104–118
Kalinowska MB, Rowiński PM (2008) Numerical solutions of two-dimensional mass transport equation in flowing surface waters. In: Monographic volume, publications of the institute of geophysics E-8(404). Polish Academy of Sciences, Warsaw
Kalinowska MB, Rowiński PM (2012) Uncertainty in computations of the spread of warm water in a river—lessons from environmental impact assessment case study. Hydrol Earth Syst Sci 16:4177–4190
Kalinowska MB, Rowiński PM (2014) Modeling of the spread of thermal pollution in rivers with limited data. In: River Flow 2014. Lausanne, Switzerland
Kalinowska MB, Rowiński PM, Kubrak J, Mirosław-Swiątek D (2012) Scenarios of the spread of a waste heat discharge in a river—Vistula river case study. Acta Geophys 60:214–231
King IP (2003) RMA11—a three-dimensional finite element model for water quality in estuaries and streams, update documentation, version 4.0a. Resource Management Associates, Fairfield, 90p
Maderich V, Heling R, Bezhenar R, Brovchenko I, Jenner J, Koshebutskyy V, Kuschan A, Terletska K (2008) Development and application of 3D numerical model THREETOX to the prediction of cooling water transport and mixing in the inland and coastal waters. Hydrol Process 22:1000–1013
McCutcheon SC (1990) Water quality modeling: river transport and surface exchange, vol I. CRC Press, Boca Raton, 344p
Murray JD (2002) Mathematical biology. Springer, Berlin, 553p
Null SE, Viers JH, Deas M, Tanaka S, Mount JF (2012) Stream temperature sensitivity to climate warming in California's Sierra Nevada: impacts to coldwater habitat. Clim Change 116 (1):149–170
Rajwa A, Rowiński PM, Bialik RJ, Karpiński M (2014) Stream diurnal profiles of dissolved oxygen—case studiem. In: Book of proceedings of 3rd IAHR europe congress. Porto-Portugal
Roberts PJW (2011) Diffusers for heated water disposal from power plants. In: Proceedings of the 2011 Georgia water resources conference. University of Georgia, Georgia
Rodi W, Pavlovic R, Srivatsa S (1981) Prediction of flow and pollutant spreading in rivers. In: Fischer HB (ed) Transport models/inland and coastal waters: proceedings of a symposium on predictive ability. Academic Press, London, pp 63–111
Rowiński PM (2002) Constituent transport. In: Dooge JCI (ed) Encyclopaedia of life support systems (EOLSS), chapter: fresh surface water. Eolss Publishers Co., Oxford (developed under the auspices of the UNESCO)
Rowiński PM, Kalinowska MB (2006) Admissible and inadmissible simplifications of pollution transport equations. In: Ferreira RML, Alves ECTL, Leal JGAB, Cardoso AH (eds) River flow 2006: proceedings of the international conference on fluvial hydraulics, Lisbon, Portugal. Taylor and Francis, London, pp 199–209, 6–8 Sept 2006
Rowiński PM, Guymer I, Bielonko A, Napiórkowski JJ, Pearson J, Piotrowski A (2007) Prediction of dispersion coefficients in a small stream using artificial neural networks. In: Proceedings of 32nd congress of international association of hydraulic engineering and research. Harmonizing the Demands of Art and Nature in Hydraulics, 1–6 July 2007 (paper 297)
Rowiński PM, Guymer I, Kwiatkowski K (2008) Response to the slug injection of a tracer—large scale experiment in a natural river. Hydrol Sci J 53(6):1300–1309
Rup K (2006) Transfer processes of pollution in the environment. WNT, Warsaw (in Polish)
Rutherford JC (1994) River mixing. Wiley, Chichester

Rutherford JC, Williams BL, Hoare RA (1992) Transverse mixing and surface heat exchange in the Waikato river: a comparison of two models. NZ J Mar Freshw Res 26(3–4):435–452
Rutherford JC, Macaskill JB, Bryan LW (1993) Natural water temperature variations in the lower Waikato river, New Zealand. J Mar Freshw Res 27(1):71–85
Seo IW, Choi HJ, Song CG (2010) Two-dimensional finite element model for analysis of heat transport in river. In: Christodoulou G, Stamou A (eds) Environmental hydraulics. Taylor & Francis Group, London, pp 277–282
Sherman B (2000) Scoping options for mitigating cold water discharges from dams. CSIRO Land and Water Canberra Consultancy Report 00/21, 48p
Smith NP (2002) Observations and simulations of water-sediment heat exchange in a shallow coastal lagoon. Estuaries 25(3):483–487
Steele JG (1989) High-resolution profiles of temperature and dissolved oxygen in a river. J Hydrobiol 179(1):17–24
Szymkiewicz R (2010) Numerical modeling in open channel hydraulics. In: Water science and technology library, vol 83. Springer, Berlin
Tang H, Paik J, Sotiropoulos F, Khangaonkar T (2008) Three-dimensional numerical modeling of initial mixing of thermal discharges at real-life configurations. J Hydraul Eng 134(9):1210–1224
Theurer FD, Voos KA, Miller WJ (1984) Instream water temperature model. In: Instream flow information paper no. 16: cooperative instream flow and aquatic systems group. U.S. Fish and Wildlife Service, Fort Collins
Thomann RV, Mueller JA (1987) Principles of surface water quality modeling and control. Harper & Row, New York
Whitaker S (1983) Fundamental principles of heat transfer. Krieger, New York, p 30

[illegible]

[illegible]

[illegible]

[illegible] Stream Network Temperature Model [illegible] Water Resources Council [illegible]

[illegible]

[illegible]

[illegible]

[illegible] Wiley, New York

[illegible]

MIX
Papier aus verantwortungsvollen Quellen
Paper from responsible sources
FSC® C105338

If you have any concerns about our products,
you can contact us on
ProductSafety@springernature.com

In case Publisher is established outside the EU,
the EU authorized representative is:
Springer Nature Customer Service Center GmbH
Europaplatz 3, 69115 Heidelberg, Germany

Printed by Libri Plureos GmbH
in Hamburg, Germany